OPERATIONS MANAGEMENT

A Systems Approach

Martin K. Starr

COLUMBIA UNIVERSITY

**boyd & fraser
publishing
company**

I(T)P An International Thomson Publishing Company

Danvers   Albany   Bonn   Boston   Cincinnati   Detroit   London   Madrid   Melbourne
Mexico City   New York   Paris   San Francisco   Singapore   Tokyo   Toronto   Washington

Executive Editor/Acquisitions Editor: DeVilla Williams
Project Manager: Jeanne Busemeyer, Hyde Park Publishing Services
Development Editor: Laura Ellingson
Editorial Assistant: Jason Morris
Production Editors: Jackie Bedoya/Sandra Cence
Photo Researcher: Abigail Reip
Manufacturing Coordinator: Lisa Flanagan
Director of Marketing: William J. Lisowski
Production Services: Douglas & Gayle, Ltd.
Composition: Douglas & Gayle, Ltd.
Interior Design: Rebecca Lemna, Lloyd Lemna Design
Cover Design: Rebecca Lemna, Lloyd Lemna Design
Cover Photo: Guy Motil, Westlight

For more information, contact boyd & fraser publishing company:

boyd & fraser publishing company
One Corporate Place • Ferncroft Village
Danvers, Massachusetts 01923, USA

International Thomson Publishing Europe
Berkshire House 168-173
High Holborn
London, WCIV 7AA, England

Thomas Nelson Australia
102 Dodds Street
South Melbourne 3205
Victoria, Australia

Nelson Canada
1120 Birchmount Road
Sarborough, Ontario
Canada MIK 5G4

International Thomson Editores
Campose Eliseos 385, Piso 7
Col. Polanco
11560 Mexico D.F. Mexico

International Thomson Publishing GmbH
Konigswinterer Strasse 418
53227 Bonn, Germany

International Thomson Publishing Asia
221 Henderson Road
#05-10 Henderson Building
Singapore 0315

International Thomson Publishing Japan
Hirakawacho Kyowa Building, 3F
2-2-1 Hirakawacho
Chiyoda-ku, Tokyo 102, Japan

1 2 3 4 5 6 7 8 9 10 K 9 8 7 6 5

ISBN 0-8770-9471-3 Student Edition
ISBN 0-8770-9481-0 Annotated Instructor's Edition

BRIEF CONTENTS

PART I

INTRODUCTION TO *OPERATIONS MANAGEMENT: A SYSTEMS APPROACH*

1 Operations Management (OM) and the Systems Approach *4*

2 OM's Key Role In Productivity *40*

PART II

STRATEGIC PERSPECTIVES FOR OPERATIONS MANAGEMENT

3 Forecasting Life-Cycle Stages *78*

4 Quality Management *112*

INDUSTRY PERSPECTIVE ON QUALITY:
Tom's of Maine *152*

PART III

PROCESS FLOW: DESIGN, ANALYSIS AND IMPROVEMENT

5 Process Configuration Strategies *166*

6 Process Analysis and Process Re-design *206*

7 Methods for Quality Control *248*

8 Management of OM Technology *306*

9 Planning Teamwork and Job Design *346*

10 Capacity Management *394*

11 Facilities Planning *446*

INDUSTRY PERSPECTIVE ON SERVICE:
Rosenbluth International *495*

PART IV

SYSTEMS MANAGEMENT

12 Materials Management *512*

13 Aggregate Planning *548*

14 Inventory Management *588*

15 Material Requirements Planning (MRP) *640*

16 Production Scheduling *684*

17 Cycle Time Management *732*

INDUSTRY PERSPECTIVE ON TECHNOLOGY:
AT&T Global Business Communications Systems *781*

PART V

TRANSITION MANAGEMENT: PREPARING FOR THE FUTURE

18 Rapid-Response Project Management *796*

19 Change Management *848*

INDUSTRY PERSPECTIVE ON THE ENVIRONMENT:
Saturn Corporation *898*

APPENDICES: Linear Programming *A1*

Table of Normal Distribution *A3*

Poisson Probability Distribution *A4*

SOLUTIONS *S1*

GLOSSARY *G1*

INDEX *I1*

ABOUT THE AUTHOR

MARTIN K. STARR

Martin K. Starr presently serves as President of the Production and Operations Management Society (POMS). A full professor at Columbia University, Dr. Starr is also Director of the Center for the Study of Operations at Columbia University. The Center, whose mission is to ascertain what makes companies competitive in the POM area, has studied successful companies around the world to use as benchmark models for excellence in operations.

Dr. Starr regularly teaches the operations management course at Columbia and has been a regular contributor to research which focuses on systems that coordinate all the functions of business interacting with the production area. He has been an active consultant for companies on the interaction of marketing and finance with production and operation issues. Clients include such companies as AT&T, Bristol Myers Squibb, Citibank, Digital Equipment Company, DuPont, Eastman Kodak, Grand Metropolitan, IBM, Lever Brothers, Merrill Lynch, and Young & Rubicam.

Dr. Starr began his academic career with a bachelor's degree in engineering from MIT, and went on to earn masters' degrees in operations research and industrial engineering from Columbia, and ultimately a Ph.D. in operations research and management science from Columbia. He is the author of numerous books, as well as many articles in leading journals.

CONTENTS

PART I

INTRODUCTION TO *OPERATIONS MANAGEMENT: A SYSTEMS APPROACH*

CHAPTER 1

Operations Management (OM) and the Systems Approach *4*

1.0   Explaining Operations Management *6*
    Use of Models by OM *6*
    Working Definitions of Production
      and Operations *7*

1.1   Operations Management and the Systems
    Approach *10*
    Using the Systematic—Constructive Approach *11*
    Why is the Systems Approach Required? *12*
    Defining the System *12*
    Structure of the Systems Approach *13*

1.2   Examples of the Systems Approach *14*

1.3   Differentiating between Goods
    and Services *15*

1.4   Contrasting Production Management
    and Operations Management *17*
    Information Systems are Essential for
      Manufacturing and Services *18*

1.5   The Basic OM Transformation Model:
    An Input-Output System *19*
    Costs and Revenues Associated with the
      Input-Output Model *21*

1.6   OM Input-Output Profit Model *24*
    How Has the Profit Model Changed Over Time? *25*

1.7   The Stages of P/OM Development *25*
    Special Capabilities *26*

1.8   Organizational Positions and Career
    Opportunities in OM *29*
    Career Success and Types of Processes *29*
    Operations Management Career Paths *31*
    Global Aspects of Career Paths *31*
    Supplement 1: The Decision Model *37*
    Focus on People: John Reed, Citicorp *28*
    Focus on Quality: IBM Helpware Customer
      Service *14*
    Focus on Technology: ATMs: Changing the Way
      We Bank *24*

CHAPTER 2

OM's Key Role in Productivity *40*

2.0   Productivity—A Major OM Issue *42*
    Operational Measures of the Organization's
      Productivity *46*
    Productivity is a Systems Measure *47*
    Productivity is a Global Systems Measure *48*
    Bureaucracy Inhibits Flexibility
      and Productivity *49*

2.1   Productivity and Price-Demand Elasticity *51*
    Elasticity Relationship *51*
    Elasticity of Quality *55*

2.2   Economies of Scale and the Division
    of Labor *57*

2.3   The History of the Improvement of OM
    Transformations *58*
    Discussion of the History of P/OM *60*
    Supplement 2: Decision Trees *70*
    Focus on People: Al Scott, Wilson Sporting
      Goods *49*
    Focus on Quality: McKesson Corporation *46*
    Focus on Technology: Object-Oriented
      Programming *55*

PART II

STRATEGIC PERSPECTIVES FOR OPERATIONS MANAGEMENT

CHAPTER 3

Forecasting Life-Cycle Stages 78

3.0 Strategies for Operations Management 80
3.1 Participants in Planning Strategies 81
  OM's Role in Developing Strategies 81
  Who Carries Out the Strategies? 82
  Adopting "Best Practice" 84
  OM is a Scarce Resource 85
3.2 Understanding Life Cycle Stages for OM Action 86
  Introduction and Growth of the New Product 86
  Maturation and Decline of the New Product 86
  Protection of Established (Mature) Products 88
3.3 Forecasting Life Cycle Stages for OM Action 89
  Perspectives on Forecasting 90
  Extrapolation of Cycles, Trend Lines and Step Functions 91
  Time Series Analysis 92
  Correlation 93
  Historical Forecasting with a Seasonal Cycle 93
3.4 Forecasting Methods 95
  Base Series Modification of Historical Forecasts 95
  Moving Averages for Short-term Trends 96
  Weighted Moving Averages (WMA) 97
  Use of Regression Analysis 98
  Correlation Coefficient 100
  Coefficient of Determination 101
  Pooling Information and Multiple Forecasts 102
  The Delphi Method 102
  Comparative Forecasting Errors for Different Forecasting Methods 102
  Keeping a Record of Past Errors 104
  Supplement 3: Exponential Smoothing 109
  Focus on People: Doug Smith, Computervision 85
  Focus on Quality: Westinghouse Commercial Nuclear Fuel Division 94
  Focus on Technology: Tracking the Power Rangers 89

CHAPTER 4

Quality Management 112

4.0 Why is Better Quality Important? 114
4.1 Why is Total Quality Management Important? 114
4.2 Two Definitions for Quality 116
4.3 The Dimensions of Quality 118
  Models of Quality 119
  Quality Dimension—Purpose Utilities 121
  Quality Dimension—Other "ilities" 121
  Definition of Failure—Critical to Quality Evaluation 123
  Warranty Policies 123
  The Service Function—Repairability 124
  Nonfunctional Quality Dimension—Human Factors 124
  The Variety Dimension 125
4.4 Setting International Producer Standards 125
  U.S. Quality Standards 125
  International ISO 9000 Standards 126
  Japanese Quality Standards 129
  Database of Quality Standards 130
4.5 The Costs of Quality 130
4.6 The Costs of Prevention 130
4.7 The Costs of Appraisal (Inspection) 132
4.8 The Costs of Failure 133
4.9 Quality and the Input-Output Model 133
4.10 Controlling Output Quality 134
4.11 TQM is the Systems Management of Quality 135
4.12 Mapping Quality Systems 135
  House of Quality (HOQ) 137
  Quality Function Deployment (QFD) 138
  Teamwork for Quality 140
4.13 Industry Seeks Recognition 140
  The Deming Prize 141
  The Malcolm Baldrige National Quality Award 142
  Supplement 4: A Scoring Model for the House of Quality 149
  Focus on People: Curt Reimann, Director Malcolm Baldrige National Quality Award 115
  Focus on Quality: Malcolm Baldrige Award-Winning Lessons 143
  Focus on Technology: Help Systems Inc. Wins ISO 9001 127
  INDUSTRY PERSPECTIVE ON QUALITY:
  Tom's of Maine 152

PART III

PROCESS FLOW: DESIGN, ANALYSIS AND IMPROVEMENT

CHAPTER 5

Process Configuration Strategies *166*

....

5.0   What is a Process? *168*
    Work Configurations Defined *168*
    Process Simulations *170*
    Classifying the Process *170*
    Charting the Process *171*
    Choosing the Process *171*
    Means and Ends *171*

5.1   Strategic Aspects of Processes *172*

5.2   Types of Process Flows—Synthetic Processes *173*

5.3   Types of Process Flows—Analytic Processes *176*

5.4   Sourcing for Synthetic Processes *178*

5.5   Sourcing for Analytic Processes *179*

5.6   Marketing Differences *181*
    Marketing Products of Analytic Processes *181*
    Best Practice *182*

5.7   Mixed Analytic and Synthetic Processes *182*
    The OPT System's V-A-T *184*

5.8   Work Process Configurations *184*

5.9   The Custom Shop *184*

5.10   The Job Shop *185*

5.11   The Flow Shop *187*
    The Benefits and Constraints of Flow Shops *190*

5.12   Economies of Scale—Decreasing the Cost per Unit *190*

5.13   The Intermittent Flow Shop *192*

5.14   Flexible Process Systems *193*

5.15   Moving from One Work Configuration Stage to Another *195*
    Expanded Matrix of Process Types *196*

5.16   The Extended Diagonal *198*
    Adding the Custom Shop *198*
    The Role of FMS *198*
    Applied to Services *198*

5.17   Aggregating Demand *199*
    Supplement 5: How Purchasing Agents Use Hedging *204*
    Focus on People: Jeffrey Bleustein, Harley-Davidson Inc. *191*

Focus on Quality: Boutwell Owens and Co., Inc. *172*

Focus on Technology: BiCMOS Technology at National Semiconductor *183*

CHAPTER 6

Process Analysis and Process Redesign *206*

....

6.0   Process Improvement and Adaptation *208*

6.1   Process Flow Charts *209*

6.2   Background for Using Process Analysis *211*
    Process Management is a Line Function *211*
    Demand is Greater than Supply *212*
    Seven Points that Deserve Consideration for Process Flow Design *212*
    The Purpose of Process Charts *213*
    The Use of Process Charts for Improvements *214*
    New Orders *216*

6.3   Process Charts with Sequenced Activity Symbols *216*
    Computer-Aided Process Planning (CAPP) *220*
    Creating Process Charts *220*
    Re-engineering is Starting from Scratch in Redesigning Work Processes *221*

6.4   Process Layout Charts *223*

6.5   Analysis of Operations *225*
    Value Analysis for Purchased Parts or Services *226*
    Other Process Activities *227*

6.6   Information Processing *227*

6.7   Process Changeovers (Set-ups) *230*
    Single-minute Exchange of Dies (SMED) *231*

6.8   Machine-Operator Charts: Process Synchronization *234*

6.9   Total Productive Maintenance (TPM) *236*
    The Preventive Maintenance Strategy *239*
    The Remedial Maintenance Strategy *239*
    Zero Breakdowns *240*
    TPM for the Flow Shop *240*
    Supplement 6: A Total Productive Maintenance Model for Determining the Appropriate Level of Preventive Maintenance to be Used *245*
    Focus on People: Motorola Production Manager *213*
    Focus on Quality: Toyota Motor Manufacturing U.S.A. Inc. *215*
    Focus on Technology: IBM's FlowMark—Workflow Management Technology *233*

CHAPTER 7

Methods for Quality Control *248*

. . . .

7.0   The Attitude and Mind-Set for Quality
       Achievement *250*
       The Zero-Error Mind-set *250*
       The Olympic Perspective *251*
       Compensating for Defectives *251*
7.1   Quality Control (QC) Methodology *251*
7.2   Introduction to Seven Classical Process Quality
       Control Methods *253*
7.3   Method 1—Data Check Sheets (DCs) *255*
7.4   Method 2—Bar Charts *256*
7.5   Method 3—Histograms *256*
       Auto Finishes—Quality on the Line *257*
7.6   Method 4—Pareto Analysis *258*
       The 20:80 Rule *259*
       Pareto Procedures *259*
7.7   Method 5—Cause and Effect Charts
       (Ishikawa/fishbone) *261*
       A Casual Analysis *261*
       Scatter Diagrams *263*
7.8   Method 6—Control Charts for Statistical
       Process Control *265*
       Chance Causes are Inherent Systems Causes *266*
       Assignable Causes are Identifiable Systems
          Causes *267*
7.9   Method 6—$\bar{x}$ Control Chart Fundamentals *267*
       The Chocolate Truffle Factory *268*
       Upper and Lower Control Limits
          (UCL and LCL) *269*
       How to Interpret Control Limits *274*
7.10 Method 6—Control Charts for Ranges—
       R Charts *274*
7.11 Method 6—Control Charts for Attributes—
       p Charts *275*
       c Charts for Number of Defects Per Part *278*
7.12 Method 7—Analysis of Statistical Runs *280*
       Start-ups in General *281*
7.13 The Inspection Alternative *282*
       100% Inspection, or Sampling? *282*
       Technological and Organizational Alternatives *282*

7.14 Acceptance Sampling (AS) *283*
       Acceptance Sampling Terminology *284*
       O.C. Curves and Proportional Sampling *285*
       Negotiation between Supplier and Buyer *287*
       Type I and Type II Errors *288*
       The Cost of Inspection *288*
       Average Outgoing Quality Limits (AOQL) *289*
       Multiple-Sampling Plans *290*
7.15 Binomial Distribution *291*
7.16 Poisson Distribution *294*
       Supplement 7: Hypergeometric OC Curves *301*
       Focus on People: W. Edwards Deming,
          Father of Quality *254*
       Focus on Quality: Copley Pharmaceutical Inc. *266*
       Focus on Technology: U.S. Postal Service Uses
          Optical Scanning to Speed Mail Sorting *280*

CHAPTER 8

Management of OM Technology *306*

. . . .

8.0   Technology Component of
       Transformation *308*
8.1   Management of The Technology
       Component *308*
8.2   The Technology Trap *309*
8.3   Technology Timing *310*
8.4   The Art and Science of Technology *311*
8.5   The Development of Glass Processing
       Technologies *315*
       Technological Innovations—Product
          and Process *317*
       Supply Chain Technology *320*
       Replacement vs. Original Market Technology *323*
8.6   The Technology of Packaging
       and Delivery *324*
       Systems Analysis *324*
8.7   The Technology of Testing *325*
       Testing by Simulation *326*
8.8   Quality Control Technology *327*
8.9   Technology of Design for Manufacturing
       (DFM) and Design for Assembly (DFA) *328*
       Changeover Technologies *331*

8.10 The New Technologies—Cybernetics *332*
  Flexibility—FMS, FOS, FSS and FPS *332*
  Expert Systems *333*
  Cyborgs and Robots *336*
  The Technology of Computer-User Interfaces *337*
  Other New Technology Developments—Materials
    and Miniaturization *338*
  Supplement 8: The Poka-Yoke System *342*
  Focus on People: Vickie Eckert, Megatest *322*
  Focus on Quality: Elkum Metals
    Expert System *331*
  Focus on Technology: Neural network programming
    Mellon Bank & LBS Capital Management *314*

CHAPTER 9

Planning Teamwork and Job Design *346*

. . . .

9.0 Training and Teamwork *348*

9.1 Training and Technology *350*

9.2 Operations Managers and Workers *351*

9.3 Present Training Methods *352*

9.4 New Training Methods *352*
  Action Learning *352*
  Team Learning *352*
  Computer-Based Training (CBT) *354*
  Distance Learning *354*

9.5 Performance Evaluation (PE) *355*

9.6 Using Production and Cost Standards *357*
  Activity-Based Cost (ABC) Systems
    and Job Design *358*

9.7 How Time Standards are Set—Job
  Observation *361*
  Cycle Time Reduction *361*
  Computers Substitute for Stopwatches *362*
  Shortest Common-Cycle Time *366*
  Productivity Standards *367*
  Training for Job Observation *368*
  Setting Wages Based on Normal Time *368*
  What is Normal Time? *368*
  Learning Curve Effects *369*
  Job Observation Corrections *369*
  Expected Productivity and Cost of the
    Well-Designed Job *371*
  Job Observation Sample Size *371*

9.8 Work Sampling and Job Design *373*
  Work Sample Size *374*

9.9 The Study of Time and Motion—Synthetic
  Time Standards *375*
  Evaluation of Applicability *376*

9.10 Managing the Work System *379*
  Job Design, Evaluation, and Improvement *380*
  Wage Determination *380*
  Work Simplification and Job Design *381*
  Job Design and Enrichment *382*
  The Motivation Factor *383*
  Supplement 9: Ergonomics or Human Factors *389*
  Focus on People: Nancie O'Neill, Westin Hotels
    & Resorts *360*
  Focus on Quality: TQM at Martin Marietta *379*
  Focus on Technology: India's Programming
    Industry *370*

CHAPTER 10

Capacity Management *394*

. . . .

10.0 Definitions of Capacity *396*
  Standard Hours *397*
  Maximum Rated Capacities *398*
  Difficult to Store Service Capacity *400*
  Backordering Augments Service Capacity *400*

10.1 Peak and Off-Peak Demand *400*
  Qualitative Aspects of Capacity *401*
  Maximum Capacity—100% Rating *402*
  Capricious Demand *404*
  Capacity Requirements Planning (CRP) *404*

10.2 Upscaling and Downsizing *405*

10.3 Bottlenecks and Capacity *407*

10.4 Delay Deteriorates Performance in the
  Supply Chain *409*
  Contingency Planning for Capacity Crises *410*
  A Supply Chain Game—Better Beer Company *411*
  Cost Drivers for Capacity Planning *414*
  Forecasting and Capacity Planning *415*

10.5 Capacity and Breakeven Costs
  and Revenues *416*
  Interfunctional Breakeven Capacity Planning *418*

10.6 Breakeven Chart Construction *419*
Three Breakeven Lines *420*
Profit, Loss and Breakeven *420*

10.7 Analysis of the Linear Breakeven Model *421*
Linear Breakeven Equations *424*
Breakeven Capacity Related to Work
Configurations *424*
Capacity and Margin Contribution *425*

10.8 Capacity and Linear Programming *425*

10.9 Modular Production (MP) and Group
Technology (GT) *427*
Group Technology (GT) Families *431*

10.10 Learning Increases Capacity *431*
The Learning Curve *432*
Service Learning *434*

10.11 Capacity and the Systems Approach *435*
Supplement 10: The Breakeven Decision
Model *441*
Focus on People: Scott Haag, Cinergy Corp. *430*
Focus on Quality: Grand Island Contract
Carriers *409*
Focus on Technology: Bellcore TEC, New Era
of Telephony *417*

CHAPTER 11

Facilities Planning *446*

11.0 Facilities Planning *448*
Who Does Facility Planning? *451*

11.1 Models for Facility Decisions *451*

11.2 Location Decisions—Qualitative Factors *452*
Location to Enhance Service Contact *452*
Just-in-Time Orientation *453*
Location Factors *453*

11.3 Location Decisions Using the Transportation
Model *454*
Suppliers to Factory—Shipping Costs *454*
Factory to Customers—Shipping Costs *455*
Obtaining Total Transport Costs *458*
Obtaining Minimum Total Transport Costs *458*
Testing Unit Changes *459*
Northwest Corner Method (NWC) *460*
Allocation Rule: $M + N{-}1$ *461*
Plant Cost Differentials *462*
Maximizing Profit Including Market
Differentials *462*
Flexibility of the Transportation Model *464*
Limitations of the Transportation Model *464*

11.4 Structure and Site Selection *466*
Work Configurations Affect Structure
Selection *466*
Facility Factors *467*
Rent, Buy, or Build?—Cost Determinants *467*

11.5 Facility Selection Using Scoring Models *468*
Intangible Factor Costs *469*
Developing the Scoring Model *470*
Weighting Scores *472*
Multiple Decision-Makers *472*
Solving the Scoring Model *473*

11.6 Equipment Selection *474*
Design for Manufacturing—Variance
and Volume *475*
Equipment Selection and Layout Interactions *476*
Life Cycle Stages *476*

11.7 Layout of the Workplace and Job Design *477*
Opportunity Costs for Layout Improvements *478*
Sensitivity Analysis *479*

11.8 Layout Types *480*

11.9 Layout Models *482*
Layout Criteria *482*
Floor Plan Models *483*
Layout Load Models *483*
Flow Costs *487*
Heuristics to Improve Layout *488*
Focus on People: Jane Olson, FACS *450*
Focus on Quality: Disney's America *465*
Focus on Technology: Citicorp Technical
Center *477*
INDUSTRY PERSPECTIVE ON SERVICE:
Rosenbluth International *495*

PART IV

SYSTEMS MANAGEMENT

CHAPTER 12

Materials Management *512*

12.0 What Is Materials Management? *514*
The Systems Perspective for Materials
Management *514*
Taxonomy of Materials *515*
Cost Leverage of Savings in Materials *518*
The Penalties of Errors in Ordering Materials *519*

12.1 Differences in Materials Management by
Work Configurations *520*

12.2 The Materials Management Information System *523*
  Global Information System—Centralized or Decentralized *523*
  The Systems Communication Flows *524*

12.3 The Purchasing Function *525*
  The 21st Century Learning Organization *525*
  Buyers' Risks *525*
  Turnover and Days of Inventory—Crucial Measures of Performance *526*
  Turnover Examples *527*
  Purchasing Agents *527*
  The Ethics of Purchasing *529*

12.4 Receiving, Inspection and Storage *531*

12.5 Requiring Bids Before Purchase *532*

12.6 Materials Management of Critical Parts *533*
  The Complex Machine—How Many Parts to Carry *534*
  The Static Inventory Problem *535*

12.7 Value Analysis (Alternative Materials Analysis) *536*
  Methods Analysis *536*

12.8 ABC Classification—The Systems Context *539*
  Material Criticality *539*
  Material Dollar Volume *540*

12.9 Certification of Suppliers *541*
  Focus on People: Wendell Kelly and Dennis Garrett, QualitiCare Inc. *521*
  Focus on Quality: Eastman Chemical Company *530*
  Focus on Technology: Inco Ltd.'s "Vendor Rationalization Process" *538*

CHAPTER 13

Aggregate Planning *548*

13.0 What Is Aggregate Planning? (AP) *550*
  Aggregation of Units *551*
  Standard Units of Work *553*
  The Importance of Forecasts *556*
  Basic Aggregate Planning Model *559*

13.1 Three Aggregate Planning Policies *559*
  Pattern A—Supply is Constant *564*
  Pattern B—Supply Chases Demand *565*
  Pattern C_1—Increase Workforce Size by $\pm$ One *566*
  The Cost Structure *568*

13.2 Linear Programming (LP) for Aggregate Planning (AP) *569*

13.3 Transportation Model (TM) for Aggregate Planning (AP) *574*
  Network and LP Solutions *576*

13.4 Nonlinear Cost Model for Aggregate Planning *577*
  Focus on People: Ken Bogard, Registrar, Miami University *558*
  Focus on Quality: 1996 Centennial Olympic Games *573*
  Focus on Technology: Planning Services in the Healthcare Industry *570*

CHAPTER 14

Inventory Management *588*

14.0 Types of Inventory Situations *590*
  Static versus Dynamic Inventory Models *591*
  Make or Buy Decisions—Outside or Self-supplier *592*
  Type of Demand Distribution—Certainty, Risk, and Uncertainty *593*
  Stability of Demand Distribution—Fixed or Varying *594*
  Demand Continuity—Smoothly Continuous or Lumpy *594*
  Lead-Time Distributions—Fixed or Varying *594*
  Independent or dependent demand *594*

14.1 Costs of Inventory *595*
  Costs of Ordering *596*
  Costs of Setups and Changeovers *596*
  Costs of Carrying Inventory *597*
  Determination of Carrying Cost *599*
  Costs of Discounts *599*
  Out-of-Stock Costs *600*
  Costs of Running the Inventory System *600*
  Other Costs *601*

14.2 Differentiation of Inventory Costs by Process Type *601*
  Flow Shop *601*
  Job Shop *602*
  Project *603*
  Flexible Process Systems (FPS) including FMS *603*

14.3 Order Point Policies (OPP) *603*

14.4 Economic Order Quantity (EOQ) Models—Batch Delivery *605*
  Total Variable Cost *608*
  An Application of EOQ *610*

14.5 Economic Lot Size (ELS) Model *612*

The Intermittent Flow Shop (IFS) Model *614*

An Application of ELS *617*

From EOQ through ELS to Continuous Production *618*

Lead Time Determination *619*

Expediting to Control Lead Time *621*

Lead-Time Variability *622*

14.6 Perpetual Inventory Systems *623*

Reorder Point and Buffer Stock Calculations *624*

Imputing Stock-Outage Costs *625*

Operating the Perpetual Inventory System *626*

Two-Bin Perpetual Inventory Control System *627*

14.7 Periodic Inventory Systems *628*

14.8 Quantity Discount Model *630*

Focus on People: Les Waltman, GE Aircraft Engines Procurement Division *620*

Focus on Quality: Auto Industry Outsourcing *595*

Focus on Technology: Baird Information Systems *605*

CHAPTER 15

Material Requirements Planning (MRP) *640*

15.0 The Background for MRP *643*

15.1 Dependent Demand Systems Characterize MRP *643*

Forecasting Considerations *645*

15.2 The Master Productions Schedule (MPS) *646*

Total Manufacturing Planning and Control System (MPCS) *648*

MPS: Inputs and Outputs *650*

Changes in Order Promising *651*

15.3 The Bill of Material (BOM) *653*

Explosion of Parts *655*

15.4 Operation of the MRP *655*

Coding and Low-level Coding *656*

15.5 MRP Basic Calculations and Concepts *658*

15.6 MRP in Action *660*

Scenario 1 *661*

Scenario 2 *662*

Scenario 3 *662*

Scenario 4 *664*

15.7 Lot Sizing *664*

15.8 Updating *665*

15.9 Capacity Requirements Planning (CRP) *666*

Rough-Cut Planning *668*

15.10 Distribution Resource Planning (DRP) *670*

15.11 Weaknesses of MRP *670*

15.12 Closed-Loop MRP *671*

15.13 Strengths of MRP II (Manufacturing Resource Planning) *673*

Supplement 15: Part-Period Lot Sizing Policy *680*

Focus on People: Glend Paquin, Mitel Corporation *669*

Focus on Quality: Valendrawers Inc. *659*

Focus on Technology: Software Systems Inc. *651*

CHAPTER 16

Production Scheduling *684*

16.0 Production Scheduling: Overview *686*

16.1 Loading *687*

Plan to Load Real Jobs—Not Forecasts *687*

16.2 Gantt Load Charts *688*

16.3 The Assignment Method for Loading *691*

Column Subtraction—Best Job at a Specific Facility *693*

Row Subtraction—Best Facility for a Specific Job *694*

16.4 The Transportation Method for Shop Loading *696*

16.5 Sequencing Operations *702*

First-In, First-Out (FIFO) Sequence Rule *702*

16.6 Gantt Layout Charts *703*

Evaluatory Criteria *705*

16.7 The SPT Rule for n Jobs and $m = 1$ Facilities *707*

16.8 Priority-Modified SPT Rules: Job Importance—Due Date—Lateness— Other Priorities *712*

Critical Ratio for Due Dates *712*

16.9 Sequencing with More Than One Facility ($m > 1$) *713*

16.10 Finite Scheduling of Bottlenecks—OPT Concept *715*

Flow Shops and Continuous Processes Do Not Have Bottlenecks *715*

Queue Control *717*

Synchronized Manufacturing *718*

Changing the Capacity of the Bottleneck *720*

Focus on People: Eliyahu Goldratt *716*

Focus on Quality: Digital's "Factory Scheduler" *795*

Focus on Technology: Red Pepper Software's Response Agents *708*

CHAPTER 17

Cycle Time Management *732*

. . . .

17.0 Cycle Time Management *732*
  Time-Based Management and Synchronization
    of Flows *735*
17.1 Line-Balancing Methods *736*
  Basic Concepts for Line Balancing *739*
  The Precedence Diagram of the Serialized
    Process *743*
  Cycle Time and the Number of Stations *745*
  Constraints on Cycle Time *746*
  Developing Station Layouts *749*
  Perfect Balance with Zero Balance Delay *751*
  Rest and Delay *753*
17.2 Heuristic Line Balancing *753*
  Kilbridge and Wester's (K & W) Heuristic *754*
  Other Heuristics *757*
17.3 Stochastic Line Balancing *759*
  Queuing Aspects of Line Balancing *760*
  Single-Channel Equations *761*
  Multiple-Channel Equations *764*
17.4 Just-In-Time Delivery *765*
17.5 Push Versus Pull Process Discipline *768*
  Supplement 17: Simulation of Queuing Models *776*
  Focus on People: Owner, Mail-Order
    Company *742*
  Focus on Quality: Boeing Co. *754*
  Focus on Technology: Quality Control Color *766*
  INDUSTRY PERSPECTIVE ON TECHNOLOGY:
    AT&T Global Business Communications *781*

PART V

TRANSITION MANAGEMENT: PREPARING FOR THE FUTURE

CHAPTER 18

Rapid-Response Project Management *796*

. . . .

18.0 Defining Projects *798*
18.1 Managing Projects *800*
  Project Managers as Leaders *800*
18.2 Types of Projects *801*
18.3 Basic Rules for Managing Projects *801*
18.4 Gantt Project-Planning Charts *802*

18.5 Critical-Path Methods *805*
  Constructing PERT Networks *806*
  Deterministic Estimates *806*
  Activities Labeled on Nodes (AON) *807*
  Activities Labeled on Arrows (AOA) *808*
  Activity Cycles *810*
18.6 Critical-Path Computations *811*
  Forward Pass *813*
  Backward Pass *818*
  Slack Is Allowable Slippage *820*
18.7 Estimates of Time and Cost *823*
18.8 Distribution of Project-Completion Times *827*
  The Effect of Estimated Variance *828*
  Why Is Completion Time So Important? *829*
  Progress Reports and Schedule Control *829*
18.9 Designing Projects *829*
  Resource Leveling *830*
18.10 Cost and Time Trade-Offs *832*
  PERT/Cost/Time(and Quality) *833*
18.11 Rapid-Response Time-Based Management
    (TBM) of Projects *837*
  Concurrent Engineering (CE) *838*
  Focus on People: Tom Walker, boyd & fraser *824*
  Focus on Quality: Quality Aspects of Project
    Management *832*
  Focus on Technology: Machine Tool
    Manufacture *812*

CHAPTER 19

Change Management *848*

. . . .

19.0 Adaptation to External Changes *851*
19.1 OM's Part in Change Management *853*
19.2 Factors That Allow Organizational
    Change *854*
19.3 OM's Benchmarking Role in Change
    Management *855*
19.4 Continuous Improvement (CI) for Change
    Management *857*
  Planning Transitions *859*
19.5 Re-Engineering (REE) for Change
    Management *859*
  Oticon Holding S/A *866*
  W. L. Gore & Associates, Inc. *867*
  Continuous Project Development Model *868*
19.6 Greening the Environment *870*
  Waterjetting *871*
  Design for Disassembly (DFD) *871*
  Recycling Concrete in Place *874*

19.7 Supporting Ethics *874*
    Quality Integrity *875*
    Hazard Analysis Critical Control Point (HACCP)
       System *875*
    Food & Drug Testing *879*
19.8 Readying the Organization for Future
Technology *880*
    Mobile Robots *880*
    Service Robots *882*
    The Virtual Office *883*
19.9 P/OM's Responsibility for Going Global
Successfully *885*
    Agribusiness–Consulting–Financial
      Services–Machinery *885*
    ISO 14000/14001 *886*
    The Olympics as a Systems Symbol *886*
    Supplement 19: NPV Net Present Value Model
      Applied to the Make or Buy Problem *891*
    Focus on People: Lars Kolind, Oticon *858*
    Focus on Quality: Gensym Corp. *878*
    Focus on Technology: Recycling Technology *873*
    INDUSTRY PERSPECTIVE ON THE
      ENVIRONMENT: Saturn Corporation *898*

APPENDICES: Linear Programming *A1*

Table of Normal Distribution *A3*

Poisson Probability Distribution *A4*

SOLUTIONS *S1*

GLOSSARY *G1*

INDEX *I1*

PREFACE

Operations Management: A Systems Approach, a fully integrated instructional support package, explains in simple terms the many methods and techniques important for the successful competitive practice of Operations Management (OM). The fresh concepts and approaches of the 1990s are the foundation of functional business practices and build on industrial engineering concepts and operations research methodology. Coverage reflects standards set forth by the American Production and Inventory Control Society (APICS) and the American Society for Quality Control (ASQC). Coverage also conforms with AACSB standards, which call for accredited schools to include the OM course in their curricula.

The Course

The text is targeted to the Operations Management course required of most business majors. The dynamics of the changing business environment have necessarily and profoundly altered the content of the OM course. A lot of the "old" methods and practices are changing before the very eyes of those who work in the field. *Operations Management: A Systems Approach* has been developed through extensive research to reflect both the new practices and traditional concepts and methods that instructors of Operations Management feel are important for students to know.

Systems Approach

The systems approach integrates OM decisions with the functional areas of business whose challenge is to make the firm perform as a team. The systems approach followed by this text is systematic and constructive, combining the methods of analysis and synthesis. This approach leads to better decisions and provides better problem-solving skills for any complex situation.

Organization

The text is comprised of five parts as illustrated below. This organization allows instructors to easily incorporate systems thinking and emerging OM practices into their course.

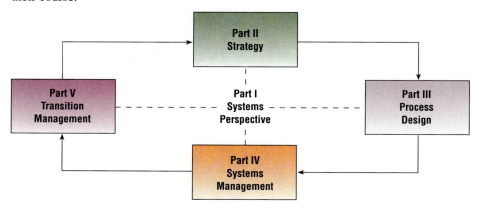

Real-World Themes

Quality concepts such as continuous improvement, which pervade the cultures of all excellent companies, are woven into the fabric of each chapter. Two chapters are devoted specifically to quality. Chapter 4 develops the total quality management (TQM) environment with emphasis on the Malcolm Baldrige National Quality Award and ISO9000 certification criteria. Chapter 7 covers statistical quality control methods. Application boxes on quality in every chapter showcase companies striving to create and maintain quality in their operations.

Technological changes are forcing companies accustomed to the old methods to either adapt or disappear. A major focus of this text is the impact of information systems and their ability to create "smart" operating processes. Each chapter's "technology box" introduces students to worldwide manufacturing and service applications (such as expert systems, object-oriented programming, and electronic data interchange). Leading commercial software packages are referenced to familiarize students with the control and execution of high-technology processes.

Globalization of business, a major emphasis of the AACSB guidelines, is addressed throughout the text in examples, application boxes, and cases to reflect the increasingly global nature of business. Among the many global topics explored in the text are global telecommunications reach, global logistics systems, and global impact on the supply chain.

Service and manufacturing applications are balanced throughout the text in recognition of the service industry's increasing importance in the global economy and manufacturing's growing awareness of the importance of providing exceptional service to customers.

Teamwork, a major component of all TQM programs, is emphasized as an essential element for success in all business operations. Students gain the understanding that working in business today means working as part of a team. Team-oriented classroom exercises are included in the instructor's manual to reinforce this important theme.

Environmental initiatives and awareness are stressed as a major operational responsibility. Coverage of the topic culminates in Chapter 19 on Change Management. The supporting Saturn case focuses on the environmental initiatives taken by the auto manufacturer to ensure recyclability of car components and preservation of the environment.

Problem solving methods introduced in the text present the mathematical foundations so that students know how and why they work. Most of the methods and problems presented can be performed by computer using commercial software programs. The Instructor's Manual contains computer output for selected problem solutions. The text is supported by—and can be packaged with—a number of commercial software packages.

"I think Operations Management: A Systems Approach will prove to be an excellent learning guide for any student in the OM arena. It is the most up-to-date text I have read, plus it incorporates the all-important global perspective...which is vital in today's business world."

William Strassen, *California Polytechnic State University*

"Dr. Starr truly takes essentially dull, pedantic material and through his insight and interspersed 'war stories,' makes the stuff interesting, very interesting."

Gus Widhelm, *University of Maryland at College Park*

THE TEXT SYSTEM

Pedagogy

Learning Objectives for each chapter highlight important concepts. Using this tool, students can assess their mastery of the full range of concepts and skills covered in each chapter.

Chapter Outlines show topics and organization at a glance. The outline functions as an overview which helps students place each section into the context of the entire chapter as they read.

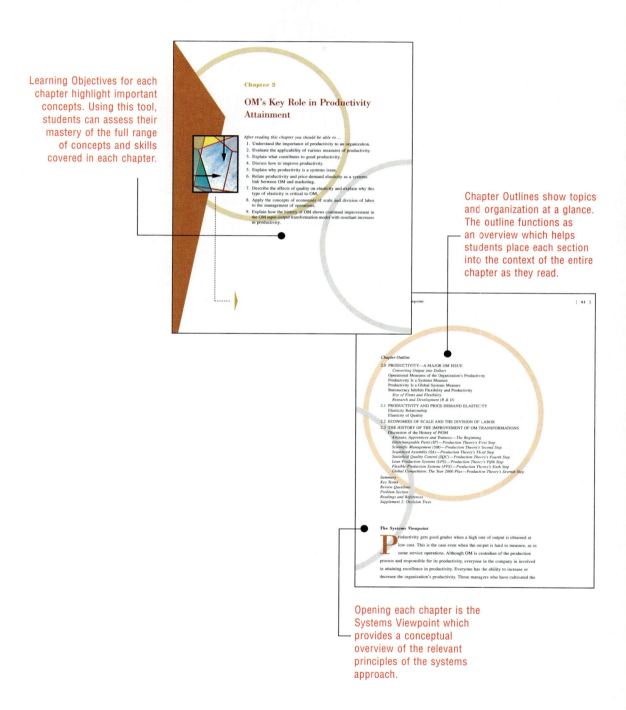

Opening each chapter is the Systems Viewpoint which provides a conceptual overview of the relevant principles of the systems approach.

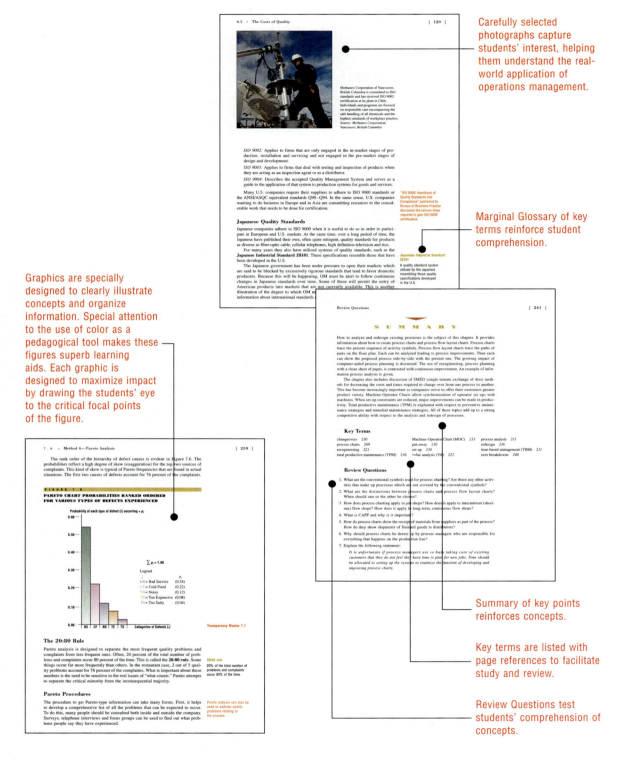

Carefully selected photographs capture students' interest, helping them understand the real-world application of operations management.

Marginal Glossary of key terms reinforce student comprehension.

Graphics are specially designed to clearly illustrate concepts and organize information. Special attention to the use of color as a pedagogical tool makes these figures superb learning aids. Each graphic is designed to maximize impact by drawing the students' eye to the critical focal points of the figure.

Summary of key points reinforces concepts.

Key terms are listed with page references to facilitate study and review.

Review Questions test students' comprehension of concepts.

Graded problems provide an opportunity for students to practice problem-solving skills covered in the chapter. Selected problems in each chapter can be solved with software.

Answers to odd problems provided at the end of the book offer feedback for students as they work through homework problems and prepare for exams.

Readings and References provide opportunity for further exploration of topics.

Notes for each chapter reflect both current research and historical foundations, both in operations management.

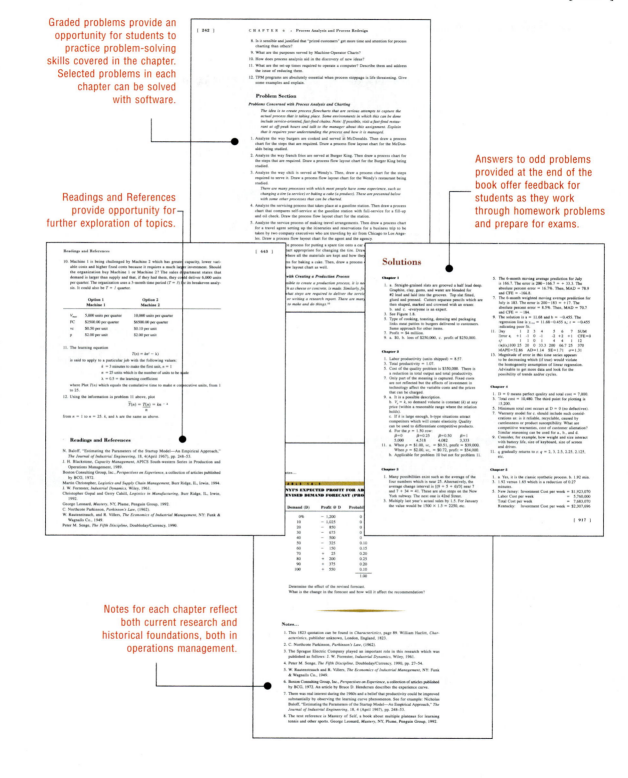

Real-World Examples

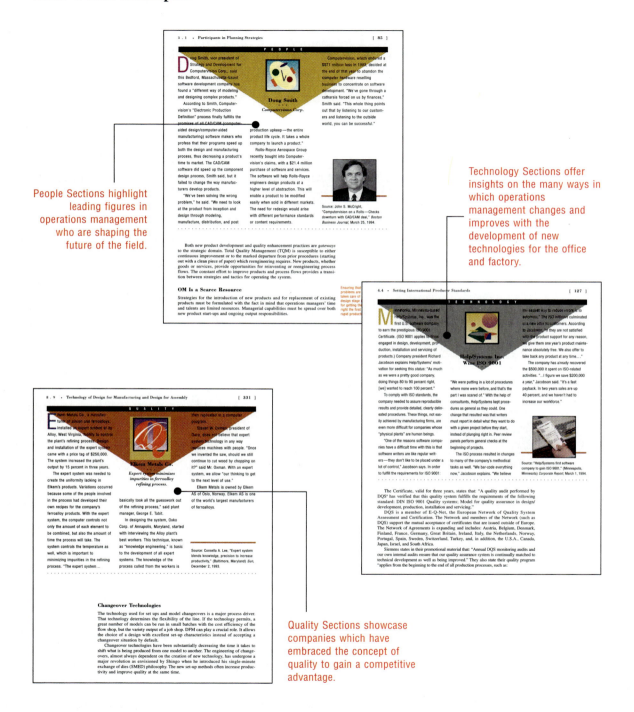

People Sections highlight leading figures in operations management who are shaping the future of the field.

Technology Sections offer insights on the many ways in which operations management changes and improves with the development of new technologies for the office and factory.

Quality Sections showcase companies which have embraced the concept of quality to gain a competitive advantage.

Hundreds of real-world examples from manufacturing and service organizations give students practical insight into the operations of contemporary businesses. Application boxes and cases provide additional insight into business operations today.

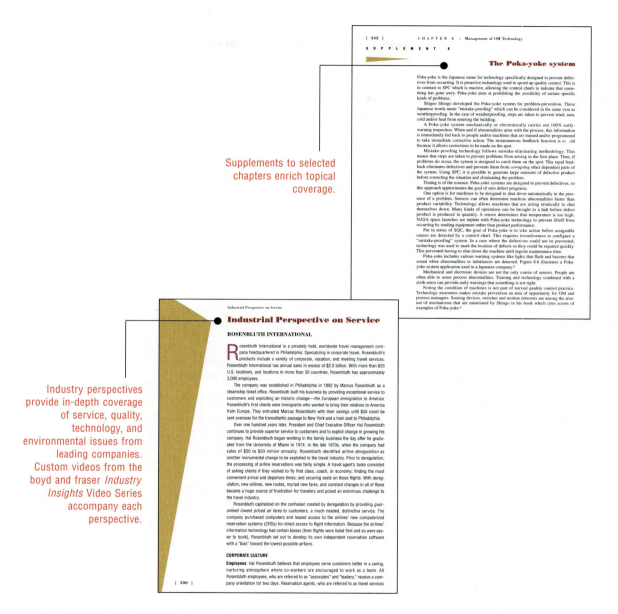

Supplements to selected chapters enrich topical coverage.

Industry perspectives provide in-depth coverage of service, quality, technology, and environmental issues from leading companies. Custom videos from the boyd and fraser *Industry Insights* Video Series accompany each perspective.

INDUSTRY PERSPECTIVE ON QUALITY— *Tom's of Maine*—Explores the quality culture pervading this manufacturer of natural personal care products with a special focus on the process of making toothpaste.

INDUSTRY PERSPECTIVE ON SERVICE— *Rosenbluth International*— An unprecedented look into the inner workings of a high-tech service operation with a focus on the complex automated airline reservation process.

INDUSTRY PERSPECTIVE ON TECHNOLOGY — *AT&T Global Business Communications Systems* — Examines the state-of-the-art "smart" material logistics management system developed at AT&T's Little Rock Repair and Distribution Center.

INDUSTRY PERSPECTIVE ON THE ENVIRONMENT — *Saturn Corporation* — A study of Saturn's environmental initiatives for the automobile building and dismantling process.

PACKAGE

The Annotated Instructor's Edition, with annotations written by Narayan Raman of the University of Illinois, features more than 300 annotations placed throughout the text. These annotations offer teaching tips, additional. examples, suggested readings, and other lecture aids. ISBN 0-87709-481-0

The Instructor's Manual, prepared by L. Wayne Shell of Nicholls State University and Marilyn Helms of the University of Tennessee-Chattanooga, offers chapter outlines and objectives, transparency masters of key text art, notes on using the cases and videos in class, critical thinking questions to accompany the special sections on People, Technology, and Quality, and much more. ISBN 0-87709-473-X

The Solutions Manual, written by Gus Widhelm of the University of Maryland, offers complete solutions to all end-of-chapter problems, with computer printouts for selected problems solved with software. ISBN 0-87709-476-4

A printed **Test Bank,** written by Generao Gutierrez of the University of Texas at Austin, offers more than 1800 multiple choice, true/false, and short answer questions and problems. These high-quality questions, with answers and page references to the text, provide instructors the flexibility to tailor their exams to meet their specific course needs. ISBN 0-87709-474-8

The test bank is also available in computerized form. The **MicroSWAT DOS** version of the test bank, a full-featured testing program, allows instructors to customize exams to meet the objectives of their course. ISBN 0-7895-0705-6

EXPTest software provides the test bank in the user-friendly Windows environment. Easy to learn and to use, this program offers the instructor maximum flexibility in exam preparation. ISBN 0-7895-0707-2

New **NetTest** software allows instructors to give on-line exams in computer labs. The program offers several exciting features including immediate evaluation and grading and the option to present test questions in several different sequences to discourage cheating. ISBN 0-7895-0706-4

boyd & fraser Online—**"Office Hours with Dr. Starr"** provides a hands-on opportunity for professors and students to interact with Dr. Starr. For more information, contact us via the Internet at info@bf.com.

CD-ROM—This multimedia ancillary is the first of its kind for the Operations Management course. The CD contains an innovative multimedia lecture presentation outline utilizing Microsoft Power Point™; digitized versions of graphics, photos, and artwork from the text; key terms and definitions; additional video to complement the four FOCUS cases; and CNBC business network clips that tie into text topics and concepts. ISBN 0-87709-482-9

Videos—Custom videos from the boyd and fraser video series accompany each of the cases in the text, providing in-depth coverage of leading corporations. ISBN 0-87709-477-2

The **Study Guide,** developed by Vincent Callluzzo of Iona College, features solved problems to help students understand models presented in class. Chapter outlines, objectives, and overviews help students comprehend the key concepts of the chapters. Multiple choice and true/false questions and problems allow students to quiz themselves in preparation for exams. ISBN 0-87709-472-1

Packaging Options

Software

LINGO is a mathematical modeling language that provides an environment in which you can develop, run, and modify mathematical models in the areas of distribution, marketing, and finance. Contains a direct solver, a simultaneous linear solver/optimizer, and a simultaneous nonlinear solver/optimizer. Retains your comments when you save a model. Easy to understand programming language. ISBN-0-7895-0550-9

LINDO Systems Suite, which offers student versions of Whats*Best!*, LINDO, and LINGO at a very competitive price, provides maximum flexibility for problem solving.

XCELL+ Factory Modeling System, Release 4.0, Educational Version, Third Edition is a menu-driven, graphics-oriented software package that makes it easy to model manufacturing processes without the need to master a specialized programming language. In this release, the power of the educational version has been greatly increased and many features have been added to make XCELL+ easy to learn and use. ISBN 0-89426-183-5

Other packing options include the full range of titles offered by boyd & fraser. Recommended titles include...

Creative Problem Solving and Opportunity Solving by J. Daniel Couger, ISBN 0-87709-752-6

Total Quality Management: A Survey of its Important Aspects by C. Carl Pegels, ISBN 0-87709-274-5

Cases in Total Quality Management by Jay Heizer and Jay Nathan, ISBN 0-87709-274-5

Productivity in the Office and the Factory by Paul Gray and Jaak Jurison, ISBN 0-87709-751-8

Selecting a Manufacturing Enterprise Software Package by Nathan Hollander, ISBN 0-7895-0332-8

LINDO: An Optimization Modeling System, Text and Software, Fifth Edition by Linus Schrage, ISBN 0-7895-0151-1 (text only), 0-7895-0153-8 (text/PC Disk), 0-7895-0152-x (text/Mac Disk)

Cases in Operations Management: Using XCELL Factory Modeling System, by Thomas, Edwards, and McClain ISBN 0-89426-112-6

Of course, the study guide and solutions manual are also available for bundling. Please see your local International Thomson Publishing representative or call 800-225-3782 for more information. You can also reach boyd & fraser via e-mail at info@bf.com.

Acknowledgments

Incredible teamwork with a group of highly talented professionals characterized every stage of this project. I want to thank DeVilla Williams (Senior Acquisitions Editor) for her total support and organization of the project team. I must also thank DeVilla for her vision, which led her to the decision that the time had come to replace Operations Management books containing worn-out ideas from a by-gone era with a new book having a systems theme and a global perspective.

Thanks also to DeVilla for appointing Jeanne R. Busemeyer of Hyde Park Publishing Services (Project Manager) to write the cases and to complete the people, quality, and technology boxes. This work adds immeasurably to the value of the book. Jeanne dedicated herself to getting interviews with the right people, visiting every case site, and meeting with anyone who could bring the cases to life—which they do because she is a superb writer. Jeanne continuously gave me smart editorial guidance for which I am grateful.

At boyd & fraser, teamwork is the key to its success. Jackie Bedoya (Production Editor) knows how to make her system function. Jackie communicates status and coordinates all of the vital activities associated with the production of this book for which I have great appreciation. Jackie epitomizes the systems-oriented production manager that this text describes as a model to emulate. A special thanks to Sandra Cence, who assisted in the final stages of production.

Deserved thanks also go to Abigail Reip (Photo Research), Laura Ellingson (Development Editor), Jason Morris (Editorial Assistant), Bill Lisowski (Director of Marketing), Tom Walker (President and CEO), and Charles R. Flynn and Nancy Lamm (Copy editing and Proofreading).

Abby selected the photographs in each chapter, as well as started the book off on the right foot. Laura has had the awesome task of organizing and coordinating the supplementary materials, which are considerable. Jason has helped out in the editing process and managed the review process, which has been sizeable and constant throughout the book. Bill has provided the marketing attention that every good product requires. Everyone has done all sorts of other things as needed, which is the hallmark of a great team.

Also, I thank Polly Starr for phenomenal research work that had to be done to make this book state of the art. With the two Starrs the team was nine, enough to hit a home run.

The following people have been helpful in ensuring that the text reflects the most current coverage possible: Becky Burroughs, United Technologies, Pratt & Whitney Waterjet Systems; Heidi H. Cofran, W. L. Gore & Associates, Inc.; Gay Engelberger, Transition Research Corporation (TRC); John T. Flanagan, Department of Management, University of Wollongong, Australia; Professor Robert I. Goldberg, St. Francis College and Executive Director, Center for Packaging Education, Inc.; Prof. Charles H. Goodspeed, Department of Civil Engineering, University of New Hampshire; Jennifer Graham, Saturn Corporation; Peter Hahn, President, Oticon, Inc.; Alexander Houtzeel, President, HMS Software, Inc.; David C. Kirkpatrick, Member of U.S. Technical Assistance Group ISO TC 207; James R. Koelesch, Society of Manufacturing Engineers; Raymond L. Landes, Guardian Glass Corp.; Kathleen Lydon, American Iron and Steel Institute; Nancy J. Mauter, Society of Manufacturing Engineers; Ronald A. McMaster, glasstech, inc.; Glynda McWhorter, Harvard Business School Publishing; G. T. (Tom) Minick, Automotive Replacement Glass Products, PPG; Gary Neilson, Booz-Allen & Hamilton, Inc.; Barry R. O'Connor, President and CEO, AGS Management Systems, Inc.; Greg E. Placy, Highway Design, New Hampshire Department of Transportation; Michael G. Pollack, U. S. AGR Operations, Libbey Owens Ford Co. (LOF); Sidney J. Railson, DuPont Packaging and Industrial Polymers; John J. Resslar, Saturn Corp.; Ann R. Smith, ABB Flexible Automation, Inc.; Gail R. Starr, Steuben; Sara Stephens, American Concrete Pavement Association; Len J. Stolk, Ford Carlite Technology Group; Dr. David A. Summers, High Pressure Waterjet Laboratory, University of Missouri; Roger C. Tollefsen, Executive Director, New York's Seafood Council; Prof. Donald Uhlmann, Department of Materials Science, University of Arizona; Dr. Carl F. R. Weiman, Transition Research Corporation (TRC); Dr. William Wenger, Human Factors Dept., Dreyfuss Associates; Dr. George S. Wilkins, Chrysler Component Operations, Glass Division; Robert Zedik, Chocolate Manufacturers Association.

boyd & fraser would like to thank Warren Zaretsky and Elizabeth Agnese of Zaretsky & Associates, producer of the case videos, whose professionalism, creativity and wit helped us develop unparalleled leading-edge videos and multimedia enhancements for our package. Thanks also to our contacts at the "host" case companies including Matt Chappell and Patty Grennan at Tom's of Maine; Colleen McGuffin and Liz Joseph of Rosenbluth International; Don Stuart at AT&T; and Jennifer Graham at Saturn Corporation.

Many reviewers helped shape the direction of the text and helped ensure consistency and accuracy of the material. Special thanks to:

Antonio Arreola-Risa
Texas A&M University

James Buffington
Indiana State University

Arnold Buss
Naval Postgraduate School

Tim Butler
Wayne State University

John Buzacott
York University

Vincent Calluzzo
Iona College

Sharad Chitgopekar
Illinois State University

Lewis Coopersmith
Rider College

Dinesh Dave
Appalachian State University

Nancy Delaney
Suffolk University

Richard Discenza
University of Colorado - Colorado Springs

Peter Duchessi
S.U.N.Y. - Albany

Ellen Dumond
California State - Fullerton

Hank Giclas
University of Denver

Ed Gillenwater
University of Mississippi

Peter Haug
Western Washington University

Janelle Heineke
Boston University

Marilyn Helms
University of Tennessee - Chattanooga

David Ho
Oklahoma State University

Mark Ippolito
Indiana University

Marilyn Jones
Winthrop University

Christos Koulamas
Florida International University

Ching-Chung Kuo
Pennsylvania State - Harrisburg

Jooh Lee
Rowan College of New Jersey

Terry Nels Lee
Brigham Young University

Lance Matheson
Virginia Tech

Jerrold May
University of Pittsburgh

Taeho Park
San Jose State University

Richard Peschke
Moorhead State University

Michael Peters
Middle Tennessee State University

Narayan Raman
University of Illinois

Tracy Rishel
Susquehanna University

Dan Rinks
Louisiana State University

Andrew Ruppel
University of Virginia

Gary Salegna
Illinois State University

Bob Schlesinger
San Diego State University

Peter Stonebraker
Northeastern Illinois University

William Strassen
California State Polytechnic University

Sam Taylor
University of Wyoming

Jay M. Teets
Virginia Polytechnic Institute

Timothy Vaughn
Northern Illinois University

James Vigen
California State - Bakersfield

Reino Warren
Eastern Michigan University

Gus Widhelm
University of Maryland - College Park

Fredrick Williams
University of North Texas

Steve Yourstone
University of New Mexico

We wish to thank the following professors for their valuable contribution and assistance with the Marketing Research Survey for *Operations Management: A Systems Approach.*

Dennis K. Agboh
Morgan State University

Ajay K. Aggarwal
Millsaps College

Sal R. Agnihothri
State University of New York at Binghamton

Yasemin Aksoy
University of Florida

Edward Alban
Savannah State College

F. David Alexander
Angelo State University

Benigno Alicea
University of Puerto Rico Humacao University College

Stephen Allen
Northeast Missouri State University

David Allen
Northern Michigan University

Jerry Allison
University of Central Oklahoma

Pesi Amaria
Salem State College

Henry Amato
University of Nevada, Reno

M. Ruhul Amin
Bloomsburg University of Pennsylvania

Gordon Amsler
Troy State University

Wendell L. Anderson
St. Mary's University

Professor Mehar
University of Wisconsin-Stout

Antonio Arreola-Risa
Texas A&M University

Ronald G. Askin
University of Arizona

J. Brian Atwater
University of Colorado at Colorado Springs

Turgit Aykin
Rutgers The State University of New Jersey Newark Campus

Sunil Babbar
Kansas State University

Professor Babu
Seton Hall University

Nagraj Balakrishnan
Tulane University

K.N. Balasubramanian
California State Polytechnic University-San Luis Obispo

Raymond Balcerzak
Ferris State University

Ed Ballantyne
Northwest Missouri State University

Michael Ballot
University of the Pacific

Avijit Banerjee
Drexel University

Warren Barce
Montana State University-Northern

Jonathan F. Bard
University of Texas-Austin

Frank C. Barnes
University of North Carolina at Charlotte

Richard S. Barr
Southern Methodist University

Dr. Richard Barrett
Elmira College

Lawrence T. Bauer
Dowling College

Ali R. Behnezhad
California State University-Northridge

Saifallah Benjaafar
University of Minnesota

Harold P. Benson
University of Florida

Stephen Berry
University of South Carolina-Spartanburg

Joseph R. Biggs
California State Polytechnic University-San Luis Obispo

Charles Bimmerle
University of North Texas

D. Bischak
University of Alaska Fairbanks

William Blackerby
Siena Heights College

Paul M. Bobrowski
Syracuse University

Francis D. Booth
University of Missouri-Kansas City

Elizabeth Booth
Louisiana State University

Professor Boyd
University of Georgia

S. Bradley
Roosevelt University

Francis J. Brewerton
University of Texas-Pan American

Professor Brodzinski
Xavier University

James E. Brooks
York Technical College

Stan Bullington
Mississippi State University

John Burnham
Tennessee Technological University

John K. Burns
Rowan-Cabarrus Community College

Timothy H. Burwell
Appalachian State University

Arnold H. Buss
Washington University

Robert S. Bussom
North Carolina Wesleyan College

Darlene P. Butler
University of Arkansas

Dr. David T. Cadden
Quinnipiac College

Alec Calamidas
City University of New York Bernard M.

Baruch College

James F. Campbell
University of Missouri-Saint Louis

Professor Canel
University of North Carolina at Wilmington

Anthony Casey
University of Dayton

Yih-Long Chang
Georgia Institute of Technology

Sohail S. Chaudhry
Villanova University

Ronald G. Cheek
University of New Orleans

Wun-Hwa Chen
Washington State University

Injazz Chen
Cleveland State University

Bintong Chen
Washington State University

Milton M. Chen
San Diego State University

Ken Cherney
Wisconsin Lutheran College

Dilip Chhajed
University of Illinois at Urbana-Champaign

Wen-Cuyuan Chiang
University of Tulsa

Ying Chien
University of Scranton

Dr. Louis Chin
Bentley College

Shared Chitgopekar
Illinois State University

M. Chowdhury
Morgan State University

Orinda B. Christopher
James Madison University

Hsing-Wei Chu
Lamar University-Beaumont

Chao-Asien Chu
Iowa State University

Gary Cleveland
The University of Montana

Jill M. Clough
University of Wisconsin - Platteville

Jose F. Colon
University of Puerto Rico Mayaguez

Carole A. Congram
Bentley College

Robert Conti
Bryant College

Professor William Corney
University of Nevada- Las Vegas

William J. Cosgrove
California State Polytechnic University-Pomona

Leon D. Cox
New Mexico State University

Richard Crandall
Appalachian State University

Professor Culbertson
University of Wisconsin - Milwaukee

Prof. D'Itri
Pennsylvania State University

Dr. Kelwyn D'Souza
Hampton University

Maqboot Dada
Purdue University

Madison M. Daily
University of Missouri-Rolla

David D. Dannenbring
City University of New York Bernard M. Baruch College

Anthony Dastoli
Youngstown State University

Pierre A. David
Baldwin-Wallace College

R. Stephen Davis
Northwestern College

Samuel G. Davis
Pennsylvania State University

Elvis E. Deal
University of Houston

Dr. Richard H. Deane
Georgia State University

Frank P. DeCaro
Georgian Court College

Robert W. Dempsey
University of Missouri-Saint Louis

Juan Diaz
Saint Cloud State University

Clayton R. Diez
University of North Dakota

Karen Donohue
University of Pennsylvania

Shad Dowlatshahi
University of Texas at El Paso

Barbara Downey
University of Missouri-Columbia

Frank L. DuBois
The American University

Paul F. DuMont
Walsh University

Dr. Steve Dunn
Idaho State University

Maling Ebrahimpour
University of Rhode Island

Donald R. Edwards
Baylor University

Mostafa El Agizy
California State Polytechnic University-Pomona

Mohammad K. El-Najdawi
Villanova University

Professor Abdel El-Shaieb
San Jose State University

Salah E. Elmaghraby
North Carolina State University

Robert Emerson
State University of New York at Binghamton

Michael S. Epelman
Bentley College

Lawrence P. Ettkin
University of Tennessee at Chattanooga

Carleton S. Everett
Des Moines Area Community College

Randall Ewing
Ohio Northern University

Walter Fabrycky
Virginia Polytechnic Institute & State University

Kamvar Farahbod
Cal State U San Bernardino

Kasra Ferdows
Georgetown University

Bruce G. Ferrin
Arizona State University

D. Filev
Iona College

Paul R. Finch
New Mexico State University

Ross Fink
Bradley University

Warren W. Fisher
Stephen F. Austin State University

Kathy Fitzpatrick
Appalachian State University

James Fitzsimmons
University of Texas at Austin

Benito Flores
Texas A&M University

A. Dale Flowers
Case Western Reserve University

Barbara B. Flynn
Iowa State University

Suan Tong Foo
Marquette University

Thomas Foster
Boise State University

Oscar S. Fowler
University of Tennessee, Knoxville

Nelson M. Fraiman
Columbia University

James Fremder
Lima Technical College

Dr. Terri L. Friel
Eastern Kentucky University

Phillip Fry
Boise State University

Lissa Galbraith
Florida Agricultural and Mechancial University

Li-Lian Gao
Hofstra University

Stanley C. Gardiner
Auburn University

Leslie Gardner
University of Indianapolis

Vidyaranya Gargeya
University of North Carolina at Greensboro

William C. Giauque
Brigham Young University

Manton C. Gibbs
Indiana University of Pennsylvania

Kingsley Gnanendran
University of Scranton

Merbil Gonzalez
University of Puerto Rico Mayaguez

Juan J. Gonzalez
University of Texas at San Antonio

Franklin E. Grange
University of Colorado at Denver

Ron Green
East Tennessee State University

Barry Griffin
Fayetteville State University

James Gross
University of Wisconsin-Osh Kosh

Faruk Guder
Loyola University

Alfred L. Guiffrida
Canisius College

Thomas R. Gulledge
George Mason University

S.K. Gupta
Florida International University

Jatinder N.D. Gupta
Ball State University

Chan K. Hahn
Bowling Green State University

Professor Hamadeh
Cleveland State University

William C. Hamilton
Columbus College

Professor Hancock
University of Wisconsin-Whitewater

Ray Hansen
University of Wisconsin-Stout

Roland S. Hanson
Georgia Southern University

Craig C. Harms
University of North Florida

Dr. Michael D. Harper
University of Colorado

Dr. William L. Harris
Loyola College in Maryland

William Hartley
State University of New York College at Fredonia

Ronald B. Heady
University of Southwestern Lousiana

Dave W. Hedrick
Wake Technical Community College

Dr. George Heinrich
Wichita State University

F. Theodore Helmer
Northern Arizona University

Dr. Marilyn M. Helms
University of Tennessee at Chattanooga

Ramadan Hemaida
University of Southern Indiana

Sunderesh Heragu
Rensselaer Polytechnic Institute

Armando Heredia
Bryant College

Andy Herrity
Azusa Pacific University

Marek Hessel
Fordham University

Jeffrey E. Heyl
DePaul University

Leslie Hiraoka
Kean College of New Jersey

Johnny C. Ho
Northeast Missouri State University

John Hooker
Carnegie Mellon University

Norm Hope
Tabor College

Stephen Horner
Bethany College

Hassan Hosseini
American Graduate School of International Management-Thunderbird Campus

Mahmoud Hosseini
Grambling State University

Daniel G. Hotard
Southeastern Louisiana University

Edward M. Hufft
Indiana University Southeast

Neil J. Humphreys
Longwood College

Tim C. Ireland
Oklahoma State University

R. Issa
Western Connecticut State University

Gene Iverson
University of South Dakota

Mahyar Izadi
Eastern Illinois University

Lewis W. Jacobs
University of Wyoming

F. Robert Jacobs
Indiana University at Bloomington

Russell E. Jacobson
University of Wisconsin - Whitewater

Prafulla N. Joglekar
LaSalle University

T. Johns
Clarion University of Pennsylvania

David Johnson
St. Paulís College

George Johnson
Sonoma State University

George Johnson
Rochester Institute of Technology

Mary Jones
Mississippi State University

John Kachurick
College of Misericordia

Bharat K. Kaku
University of Maryland

Ten Kao
Kean College of New Jersey

Corinne Karuppan
Southwest Missouri State University

Rammohan R. Kasuganti
Youngstown State University

Peter Kelle
Lousiana State University

Deborah L. Kellog
University of Colorado at Denver

Walter F. Kelly
Wake Technical Community College

Bruce W. Ketler
Grove City College

Dr. Basheer Khumawala
University of Houston

DaeSoo Kim
Marquette University

Michael W. Kim
University of California-Riverside

Barry E. King
Butler University

Michael Klimesh
Upper Iowa University

David Knopp
Eureka College

Jannette G. Knowles
Long Island University-C. W. Post Campus

Bulent Kobu
University of Massachusetts-Dartmouth

Gary A. Kochenberger
University of Colorado at Denver

Christos P. Koulamas
Florida International University

Albert Kowalewski
Point Park College

T. Kratzer
Roosevelt University

Dennis Kroll
Bradley University

Ching-Chung Kuo
Pennsylvania State University at Harrisburg

Thomas Lacksonen
Ohio University

Manuel Laguna
University of Colorado at Boulder

Hamo Lalehzarian
Califprnia State University-Fresno

Charles Lamberton
South Dakota State University

Harold P.Langford
University Southwestern Lousisiana

Nikiforos T. Laopodis
Wesley College

Ronald Lau
University of South Dakota

Hon-Shiang Lau
Oklahoma State University

Thomas Lavender
Catawba Valley Community College

Louis LeBlanc
University of Arkansas at Little Rock

Jooh Lee
Rowan College of New Jersey

Choong Y. Lee
Pittsburg State University

Patrick S. Lee
La Salle University

JP Leschke
University of Virginia

William Lesso
University of Texas at Austin

Dr. Joshua B. Levy
University of Massachusetts-Lowell

David A. Lewis
University of Massachusetts-Lowell

Dr. Ed. Lewis
Belmont University

Cheng Li
California State University-Los Angeles

Dr. A. Lightner
D'Youville College

Binshan Lin
Louisiana State University in Shreveport

Wayne C. Linhardt
Lincoln University

Theodore J. Lloyd
Eastern Kentucky University

Henry T. Loehr
Belmont Abbey College

Vahid Lofti
University of Michigan-Flint

Dean Long
University Wisconsin Stout

Thomas Lucy-Bouler
Auburn University at Montgomery

Raymond P. Lutz
University of Texas at Dallas

Kenneth MacLeod
East Carolina University

Christian Madu
Pace University of New York

Timothy M. Magee
University of Colorado at Boulder

James Mahoney
Central Wesleyan College

Manoj K. Malhotra
University of South Carolina-Columbia

Tomislav Mandakovic
Florida International University

Dinesh Manocha
Slippery Rock University of Pennsylvania

John E. Mansuy
Wheeling Jesuit College

Jo-Mae Maris
Northern Arizona University

Jack Martin
Pennsylvania State University at Eric-Behrend College

Ann E. Marucheck
University of North Carolina at Chapel Hill

Lance Matheson
Virginia Polytechnic Institute & State University

William L. Maxwell
Cornell University

Kristina E. Mazurak
Albertson College

Cynthia S. McCahon
Kansas State University

Donald McCarty
University of Pittsburgh Johnstown

Dr. Michael B. McCormick
Jacksonville State University

Prof. McGaughey
Arkansas Tech University

John McGill
Trenton State College

Timothy J. McGrath
State University of New York College at Plattsburgh

Craig McLanahan
Salem State College

Bruce McLaren
Indiana State University

Curtis P. McLaughlin
University of North Carolina at Chapel Hill

Daniel E. McNamara
University of Saint Thomas

Dr. Howard R. Mead, Jr.
Virginia Commonwealth University

Eugene Melan
Marist College

John Meredith
University of Mary Hardin Baylor

Bradley C. Meyer
Drake University

Diliop Mirchandani
Rowan College of New Jersey

Holmes Miller
Muhlenberg College

David M. Miller
The University of Alabama

Edward S. Mills
Kendall College

Professor Minaie
The Ohio State University

J.R. Minifie
Southwest Texas State University

Kossuth M. Mitchell
Alice Lloyd College

George Mitchell
Southwest State University

Hamid Mohammadi
Saint Xavier University

Professor Mohan
City University of New York Bernard M. Baruch College

James Monahan
University of Massachusetts-Lowell

Laura E. Moody
Mercer University

Thomas P. Moore
Naval Postgraduate School

Graham K. Morbey
University of Massachusetts

Thomas Morton
Carnegie-Mellon University

Joseph Mosca
Monmouth College

Ronnie Moss
Northwest Missouri State University

Jaideep C. Motwani
Grand Valley State University

Robert L. Mullen
Southern Connecticut State University

John P. Mullen
New Mexico State University

Frederic Murphy
Temple University

K.G. Murty
University of Michigan

Bijayananda Naik
University of South Dakota

Behnam Nakhai
Millersville University of Pennsylvania

Paul Nelson
Michigan Technological University

Carl W. Nelson
Northeastern University

Roy L. Nersesian
Monmouth College

Richard Newman
Rockhurst College

Aldo R. Neyman
Tri-State University

Shimon Y. Nof
Purdue University

Dr. Mark Nowak
California University of Pennsylvania

Henry Nuttle
North Carolina State University

Robert L. Nydick
Villanova University

Luis G. Occeua
University of Missouri-Columbia

Felix Offodile
Kent State University

Dolun Oksoy
Alfred University

Alan Oppenheim
Montclair State College

Richard J. Penlesky
Bowling Green State University

Taeho Park
San Jose State University

R. Gary Parker
Georgia Institute of Technology

Faramarz Parsa
West Georgia College

Jayprakash G. Patankar
The University of Akron

B. Eddy Patuwo
Kent State University

James W. Pearce
Western Carolina University

David Pentico
Duquesne University

Charles Perkins
Missouri Western State College

Richard Perle
Loyola Marymount University

Myron I. Peskin
Iona College

Charles Peterson
Elon College

Patrick R. Phillipoom
University of South Carolina-Columbia

William E. Pinney
University Texas at Arlington

Selwyn Piramuthu
University of Florida

Paul Pittman
Indiana University Southeast

John Platt
Lima Technical College

Gerhard Plenert
Brigham Young University

Michael F. Pohlen
University of Delaware

Joseph D. Pope
University of Texas at El Paso

Willard Price
University of the Pacific

Zbigniew H. Przasnyski
Loyola Marymount University

David Pyke
Dartmouth College

Fred Raafat
San Diego State University

Srikant Raghavan
Lawrence Technological University

M. Shakil-Ur Rahman
Frostburg State University

Jayant Rajgopal
University of Pittsburgh

Narayan Raman
University of Illinois at Urbana-Champaign

Ranga V. Ramasesh
Texas Christian University

Mohan P. Rao
Salisbury State University

Ashok Rao
Babson College

Amitabh S. Raturi
University of Cincinnati

Emil Rau
Hofstra University

Russell Reddoch
University of Arkansas at Little Rock

John Reed
Clarion University

Richard A. Reid
University of New Mexico

Dan Reid
University of New Hampshire

James W. Rice
University of Wisconsin-Oshkosh

Dr. Anthony Rizzi
Central Missouri State University

P. Robinson
Texas A & M University

David Ronen
University of Missouri-Saint Louis

L. Drew Rosen
University of North Carolina at Wilmington

Edward Rosenthal
Temple University

Leonard E. Ross
California State Polytechnic University-Pomona

Robin O. Roundy
Cornell University

B. Roychoudhury
Mankato State University

Paul Rubin
Michigan State University

Narendra K. Rustagi
Howard University

Carl R. Ruthstrom
University of Houston-Downtown

Masud Salimian
Morgan State University

Robert Saltzman
San Francisco State University

James M. Salvate
California State Polytechnic University-Pomona

J. Sandvig
Indiana University

John P. Schenck
Northern State University

Charles P. Schmidt
University of Alabama

Carl Schultz
University of New Mexico

Professor Schvaneveldt
Weber State University

Professor Miquel
University of Puerto Rico

Jose A. Sepulveda
University of Central Florida

Edward Sewell
Washington University

John F. Seydel
University of Mississippi

Robert Seyfarth
Lock Haven University of Pennsylvania

Vijay Shah
West Virginia University at Parkersburg

M. Kim Sharp
Iowa State University

Lloyd Shell
Nicholls State University

A. Kimbrough Sherman
Loyola College in Maryland

William B. Sherrard
San Diego State University

Charles Sherwood
California State University-Fresno

Chwen Sheu
Kansas State University

Dr. Margaret F. Shipley
University of Houston-Downtown

Michael Shurden
Lander University

Dr. Sue P. Siferd
Arizona State University

John Simon
University of Notre Dame

Michael C. Simone
Delaware Valley College

Diptendu Sinha
University of Notre Dame

Donald G. Sluti
University of Nebraska at Kearney

Tim Smunt
University of Illinois at Urbana-Champaign

Ramesh G. Soni
Indiana University of Pennsylvania

Victor E. Sower
Sam Houston State University

Charles R. Sox
Auburn University

Michael E. Spangler
Texas A&M University

Stuart Spero
Nebraska Wesleyan University

Barry Spraggins
University of Nevada, Reno

Daniel C. Steele
University of South Carolina-Columbia

John Steelquist
Chaminade University of Honolulu

Herman Stein
Bellarmine College

Harold Steudel
University of Wisconsin-Madison

Scott P. Stevens
James Madison University

William Strassen
California State Polytechnic University-
Pomona

M. Sullivan
University of Wisconsin-Stevens Point

Donna C. S. Summers
University of Dayton

Minghe Sun
University of Texas at San Antonio

R. Meenakshi Sundaram
Tennessee Technological University

Paul Sweeney
University of Michigan-Ann Arbor

Kambiz Tabibzadeh
Eastern Kentucky University

Suresh Tadisina
Southern Illinois University at Carbondale

Rajesh Tahiliani
University of Texas at El Paso

Shahram Taj
University of Detroit Mercy

William J. Tallon
Northern Illinois University

Assad Tavakoli
Fayetteville State University

Richard W. Taylor
The University of Akron

Ann Theis
Adrian College

Robin Thomas
Montana State University

John Thompson
California University of Pennsylvania

Kathleen A. Tini
Delaware Technological and Community
College

Frank Tomassi
Johnson & Wales University

Milton Torres
Florida International University

Mark Treleven
John Carroll University

RJ Trent
Lehigh University

Marvin Troutt
Southern Illinois University at Carbondale

William H. Turnquist
Central Washington University

M. Tzur
University of Pennsylvania

Benny Udemgba
Alcorn State University

Onur Ulgen
University Michigan-Dearborn

Timothy L. Urban
University of Tulsa

John Usher
Mississippi State University

Evert Vander Heide
Calvin College

Emre Veral
City University of New York
Bernard M. Baruch College

Jennifer Videtto
University of Texas at Tyler

Dr. Michael Vineyard
Memphis State University

Marwan Wafa
University of Southern Indiana

B. Wagner
Michigan State University

Dr. Avinash Waikar
Southeastern Louisiana University

Burley Walker
University of Texas at Arlington

B.J. Wall
Middle Tenn State University

Chaing Wang
California State University-Sacramento

Peter T. Ward
The Ohio State University

John B. Washbush
University Wisconsin-Whitewater

Edward A. Wasil
The American University Washington

Kerr F. Watson
Virginia Intermont College

M. Way
Indiana University at Bloomington

Lawrence R. Weatherford
University of Wyoming

Scott Webster
University of Wisconsin-Madison

Professor Jerry Wei
University of Notre Dame

Marek Wermus
Old Dominion University

David West
Bryant College

Terry Wharton
Oakland University

Larry R. White
Aurora University

Gregory White
Southern Illinois University Carbondale

Edna M. White
Florida Atlantic University

Michael J. White
University of Southern Florida

C. David Wieters
New Mexico State University

John R. Willems
Eastern Illinois University

R. Williams
Abilene Christian University

Dr. T.H. Willis
Louisiana Tech. University

James Wilmer
Northern Kentucky University

George Wilson
Lehigh University

Richard Wilson
Fort Valley State College

Bruce Winston
Regent University

Joel D. Wisner
University of Nevada-Las Vegas

Richard Withycombe
The University of Montana

Keith L. Witwer
Minot State University

Craig Wood
University of New Hampshire

Jimmie D. Woods
Hartford Graduate Center

Haw-Jan Wu
Whittier College

Nesa Wu
Eastern Michigan University

Jiaqin Yang
University of North Dakota

Ira Yermish
St. Josephís University

Jon Young
Thomas College

John Zaner
University of Southern Maine

Eitan Zemel
Northwestern University

Xiande Zhao
Hampton University

Zhiwei Zhu
University Southwestern Lousiana

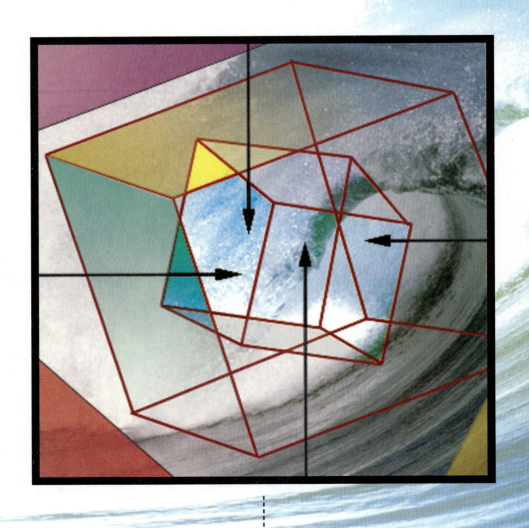

THE OVERVIEW AND INTRODUCTION TO OPERATIONS MANAGEMENT

A Systems Approach

Chapter 1 defines operations management (OM) and explains how this management field is applicable to both manufacturing and services. It also elaborates on the advantages of the systems perspective which links operations management to all other managerial functions in the organization.

The book addresses students of all management functions who need to know why OM is an important organizational partner for the achievement of competitive success. Chapter 1 provides a description of typical operations management positions, so that managers of all functions can understand the various ways in which they can expect to interact with operations managers.

For students who would like to know more about OM as a potential concentration, Chapter 1 provides career path information and career choice guidance. OM planning and decision making offer exciting competitive leverage for those organizations that know how to use it.

Chapter 2 immediately illustrates this point. OM is the function responsible for creating and running highly productive systems. The history of operations management has been to continually invent and develop methods that improve productivity.

Chapter 1

Operations Management (OM) and the Systems Approach

After reading this chapter you should be able to...

1. Define and explain operations management and contrast it with production management.
2. Explain the categories of the systems approach, what it means and why it is important to OM.
3. Detail the systems approach that is used by OM.
4. Understand how OM—using the systems approach—increases the competitive effectiveness of the organization.
5. Understand why this text is titled: *Operations Management: A Systems Approach.*
6. Distinguish between the application of P/OM to manufacturing and services.
7. Explain how special OM capabilities provide competitive advantages.
8. Relate information systems to the distinction between production and operations.
9. Explain how the input-output model defines production and operations.
10. Describe the stages of development of companies with respect to OM and P/OM.
11. Discuss positions in OM that exist in the organization and career success in OM as a function of process types.
12. Explain the effects of globalization on OM careers.

Chapter Outline

1.0 EXPLAINING OPERATIONS MANAGEMENT
Use of Models by OM
Working Definitions of Production and Operations
Defining Operations
Defining Operations Management (OM)
Manufacturing Applications of OM
Service Applications of OM

1.1 OPERATIONS MANAGEMENT AND THE SYSTEMS APPROACH
Using the Systematic—Constructive Approach
Why Is the Systems Approach Required?
Defining the System
Structure of the Systems Approach

1.2 EXAMPLES OF THE SYSTEMS APPROACH

1.3 DIFFERENTIATING BETWEEN GOODS AND SERVICES

1.4 CONTRASTING PRODUCTION MANAGEMENT AND OPERATIONS
MANAGEMENT
Information Systems Are Essential for Manufacturing and Services

1.5 THE BASIC OM TRANSFORMATION MODEL: AN INPUT-OUTPUT SYSTEM
Costs and Revenues Associated with the Input-Output Model
Inputs Associated with Variable (or Direct) Costs
Transformations Associated with Fixed (or Indirect) Costs
Outputs Associated with Revenues and Profits

1.6 OM INPUT-OUTPUT PROFIT MODEL
How Has the Profit Model Changed Over Time?

1.7 THE STAGES OF P/OM DEVELOPMENT
Special Capabilities

1.8 ORGANIZATIONAL POSITIONS AND CAREER OPPORTUNITIES IN OM
Career Success and Types of Processes
Operations Management Career Paths
Global Aspects of Career Paths
Manager of Production, Manufacturing (or Service) Operations
Department Supervisor
Inventory Manager or Materials Manager
Director of Quality
Production Manager/Consultant (Internal or External)
Performance Improvement Manager

Summary
Key Terms
Review Questions
Problem Section
Readings and References
Supplement 1: Decision Models

The Systems Viewpoint

This chapter explains operations management (OM). If the part that operations managers play is to be effective, it should be systems-based. Compare the operations manager to the coach of a sports team. What is the coach's job for a baseball, football, basketball or soccer team? It is to guide the team to achieve competitive excellence. The coach knows that to make the team win requires coordinating the contributions of the individual players. Winning takes teamwork and the coach tries to develop that cooperative ability. Teamwork skills require a systems viewpoint.

Assume that the game is business and that players' positions are known as marketing, finance, and operations. The successful coach emphasizes coordination of these functions. Managers of all functional areas need to understand OM and OM managers need to understand areas that interact with their own. To understand global competitors requires understanding the international character of the operations management system. That is why this book directs all OM managers to focus on the use of the systems approach.

1.0

EXPLAINING OPERATIONS MANAGEMENT

Operations management is the work function that makes goods and provides services. Because it provides what others sell, finance and account for, it is an indisputable partner in the business.

This book will familiarize you with the language and abbreviations used by operations managers—such as writing OM for operations management. Important OM terms are explained in the text and also presented with their definitions in the page margins.

The OM language describes *methods, tools, procedures, goals* and *concepts* that relate to the *management of people, materials, energy, information,* and *technology.* Operations managers learn how to study a process by observing it and mapping its flow; from that platform, its performance can be improved. OM allows the state of a production process to be assessed.

Use of Models by OM

The Greyhound bus driver is an operations manager assessing highway driving conditions. The driver knows how rain slows velocity (v) which cuts down miles that can be driven per day (m). The manager in charge of operating the fleet of buses could describe this relationship as follows: $m = vt$ where m is the driver's output in miles driven per 8-hour day; v is the velocity, measured in miles per hour and $t = 8$ hours.

This method of quantitative description is often used by OM to build a model— *a representation of the real situation.* The model permits OM to test the effect of different t's and v's. A general quantitative model that describes output is $O = pt$

where O is output per day. O changes as a function of the production rate per hour (p) and the length of time worked (t). OM develops models to describe productivity (p) as a function of scheduling, training, technology and capacity.

There are various OM models used to make equipment selection, workforce and production scheduling, quality control, inventory, distribution, plant location, capacity, maintenance and transportation decisions, among others. Decision models organize the elements of a problem into actions that can be taken, forecasts of things that can happen which will affect the results, and thereby, the relative likelihood of the various outcomes occurring. Thus, decision models organize all of the vital elements in a systematic way.

Working Definitions of Production and Operations

The generic or collective definition of **operations** emphasizes *rational design, careful control and the systematic approach* that characterizes the methodology of OM.

Operations
Purposeful actions methodically conducted as a work plan to achieve practical ends.

Defining Operations

Operations are purposeful actions (or activities) methodically done as part of a plan of work by a process that is designed to achieve practical ends.

Defining Operations Management (OM)

Operations management is the systematic planning and control of operations. This definition implies that management is needed to insure actions are *purposeful—* designed to achieve practical ends so goals and targets are required.

Operations management
Systematic planning and control of operations; a methodical, purposeful process.

Operations management (OM) makes sure the work is done *methodically—* characterized by method and order. The fact that a *process* is used suggests the presence of management to install a procedure for working systematically.

Operations used by manufacturing include assembly phases such as fitting components and joining pieces together.
Source: The Document Company, Xerox

Operations management is responsible for a *plan of work*—a thoughtful progression from one step to another. Plans require details for accomplishing work. *Practical ends* are not realized without operations management which is able to provide public services, gain market shares, or make profits.

Operations management uses *methodology which consists of procedures, rules of thumb, and algorithms for analyzing situations and setting policies*. They apply to many different kinds of service and manufacturing processes. In brief, operations management consists of: *scheduling work, assigning resources including people and equipment, managing inventories, assuring quality standards, process management including capacity decisions, maintenance policies, and equipment selection*. The functions of OM overlap and interact. They are driven by demand from the market where OM manages supply.

Manufacturing Applications of OM

The operations used by manufacturing can be represented by many different verb and object phrases that describe doing things to various materials such as: pressing metal, cutting paper, sewing clothes, sawing wood, drilling metal, sandblasting glass, forming plastics, shaping clay, heat-treating materials, soldering contacts, weaving fabric, blending fuels, filling cans, and extruding wires. Similarly, there are a variety of assembly phrases such as: snapping parts, gluing sheets, fitting components and joining pieces together. This kind of work is done in factories.

E X A M P L E S :

Autos, planes, refrigerators and light bulbs are made in factories. On the other hand, fast food chains like McDonald's and Burger King view the assembly of sandwiches from meat, buns and condiments as a manufacturing application that demands the highest levels of quality for cleanliness and consistency. Also, costs must be kept low enough so that a profit can be made at a price that attracts customers. The price of fast food products, said to be highly elastic, strongly affects the demand volume. The concept of manufacturing sandwiches is a departure from traditional restaurant operations. It provides an OM basis for requiring methodical performance to achieve practical goals.

Service Applications of OM

Service operations in the office environment are quite familiar, i.e., filing documents, typing input for the word processor, and answering the phone. There are similar lists of verbs and objects that apply to jobs done in banks, hospitals and schools; grant a loan, take an x-ray, teach a class, are a few examples. Movies are one of the biggest export products of the U.S.A. Operations management for entertainment and sports provides great opportunities. Administration of the law is a major industry in the U.S. that requires operations management. In recent years, law firms have begun to understand the importance of productivity management and quality improvement.

Jobs in the service industries are not as well paid as jobs in manufacturing. This is because the rational design of service industry operations is well behind that of manufacturing. In the earliest years of manufacturing, many employees were paid small amounts of money. The kinds of operations that each person could do made only minor contributions to the profitability of the enterprise. Successful firms integrated the minor contributions of the many. As manufacturing operations and the technology behind them became better understood and the process was ratio-nalized, fewer people were needed because each could make greater contributions.

The service industries are in the process of rationalizing operations so that indi-vidual contributions will become increasingly valuable in the same way that occurred in manufacturing. Consider the labor-intensive environment of check processing in a bank. It is reminiscent of the labor-intensive character of work that dominated manufacturing from the turn of the twentieth century through the mid-twentieth century, and still exists in various places in the world.

E X A M P L E S :

If one has worked for UPS, RPS, Federal Express, DHL or the post office they will be able to list the various service operations related to delivering mail and packages. If one has worked for the IRS, they will have another set of job descriptions to define the specific operations that characterize tax collection activities of the federal government. Also, if one has worked for The Gap, Eddie Bauer, The Limited, Wal-Mart, The Sharper Image, K-Mart, Caldors, Sears, or other retail operators, they will be able to define processes that are pertinent to merchandise selec-tion and pricing, outsourcing, distribution logistics, display and store retailing in general.

Successful mail-order companies like Lands' End, COMB, Sporty's, L.L. Bean, Victoria's Secret, Norm Thompson, and Spiegel have become some of the best examples of entrepre-neurial firms that have mastered the operational advantages of smart logistics. Distribution, in retail and mail order, is a production process that lends itself to all of the benefits that excel-lent information systems and new technology can bestow. The credit card business combines many aspects of service functions. Master Card, Visa and American Express are totally depen-dent upon smart operations management to provide profit margins.

Service sector businesses like McDonald's Restaurants have manufacturing applications such as assembly processes for making sandwiches.
Source: McDonald's Corporation

1.1

OPERATIONS MANAGEMENT AND THE SYSTEMS APPROACH

There are two approaches that OM can use:

1. the functional field approach
2. the systems approach

Functional field approach

A tactical approach concentrating on specifics to achieve an end goal.

With the **functional field approach,** operations management is explained with minimum reference to other parts of the business—such as marketing and finance. The functional field approach concentrates on the specific tasks that must be done to make the product or deliver the service. It is tactical, not strategic.

A typical organization chart—but with details only for OM—is shown in Figure 1.1. The OM department is headed by a senior vice president of operations who has the plant manager as well as staff heads reporting to him or her. The chart could also detail the marketing, finance, human resources, R & D, and other functions.

FIGURE 1.1

TRADITIONAL ORGANIZATION CHART

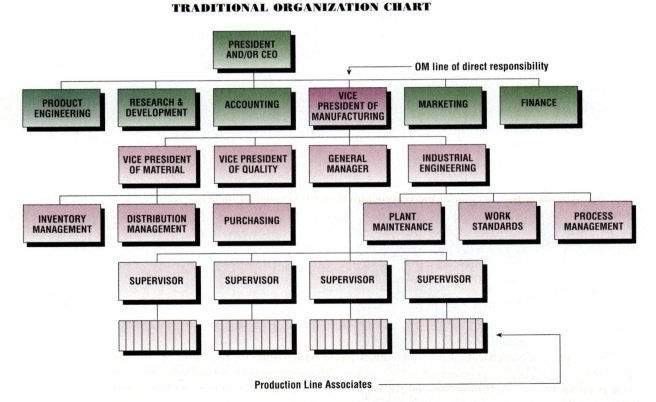

Traditional organization chart shows self-contained functional areas.

There are no lines connecting people in the other functional areas to people in OM. The only connection is at the president's level. Within the OM area, there are a limited number of connections and these are hierarchically structured. The traditional organization chart does not reflect the systems approach wherein anyone can talk to anyone else if they are part of the problem or the solution. Teamwork is difficult to achieve with self-contained functions.

The **systems approach** integrates OM decisions with those of all other business functions. *The challenge is to make the firm perform as a team.* The systems approach entails having all participants cooperate in solving problems that require mutual involvement. It begins with strategic planning and moves to tactical accomplishments.

Heiner Müller-Merbach has recently provided a useful way of describing the systems approach as a combination of concepts.[1] Figure 1.2 depicts the various meanings that he has associated with the systems approach.

Systems approach

Integrates OM decisions with business functions as a team approach to problem resolution.

FIGURE 1.2

TAXONOMY OF THE SYSTEMS APPROACH

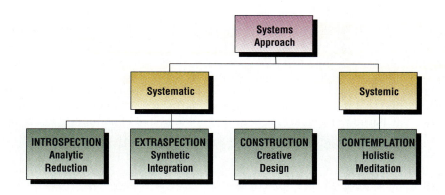

"The systems approach focuses the consideration of wholes and of their relations to their parts. The systems approach is necessarily comprehensive, holistic and interdisciplinary. However, there are several types of systems approaches around, quite different from each other and competing with one another,"[2] Müller-Merbach noted.

Using the Systematic—Constructive Approach

The systematic systems approach is considered the Western tradition whereas the systemic systems approach is characteristic of Eastern philosophies. The systems approach this book uses is systematic and constructive.

The systems approach called *introspection* is based on the analytic reduction of systems into their parts, which is characteristic of the sciences. The systems approach called *extraspection* is characteristic of philosophy and the humanities and strives to integrate objects and ideas into higher-order systems using synthesis. The field of General Systems is closely associated with this effort to develop meta-systems of knowledge.

Say that a desktop computer stops working. Following introspection it is opened up and taken apart. By means of analysis, components are tested to find the cause of the trouble.

Synthesis is required to reassemble the computer. Using extraspection, perhaps a better overall configuration can be found.

Combine analysis and synthesis to obtain the third systems approach, called *construction*. It is "characteristic of the engineering sciences and their creative design of systems for practical purposes."[3] Creative design that uses both analysis and synthesis is the systems approach described in this book.

Why Is the Systems Approach Required?

The systems approach is needed because it produces better solutions than other approaches—including the functional field approach. It leads to better decisions and provides better problem-solving for complex situations, enabling those that use it to be more successful.

Think of the systems approach in terms of the sports team. If the players are coordinated by communication and training, they play a better game. Similarly, in business, those that use the systems approach are the leading competitors in every industry.

Defining the System

Elements that qualify to be part of a system are those that have a direct or indirect impact on the problem, or its solution; on the plan, or the decision. Thus, an **OM system** is everything that affects process planning, capacity decisions, quality standards, inventory levels, and production schedules.

Figure 1.3 is a symbolic picture of a system. The shape encloses all factors that have an effect on the purposes and goals of the system. Anything outside of that system is believed to be independent of the problems and opportunities inside of it.

OM system

Incorporates all factors with an effect on the purposes and goals of the system.

FIGURE 1.3

REPRESENTATION OF A SYSTEM

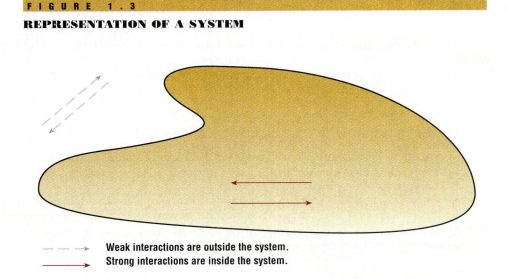

- - - → Weak interactions are outside the system.
——→ Strong interactions are inside the system.

Figure 1.4 shows an organization chart with the systems shape mapped across it to reflect the fact that the problems and opportunities include various people and functions. The problem map cuts across OM as well as other functions.

FIGURE 1.4

SYSTEM MAPPING ACROSS TRADITIONAL ORGANIZATION CHART

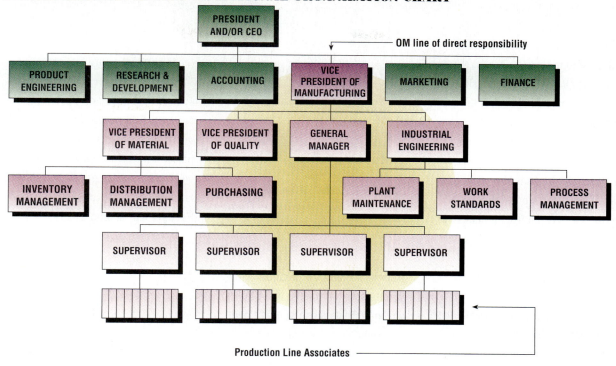

This is a symbolic representation of the fact that some people in departments covering the yellow area are involved with that area. The main responsibility falls on the general manager, plant maintenance, and a supervisor.

The key is to identify all the main players and elements that interact together to create the system in which the real problem resides. Even though the problem solution may be assigned to the operations management team, the resolution requires cooperation of all the organizational participants in the problem.

Structure of the Systems Approach

1. The systems approach requires identification of all the elements related to purposes and goals. The question to be answered is: what accounts for the attainment of the goals? The visual concept depicted in Figure 1.3 is one way to answer the question. Another way is to use a math model which shows what accounts for the performance of the systems and the attainment of its goals. The equation, given below, is read, the goals y_i are a function of all relevant factors x_j and t_j. That is:

$$\{y_i\} = f \{x_1, x_2, \ldots x_j, \ldots t_1, t_2, \ldots t_j\}$$

2. The systems approach requires control of timing. It is to be noted that the equation above includes measures of time (t_j) as an important systems parameter. By recording conditions over time it is possible to measure rates of change—which are also important systems descriptors. Timing is critical to the performance of symphony orchestras, sports teams and business organizations. To achieve synchronization of functions, the systems approach is required.

3. The systems approach requires teamwork. Coordination of all participants is essential. Designing a productive process for making hamburgers, VCRs, delivering packages or servicing cars, requires the cooperation among all members of the system. To operate a process properly, it is necessary for all to have the systems perspective. In particular, the attainment of quality requires a dedicated team effort. A mistake by anyone involved in gaining the perception of quality is like the weak link of a chain which causes the whole system to fail.

International Business Machines Corporation
. . .

IBM offers HelpWare customer service as differentiation from PC clone makers.

In an effort to regain market share in the personal computer (PC) market, IBM has implemented a new customer service program it calls HelpWare. This emphasis on customer service is part of the computer giant's strategy to differentiate itself with superior service and support from clone makers who operate on much lower margins. IBM's worldwide market share for PCs fell from a high of 37 percent in 1984 to a low of 13 percent in 1992.

The HelpWare program is made up of three segments: (1) "HelpLearn," a nationwide education program with centers led by IBM-licensed and certified trainers offering programs on topics from multimedia to IBM's OS/2 operating system; (2) "HelpBuy" finance and leasing programs that offer financing from IBM Credit Corporation, with credit approval granted in 24 hours via an 800 telephone number; and (3) "HelpCenter," a "clearinghouse" for all questions related to IBM PCs and OS/2.

The main purpose of HelpCenter is to "serve as a focal point for those lost in the IBM maze,"

according to a *ComputerWorld* article. All previous customer service programs have been rolled under the HelpWare umbrella. Larry Deaton, IBM's director of personal systems services and customer support, said customers with technical problems will not be turned away. The center will have electronic and telephone ties to IBM laboratories in Boca Raton, FL, and Raleigh, NC, among others.

Sources: Carol Hildebrand, "IBM's HelpWare draws polite applause," and "IBM to offer PC help," *ComputerWorld*, March 23, 1992, Vol. 26., pp. 8 and 12.

1.2

EXAMPLES OF THE SYSTEMS APPROACH

Managing a sports team is an excellent example of a purposeful effort to manage by the systems approach. Another good example is a symphony orchestra. The conductor of the orchestra makes certain that all participants are synchronized. If the violins, woodwinds and brass treat their participation as if they were separate functional fields, bedlam would result. Everyone looks to the conductor to keep the components of the system related and balanced.

The third example is a business. Operations makes the product and/or delivers the service. The systems approach coordinates the business-unit team so that supply and demand can be balanced and the use of resources is optimal.

Finally, to complete a jigsaw puzzle also requires a systems perspective. This is the vision needed to relate the clues of many interdependent elements. Difficult puzzles, with pieces that are cut to look alike, provide little color differentiation and few hints concerning the outlines of objects.

The systems approach requires teamwork. At Xerox, operations manufactures the product and provides repair and maintenance service to the customer.
Source: The Document Company, Xerox

Puzzles become geometrically harder as the number of pieces increases. As a system becomes more complex, fathoming its structure and understanding how it functions becomes more difficult. Operations management problems are similarly composed of complex subsystems which require interfunctional communication to uncover the patterns that relate the subsystems to the whole system.

1.3

DIFFERENTIATING BETWEEN GOODS AND SERVICES

There is less difference than similarity between OM in **manufacturing** and OM in **service** organizations. Manufacturing is the fabrication and assembly of goods, whereas services generate revenues either independently of goods, or to help the user of those goods. Banking, transportation, health care, and entertainment are all services. They change the customer's location, financial condition, and sense of well-being. Increasingly manufacturers recognize the importance of servicing customers and service systems recognize the value of using manufacturing capabilities.

The methodology of OM was first developed by and for manufacturing, but it can be applied to services. Service industries involve an increasing percent of the workforce. Thus, more attention needs to be directed toward achieving coherent and efficient operations for services.

Similarities between services and manufacturing can be noted when service operations are based upon repetitive steps in information processing. Almost identical methods apply with respect to production scheduling, job design and design of the workplace, process configurations, and quality achievement. High volume repetitive operations on physical items (i.e., for fast foods or blood testing) is production whether they belong to manufacturing or services. Similar analogies can be made for lower volumes of production and services delivered.

Manufacturing
Fabrication and assembly of goods.

Service
Generates revenues either independently of goods, or to help the user of those goods.

A significant difference between the provision of services and manufacturing occurs because of inventory. When the toll collector is idle there is no way to build up inventory that can be used when the traffic increases.
Source: HNTB Corporation

The *similarities stop and significant differences occur when the operations involve contact between people.* Person-to-person activities which require transfer of information and/or treatments offered by one to another are difficult to schedule; activity times vary more than with machines. Human-to-human interactions involve many more intangibles than interactions between people and machines. The contact aspect of services requires different methods for analysis and synthesis than are needed for manufacturing systems.

At the same time, care should be taken to avoid stereotyping services as being all too human and, therefore, difficult to control for quality and productivity. It does a disservice to services to consider them quixotic or flawed by humanism, while manufacturing is admired for its elegant, efficient technological component. A most respected thinker has written, "Until we think of service in more positive and encompassing terms, until it is enthusiastically viewed as manufacturing in the field, receptive to the same kind of technological approaches that are used in the factory, the results are likely to be...costly and idiosyncratic."[4]

The point to make is that services that are presently rendered in an inherently inefficient way often can be transformed into rational repeatable activities that emulate the best of manufacturing environments. However, often is not always. Some services are not amenable to the concept of manufacturing in the field. One would be fearful if the services rendered by a doctor were based on a repetitive manufacturing model. At the same time, many aspects of open heart surgery are the better for such systematization. The same can be said for blood testing, taking X-rays, and other repetitious aspects of the health care business. Juxtaposed to this, some products, such as art work, are epitomized by being custom-made.

Another significant difference between the provision of services and manufacturing occurs because of inventory. It is not considered possible, usually, to stock services. For example, when the machine repair person is idle there is no way to build up an inventory of repair hours that can be used when two machines go down at the same time. In most service businesses this is one of the great waste factors.

On the other hand, many companies are using automated systems to provide custom-tailored information such as stock, bond and mutual fund quotes for anyone knowing the symbols. Phone-call requests for product information are answered by a digitized voice that instructs the caller to use a Touch-Tone phone to input the product of interest and his or her fax number. The appropriate fax is automatically transmitted within a minute. This entire service transaction, without human intervention, is becoming increasingly common and epitomizes an automated manufacturing process.

1.4

CONTRASTING PRODUCTION MANAGEMENT AND OPERATIONS MANAGEMENT

What is the difference between *production* management and *operations* management? **Production** is an old and venerable term used by engineers, economists, entrepreneurs and managers to describe physical work both in homes and in factories to produce a material product.

Traditionally, **production management** involves planning and decision making for the manufacture of goods such as airplanes, cars, tires, soup, paper, soap, cereal, shoes, shirts, skirts, sweaters, dentifrice, clothing, furniture, and the parts that go into many of these products. Parts and components manufacture is a large portion of finished goods manufacture. Production managers are responsible for food, clothing and shelter, as well as luxuries.

Operations management is a more recent term associated with services performed by organizations such as banks, insurance companies, fast food servers, and airlines. Government jobs are also in the services. Health care providers, including hospitals and schools, belong in the services category. It is not surprising that there has been a rapid growth of service jobs in the U.S. economy. The present ratio of service jobs to manufacturing jobs is nearly four to one, compared to an approximate one-to-one ratio in the 1950s. As a result, the number of people that are now engaged in operations is far larger (and still growing) than the number of people that work in the production departments of manufacturing firms.

Manufacturers have come to view service to the customer as part of the quality of their products (i.e., repairing defective product as well as providing regularly scheduled maintenance). Auto manufacturers learned a great deal from Acura when Honda launched that division with a service mission that eclipsed anything in auto service that preceded it. One of the best automotive service operations before Acura was Honda itself. Xerox established strict guidelines for the maximum allowable downtimes that would be tolerated for their copying machines. Until the 1990s, IBM provided no service to its small customers. After a serious fall from grace, IBM changed their policy, and became a full-service company to all customers with their HelpWare support run by their Personal Systems HelpCenter.

In this transition decade, preventive maintenance targets generating customer satisfaction instead of additional revenues. Many manufacturers have established toll free 800 telephone numbers permitting their customers to call without charge for information about the products they make, and to register complaints. Plant tours educate customers and convey information about quality efforts with the intention of attracting new customers and building loyalty.

Production
Physical work which produces a material product.

Production management
Planning and decision making for the manufacture of goods.

Operations management
Associated with services performed by organizations; but also with manufactured goods.

Manufacturers like Honda have come to view service to the customer as part of the quality of their products.
Source: Honda of America Mfg., Inc.

Information Systems Are Essential for Manufacturing and Services

The growing recognition of the importance of the service function in manufacturing has broadened the situations to which the term operations is applied. Manufacturers have become more comfortable with the notion that they must cater to the customer's service requirements. **Information systems** provide the necessary data about customer needs so that operations management can supply the required services.

Information systems

Provide customer data for operations management to supply required services.

Both services and manufacturing are increasingly responsive to—and controlled by—information systems. Therefore, knowledge of computers, computer programming, networking and telecommunications is essential in both the manufacturing and service environment.

Schools of business include both goods and services under the term "operations" whereas industrial engineering departments are still inclined to teach "production" courses. Nevertheless, there is inevitable convergence of both to an information-dominated workplace. *Operations is the familiar management term for an information systems environment,* so the word "operations" fits nicely.

Programming and maintenance (both service functions) have become increasingly important to manufacturing. Further, the relevance of service to customers increasingly is viewed as a part of the total package that the manufacturer must deliver. Manufacturing joins such distinguished service industries as transportation, banking, entertainment, education and health care. In that regard, note the following trends for manufacturing:

1. The labor component (the input of blue collar workers) has been decreasing as a percent of the cost of goods at an accelerating rate for over 50 years.
2. The technological component as a percent of the cost of goods has been increasing for many years. In the past 20 years, this effect has become multiplicative, with computers controlling sophisticated and costly equipment across vast distances via satellites and networks.

3. As information systems play a larger part in manufacturing, highly-trained computer programmers (sometimes called gold collar workers) and white collar supervisors add to growing sales and administrative (overhead) costs which have to be partitioned into the cost of goods. These costs are an increasing percent of the cost of goods. Traditional methods for assigning these costs can lead to detrimental OM decisions. New accounting methods, called activity-based costing (ABC), should be used to improve overhead accounting. Operations managers need to discuss these issues with their colleagues from accounting.

4. The systems approach requires communication between functions and the sharing of what used to be (and still are, in many traditional firms) mutually exclusive databases. The databases of marketing and sales, OM, R & D, engineering, and finance are cross-linked when advantageous. That sharing is crucial to enabling the systems approach to work. There are many examples of both manufacturing and service industries where shared databases have been installed and utilized successfully.

5. The technology of the twentieth century is moving rapidly into retirement along with a lot of executives who grew up with its characteristics. It's a new ball game with new players who feel free to deal with the distinction between services and manufacturing as well as between operations and production in their own way.

Practitioners now have one foot in the twentieth century while the second foot is poised to step into the twenty-first. It is a good bet that the taxonomy of the twenty-first century will categorize production as a subheading under operations, and services will be an integral part of manufacturing.

When a discussion applies equally well to both manufacturing and services, it is often referred to as P/OM. As explained above, it is increasingly common to call it OM. In this text, OM will be used to describe both manufacturing and services. P/OM will be used only in situations where the manufacturing component is critical and conclusions are likely to be different from those applicable to service systems.

1.5

THE BASIC OM TRANSFORMATION MODEL: AN INPUT-OUTPUT SYSTEM

All operations management and production systems involve **transformation.** Alteration of materials and components adds value and changes them into goods that customers want to own. The raw materials and components before transformation could not be used—and therefore had no utility—for the customer. Service conversions have customer utility even though no transfer of goods takes place. The conversion is locational or related to the customer's state of well-being.

Transformation

Added-value alteration of materials and components into desired goods.

The manufacturing transformation of raw materials into finished goods is successful if customers are willing to pay more for the goods than it costs to make them. Consider what has to be done to make a product. The raw materials for glass, steel, food, and paper have no utility without technological transformations. New processes are constantly being invented for improving the transformations and the products that can be obtained from them.

The same transformation rules apply to services. The conversion is successful if customers are willing to pay more for the services than it costs to provide them. To illustrate a service transformation, consider an information system in a bank. A check that is issued results in the electronic transfer of funds (ETF) from the paying account to the paid account, clearly an input-output transformation.

Another information transformation is to take raw data and turn it into averages and standard deviations. The latter is characteristic of the operations aspect of market research. As another service example, consider the transformation that is at the heart of the airline business—moving people from one place (input) to another (output) for profit. Other inputs are fuel, food, and the attention of the flight attendants.

Figure 1.5 presents a picture of a generalized input-output transformation model. Generalized means that it is a standard or generic form which could be applied to any system where conversions are taking place. Inputs are fed into the transformation box, which represents the process. The "process" can include many subprocesses. If the process is to make a sandwich, then important subprocesses include cooking the hamburger and toasting the buns.

FIGURE   1 . 5

INPUT-OUTPUT TRANSFORMATION MODEL

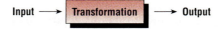

Input-output is the basic OM model.

The inputs are combined by the process, resulting in the production of units of goods or the creation of types of services. The transformed units emerge from the facility (factory, office, etc.) at a given rate. Time needed to carry out the transformation determines the production rate. The transformed inputs emerge as outputs to be sold or used beneficially.

The transformation model depicts work being done. This work involves the use of resources made up of people, materials, energy, and machines to achieve transformations. Figure 1.6 illustrates the expanded version of the input-output transformation model. There are many boxes now within the transformation grid. Each box represents operations that generate the product line, which can be goods or services. Productive OM systems have well designed transformations.

FIGURE   1 . 6

EXPANDED INPUT-OUTPUT OM MODEL

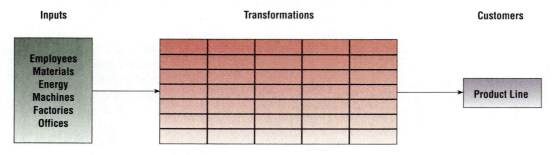

Transformations are being accomplished when people are served chili at Wendy's, or when they give blood to the Red Cross, have their teeth cleaned by the dentist, or visit Disney World to be entertained. A travel agency will have secured the necessary reservations and tickets for the customer's flight to Orlando, Florida, and for the hotel. The travel agent designs the trip and fits it to the customer's specifications concerning dates and costs using reservation and other information systems to complete all necessary transactions. When the desired outputs are fully specified, the transformations can be planned, along with the inputs, and the plan can be carried through to completion. The culmination of the transformation process constitutes the desired output—a visit with Mickey Mouse.

Costs and Revenues Associated with the Input-Output Model

Cost management is a key function associated with all aspects of OM. A major portion of the cost of goods or services originates with operations. Figure 1.7 is meant to illustrate how costs are related to the I/O model. Controlling costs is of prime concern to all managers.

For the most part, costs are readily categorized into variable costs and fixed costs. Generally, costs are considered to be easily measured, although the treatment of overhead costs is subject to debate. Also, a variety of accounting methodologies exist. The differences between them are not trivial because they can impact OM decisions in significantly disparate ways. OM and accounting coexist in the same system, and they are interdependent when the measurement of costs interacts with OM decision making. Quality, another key criteria associated with all aspects of OM, interacts with costs in a variety of ways, as do productivity, timeliness of delivery, and styles and sizes of products and services.

FIGURE 1.7

INPUT-OUTPUT OM MODEL: COST AND REVENUE STRUCTURE

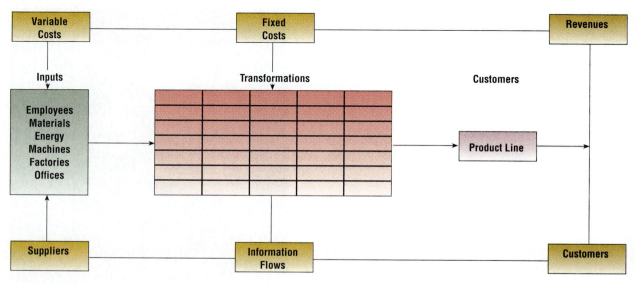

Inputs Associated with Variable (or Direct) Costs

The input components of the transformation model which apply to an airline transportation process include fuel, food, crew pay, and other costs. Variable operating costs increase as there are more flights flown and more people flying. Variable costs are also called direct costs because they can be applied directly, without ambiguity, to each unit that is processed.

The same reasoning applies to a manufacturing example. The variable costs for the inputs include labor, energy, and all of the materials purchased from suppliers and used to make the product. Materials include raw materials, subassemblies, semi-finished materials, and components. The more finished the purchased materials, the less work that has to be done by the purchaser (i.e., less value can be added by the purchaser).

Transformations Associated with Fixed (or Indirect) Costs

When Delta or American Airlines buy aircraft from Airbus Industries or Boeing, the airline increases its already substantial fixed cost investment in planes. Fixed costs are also called indirect costs because they are part of overhead and must be allocated to units of output by some formula.

Often *the charge per year—called the depreciation—is calculated by dividing the cost of the investment by the number of years in the estimated lifetime of the investment.* For example, a 30 million dollar aircraft with a 15-year lifetime would generate two million dollars of depreciation per year. This is called straight-line depreciation because the amount per year does not change. There still remains the question of how to allocate a portion of the two million dollars as an applicable charge for a particular passenger flying from Milan to New York on that aircraft. Determining the appropriate fixed costs to be charged to each job, unit made or passenger mile flown is a joint responsibility of OM and accounting.

An alternative approach for adding capacity without increasing investment is to lease instead of buy aircraft. This shifts the financial burden of buying capacity to the variable input cost mode. The same distinction applies when a manufacturer

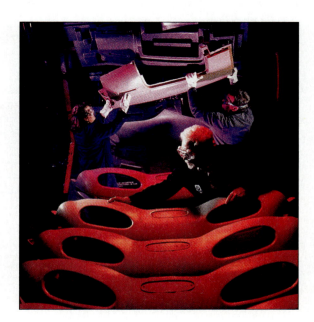

Plastic components purchased from outside suppliers by automobile manufacturers constitute variable costs for inputs in manufacturing. Source: The Dow Chemical Company

analyzes the net long-term economic value of renting a factory building instead of either building or buying a facility. Leasing a facility shifts the fixed costs of building acquisition to the variable costs associated with inputs.

Delta and American also have investments in maintenance facilities, airport terminals, training and education systems, as well as in their workers and management. The payments that airlines make to support the operations of airports generally are fixed and not variable costs. Airports—like factories—are major fixed cost facilities; treat them as fixed costs because the same expenses must be met no matter how many flights depart or arrive there. However, if part of the airport charges are based on the number of flights an airline makes, then both fixed and variable (input) costs must be considered. As for investments in training and management, companies use consultants as a way to keep down their fixed costs. For similar reasons, they often lease transportation.

Outputs Associated with Revenues and Profits

Passengers pay the airline for transportation. The number of passengers (units) that are transported (processed) by the airline is the output (sometimes called "throughput" to emphasize the output rate) of the system. It usually is measured by passenger miles flown system wide (or between two points) in a period of time. Throughput is managed to balance supply (capacity) and demand. The demand level for transportation between any two points is related to marketing factors, not the least of which is the price for a roundtrip ticket.

All of the airlines do not charge the same amount for a roundtrip ticket. By adjusting price, airlines often can affect the percent occupancy of their flights. Such *marketing decisions* are part of the total system that affects operations. They typify the need for systems coordination to relate OM with the other functional areas within the framework of the transformation model. Southwest Airlines has used efficient operations to maintain low costs which allow low prices to be charged. This has made Southwest Airlines uniquely profitable.

The manufacturer can measure output in terms of the number of units of each kind of product it produces. Because there may be many varieties such as sizes and flavors, it is usual OM practice to aggregate the output into some common unit such as standard units of toothpaste produced. Both forms of information would reflect the variability of demand, but the aggregate measure much less so than the detailed product reports. Depending upon the demand levels, the marketing department could stimulate sales by dropping prices or raise prices to slow down demand that is exceeding supply capabilities.

If lower prices are effective in generating new business, then the demand is said to be elastic to price. When there is price elasticity, operations must keep costs down so that the advantage of low prices can be obtained. Marketing ultimately controls the volume of business that production must process. Financial planning has determined the operating capacity for peak demand. This in turn, translates back to the amount of inputs that need to be purchased to meet demand.

The systems perspective is required to make sure that all participants are connected directly to the revenue generating capacity of the OM I/O system. The information system helps to foster the process of keeping connected. Many kinds of data are regularly transmitted between participants. For example, information about what is selling—and what is in stock—leads to production scheduling decisions. It also leads to initiatives by the sales department. The levels of inventory are perpetually examined to make sure that no stock outages occur, and care is taken to keep track of what is in the finished goods inventory.

T E C H N O L O G Y

A major revolution in banking operations began in the 1970s with the advent of the Automated Teller Machine (ATM). Consumers primarily use ATMs for basic banking transactions, such as cash withdrawals and verification of account balances. The banking industry wants to expand the use of ATMs because automated transactions create significant savings in time and labor costs. In addition, banks recognize income from ATM usage fees. Consumers seem willing to pay for the convenience of automated banking.

To alleviate consumer concerns about security and machine accuracy, ATMs are being designed with enhanced security features. These features should increase their usage for basic transactions, including deposits. More banks will introduce fully automated centers where

**ATMs:
Changing the Way
We Bank**

consumers can perform basic cash transactions as well as apply for loans and open new accounts. Banks can achieve considerable cost savings from "fully automated banking," particularly for low-volume branches, by not having to provide staff at these locations.

ATM manufacturers are being pushed to build machines with options beyond banking. Technology is being developed for ATMs that can store, retrieve, and update records

for businesses. Existing ATMs can now be upgraded to accept AT&T Smart Cards. Resembling credit cards, the Smart Cards hold customer account information. The customer can purchase electronic tickets for airlines, sporting events, plays or concerts; have the tickets stored on the card; then access them at the airport, stadium, theater, or concert hall.

Sources: "AT&T and NCR offer Smart Card upgrade kits for bank machines," PR Newswire, November 17, 1993, and The American Banker/Ernst & Young 1993 technology survey conducted by Tower Group.

1.6

OM INPUT-OUTPUT PROFIT MODEL

The purpose of this section is to link the equation for profit—which is familiar to people in the business field—to the input-output transformation model.

I/O profit model

Derived from the costs and revenues of the traditional model based on a specific time period.

The **I/O profit model** is derived from the costs and revenues shown in Figure 1.7. The model assigns the costs and revenues of the traditional equation of profit to the inputs, the outputs, and the transformation process—all based on a specific period of time (t).

In Equation 1.0, Total Costs, TC, are subtracted from Revenue, R, yielding Profit, P.

$$P = R - TC$$ **Equation 1.0**

In Equation 1.1, Total Costs, TC, equal the sum of Total Fixed Costs, FC, and Total Variable Costs, $vc(V)$. The latter is the variable costs per unit (vc) multiplied by (V) which is volume in production units for the time period (t).

$$TC = FC + vc(V) \qquad \text{Equation 1.1}$$

Equation 1.2 represents the calculation of Revenue where price per unit, p is multiplied by V, sold in (t). $p(V)$ equals Revenue.

$$R = p(V) \qquad \text{Equation 1.2}$$

$$P = p(V) - FC - vc(V) = (p - vc)V - FC \qquad \text{Equation 1.3}$$

This equation can be used in a variety of ways. How do the variables relate to each other? For example, how does demand (V) relate to the price charged? What is the effect of FC on vc?

How Has the Profit Model Changed Over Time?

Nothing in the structure of the profit model has changed over time; the equations are the same. What has changed is the technology that the fixed costs can buy, and in turn, that investment affects the variable costs. In addition, there have been major changes in the knowledge base about the productivity of processes and the quality they can deliver. Knowledge has been developed which affects the volume of output that the fixed costs can deliver.

It is the production system (or the system of operations) which is embedded in the profit model that has changed a great deal. The performance of the input-output transformation system has been altered, resulting in increased productivity of the system. The architecture of the operating system has been changing for many years, but recently at an accelerated rate.

1.7

THE STAGES OF P/OM DEVELOPMENT

The profit model operates differently according to the stage of development of the company's input-output operating system. Therefore, the issue of stage is the next point to address. The *stage* reflects the degree to which a company's activities are planned for in a coordinated way. Thus, stage determines company effectiveness (ability to do the right thing) and efficiency (ability to do the thing right). As a company improves its stage of operations, it is expected that its profitability will increase. However, it is necessary to relate the company's stage of development to that of its competitors.

Each company's input-output profit model indirectly and directly reflects the impact of the competitors' input-output models. The cost structures, prices, volumes and profit margins reflect the influence and extent of the competition. If the competitors all are at the same stage and one of them starts to move to a higher stage, the expected result is that the advancing company will gain market share, while the remaining companies experience a decrease in volume. This then gets translated into higher variable costs and lower profit margins. The OMs should be on top of the competitive analysis which means that all participants fully understand the stages of development of all competing companies.

How a company manages its profit model provides insight into the role that P/OM can play in a company. Decisions concerning capacity and the resulting economies of scale capture only a portion of the story; the extent to which new technology is utilized to offset high variable labor costs also plays a part. The development stage relates as well to the management of the throughput rate, quality attainment and variety achievement.

Stage I companies operate on the premise that there is no competitive advantage to be gained by changing the production process. Therefore, the process usually is relatively unrefined and often out-of-date. Management, seeing no leverage in processes, pays minimal attention to production and operations. Stage I firms squeak by on quality. Because the competitors are not much better, everyone appears indifferent. In such firms, management and its resources are insufficient to do more than keep up with demand. Survival occurs only when and if the competition is in the same boat. Marginal firms do not survive long in a market that is dominated by higher-stage companies.

Special Capabilities

At the other end of the spectrum, there are companies that use operations to gain unique basic advantages through the development of special capabilities. The concept of competing on capabilities is clearly formulated by Stalk, Evans and Shulman.[5] The advantages gained can vary from speeding up distribution, to tactical superiority in managing inventories, or product preeminence obtained from total quality management (TQM).

Stage IV companies practice *continuous improvement (CI) which means that they persistently remove waste*. They aggressively seek to innovate in their unswerving pursuit of quality. Stage IV companies have a high level of basic advantages that are unique to them, whereas Stage I companies have virtually none.

This approach is based on work done by Wheelwright and Hayes (W&H) cited in their article in the *Harvard Business Review*.[6] They have created a framework applicable to manufacturing firms (P of P/OM). This approach also utilizes work by Chase and Hayes (C&H) detailed in their article in the *Sloan Management Review*.[7] Their framework is applicable to service firms (OM of P/OM). This article puts together the concept of stages of development for both manufacturing and service firms.

Stage IV companies aggressively seek to innovate and pursue quality. Hewlett-Packard uses self-directed teams of highly-focused engineers to maintain their leadership in inkjet printing.
Source: Hewlett-Packard Company

1. A Stage I company is centered on meeting shipment quotas and providing service when requested. C&H call it "available for service," and cite as an example, a government agency. A Stage I company has no planning horizon and is predisposed to be indifferent to P/OM goals. It is reactive to orders and has no quality agenda. Worker control is stressed. The company is not conscious of special capabilities for itself or for its competitors. W&H describe such firms as being internally neutral, which means that top management does not consider P/OM as able to promote competitive advantage and, therefore, P/OM is kept in neutral gear.

2. A Stage II company manages traditional P/OM processes and has a relatively short-term planning horizon. It makes efforts to secure orders and to meet customers' service desires. The primary goal of Stage II companies is to control costs. Quality tends to be defined as products or services that are not worse than some standard. These companies consider the most important advantages to be derived from economies of scale which means that as output volume increases, costs go down. W&H describe such firms as being externally neutral; they strive to have parity in P/OM matters with the competition.

3. A Stage III company installs and manages manufacturing and service processes that are equivalent to those used by the leading companies. C&H describe this as "distinctive service competence." A Stage III company makes efforts to emulate the special capabilities of the best companies. Quality and productivity improvement programs are utilized in an effort to be as good as the best. Stage III firms have a relatively long-term planning horizon supported by a detailed P/OM strategy. W&H describe such firms as being internally supportive, meaning that P/OM activities support the Stage III company's competitive position.

4. A Stage IV company is a P/OM innovator. It has both short- and long-term planning horizons that are integrated. P/OM is part of the top management strategy team because the production processes are held to be a source of unique advantage gained through special capabilities, as are product and service design. **Project management** is a P/OM responsibility which offers significant advantage to those in OM who know how to use project management methods to innovate quickly and successfully for competitive advantage. W&H describe such firms as externally supportive which means that competitive strategy "rests to a significant degree on a company's manufacturing capability." C&H conclude that Stage IV firms offer services that "raise customer expectations." Stage IV firms utilize the systems approach to integrate service and manufacturing activities.

Project management
A P/OM responsibility which provides a significant competitive advantage for P/OM.

Figure 1.8 provides a useful way to represent the classification system for the manufacturing and/or service system stages of P/OM development.

FIGURE 1.8

STAGES OF P/OM DEVELOPMENT

	Internally	Externally
Neutral	Stage I	Stage II
Supportive	Stage III	Stage IV

P E O P L E

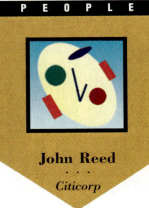

John Reed
• • •
Citicorp

Citicorp CEO John Reed faced red ink, bank regulator scrutiny, a bloated bureaucracy, and a crippling corporate culture in the early 1990s. A practitioner of broad strategic thinking and an ivory-tower management style, Reed confronted the fact that he and his organization needed to change in order to survive and grow. An engineer by training, he toured various manufacturing companies, including General Electric Co. and Ford Motor Co., looking for ideas on how to apply their successful reengineering strategies to Citicorp's operations. To help implement these strategies, Reed hired Christopher Steffan, who was instrumental in developing successful cross-functional teams at Chrysler Corporation in the early 1980s.

After cutting the workforce by 15%, cutting operating expenses by 12% and consolidating the number of data centers from 240 to 60, Reed brought earnings back into the black, posting a $722 million profit in 1992. He continues working to regain the trust of bank regulators through tighter financial controls, minimizing risky loans, and mandating an additional $1.5 billion in capital reserves.

Reed addressed the bloated bureaucracy through reorganizing top management into a management committee and 15 line officers (emulating the structure at Exxon Corp.) and setting up task forces across business lines (like those at Ford and other manufacturing companies).

To change the culture at Citicorp, Reed began by changing his own behavior. He is becoming more "hands-on" as he involves himself in many more details concerning day-to-day operations. Despite Reed's personal evolution, changing the culture at Citicorp will not be done overnight.

Source: Steven Lipin, "A New Vision: Citicorp Chief Reed, Once a Big Thinker, Gets Down to Basics," *Wall Street Journal,* June 25, 1993.

Reengineering

Starting from scratch to redesign a system; way to jump stages.

It takes a lot of work to progress through successive stages. It is unlikely that an existing P/OM organization can skip a stage. Only with total reorganization is it possible for a Stage I company to become a Stage III or IV. **Reengineering** *which is defined as starting from scratch to redesign a system,* is an appealing way to circumvent bureaucratic arthritis and jump stages. However, it is costly and, if not done right, has a high risk of failure. Mastery of this OM text can lower that risk.

1.8

ORGANIZATIONAL POSITIONS AND CAREER OPPORTUNITIES IN OM

To qualify for an operations management job, there are several reasons it helps to have an undergraduate degree in business or an MBA. First, the degree aside, OM has much real content. There are many concepts to learn and a special OM language to master. Being comfortable with technology also is helpful. Understanding detailed technology remains an engineering specialty. The specific course of study for operations is demanding; there are entirely new phenomenon to learn about.

A single, introductory course in OM will not suffice. The system's perspective, instrumental for success, requires knowledge of various business functions including marketing, finance, accounting and human resources management. The business school is becoming a proponent of the systems approach. The ideal academic situation for an OM career is a concentration or major in OM. However, many graduates from other concentrations, with some exposure to OM, find themselves attracted to OM jobs.

Career Success and Types of Processes

It is essential when talking about careers in OM to recognize that one of the *major differences in OM jobs relates to the kinds of processes* that are involved in the transformation of inputs into outputs. This means such things as the continuity of processing, the number of units processed at one time, the volume of throughput between set ups, and the degree of repetition of the operations.

It can be seen from the historical development of OM that manufacturing started with custom work, which in many ways resembles an artist at work. Consider the shoemaker who fits and makes the entire shoe for each customer. Most often, the left shoe differs from the right shoe. However, for store-bought shoes, such attention to fitting the customer is not possible.

Services often are of the custom variety. The medical doctor sees one patient at a time, and treats that patient as warranted. Service processes can prosper by making them more like manufacturing. In time, manufacturing learned how to process small batches efficiently. Some service systems, like elevators, lend themselves to batch processes.

Continuous flow processes were developed by a variety of industries, including chemical processors, refineries and auto assembly manufacturers. Fast food chains emulated this kind of process to handle a continuous flow of information and to assemble hamburgers. Until the late 1970s, there were basically three different ways to get work done. A fourth was added when computers began to change the way processes were achieved. The four categories are:

- Project—each project is a unique process, done once, like launching a new product, building a plant or writing a book. Both service providers and manufacturers need to know how to plan and complete projects which are associated with the evolving goals of "temporary" organizations. Projects appeal to people who prefer nonrepetitive, constantly evolving, creative challenges. Projects do not attract people who opt for a stable environment and the security of fixed goals—associated with the flow shop. There is a unique profile of people who prefer the project environment to other process types, and who excel in that milieu.

- **Batch processing**—facilities are set up and n units are made or processed at a time. Then the facility is reset for another job. *When $n = 1$, or a very few, it is called custom work*, and it is done in a custom shop. When n is more than a few, and the work is done in batches, it is called a job shop. The average batch size in job shops is 50. The work arrangement ceases to be a job shop when the work is done in serial flow shop fashion. With the job shop, many different kinds of goods and/or services can be processed. As the batch size gets larger for manufacturing or services, more effort is warranted to make the process efficient and to convert it to a serialized production system. Job shops, with their batch production systems, appeal to people who prefer repetitive assignments within a relatively hectic environment. The job shop generally involves a lot of people interactions and negotiations. The tempo of batch production is related to the number of set ups, cleanups, and changeovers.

- **Flow shop processing**—as the batch size increases so that production can be serialized, either continuously or intermittently, it is rational for both manufacturing and services to pre-engineer the system. This means that balanced flow is designed for the process before it is ever run. It is expected that variable costs will decrease as the fixed cost investments increase. Continuous process systems require a great deal of planning and investment. Flow shops run the gamut from crude set ups arranged to run for short periods of times (such as days or weeks) to continuous process systems that have been carefully designed and pre-engineered for automation. The more automated processes appeal to people who like a controlled, stable and well-planned system.

- **Flexible (programmable) processing** systems—in the 1980s, a new process category began to emerge which continues to grow faster than any other P/OM segment. Flexibility is derived from the combination of computers controlling machines, making this option the hi-tech career choice. People who enjoy working with computers prefer these technologically-based environments. There are two aspects to this attraction. First is the application of the technology to do the work, and second is the programming of the computers to instruct and control the equipment that does the work. Associated with the adoption of the new technology is much experimentation. An open attitude to learning is essential because the systems are continuously changing and require high levels of adaptability.

There are specific categories of people who prefer working in different process surroundings. There are also people who prefer either manufacturing or services; these issues usually are more important than type of industry preferences. For example, airplanes and automobiles are both assembly-oriented industries. Real advantages often accrue to both industries from employee cross-hiring. However, industrial preferences do play a part in career choice.

The same should be said about preferences for different kinds of service industries. A person having expertise in the hotel business is likely to gravitate between resorts, hotels, and restaurants with a great deal of crossover between them. The Ritz Carlton Corporation has made some remarkable competitive strides with respect to the quality of hotel service that can be applied broadly to this class of service. Club Med, which represents the best of the resort industry, has a very strong—transferable—OM orientation.[8] Media and entertainment are two other service areas with a strong draw on career selection.

Certain industries and services have intense regionality, i.e., Florida, Hawaii, and the Caribbean for the resort business; Michigan and Tennessee for automotive activities; New York (New York City) and Amsterdam for diamonds; other industries like McDonald's and Wendy's, Mr. Goodwrench, The Gap and Century Theaters are widely spread over the map. People may have strong regional preferences which could influence the industries with which they choose to be affiliated. Factors such as these need to be taken into account for job planning.

Operations Management Career Paths

There are many different kinds of careers in OM. No one can describe all of them because the scenarios are limited only by the imagination and the fact that truth is often stranger than fiction. Twelve career paths showing the difference between line and staff positions are provided in the following list. *Line* position means responsibility for producing the product or service. The term comes from working on the production line. *Staff* position by definition is not line, but supportive of the line. Staff positions are responsible for providing advice on such topics as costs and quality, inventories and work schedules. Titles for both line and staff positions vary in different companies, and specific details of responsibilities would differ for each position. However, the generalities regarding accountability remain the same.

Knowing about career paths provides a useful perspective for a person starting the study of OM. It should be noted, however, that the field is dynamic and changing. It is involved in organizational experimentation with teamwork and the systems approach. The use of multifunctional teams is increasing and likely to spawn new kinds of positions and career opportunities.

Global Aspects of Career Paths

Among the vast possibilities of exciting new careers are managers of global OM support networks which need to connect and synchronize factories and service systems located all over the world. OM is an international endeavor. With the North American Free Trade Agreement (NAFTA) accepted and the General Agreement on Tarrifs and Trade (GATT) being implemented, the European Union (EU) is a giant market for goods and services as well as a new environment for manufacturing and service operations. The Pacific Rim has come alive with manufacturing and great new markets are opening up in Southeast Asia. The agreement to create a free trade zone for the Americas stretching from Alaska to Argentina (34 countries including the U.S.A.) is yet another indicator of the internationalization of operations management. Suppliers from everywhere will be competing in the global market. A career in OM will involve much travel and an equally large amount of global communication. All of the careers that are listed will require the ability to coordinate and synchronize systems on a global scale.

Manufacturing jobs are followed by a list of service professions. Both include line and staff positions. Titles can only approximate the progression from top management through middle management to first-level management. For services there is an even greater problem to get titles that fairly represent progression through the management hierarchy.

OM Careers in Manufacturing–Line

1. Corporate Vice President of Manufacturing
2. Divisional Manager of Production
3. Plant Manager

4. Vice President of Materials Management
5. Project Manager of Transitions
6. Department Foreman or Forelady, Department Supervisor

OM Careers in Manufacturing–Staff

7. Director of Quality
8. Inventory Manager, Materials Manager or Purchasing Agent
9. Production Schedule Controller
10. Project Manager/Consultant (Internal or External)
11. Performance Improvement Manager
12. Methods Analyst

OM Careers in Services–Line

1. Corporate Vice President of Operations
2. Divisional Manager of Operations
3. Administrative Head
4. Department Manager or Supervisor
5. Facilities Manager
6. Branch Manager or Store Manager

OM Careers in Services–Staff

7. Quality Supervisor
8. Materials Manager or Purchasing Agent
9. Project Manager/Consultant (Internal or External)
10. Staff Schedule Controller
11. Performance Improvement Manager
12. Systems or Methods Analyst

By explaining some of the titles in more detail, it is possible to provide quick insight into various aspects of the P/OM field. Also, it further clarifies differences between manufacturing and service systems.

Manager of Production, Manufacturing (or Service) Operations

The manager of operations in services and the production manager in a manufacturing plant are in line positions, meaning that they are responsible for the inputs, the outputs and the transformation process. They work with technology and a hierarchy of people to oversee the production of the product. As a rule, they report to the corporate vice president who, in turn, reports to the president of the company. Middle managers and selected staff functions report to them. This is a high-level position, but the character of the job depends upon the stage of the company. In Stage IV firms, this person will regularly be invited to lunch in the boardroom because of top management's keen interest in operations.

Department Supervisor

The supervisor's position is a line job. In manufacturing, the supervisor, who is often called the foreman (or forelady or foreperson) typically oversees some part of the production process. In service operations, the title of supervisor commonly describes the person responsible for some specific function such as reservations and ticket processing, or insurance premium collection. The workers look to the foreman or supervisor for help with problems that arise. Fewer firms today consider the supervisor a police officer who makes sure that everyone is doing their job. To signal this change, many firms are calling both workers on the line and their supervisors "associates." Japanese companies in the U.S.A. started using the term "associate" to describe workers as well as supervisors with the idea of empowering employees. Similar practices can now be found among companies all over the world.

In many factories supervisory jobs are held by people who are promoted from the shop floor. The same applies to service operations. At the same time, some companies like to place a new business graduate into a foreman's job so they can learn the ropes. It is not unusual for a foreman to become the operations manager, but in many companies an engineering degree for manufacturing and a business degree for services are considered desirable.

Inventory Manager or Materials Manager

The inventory or materials manager holds a staff position which is accountable for controlling the flow of input materials to the line. The function of this job is to determine when and how much to order, and how much stock to keep on hand. There are myriad titles associated with these jobs. Many manufacturing and some service firms have vice presidents of materials management because the cost of materials as a percent of the total cost of goods sold is high. In service firms like McDonald's and American Airlines materials are similarly important.

Director of Quality

There are a great number of jobs in the quality assurance area and even more titles. Most of these jobs are staff positions which range from auditing quality levels to doing statistical analyses for control charts. The director of quality or quality manager (who could be a vice president) is in charge of the various quality activities that are going on in the firm. In some companies, line workers have been given quality responsibilities so it is possible to find supervisors with quality team assignments. Quality adjustments to inputs, including vendors, and to the transformation process are common. Quality positions usually focus on improving the quality of outputs by inspecting for defects, preventing them from happening, and correcting their causes. Quality management is just as important in service firms as it is in firms manufacturing goods, but it is more elusive and, therefore, challenging to deal with it.

Project Manager/Consultant (Internal or External)

There are important OM jobs relating to projects which can include the development of new products and services as well as the processes to make and deliver them. Construction of a refinery, putting the space station into orbit, writing this book— all are projects. Consultants, both internal and external, are usually engaged in project management. External consultants are employees of a consulting firm. P/OM is an excellent entree for a consulting career. Internal consulting applies only to the company for which the employee works (i.e., GE and Bristol Meyers are two companies that have very successful records using their own internal consultants).

During these turbulent times, transition management is used with various scenarios such as downsizing, turnarounds, and business process redesign. Companies have created positions which indicate responsibility for some form of transition in the job title. For example, a project manager of transitions is in charge of down-sizing, or rightsizing the company, turnarounds—restoring a company that is in trouble to good financial health, and reengineering—starting from scratch to redesign the firm. Various new job titles have appeared to indicate that the management of change is in place.

Performance Improvement Manager

Continuous improvement (CI) is a new item on the agenda of many companies. There are a variety of jobs that signify a company's interest in pursuing CI goals. The titles of such positions are more varied than most others in the OM field. For example, PIPs are Productivity Improvement Programs and some firms have appointed managers to head up these initiatives. In general, performance improvement managers are hybrids. They work on improving processes by means of total quality management (TQM), which includes waste reduction and value-added augmentation. They are likely to be responsible for coordinating supplier activities to achieve just-in-time (JIT) deliveries. Fast-paced project development is another responsibility. Another new title is based on variants of the words for best-in-class benchmarking which means comparing the system with the best one. In this category, as well as with the project manager/consultant above, there exist reengineering titles which represent new opportunities for OM.

SUMMARY

The chapter begins with an explanation of operations management (OM) and introduces the use of models by operations managers. Next, the systems approach is defined as systematic and constructive methodology using a systems taxonomy. The importance of the systems approach as employed by OM is detailed and explored with examples. The application of OM to manufacturing of goods has similarities and differences compared to its application to services. In that context, production management is compared to operations management. It is explained how information systems play a vital role in both manufacturing and services. The basic model of OM—the input-output transformation model—is developed, along with the costs and revenue that are associated with its use. This leads to a discussion of the input-output profit model and how it has changed over time. The stages of P/OM development are introduced with attention paid to special capabilities derived from OM that allow a company to reach highly competitive stages of development. This leads to deliberation about the kind of OM positions that might be encountered by another manager in the organization, as well as the type of careers that OM offers.

▶ Key Terms

functional field approach *10*	information systems *18*	I/O profit model *24*
manufacturing *15*	OM system *12*	operations *7*
operations management *17*	production *17*	production management *17*
project management *27*	reengineering *28*	service *15*
systems approach *11*	transformation *19*	

▶ Review Questions

1. What is operations management? Define OM. Compare it to P/OM. Distinguish between production and operations management.

2. What are the differentiating characteristics of services as compared to those of manufacturing? Illustrate the distinctive aspects of each by naming industries and citing specific companies that represent each.

3. To what category of P/OM does hotel management belong? What fixed and variable costs are appropriate for this industry?

4. To what category of P/OM does agricultural management belong? What fixed and variable costs are appropriate for this industry?

5. How do information systems relate to operations management?

6. Why is the systems approach essential to assure real participation of OM in a firm? What does the systems approach have to do with company strategy?

7. What types of costs are usually associated with inputs? Give specific examples from the education field.

8. What costs are identified with the equipment that provide transformations? Give specific examples from the fields of manufacturing and transportation.

9. How do outputs convert into dollars of revenue? How are these dollars related to fixed and variable costs as described in Questions 7 and 8?

10. Explain the stages of P/OM development and try to identify some companies that might be representative of each stage.

11. What changes have occurred over time in the following profit model?

$$p = (p - vc)V - FC$$

12. Describe career paths of P/OM for both manufacturing and service systems.

▶ Problem Section

1. Develop the transformation process to:
   a. produce #2 lead pencils
   b. bake bread
   c. complete this homework assignment

2. Draw the input-output model that could be used to run a hotel. Label the variable costs for labor and materials, and detail those costs. What are the fixed costs for this model?

3. What are the headings for the columns and the titles for the rows? The stages refer to the development level of the P/OM function.

	Stage I	Stage II
	Stage III	Stage IV

4. Draw an input-output model for a Burger King fast food outlet. Label as many of the specific and detailed inputs and outputs as you can. (Use Figure 1.7 as a model of categories to include.) What transformations link the inputs with the outputs? (It may be

helpful to visit a Burger King. Observe what is going on and if possible talk to the manager about how they cook and assemble their food.)

5. Repeat what you have done in Problem 4 for a McDonald's fast food outlet. (You may want to visit a McDonald's. As above, observe what you can and talk to the manager if possible.) Then, make a detailed comparison between Burger King and McDonald's.

6. A manufacturing plant and equipment cost 120 million dollars and are estimated to have a lifetime of twenty years. Straight-line depreciation is to be used. Additional fixed costs per year are 4 million dollars. Variable costs are $1.50 and price is set at $2.50. What will annual profit be if the annual volume is:
   a. 10 million units?   b. 5 million units?
   c. 20 million units?   d. 8 million units?

7. A manufacturing plant and equipment cost 120 million dollars and are estimated to have a lifetime of twenty years. Straight-line depreciation is to be used. Additional fixed costs per year are 4 million dollars. Variable costs are $2.50 and price is set at $3.50. What will annual profit be if the annual volume is 10 million units?

8. A manufacturing plant and equipment cost 180 million dollars and are estimated to have a lifetime of thirty years. Straight-line depreciation is to be used. Additional fixed costs per year are 4 million dollars. Variable costs are $1.50 and price is set at $2.50. What will annual profit be if the annual volume is 10 million units?

9. A service center has installed a new computer system with local area networking at a cost of 1.6 million dollars. The system is expected to serve for eight years, and straight-line depreciation is acceptable. There are additional fixed costs of $300,000 per year. This service repair center charges each customer a flat fee of $30. The variable costs are $20. What will profit be if the annual volume is:
   a. 50,000 units?
   b. 25,000 units?
   c. 75,000 units?

10. Four companies are described below. Characterize each in terms of the stage of P/OM development that seems appropriate as described in Chapter Section 1.7. Classify each situation and explain the reason for the classification used.
   a. This manufacturer pays little attention to the quality of the product. The owner is convinced that customers are plentiful but not loyal. Hard sell is stressed.
   b. This service organization tries to keep up with its competitors by copying everything they do as soon as possible. The president believes that development costs are saved and that being a fast imitator results in a competitive advantage.
   c. This manufacturer is constantly working at being as good as the best of the competitors. They have improved quality repeatedly while holding costs constant. The production manager participates in strategy formulation.
   d. This service organization aims at global leadership based on the finest operations in the world. The service is constantly being improved which gives the organization a proactive leadership role in its industry.

▶ Readings and References

Abernathy, William J., K.B. Clark, and A.M. Kantrow, *Industrial Renaissance,* New York: Basic Books, 1983.

Chase, Richard B., and Robert H. Hayes, "Beefing-Up Operations in Service Firms," *Sloan Management Review,* Fall, 1991, pp. 17-28.

Horovitz, Jacques, *Winning Ways: Achieving Zero-Defect Service,* Productivity Press, 1990.

Levitt, Theodore, "Production-line Approach to Service," *Harvard Business Review,* Sept.-Oct., 1972.

Müller-Merbach, Heiner, "A System of Systems Approaches," *Interfaces,* Vol. 24, No. 4, July-August, 1994, pp. 16-25.

Parkinson, C. Northcote, *Parkinson's Law,* Boston: Houghton Mifflin Company, 1957.

Peters, Tom and R. H. Waterman, Jr., *In Search of Excellence,* New York: Basic Books, 1983.

Stalk, George, Philip Evans, and Lawrence E. Shulman, "Competing on Capabilities: The New Rules of Corporate Strategy," *Harvard Business Review,* March-April, 1992, pp. 57-69.

Wheelwright, Steven C., and Robert H. Hayes, "Competing Through Manufacturing," *Harvard Business Review,* Jan.-Feb., 1985, pp. 99-109.

S U P P L E M E N T   1

Decision Models

A Distribution Problem—How to Ship—Plane or Train?

A decision matrix arranges outcome data into rows and columns. The following matrix presents a decision problem where the rows are shipping "strategies" (plane or train) and the columns are "states of nature"—so called because they are events not under the decision-maker's control. In this case, the states of nature are weather conditions.

Outcomes are the entries in the cells of the matrix. They will reflect the fact that planes will be delayed if there is fog, while trains are not affected by the weather. The buyer has promised the shipper a bonus if the delivery can be early. Conversely, the buyer sets a penalty if the delivery is delayed. The decision problem is organized by the matrix as follows:

Probabilities	Low	High
States of Nature	Fog (F)	Clear (C)
Strategy 1: Plane	delayed	early
Strategy 2: Train	on time	on time

(Note that at this time of year, there is a low probability of fog, *F*, and a high probability of clear weather, *C*.)

Before putting numbers in this matrix, look at the symbolic representation of the following decision matrix. It shows outcomes as $O(i,j)$ where i represents different strategies (rows), and j represents different states of nature (columns). Also, the probabilities of the states of nature—obtained by some forecasting method—are represented by $p(j)$'s, j = either *F* or *C*.

Probabilities	p(F)	p(C)
States of Nature	Fog	Clear
Strategy 1: Plane	O(P,F)	O(P,C)
Strategy 2: Train	O(T,F)	O(T,C)

Calculating the Numerical Outcomes or Payoffs

Assume the set of values (below) describes the profit that the shipper gets if the delivery is early, late, or on time. Weather permitting, the flight arrives on time and the company makes $4,500. If the flight is delayed, the firm can still earn $1,500. Train delivery will earn $3,000 either way because the train arrival time will not be subject to weather. In addition, shipment by plane costs $800 and by train $300. These costs will be subtracted from the profits. Often, the payoff portion of the matrix is calculated by formulas or observations. The payoff matrix for each outcome is shown below with probabilities of 0.1 and 0.9 obtained from weather forecasts. (Note: The sum of the probabilities must always be equal to one.)

Probabilities			0.1	0.9
States of Nature			Fog	Clear
		Cost		
Strategy 1: Plane		− 800	+ $1,500	+ $4,500
Strategy 2: Train		− 300	+ $3,000	+ $3,000

After subtraction:

Probabilities		0.1	0.9
States of Nature		Fog	Clear
Strategy 1: Plane		+ $ 700	+ $3,700
Strategy 2: Train		+ $2,700	+ $2,700

Calculating the Expected Values (EV)

An expected value is a measure of the average outcome. There is an expected value for each strategy. It is obtained by multiplying the probabilities for each state of nature by the outcome in that column and adding the products across all of the columns. The computing formulas are shown below:

$$EV(\text{Plane}) = p(F) \times O(P,F) + p(C) \times O(P,C)$$

$$EV(\text{Train}) = p(F) \times O(T,F) + p(C) \times O(T,C)$$

Now using numbers:

$$EV(\text{Plane}) = 0.1 \times 700 + 0.9 \times 3700 = 3400*$$

$$EV(\text{Train}) = 0.1 \times 2700 + 0.9 \times 2700 = 2700$$

The decision—noted by an asterisk—is to ship by plane because that strategy has an expected value of $3,400 whereas the expected value for shipping by train is $EV(\text{Train}) = \$2,700$.

Problem for Supplement 1

1. The forecast of 0.1 and 0.9 is based on last year's weather. When the past five years of weather data is consulted, the following results are obtained.

Year	1	2	3	4	5
P(Fog)	0.1	0.2	0.3	0.2	0.5

Determine for each year the best strategy. Then decide how to use all of the weather data to reach the best possible decision.

Notes...

1. Heiner Müller-Merbach, "A System of Systems Approaches," *Interfaces*, Vol. 24, No. 4, July-August 1994, pp. 16-25.

2. Ibid, pp. 16-17.

3. Ibid, p. 17.

4. Theodore Levitt, "Production-line Approach to Service," *Harvard Business Review*, Sept.-Oct. 1972. Although this article is old, it is still totally relevant. It is considered a classic. Our interpretation of "idiosyncratic" is eccentric and erratic.

5. George Stalk, Philip Evans, and Lawrence E. Shulman, "Competing on Capabilities: The New Rules of Corporate Strategy," *Harvard Business Review*, March-April 1992, pp. 57-69.

6. Steven C. Wheelwright, and Robert H. Hayes, "Competing Through Manufacturing," *Harvard Business Review*, Jan.-Feb. 1985, pp. 99-109.

7. Richard B. Chase, and Robert H. Hayes, "Beefing-Up Operations in Service Firms," *Sloan Management Review*, Fall 1991, pp. 17-28.

8. Jacques Horovitz, *Winning Ways: Achieving Zero-Defect Service*, Productivity Press, 1990.

Chapter 2

OM's Key Role in Productivity Attainment

After reading this chapter you should be able to...

1. Understand the importance of productivity to an organization.
2. Evaluate the applicability of various measures of productivity.
3. Explain what contributes to good productivity.
4. Discuss how to improve productivity.
5. Explain why productivity is a systems issue.
6. Relate productivity and price-demand elasticity as a systems link between OM and marketing.
7. Describe the effects of quality on elasticity and explain why this type of elasticity is critical to OM.
8. Apply the concepts of economies of scale and division of labor to the management of operations.
9. Explain how the history of OM shows continual improvement in the OM input-output transformation model with resultant increases in productivity.

Chapter Outline

2.0 PRODUCTIVITY—A MAJOR OM ISSUE
  Converting Output into Dollars
  Operational Measures of the Organization's Productivity
  Productivity Is a Systems Measure
  Productivity Is a Global Systems Measure
  Bureaucracy Inhibits Flexibility and Productivity
  Size of Firms and Flexibility
  Research and Development (R & D)

2.1 PRODUCTIVITY AND PRICE-DEMAND ELASTICITY
  Elasticity Relationship
  Elasticity of Quality

2.2 ECONOMIES OF SCALE AND THE DIVISION OF LABOR

2.3 THE HISTORY OF THE IMPROVEMENT OF OM TRANSFORMATIONS
  Discussion of the History of P/OM
  Artisans, Apprentices and Trainees—The Beginning
  Interchangeable Parts (IP)—Production Theory's First Step
  Scientific Management (SM)—Production Theory's Second Step
  Sequenced Assembly (SA)—Production Theory's Third Step
  Statistical Quality Control (SQC)—Production Theory's Fourth Step
  Lean Production Systems (LPS)—Production Theory's Fifth Step
  Flexible Production Systems (FPS)—Production Theory's Sixth Step
  Global Competition: The Year 2000 Plus—Production Theory's Seventh Step

Summary
Key Terms
Review Questions
Problem Section
Readings and References
Supplement 2: Decision Trees

The Systems Viewpoint

Productivity gets good grades when a high rate of output is obtained at low cost. This is the case even when the output is hard to measure, as in some service operations. Although OM is custodian of the production process and responsible for its productivity, everyone in the company in involved in attaining excellence in productivity. Everyone has the ability to increase or decrease the organization's productivity. Those managers who have cultivated the systems point of view recognize that *good productivity is contagious*; so is poor productivity. Employees sense whether the company culture promotes high or low productivity, and they respond in accordance with the cultural norm. This makes the productivity condition a contagion factor with systems-wide implications.

Another strong systems-type interaction relates to price-demand elasticity which links the price charged and the volume which can be sold. Competitive pressures to reduce prices lead to demands that OM improve productivity to decrease costs. Also, business lost because of price competition decreases volume, which reduces capacity utilization, leading to larger overhead charges per unit. Reductions in discounts that are based on volume increase variable costs per unit.

From a broader point of view, OM history shows that the overall trend in global economies is to increase productivity. Productivity growth has reflected the impact of a continuous stream of developments in technology and operations management methodology.

2.0

PRODUCTIVITY—A MAJOR OM ISSUE

Productivity is a critical business variable which directly impacts the "bottom line"; improved productivity raises net profits. OM is responsible for the productivity of the process. This is such a critical factor in a company's overall productivity that excellence in productivity achievement is a major OM issue.

Productivity measures the performance of the organization's processes for doing work. The American Production and Inventory Control Society (APICS) definition follows, but other definitions also will be given. Productivity is an important way of grading how well OM and the rest of the system are doing. *Productivity is a score like RBIs* (runs batted in) or *ROI* (return on investment).

Productivity Overall measure of the ability to produce a good or a service; OM views productivity as the ratio of output over input.

Productivity is defined as: "An overall measure of the ability to produce a good or a service. It is the actual output of production compared to the actual input of resources. Productivity is a relative measure across time or against common entities. In the production literature, attempts have been made to define total productivity where the effects of labor and capital are combined and divided into the output."[1]

Operations management views *the measurement of productivity as essential for assessing the performance of the organization's productive capacity over time and in comparison to the competition.* Operations management, from a production point of view, views productivity as the ratio of output over input. When outputs are high and inputs are low, the system is said to be efficient.

$$\text{Productivity} = \frac{\text{Outputs}}{\text{Inputs}} = \text{Measure of Production Efficiency}$$

The productivity measure compares the quantity of goods or services produced in a period of time (*t*), and the quantity of resources employed in turning out these goods or services in the same period of time (*t*).[2]

Productivity measures can be common-sense ratios, like the number of sweatshirts embossed per hour.

This measure is relatively easy to accomplish for physical goods. It is more diffi-cult to find appropriate measures for some services. Measuring the outputs for educa-tion, health care and creative knowledge work provide some instances of intangibles that are highly valued, but elusive to calculate. The effort has to be made to appraise the value of these outputs in a standardized way in order to provide a benchmark (or standard) for measurement.

Sales perceives productivity as high customer sales volume (called an effective marketing system) accompanied by low producer costs (called an efficient producer system). Thus:

$$\text{Productivity} = \frac{\text{Effectiveness}}{\text{Efficiency}} = \frac{\text{High Customer Sales Volume}}{\text{Low Producer Expenses}}$$

From a systems point of view, the inclusion of sales provides a correct measure of productivity. It is hardly productive to make a lot of product which is not sold, even if the cost of making it is low.

At the same time, OM employs productivity measures to assess how well the production system is functioning. The kinds of questions that are being addressed are: how many resources are consumed to produce the output, and, how many units of output can be made with a fixed amount of capacity?

Both ways of viewing productivity have benefits. They stem from different inter-ests that need to be shared. The best interests of the company are served by merg-ing what is learned about OM's efficiency and sales effectiveness.

Addressing the specifics of measuring productivity: productivity is a ratio of output units produced to input resources expended per unit of time (t). For exam-ple, productivity could be measured as units of output per dollar of labor in a period of time. That would provide a measure of labor productivity. The dimen-sions for output and input are as follows:

output = units/hour, and input = dollars/hour = hourly wages

$$\text{productivity } (t) = \frac{\text{output}}{\text{input}} = \frac{\text{units of output}}{\text{dollars of labor}}$$

This relationship measures how many dollars of labor resources are required to achieve the output rate, and indicates how much output is being obtained for each dollar spent.

Converting Output into Dollars

Another way of stating the measure of productivity is to put a dollar value on the output volume per unit of time. The input cost is already stated in dollars per hour. The advantage of this method is that it can deal with the productivity of the system across different kinds of units; for example, the productivity of a paint company that puts paints of many colors in cans of many sizes could be measured. This approach is used for national accounting of productivity where there are so many different kinds of units to be included (i.e., furniture, clothing, and foods). Dollars can standardize the output measure across diverse categories.

Labor productivity

Partial measure of productivity; ratio of sales dollars to labor cost dollars.

Labor productivity, often measured as the ratio of sales dollars to labor cost dollars, is called a "partial measure of productivity."

$$\text{labor productivity } (t) = \frac{\text{output}}{\text{input}} = \frac{\text{sales in dollars}}{\text{cost of labor dollars}}$$

Multifactor productivity

Total productivity; difference between change in output and change in labor and capital inputs engaged in the production of the output.

The U.S. Labor Department measures **multifactor productivity** (also called total productivity) on a regular basis to determine whether the U.S. industrial base is improving its competitive position in the world.

> "*A change in multifactor productivity reflects the difference between the change in output (production of goods and services) and the change in labor and capital inputs engaged in the production of the output. Multifactor productivity does not measure the specific contributions of labor, capital, or any other factor of production. Instead, it reflects the joint effects of many factors, including new technology, economies of scale, managerial skill, and change in the organization of production.*"[3]

Multifactor productivity is measured below in dollars as the ratio obtained when the costs for all the resources used are divided into the total output of goods and services.

$$\text{multifactor productivity } (t) = \frac{\text{all outputs of goods and services}}{\text{total input resources expended}}$$

Table 2.1 shows the average annual percent changes in multifactor productivity for the U.S.A., over different periods of time.[4]

T A B L E   2 . 1

AVERAGE ANNUAL PERCENT CHANGES IN MULTIFACTOR PRODUCTIVITY

Years	1948-73	1973-79	1979-90	1990-93
Multifactor Productivity	1.9	0.3	−0.1	0.2

U.S. productivity has begun to increase again after many years of declining growth rates, and even some years of negative growth (average annual shrinkage of −0.1 from 1979 to 1990). Although the U.S. has experienced declining multifactor manufacturing productivity growth, the recent experience of other countries, such as Japan, has been similar. Further, the trend has been reversing for the U.S.A.

The renewal of U.S. productivity growth is not surprising. Great efforts have been made in the manufacturing sectors to improve efficiency. Subjects such as total quality management (TQM), reengineering, and turnarounds, discussed in the general press, are not fads. They herald solid accomplishments, and more developments yet to come.

However, it is important to separate the manufacturing component of the total productivity measure from the service component. *Measures of service productivity are much lower than those of manufacturing* and have not been posting notable improvements around the world. There have been substantial investments in computers and telecommunications within the service sectors, but only recently are there real glimmers of improvement. For the time being, no broad-based changes in productivity can be reported. This will change as OM managers learn how to apply the new technologies to service systems. This productivity comparison constitutes a significant difference that presently exists between services and manufacturing.

There are many ways to measure productivity in line with other purposes. The equations below measure **capital productivity** in two different ways. Capital productivity is a partial productivity due to invested capital. First, there is the number of units of output per dollar of invested capital.

Capital productivity

Partial measures of productivity are the number of units of output per dollar of invested capital. Also, a ratio based on dollars of output per dollar of invested capital.

$$\text{capital productivity } (t) = \frac{\text{output}}{\text{input}} = \frac{\text{units of output}}{\text{dollars of capital}}$$

The second measure of capital productivity is a pure ratio based on dollars of output per dollar of invested capital.

$$\text{capital productivity } (t) = \frac{\text{dollar value of output units}}{\text{dollars for capital resources}}$$

Such measures of capital productivity might help organizations in the service sectors to address the value of returns on investments in computers and telecommunications. By extending this reasoning it is possible to develop other partial measures of productivity with respect to areas such as energy expended, space utilized, and materials consumed.

Students and practitioners of business know that productivity improvements translate into more profits and greater profitability for organizations, and improves the state of the economy. In the same vein, economists believe that productivity improvements translate into higher standards of living and greater prosperity. Productivity measures get factored into inflation calculations as well as other economic scenarios. Increases in productivity are generally regarded as a means of checking inflationary trends.

This is because by working smarter—not harder—profit margins increase and there is more money available to invest in other business opportunities. The shortage of capital drives interest rates up. However, if there are more jobs available than people to fill them, labor becomes scarce and rising salaries begin to fuel the expansion of inflation. **Relative productivity** compares the performance of competitive processes for which OM is accountable. The contribution of OM to the well-being of the national economy should be recognized.

Relative productivity

Compares the performance of competitive processes for which OM is accountable.

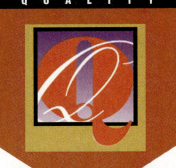

McKesson Corporation
• • •

Quality in the distribution business means that orders are filled quickly and accurately.

McKesson Corporation dominates the fast-growing $22 billion market for the distribution of pharmaceuticals and personal care products. The 160-year-old giant, which has its roots in selling medicines to seagoing clipper ship captains, remains number one because of the quality service it continues to provide its customers. And *quality* in the distribution business means that orders are filled quickly and accurately.

"Quality costs less," says Thomas Simone, president of McKesson. "When we get out a complete order, without mistakes, with as few shortages as we can, and that's clean and neat and delivered on time, the customer's satisfied. That literally will make the difference over time between a customer choosing to remain with us, or choosing to go somewhere else."

Despite the trend toward regional distribution of these products, McKesson serves the entire nation from a computer and communications center near Sacramento, California. Using leading-edge distribution technology, the center can process 30,000 orders per day.

The order-filling process begins when the retailer places the order with a handheld computer while walking the store's aisles. When the retailer discovers inventory that needs to be replaced, the information is entered into the computer. The retailer then sends the order information to McKesson's mainframe IBM computer in Sacramento—via modem and an AT&T satellite substation. The computer routes the order to one of 45 distribution centers around the country. Once the center receives an order, it is processed in less than an hour for next-day delivery. The order is filled by automation in the same sequence as placed. So when the bins are delivered to the retailer, the merchandise is organized by aisle, allowing for ease in restocking the shelves. That's quality service!

McKesson counts among its satisfied customers both large and small retailers, from Wal-Mart to mom-and-pop drugstores. For these smaller, independent retailers, McKesson also offers membership in a "voluntary chain" or buying group that affords economies in purchasing clout, sales promotions, and other services.

Sources: Barnaby J. Feder, "McKesson: No. 1 but a Doze on Wall Street," *New York Times*, March 17, 1991, and John Karolefski, "McKesson Strengthens through Partnerships," *Supermarket News*, September 9, 1994, Vol. 41, No. 27, p. 37.

Operational Measures of the Organization's Productivity

Productivity measures can be common-sense ratios such as the value of pieces made in a factory divided by the cost of making them, or the number of documents produced by the typing pool divided by the number of people doing word processing. Such operational measures of productivity are valuable to companies that are focused on continuous improvement.

A productivity measure used by restaurants is the daily dollars generated per table. Airlines measure plane occupancy per flight and average occupancy per route, as well as for all routes flown. Department stores use sales dollars per square

The productivity of a company is the composite of the contributions of the individual functions. Here, U.S. produced tires are mounted on rims in Hermosillo, Mexico and delivered to a nearby U.S. automaker's assembly plant.
Source: The Goodyear Tire and Rubber Company

foot of space. Mail-order companies measure sales dollars for categories (i.e., fashion, toys, and luggage) by type of illustration (i.e., color and size) on a percent of page basis.

Many companies measure the productivity of their complaint departments by the ratio of the number of complaints dealt with per day divided by the number of complaint handlers.

Formulating appropriate productivity measures to capture the effectiveness of operations is an OM challenge. It takes insight and creativity to measure what really matters in performance and what really can be controlled and corrected.

Productivity Is a Systems Measure

Every function in the company that has some measurable accomplishment can be evaluated with respect to productivity. The productivity of the company is the composite of the contributions of the individual functions.

Volume of output sold is a measure of the productivity of the sales department. Cost of goods sold is a measure of the productivity of the process designers, the R & D department, and the operations managers. It also is a measure of the purchasing department's ability to find the best materials obtained at the lowest possible costs and highest qualities.

Output delivered to the customer is a measure of the ability of the distribution system to be on time with undamaged delivery. Output delivered is subject to warranties. Does the company listen to the "voice of the customers" with respect to difficulties in repairing or returning defectives? If not, future sales productivity can be impaired. Further, warranties exercised represent productivity decreases.

Productivity issues are woven into all parts of the supplier-producer-customer chain and the input-output transformation process. This further explains why productivity is a systems issue. The supplier-producer-customer chain has a global reach. Therefore, the next section explores the impact of international factors on productivity.

Productivity Is a Global Systems Measure

Although much knowledge exists about production systems and operations management, there have been and continue to be serious productivity problems in the world. These problems have afflicted many developing countries where capital to invest in new technology is scarce, and technical knowledge and training are lacking. There also have been productivity problems in industrialized countries where early productivity growth was not sustainable, as in the U.S.

Japan's phenomenal productivity growth rates of the 1980s could not be maintained. Nevertheless, Japanese productivity in a variety of industries continues to be formidable. Thus, with respect to auto parts, "On average, the plants in Japan were 18 percent more productive than ones in the United States, and 35 percent more productive than ones in Europe."[5] Further, "Japanese parts makers surveyed increased their productivity by almost 38 percent from 1992 to 1994; the American companies made gains only in the mid-20s."[6]

Such data needs to be questioned for accuracy. Although the basic trends are generally accepted, it is unsatisfactory to measure manufacturing sector productivity as output per person directly employed. Productivity improvements can be registered by companies that reduce the number of direct employees, increase the use of part-time help, rely more on subcontracting of parts, and employ outside maintenance companies, among other things. It is to be noted that Japanese firms typically rely heavily on their suppliers and part-time help. Other aspects of Japanese manufacturing methods logically account for world-class productivity accomplishments. Nevertheless, the phenomenal Japanese productivity records have diminished substantially in recent years.

Other leading industrial nations have experienced productivity declines. Many hypotheses have been offered to explain the inability to sustain stable productivity growth in this turbulent era of new technological development. One is that the productivity of industrial nations is converging.[7] Another is that old technology gets used up and it is difficult to switch to the new technologies and make them profitable.[8] It also takes smarter management.

Competitors keep jockeying and leap-frogging each other with the adoption of new technologies that initially reduce productivity. There are various causes for the reduction. The new technology is imposed on old processes by employees who lack training and experience with the new technology. Product life is short. This allows little time to enjoy the advantages of new technology applied to evanescent and even obsolescent product lines. Furthermore, the global economies have been buffeted by recessionary economic conditions which penalize high volume export industries such as those of Japan and the U.S.

Increasingly, companies buy from suppliers located around the world and sell in markets that are equally dispersed. Production facilities, including fabrication, assembly, chemical and drug processes, and service facilities are located globally. Therefore, productivity performance becomes the result of international interactions. If productivity is being measured in dollars or local currencies, exchange rate problems can distort the picture. On the other hand, exchange rate imbalances cause gains and losses that relate to the productivity of investments. These are reflected in ROI (return on investment) measures.

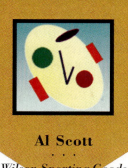

Al Scott
. . .
Wilson Sporting Goods

When Al Scott took over as plant manager of the Wilson Sporting Goods golf ball plant in Humboldt, Tennessee in 1985, he found low morale and low productivity. "It was either turn it around or lose it," Scott said.

His guiding principles in turning the plant around, less familiar in 1985 than now, were continuous improvement, associate involvement, total quality management, just-in-time methods for waste elimination, and a focus on lowest total cost manufacturing. Since implementing these programs, productivity has increased 177 percent and market share for Wilson golf balls has increased from 2 to 19 percent. Wilson sells golf balls under the Ultra and ProStaff brands.

In the manufacture of golf balls, "We need a production lead time of three to four days," Scott says. "Our goal is four hours from rubber mix to pack, which will give us a big competitive advantage." To this end, the company has begun making its own rubber in a new $8.5 million rubber mix facility in Humboldt.

Scott is a strong believer in the team approach. Humboldt employees are called "associates," and supervisors are called "coaches." A coach's shirt and terminology were first given to only exempt salaried employees. The second phase was to include all salaried employees as coaches. In the third phase, which is now complete, everyone is referred to as a coach. The principle is that each coach can actively influence others on the team! Collectively the staff at the plant is known as "team Wilson."

Forty smaller teams meet for an hour each week to discuss various projects to improve productivity and cut costs. Scott encourages full participation in teams. "When we started, we wanted the teams to be voluntary," he said. "Now we have changed the wording in job descriptions to make the team activities as important as the individual job."

Wilson Sporting Goods' Humboldt facility has achieved recognition for its efforts, winning the *Industry Week* "best plant award," the Shingo Prize for Excellence in Manufacturing, and Clemson University's 21st Century Organizational Excellence Award.

Source: Laurel Campbell, "Focus on Quality helps Wilson expand," *The Commercial Appeal*, April 24, 1994.

Bureaucracy Inhibits Flexibility and Productivity

Another factor that accounts for poor productivity is the great inhibitor of flexibility, bureaucracy. Bureaucratic systems are rampant worldwide. North America and Europe have more than their share, and they are equally prevalent in Asia, Latin America, and Africa.

Flexibility is related to productivity in a number of ways. Conditions change and the ability to adapt to new situations is measured by flexibility. New technology and the need to be global are among the most important changes in conditions that require flexibility. Product life is shorter and the need to modify product designs requires flexibility. The productivity advantages associated with producing large volumes of identical units is being replaced by methods that permit smaller volumes to be produced having greater variety. Design variations are being used for different countries and even for various regions of the same country.

Bureaucratic organizations are dedicated to resisting change. What every operations manager should know is that bureaucracy is, by intent, the protector and champion of the status quo. Bureaucracy is the opponent of operational change. When bureaucratic constraints are removed, often by decentralization, and more recently by reengineering, afflicted organizations regain some ability to rebound.

Japanese organizations which exhibited great resilience when they first started their major export drives, also began to succumb to the problems of age and success. Age is correlated with circulatory insufficiencies in human beings and a lack of communication in organizations. This lack also is associated with zero empowerment of employees to do what makes sense instead of what the rule book dictates.

Success leads to complacency and arrogance—even though it doesn't have to. Bureaucratic organizations are very successful at inhibiting innovation and change. It remains to be seen how successful organizations worldwide will be in learning to counter these inhibitors.

Size of Firms and Flexibility

It is worth noting that small- and medium-sized firms, and new businesses as well, tend to exhibit greater flexibility and adaptability to change than large, centralized organizations. That is why under stress, Digital Equipment, Ford, GM, IBM, Sears, and other giant corporations used different forms of decentralization to improve their chances of recovering market preeminence. Organizational awareness of the need for flexibility has surfaced, but solutions for big bureaucratic companies have been elusive. The Iacocca Institute at Lehigh University in the U.S. supports research on "agile manufacturing" which recognizes the extent of the problem.

Small- and medium-sized firms are organizations with about 300 people. That number has been suggested by various managers and there also is a great deal of unanimity that the upper limit should be no more than 500.[9] Given the usual proportions of administrative personnel and those of other functions, this suggests that a sensible limit for the size of an efficient production system is in the neighborhood of 100 to 200 people. By using divisional structures, it is reasonable to assume that a number of relatively autonomous divisions of sensible size can be related within the firm.

Research and Development (R & D)

The characterization of age and size also applies to the productivity of R & D departments. Some of the largest expenditures for R & D amounting to billions of dollars have been highly unproductive. IBM and GM spent multiple billions of dollars over many years while their fortunes declined.

There is ample evidence that small research budgets produce the most impressive results. The employees of a small start-up company are in constant communication. Large research units have all of the problems that previously have been mentioned in connection with bureaucracy. Smaller dynamic organizations utilize the systems approach without having to consciously decide to do so. Everyone knows everyone else and they talk regularly. In many cases, the employees achieve a high degree of interfunctional coordination.

The productivity of R & D efforts is a legitimate concern of OM. Applied research and development fosters the next wave of new products and services. OM should be part of the team that carries the ball from start to finish for actual new product implementation. The way in which this is done is the concern of OM. Poor productivity in R & D translates into inferior processes, difficult operations, poor quality, high costs, and low profit margins. The rules for achieving high levels of productivity in research and development are similar to those which apply to good project management, which is an OM responsibility.

2.1

PRODUCTIVITY AND PRICE-DEMAND ELASTICITY

When a company is competing on price, it means that it will lose sales if a competitor offers a lower price which it cannot match. Everyone in the company looks to OM at this point. The CEO or the president asks for increased productivity which translates into greater output volume at the same or lower total costs. It is assumed that quality will remain unchanged.

To ask for increased productivity is a special way of asking for lower costs. Unions often take it to mean work faster for the same pay which makes them reluctant to participate in productivity improvement. Speeding up production can compromise quality. Operations management should try to avoid supporting productivity increases gained in this way; the improvement is temporary, at best. Other ways of obtaining lower costs include the use of cheaper components and raw materials which lowers quality.

The CEO had something else in mind. When requesting increased productivity, the CEO meant using technology and good OM methods to improve the process, not lower the quality.

The CEO's call for increased productivity is in response to competitive bids. Decreasing quality to match lower prices is not a way to hold on to customers. Improved productivity, if it is to translate into greater customer satisfaction and loyalty, must come from working smarter, not harder. This means improving productivity by means other than asking people to work faster, which usually degrades quality.

This highlights a strong functional interaction between marketing and OM. The managers of these areas are associates working together to manage the effects of price-demand elasticity (defined in the following section) on production costs and on meeting quality standards.

Elasticity Relationship

Elasticity is a rate-of-change measure that expresses the degree to which demand grows or shrinks in response to a price change. In Equation 2.0, elasticity is expressed as a function that relates changes in price with changes in demand volume. Thus:

$$V_p = f(p, \beta),\ 1 \geq \beta \geq 0$$ **Equation 2.0**

where V_p is the demand volume associated with price p, and the elasticity coefficient, β. The degree of elasticity increases as beta moves to one. Demand is inelastic when beta equals 0.

Elasticity

Rate-of-change measure expressing degree to which demand grows or shrinks in response to a price change.

A product with high elasticity experiences large decreases in demand when there are price increases, whereas, a product with low elasticity experiences small decreases in demand with the same degree of price increases. Low elasticity, called inelasticity, means that demand levels are relatively insensitive to price changes.

Marketing managers frequently ask market researchers about price elasticity of products or services to determine how fast demand falls off as price is increased. Figure 2.1 shows a highly elastic situation ($\beta = 1$) where the demand volume changes greatly with price changes. In the same figure, an inelastic case is shown ($\beta = 0$) where demand volume does not change at all with price changes. The midway case ($\beta = 0.5$) falls between the other two. The equation used for these lines is

$$V_p = kp^{-\beta}$$

where k is a constant that reflects demand volume levels.

FIGURE 2.1

PRICE ELASTICITY FOR BETA = 0, 1, 0.5, AND R = 1

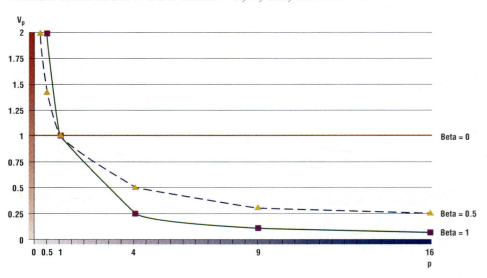

Perfect inelasticity—when demand does not change, no matter what the price—is an accurate description of the situation when an industrial customer is dependent on one supplier for special materials. Most customers try to get out of such a constraining situation.

Elasticity is a complex relationship. The rate of change between price and demand is not always smooth and regular. There can be kinks in the line or curve. These occur, for example, when an increase in price causes demand to increase, which might happen when price becomes high enough to have "snob appeal," which opens a new market. Despite difficulties, it is important to measure elasticity thereby relating price and volume.

The elasticity-productivity tie between operations management and marketing is:

1. Demand volume falls as price rises, but this is relative to what prices competitors charge. When a competitor lowers his/her prices, that is equivalent to a price increase for the customer that stays with a supplier who does not lower his/her prices.

2. To be competitive it is often necessary to find ways to match price decreases offered by competitors. This is a price-demand volume elasticity issue which assumes that quality remains unchanged.
3. If marketing lowers the price, then the profit margin will decrease in accordance with the formula $(p - vc)V$ where $(p)V$ is revenue and $(vc)V$ is total variable cost.
4. OM is asked to find a way to decrease total variable cost without degrading quality.
5. The only way to do this is by means of technology-based or methodology-based productivity improvements.
6. Marketing tries to control demand volume through pricing.
7. OM tries to match supply to demand through production scheduling and capacity planning. Marketing and OM should work together combining their interactions—the systems approach.

To understand these issues in a concrete way, consider the productivity problem of the Market Research Store (MRS). Service industries are especially concerned about productivity because so many are labor intensive and the cost of labor in the U.S. is relatively high. The Market Research Store is a market research firm that processes information.

There are many small- and medium-sized firms that perform the market research function for large companies and advertising agencies. Small- and medium-sized market research businesses have an advantage over large competitors because of their low overhead. Entry into the field does not require a large investment and labor costs are the main operating expense. Consequently, low labor costs, but not unskilled labor, are sought. Often this means making use of part-time people or training people who are willing to work at lower rates because their skills are being improved. High productivity is not associated with part-time labor, except for special groups of people such as married women with children and the elderly. The source of the labor pool is crucial.

Creation of questionnaires and design of focus groups to collect data is knowledge work. Because it entails substantial skills that would be costly if bought outside, it is done by the owners. This ability is often the most valuable asset of the firm.

The input-output transformation model provides a basis for understanding the pressure to achieve high levels of labor productivity. The main input for market research activities is data. It has to be collected and put into the computer. The major input activity is labor intensive with people sitting at computers, typing information into a program that can calculate averages and prepare standard-format reports. The same data may be entered twice in order to check for consistency between data-entry operators. The input costs are variable because they increase with the amount of data that must be processed. In addition, purchased materials such as computer paper, printer cartridges, and electricity add to the variable costs.

The outputs of the Market Research Store are graphs, charts and tables, as well as other forms of aggregated data, and statistical reports, such as regression analyses. These would be studied by market research analysts to spot trends, determine conditions, and write reports about market performance, and new product comparison tests, among other things. The volume of output can be measured in a variety of ways that relate to processing time.

Output rates also can be treated in terms of revenues generated. Both the number of units processed and the revenue obtained are demand-driven variables. When output falls because demand slacks off, the sales force contacts potential clients and offers them market research studies at discounted prices. The lower prices are somewhat effective in bringing in new business because demand is at least partially elastic to price. Marketing has some control over the volume of business that production must process.

The Market Research Store has a minimum capital investment in computers and printers. Its owners have no intention of trying to attract additional investment capital. Its computers and printers are networked in a topsy-turvy fashion. When heavy loads occur, the MRS rents additional equipment. It owns the software to do statistical analyses, graphics, and desktop publishing—all of which are needed for their clients' reports. The owners of the MRS are so busy processing clients' reports and dealing with processing problems that they have little time to determine if other types of computers, different network configurations and alternative software packages would be better for them.

The problem is that one of their long-term competitors has invested in developing networks to connect homes and offices with laptop computers. This allows part-time employees to sign out a laptop computer from the stockroom and work from home at labor rates that are more favorable than MRS can offer. The result is that the competitor is regularly underbidding MRS with many of its oldest clients. The volume of work is dropping off and the owners of MRS are perplexed. They do not have the money to support the technological expansion that is required to compete with their market research adversary. Somehow they need to increase the productivity of their operations because the market is sufficiently price elastic to put them in trouble.

The owners of MRS hope that there are better methods for doing work using essentially the same equipment they have used in the past, and leasing some additional equipment if that would give them an advantage. At first, they competed fiercely, dropping prices to meet their competition. Unfortunately, they began losing money on jobs which limited their aggressive stance. Are there operations management methods which can return the MRS to a competitive position? Because the owners are excellent in market research questionnaire development, response analysis and interpretation, and report writing, it may be presumed that they were not familiar with the methodology of OM. If that is the case, they have a good chance of saving a substantial amount of money by hiring an operations manager.

The MRS can obtain great productivity improvements by using better production methods, available those who have studied the field of OM. Further, certain kinds of market research jobs are more profitable than others. OM and marketing can work together to ascertain which jobs are best for profits and to make every effort to sell those jobs. In addition, some customers and types of jobs are less price elastic than others. For example, customers are willing to pay more for situations that require precision. This is equivalent to paying for higher quality.

By specializing in certain kinds of jobs and providing the highest quality reports, MRS can improve its competitive position over time. Also, by using teamwork, MRS has a good chance to pull out of its difficulties and start planning for a better tomorrow. This will include the use of up-to-date technology. The MRS should consider the advisability of following their competitor's lead with part-time employees working at home using laptop computers networked to the central office.

Summing up, an OM approach to improving productivity can be found for this company. It must address making shifts in the products that MRS sells and making changes in their processes. MRS will gradually learn to use new technology to their advantage through training its employees to perform with excellence in the new environment.

TECHNOLOGY

Object-Oriented Programming

To increase productivity of computer software programmers, major U.S. companies are turning to object-oriented programming (OOP). This technological innovation allows software to be built using "objects" or modules of computer code to control specific functions. These modules are then reused in new applications. Object-oriented programmers may assemble a number of objects, then customize as necessary, in creating new applications.

Advantages of OOP include the ability of the software to work within an existing system, increased speed with which new programs can be developed, and higher quality applications. Experts say that object-oriented programs can be written in one-half to one-fifth the time it takes using traditional programming methods. According to industry analysts, the high cost of OOP is a disadvantage of this technology at present, but the cost should be driven down as more programmers are trained in the area and as orders for OOP increase.

One of the first object-oriented applications specific to production and operations management was the Protean Maintenance System, developed by Marcam Corp. of Newton, Massachusetts. This system includes comprehensive maintenance and inventory applications which can interface with a company's existing purchasing and resource planning systems. Marcam plans to release additional Protean products over the months and years ahead including applications for logistics, financials, procurement, and costing, as well as additional production and maintenance modules.

"We are very excited about the development productivity that object-oriented programming is affording us in terms of creating value for our customers," said Paul Margolis, president and CEO at Marcam. He added that the shipment of the Protean Maintenance system in June 1994 validates the company's "aggressive delivery plans for object technology applications over the coming years."

Marcam Corp. has 35 offices worldwide and is represented in more than 60 countries. The company's customers include Coca-Cola USA, GAF, Gillette, Kraft General Foods, and Westinghouse, among many others.

Sources: David Ranii, "The latest object of their affection," *News and Observer,* October 2, 1994, and "Marcam delivers object technology applications as promised," Business Wire, June 27, 1994.

Elasticity of Quality

The volume of goods or services sold as a function of price is the traditional focus of elasticity analysis. Years ago that simple model may have sufficed, but it no longer is valid. This subject is of mutual concern to OM and marketing.

Customers in the marketplace take both price and quality into account. Customers' quality expectations usually override price considerations. This also applies to commercial and industrial customers.

Using new technology such as laptop computers allows people to work virtually anywhere, thereby improving productivity.
Source: Motorola

A product or service that has special qualities is said to have uniqueness which means that other competitive products have a lower degree of substitutability. For example, a product with a special feature, or a service rendered by a well-liked person, have a competitive advantage. The competitive analysis would show that they are not competing head-on, as if they were identical, perfect substitutes for each other. Perceived high quality renders a product less vulnerable to substitution which means that the product is less quality elastic.

How can the effects of quality on demand levels be determined? This is equivalent to asking how to determine the quality-demand volume elasticity. There are ways in which market research can approach this issue that are similar to the way price-demand volume elasticity is determined. Essentially, it is necessary to establish how much extra money customers would be willing to pay for superior quality or for an added quality feature.

By noting the distribution of the additional amounts of money that people would pay for superior quality or an added feature, it is possible to quantify the effects of quality and price on demand elasticity. In effect, two price elasticity studies are conducted. The first study is done without the added quality. The second study is done with it. Throughout this discussion, it should be kept in mind that *the achievement of quality standards is a direct responsibility of OM*.

Table 2.2 indicates that the demand elasticity in the second study should be less than in the first one. This is because the added quality in the second study makes alternative products or services less substitutable.

TABLE 2.2

COMPARING ELASTICITIES AND BETAS WITH HIGHER OR LOWER QUALITY LEVELS

Study	Quality	Beta Values	Elasticity
1	lower	closer to 1	more elastic
2	higher	closer to 0	less elastic

Although this is a book about OM and not market research, these functions are highly interdependent. Market research enables OM to determine the kind of connections that link quality, price and demand elasticity in the customer's mind.

These factors relate design and process decisions with the financial choices that are available to the firm. The system interaction includes the fact that quality varies with the kind of equipment that is used, and the amount of training that the employees receive.

2.2

ECONOMIES OF SCALE AND THE DIVISION OF LABOR

There are two principles that OM must be schooled in to take advantage of productive opportunities: economies of scale and the division of labor. **Economies of scale** are reductions in variable costs directly related to increasing volumes of production output.

Economies of scale
Reductions in variable costs directly related to increasing volumes of production output.

Economies of scale are driven by volume (V). Scale, as used here, is another word for volume. The equation for total variable costs is both an OM and marketing responsibility.

$$Total\ Variable\ Costs = vc(V)$$

V is a function of the total market size that exists, the number of competitors and their shares, and the controllable variables of the organization's price and quality. While marketing and OM work together on the consequences of V, OM is working on the reduction of vc—which is also a function of V—with no loss in quality.

Materials and labor are an important component of the variable per unit cost, vc. The design of the product or service determines what materials are needed. It is common knowledge that greater purchase volumes generally are rewarded with discounts. That is only one of the interactions of vc and V. The machines that can be used for high volume outputs are significantly faster than the kind that are economic for low volume outputs. High volumes can sustain pre-engineering and improvement studies of the interactions between the design of jobs and the processes used, while low volumes cannot. High volumes generate learning about how to do the job better; low volumes do not.

The design of jobs determines the amount of labor and the skill levels that are required. The responsibility for low unit variable costs leads OM to want high volumes so that it can take advantage of the resulting economies of scale.

For many reasons, including the material discounts previously discussed and a general learning effect, variable costs per unit decrease as volume increases. This result, called the "economies of scale," is quite similar to the Boston Consulting Group's (BCG) "Experience Curve" which yields a 20 to 30 percent decrease in per unit costs with each doubling of the volume.[10] It is reasonable to consider the "doubling of volume" as a surrogate for the "doubling of experience," as in BCG terminology.

An approximation of this relationship is shown in Figure 2.2.

FIGURE 2.2

ECONOMIES OF SCALE WHERE PER UNIT VARIABLE
COSTS DECREASE BETWEEN 20 TO 30 PERCENT
WITH INCREASING VOLUME

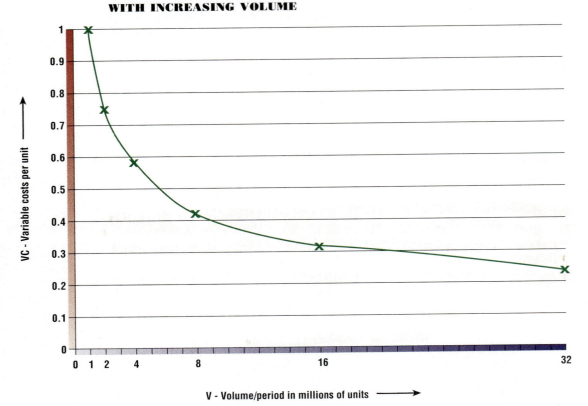

A coincident concept, proposed by Adam Smith (a Scottish economist) in the 1700s, was for the division of labor.[11] Labor was to be divided into specialized activities that could be honed to ever-greater skill levels. This notion follows from the theory that "practice makes perfect." To make division of labor worthwhile, the volume of production must be sufficient. Adam Smith said, "The division of labor depends on the extent of the market." With a large enough volume, activities could be segmented, and serialized process flows could be developed. Division of labor appears in the history section which follows.

2.3

THE HISTORY OF THE IMPROVEMENT OF OM TRANSFORMATIONS

Literacy in OM requires an understanding of how the OM field has developed with respect to the transformation process and, thereby, productivity, quality and variety. The stages of history have moved production and operations capabilities from custom work through high-speed continuous output systems.

Attention shifts from custom crafts which are art-based to the theory of production which has evolved over time. This theory consists of six steps and a potential seventh one. There is an emphasis on manufacturing because the theory evolved from the production of goods, but it lends itself to service operations.

Figure 2.3 depicts the history in a timeline chart. Dates mentioned are approximate because it is not possible to pinpoint exactly when each contribution was made.

FIGURE 2.3

THE HISTO-MAP: A TIMELINE OF OM DEVELOPMENTS

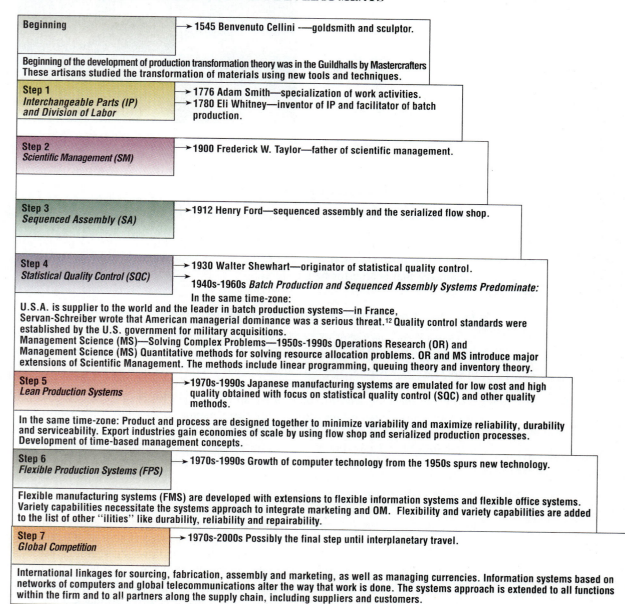

Beginning
→ 1545 Benvenuto Cellini -—goldsmith and sculptor.

Beginning of the development of production transformation theory was in the Guildhalls by Mastercrafters These artisans studied the transformation of materials using new tools and techniques.

Step 1
Interchangeable Parts (IP) and Division of Labor
→ 1776 Adam Smith—specialization of work activities.
→ 1780 Eli Whitney—inventor of IP and facilitator of batch production.

Step 2
Scientific Management (SM)
→ 1900 Frederick W. Taylor—father of scientific management.

Step 3
Sequenced Assembly (SA)
→ 1912 Henry Ford—sequenced assembly and the serialized flow shop.

Step 4
Statistical Quality Control (SQC)
→ 1930 Walter Shewhart—originator of statistical quality control.
1940s-1960s *Batch Production and Sequenced Assembly Systems Predominate:*
In the same time-zone:
U.S.A. is supplier to the world and the leader in batch production systems—in France, Servan-Schreiber wrote that American managerial dominance was a serious threat.[12] Quality control standards were established by the U.S. government for military acquisitions.
Management Science (MS)—Solving Complex Problems—1950s-1990s Operations Research (OR) and Management Science (MS) Quantitative methods for solving resource allocation problems. OR and MS introduce major extensions of Scientific Management. The methods include linear programming, queuing theory and inventory theory.

Step 5
Lean Production Systems
→ 1970s-1990s Japanese manufacturing systems are emulated for low cost and high quality obtained with focus on statistical quality control (SQC) and other quality methods.
In the same time-zone: Product and process are designed together to minimize variability and maximize reliability, durability and serviceability. Export industries gain economies of scale by using flow shop and serialized production processes. Development of time-based management concepts.

Step 6
Flexible Production Systems (FPS)
→ 1970s-1990s Growth of computer technology from the 1950s spurs new technology.
Flexible manufacturing systems (FMS) are developed with extensions to flexible information systems and flexible office systems. Variety capabilities necessitate the systems approach to integrate marketing and OM. Flexibility and variety capabilities are added to the list of other "ilities" like durability, reliability and repairability.

Step 7
Global Competition
→ 1970s-2000s Possibly the final step until interplanetary travel.
International linkages for sourcing, fabrication, assembly and marketing, as well as managing currencies. Information systems based on networks of computers and global telecommunications alter the way that work is done. The systems approach is extended to all functions within the firm and to all partners along the supply chain, including suppliers and customers.

The flexibility concept joins computers and equipment, revising the way machines operate. Source: ©1993. Westvaco Corporation; Photographer—David Pollack

Discussion of the History of P/OM

The capability of P/OM processes to deliver goods and services has changed in steps or stages over time. The study of the history of OM production transformation processes allows us to determine which events triggered these stages of production theory. The ultimate goal is to learn the theory, understand it, and possess the advantages that accrue for being literate.

Artisans, Apprentices and Trainees—The Beginning

The Renaissance period (1300s–1600s) signaled a surge of intellectual and productive vitality in Europe. That surge swept away the dark ages and fostered accomplishments in the arts and sciences centered around artisans, apprentices and craft guilds. Production transformations were by hand and output volumes were very small.

Before the Industrial Revolution began (around 1770) craft guilds emphasized pride of workmanship and training for basic manual operations with appropriate handtools. The shoemakers' children learned from their fathers and mothers. Process techniques were handed down from generation to generation.

From a transformation point of view, this was good management of the labor inputs. The use of apprentices improved productivity in the artisans shops because the less skilled (and lower paid) apprentices did much of the preliminary work. This freed the master craftsmen to devote their time to the activities requiring higher skills. On-the-job training produced a continuous stream of greater skills.

Apprenticeship still has significance for many service functions. Great chefs almost always are the pupils of great chefs. The formula would seem to reside in the balance of art and science. When the important knowledge resides in the minds and hands of skilled workers, then the percent of art is high and the percent of science is low. Over time, this percentage has shifted in manufacturing so that engineering, technology and computer programming play an increasing role.

The art element is disappearing which means that the valuable know-how now resides mostly in the minds, not in the hands of the critical workers. It used to be that the tool and die department was crucial to the success of metal-working companies and the best die makers were considered artists. (Tools and dies are the shape formers in the metal-working businesses.)

Now, CAD and CAM (discussed below) are primarily science and the old industrial arts are giving way to the new programming arts. This also is happening in service industries, and is an effect that can be expected to accelerate in the future.

Interchangeable Parts (IP)—Production Theory's First Step

Eli Whitney invented the concept of interchangeable parts for the fabrication of rifles around 1780, which coincides with the dates usually given for the beginning of the Industrial Revolution. The notion of interchangeable parts was the catalyst around which new methods for production transformation began to develop. These methods spawned and supported the Industrial Revolution.

Whitney was not the sole inventor of interchangeable parts. In France, Nicolas LeBlanc had invented the same OM concept. Neither Whitney nor LeBlanc knew about each other's ideas. Whitney obtained a United States government contract for "ten thousand stand of arms." The contract was awarded because of his newly developed production capabilities.

The concept of **interchangeable parts** is defined as follows: it allows batches of parts to be made, any one of which will fit into the assembled product. For example, headlights, fenders, tires and windshield wiper blades are not specially made for each car. One 60-watt bulb is like another and does not have to be fitted to each socket. The reason that the parts are interchangeable is that each one falls within the design tolerances.

Interchangeable parts
Concept that allows batches of parts to be made, any one of which will fit into the assembled product.

Machines that could produce parts to conform to the designer's tolerances were the keystone. Hand labor, better suited to custom work, began to be replaced by machinery. The effects of this change hastened the Industrial Revolution. Within a short time IP was an accepted part of the production transformation process being applied to the manufacture of rifles, sewing machines, clocks, and other products.

In 1776, Adam Smith saw that the use of the division of labor as a means of increasing productivity was market volume dependent. The pin factory that he studied had sufficient production volume to warrant specialization. The production transformation process was revolutionized—combining worker specialization with interchangeable parts changed all of the productivity standards. Expectations were raised to new levels.

Scientific Management (SM)—Production Theory's Second Step

Frederick Winslow Taylor (1865-1915) introduced **scientific management**, the numerical measurement and analysis of the way work should be done. One of his landmark studies dealt with the speed and feed rates of tools and materials for metal cutting. Other studies focused on how to lay bricks and how to move iron castings. Taylor's testimony at hearings concerning the setting of rational railroad fees for shipments in interstate commerce brought national prominence to his analytic methodology.

Scientific management
Numerical measurement and analysis of the way work should be done.

This step in production theory added the idea that the transformation processes could be improved by studying and simplifying operations. This view required rationalizing the job, the workplace and the workers. Strangely, it had been over-looked until the turn of the twentieth century. Finding economies of motion and putting materials near at hand were the kinds of improvement that Taylor addressed. In this step, the workplace and the design of the job were enhanced to improve the productivity of the transformation process. These industrial engineering ideas, often called "methods engineering," have proved to be as useful for service applications as for manufacturing.

Called the father of "scientific management," Taylor was one of the key progenitors of industrial engineering (IE). There were others as well. Associated with this era are Henry L. Gantt, Frank and Lillian Gilbreth, and other individuals who were attempting to develop a theory of managing workers and technology in the U.S. Henri Fayol was developing similar management theories in France.

Taylor started the process of systematizing all of the elements that are part of the manufacturing system. The same industrial engineering techniques developed by Taylor are still used by banks, insurance companies and investment houses as well as by truckers, airlines and manufacturers. They are well-suited for repetitive operations such as those which characterize fast food chains and information processing systems. Industrial engineering methods are recognizable as the forerunners of techniques which are presently applied in the search for continuous improvement.

Taylor and his associates developed principles and practices that led to a present-day backlash. He and other contributors to scientific management have been accused of dehumanizing the worker in pursuit of efficiency. Today, the accusation might hold, but it did not when judged by the value system of the early 1900s.

The criticism does not damage the case for benefits that can be derived from using the industrial engineering approach which grew out of scientific management. Work simplification and methods engineering are both IE techniques for making jobs better for workers.

Sequenced Assembly (SA)—Production Theory's Third Step

Sequenced assembly

Allows assembly to be a continuous flow shop process; timing must be perfect.

In 1912, Henry Ford developed **sequenced assembly** which allows assembly to be a continuous flow shop process. Timing must be perfect so that what is needed for assembly arrives on time. Ford developed the sequenced assembly process as a continuous flow production line for automobiles, changing the pace from batch to continuous sequenced assembly.

The serialized flow shop was born. The key was learning to achieve synchronization and control of the process flows. The moving assembly line required a high level of component interchangeability. Ford succeeded in achieving complete synchronization of the process flows.

By means of the principles of interchangeability, division of labor and flow synchronization, Ford altered the production transformation process. He changed the perception of productivity standards and goals in a conclusive way. In so doing, he built an industrial empire which helped the U.S. become the world leader in productivity. The U.S. continues to maintain its lead, although other nations—especially those considered to be less developed countries—have been improving their productivity consistently.

Ford's contribution to production theory and to the revision of the transformation process had a major impact on the Japanese automobile industry. It also affected other industries of many kinds all over the world. There was a new rhythm to the transformation process.

Contrast U.S. and Japanese production processes in the 1980s. The major portion of production and operations activities in the U.S. utilized batch processes. Batch work with small lots does not lend itself to the kind of synchronization that applies to the automobile industry or the continuous flows of chemical processes.

When the Japanese export industry began to compete aggressively in global markets, they chose to shun batch-type production systems. Instead they elected to specialize in high-volume, serialized flow shops which extended the concept and application of assembly synchronization to manufacturing and assembly systems.

Statistical Quality Control (SQC) — Production Theory's Fourth Step

Interchangeable parts required manufacturing methods that made batches of parts conforming to tolerance limits. Shewhart developed the theory of **statistical quality control (SQC)** which enabled manufacturing to design and control processes which could achieve these objectives. SQC was focused on the producer's ability to control the variability of the process that was making the parts that had to fit within the specified tolerance limits. For the first time, the output of the transformation process could be stabilized and controlled. This was a major contribution to production theory.

Statistical quality control (SQC)
Producer's ability to control the process's variability of part qualities within specified tolerance limits.

Walter Shewhart's major work was published in 1930 describing his concepts about why SQC works and how to apply it.[13] Deming and Juran also participated in the development of SQC theory and later on played a crucial role in its implementation and dissemination.[14]

The U.S. was the first country that consistently used SQC, which it did through the 1940s and the early 1950s, but by 1960 the majority of SQC users were in Japan. U.S. organizations reported that they had dropped SQC to make cost reductions. Quality was considered good enough to replace costly staff departments with inspectors at the end of the production line. By the 1980s, however, under great competitive pressure from quality-driven Japanese organizations, many U.S. companies restored SQC and enhanced it with broader concepts into total quality management (TQM) activities.

Organizations like Motorola and Xerox are considered to be pioneers leading the development of TQM within the framework of the systems approach. The total quality management approach applied to the production transformation system integrates the goals of productivity and quality. It represents a major step forward in the theory of production and an organizational feat to have gained broad acceptance at all levels.

The next three steps are in formative stages and their impact on productivity cannot be fully evaluated at this time.

Lean Production Systems (LPS) — Production Theory's Fifth Step

During the 1970s–1990s, Japanese organizations spearheaded by Toyota, Mazda, Nissan, Honda, and Mitsubishi developed a new kind of production methodology called **lean production systems**. These systems combine a deep understanding of quality with a desire to be fastest and a fanatical distaste for all kinds of waste.

Lean production systems
Combine quality with speed and waste avoidance.

Time wasted was singled out. Every effort was made to use pre-engineering of products and process design to maximize quality achievements, minimize variability, and do it all as rapidly as possible. Part of being lean was being fast—in production. The Japanese were not fast in reaching decisions. For the most part, the export industries became lean producers with high-output volume targets, minimum cycle times and rapid new product development.

By introducing time management and goals for short cycle times and rapid project development, other new factors were introduced into the production transformation system. The timing of transformations rose to a new level of importance, and, secondly, rapid project management began to mean that the transformation process could be changed from doing one thing to another very quickly.

The Japanese auto industry has been leading in the development of lean and fast production systems. Toyota's production planners, who were architects of the revised production system, stated that Toyota's ideas were a continuation of the concepts that Henry Ford had been developing.[15] In Europe, the notion of leanness was directly associated with speeding up cycle times and project development times. Half-time systems were advocated by Saab, meaning the goal was to cut in half the time presently required to do anything.

Most U.S. organizations are trying out some aspects of lean and rapid manufacturing methods. Focusing on TQM and JIT, Motorola's management set the goal of reducing defectives to less than four defects per million parts.[16] Having demonstrated that a near-zero defects objective is attainable, Motorola now has set another radical target which is to reduce existing cycle times by 90 percent. This means that a part which presently takes ten minutes to produce eventually will be made in one minute instead of five minutes as dictated by the half-time objective previously mentioned.

Highlighting reduced time objectives brings new features to the production transformation process.

Flexible Production Systems (FPS)—Production Theory's Sixth Step

Computer technology and production machinery technology are two distinct areas of technological development. When these technologies are combined, opportunities develop for computers to instruct and control machines instead of needing hands-on control by human operators.

These developments were quite primitive in the 1960s and 1970s, but they are becoming more sophisticated and reliable now. In the mid-1990s, few factories in the world do not employ some form of computer-controlled machinery. Nevertheless, flexible manufacturing equipment still must be classed as being experimental. The sixth step is predicted by many, rejected by some. The real issue is how long it will take to come.

Flexible manufacturing systems (FMS)

Produces a high variety of outputs in small quantities at low cost; components include computer-controlled system changeover and machine software interfaces.

The initial thrust was in manufacturing. **Flexible manufacturing systems (FMS)** are designed to produce a high variety of outputs at low cost. Computer-controlled changeover is engineered into the system. Instead of human hands changing machine settings, electronics and mechanics provide the interface between the computer and machinery. They are flexible, allowing fast and inexpensive changeovers. The equipment can be programmed to move from one product set up to another product set up in an instant.

Computer-aided design (CAD)

Design software which communicates specification to CAM software which translates, instructs, and controls production machinery.

Another characteristic of FMS is design and machine software that talk to each other. **Computer-aided design (CAD)** software does the new design drawings and is able to test the characteristics, such as strength. CAD can communicate the

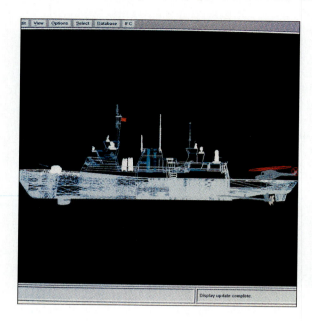

CAD terminals communicate design specifications for the construction of ships and are linked to an integrated computer-aided manufacturing (CAM) production network throughout the shipyard.
Source: Litton Industries, Inc.

design specifications to software that translates, instructs, and controls the production machinery; this software is called **computer-aided manufacturing**. CAD and CAM work together to determine the feasibility of manufacturing the new design and suggest improved design alternatives.

CAD/CAM technology is used to design and manufacture many different products such as semi-conductors, automobile grills and aircraft parts. John Deere has invested millions of dollars in the creation of CAD/CAM systems for the manufacture of tractors. Boeing uses CAD/CAM for all of the parts of its 777. Design has moved from the draftsman's table to the computer.

The flexibility concept joins computers and equipment of many other kinds, including assembly-line processes and office machines. It also is possible and often desirable to include human beings in the network. Flexibility can be applied to information systems—flexible information systems (FIS) and to flexible office systems (FOS) as well as to FMS.

The capability to produce small numbers of many varieties includes extensive application of systems thinking to integrate marketing needs and OM scheduling abilities. Marketing forecasts and OM decides what to make, but it is constrained by the FMS menu which was predetermined at the planning stage.

The need to produce increased variety is market-driven. The transformation process has to be able to change over from making one thing to another, quickly and inexpensively. A great deal of effort is going into altering the transformation process so it can deal with the goal of increased variety. Technology and methodology must enable nearly instant set ups and changeovers from one model to another to satisfy the market demands. The payoff will be increased productivity of the joint production-marketing system, as exemplified by customized jeans, described below.

Levi Strauss has put the customer directly in touch with the factory. "Sales clerks at an original Levi's store can use a personal computer and the customer's vital statistics to create what amounts to a digital blue jeans blueprint. When transmitted electronically to a Levi's factory in Tennessee, this computer file instructs a robotic tailor to cut a bolt of denim precisely to the woman's measurements."[17]

Computer-aided manufacturing (CAM)

Software used together with CAD software to determine the feasibility of manufacturing a new design or suggest alternatives.

There is a dedicated strategic effort to achieve competitiveness through flexibility in America. It combines the goals of leanness (speed) and flexibility and is known as Agile Manufacturing Enterprises.[18] This name is meant to emphasize the ability to react quickly, and with flexibility. The agile organization is supposed to be alert and nimble, keen and lithe. Bureaucratic organizations cannot qualify.

Global Competition: The Year 2000 Plus— Production Theory's Seventh Step

It is conjectured that in the future the transformation process will continue increasing in complexity and productivity. On a worldwide scale, a broad range of goods should be within the spending capabilities of more people. More management will be needed to plan and control the systems. More operations managers will be required with far fewer workers on the production line. More people will be sharing services— such as education and health care—that are mutually rewarding. Onerous service tasks will be relegated to service robots. Hopefully, people will have more time to spend their money as they wish.

The input-output production transformation model will be internationalized. There will be global competition at every link in the supply chain. International sourcing, fabrication, assembly, distribution and marketing will prevail. The costs of the inputs and the values of the outputs will be affected by dozens, if not hundreds, of different currencies. Managing currencies will be part of the transformation process.

Information systems will be based on international networks of computers. Global telecommunication systems will transmit conversations which are spoken in eighty different languages. Translation will be accomplished by language computers.

The systems approach is extended to all functions within the organization and all partners along the supply chain, including suppliers and customers. Production transformation processes will require the best of management while decreasing burdensome labor components. Substantial productivity increases will be required.

SUMMARY

Productivity is a critical business variable because it strongly impacts the bottom line. Because it is so important, productivity is a major OM issue. There are many ways to measure productivity, including those used to describe the national economy. The measures can be made in terms of labor productivity, capital productivity, or both. For business units, the measures relate more directly to the kind of activities the company does. These are the operational measures of productivity used by an organization. Productivity measures should be adjusted to capture the relevant system—so the systems approach applies.

Bureaucracy inhibits flexibility and, therefore, constrains productivity improvement. Small firms are less bureaucratic than large ones and, therefore, are more flexible and able to pursue initiatives for greater productivity. Research and Development (R & D) organizations are related to OM and should be studied in terms of their productivity, which is usually better for smaller organizations.

Productivity and price-demand elasticity are interdependent. The importance to OM of the elasticity relationship for both price and quality are explained. Then, economies of scale and the division of labor are linked to productivity. Finally, the history of the improvement of OM input-output transformations and resultant productivity increases is presented in seven steps.

▶ **Key Terms**

capital productivity *45*

economies of scale *57*

interchangeable parts *61*

multifactor productivity *44*

sequenced assembly *62*

computer-aided design (CAD) *64*

elasticity *51*

labor productivity *44*

relative productivity *45*

statistical quality control (SQC) *63*

computer-aided manufacturing (CAM) *65*

flexible manufacturing systems (FMS) *64*

lean production systems *63*

scientific management *61*

▶ **Review Questions**

1. Why is productivity measurement important to operations management?

2. What is the importance of productivity measurement to marketing management?

3. What role does the systems approach play with respect to productivity measurement?

4. What is good and bad about bureaucracy with respect to productivity? One hundred years ago it was considered a major organizational advance over what preceded it. Why is it now considered an impediment?

5. Explain how the division of labor might help the Market Research Store put together a report for a client.

6. What are the six historical steps in the development of production theory? Is the suggested seventh step likely to have an impact on the future of operations management?

7. Explain the concept of interchangeable parts for:
    a. a vacuum cleaner
    b. a jigsaw puzzle
    c. a flashlight

8. Describe what is meant by lean (or agile) production systems.

9. How does the global viewpoint create any new problems or opportunities for operations management?

10. What is the relationship between productivity and elasticity of price and demand volume?

11. What is the relationship between productivity and elasticity of quality and demand volume? Is price implicitly included?

12. Explain what economies of scale are and how they relate to productivity and to the systems approach.

▶ **Problem Section**

Data for Problems 1-6: In a one year period, the Productive Components Corporation (known as PCCorp) has shipped units worth $1,200,000 to its customers. It produced units worth $250,000 for finished goods (FG) inventory. PCCorp has $50,000 of work-in-process (WIP) units. During the same period of time, PCCorp had a labor bill of $140,000.

Its capital expenses for the year are calculated to be $430,000. Materials were purchased costing $530,000. Energy expenses were $225,000 and miscellaneous expenses were estimated to be $75,000. All figures are annual.

1. Calculate the labor productivity for PCCorp with respect to units shipped.

2. Calculate the multifactor productivity composed of labor and capital for units shipped plus finished goods for PCCorp.

3. What is PCCorp's total productivity?

4. What is PCCorp's capital productivity?

5. The value of the units shipped must be reduced because $350,000 worth of them have been returned as defective. Of the finished goods units (FG), $150,000 have been found to be defective. All work-in-process (WIP) units are within tolerances. Rework on the defective units reduces their value by 70 percent. What is the cost of the quality problem that has surfaced? Discuss what this means in qualitative terms regarding productivity.

6. Calculate the multifactor productivity composed of capital, materials and energy consumed for the total reworked output, which consists of units shipped plus FG plus WIP.

7. Productivity relates the value of outputs to the cost of inputs. Good productivity is shown by high values; improved productivity by higher values over time.

Does the following equation capture the meaning of productivity?

$$\text{Productivity} = \frac{pV}{(vc)V}$$

where pV = price per unit × Volume per year, and vc is the variable cost per unit.

8. Because V is in both the numerator and the denominator of the productivity equation,

$$\text{Productivity} = \frac{pV}{(vc)V}$$

reduce the equation as follows, and interpret the results.

$$\text{Productivity} = p/(vc)$$

Is productivity well described by the ratio of price per unit to variable cost per unit?

9. Elasticity describes how V changes as a function of p.

a. Does the following equation capture the meaning of elasticity?

$$V_p = kp^{-\beta}$$

As in the text, elasticity of demand volume V_p is a function of a price, p. The coefficient of elasticity is beta, which ranges from zero to one, and k is a constant.

b. Explain the following statement:

If beta equals zero—indicating perfect inelasticity—the constant, k has a value equal to the demand volume for the product or service which is irrespective of price.

c. Why is it critical for OM to make every effort to continually raise the quality standards for the product described in Problem 9(b) above?

d. Fill in the table below for the elasticity equation:

$$V_p = kp^{-\beta}$$

where $k = 5,000$ and $\beta = 0, 1, 0.5$, and 0.25.

p	V_p ($\beta=0$)	V_p ($\beta=0.25$)	V_p ($\beta=0.5$)	V_p ($\beta=1$)
1.00				
1.50				
2.00				
2.50				

10. Link productivity, elasticity and profit (P) using the typical profit equation:

$$P = (p - vc)V - FC$$

where: p = \$1.00, vc = \$0.50 and fixed costs (FC) = \$10,000.
Also, k = 100,000 units per year when β = 1.0 and p = \$1.00.

a. Describe what occurs if the price is raised to \$2.00.
b. Assume that increasing investments in technology improves productivity, such that by increasing FC to \$15,000, vc is reduced to \$0.40. Does this investment make sense? Do the necessary calculations? Explain.

11. For Problem 10, assume that the equation for the economies of scale is:

$$vc_V = KV^{-\alpha}$$

where α = 0.5 and K = 160.
Also, vc_V is the per unit variable cost for volume V, K is a constant and alpha is the co-efficient for economies of scale.

12. Explain how the division of labor might increase the size of the alpha coefficient (in Problem 11). How might this affect the profit equation?

▶ Readings and References

William J. Baumol, Sue Anne Blackman and Edward N. Wolff, *Productivity and American Leadership: The Long View*, MIT Press, Cambridge, MA, 1991.

W. E. Deming, *Out of the Crisis*, MIT Center for Advanced Engineering Study, 1986.

Soloman Fabricant, "A Primer on Productivity," NY: Random House, 1969.

Jay W. Forrester, "Changing Economic Patterns," *Technology Review*, August/September, 1978.

J.M. Juran, and F.M. Gryna, Jr., *Quality Planning and Analysis*, NY, McGraw-Hill, 2nd Ed., 1980.

Taiichi Ohno, *Toyota Production System: Beyond Management of Large Scale Production*, Diamond Publishing Co. Ltd., 1978.

Jean-Jacques Servan-Schreiber, *The American Challenge*, New York: Atheneum, 1968.

W. A. Shewhart, *Statistical Method from the Viewpoint of Quality Control*, Washington, DC, The Department of Agriculture, 1939.

Adam Smith, *An Enquiry into the Nature and Causes of the Wealth of Nations,* 1776.

Andrew Britton, Economic Commission for Europe Discussion Papers, Volume 1, No.1, 1991, *Economic Growth in the Market Economies 1950–2000*, United Nations, NY, 1991, (GE.91-23315).

Decision Trees

Decision trees (DTs) provide another method for representing decision problems. There are certain kinds of problems that DTs can organize which are very cumbersome in matrix form. It is a toss-up, however, as to which is preferred.

The decision problem presented in Supplement 1 can be modeled with decision trees. Later on, a productivity decision example will be given where decision trees are much preferred to matrices, and by extension, where they must be used. First, the decision tree is drawn in Figure 2.4 to represent the shipper's situation.

F I G U R E   2 . 4

DECISION TREE FOR THE SHIPPING DECISION

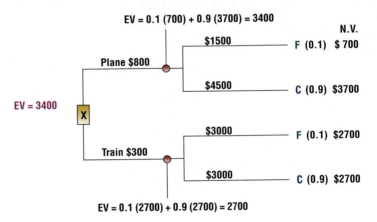

Legend:
N.V. = Net Value
F   = Fog
C   = Clear

Below are the rules for using DTs. It is helpful to observe how the problem is translated from decision-matrix form into a decision tree.

1. Start with the various strategies emanating from the first decision box. Some trees have many decision boxes (see above) although this tree has only one. Also, there can be many lines that fan out from any decision box. This tree has only two.
2. Label the strategies with their costs. At the end of each strategy line there is a chance-event circle. From that circle emanate the states of nature that can occur. There is a line for every state of nature and each is labeled with the appropriate probability and revenue. By tracing out the costs and revenues encountered along each branch of the tree, a unique endpoint for profits is derived. Label the net value (N.V.) of each endpoint.
3. Work backwards through the tree. Wherever two or more states of nature merge at a chance-event circle, obtain the expected value for profit at that circle and mark it.
4. Continue working backwards through the tree. Whenever two or more strategies (such as by plane or by train) merge at a decision-point box, *carry the largest expected value to that box.* For this tree, the choice is between 3400 and 2700. Because the former is larger, and this tree is profit oriented, label the decision-point box with *EV* = 3400. If the decision problem is not in terms of profits, then use net costs and expected costs.

5. When the decision tree has a sequence of strategies (see the next example) then, working backwards, follow that sequence of strategies which accounts for the largest *EV* for profit. This is the *EV* that has been carried to the first decision-point box of the network. For the preceding tree, the strategy ship by plane, which is the top branch of the tree is the winner.

Decision Sequences of Greater Length—A Game of Chance

Decision problems that involve a sequence of alternative strategic choices are particularly well suited to decision tree organization for solution. When there are many sequences of the type "if this, then that," the decision tree is the only sensible way to represent such complexity.

Start with a simple example. Assume the following game of chance:

1. If *X* chooses to play the first game, called *A*, and wins, then *X* can play the second game, called *B*. *X* does not have to play *B* or *A*.
2. To play *A* or *B* costs $50.
3. If *X* wins *A*, *X* receives $50. There is an 80 percent probability of winning *A*. If *X* wins the second game, *B*, *X* receives $200. There is a 60 percent probability of *X* winning *B*. What should *X* do?

Figure 2.5 shows that the expected value for trying to play both games is $46. There is a 20 percent chance that *X* will lose the first game and be out the $50. However, there is an 80 percent chance that *X* will win the first game which means that *X* just breaks even. *X* must go on to play the second game.

FIGURE 2.5

DECISION TREES FOR GAME

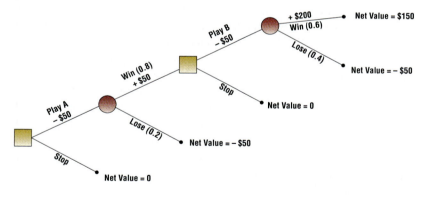

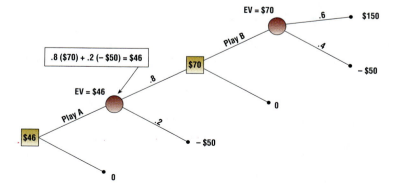

Decision-Matrix Equivalent

When the tree in Figure 2.5 is converted into its decision-matrix form (see Table 2.3 below) it is possible to appreciate the economy of representation of the tree approach for large, multi-stage decision problems.

TABLE 2 . 3

DECISION MATRIX FOR TWO-STAGE PROBLEM

	Probabilities			
	0.48	0.32	0.20	EV
	Win A	Win A	Lose A	
	Win B	Lose B	Stop	
Play A	0	0	-$50	-$10
Play A and B	$150	-$50	-$50	$46
Do not play	0	0	0	0

In the problem below, a productivity decision problem is posed which is best resolved by the decision-tree method.

Problems for Supplement 2

1. The Pin Factory can buy a new machine—called M1—which costs $40,000. There is an 80 percent chance that sales will get an order for which this machine can be used with significant productivity advantages. If that order does not materialize, then the machine will not be used for the foreseeable future. Normally, one would wait to see if the order was received, but that cannot be done in this case because the M1 has a three-month lead time. If M1 is purchased today, it will be online just in time.

   If the order is received and the machine is online, then the Pin Factory will earn $50,000. It also will be given the opportunity to bid on a much bigger job. The bigger job requires buying four more M1 machines which will be delivered in time to be online for the new job, if the Pin Factory wins the bid. The management team estimates that the probability of winning the bid is 60 percent. The revenue from getting the new order will be $300,000. Should the Pin Factory decide to try for these orders?

2. American Auto Supply (AAS) has been given an opportunity to become a certified supplier by one of the major auto companies in Detroit. They have been offered the following proposition:

   Stage 1 Submit engine-mount parts from a sample mold. If they pass, there will be an order for 100,000 engine-mount parts, and a chance to:

   Stage 2 Submit samples for shock-absorber assemblies. If these pass lab tests, there will be an order for 100,000 shock-absorber assemblies, and a chance to:

   Stage 3 Submit samples for bumper assemblies. If these pass the tests, there will be an order for 100,000 bumper assemblies.

   The deadline is tight for accomplishment of each stage.

   The conditions are that AAS pays for the molds and all costs associated with making sample parts. There is no reimbursement for tooling-up for parts that fail to pass the test. Further, engineering has stated that these parts are difficult to make to meet the specifications. The chances of success by the deadline are estimated to be 50 percent for the engine-mount and bumper assemblies. There is a 40 percent chance of failure for the shock-absorbers assemblies.

Total direct costs for 100 samples are as follows:

Engine-mounts – $ 8,000 (0.20)
Bumpers – $16,000 (0.40)
Shock-absorbers – $10,000 (0.20)

The profits per unit *if awarded the contract* for 100,000 units are listed next to the total direct costs within parentheses.
Should AAS accept the challenge?

3. Because the deadline is so tight, American Auto Supply has asked for changes in the proposition to give them more latitude in preparing the samples. The changes are in Stages 2 and 3, as follows:

Stage 2 Submit samples for either the shock-absorber assemblies or the bumper assemblies, not both. If the second part passes the lab tests, there will be an order for 100,000 units. Then AAS can go on to qualify for the third part. Therefore, Stage 3 would become:

Stage 3 Submit samples for the third part which is either the bumper assembly or the shock-absorber assembly. If these pass the tests, there will be an order for 100,000 units of that assembly.

Notes...

1. Jon F. Cox, John H. Blackstone, and Michael S. Spencer, *APICS Dictionary*, 7th Ed., APICS Educational and Research Foundation, 1992.

2. Soloman Fabricant, *A Primer on Productivity* (New York: Random House, 1969), p. 3. This definition adds "in a period of time (*t*)," to Fabricant's definition. He is considered to be the "father of productivity measurement."

3. The U.S. Department of Labor, Bureau of Labor Statistics, *News*, USDL 94-327, July 7, 1994. The Bureau of Labor Statistics public database, LABSTAT, is available on Internet.

4. The U.S. Department of Labor, Bureau of Labor Statistics, Press Release, Feb., 1995.

5. *New York Times*, November 5, 1994, p. D19.

6. Ibid., but this quote is attributed to Donald P. Wingard, managing partner for Andersen's Americas Automotive Practice.

7. William J. Baumol, Sue Anne Blackman, and Edward N. Wolff, *Productivity and American Leadership: The Long View*, MIT Press, Cambridge, MA, 1991.

8. Jay W. Forrester, "Changing Economic Patterns," *Technology Review*, August/September, 1978.

9. Cited by the study, "Foreign-Affiliated Firms in America," conducted by the Center for the Study of Operations at Columbia University, 1991.

10. Boston Consulting Group, *On Experience*, was published as part of the *Perspectives* series by the company in 1968, with a third printing and preface by Bruce Henderson, then president of BCG, in 1972.

11. Adam Smith, *An Enquiry into the Nature and Causes of the Wealth of Nations*, 1776, (a five volume treatise).

12. Jean-Jacques Servan-Schreiber, *The American Challenge*, New York: Atheneum, 1968.

13. W.A. Shewhart, *Statistical Method from the Viewpoint of Quality Control*, Washington, DC, The Department of Agriculture, 1939.

14. W. E. Deming, *Out of the Crisis*, MIT Center for Advanced Engineering Study, 1986. J.M. Juran, and F.M. Gryna, Jr., *Quality Planning and Analysis*, NY, McGraw-Hill, 2nd Ed., 1980.

15. T. Ohno, *Toyota Production System: Beyond Management of Large Scale Production*, Diamond Publishing Co. Ltd., 1978.

16. Called the six sigma program, it strives for 3.4 defects per million parts.

17. *New York Times*, Nov. 8, 1994, p. 1.

18. The agile manufacturing strategy and recommendations for future action were developed by an industry-led team, partially funded by the Office of the Secretary of Defense Manufacturing Technology (MANTECH) Program and facilitated by the Iacocca Institute at Lehigh University. The study resulted in the publication of two volumes entitled, "21st Century Manufacturing Enterprise Strategy: An Industry Led View of Agile Manufacturing," dated November, 1991 and released by Agility Manufacturing Enterprise Forum at Lehigh University.

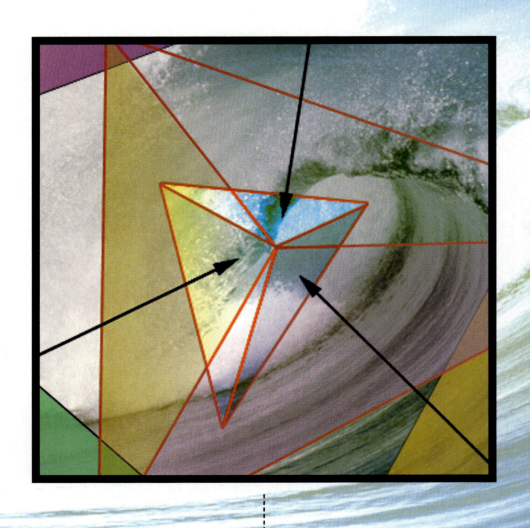

STRATEGIC PERSPECTIVES FOR OPERATIONS MANAGMENT

Part II

In Part II, strategy is defined. OM's participation with other functions in developing strategy is explained. OM's role in carrying out strategy through tactics is described. The focus moves from new products (goods and services) in Chapter 3, to quality attainment in Chapter 4. These two chapters which comprise Part II provide the framework for understanding how operations management participates in strategic planning.

Chapter 3 explores strategic planning for new and existing products—referred to as "life cycle planning." This involves understanding the interactions of OM with marketing. Scoring models to rate idea and design acceptability, testing methods to evaluate consumer acceptance, and forecasting models to estimate demand (in terms of sales and market share), based on market research data are basic to this aspect of strategy formulation.

Chapter 4 examines the role quality plays in strategic planning. This involves answering such questions as: how should strategic planners set the right quality goals and what must operations managers do to attain the goals that have been set? Understanding Total Quality Management (TQM) is necessary to get the answers. TQM is explained as a systems concept that cuts across functional areas.

Chapter 3

OM Strategies: Forecasting Life Cycle Stages

After reading this chapter you should be able to...

1. Clarify the distinction between strategy and tactics.
2. Detail how OM can contribute to the strategy planning team.
3. Explain what "Best Practice" means to OM.
4. Define life cycle stages and why they are important.
5. Forecast relevant information for production scheduling.
6. Teach others about choosing forecasting techniques.
7. Contrast various forecasting methods.
8. Do a linear regression analysis.
9. Show how to evaluate forecast errors.
10. Explain why it is important to maintain forecasting history.

Chapter Outline

3.0 STRATEGIES FOR OPERATIONS MANAGEMENT

3.1 PARTICIPANTS IN PLANNING STRATEGIES
OM's Role in Developing Strategies
Who Carries Out the Strategies?
Adopting "Best Practice"
OM Is a Scarce Resource

3.2 UNDERSTANDING LIFE CYCLE STAGES FOR OM ACTION
Introduction and Growth of the New Product
Maturation and Decline of the New Product
Protection of Established (Mature) Products

3.3 FORECASTING LIFE CYCLE STAGES FOR OM ACTION
Perspectives on Forecasting
Extrapolation of Cycles, Trend Lines and Step Functions
Time Series Analysis
Correlation
Historical Forecasting with a Seasonal Cycle

3.4 FORECASTING METHODS
Base Series Modification of Historical Forecasts
Moving Averages for Short-Term Trends
Weighted Moving Averages (WMA)
Use of Regression Analysis
Correlation Coefficient
Coefficient of Determination
Pooling Information and Multiple Forecasts
The Delphi Method
Comparative Forecasting Errors for Different Forecasting Methods
Keeping a Record of Past Errors

Summary
Key Terms
Review Questions
Problem Section
Readings and References
Supplement 3: Forecasts Using Exponential Smoothing Methodology

The Systems Viewpoint

Strategic planning involves teamwork and OM is part of that team. The hub and spoke diagram in Figure 3.1 which follows puts strategy planning at the center of the system and all of the business functions at the end of the spokes. The coordinating systems mechanism is strategy. The OM function sets its tactics accordingly, to pull together the various functions. There is a need for coordinated forecasting. In some companies, the forecast for market demands sets the agenda for how the entire company will commit its resources. In other companies, the forecast is the platform for strategic planning which is aimed at modifying the forecast to better fit production capabilities and maximize profits.

3.0
STRATEGIES FOR OPERATIONS MANAGEMENT

Strategy

Comprehensive and overall planning for the organization's future; a continuous, ongoing process formulated by top management and appropriate specialists.

Strategy is the comprehensive and overall planning for the organization's future. It has to be product-oriented and marketing-aware in order to take the customer and the competition into account, and process-oriented and OM-oriented to deliver the product that customers want. It needs to be systems-oriented to coordinate marketing and OM. This entails broad-based, cooperative goal setting followed by planning to achieve four goals:

1. survival of the organization under the stress of technological change
2. adaptation to a global business environment
3. conversion from a reactive mode to a proactive mode
4. profit maximization

Successful strategies are required for the organization's effective pursuit of goals.

Many companies are divided into strategic business units (SBUs) in order to help them define clear, strong and effective strategies. Each strategic business unit develops plans centered on its own set of products, and each can be compared to similar business units with respect to process performance. Comparisons also can be made with OM and marketing effectiveness. The systems perspective is essential to the competitive advantage of successful SBUs.

Illustrating the distinction between strategies and tactics helps to further define strategy: *Do the right "thing"* (strategic choice), *before doing the "thing" right* (tactical choice)! As an example, consider the company that tried to make better buggy whips in order to fend off the competition of Henry Ford's Model T. That company would have been well advised to work on an electric starter because the Model T engine had to be hand-cranked to start it.

Another analogy that is used to help differentiate between strategies and tactics is: *Know where you want to go* (the strategic choice–a goal), *before deciding on the best way to get there* (the tactical choice–the means to an end). This is equivalent to stating that the goals set by strategies are like destinations. The decision to fly, drive or take a train is a tactical decision which cannot be made until the origin and destination are known.

3.1

PARTICIPANTS IN PLANNING STRATEGIES

Because strategies are the future plans for the company, they must be made by those who know enough about the business to do such planning. Usually, therefore, strategies will be formulated by those at the top of the organization, aided by those with special knowledge who may be located anywhere in or out of the company. The cast of characters involved in strategic planning usually is composed of a top management group, line and staff managers, and various specialists in relevant fields. The participants change as conditions are altered.

Strategy development is a continuous and ongoing process, because strategies must be revised as conditions change. It is a matter of survival through adaptation. A new competitive product entry, an improvement in the competitor's delivery system, a change in the law, or in economic conditions, or a new technological development that enhances the competitor's process often will call for modification of prior strategies. Therefore, the nature of the changes that are occurring will determine which specialists may be called upon. Those responsible for strategic planning must know enough to seek out the appropriate specialists as conditions warrant.

OM's Role in Developing Strategies

From the systems point of view, only in small companies can one person know enough about the business to do strategic planning by themself. As firms get larger, no one person can handle all of the activities that need to be done.

In larger firms, the systems perspective dictates that representatives of all of the functional departments that define "knowing enough about the business" play a part in strategy development. In Figure 3.1, no one person holds all the cards.

F I G U R E   3 . 1

HUB AND SPOKE DIAGRAM

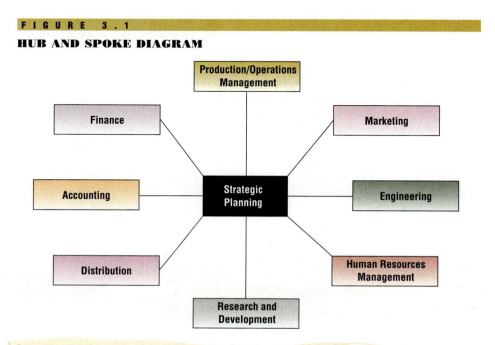

Strategy planning is the hub and business functions are at the end of the spokes.

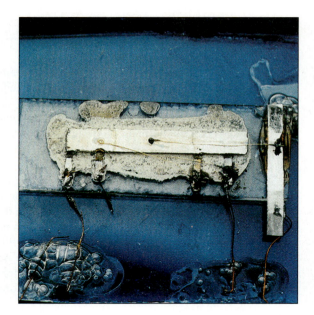

New technological developments
can change a company's strategy
(first integrated circuit, 1958).
Source: Texas Instruments

There are some additional differences between organizations with respect to who participates in planning the strategies for products and processes. If the technology used for the process is recent and complex (new and high-tech), then operations managers are more likely to be part of the strategic planning team, or at least consulted by them. Such proactive participation applies to both manufacturing and service industries.

Chembank's operations managers participated in strategy formulation for a variety of new products. They were the only part of the bank that was on top of the technology for new banking products using computerized voice-responding systems for account information inquiries, and PCs for home-banking. Similarly, in General Electric's smart manufacturing plants, OM is coordinated with the other functions by means of a common understanding of strategic goals. In smart plants, OM is part of the strategic planning team.

If OM knowhow is not properly represented, then strategic planning is operating at a limited or damaged level. The same is true if other functional areas are not represented. The benefits of the systems approach are missing and the organization is operating at a disadvantage with respect to survival under technological stress, adaptation to a global environment and conversion from reactive to proactive competitive positions.

Who Carries Out the Strategies?

Operations managers are always involved in carrying out strategic plans, even when they are not part of the strategic planning team. Operations is where the work is done that creates the product which the customer buys. This applies to the operations room of banks and airlines and the production department of the manufacturer.

Independent of who formulates company strategy, decisions that stem from that strategy are carried into action by tactics. **Tactics** are the procedures used by the line managers of the organization to carry out the strategic plans. Tactics are the normal operating mode of line managers of the organization, and include production scheduling, materials planning and quality control.

The terms "strategy" and "tactics" have a military origin. Strategy is directed at winning the "war" while tactics are designed to win the "battles." The generals decide what battles have to be fought, when, where, and with what. The line officers carry out the mission determining the "hows."

Use of the terms strategy and tactics is common in companies where line management is recognized as being an OM responsibility. Overseeing production workers who are on the line, making goods and delivering services, is only part of the job. Line managers must be aware of the company's goals and objectives to be an effective part of the strategy team.

Strategic planners determine the *what, when,* and *where* of new products. OM is then responsible for *how* to make and deliver these new products. New product decisions require planning future capabilities and capacities. Strategies also are concerned with the production and replacement of existing products at given costs and levels of investment. Whether or not the OMs participate in the development of such strategic plans, they are administered by operations managers.

In organizations that are strongly integrated, operations managers participate in planning the evolution of both new and existing products. That evolution over time is called the "product life cycles." Plans for the life cycles—called "life cycle strategies"—set in motion the designing of the input-transformation-output systems which are the basis of production. Life cycle strategies will be further defined in this chapter.

Tactics

Procedures used by the organization's line managers to carry out strategic plans; include production scheduling, materials planning, and quality control.

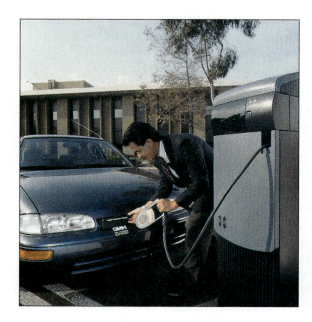

Because of pollution concerns and emissions controls, many auto companies have developed electric cars.
Source: GM Hughes Corporation

Organizations that include OM knowhow in strategic planning for both product and process life cycles avoid suboptimization, defined as doing less well than would have been possible if the planner had known how to best optimize the plan. When the best strategy for a situation is not chosen, because of a lack of information—which was obtainable, but was not sought—suboptimization occurs. The systems approach protects against serious suboptimization.

Three kinds of advantages occur when OMs are part of the strategy team.

- Advantage 1 is that OM knowledge about existing technology and methodology is beneficial in the design of strategy.
- Advantage 2 is that OM can evaluate new technology and assess its impact on operating methodology (such as scheduling work).
- Advantage 3 is that knowledge of company-wide strategies can lead to tactical benefits.

Overall, it is a good idea for line operators to understand the company strategy. Communications from top management with OM are poor in Stage I organizations. In Stage II organizations, OM understands the goals, but does not participate in formulating them. Stage III organizations encourage suggestions from OM about the production and operations factors that can improve company strategies and affect the tactical capabilities of the organization.

Synergy is the action of separate agents that produces a greater effect than the sum of their individual actions. One plus one makes three. **Syzygy,** besides being the only word in English that has no vowels and three y's, comes from a Greek word for conjunction. It is the point when maximum tidal pull is exerted because the planets and moon have lined up to support each other's gravitational forces. Stage IV organizations capitalize on synergy and syzygy, by encouraging communication about all matters relating to strategic planning. Stage IV organizations achieve the teamwork ideals of the systems approach.

Stage I companies are moving to Stage II and higher as most American organizations have come to realize that important strategic leverage is gained through the participation of OM. It is an exciting time for operations managers because in almost every organization OM is being integrated into the corporate planning fabric. Successful new competitors are not Stage I companies. Often from abroad, they have fully-coordinated strategic planning teams.

Synergy

Action of separate agents that produces a greater effect than the sum of their individual actions.

Syzygy

Teamwork is pulling together with the analogy of maximum tidal pull exerted because the planets and the moon have lined up to support each other's gravitational forces.

Adopting "Best Practice"

Being "in the know" prevents OMs from inadvertently working against the company's success. Yet, what is to be done if the stage of the organization, or the politics of the organization, or the chemistry of the managers, causes policies to be followed which do not place the operations managers onto the strategy teams?

By working actively (as compared to working passively) in the realm of:

1. new product development, and
2. quality enhancement practices,

OM can get involved with the strategic domain, without ever having been appointed to a strategic planning committee.

Passive participation of OM is characterized by doing just what is asked and nothing more. Active participation involves learning about what is being done elsewhere and then trying to get *Best Practice* adopted whenever possible. OM must play an instrumental role at both the strategic and tactical levels to determine what Best Practice is regionally, especially with competitors, and ultimately, anywhere in the world.

P E O P L E

Doug Smith
. . .
Computervision Corp.

Doug Smith, vice president of Strategy and Development for Computervision Corp., said this Bedford, Massachusetts-based software development company has found a "different way of modeling and designing complex products."

According to Smith, Computervision's "Electronic Production Definition" process finally fulfills the promises of all CAD/CAM (computer-aided design/computer-aided manufacturing) software makers who profess that their programs speed up both the design and manufacturing process, thus decreasing a product's time to market. The CAD/CAM software did speed up the component design process, Smith said, but it failed to change the way manufacturers develop products.

"We've been solving the wrong problem," he said. "We need to look at the product from inception and design through modeling, manufacture, distribution, and post

production upkeep—the entire product life cycle. It takes a whole company to launch a product."

Rolls-Royce Aerospace Group recently bought into Computervision's claims, with a $21.4 million purchase of software and services. The software will help Rolls-Royce engineers design products at a higher level of abstraction. This will enable a product to be modified easily when sold in different markets. The need for redesign would arise with different performance standards or content requirements.

Computervision, which endured a $571 million loss in 1993, decided at the end of that year to abandon the computer hardware reselling business to concentrate on software development. "We've gone through a catharsis forced on us by finances," Smith said. "This whole thing points out that by listening to our customers and listening to the outside world, you can be successful."

Source: John S. McCright, "Computervision on a Rolls—Checks downturn with CAD/CAM deal," *Boston Business Journal,* March 25, 1994.

Both new product development and quality enhancement practices are *gateways* to the strategic domain. Total Quality Management (TQM) is susceptible to either continuous improvement or to the marked departure from prior procedures (starting out with a clean piece of paper) which reengineering requires. New products, whether goods or services, provide opportunities for reinventing or reengineering process flows. The constant effort to improve products and process flows provides a transition between strategies and tactics for operating the system.

OM Is a Scarce Resource

Strategies for the introduction of new products and for replacement of existing products must be formulated with the fact in mind that operations managers' time and talents are limited resources. Managerial capabilities must be spread over both new product start-ups and ongoing output responsibilities.

Because OM is a scarce resource its personnel should be used wisely. This means bringing OM knowhow to bear as early as possible—in the strategy formulation process. By doing so, it is possible to avoid squandering limited OM time and talent for remedial "fire-fighting" and "catch-up" later on.

3.2

UNDERSTANDING LIFE CYCLE STAGES FOR OM ACTION

Throughout the company, the planning function marches to the drum beat of life cycle stages. Operations managers need to be aware of the timing and stages that drive the project management schedules of new products, as well as the production and delivery schedules of the company's mature products.

Cycles

Composed of four discrete stages: introduction, growth, maturation, and decline.

Cycles are composed of four stages which appear in a regular way over time, and which all products and services go through:

1. introduction to the market
2. growth of volume and share
3. maturation, which is equilibrium, and
4. decline, leading to restaging or withdrawal.

These life cycle periods are discrete stages in each product's life which need to be understood in order to manage that product. Marketing is responsible for using different pricing, advertising and promotion activities during appropriate stages. OM is responsible for intelligent management of the transformation system which changes in various ways according to the life cycle stage. The changes or transitions between stages require knowing how to adjust the production system's capabilities.

Introduction and Growth of the New Product

There are two main phases of life cycle stages. The first consists of the introduction and growth of the product. The "idea" for the product and its development precedes the introduction. The entire team works on ascertaining the marketing feasibility of the idea, as well as the feasibility of making it and delivering it. R & D may have made sample product so that market research can test its customer acceptability. When it is approved, OM and engineering swing into action to create the production system that can make and/or assemble it. During this process of bringing the idea to reality, there are many make or buy analyses.

None of this is easy. It takes a lot of effort and attention to detail. There is much time and talent needed to conceptualize the product, design its specifics, organize the process for making it, cost it out, pilot test it, and so forth. When the product is accepted, it is released for production and marketing. All of this takes place in the introductory stage.

Maturation and Decline of the New Product

When the new product or service stops growing, it is considered mature. This means that its volume is stabilized at the saturation level for that brand. The competitors have divided up the market and only extraordinary events, such as a strike at a competitor's plant, are able to shift shares and volumes.

Previously OM had to deal with producing larger and larger quantities of product. Now the input-output relationship reaches equilibrium. Marketing takes specific actions during this phase to maintain the product's share of the market. Prices often are lowered. Coordination between OM and sales is essential to meet delivery schedules on time. Finally, the product begins to lose share, volume drops and, depending on the strategy, the product is either restaged or terminated.

It is expected that a new product will have been introduced previously which has grown to a reasonable extent. It is the replacement for the other product and has been assigned the capacity of the product that has been replaced. The cycle of the new product is similar to that of the one that is replaced in that it goes through introduction, growth, maturation and decline. Figure 3.2 illustrates the curve of the evolution of life cycle stages.

FIGURE 3.2

EVOLUTION OF LIFE CYCLE STAGES

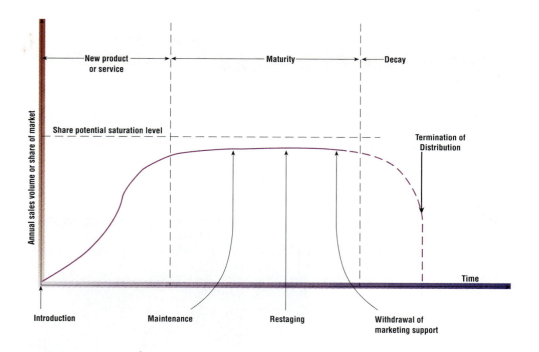

Consider the revenue generating lifetime of the new product. During the introduction, expenditures usually are high to bring the new product to the distributors, retailers and customers. The cash flow is negative. More is being spent on the introduction and promotion of growth than product revenue returns. Later, when the customer base has been established, net revenue increases, and expenditure decreases. If the product enjoys a loyal following, promotions of any kind may be unwise. Eventually, competitive moves are likely to lead to price cuts and the withdrawal of the product from the market.

Product life cycles have been speeding up which means that growth has to occur faster. The product does not stay in the mature stage as long and has to be replaced more often. More new product introductions are required to replace waning products. The challenge for OM is the rapidity of adjustment to all of the life cycle stage changes and to go through them frequently.

OM thinking is crucial to each of the four stages. The new design that is about to enter production has been finalized. There still are many questions to be answered. How much product must be made for distribution at start-up, and then, to keep the supply lines filled to meet the demand during growth? How much capacity is to be used at each of the stages? How much training will the evolving work configurations require? Make or buy decisions can change. What should be bought at first, at the lower volumes, might be better made in the plant, at the higher volumes that are associated with maturity.

Protection of Established (Mature) Products

Organizations that have been successful with their new product introductions can bank on having established products or services which generate cash flow. The "bank" is not as good as it used to be though, because the competitive rate of new product introductions has increased markedly in recent years. OM requires the cash to pay labor and buy materials from suppliers. Generally, OM spends more cash than any other department in the company. These cash accounts for operating expenses may total 70 percent of all cash spent by the company.

Because of global organizations, there are more competitors in the game, and because there are more competitors, each organization competes harder. Increased competition has led to higher levels of market volatility. Nevertheless, for most product categories, there is still plenty of opportunity to benefit from the mature product life cycle stage. OM benefits from the security of established products because they have stable material and labor skill requirements. They also provide the opportunity to improve the productivity of these processes.

To compete in the global market, banks use state-of-the-art technologies and advanced processes to push quality up to levels competitors find difficult to match.
Source: Citibank

TECHNOLOGY

Tracking the Power Rangers

A toy buyer at Target Stores headquarters in Minneapolis, tracking sales on his computer in August 1993, noticed unusually high sales activity for the new and relatively unknown Mighty Morphin Power Rangers.

No one in the office could interpret the sales activity. Target buyers decided to delay ordering more inventory to see how the toys sold after the related television show aired in September. Other toy retailers did the same, and the delay proved costly. By October Power Rangers were in short supply and the Japanese manufacturer, Bandai Co., expected to ship about 600,000 of the action figures by Christmas, far below the estimated demand of 12 million.

Historically retailers ordered merchandise inventories based on what they thought they could sell. Retail sales forecasting is a very inexact science because it depends on the retailer's ability to predict consumer tastes and demand. Overstocking of merchandise is extremely costly, and nowhere is this more apparent than in the toy industry. In the 1980s, excess

inventories contributed to the bankruptcies of several toy makers, including Coleco Industries Inc. (producer of Cabbage Patch Kids) and Worlds of Wonder Inc. (maker of Teddy Ruxpin).

Retailers, who often had to sell toy manufacturers' "mistakes" at a loss, became more conservative in their inventory policies. Using electronic tracking systems for sales and inventories, retailers began ordering only what they needed to replace products just as they are sold. This approach, called "just in time" or JIT, was introduced in retailing on a large scale by such companies as Toys "Я" Us and Wal-Mart. JIT reduces carrying costs of inventory and the probability of stocking products that customers won't buy.

But JIT has a downside for both retailers and manufacturers— it doesn't give them time to take advantage of unexpected heavy demand for a product such as the Mighty Morphin Power Rangers. Despite such pitfalls, large retailers remain proponents of JIT. "The later you flow the goods through the channels, the lower you keep your inventory and the lower your markdowns," says Michael Goldstein, vice chairman of Toys "Я" Us. Further, electronic tracking keeps the company "battle station ready." From computer command posts "we know what's on our trucks, where they're headed, and who needs new supplies."

Source: Joseph Pereira, "Toy Industry Finds It's Harder and Harder to Pick the Winners," *Wall Street Journal*, Tuesday, Dec. 21, 1993, pp. A1 and A5.

- -

3.3

FORECASTING LIFE CYCLE STAGES FOR OM ACTION

Life cycle stages provide a classification for understanding the kinds of trends that can be expected in demand. During the introduction, the demand is led by the desire to "fill the pipe line." This means getting product into the stores or into warehouses or wherever it must be to supply the customers.

When growth starts to occur, there is a trend line of increasing sales. The trick is to estimate how fast demand will increase over time, and for how long a period growth will continue. There are good methods used by market research to find out what is happening. This information is fed back through sales to OM for purposes of production scheduling.

Perspectives on Forecasting

An umbrella manufacturer is preparing production and distribution schedules. There is a sign on the wall: "Where will it rain next week?" When it rains, umbrella sales pick up. Meteorological models have improved in recent years so that fairly good weather forecasts can be made. Many businesses, especially when scheduling outdoor events, hire weather forecasters who have a good track record.

Forecasting is not used to talk about a hand of cards or how a roulette wheel performs. Those are statistical phenomena for which probabilities are known. Las Vegas and Atlantic City have built their gambling casino profits on the laws of probability—but customers still have to come and play. Customer attendance is not a known probability.

Forecasting

To calculate or predict events. Types include: base series modification, moving averages, weighted moving averages (WMA), regression analysis, correlation, multiple, Delphi method, exponential smoothing, and the use of forecasting errors.

Forecasting is used to foretell sales demand volume even though the probabilities have never been formally studied. Business people often use their sense of what is happening to reach decisions that might be better made if someone had kept a record of what had taken place already. There is often some empirical basis for estimating what is likely to happen in the future.

Marketing models for predicting sales (lacking a contract) deal with levels of uncertainty that make forecasts of demand volumes, market shares and revenues difficult, but not irrational. One of the best sources of information about the future is the past. For new products, there is no past, and so other methods can be tried.

The focus is on using existing data to develop forecasts. The Rivet and Nail Factory has to forecast sales of products to develop departmental schedules for the next production period. The Mail Order Company has to forecast demand in order to have the right number of trained agents and operators in place. Ford Motor Company has to forecast car sales so that dealer stocks are of reasonable size for every model.

In every sales forecasting situation, the volatility of demand will determine how likely it is that a good forecast can be made. Stable patterns that persist for a long period of time make company forecasters confident that a credible job of forecasting can be done. Shaky estimates make company forecasters uneasy so they search for new factors to correlate with the demand system.

Sales patterns are becoming less stable with increasing competition and information. Food sales that once were stable now are affected by medical reports about the food's effect on health. Auto and home sales are among many products that are strongly influenced by interest rates which fluctuate more in a global environment. Exports and imports are sales that move around the globe in response to currency fluctuations which are increasingly unstable. Consequently, better forecasting methods need to be used by the companies that are affected by increased volatility.

How well can one forecast the future? The answer will depend on the stability of the pattern of the time series for the events being studied. The underlying pattern can be hard to find, but not impossible. Even if a pattern is found, the question remains, how long will it persist? When will it change? Those willing to forecast accept that it is a challenge.

Extrapolation of Cycles, Trend Lines and Step Functions

Extrapolation is the process of moving from observed data (past and present) to the unknown values of future points. The extrapolation of time series is one of the main functions of forecasting.

A **time series** is a stream of data that represents the past measurements. Each event is time-tagged so that it is known where it is located in the series of data. Figure 3.3 shows a linear trend extrapolated by extending a line drawn between the last two observations. Figure 3.4 shows an erratic pattern of peaks and valleys which represents a somewhat cyclical pattern of monthly sales. Three forecasts had been made for those sales by extrapolating the cyclical patterns of prior months' sales using different techniques.

Extrapolation

Process of moving from observed data (past and present) to unknown values of future points; main function of forecasting.

Time series

Stream of data that represents past measurements.

FIGURE 3.3

EXTRAPOLATION OF TREND LINE

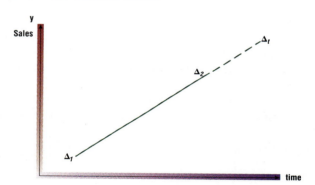

FIGURE 3.4

CYCLICAL PATTERN EXTRAPOLATION
SALES PATTERN FORECASTS

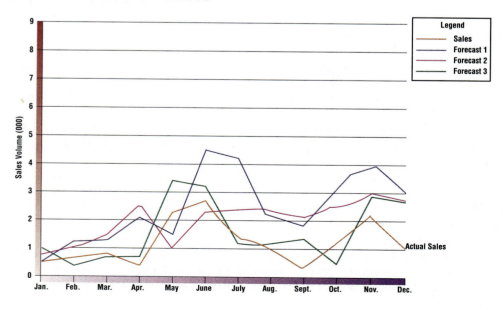

Extrapolations of time series can be based on:

1. cyclical wave patterns,
2. trend lines,
3. step functions, and
4. combinations of any or all of these patterns.

Step functions look like stairs—going up or going down. It is hoped that the forecast can predict when the next step will occur, and how far up or down it goes. Short-term cycles can be piggy-backed on long-term cycles. All kinds of combinations are possible. The key point is that cycles, trends and steps are the basic pallet for the development of forecasting models.

Time series data also can reflect erratic bursts called impulses. If these spikes appear from time to time and some pattern can be associated with them, they could be added to the list of what can be extrapolated or projected into the future.

Time Series Analysis

It is always useful to try to find causal links between well-known cycles—such as the seasons—and demand patterns that are being tracked. Thus, there are seasons for resort hotels and home heating oil. Cycles are one form of time series that is often used by business to forecast other events related to that time series. There are other categories of time series that are useful to know about. *Time series analysis deals with using knowledge about cycles, trends and averages to forecast future events.*

▼ **Time series analysis**

Uses statistical methods including trend and cycle analysis to predict future values based on past history.

The term **time series analysis** deserves explanation. Time series analysis is "analysis of any variable classified by time, in which the values of the variable are functions of the time periods."[1]

The time series consists of data recorded at different time periods such as weekly or daily for the variable which could be units produced or demands received. Forecasters attempt to predict the next value or set of values that will occur at a future time. In time series analysis, external causes are not brought into the picture. The pattern of the series is considered to be time-dependent, which is why the APICS definition states that "the values of the variables are functions of the time periods."

Using symbols, a forecast is to be made from some series of numbers x_1, x_2, ... x_n. These numbers might be sales figures for the past year. The question is: What will be next year's sales figures? From the forecasted series, production schedules will be made, orders will be placed for inventory to be purchased, workers will be trained or downsized, rented space will be increased or decreased. Many other decisions need to be based on forecasts.

Time series analysis uses statistical methods, including trend and cycle analysis, to predict future values based on the history of sales or whatever the time series describes.

If the time series shows a linear trend, as in Figure 3.3, there is a constant rate of change for increasing the size of the numbers. Linear decreases also can occur. Nonlinear trend lines occur where the rate of change is geometric. In such cases, it is possible to extrapolate the curve by eye, but also, the rate of change can be calculated, and projections can be made mathematically.

Similar reasoning applies to step functions. If they occur regularly, a good estimate can be made of the period between steps and the timing of future steps. Step functions are characteristic of systems that can only change in given quantities. Thus, if sales must be made in lots of 100 units, then each change in demand will occur in hundred units. The sales for January could be 400, 500, or 600. It could step up or down, or stay the same as December sales which were 500.

Correlation

When a series of x numbers (such as the monthly sales of 35mm cameras over a period of years) is causally connected with the series of y numbers (the monthly sales of 35mm film), then it is beneficial to collect the information about x in order to forecast y. In this case, the direction of causality is from x to y since cameras sold create the demand. Thus:

$$y = f(x)$$

To forecast the y series of numbers $y_1, y_2, \ldots y_n$, collect the x series of numbers and modify each element of x to produce the equivalent value of y. Thus:

$$y_i = f(x_i)$$

This relationship is not perfect. Therefore, y is said to be correlated with x. A correlation coefficient (r) can be calculated to measure the extent of *correlation* of x with y, where r can take on values from -1 to $+1$. At -1 x and y are perfectly correlated—going in opposite directions. As x gets large, y gets small and vice versa. When $r = 0$, there is no correlation and when $r = +1$, x and y are perfectly correlated going in the same direction. Note that for the historical forecast, if each month of the time series is identical with the one 12 months before, then the auto-correlation coefficient would be $+1$. The equation for correlation coefficients is given below with the material on linear regression.

Historical Forecasting with a Seasonal Cycle

The historical forecast is based on the assumption that what happened last year (or last month, etc.) will happen again. The pattern is expected to repeat itself within the time period. This method works if a stable pattern (which is often seasonal) exists. The hotel and resort business is typically involved with historical forecasts, as is the agriculture business. In each case, special circumstances can arise which lead to the desire to modify the historical forecast. In the case of hotels and resorts, the state of the economy can modify the ups-and-downs of occupancy rates. Agricultural pursuits are modified by rainfalls and temperature fluctuations (see modifications to the historical forecast, below).

When last year's monthly sales values are used to predict the next year's monthly sales values, the method is based on using past history to forecast. The same concept applies to weekly or daily sales which can be forecast on a semiannual, quarterly, or monthly basis. Table 3.1 shows the historical forecast technique applied to data which exhibit a seasonal cycle. No modification has been made for an overall change in annual sales, which is the basis of the historical forecast.

After Table 3.1, the method for making a modified historical forecast is presented. It should be pointed out that in practice the greatest percent of all forecasts are made using the nonmodified historical method.

When stable, cycles can provide insights which are very important for OM tactical planning. When they work, historical cycles allow OM to excel at capacity planning and production scheduling for mature manufactured products and services. The historical forecast is not appropriate for a new product introduction—unless there is a close tie to some other product that already is on the market.

Q U A L I T Y

Westinghouse Commercial Nuclear Fuel Division

Building a quality culture asks employees to do the "right things right the first time."

When electric utilities operating nuclear power plants install fuel-rod assemblies made by the Westinghouse Electric Corp.'s Commercial Nuclear Fuel Division (CNFD), they can be 99.995 percent sure that each of the thousands of rods supplied will perform flawlessly.

Westinghouse's CNFD, whose objective is to be recognized as the world's highest quality supplier of commercial nuclear fuel, is building a quality culture that asks employees to do "the right things right the first time." This philosophy considers every employee action a quality initiative.

Customer satisfaction is CNFD's guiding principle, whether the customer is the ultimate consumer of nuclear-generated energy or the next employee in the process. CNFD maintains almost daily contact with its utility customers and regularly collects data to evaluate the performance of its fuel assemblies. Customer service plans created for each client are reviewed jointly each quarter.

In strategic planning, top management develops formal quality initiatives and "Pulse Points" that are deemed most critical to improving performance and customer satisfaction. The unique Pulse Points system tracks improvements in over 60 key performance areas identified with statistical techniques and other evaluative tools. This system helps CNFD set measurable goals within each unit, down to the jobs of hourly workers. Pulse Point trends are reviewed each month in a teleconference that includes top management at each division site.

The CNFD employs approximately 2,000 people at three sites, the Specialty Metals Plant near Pittsburgh, Pennsylvania; the Columbia, South Carolina, plant that manufactures fuel-rod assemblies; and its headquarters for Operations and Nuclear Engineering in Monroeville, Pennsylvania. The CNFD supplies about 40 percent of the U.S. market for fuel-rod assemblies and about 20 percent of the world market. Westinghouse's CNFD won one of the first Malcolm Baldrige National Quality Awards in 1988.

Source: Malcolm Baldrige National Quality Award, *Profiles of Winners,* 1988-1993.

	ACTUAL Sales	FORECAST of Sales
Month	**Last Year**	**Next Year**
January	1500	1500
February	1600	1600
March	1800	1800
April	2000	2000
May	2300	2300
June	2500	2500
July	2350	2350
August	2100	2100
September	1850	1850
October	1650	1650
November	1550	1550
December	1400*	1400

T A B L E   3 . 1

*Note: The December estimate of 1400 units is based on the prior year's sales.

Historical cycles allow OM to
excel at capacity planning and
production scheduling for mature
manufactured products.
Source: Honda of America, Inc.

3.4

FORECASTING METHODS

This book discusses the following forecasting methods:

1. base series modification of historical forecasts
2. moving averages for trends
3. weighted moving averages
4. regression analysis and correlation
5. use of multiple forecasts
6. the Delphi method
7. means of comparing forecasts using forecasting errors
8. exponential smoothing

   Mathematical equations will be used for various kinds of forecasting. It is important to stress that the use of equations does not make a forecast "truth." Also, a
great deal of good forecasting can be done without mathematics. Further, with or
without mathematics, no forecast is ever guaranteed.

Base Series Modification of Historical Forecasts

If the pattern remains fixed, but the demand level has increased overall then a base
series modification can be used. Assume that in 1999, the quarterly demands were
10, 30, 20, 40. This gives a yearly demand of 100 units. Further, assume that in
2000 the yearly demand is expected to increase to 120 units. Then the quarterly forecasts would be adjusted:

$$120 \ (10/100) = 12$$
$$120 \ (30/100) = 36$$
$$120 \ (20/100) = 24$$
$$120 \ (40/100) = 48$$

The adjusted quarterly demands total 120 units. The cyclical patterns are matched. The next year, assuming the pattern continues and the base level increases to 150, the time series would be 15, 45, 30, 60 with the sum of 150.

Moving Averages for Short-Term Trends

Moving averages might be used to extrapolate next events if the following occurs:

1. if there is no discernible cyclical pattern, and
2. if the system appears to be generating a series of values such that the *last set of values provides the best estimate* of what will be the next value,
3. then, there is forecasting momentum in the time series' movements.

A decision must be made concerning how many readings should be included in the moving average set (N). Is it better to use two, three, four, or more, periods? The quest is to find the optimal number of prior periods to include in the moving average series. How far back to go depends on the speed with which the series changes, and the recency of events that tend to determine the future. "Very recent" means few values; "not so recent" means more than a few values. That is a workable rule of thumb.

When the *magnitude* of the trend is great, and the pattern is very *consistent,* then, the fewer the number of periods in the set the better. If the trend is gradual (up or down), and if fluctuations around the average are common, then having more periods of time in the set is better than having too few.

A moving average supplies a forecast of future values based on recent past history. The most appropriate (often the last, and latest) N consecutive values, which are observations of actual events such as daily demand, are recorded. These data must be updated to maintain the most recent N values. Then, using the last, and the latest N values:

$\hat{x}_t$ = the forecast value of x in period t
   (the hat sitting on top of x signifies it is the forecast)

x_t = the actual value of x observed in period t

The computing equations are:

$$\hat{x}_{t+1} = \frac{(x_t + x_{t-1} + \ldots + x_{t-N+1})}{N}$$

Equation 3.0

and:

$$\hat{x}_{t+2} = \frac{\hat{x}_{t+1} + (x_{t+1} - x_{t-N+1})}{N}$$

Equation 3.1

Each next period's forecast is updated from the last period's forecast by dropping the value of the oldest period of the series x_{t-N+1} and adding the latest observation, which is x_{t+1}. This yields the Equation 3.1 (above) for $\hat{x}_{t+2}$.

In this way, new trends are taken into account, and old information is removed from the system. The moving average method applies to any time period: hours, days, weeks, months, or years.

As an example, consider a four-period monthly moving average in Table 3.2.

TABLE 3 . 2

Month	Actual x_t	Forecast $\hat{x}_t$	Error $(x_t - \hat{x}_t)$
1	10		
2	30		
3	40		
4	40		
5	50	30	+20
6	60	40	+20
7		47.5	

Four values of x_t are required before a forecast $\hat{x}_t$ can be made and before the error can be calculated.

The first forecasted value is 30. It is derived as follows:

$$\hat{x}_5 = (40 + 40 + 30 + 10)/4 = 30$$

When the actual demand occurs, it turns out to be 50. The error of the forecast is +20. The forecast has fallen short by 20 units.

The sixth period forecast using Equation 3.1 is:

$$30 + (50 - 10)/4 = 40$$

It also could be calculated using Equation 3.0:

$$(50 + 40 + 40 + 30)/4 = 40$$

If the forecast of 40 is incorrect, and the actual result is 60, then the following period forecast is 47.5.

The moving average is slowly increasing with the trend in the actual values of x. Being slow is acceptable if there is a gradual trend with some variability in successive values. In this case, the OMs might wish for a faster response in order to schedule production closer to actual. Also, if there is a turn in the trend line, this method will follow the turn, but slowly. Some things can be done to speed things up.

The situation can be addressed in a number of ways. One approach is to decrease the number of observations N in the moving average. This will make the forecasts more responsive to recent events. Try it out with $N = 2$ and see for yourself that the seventh week forecast will be 55 instead of 47.5 as in the example above. Another approach is to use weighted moving averages.

Weighted Moving Averages (WMA)

A way to make forecasts more responsive to the most recent actual occurrences is to use weighted moving averages. The new forecast equation is:

$$\hat{x}_{t+1} = w_t x_t + w_{t-1} x_{t-1} + \ldots + w_{t-N+1} x_{t-N+1} \qquad \textbf{Equation 3.2}$$

$$w_t = \text{weights for periods } t$$

Let the sum of the weights w_t be 1; thus, $\sum\limits_{t-N+1}^{t} w_t = 1$

Note that when all weights are equal, the weighted moving average is the same as the moving average.

Thus:

$$w_t = w_{t-1} = \ldots = w_{t-N+1} = \frac{1}{N} \quad \forall t \; (\forall t = \text{for all t})$$

and:

$$\hat{x}_{t+1} = \frac{1}{N}x_t + \frac{1}{N}x_{t-1} + \ldots + \frac{1}{N}x_{t-N+1} = \text{Equation 3.0}$$

Choosing weights which allow the latest period to have the greatest impact, and later periods decreasing importance, the equation is derived:

$$\hat{x}_{t+1} = 0.4x_t + 0.3x_{t-1} + 0.2x_{t-2} + 0.1x_{t-3}$$

where the weights: 0.4, 0.3, 0.2, and 0.1, sum to one.

The biggest weights are assigned to the most recent events when there is a continuing trend. This would not be applicable where the values cycle. Thus, Thursday's sales are not the best predictors of Friday's sales—when sales on Fridays are more like sales on other Fridays than any other day.

In a rapidly changing system w_t may be much greater ($\ggg$) than w_{t-1}. Thus, $w_t \ggg w_{t-1} \ggg w_{t-2} \ldots$ etc. This is equivalent to reducing the size of N, which was discussed in the section on moving averages.

As has been noted, with the previous method of moving averages, all the w_t were treated as being equal. Then, using Table 3.2 above for $t = 5$, which requires a forecast for $t + 1 = 6$:

$$\hat{x}_{t+1} = \hat{x}_6 = 0.25(50) + 0.25(40) + 0.25(40) + 0.25(30) = 40$$

Using the weights (.4, .3, .2, .1) and the numbers from the previous example of Table 3.2, the result is obtained:

$$\hat{x}_{t+1} = .4(50) + .3(40) + .2(40) + .1(30) = 20 + 12 + 8 + 3 = 43$$

The weighted system is responding more rapidly to the possible trend. It has raised the forecast to 43 as compared to the unweighted moving average of 40. The same effect is seen for the next period. Introducing the sixth period's actual value of 60, the new forecast for the seventh period has a better reaction rate, rising to 51 from 47.5.

$$\hat{x}_7 = .4(60) + .3(50) + .2(40) + .1(40) = 24 + 15 + 8 + 4 = 51$$

Thus, weighted moving averages can track strong trends more accurately than unweighted moving averages. It should be noted that the procedures can easily be programmed for computers.

Use of Regression Analysis

Another valuable method that assists forecasting is regression analysis. This method is useful when trying to establish a relationship between two sets of numbers that are time series. The numbers can be simultaneous so that $x(t = 1)$ and $y(t = 1)$ are considered correlate pairs. In more general terms, these pairs would be $x(t = k)$ and $y(t = k)$. On the other hand, one time series can lead another. Thus, when $x(t = k)$ and $y(t = k + r)$, the time series y leads the time series x by r periods. It also is said that the time series x lags the time series y by r periods. Whichever way it is stated, it means that future values of y can be predicted on the basis of actual observations of the values of x—which are made r time units before the forecast interval.

When outcomes are to be forecast, the knowledge that one series leads another can be valuable information. The pairs of points x_t, y_{t+r} ($t = now$) can be plotted, and a regression line can be calculated as described below and as drawn in Figure 3.5. It is then possible to forecast a sequence of future outcomes by extrapolating the regression line to predict future points.

Consider the fact that sales of aftermarket tires—replacement tires, not tires that come with a brand new car—are related to new car sales as well as used cars that are on the road. To keep this illustration simple, assume that the sale of light truck replacement tires is well forecast by the sales of new light trucks *five years earlier*.

For real situations, used truck tires might have to be considered, and they would have different lead times depending upon their age. The number of trucks of different ages that are on the road would be part of the picture.

Regression analysis uses the least-squares method to estimate the value of future outcomes, y_{t+k}, if a leading factor x_t can be found. Causality between y_{t+k} and x_t is not assumed. A causal factor that is common to both x and y and which operates as an unknown link may be responsible for whatever relationship is found. Least-squares means that the analysis is based on minimizing the sum of the squared deviations of the points from the regression line.

Tire sales are to be forecast from a time series of light truck sales. These kind of data are shown in Table 3.3.

TABLE 3.3

Year t	Light Truck Sales x_t in Year t (in millions)	Year $t + 5$	Light Truck Tire Sales y_{t+5} in Year $t + 5$ (in millions)
1	6	6	4
2	8	7	6
3	12	8	10
4	8	9	10
5	16	10	12

It has been assumed that the strongest relationship between light truck sales and light truck tire sales in the aftermarket is when the two time series are five years apart.

It is appropriate to fit a least-squares line to these data. This line will be an estimate of the way in which truck sales affect tire sales five years later. For this example, a linear trend is assumed. More complex equations can be used for nonlinear trends. Once the line is determined, it can be used to extrapolate future tire sales.

The model uses what are called *normal equations* (based on the assumption of a linear relationship between x_t and y_{t+5} to derive the least-squares line. The least-squares line minimizes the total variance of the distances of the observed points from that line. The normal Equations 3.3 and 3.4 achieve this objective.

$$\sum_{t=1}^{t=N} y_{t+5} = aN + b\sum_{t=1}^{t=N} x_t \qquad \textbf{Equation 3.3}$$

$$Y = aN + bX \qquad \textbf{(simplified)}$$

(see the last row of Table 3.4)

$$\sum_{t=1}^{t=N} x_t y_{t+5} = a\sum_{t=1}^{t=N} x_t + b\sum_{t=1}^{t=N} x_t^2 \qquad \text{Equation 3.4}$$

$$XY = aX + bX^2 \qquad \text{(simplified)}$$

(see the last row of Table 3.4)

For our example, $N = 5$, which is the total number of pairs of x_t and y_{t+5}. Table 3.4 presents the data needed for this analysis.

t	x_t	y_{t+5}	$x_t y_{t+5}$	x_t^2	$y^2_{(t+5)}$
1	6	4	24	36	16
2	8	6	48	64	36
3	12	10	120	144	100
4	8	10	80	64	100
5	16	12	192	256	144
SUMS	50	42	464	564	396
	$X = 50$	$Y = 42$	$XY = 464$	$X^2 = 564$	$Y^2 = 396$

TABLE 3.4

The Equations, with numbers derived from Table 3.4 are:

$$42 = a(5) + b(50) \qquad \text{using Equation 3.3}$$

$$464 = a(50) + b(564) \qquad \text{using Equation 3.4}$$

Solving for a and b, obtain: $a = 61/40$; $b = 11/16$. These values are introduced in the least-squares line:

$$y_{t+5} = a + bx_t = \frac{61}{40} + \frac{11}{16} x_t \qquad \text{Equation 3.5}$$

which simplifies to:

$$y_{t+5} = a + bx_t = 1.53 + 0.69 \, x_t \qquad \text{Equation 3.6}$$

This least-squares line and the actual scatter of values around it are shown in Figure 3.5.

To show how to use this line, assume that when $t = 6$, $x_t = 10$. The regression line indicates that in year 11, $y_{t+5} = 8.43$ which means that 8,430,000 tires would be the sales forecast. Next, assume that when $t = 7$, $x_t = 20$. The regression line indicates that $y_{t+5} = 15.33$ million tires would be the sales forecast for year 12. This line would be used for forecasting only if the observed data appear to fit it well. These data do fit well and the correlation coefficient determined below supports the visual confirmation of a good fit.

Correlation Coefficient

The correlation coefficient is:

$$r = \frac{n\sum_{t=1}^{t=N} xy - \sum_{t=1}^{t=N} x \sum_{t=1}^{t=N} y}{\sqrt{\left[n\sum_{t=1}^{t=N} x^2 - \left(\sum_{t=1}^{t=N} x\right)^2\right]\left[n\sum_{t=1}^{t=N} y^2 - \left(\sum_{t=1}^{t=N} y\right)^2\right]}}$$

which equals:

$$\frac{(5)(464) - (50)(42)}{\sqrt{[5(564) - (50)^2][5(396) - (42)^2]}} = 0.84$$

There is a strong positive correlation.

Coefficient of Determination

The coefficient of determination (r^2) is a measure of the variability that is accounted for by the regression line for the dependent variable, in this case, y_{t+5}.

$$r^2 = \frac{a\sum y + b\sum xy - \frac{\left(\sum y\right)^2}{n}}{\sum y^2 - \frac{\left(\sum y\right)^2}{n}}$$

$$= \frac{1.53(42) + 0.69(464) - \frac{(42)^2}{5}}{396 - \frac{(42)^2}{5}} = \frac{32}{43} = 0.74$$

The coefficient of determination always falls between 0 and 1.
The x values account for a majority of the variability in the y values.

F I G U R E   3 . 5

**SCATTER DIAGRAM USING LEAST-SQUARES METHOD
FOR REGRESSION ANALYSIS**

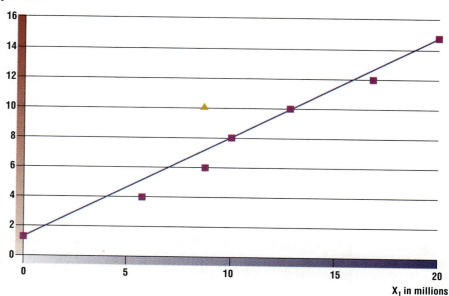

Y_{t+5} in millions

X_t in millions

Pooling Information and Multiple Forecasts

Methods for pooling information to provide stronger forecasts should be explored. It is critical that all parties share their forecasts as much as possible, and try to find ways to combine them. Usually, stronger forecasts can be obtained if both data and experience are pooled.

One of the keys to success in combining forecasts is trial and error. What seems to work is retained and what fails is discarded. As an example of pooling, the results of the regression analysis just completed, could be augmented by a Delphi-type estimation (described below) of light truck tire sales for the next year. The Delphi participants might be the regional sales managers.

Formal methods can be used to evaluate how well different forecasting techniques are doing. At each period that method which did best the last time is chosen for the forecast that will be followed. Forecasts are still derived from the alternative methods, but they are recorded and not followed. When the actual demand results are known, the various forecasting methods are evaluated again and the one that is most successful is chosen to make the prediction for the next period.

Averaging of forecast results also is used. The results of taking forecasts from more than one method and averaging these results to predict demand has been successful in circumstances where choosing the best method (as described in the paragraphs above) produces frequent alteration of the chosen method.

The Delphi Method

Delphi is a forecasting method that relies on expert estimation of future events. In one of its forms, the experts submit their opinions to a single individual who is the only one that knows who the participants are and what they have to say. The person who is the Delphi manager combines the opinions into a report which, while protecting anonymity, is then disseminated to all participants. The participants are asked whether they wish to reevaluate and alter their previous opinions in the face of the body of opinion of their colleagues. Gradually, the group is supposed to move toward consensus. If it does not, at the least, a set of different possibilities can be presented to management.

The regression results could be shown to the Delphi panel of experts with the questions: "Do you think sales will be higher, lower, or the same as the regression results, and why?" Consensus might lead to modification of the managers' targets that were based on the regression results. Why not let these "experts" talk to each other and discuss their opinions? If one of the experts is the CEO, or a Nobel prize winner, the dialogue might not be unbiased. People with greater debating capabilities are not necessarily people with more insight. The Delphi method is meant to put all participants on an equal footing with respect to getting their ideas heard.

Comparative Forecasting Errors for Different Forecasting Methods

All forecasting errors $x_t - \hat{x}_t = \epsilon_t$ are based on comparing x_t the actual demand for time period t, with $\hat{x}_t$ the forecast demand for that same period.

Two kinds of forecasting errors can occur. First, the actual demand is greater than the forecast, which is an underestimate although ϵ_t is positive. Second, the actual demand is less than the forecast, an overestimate although ϵ_t is negative.

To choose a forecasting method, it is necessary to be able to compare the errors that each method generated. There are various measures of error that are used. The choice is dependent upon the situation and what kind of comparison is wanted. Thus, absolute measures are used to count all errors conservatively.

Absolute measures are signified by open brackets $|\epsilon_t|$ which mean that positive and negative errors are treated the same way. The reason for doing this is to prevent positive and negative errors from canceling each other out.

The most common measure is called mean absolute deviation (MAD). It is calculated by taking the sum of the absolute measures of the errors and dividing that sum by the number of observations.

Also, quite common is the mean squared error (MSE) measure. It is calculated by squaring all of the error terms and adding them up. Then, this sum is divided by the number of observations. Because squares of both positive and negative errors become positive numbers, MSE does not allow the two kinds of errors to cancel out each other. However, the squares of large errors become significantly larger than the squares of small errors. Therefore, MSE magnifies large errors. MAD treats all errors linearly, and is a more conservative measure.

Cumulative forecast error (CFE) is simply the sum of all the error terms and is useful when overestimation errors do cancel out underestimation errors. As an example, if the error term represented inventory shortages when $x_t > \hat{x}_t$ and stood for stock on-hand when $x_t < \hat{x}_t$, assuming that stock on-hand could be used to fulfill shortages, the CFE would represent the net inventory situation.

Sometimes the standard deviation of the error distribution, σ_ϵ, is used. It is the square root of the MSE. Another method for comparing errors is called mean absolute percent error (MAPE). It is the result of dividing each absolute error term by its respective actual demand term. This gives a fraction which describes how large the error is as compared to the demand. The fraction is multiplied by 100 to convert to percents, and all of the terms are summed and divided by the number of observations as shown in Table 3.5.

This table examines one forecasting method with observations taken over $n = 4$ weeks. All are row sums—taken across columns.

TABLE 3.5

TABLE OF COMPUTATIONS FOR COMPARING FORECAST ERRORS

					Sum
Week Number, t	1	2	3	4	
Actual Demand, x_t	3	8	5	7	
Forecast, $\hat{x}_t$	6	4	9	6	
Error, $x_t - \hat{x}_t = \epsilon_t$	−3	+4	−4	+1	−2 CFE
Error Squared, ϵ_t^2	9	16	16	1	42
Absolute Error, $|\epsilon_t|$	3	4	4	1	12
Absolute Percent Error, $(|\epsilon_t| / x_t)\,100$	100	50	80	14.3	244.3

Each of the forecast error measures is presented in the following equations:

$$\text{MAPE: } \sum_{t=1}^{t=N} (|\epsilon_t|/x_t)100 \div n = 244.3/4 = 61.08$$

MAPE reveals poor forecasts. Note that for week one, the absolute percent error was 100, and the next two weeks are large percentages as well. Because the fourth week was reduced to 14.3, perhaps the system is stabilizing.

$$\text{CFE: } \sum_{t=1}^{t=N} \epsilon_t = -2$$

as shown in the sum column of Table 3.5.

This minus CFE measure may reveal a bias toward overestimation. The sample is too small to tell.

$$\text{MAD: } \sum_{t=1}^{t=N} |\epsilon_t| \div n = 12/4 = 3.0$$

This is the most conservative estimate of error. Without knowing the size of demand, these 3 units of error might seem very small. They are not, and the MAD measure of 3 is not insignificant because the average demand is 23/4 = 5.75. MAD is more than 50 percent of the average demand.

$$\text{MSE: } \sum_{t=1}^{t=N} \epsilon_t^2 \div n = 42/4 = 10.5$$

As expected, the MSE measure of 10.5 units is large. It is more than three times MAD and 1.8 times the average demand of 5.75.

MSE was magnified by the fact that the error values of 3, 4, and 4 became 9, 16, and 16.

Standard Deviation: $\sigma\epsilon = \sqrt{MSE} = \sqrt{\sum_{t=1}^{t=N} \frac{\epsilon_t^2}{n}} = \sqrt{10.5} = 3.24$

This standard deviation of the error distribution is closely approximated by MAD. That is because the square root of the MSE removes the magnification of error that is inherent in MSE. It should be noted that MSE is a preferred measure when larger errors have disproportionate penalty as compared to small errors. This would be the case when the penalties are proportional to the square of the errors.

Special new measures can be devised when special circumstances exist. Thus, if underestimation results in stock outages which lead to severe penalties whereas overstock carries a small charge, it is necessary to adjust the comparative measures that are used to evaluate forecasting systems.

Keeping a Record of Past Errors

Managers change forecasts that are derived by numerical methods because they know about other factors that are not included in the calculations. This results in modifications of forecasts which can be called predictions and estimates.

It is, therefore, wise to maintain a *history of all forecasting errors*, $x_t - \hat{x}_t = \epsilon_t$. This should be done for all methods that are used, and for all personnel that are making the forecasts, predictions and estimates. Some people are good at forecasting and others are not. Also, some people are good at forecasting under certain circumstances and not good in others.

Companies gain major advantages by finding out who can make good estimates under what circumstances. Further, there are situations where people can learn to make better forecasts, predictions and estimates as a result of feedback about how well they have done in the past. When a record is not kept, all such advantage potentials are lost.

Mail-order companies need to forecast demand in order to have the right number of trained agents and operators in place.
Source: Lands' End, Inc.

S U M M A R Y

Strategic planning requires teamwork and OM should be part of the team. The chapter discusses OM's role in helping to develop strategies and its assignment to carry them out. OM is responsible for seeking out and adopting "Best Practice." Life cycle stages are developed. This includes the introduction and growth of new products, followed by their maturation and decline. OM is expected to protect established products. Forecasting is critical for dealing to advantage with the life cycle stages. Time series analysis is treated with extrapolation and correlation. Important forecasting methods and measures are covered, including moving averages, weighted moving averages, regression analysis, correlation coefficients, coefficients of determination, multiple forecasts, and the Delphi method. Also, the means for evaluating the different forecasting methods by comparing the kinds of errors they generate is explained. A supplement on the exponential smoothing method is provided.

▶ Key Terms

cycles   *86*   extrapolation   *91*   forecasting   *90*
strategy   *80*   synergy   *84*   syzygy   *84*
tactics   *83*   time series   *91*   time series analysis   *92*

▶ Review Questions

1. Contrast strategies and tactics from an OM perspective.

2. What is suboptimization and why is it a concern of OM?

3. What is meant by the statement "OM is a scarce resource"?

4. What is a time series? How would it be used to make predictions about emerging technological developments?

5. Is there a basis for predicting periods of prosperity and of economic slowdowns on the
basis of cycles?

6. What are life cycle stages? Why does the concept of life cycle stages unite P/OM with
other members of the organization in using the systems approach?

7. Detail the life cycle stages and explain them.

8. What is extrapolation and how is it used?

9. What is correlation? What is autocorrelation? Distinguish between them.

10. What is an historical forecast? When is it used?

11. Why are long life cycle stages considered desirable? Does this apply to start-up and
growth as well as the mature stage?

12. How does the hub and spoke diagram for a large company differ from that of a small
company? Draw the contrast with the classic organization chart. Relate both of these
types of diagrams to organizational structure and infrastructure.

13. What is Best Practice when applied to a business system?

14. When are multiple forecasts desirable?

▶ Problem Section

1. If a series of numbers are { 14, 23, 28, 34 } what is the best estimate for the next
value? What method can be used?

2. A time series for actual sales figures is given in the table below.

Month	Actual x_t	Forecast $\hat{x}_t$	Error $\epsilon_t = (x_t - \hat{x}_t)$
1	10		
2	30		
3	40		
4	40		
5	50		
6	60		
7	50		
8	40		
9			

a. Use the moving average method with $N = 3$ to develop forecasts for months four
through nine. Calculate the error terms, ϵ_t.

b. Employ weights of 0.5, 0.3, and 0.2 to obtain weighted moving average forecasts for the
same time period. Calculate the error terms, ϵ_t. Compare the results in parts 2(a) and 2(b).

3. It has been stated that sales for next year will be 50 percent greater than this year, but
the historical pattern will continue to hold. Complete next year's forecast.

Month	ACTUAL Sales Last year	FORECAST of Sales Next year
January	1500	
February	1600	
March	1800	
April	2000	
May	2300	
June	2500	
July	2350	
August	2100	
September	1850	
October	1650	
November	1550	
December	1400*	

*Note: This December estimate of 1400 units is based on the prior year's sales.

4. The Highway Department is considering using a 6-month moving average to forecast crew hours needed to repair roads month by month for the coming year. To test whether this method is accurate, use last year's data given below to predict the last six months' crew hours to compare with the observed values. Start with the average of January through June to compare with the actual value of 200 for July and so on through December. Remember that the prediction for the next month is the average of the ACTUAL values of the previous six, not the predicted values!

Month	Crew Hours	Month	Crew Hours
January	110	July	200
February	120	August	220
March	140	September	280
April	180	October	120
May	250	November	100
June	200	December	80

5. In reference to Problem 4, prepare an analysis of the error in using the 6-month moving predictor.

6. Referring back to Problems 4 and 5, use a 3-period moving average to forecast the last six months and compare directly to the 6-month forecast developed there. Here, the first forecast is the average of April, May, and June to compare with the actual value of 200 for July and so on through December.

7. Referring back to Problems 4, 5, and 6, use a 6-period weighted moving average to forecast the last six months and compare directly to the 6-month and 3-month forecasts developed there. Use the weights 0.1, 0.1, 0.1, 0.2, .0.2, and 0.3 with the 0.3 weighting the most recent datum point.

8. The following table presents data concerning the sales of minivans for 10 years. It is believed that the demand for minivan replacement tires is highly correlated with the sales figures three years earlier.

Year t	Minivan Sales x_t in Year t (in millions)	Minivan Tire Sales y_t in Year t (in millions)
1	10	4
2	12	6
3	11	7
4	9	5
5	10	8
6	12	7
7	10	5
8	9	7
9	8	8
10	7	6

Do an analysis of x_t and y_{t+k}, where $k = 3$ years, using the least-squares method and linear regression. Note that seven years of data will be used to prepare the regression line. The forecast for tires will go from year 11 through year 13.

9. Using the information given in Problem 8, it is now stated that there is another person in the company who feels that $k = 4$ should be tried. Do it and compare the results. Note that six years of data will be used to prepare the regression line. The forecast for tires will go from year 11 through year 14.

10. What does autocorrelation have to do with the methodology used in Problem 8?

11. This table presents the results of forecasting using moving averages. Comparison is made of actual demand and the forecast for a period of seven days.

Day Number, t	1	2	3	4	5	6	7
Actual Demand, x_t	4	5	2	3	1	3	4
Forecast, $\hat{x}_t$	3	6	2	4	3	1	3
Error, $x_t - \hat{x}_t = \epsilon_t$							
Error Squared, ϵ_t^2							
Absolute Error, $|\epsilon_t|$							
Absolute Percent Error, $(|\epsilon|/x_t)100$							

Complete this table and then work out each of the forecast error evaluation methods that are listed below.

MAPE: $\sum_{t=1}^{t=N} (|\epsilon_t|/x_t)100 \div n$

CFE: $\sum_{t=1}^{t=N} \epsilon_t$

MAD: $\sum_{t=1}^{t=N} |\epsilon_t| \div n$

MSE: $\sum_{t=1}^{t=N} \epsilon_t^2 \div n$

Standard Deviation: $\sigma_\epsilon = \sqrt{\text{MSE}} = \sqrt{\sum_{t=1}^{t=N} \epsilon_t^2}$

Discuss the results comparing the various evaluation methods.

12. The time series for ϵ_t is as follows:

$$\{ +600, -300, +500, +100, -50, +60, -40, +30, -50, +20 \}$$

Measure MAD, MSE, CFE, and $\sigma\epsilon$.

13. The time series for ϵ_t is as follows:

$$\{ +600, -300, +500, +100, -50, +60, -40, +30, -50, +20 \}$$

Analyze this time series in such a way as to shed additional light on what seems to be happening.

14. It should be noted that MSE is preferred to MAD when larger errors have disproportionate penalties as compared to small errors. This would be the case when the penalties are proportional to the square of the errors.

Two forecasting methods are used. The first method gives the error time series for ϵ_t as follows:

$$\{ +600, -300, +500, +300, -500 \}$$

The second gives this time series:

$$\{ +100, -200, +800, +100, -100 \}$$

Compare the two time series using the various forecast error evaluation methods. Discuss the results in light of the statement concerning penalties.

▶ Readings and References

G. E. P. Box, and G. M. Jenkins, *Time Series Analysis: Forecasting and Control,* San Francisco, CA, Holden-Day, 1970.

S. A. DeLurgio, and C. D. Bhame, *Forecasting Systems for Operations Management,* Burr Ridge, IL, Irwin, 1991.

P. Drucker, *Managing in Turbulent Times,* London, Pan Books, 1981.

I. Magaziner, and R. Reich, *Minding America's Business,* NY, Vintage Books, 1982.

S. Makridakis, S. C. Wheelwright, and V. E. McGee, *Forecasting: Methods and Applications,* NY, John Wiley & Sons, 1983.

S U P P L E M E N T 3

Forecasts Using Exponential Smoothing Methodology

for short term produ. planning.

Often, the use of the exponential smoothing method for forecasting is preferred to the weighted moving average method. Like the weighted moving average (WMA) method, exponential smoothing calculates an average demand—usually giving more weight to recent actual demand values than to the older ones.

Exponential smoothing is a simpler method, requiring fewer calculations than WMA, which needs N weights and N periods of data for each forecast estimate. Exponential smoothing needs only three pieces of data, as listed below. Also, it can be more effective because only one weight, alpha(α)—called the smoothing constant—has to be chosen. This makes it easier to experiment with past data to see which value of alpha provides the least forecast error.

Exponential smoothing has been found to be more effective in a variety of situations. Many forecasting and control systems employ exponential smoothing because it works better than the older methods of moving averages and WMA. It has proved effective for diverse applications. Fighter aircraft use exponential smoothing to aim their guns at moving targets. In effect, they forecast the location of enemy jets during flying missions. This application shows how fast the exponential smoothing method can track a consistent, but dynamically changing pattern.

Manufacturers use exponential smoothing to forecast demand levels which experience the same kind of nonrandom but volatile shifts from time to time. In such cases, the situation is dynamically changing, but the recent past has the most information about the near future. Exponential smoothing can catch these shifts and make rapid adjustments to inventory levels. Other manufacturers use it because it requires less computational work and is readily understood.

Exponential smoothing methodology remembers the last estimate of the average value of demand and combines it with the most recent observed, actual value to form a new estimated average. As in prior materials, hats over the x's stand for forecast values of demand; t is now and $t + 1$ is the next period for which the forecast is to be made:

$\hat{x}_{t+1}$ is the new forecast of demand to be made for period $t + 1$.

$\hat{x}_t$ is the last forecast made—it was for period t, which is now.

x_t is the actual result observed for the period t, which is now.

All x_t values—no hats—are recorded values of actual demand.

Equation 3.7 below forecasts demand in period $t + 1$ as an alpha-based average of the actual demand in the last period t, and the demand that had been forecasted for that period, t.

$$\hat{x}_{t+1} = \alpha x_t + (1 - \alpha)\hat{x}_t \qquad \textbf{Equation 3.7}$$

The preceding period's equation would then have been:

$$\hat{x}_t = \alpha x_{t-1} + (1 - \alpha)\hat{x}_{t-1} \qquad \textbf{Equation 3.8}$$

Substituting Equation 3.10 into Equation 3.7 yields Equation 3.9:

$$\hat{x}_{t+1} = \alpha x_t + (1 - \alpha)[\alpha x_{t-1} + (1 - \alpha)\hat{x}_{t-1}] \qquad \textbf{Equation 3.9}$$

$$\hat{x}_{t+1} = \alpha x_t + \alpha(1 - \alpha) x_{t-1} + (1 - \alpha)^2 \hat{x}_{t-1} \qquad \textbf{(simplified)}$$

Substituting Equation 3.10 into Equation 3.9, and continuing in this fashion, yields Equation 3.11—which is the generalized form for the forecast of demand for period $t + 1$:

$$\hat{x}_{t-1} = \alpha x_{t-2} + (1 - \alpha)\hat{x}_{t-2} \qquad \textbf{Equation 3.10}$$

$$\hat{x}_{t+1} = \alpha x_t + \alpha(1 - \alpha) x_{t-1} + \alpha(1 - \alpha)^2 x_{t-2} + \alpha(1 - \alpha)^3 x_{t-3} + \ldots \qquad \textbf{Equation 3.11}$$

The response rate of the exponential smoothing model is a function of the alpha value that is used. When alpha is large, the actual demand in the last period is given great weight. The forecast results are markedly affected by the alpha that is used, as is shown in Table 3.6.

Table 3.6 employs the same numbers that were used for the moving average and weighted moving average calculations. The table shows the effect of α on the forecast for $\hat{x}_{t+1}$.

The exponential smoothing updating system requires only one operation. Thus, if $\alpha = 0.100$, and the actual result is found to be $x_t = 45$, then (from Table 3.6) $\hat{x}_{t+1} = 42$. Then the next period forecast will be:

$$\hat{x}_{t+1} = 0.100(45) + (0.900)42 = 42.3$$

Small values of alpha are used for stable systems where there is, at most, some small amount of random fluctuation. Large values of alpha are used for changing and evolving systems where much credence can only be placed on the last observation. New products, as they move through their life cycle stages, start with a large alpha value which gradually diminishes as the product enters its mature stage. For most production scheduling systems in both the job shop and the flow shop, alpha is kept small, in the neighborhood of 0.050 to 0.150, to decrease the systems response to random fluctuations.

TABLE 3.6

EXPONENTIAL SMOOTHING WITH A RANGE OF ALPHAS

$\hat{x}_{t-1}$	α	
$0.000(60) + 1.000(40) = 40$	0.000	
$0.100(60) + 0.900(40) = 42$	0.100	
$0.200(60) + 0.800(40) = 44$	0.200	
$0.300(60) + 0.700(40) = 46$	0.300	
$0.375(60) + 0.625(40) = 47.5$	0.375	← moving average result
$0.400(60) + 0.600(40) = 40$	0.400	
$0.500(60) + 0.500(40) = 50$	0.500	
$0.550(60) + 0.450(40) = 51$	0.550	← weighted moving average result
$0.600(60) + 0.400(40) = 52$	0.600	
$0.700(60) + 0.300(40) = 54$	0.700	
$0.800(60) + 0.200(40) = 56$	0.800	
$0.900(60) + 0.100(40) = 58$	0.900	
$1.000(60) + 0.000(40) = 60$	1.000	

Problems for Supplement 3

1. Using the data in Problem 2 above, apply exponential smoothing to develop a forecast comparable to what is asked for in parts A and B. Note that exponential smoothing goes back to the first prediction, so assume that the forecast *for t = 1 is 20 and sequentially calculate* all forecasts through $t = 9$. However, only analyze the errors for $t = 4, \ldots, 8$ for direct comparison with the 3-period moving averages calculated in parts a and b of Problem 2. Run for α of 0.3 and 0.7 and discuss the benefits of each kind of approach.

2. Simulate the effect of using $\alpha = 0.05$ on the following set of very seasonal data:

Month	Sales	Month	Sales
1	500	7	10,000
2	800	8	7,000
3	1,200	9	6,000
4	2,000	10	1,000
5	4,000	11	500

Notes...

1. Jon F. Cox, John H. Blackstone, and Michael S. Spencer, *APICS Dictionary,* 7th Ed., APICS Educational and Research Foundation, 1992.

Chapter 4

Quality Management

After reading this chapter you should be able to . . .

1. Distinguish between producer's and consumer's quality concepts.
2. Analyze quality in terms of its many dimensions.
3. Explain how quality is measured for both tangible and intangible quality dimensions.
4. Detail the things to consider when developing a rational warranty policy.
5. Discuss ISO 9000 standards in an international context.
6. Address and apply the costs of quality.
7. Describe the control monitor feedback model.
8. Relate quality mapping and the House of Quality (HOQ).
9. Explain quality prizes being given worldwide.

Chapter Outline

4.0   WHY IS BETTER QUALITY IMPORTANT?

4.1   WHY IS TOTAL QUALITY MANAGEMENT IMPORTANT?

4.2   TWO DEFINITIONS FOR QUALITY

4.3   THE DIMENSIONS OF QUALITY
Models of Quality
Quality Dimension—Purpose Utilities
Quality Dimension—Other "ilities"
Definition of Failure—Critical to Quality Evaluation
Warranty Policies
The Service Function—Repairability
Nonfunctional Quality Dimension—Human Factors
The Variety Dimension

4.4   SETTING INTERNATIONAL PRODUCER STANDARDS
        U.S. Quality Standards
        International ISO 9000 Standards
        Japanese Quality Standards
        Database of Quality Standards

4.5   THE COSTS OF QUALITY

4.6   THE COST OF PREVENTION

4.7   THE COST OF APPRAISAL (INSPECTION)

4.8   THE COST OF FAILURE

4.9   QUALITY AND THE INPUT-OUTPUT MODEL

4.10   CONTROLLING OUTPUT QUALITY

4.11   TQM IS THE SYSTEMS MANAGEMENT OF QUALITY

4.12   MAPPING QUALITY SYSTEMS
        House of Quality (HOQ)
        Quality Function Deployment (QFD)
        Teamwork for Quality

4.13   INDUSTRY SEEKS RECOGNITION
        The Deming Prize
        The Malcolm Baldrige National Quality Award

Summary
Key Terms
Review Questions
Problem Section
Readings and References
Supplement 4: A Scoring Model for the House of Quality
Industry Perspective on Quality—Tom's of Maine

The Systems Viewpoint

Everyone in the company contributes to the quality of the product. In recognition of this fact, excellent companies train everyone in quality. It applies to every aspect of manufacturing, service for the product, administrative support and office management. The systems viewpoint is inherent in good quality because it links what customers want to the production transformation process that makes and delivers the goods. Quality goals are determined by the voice of the customer and OM tries to keep on target. The approach is: to look into each place where quality could be improved, analyze the pieces and develop quality standards, and then synchronize the system to produce quality products.

4.0

WHY IS BETTER QUALITY IMPORTANT?

Quality is one of the most important variables in the business system. When it comes to gaining competitive advantage, quality has tremendous leverage with new and old customers alike.

Products having better quality have a marketplace advantage. Because customers prefer to buy products they perceive to be of better quality, they often are willing to pay more for the superior qualities. Products perceived to be of lower quality are less likely to be substituted for products of better quality. Therefore, lower quality is equivalent to higher elasticity which means that price increases drop demand rates faster. Better quality is equivalent to more protection against competitors that compete on price.

Better quality saves money for the company by decreasing defectives that must be scrapped or reworked. This can be a very big savings because a great deal of money and time has been used to add value to the product, which then fails inspection and has to be scrapped. With so much value added, rework is the only reasonable way to go. In turn, rework expenses often are considerable. Further, the company with a better product can afford to offer a superior warranty which has marketing attraction for customers. Even with the superior warranty policy, the better product leads to less frequent rework and replacement.

Better quality increases customer loyalty. An investment in quality can be called a customer holding expense. The greatest part of marketing expenditures is to attract new customers, called the switching expense. Retaining current customers makes good sense. The trade-off between the cost of getting new customers and the additional price paid for better quality which holds onto existing customers favors the investment in quality.

Alienating existing customers with product weaknesses and failures has a host of other side effects which are undesirable. Word spreads quickly about product failures. There are a growing number of consumer protection publications. Government agencies pursue other aspects of consumer protection using various means to remove defective products from stores. Large-scale auto callbacks are given prime time coverage on TV.

Better product quality obtained at a reasonable price generally goes hand-in-glove with growth in market shares and increases in revenues. Sometimes quality improvements result in decreases in court costs for claims of harm caused by malfunctions, and other types of liability. Another advantage of better quality is that people who work in companies that really have better quality enjoy an environment of higher morale.

4.1

WHY IS TOTAL QUALITY MANAGEMENT IMPORTANT?

Total Quality Management (TQM)

Called "The Prime Directive." TQM must be a company-wide systems approach to quality.

TQM, which stands for **Total Quality Management,** is called "The Prime Directive" in this book. What is meant by this? The "prime directive" denotes a critically important and overriding rule that cannot be compromised. *If quality management is to achieve the best results, then TQM is the procedure of choice.*

Translating this into quality-specific terms, the prime directive for TQM is that the systems approach to quality must be used company-wide. Thus, as an unequivocal rule:

**TQM is a management strategy that requires
a company-wide systems approach to quality.**

To Curt W. Reimann, director of the Malcolm Baldrige National Quality Award, the major objective of the Baldrige Award is to "promote change at the national level, by way of a body of knowledge about quality that connects process and results." Reimann says the Baldrige Award serves to promote quality awareness, to recognize quality achievement of U.S. businesses, and to publicize successful quality strategies. Reimann, who is also director for quality programs at the National Institute of Standards and Technology (NIST), has directed the Baldrige Award since its inception.

Reimann has been instrumental in developing and annually revising the Award examination, which operationalizes the basic concepts of total quality management and makes performance assessible. Reimann says that "Through the Baldrige Award criteria, we are attempting to create an integrated organizational management discipline, built around the factors that determine competitive success. Product and service

Curt W. Reimann
· · ·
Director, Malcolm Baldrige
National Quality Award

quality are among these factors, but by no means are they the only factors. The discipline must integrate all aspects of performance—productivity, speed, innovation, market knowledge, workforce development, and many others."

The quality-improvement thrust of the Baldrige Award "goes well beyond the contest itself," according to Reimann. "To me, the contest 'pays the rent' for everything and gets most of the attention." But the real purpose of the Baldrige Award is information transfer and its affect on all U.S. institutions.

Reimann supports the inclusion of the Baldrige Award concepts and criteria

in business school curricula. "From the point of view of national competitiveness, the teaching of the Baldrige Award principles in business schools represents a very important development— one that could influence the next generation of business leaders."

A chemist by training, Reimann has a Ph.D. from the University of Michigan and was a scientific researcher and science manager in chemistry at NIST before becoming involved with the Baldrige Award.

Sources: Curt W. Reimann, "The Baldrige Award: Leading the Way in Quality Initiatives," *Quality Progress,* July, 1989; Scott Madison Paton, "An Interview with Curt W. Reimann, Director, Malcolm Baldrige National Quality Award," *Quality Digest,* 1991; and conversations and correspondence with Dr. Reimann.

The attainment of TQM is the critical precursor of excellence in quality. To begin with it is an organizational matter. Everyone in the organization has one or more roles that are related to the attainment of quality. People must be able to communicate with each other in the common pursuit of quality for TQM to operate. TQM puts the entire system in perspective.

TQM enables the systems approach to operate effectively. This is because everyone in the company knows the goals. They also know they share the same objectives for attaining quality. Like the orchestra that needs to play in synchronization to achieve quality sounds, the company needs to synchronize its players to achieve company-wide quality. Ignore or discard any one of the elements and the quality chain is as weak as the weakest link. That, too, is the philosophy of TQM.

Some people do not like the name TQM, and they think that it has been overblown or overused. However, everyone agrees that what TQM stands for is here to stay. There is so much business publication that the name for any concept is likely to get used up, and misused too, in a very short time. It is, therefore, necessary to examine the concept of TQM carefully to see what makes it so important and successful that business and academia alike are proponents of the concepts.

Those who practice the systems approach know that this is a sound foundation for the attainment of superior quality. If the practice of TQM is done right, it can put a faltering company back on track. If done badly, TQM can alienate those who put themselves on the line to support it. The label TQM assures no one of anything. It is essential to read the ingredients on the label of this system of practice. All TQM's are not alike.

4.2
TWO DEFINITIONS FOR QUALITY

Before the correct and sound approach to TQM can be developed, it is necessary to define quality. The word quality is used in two different ways.

1. The producer's view—qualities are properties of things without distinction about being good or bad. Thus, the product conformed in all regards to the company's quality specifications.
2. The consumer's view—qualities relate to the "degree of excellence" of things. Thus, consumers expect and deserve a quality product.

The two definitions of quality are held and exercised by different interests. One person can have both points of view, depending on what hat they are wearing at the time. Everyone has an intrinsic sense of what quality means to them as a consumer. They want the best that can be had so they are always judging the "degree of excellence."

As a producer, it is often necessary to compromise and set quality standards that are not the best in their class. Inexpensive automobiles do not have the same set of qualities as the expensive ones. There are different markets for town cars and compacts. It is the producer's problem to choose appropriate quality standards so that the market judges them to be excellent for that price class.

Producers view the quality of golf balls by their bounce and durability.
Source: The Goodyear Tire and Rubber Company

Producers strive to balance the market forces for high quality with consumers' cost preferences and the company's production capabilities. Figure 4.1 illustrates a common economic concept of the optimal quality level.

COST OF QUALITY AND DOLLAR VOLUME OF SALES YIELD MAXIMUM PROFIT AS A FUNCTION OF THE QUALITY LEVEL

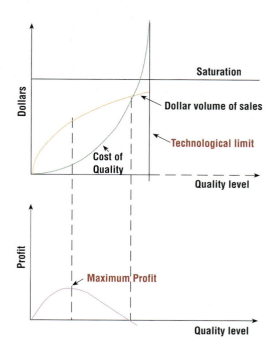

There are valid exceptions to the relationships that Figure 4.1 depicts. The cost of quality does not necessarily rise as the quality level increases. The fact that many times quality can be improved without spending any money will be discussed at a later point. There also is the issue of how improved quality is obtained allowing that money is spent; is it on training, technology or both?

According to Figure 4.1, there is a limit to how much quality can be improved. This concept can be faulted as not taking scientific possibilities and the ingenuity of creative people into account. There also is a question about the dollar volume of sales approaching saturation with respect to improvements in the quality level. Given all of those issues, it is difficult to know where the point of maximum profit with respect to quality level really will occur. Nevertheless, the quality level standards must be set, and that should be done with the notion of maximum profit in mind.

Some producers may be born with an innate sense of how to set quality standards so as to best balance the costs and benefits. Most producers have to learn how to deal with such issues. This is an essential part of what the study of OM is about. The lessons involve developing models about quality that increase awareness of the categories of quality and the means of improvement.

Consumers view seafood by the degree
of freshness.
Source: Legal Sea Foods

4.3

THE DIMENSIONS OF QUALITY

Dimensions of quality

Descriptors examined to
determine a product's quality.

The **dimensions of quality** are the descriptors that must be examined to determine the quality of a product. The wine business is an enormous industry on a global scale. How do the presidents of wine companies characterize the quality of their products? They measure their production output by bouquet, color and taste plus chemical analyses. Cars are evaluated, within categories of cost, by their power, safety features, capacities, fuel efficiencies, and style, among other things. Appearance and style dimensions, so often important, are difficult to rate. Experts at ladies' fashion shows are at no loss for words, and yet, this multimillion dollar industry is only able to explain successes and failures after the fact.

When rating the quality of cities to live in, the authors of such research base their studies on dimensions that define the quality of life. The complex set of criteria includes amount of crime, cost of living, job availability, transportation, and other factors. Special needs emerge such as the quality of schools for families with children.[1]

The starting point for managing quality is to define the set of quality dimensions, recognizing the special needs of market niches and segments. Not everyone will agree about what should be on the list, nor about the importance of the dimensions that are on the list. Individuality accounts for different perceptions about the quality of the product with respect to its various elements.

A small sample of customers were asked what qualities could be improved in the service they received from their bank. Their answers included: length of time waiting for tellers should be decreased, availability of officers for special services should be increased, bank should be open longer hours, rates on interest-bearing accounts should be increased, and service charges should be dropped. Also, the turnover in personnel should be decreased.

An even smaller group of bank officers was asked to define the quality of the services their bank offered or should offer. Their answers show what a different perspective prevails between customers and producers, thus: the number of different products that the bank offers (CDs, checking accounts, passbooks, mortgages, loans, investment services) should be large, the average speed of tellers per transaction should be small, the variability of teller's times should be decreased, the percentage of times that customers wait longer than five minutes should be small, and the average length of waiting lines for tellers should be about three.

What could bring these two groups closer together? The customers describe quality in terms of excellence. The bankers describe quality in terms of how well they meet the standards that have been established by their banks. They also talk about changing the standards. The balance between these two positions is related to the costs and benefits of providing more of what the customer wants.

Models of Quality

This section presents models of quality in the form of lists. The lists are generic categories of quality. They will enable producers and consumers to check off and define all of the dimensions of quality that are applicable to their products.

This set of categories was derived by David Garvin, a Harvard professor who has specialized in quality issues. His eight dimensions for categorizing quality are shown in Table 4.1.[2]

TABLE 4.1

THE EIGHT QUALITY DIMENSIONS SUGGESTED BY DAVID GARVIN

1. Performance
2. Features
3. Reliability
4. Conformance
5. Durability
6. Serviceability
7. Aesthetics
8. Perceived Quality

The discussion about Table 4.1 uses the automobile as an example. The qualities of this manufactured product are widely known. The eight quality dimensions are applicable to services as well, such as at the bank or even Club Med. Describing qualities provides excellent practice in taking the first steps that are essential to install a quality program.

The Performance dimension relates to the quality of the fundamental purpose for which the product is purchased. How well does the car do what it is supposed to do?

The Features dimension refers to product capabilities not considered to be part of normal performance expectations. These might be air-conditioning, stereo, Antilock Brake System and air bags.

The Reliability dimension relates to performance that can be depended upon with a high level of assurance. The car starts and it drives. It does not break down, but the windshield wipers do not work. How that car measures up for reliability is an issue to be discussed.

The Conformance dimension alludes to the degree to which the measured production qualities correspond to the design quality standards that have been specified. This is definitely the producer's concern and not the consumer's. OM is responsible for meeting the conformance dimensions of quality. They are the quality standards.

The Durability dimension deals with how well the product endures in the face of use and stress. Some cars are on the road with well over 100,000 miles. Those cars deserve a quality medal for high durability. There is a relationship between reliability and durability for autos, but it needs to be analyzed.

The Serviceability dimension is related to how often, how difficult, and how costly it is to service and repair the product. Serviceability for cars involves the combination of preventive and remedial maintenance. This quality dimension applies to services for manufacturing products and the products of service firms.

The Aesthetics dimension refers to the appearance of the product. For autos, design styling counts. Industrial designers and operations managers form a powerful team that epitomizes the systems approach. Together, as a team they consider all of the linked factors which relate to manufacturability in addition to the eight quality dimensions.

The Perceived Quality dimension relates to the customers' perceptions of the product's quality. This dimension integrates the prior seven dimensions with the customers' values for them. Market research is one of the most important means for determining the customers' perceptions. The blue book, which is the used car dealers' Bible for determining the trade-in values of automobiles, reflects perceived quality.

Table 4.2 presents a different and more detailed categorization of quality dimensions. It emphasizes the difference between tangible and intangible attributes. Because it is more detailed, it shows the weakness of attributing quality to one or two dimensions. Quality resides in the totality—which is the essence of the systems approach.

TABLE 4 . 2

AN EXTENDED TAXONOMY OF QUALITY DIMENSIONS

I. Functional Qualities (the "ilities")
  1. Utility of main purpose or performance
  2. Dependability of function
    a. Conformance to standards
      from day one, and on, over time
    b. Reliability
      does it function properly over time?
    c. Durability
      how long does it function with heavy use?
      how much wear (stress) can it take?
      functional deterioration with use or time?
    d. Failure characteristics
      expected lifetime
    e. Maintainability–Serviceability I
      cost of preventive maintenance (PM)
      frequency and time required for PM
    f. Repairability–Serviceability II
      quality—as good as new?
      cost and time for remedial maintenance
    g. Guarantees and warranties
  3. Human Factors using the product
    a. Safety—can be tested
    b. Comfort—market researchable
    c. Convenience—market researchable

II. Conceptual (Nonfunctional) Qualities
   1. Style
   2. Appearance
   3. Self-image of user: related to price and prestige
   4. Timeliness of design
   5. Variety

For different individuals, certain dimensions are more important than others. Overall, it should be evident that quality definition is enhanced by requiring a systems perspective.

Quality Dimension—Purpose Utilities

Purpose utilities are related to the specific, functional classes of use, i.e., soap to wash the face, tires to keep the wheels rolling, and credit cards for purchases without cash. Often, purposes seem to be clear, but market research is needed to explore alternative uses for a product. When awareness of alternative uses develops, then quality standards may have to be developed to assess the product's performance in comparison with that of competitors.

Many customers consider shower soap to be different from face soap. Snow tires have a special market niche with their own quality dimensions. The point about "alternative uses" is that it is necessary to explore in-depth the way customers perceive quality in order to properly deal with the quality dimensions. Only then can OM make sure that conformance to the correct set of standards is taking place.

Frequently, there are regional differences in quality perceptions. Hard water areas have their own special quality dimensions for soap. Snow tires have little impact in the South. Such effects are amplified in the international marketplace. This issue of alternative uses (which can run counter to conventional marketing wisdom) creates a critical marketing-production situation. OM has one set of standards for region A and another for region B.

Quality definition is not simple. The measurement of how well a product or service performs its conforming functions is essential if standards are to be set up. There is no point in measuring the wrong things. When the standards are established to everyone's satisfaction, then OM can make the right products, test them and improve the production process.

Assumptions must be made about the way in which the measurable, physical factors relate to the consumers' evaluations of the "ilities" of the product. The "ilities" include: utility, dependability, reliability, durability, serviceability, maintainability, repairability, and warrantability. Manufacturability, which includes assembly, should be added to this list.

Quality Dimensions—Other "ilities"

There are different ways for describing dependability. Durability refers to the ability to resist wear, stress and decay. Reliability refers to the ability to continue to provide the functional utility attributes—within conformance standards or limits—over a given period of time. The width of the limits represents an important aspect of the definition of quality for the design.

To understand these limits, consider the following transactions between designers and operations managers.

1. Telephone cords have 4-pin modular connectors at each end. One connector is plugged into the modular wall jack and the other is plugged into the telephone.

The 4-pins of the connectors are gold plated to insure adequate conductance. OM will have to oversee the plating and can cost it out in terms of both materials and labor. R & D and the designers decide how thick the plating should be for specifications furnished by marketing and market research concerning how many insertions and removals the connectors should be able to withstand.

This durability specification used to be 10,000 insertions. That has been reduced significantly. It is not believed that this compromises the way the product is viewed by the customer. From a production point of view, productivity rises and costs are reduced. There is more time for OM as a scarce resource to do other things that need attention.

2. Some light bulbs are made to have long lifetimes. The materials used in such bulbs are different from those in the regular lifetime bulbs and manufacturing processes are different as well. The durability dimension of each kind of bulb is involved.

With respect to durability, parts of a product will become worn with use, whereas other parts will deteriorate with age, independent of use. Most often, both kinds of forces are at work. Generally, the expected performance will increasingly deviate from the initial design standard over a period of time. That is what happens to light bulbs; as they age, they become increasingly more dim.

They would fail a test that measures durability of the output in lumens of light. The light bulb output is decaying and slowly drifting to total failure, but it may be quite a while before it gets there. OM has to develop standards and testing procedures, to determine how many hours of acceptable output the light bulbs it is producing have.

Drift-decay quality phenomenon

A product's performance can gradually diminish as a result of age and wear.

This **drift-decay quality phenomenon** in which performance of a product gradually diminishes is characteristic of a great many functional attributes of mechanical, chemical, and electrical products. An electric light bulb, when it is first used, will generate some given number of lumens. A typical 75-watt bulb starts its life with about 1170 lumens of light. It ages as a function of both hours of use and the number of times that it is turned on and off.

OM needs to work with R & D and engineering to test durability of outputs. Materials used and characteristics of the process play an important part in product performance. For many products such as food, time spent as work in process and in final storage can cut down on the useful life of the product in the consumer's hands.

Table 4.3 presents a chart which describes how the measurements for all of the dimensions in the quality tables can be obtained. The distinction between manufacturing and services need not be drawn because the products of both types of producers have many tangible and intangible elements.

T A B L E 4 . 3

HOW QUALITY IS DEFINED AND MEASURED

	Quality Dimensions	
	Tangible	**Intangible**
Conformance to Specifications	SQC/TQC Zero Defects	Arbitrary Critiques By Experts
Consumer Perceptions	Market Research	Market Research
Management Evaluations	Competitive Comparisons	Competitive Comparisons

Definition of Failure—Critical to Quality Evaluation

Failure occurs when a product ceases to perform in an acceptable fashion. Accordingly, sometimes failure can be judged objectively. Sometimes it is in the mind of the customer. With respect to the objective kind of failure, there are engineering as well as logical conventions that define physical failure. The car does not start. The light bulb is burned out. However, if the light source is considered "failed" after its output falls below a certain threshold, then OM needs to include that fact in its quality standards.

A certain percent of light bulbs fail at start-up. Failure probabilities follow a U-shaped curve which is known as the Weibull distribution. Customers who buy a new bulb, insert it and see it fail are unlikely to write a letter of complaint. The store where it was bought could question the authenticity of the claim. There is ample room for customer hard feelings about the quality of a simple product like a light bulb.

OM needs to address the questions: Why do these early failures occur? and Is there something in the production process that could be changed? Marketing needs to develop a policy for the early failures. OM and marketing working together might be able to remove the commodity stigma that is associated with light bulbs; that all light bulbs are alike. The same commodity stigma applies to gasoline, aspirin, vodka, detergents and salt.

Operations managers should be fully aware of how they can influence the failure rates and characteristics of the products which they are producing. They also should be working closely with R & D and engineering design to develop new production capabilities that can increase the product's expected lifetime or **mean time between failures (MTBF).** MTBF is a measure frequently used to discuss reliability. In the case of light bulbs, MTBF is often as low as 750 hours, which can require changing bulbs every month. With marketing's help, OM should be fully aware of the MTBF rates and other failure characteristics of competitors' products.

Mean time between failures (MTBF)

A measurement of the product's expected lifetime.

There are many reasons why failure and reliability, as definitions of quality, play an extremely important role. Some types of failure are life threatening while others are not. This consideration plays an important part in the costs of protection through insurance and litigation. Every effort must be made to insure safety and to document that this effort has been honest. The courts of law expect it.

Some types of failure do not permit repair, whereas others do. The definition and specification of quality also should be concerned with the ease of maintenance and the cost of replacement parts. These factors affect the consumer's judgment of quality.

Warranty Policies

A product warranty is a guarantee by the producer to protect the customer from various forms of product failure. The specifics are spelled out in contractual fashion. Thus, it is typical to state for how long a time, and for how much use, the product is covered. The conditions of use are generally stated.

What is the basis for a sensible warranty policy? First, there are the competitive marketing requirements. If Chrysler guarantees 50,000 miles and 5 years, what should Ford offer? Does it make sense to offer a warranty policy that costs more than it gains? Warranties involve interactions between products and promises made by production and marketing. The systems approach is essential to make certain that costs and gains are, at least, balanced.

Second, the operational reliability of the product is critical. How many cars will fail, in what ways, after how much time? Given the schedule of failures by time and type, it is possible to figure out how much a blanket, or partial, warranty will cost the company. It is feasible to determine how many customers will be left without company support. These customers will not be repeat purchasers of that brand of auto in the future. Policies of partial warranties to cover different components under varying circumstances also can be formulated in terms of costs and ultimate customer dissatisfaction. OM will be instrumental in helping to determine a rational warranty policy.

Management must know the reliability of its products and services to come up with a rational specification of guarantees and warranty periods. An important systems dialogue must take place to set the terms of warranty coverage and period. The essence of this thinking should be founded on the knowledge of the product and the process. That is the responsibility of the OM and quality control functions within the organization.

The Service Function—Repairability

The Repairability (and/or maintainability) quality dimension requires that a service function be developed and operated. Speed of service is an attribute that usually is considered very important. A service policy is an agreement between the company and its customers concerning how fast service will be rendered, what service steps will be taken, and what charges will be borne by the customer (warranty contract).

Nikon has a service policy of repairing the entire camera and not just the parts that are responsible for the immediate cause of failure. Their service policy includes furnishing an estimate by mail or phone before beginning work. Service policies are taken very seriously by customers who require service. Organizations distinguish themselves from one another by the care that they show and the fairness of their service policy.

Another service issue is the use of preventive versus remedial service. In the 1960s autos received preventive maintenance every 3000 miles. Typically, the service interval is now 7500 miles or more. The preventive maintenance advantage from an OM perspective is that it can be scheduled. It also reduces the severity, frequency and cost of unexpected failures.

Setting up the service policy requires an understanding of the product to determine the preventive maintenance service interval and the charges for that service, as well as the cost of repair parts. In some instances, the service function is a profit center. On the other side of the coin, breakdowns result in calls for nonscheduled maintenance. They jeopardize customer loyalty and usually involve higher costs for the same procedures as preventive maintenance. A broken down car must be towed. From this point of view, the service function is a cost center with opportunities to reduce costs through careful planning.

How should the optimal maintenance function be designed? Should it be a profit center, a cost center, or can it be allowed to function on a reactive basis? OM is the only business function that can propose reasonable alternatives and cost each out in an approximate fashion. These alternatives are not likely to include the optimal service policy, but using iteration, successive service scenarios can be tested and then implemented to achieve gradual improvement over time.

Nonfunctional Quality Dimension—Human Factors

Quality management also must focus on the importance of the qualities of the human factors such as safety, comfort, and convenience. The human factors area relates equally well to the office, the factory, the process equipment, as well as products in use including such services as plane trips and taxi rides.

Nonfunctional qualities play a major role in the consumer's judgment of quality. Appearance and style are intangibles and difficult to measure. There are no unambiguous design criteria for what works. Because appearance, style and other nonfunctional qualities are intangibles, expert opinion and market research are the only way to measure and compare. Nevertheless, these dimensions and their roles are as important to the definition of quality as any that are found in the functional category.

How consumers interpret intangible qualities involves sociological and psychological dimensions of quality. Thinking along these lines is not easily associated with OM. That is why industrial designers can greatly assist engineers and designers in the development of the nonfunctional attributes of a product. It is the OM responsibility to make what the designers fashion.

The Variety Dimension

Marketing should be able to assist in the evaluation of the importance of variety of choice. **Variety** is defined as the number of product alternatives—with respect to all of the quality dimensions and cost—that the producer offers to its customers. One illustration of the quality of variety is reflected by the number of flavors that Jello produces. People like to switch between flavors even if they have favorite ones.

Variety
The number of product alternatives a producer offers customers.

Industrial consumers often want tailor-made equipment. The most expensive clothing is hand-tailored. The highest variety level stems from such customization. OM has to produce the level of variety determined by marketing strategy. However, variety has production costs that marketing has to factor into the equation. OM is the only function in the business unit that can work with marketing to raise everyone's awareness of the trade-offs and net benefits.

At the same time, variety develops loyal customers in different market niches by appealing to a wide spectrum of customer preferences. Different bank certificates of deposit appeal to different segments in the population. The various color choices of automobiles, the different styles of TVs, the variety of items on a menu, all attest to the fact that choice is a quality that must be understood by marketing and operations alike.

4.4

SETTING INTERNATIONAL PRODUCER STANDARDS

U.S. Quality Standards

The U.S. military established standards in the 1940s, during World War II, which related inspection sample sizes and observed sample values of averages and variability to the probable true values of the entire production batch.[3]

Practically speaking, almost every purchase made by the military was subject to these standards. This introduced quality standards to a broad range of private businesses in the U.S.

The American National Standards Institute (ANSI), and the American Society for Quality Control (ASQC), have jointly published a series of quality specifications—called the Q90 series—which reflect the standards of the International Standards Organization (ISO). In fact, Q90 through Q94 are Americanized versions of the international standards ISO 9000 through ISO 9004.

International ISO 9000 Standards

The International Standards Organization (ISO) is composed of national organizations that provide certification in 91 countries. They have participated in drawing up the quality standards that are widely-known as the **ISO 9000 series.**

ANSI has been the U.S. representative, while its counterpart in the European Union (EU) also has played a major role. This has led ISO 9000 to be associated with doing business with the members of the European community. In fact, if a company wants to do business in Europe, especially as a supplier, it had best be familiar with the ISO system of standards.

Figure 4.2 shows the Quality Assurance Certificate that has been awarded to the Power Transmission and Distribution Group of the High Voltage Division of the Siemens AG Company. This is one excellent company out of many—in Europe, the U.S.A., and around the world—that could have been used to illustrate the growth in compliance with the ISO standards.

ISO 9000 series

Widely-known quality standards established by ISO—the international organization for quality standardization.

F I G U R E 4 . 2

ISO 9001 QUALITY ASSURANCE CERTIFICATES

The Certificate names the specific group that "has established and is maintaining a quality system." Source: Siemens

T E C H N O L O G Y

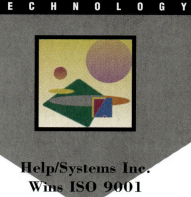

Help/Systems Inc.
Wins ISO 9001

Minnetonka, Minnesota-based Help/Systems, Inc., was the first U.S. software company to earn the prestigious ISO 9001 Certificate. (ISO 9001 applies to firms engaged in design, development, production, installation and servicing of products.) Company president Richard Jacobson explains Help/Systems' motivation for seeking this status: "As much as we were a pretty good company, doing things 80 to 90 percent right, [we] wanted to reach 100 percent."

To comply with ISO standards, the company needed to assure reproducible results and provide detailed, clearly delineated procedures. These things, not easily achieved by manufacturing firms, are even more difficult for companies whose "physical plants" are human beings.

"One of the reasons software companies have a difficult time with this is that software writers are like regular writers—they don't like to be placed under a lot of control," Jacobson says. In order to fulfill the requirements for ISO 9001:

"We were putting in a lot of procedures where none were before, and that's the part I was scared of." With the help of consultants, Help/Systems kept procedures as general as they could. One change that resulted was that writers must report in detail what they want to do with a given project before they start, instead of plunging right in. Peer review panels perform general checks at the beginning of projects.

The ISO process resulted in changes to many of the company's methodical tasks as well. "We bar-code everything now," Jacobson explains. "We believe

the easiest way to reduce errors is to automate." The ISO initiative culminated in a new offer to customers. According to Jacobson, "If they are not satisfied with the product support for any reason, we give them one year's product maintenance absolutely free. We also offer to take back any product at any time... ."

The company has already recovered the $500,000 it spent on ISO-related activities. "...I figure we save $200,000 a year," Jacobson said. "It's a fast payback. In two years sales are up 40 percent, and we haven't had to increase our workforce."

Source: "Help/Systems first software company to gain ISO 9001," (Minneapolis, Minnesota) *Corporate Report,* March 1, 1994.

The Certificate, valid for three years, states that: "A quality audit performed by DQS[4] has verified that this quality system fulfills the requirements of the following standard: DIN ISO 9001 Quality systems; Model for quality assurance in design/development, production, installation and servicing."

DQS is a member of E-Q-Net, the European Network of Quality System Assessment and Certification. The Network and members of the Network (such as DQS) support the mutual acceptance of certificates that are issued outside of Europe. The Network of Agreements is expanding and includes: Austria, Belgium, Denmark, Finland, France, Germany, Great Britain, Ireland, Italy, the Netherlands, Norway, Portugal, Spain, Sweden, Switzerland, Turkey, and, in addition, the U.S.A., Canada, Japan, Israel, and South Africa.

Siemens states in their promotional material that: "Annual DQS monitoring audits and our own internal audits ensure that our quality assurance system is continually matched to technical development as well as being improved." They also state their quality program "applies from the beginning to the end of all production processes, such as:

- Offer preparation
- Design control
- Purchasing
- Shipping
- Inspection and Testing

- Contract review
- Design
- Production
- Commissioning

- Contract processing
- Ordering
- Installation
- Post-sale service

The basis for this is an in-house quality assurance system…(based on the Canadian QA standard CSAZ299.1) which has since been continually optimized."

The ISO quality initiative is an ideal example of the seriousness of the internationalization of the quality issue. Because quality can be called a number one responsibility of OM, the need for understanding the global aspects of OM is underscored.

Which people make professional use of these standards? They are the designers, the engineers and the operations managers. Reporting to the OMs are staffers responsible for purchasing decisions, quality standards, and production scheduling. OM is a key player in doing business with outside purchasing authorities, an increasing number of whom are adhering to ISO standards.

OM is also a major league buyer of goods and services. Intelligent and thorough understanding of the ISO guidelines is essential if there is something to be made and sold. The same must be said about the purchase of materials and components. Standards are changing, too. OM must be able to keep pace with developments.

The ISO materials are written in great detail and the guideline books that contain policies, rules and principles are thick and voluminous. They apply to both manufactured and service products. Being familiar with these standards and their complete specification in the series is not a trivial pursuit.

ISO 9000: Provides overall guidance for prospective users of the ISO standards. It explains the ISO system, in general, as well as giving directions for using the other components of the ISO 9000 series as described in the following standards.

ISO 9001: Applies to firms that are engaged in the pre-market stages including design and development, and also in the in-market stages including production, installation and servicing.

Dimensions of quality in banking services can range from time spent in a teller line, to branch hours, to number of products, to turnover of branch personnel. Source: Citicorp

Methanex Corporation of Vancouver, British Columbia is committed to ISO standards and has received ISO 9002 certification at its plant in Chile. Individuals and programs are focused on responsible care encompassing the safe handling of all chemicals and the highest standards of workplace practice. Source: Methanex Corporation, Vancouver, British Columbia

ISO 9002: Applies to firms that are only engaged in the in-market stages of production, installation and servicing and not engaged in the pre-market stages of design and development.

ISO 9003: Applies to firms that deal with testing and inspection of products when they are acting as an inspection agent or as a distributor.

ISO 9004: Describes the accepted Quality Management System and serves as a guide to the application of that system to production systems for goods and services.

Many U.S. companies require their suppliers to adhere to ISO 9000 standards or the ANSI/ASQC equivalent standards Q90–Q94. In the same sense, U.S. companies wanting to do business in Europe and in Asia are committing resources to the considerable work that needs to be done for certification.

Japanese Quality Standards

Japanese companies adhere to ISO 9000 when it is useful to do so in order to participate in European and U.S. markets. At the same time, over a long period of time, the Japanese have published their own, often quite stringent, quality standards for products as diverse as fiber-optic cable, cellular telephones, high definition television and rice.

For many years they also have utilized systems of quality standards, such as the **Japanese Industrial Standard Z8101**. These specifications resemble those that have been developed in the U.S.

The Japanese government has been under pressure to open their markets which are said to be blocked by excessively rigorous standards that tend to favor domestic producers. Because this will be happening, OM must be alert to follow continuous changes in Japanese standards over time. Some of these will permit the entry of American products into markets that are not currently available. This is another illustration of the degree to which OM must be in the network to receive the latest information about international standards and specifications.

Japanese Industrial Standard Z8101

A quality standard system utilized by the Japanese resembling those quality specifications developed in the U.S.

Database of Quality Standards

A worldwide database of prices and qualities for all potential suppliers and customers provides a competitive advantage. In business environments where every factor has been carefully studied, it has become essential to know what standards can be realized for given prices. This database can be made to reflect the prevailing suppliers' prices for various quality standards. It also leads to the question of how the costs of quality are derived.

4.5

THE COSTS OF QUALITY

Costs of quality

Three basic quality costs which include prevention, appraisal, and failure.

A good approach to understanding quality is to analyze the costs associated with achieving, or failing to achieve, it. There are many shapes that cost curves can take. Therefore, nothing should be taken for granted in a specific case. However, it is useful to state some generalizations about the costs of quality. First, there are three basic **costs of quality.** These include the costs of prevention, appraisal, and failure as explained in the next three sections.

4.6

THE COST OF PREVENTION

Prevention involves the use of conscious strategies to reduce the production of defective product which by definition does not conform to agreed upon quality standards. The entire system must be designed, coordinated and controlled to prevent defectives. This includes the materials and equipment used, appropriate skills, and the correct process to deliver product conforming to standards. Presumably by spending more, the percent of defectives can be reduced. Figure 4.3 shows this kind of relationship, although the real shape of the curve would have to be determined for a specific situation.

FIGURE 4.3

PERCENT DEFECTIVES VERSUS PREVENTION COSTS

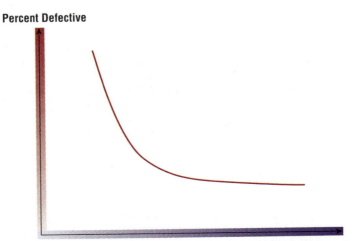

If something wrong is being done, causing defectives to occur, and the error is noticed and corrected, improving quality, then it is fair to say that quality is free. Defectives have been prevented from occurring by remedying a wrong. In service organizations as in manufacturing there are many examples of cost-free quality improvements. Philip Crosby wrote a book about quality being free[5] and established a quality school for executives that used this focus as part of its curriculum.

The cost of prevention goes up when quality standards are raised. *It is reasonable policy to raise the quality standards whenever the prior standards are being consistently met.* The costs involved in achieving these ever more stringent specifications will increase, gradually at first, and then markedly as the technological limitations of the materials and the process are approached. This is shown in Figure 4.4.

In manufacturing, the **tolerance limits** define the range of acceptable product. As an example, the lead that fits into a mechanical pencil might be specified as 0.5mm. This describes the outside diameter of the lead. No one supposes that each piece of lead that comes in a container marked 0.5mm will be exactly 0.5mm. The tolerance limit sets the standard. Each piece of lead must fall within the range 0.5mm plus or minus one twentieth of a millimeter (0.5 ± 0.05). This means that the outside diameter could be as wide as 0.55mm and as thin as 0.45mm.

Tolerance limits
Define the range of acceptable product.

FIGURE 4.4

RANGE OF QUALITY TOLERANCE LIMITS

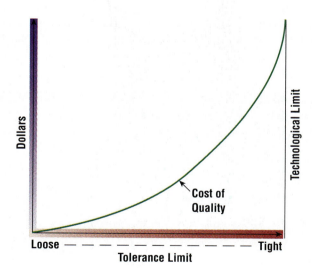

Illustrates the exponential increase in costs as the tolerance range is tightened and approaches the technological limits of the process.

As shown in Figure 4.4, as the tolerance range becomes tighter (there are more zeros after the plus and minus signs), the cost of getting all of the units produced to fall within that range becomes greater. When the designer specifies a narrower acceptable range, the existing equipment's output range is usually too wide. More output will not conform. Everything that falls outside the tolerance range limits is called defective. These defectives must either be scrapped or reworked, which is considered part of the cost of failure.

4.7

THE COST OF APPRAISAL (INSPECTION)

When product is examined to see if it conforms to the agreed upon standards, it is undergoing inspection and appraisal. Both terms are used interchangeably and have to do with the evaluation of whether or not the product conforms to the standards. Product that is not judged to conform, because it fails to fall within the tolerance limits, is sorted out. There must be a policy regarding what to do with product that does not conform, for each way that it may not meet specifications. Some types of defectives have to be scrapped, others can be reworked, sold for scrap or at a discount.

The usual way to sort out the items which do not fall within the tolerance limits is to inspect all of the items. It is possible to use **acceptance sampling methods** which, as the name indicates, consist of inspecting a sample of the production lot. If the sample fails to pass, then the entire lot is inspected and detailed. *Detailing* means removing the defectives so that every item in the lot conforms to the specifications. Therefore, when detailing, the inspector separates the bad from the good.

Figure 4.5 shows that the cost of inspection increases as the percent defective in the production lots increases. The reasons for the cost increase are that more inspectors are needed and an increased amount of detailing is necessary. If sampling is used, more samples of greater size will be taken. All other things being equal, the increase in inspection costs would be almost linear, and not too steep, as a function of increasing percent defectives.

Acceptance sampling methods

Inspection of a production lot sample. Detailing is performed to separate good from bad.

F I G U R E   4 . 5

THE THREE BASIC COSTS OF QUALITY

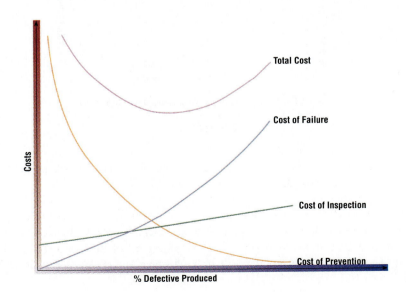

Total Cost

Cost of Failure

Cost of Inspection

Cost of Prevention

Costs

% Defective Produced

4.8

THE COST OF FAILURE

Figure 4.5 shows three costs curves, one of which is the cost of failure, which rises linearly, at first, as the percent defectives increases. This might reflect a replacement cost for failed items that are under warranty. As the percent defectives rises, product failure costs can be far more severe. The curve starts to move up geometrically. Severe costs of failures occur when customers begin to defect to competitors as a result of product failures. This lost revenue stream, often called the lost lifetime value of a customer, has to be taken into account. It can be significant.

Further, serious failures could involve liability of very large sums of money. The expense of litigation in court trials, as well as the accompanying bad publicity, have costs that are difficult to estimate. Additional costs of failures are related to **product callbacks** which require rework involving labor and material costs. Product callbacks often carry other penalties beyond the cost of repairing or replacing failed product. These include possible legal damage claims, bad publicity and the loss of customers. Curves that chart the cost of failure are likely to have steeply ascending, exponential or geometric shapes.

Figure 4.5 puts the cost of failure together with the cost of preventing defectives and the cost of inspection. The prevention curve should be read as follows: with greater expenses for equipment and training, the percent defectives decreases. This is consistent with Figure 4.3.

The three kinds of costs are added together in Figure 4.5 resulting in a *Total Cost Curve*. This curve has a U-shape because total costs are minimized at some level of percent defectives. There can be major disagreement with this proposition. Many situations exist in which for small increases in the prevention costs, the percent defectives will be dramatically reduced. There also are conditions when the cost of failure does not rise exponentially because inspection catches defectives. Therefore, the percent defectives produced are substantially higher than the percent defectives shipped.

Note that if the cost of failure rises very fast, then the indicated minimum point is pushed toward the zero-level of percent defectives. Then the total cost of quality increases geometrically as the percentage defectives rises.

The curves for the costs of quality are logical in certain circumstances; conjectural, and plain wrong in others. However, modeling quality costs remains a worthwhile endeavor.

Product callbacks

Rework involving labor and material costs as well as legal claims and customer dissatisfaction.

4.9

QUALITY AND THE INPUT-OUTPUT MODEL

To achieve specified output quality, two basic components of the input-output model must be treated.

First: the quality of input components—materials received from suppliers—must be maintained at designated levels. Also, work skills must be maintained. Often, because of employee turnover, new workers must be adequately trained. Other inputs, including energy and cash flow also must be maintained at designated levels.

Second: the transformation process must be controlled to deliver the desired output quality. This often is related to internal transfers made between departments within the company.

4.10

CONTROLLING OUTPUT QUALITY

The fundamental control model for quality is based on feedback and correction. This is the basic process by which OM adjusts the production system to conform to specifications for the product.

Figure 4.6 shows the information linkages of this system. The monitor (M) checks actual output measures against conformance standards. Deviations from standards are sent to the controller (C) which has been instructed to reduce the variation. After the controller has made its corrections, the monitor should pick up the fact that standards are being met again.

The monitor (M) measures $(q - s)$ where q is the output quality and s is the conformance standard. M sends information about deviations to the controller (C). A model of this kind should be developed for each quality dimension that counts. Many quality standards can be monitored by automated quality control equipment.

FIGURE 4.6

QUALITY CONTROL FEEDBACK MODEL

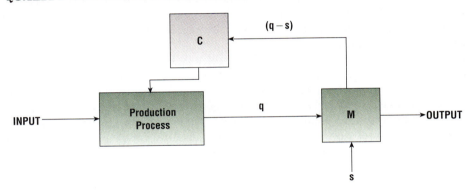

How should this job be organized? If it can be done automatically, then there is a lot of control equipment to maintain and monitor. Assume that 30 dimensions are being monitored and controlled. The cost of this control effort might be larger than the benefits. Perhaps less than 30 dimensions should be monitored.

FIGURE 4.7

QUALITY CORRECTION = f (± ε)

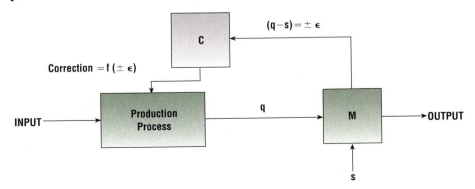

Figure 4.7 shows the difference $(q-s) = \pm \epsilon$. This is the error term which the controller (C) has been programmed to interpret and correct. The controller adjusts the production process in line with its instructions. In this case, the controller adds or subtracts white pigment to or from the mixture.

4.11

TQM IS THE SYSTEMS MANAGEMENT OF QUALITY

TQM should have been called SMQ—standing for *Systems Management of Quality*. Because it was not, every time that TQM appears, it is necessary to translate that name into systems terms. This means that partial quality concepts are not accepted. Instead, they invoke complete organizational involvement, subsequently called company-wide, in the pursuit of quality. This total systems scope means that everything and everyone that has anything to do with quality achievement (directly and indirectly) must be included in planning, designing, and then in controlling the production processes.

Because Total Quality Management requires using the systems approach to quality, all of the people, functions, factors and elements that play a relevant part in determining quality must be included in setting standards, monitoring observations, and in controlling for deviations of the process. The role of every component, ingredient and person that participates in the process, and thereby, in the quality of its product, must be understood and taken into account. This is done in planning and designing the standards for products, processes and control systems. How else can the control system correct deviations from standards?

When the input-output process is connected to the control model, as in Figures 4.6 and 4.7, a major part of the systems structure required for Total Quality Management is in place. The part that is missing is the organizational arrangement that allows for the interchange of information between the people who plan for quality and make decisions about it.

Summarizing five points that have been made:

1. There must be clear and unambiguous standards that are set and jointly held for all relevant quality dimensions.
2. Someone or something (people or machines) must serve as monitors, which entail their comparing the outputs of the process to the agreed upon standards.
3. Process managers must strive constantly to meet the standards that have been set for the products. The achievement of process standards requires that all suppliers meet specified standards.
4. Process managers and controllers must be in communication to deal with any process exceptions and deviations from the standards that trigger the controller to take action.
5. Procedures to handle deviations from quality standards will have been predetermined. Controllers, programmed to correct unacceptable deviations, are empowered by process managers to do so.

4.12

MAPPING QUALITY SYSTEMS

The systems approach to quality requires mapping the relationships between customers and those that supply them. While all functions of business should be thinking about customers, marketing is assigned the responsibility to represent them. In the

FIGURE 4.8

HOUSE OF QUALITY

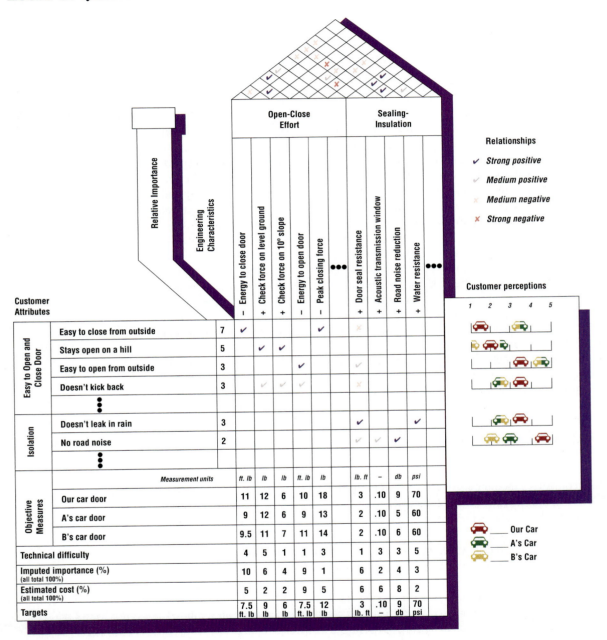

Reprinted by permission of the *Harvard Business Review*. An exhibit from "The House of Quality," by John R. Hauser and Don Clausing, May–June 1988, p. 72. Copyright ©1988 by the President and Fellows of Harvard College; all rights reserved.

same way, OM and engineering are accountable for supplying product to customers. This permits mapping the mutual concerns of sales and marketing with operations and engineering.

House of Quality (HOQ)

A well-known name for this mapping procedure is the **House of Quality (HOQ)**.[6] The "walls of the house" correspond to matrices of rows and columns. In their most basic form, the rows are lists of customer needs and the columns are lists of operations which are designed to satisfy those needs.

The "roof of the house" relates columns to columns. In one form of the House of Quality, this is equivalent to analyzing design and process interactions. Because this is difficult to explain in words, Figure 4.8 is presented. It shows a strong relationship between the energy required to close the door of a car and the energy required to open that door.

Although Figure 4.8 refers to a manufacturing example of an automobile door, the same kind of interactions can be applied to services. Examples exist for academic institutions, retail stores, investment brokers, and fast-food outlets.

The House of Quality helps to identify who should be participating in the dialogue about quality. International TechneGroup Incorporated (ITI) adapted the House of Quality to include more complete marketing information (the WHYs).[7]

The WHATs, called *customer needs,* are represented by the marketing department. The WHYs sort out the WHATs for different demographic groups and other market segments. The HOWs are the *process capabilities* and the *design factors* that are supposed to satisfy customer needs. The HOW MUCHes make comparisons between this firm's HOWs and those of competitors. There is a lot of information about the market segments, design constraints, and the competitors products. The structure is composed of interrelated matrices, as shown in Figure 4.9.

House of Quality (HOQ)

A mapping procedure which analyzes design and process interactions and can identify WHATs, HOWs, WHYs, and HOW MUCHes.

FIGURE 4.9

ITI'S HOUSE OF QUALITY

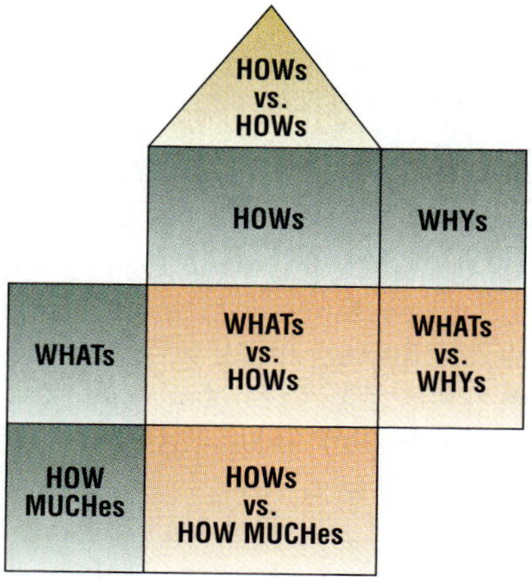

Reprinted with permission of International TechneGroup Incorporated, Milford, OH.

The matrix that results when the WHATs rows are crossed with the HOWs columns constitutes a grid of numbers. That matrix is one of four that constitute a TQM scoring model. The four are:

WHATs versus HOWs—what customers want and what the product gives

WHATs versus WHYs—different market segments have different needs

HOWs versus HOWs—physical constraints limit what a product gives

HOW MUCHes versus HOWS—comparisons between competing products

Quality Function Deployment (QFD)

As shown in Figure 4.10, it is possible to visualize a cascade of linked houses which carry customer needs (often called the voice of the customer) from design specifications to production planning. This permits operations, on the floor of the plant, or by telephone contact with the customer, to be linked with customer perceptions of quality.

LINKED HOUSES CONVEY THE CUSTOMER'S VOICE THROUGH TO MANUFACTURING

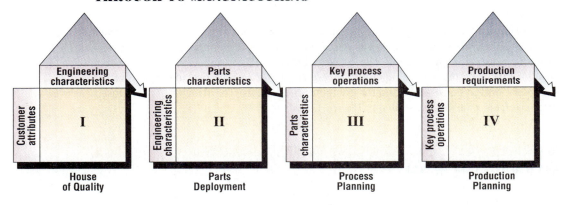

Reprinted by permission of the *Harvard Business Review*. An exhibit from "The House of Quality," by John R. Hauser and Don Clausing, May-June 1988, p. 78. Copyright ©1988 by the President and Fellows of Harvard College; all rights reserved.

This linked structure typifies the ideal realization of Quality Function Deployment (QFD). The figure conveys the idea of moving from strategies to tactics. The direct impact of tactics on the development of strategies also is revealed. If the postal sorting machine jams frequently in the house on the far right, then speedy delivery of the mail is not achievable. Customer needs and wants in the house on the far left will not be realized by this linkage. The voice of the customer has spoken, but is there anyone listening?

The systems approach requires listening and hearing in both the feedforward and the feedbackward directions. Otherwise the linkage is strictly figurative and not operational.

Quality Function Development (QFD)

Comprehensive program for quality extension company-wide.

The name specifically associated with mapping or charting the quality system throughout the organization is **Quality Function Deployment (QFD)**. QFD is a comprehensive program with a complete agenda for extending quality throughout the firm's activities.[8]

The voice of the customer has been heard. Gas service stations now provide automated credit card readers so that customers do not wait for station personnel.
Source: Ashland, Inc.

The QFD approach originated in Mitsubishi's Kobe shipyard in 1972. Toyota and its suppliers further developed the methods and their application. Many U.S. companies, both in services and manufacturing, have employed its use. The list of applications includes automakers, banks, electronic firms, retailers, schools and stock brokers. The Ford Motor Company played a leading role in the early application of QFD concepts in the U.S.A.

It is necessary to assign responsibility for each of the quality dimensions within and throughout the entire organization, as well as outside it. Each organization must determine the best deployment system, including the extension to suppliers. It also is evident that when the customer is included in the deployment pattern, there is an even greater opportunity to achieve excellence in quality. Some questions to consider: Who are the players and what training have they received? What mechanisms exist for communicating? How does the computer tie in?

Competitive systems undergo incessant change. Quality dimensions have to be reevaluated. This results in the need to constantly alter the way in which the quality function is deployed. Changes in the environment and in competitors' products necessitate upgrading the system's dynamics by constantly improving the quality function's deployment.

The procedures used in constructing the HOQ can help determine where accountability for quality resides. That is why the HOQ can affect organizational design as well as the quality of products. The basic idea is that cross-functional mapping of all elements that participate in the attainment of quality goals is a process in its own right.

Anything that affects quality must be included within the boundaries of the quality system. This is another way of saying that the map must include all factors which affect any of the quality dimensions. To be sure all relevant factors are included, it is necessary to name all of the participants that plan, design and control quality. These participants can then identify additional contacts that they make on an informal basis. Many communications that are made in the pursuit of quality are hidden.

It is a good idea to develop an *organization chart* to reflect the deployment of the quality function. This then can be studied to determine whether there are bureaucratic forces within the organization that inhibit instead of encourage and foster quality.

Even as a technique, it should be noted that the HOQ provides discipline and procedure, but its regimentation can inhibit discussion and creative thinking.

In this regard, it is important to make the House of Quality model as user-friendly as possible. This will encourage cross-functional communication. Different approaches exist for promoting interfunctional problem solving. **Action Learning** is a technique for rotating people through a variety of jobs in pursuit of broader understanding. Whatever serves to enhance communication between engineers, designers, marketers and operations is worth promoting.

Another aspect of QFD is the extent of the deployment. In Japan, there are tens of thousands of firms which are said to have total dissemination, called company-wide total quality control or CWTQC. A distinction should be drawn between CWTQC and QFD. QFD pinpoints those in the organization who have responsibility for quality. CWTQC postulates that everyone in the organization plays a part in quality attainment.

Action Learning

Technique for cross-training employees to broaden understanding.

Teamwork for Quality

Quality Circles (QCs)

Groups of workers, organized around products, who focus on quality enhancements.

CWTQC is often identified with groups of workers who meet regularly in what are sometimes called **Quality Circles (QCs)**. These groups are organized around products. Membership in a circle belongs to anyone that influences the quality of the specific products. The circle concept has been called by different names and has met with varying degrees of success.

Teamwork for productivity is enhanced when the quality perspective is brought to light. Absenteeism and turnover diminish because the team's morale is high when there is a concerted effort to produce quality products which endear customers. Pride in the job done and in the quality of the product is not a convenient fiction of human resources managers. It is a real force to reckon with and a source of energy for continuing excellence.

The negatives also must be addressed. When the team is not well-designed, if the support of top-down management is not sincere, the Quality Circle effort will backfire. Backfire means that the situation after failure is worse than it was before the effort was made. Training is important, but it must be targeted with clear goals in mind to be successful. Without training, or with poor training, the backfire is also plainly heard.

Technology interacts with quality in a variety of ways. New processes are better able to control tolerances at reasonable cost. Materials with new properties are appearing regularly. Control systems based on new technologies are able to function with increasing precision. There are optical scanners and other kinds of auto-sensory systems that can serve as inspectors. They can collect information on many dimensions and in greater detail than human inspectors. Computers analyze and chart multiple quality dimensions, spotting problems faster than ever before. Such rapid results allow quick feedback and correction.

4.13

INDUSTRY SEEKS RECOGNITION

General MacArthur's effort to put Japanese industry back on its feet after World War II was incredibly successful. There are many people, both Japanese and American, who played a part in that effort. One whose name stands out is Dr. W. Edwards Deming.

Deming espoused the notion that firms with outstanding quality products could capture markets that otherwise would not be available to them. Such firms would be able to stay in business and create new jobs. He also believed that improvements in quality benefited consumers, workers and producers alike.

Deming's 14 points, in Table 4.4, are representative of the broad systems view that he held, in which every aspect of business played a part in quality achievement.

T A B L E   4 . 4

DEMING'S 14 POINTS FOR QUALITY ACHIEVEMENT[9]

1. Create constancy of purpose toward improvement of product and service, with the aim to become competitive, stay in business, and provide jobs.
2. Adopt the new philosophy. We are in a new economic age.
   Live no longer with defective materials and poor workmanship.
3. Cease dependence on inspection. Require statistical evidence of process control from suppliers.
4. End the practice of awarding business on the basis of price. Reduce the number of suppliers.
5. Improve constantly and forever. Use statistical methods to detect the sources of problems.
6. Institute modern aids for training on the job.
7. Institute leadership. Improve supervision.
8. Drive out the fear to express ideas and report problems.
9. Break down barriers between departments.
10. Eliminate production quotas, slogans and exhortations.
11. Create work standards that account for quality.
12. Institute a training program in statistical methods.
13. Institute a program for retraining people in new skills.
14. Put everybody in the company to work to accomplish the transformation. The transformation is everybody's job. Emphasize the above 13 points every day.

The MacArthur Commission, which had been set up to deal with Japanese industry, called on Deming to help Japanese firms after World War II. The Japanese managers found Deming's ideas and his 14 points appealing. They listened to him, further developed many of his ideas, and consulted with him for years.

The Deming Prize

Since 1951, the Japanese Union of Scientists and Engineers (JUSE) has awarded the **Deming Prize** to companies from any country that have achieved outstanding quality performance. Deming, as an American statistician, advised Japanese manufacturers about various principles of quality including his work with Dr. Shewhart on statistical process control (SPC). The Japanese felt that Deming's influence was so great that they named the Prize for him.

The scope of this highly regarded Prize is twofold. It emphasizes success with both SPC and QFD. This encompasses organizational efforts and participation of many employees in such activities as quality circles. In addition, the Prize rewards companies with high standards and low defective rates.

Deming Prize
JUSE worldwide award for companies who have achieved outstanding quality performance.

Deming himself made a point of acknowledging the fact that service companies have won the award. "Service organizations have won the Deming Prize in Japan; for example, Takenaka Komuten, an architectural and construction firm, won the Deming Prize in 1979. They studied the needs of users (in offices, hospitals, factories, hotels, trains, subways)." He listed three other service firms that won the Deming Prize in successive years (1982, 1983, and 1984). Deming considered quality methodology essential for such diverse services as care of the aged and "perhaps even the U.S. mail."[10]

American companies that have won the Deming Prize are Florida Power and Light (FPL) and a Texas Instruments division located in Japan. Although FPL won the Prize in 1989, the company's quality effort was restructured in 1990 because the initiative to win and the promotional celebration thereafter had created a bureaucracy which was moving the company off the track of its real business.[11]

The Malcolm Baldrige National Quality Award

During the 1980s, U.S. companies felt the power that Japanese competitors exercised because of their devotion to quality. A variety of government initiatives were undertaken in the U.S. to spur improvement of the quality of American products. The Malcolm Baldrige National Quality Award was conceived as a program to trigger competitiveness among U.S. companies with respect to their adherence to management principles that resulted in quality products.

Baldrige Award

U.S. award for companies which lead in quality accomplishments.

One of these steps was the passage of Public Law 100-107. This was the Malcolm Baldrige[12] National Quality Improvement Act, and it was signed into law on August 20, 1987. Thereby, the U.S. Congress established the **Baldrige Award** for those companies which lead in quality accomplishments.

AT&T Consumer Communications Services was awarded the 1994 Malcolm Baldrige National Quality Award in the Service category.
Source: AT&T Archives

QUALITY

Malcolm Baldrige Award—Winning Lessons

A study of Baldrige winners in recent years reveals commonalities among the award-winning companies.

Since 1987, the Malcolm Baldrige Quality Award has been presented to more than 22 companies in recognition of the quality of their products or services. Winners include such well-known firms as AT&T Universal Card Services, American Express, Cadillac, IBM, Motorola, and Xerox, as well as many lesser-known firms. A study of Baldrige winners in recent years revealed some commonalities among the companies that appear to be crucial to their success. These commonalities translate into the following eight "quality lessons" that may help other businesses striving to become world-class competitors.

1. Develop a specific definition or "vision of quality" to guide the quality plan.
2. Enlist the support and involvement of top management from the beginning of the quality effort.
3. Focus on customer needs and establish methods for resolving customer satisfaction problems.
4. Determine the objectives to be attained, then formulate a plan of action, including key measures to be taken.
5. Train employees to use a variety of tools, including statistical process control (SPC).
6. Empower employees—give them the authority to take control and make decisions.
7. Recognize and reward employees for their accomplishments.
8. Make continuous improvement an ongoing challenge.

The last "lesson" is perhaps the most important. The journey toward quality is never ending and each increase in quality is just one step of the journey.

Source: Richard M. Hodgetts, "Quality Lessons from America's Baldrige Winners," *Business Horizons,* July-August 1994, pp. 74-78.

The funding for the Baldrige Award comes in part from the Foundation for the Malcolm Baldrige National Quality Award. Donor organizations from private enterprise have succeeded in raising funds to permanently endow the award program. The National Institute of Standards and Technology (NIST), which is part of the U.S. Department of Commerce, manages the award program. The American Society for Quality Control (ASQC) assists in that management under contract to NIST.

In the first seven years there were 22 Baldrige Awards given to companies both large and small. These companies are listed in Table 4.5. There is a special small business category; the other two categories are manufacturers and service companies. There is a maximum of two winners per category, and only companies located in the U.S. can win. Judging is done by the Board of Examiners which consists of about 270 individuals, all volunteers, most of whom are from the private sector, with some from academia.

The criteria for the Award can be characterized as being representative of TQM. This means that the criteria framework is systems-oriented. There are seven core components which span all aspects of quality and performance and add up to the total score of 1000 points.

TABLE 4.5

MALCOLM BALDRIGE AWARD-WINNING COMPANIES

Year	Company	Location	Product or Service*	
1988	Globe Metallurgical, Inc.	Cleveland, OH	(M)	iron-based metal products
1988	Motorola, Inc.	Schaumburg, IL	(M)	electronic equipment
1988	Westinghouse Commercial Nuclear Fuel Division	Pittsburgh, PA	(M)	nuclear fuel
1989	Milliken and Co.	Spartanburg, SC	(M)	textiles
1989	Xerox Corp.	Stamford, CT	(M)	business products and systems
1990	Cadillac Motor Car Co.	Detroit, MI	(M)	luxury automobiles
1990	Federal Express Corp.	Memphis, TN	(S)	express delivery service
1990	IBM Rochester	Rochester, MN	(M)	computer systems and hard disk storage devices
1990	Wallace Co., Inc.	Houston, TX	(D)	industrial pipe, valves, and fittings
1991	Marlow Industries, Inc.	Dallas, TX	(M)	thermoelectric cooling devices
1991	Solectron Corp.	San Jose, CA	(M)	printed circuit boards, systems assembly and testing services
1991	Zytec Corp.	Eden Prairie, MN	(M)	computer power supplies; repair facility
1992	AT&T Transmission Systems	Morristown, NJ	(M)	telecommunication transmission equipment
1992	AT&T Universal Card Services	Jacksonville, FL	(S)	credit and long distance calling card services
1992	Granite Rock Co.	Watsonville, CA	(M)	concrete and road treatments; also retails building materials
1992	Texas Instruments Defense Systems & Electronics Group	Dallas, TX	(M)	defense electronics equipment
1992	The Ritz-Carlton Hotel Co.	Atlanta, GA	(S)	hotel management
1993	Ames Rubber Corp.	Hamburg, NJ	(M)	rubber rollers for copiers
1993	Eastman Chemical Co.	Kingsport, TN	(M)	chemicals, fibers, plastics
1994	AT&T Consumer Communications Services	Basking Ridge, NJ	(S)	domestic and international long-distance services
1994	GTE Directories Corp.	Dallas, TX	(S)	publishing and selling advertising for telephone directories
1994	Wainwright Industries, Inc.	St. Peters, MO	(M)	stamped and machined metal products

* M = manufacturer, S = service provider, D = distributor

Source: Malcolm Baldrige National Quality Award—*Profiles of Winners*, 1988–1993.

FIGURE 4.11

BALDRIGE AWARD CRITERIA FRAMEWORK
Dynamic Relationships

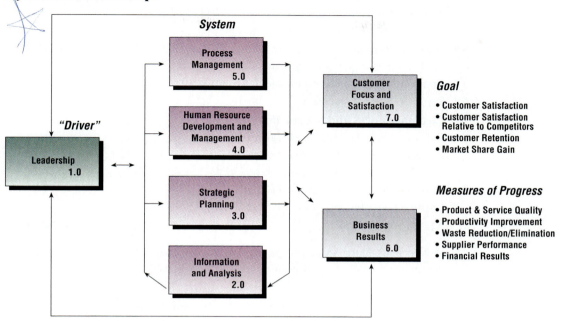

Source: Malcolm Baldrige National Quality Award—1995 Award Criteria, United States Department of Commerce, National Institute of Standards and Technology.

FIGURE 4.12

MALCOLM BALDRIGE 1995 AWARD EXAMINATION
CRITERIA CATEGORIES

1.0 Leadership . . . . **(90 pts.)**
1.1 Senior Executive Leadership (45 pts.)
1.2 Leadership System and Organization (25 pts.)
1.3 Public Responsibility and Corporate Citizenship (20 pts.)

2.0 Information and Analysis . . . . **(75 pts.)**
2.1 Management of Information and Data (20 pts.)
2.2 Competitive Comparisons and Benchmarking (15 pts.)
2.3 Analysis and Uses of Company-Level Data (40 pts.)

3.0 Strategic Planning . . . . **(55 pts.)**
3.1 Strategic Development (35 pts.)
3.2 Strategic Deployment (20 pts.)

4.0 Human Resource Development and Management . . . . **(140 pts.)**
4.1 Human Resource Planning and Evaluation (20 pts.)
4.2 High Performance Work Systems (45 pts.)
4.3 Employee Education, Training, and Development (50 pts.)
4.4 Employee Well-Being and Satisfaction (25 pts.)

5.0 Process Management . . . . **(140 pts.)**
5.1 Design and Introduction of Products and Services (40 pts.)
5.2 Process Management: Product and Service Production and Delivery (40 pts.)
5.3 Process Management: Support Services (30 pts.)
5.4 Management of Supplier Performance (30 pts.)

6.0 Business Results . . . . **(250 pts.)**
6.1 Product and Service Quality Results (75 pts.)
6.2 Company Operational and Financial Results (130 pts.)
6.3 Supplier Performance Results (45 pts.)

7.0 Customer Focus and Satisfaction . . . . **(250 pts.)**
7.1 Customer and Market Knowledge (30 pts.)
7.2 Customer Relationship Management (30 pts.)
7.3 Customer Satisfaction Determination (30 pts.)
7.4 Customer Satisfaction Results (100 pts.)
7.5 Customer Satisfaction Comparison (60 pts.)

TOTAL POINTS . . . . **(1000 pts.)**

Source: Malcolm Baldrige National Quality Award—1995 Award Criteria, United States Department of Commerce, National Institute of Standards and Technology.

Even a relatively cursory analysis of the point allocations indicates the breadth of perspective of the Baldrige approach. Note that customer focus and satisfaction is 25% of the score. The emphasis is somewhat different from the Deming Prize.

Just as Florida Power and Light experienced difficulties after winning the Deming Prize, other companies that have won the Baldrige Award have found that being a winner presents new challenges. These include being on display, going on the lecture circuit, freezing the winning combination and losing flexibility to respond to new competitive factors. There also is the winner's tendency to become bureaucratic and powerful with respect to quality, which is only successful when everyone is empowered and participating.

S U M M A R Y

The reason that better quality matters is explained. TQM—the systems management approach to quality—is the way to obtain better quality, but first quality has to be defined. There is a difference in the way that consumers and producers think about quality. Various dimensions of quality are explored. One of these is the warranty policies that companies offer. Another is the kind of service that is given. How international standards are set is discussed including U.S., international ISO 9000, Japanese and other quality standards. The costs of quality are explored, including the costs of prevention, appraisal and failure. Control of output quality is enhanced by the use of a feedback model which is explained. The use of mapping to understand quality is demonstrated using the House of Quality (HOC) and Quality Function Deployment (QFD). Quality Circles are discussed and prizes for best quality are detailed.

▶ Key Terms

acceptance sampling methods   *132*

costs of quality   *130*

drift-decay quality phenomenon   *122*

Japanese Industrial Standard Z8101   *129*

Quality Circles (QCs)   *140*

Total Quality Management (TQM)   *114*

Action Learning   *140*

Deming Prize   *141*

House of Quality (HOQ)   *137*

mean time between failures (MTBF)   *123*

Quality Function Deployment (QFD)   *138*

variety   *125*

Baldrige Award   *142*

dimensions of quality   *118*

ISO 9000 series   *126*

product callbacks   *133*

tolerance limits   *131*

▶ Review Questions

1. What is the competitive role of quality and why is it important?

2. What is TQM and why is it important?

3. What are the differences between consumer and producer definitions for quality?

4. How does the systems approach apply to the statement:

   Quality—like a chain—is betrayed by its weakest link.

5. Discuss the statement: Although the name TQM may be viewed by some as faddish, the concept is not.

6. What quality dimensions apply to the software industry?

7. Set down appropriate quality dimensions for an automobile.

8. What problem exists because quality standards can age?

9. Explain OM's relationship with market research.

10. Explain how consistency of conformity translates into low reject rates.

11. Discuss the strengths and weaknesses of Quality Circles.

    Are they similar to student cohort work groups?

12. What are the ways to reduce the cost of detailing?

13. Explain what the House of Quality does and how it works.

14. Describe how QFD functions.

15. What prize competitions exist for quality and why are these prizes given?

▶ Problem Section

1. The costs of quality are given by the following equations:

$$C_a = 100D + 1000$$

$$C_p = -50D + 6000$$

$$C_f = De^{0.5D}$$

where:

   C_a *is the cost of appraisal,*

   C_p *is the cost of prevention, and*

   C_f *is the cost of failure.*

Note that D is the percent defectives and $0 \le D \le 100$.

   Determine the total cost of quality when there are no defectives. $D = 0$ and product quality is perfect.

2. Given the information in Problem 1, determine the total cost of quality when $D = 100$. No company could stay in business with $D = 100$, as will be apprarent from the calculations.

3. Given the information in Problem 1, determine the total cost of quality when $D = 0, 10$, and 15. Plot all of the points.

4. Using the information in Problem 1, comment on the shape and position of these cost of quality curves.

5. Using the information in Problem 1, is there a minimum total cost? What is it?

6. Assume that the cost of appraisal is the same no matter what the value of D.
   a. Can this situation be explained in a rational way?
   b. With appraisal costs removed, what happens to the minimum total cost of quality?

7. Make a list of the variables that should be included in a model for a rational warranty policy for each of the products listed below. Develop a specific warranty policy for each of the four.
   a. an automobile
   b. a four-pack of light bulbs
   c. something that breaks easily
   d. something that is very unlikely to breakdown

8. How should responsibility for quality be positioned within and throughout the U.S. government? Should there be a Department of Quality?

9. Develop a House of Quality matrix for the Computer Laptop Company which manufactures and markets laptops for the mid-price range market such as students in college. Label the rows with the customers' needs and label the columns with the properties of the product. Consider customer service to be an important requirement.

10. The tolerance limits for the lead in the mechanical pencil were (0.5 ± 0.05). What tolerance limits should be specified for the tube in the pencil into which the lead is inserted? Discuss the quality problems that are related to this issue. Does it make sense to go back to the operations managers of the company that makes the lead to ask them what it would take to get tighter tolerances?

11. The quality feedback system informs the controller about the error term $(q - s) = \pm \epsilon$. The controller takes corrective action based on $f(\pm \epsilon)$. When the system is in equilibrium, $q - s = 0$ and no corrective action is warranted. Assume that the standard is $s = 2$, and the correction made by the controller is $-(1/2)(\pm \epsilon)$. The value of q, which is regularly two, has just jumped to three. Show how the quality controller corrects the production process. Note: the table below suggests a way to organize the calculations.

corrective action is taken on q ⟶

t	q	s	ϵ_t	$-(0.5)\epsilon_t$	$q - (0.5)\epsilon_t$
1	2	2	0	0	2
2	3	2	1	-0.5	$3 - (0.5)$
3	2.5				

Continue to fill out the table to $t = 5$. Explain what is happening. How would a number larger than 0.5 behave?

12. Given the information in Problem 11, assume that the standard is $s = 2$. Also, the correction made by the controller is $-(1/2)(\pm \epsilon)$. The value of q, which is regularly two, has just fallen to one. Show how the quality controller corrects the production process. Note: the following table suggests a way to organize the calculations.

corrective action is taken on q ──────┐
                  ↓

t	q	s	ϵ_t	$-(0.5)\epsilon_t$	$q-(0.5)\epsilon_t$
1	2	2	0	0	2
2	1	2	-1	0.5	$1+(0.5)$
3	1.5				

Continue to fill out the table to $t = 5$. Explain what is happening and how this differs from the situation in Problem 11. How would a number larger than 0.5 behave?

▶ **Readings and References**

Beyond Total Quality Management, Bounds, G., L. Yorks, M. Adams, G. Ranney, McGraw-Hill, Inc., 1994.

Quality Is Free (The Art of Making Quality Certain) Crosby, Philip B., NY, McGraw-Hill, 1979.

Out of the Crisis, Deming, W. Edwards, MIT Center for Advanced Engineering Study, 1986.

The New Economics for Industry, Government, Education, Deming, W. Edwards, MIT Center for Advanced Engineering Study, 1993.

Sampling Inspection Tables (2nd Edition), Dodge, Harold F. and Harry G. Romig, NY: John Wiley & Sons, Inc., 1944.

"House of Quality," Hauser, John R., and Don Clausing, *Harvard Business Review,* May-June, 1988.

Better Designs in Half the Time, King, Bob, publ. Goal/QPC, Methuen, MA, 1989.

Facilitating and Training in QFD, Marsh, S., J. W. Moran, S. Nakui, and G. Hoffherr, Publ. Goal/QPC, Methuen, MA, 1991.

S U P P L E M E N T 4

A Scoring Model for the House of Quality

A scoring application of the HOQ is shown in Figure 4.13 which rates the qualities of a gas service station.

FIGURE 4.13

RATING THE QUALITIES OF A GAS SERVICE STATION

	Importance of Need	Fast Pumping of Gas	Auto Credit Card Reader	Easy Access to Station	Limited Personnel	Total
Service Speed	5	9	8	6	3	130
Payment Ease	1	4	3	2	1	10
Helpfulness	3	5	X	X	1	18
TOTAL		64	43	32	19	158

Customer needs (rows) are assigned a measure for their relative importance (5, 1, and 3) and the process steps (columns) are rated at each intersection with respect to how well they satisfy the customer needs. A separate matrix is created for each design that is being considered.

For all numbers, the larger they are, the more important they are considered. Row values are multiplied by the weights. There is always concern about additive scoring models. If apples and oranges are being treated as if they were the same, it is because fruit is the measure of interest.

The first column of Figure 4.13 has a sum value of 64, which indicates that fast pumping is the leading quality of this station. Customers also like the automated credit card readers that let them make payments without waiting for station personnel. That column gets a 43. Easy station access is of relative importance with a sum value of 32. The limited personnel appears to handicap customer perceptions of quality. It gets only 19.

Service speed (weight of 5) is being satisfied, but helpfulness (weight of 3) and payment ease (weight of 1) are not being addressed by this service station. Helpfulness probably should be addressed first because it is more important in the mind of the customer than payment ease. There could be an interaction such that if making payments becomes easier, the customer will come to think of the station as being more helpful.

Problem for Supplement 4

The Pin Company has a supplier rating system which is based on mapping their needs for quality, fast and reliable delivery, and low prices against the capability of the suppliers to satisfy these needs.

The two matrices which follow present the data that they have collected for Suppliers X and Y. Make a comparison and choose the preferred supplier. Explain the choice.

Note: the process variables are:

1. *Own Delivery Trucks*—relates to how many trucks each supplier leases or owns.
2. *Inspected Before Delivery*—relates to the procedures used by the supplier to check the quality of outgoing shipments.
3. *Factory Stage*—relates to the kind of equipment that the supplier uses to make the products where state-of-the-art or ahead of it gets the highest rating.
4. *Percent of the Business*—refers to how many other customers each supplier has and what percentage of the supplier's total business the Pin Company represents. The best percentage is considered to be about 10%.

Supplier X

	Importance of Need	Own Delivery Trucks	Inspected Before Delivery	Factory Stage	Percent of the Business	Total
Quality	5	4	7	3	6	
Delivery	3	9	2	3	2	
Prices	4	3	5	5	7	
TOTAL						

Supplier Y

	Importance of Need	Own Delivery Trucks	Inspected Before Delivery	Factory Stage	Percent of the Business	Total
Quality	5	2	6	7	1	
Delivery	3	1	3	6	1	
Prices	4	5	2	5	8	
TOTAL						193

Notes...

1. See for example, D. Savageau, *Places Rated Almanac*, Prentice-Hall, Inc., Oct. 1993.

2. *Harvard Business Review*, November-December 1987.

3. Harold F. Dodge and Harry G. Romig, *Sampling Inspection Tables,* 2nd Ed., NY: John Wiley & Sons, Inc., 1959. These tables specified the parameters for efficient sampling plans that could be used to accept or reject production lots.

4. DQS is the German Society for the Certification of Quality Systems. Deutsche Gesellschaft zur Zertifizierung von Qualitätssicherungssystemen mbH.

5. Philip B. Crosby, *Quality Is Free (The Art of Making Quality Certain)*, NY, McGraw-Hill, 1979.

6. The best known popularization of the HOQ method is an article by John R. Hauser and Don Clausing, which appeared in the *Harvard Business Review*, May-June, 1988, pp. 63-73. A book on the subject was written by Bob King, *Better Designs in Half the Time*, Publ. Goal/QPC, Methuen, MA, 1989.

7. International TechneGroup Incorporated of Milford, Ohio has software called QFD/Capture which permits participants from all contributing functions to use a DOS, Windows, or Macintosh environment to enter data and analyze its impact on quality.

Industry Perspective on Quality

TOM'S OF MAINE

Tom's of Maine, a manufacturer of "natural" personal-care products, was founded in 1970 by Tom and Kate Chappell. The Chappells started their company after moving to Maine seeking a simpler life and a deeper connection to the land. As part of this move, they began using natural foods and products, but were not able to find the natural personal care products they wanted for themselves and their children. So they decided to make and sell their own. (Tom's of Maine considers "natural products" to be those which are minimally processed, derived from natural resources, and free of artificial colors, sweeteners, preservatives, synthetic flavors, and fragrances.) From this simple beginning in the 1970s, the company has grown to become a significant player in the mass market for personal care products. With annual sales of approximately $15 to $20 million, and 75 employees, Tom's of Maine distributes natural, environmentally-friendly products to over 20,000 food and drug stores and approximately 7,000 health food outlets.

Nestled in the picturesque New England town of Kennebunk, Maine, the company is headquartered in a renovated red-brick factory building on the Mousam river. The office area, well lit and open, bespeaks Tom's of Maine's commitment to the environment with abundant, colorful, animal-related artwork. Recycled ingredient containers, transformed into free-standing art by local students, adorn the corridors. Framed copies of the company's mission statement and posters with handwritten excerpts from customer letters also are displayed. A few miles from town, the manufacturing plant is housed in a rustic wooden structure next to a defunct railroad depot. A pleasant minty aroma pervades the inside of the facility in which all Tom's of Maine products are manufactured.

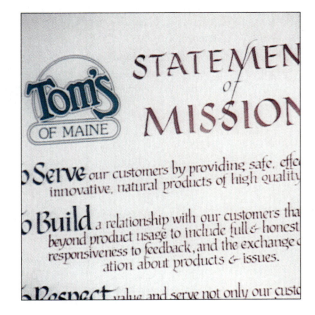

Respect for the environment and a commitment to serve their customers are just two components of Tom's of Maine Mission Statement.

Common Good Capitalism

Tom's of Maine has remained true to the Chappell's original vision to manage their business for both profit and the common good of the community, the environment, and one another. Tom's of Maine strives to integrate the corporate values of commercial success with the belief that companies have a responsibility at all times to the common good of society. Tom Chappell explains that with "common good capitalism," private aspirations or aims (such as what a business owner does with the profits from the business) are accountable to a higher good, a common good. Tom Chappell wants the community in which his company operates to be able to recognize the business as a valuable, contributing presence.[1]

Product Variety

Tom's of Maine manufactures and sells a variety of natural personal care products including flossing ribbon, alcohol-free mouthwash, anti-perspirant and deodorant, shampoo, shaving cream, and toothpaste—with and without fluoride. Sales of toothpaste account for the majority of Tom's of Maine's total revenue. We will now examine Tom's of Maine's manufacturing process for toothpaste.

Production Process for Toothpaste

Pre-production Activities

First the raw materials for the toothpaste must be received (#1). (Numbers in parentheses refer to the numbers on Exhibit 1—Production Process for Toothpaste at Tom's of Maine.) These materials include printed tubes with caps; natural ingredients (including calcium carbonate, fluoride, carrageenan, glycerin, sodium lauryl sulfate, flavors); toothpaste boxes; inner pack wrap; and shipping cartons.

Tom's of Maine sells a variety of natural personal care products.

Every incoming raw material is scrutinized against supplier specifications and United States Pharmacopoeia (USP) and Food and Drug Administration (FDA) requirements (#2). The FDA regulates ingredients that alter normal bodily functions, i.e., fluoride in toothpaste and antimicrobials in deodorant. Typical quality control (QC) tests at this stage include chemical assay, microbiology, organoleptic (appearance, smell, taste) and moisture content. QC tests normally are completed within 48 hours of receipt of a sample of the ingredient. If a problem with a one-half pound pre-shipment sample is detected, the sample is rejected, discarded, and the shipment stopped. (To avoid production-scheduling problems if a shipment is rejected, Tom's of Maine generally has a two-week supply of all materials on-hand. Lead time to order ingredients is 8 to 12 weeks.) The tubes and boxes are tested against specifications provided by the marketing department to determine that printing and colors are correct.

One problem that may arise at this stage is that the density of the calcium carbonate sometimes doesn't meet requirements. Calcium carbonate is the ingredient that dictates how much water is absorbed in the toothpaste mix, so its density must be right. Testers also must be sure that the pH rate (a scale of acidity–alkalinity) is within a narrow range of specification. If higher than the specification range, the shelf life of the toothpaste will be reduced. Shelf life is 12 to 18 months, depending on storage conditions—the colder the better!

Tom's of Maine holds one week's worth of ingredients in a transition warehouse at the plant (#3a); an off-site warehouse about a mile away is for longer-term storage (#3b). Materials stored in the off-site warehouse stay an average 40-45 days. Liquid ingredients are stored in 55-gallon drums; powdered ingredients are stored in 50-lb. bags. No supplies are used without a completed "released for use" sticker which tells the date the supply was tested and released, who analyzed it, and the quantity. This

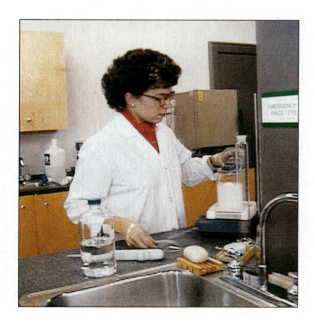

Quality control tests for incoming raw materials include chemical assay, microbiology, organoleptic (appearance, smell, taste), and moisture content.

allows Tom's of Maine to trace every ingredient back to its original tester. Ingredients in the transition warehouse are moved 20 feet by forklift into the batch preparation area on an as-needed basis (#4).

Tom's of Maine has close relationships with a few suppliers. The company works closely with suppliers to understand their processes and improve them. Target supplier flaw rate (also called the defective rate) is 0.1%. Tom's of Maine shares flaw rate information with suppliers and works with them to improve the rates. When Tom's of Maine first started buying from its tube supplier, for example, the flaw rate was 0.5%. After working closely with them to increase quality, the same supplier has achieved a 0.05% rate, and now supplies all of Tom's of Maine's toothpaste tubes.

Continuous Batch Preparation

Tom's of Maine makes one variety of toothpaste per day, mixing the ingredients in a 300-gallon vat or "change can" weighing 3,600 lbs. The toothpaste ingredients (calcium

EXHIBIT 1

PRODUCTION PROCESS FOR TOOTHPASTE AT TOM'S OF MAINE

1. Receive raw materials in 50-lb. bags and 55-gal. drums.

2. Material inspected; samples checked against supplier specs, USP & FDA requirements as they are received.

3a. Short-term storage of materials in warehouse @ plant (avg. one week).

3b. Off-site warehouse for long-term storage (40-50 days).

4. Ingredients moved 20 feet by forklift to plant floor.

5. Toothpaste slurry mixed in batches (2½ -3 hrs. in 300-gal. vat).

6. QC tests–conform to specs on pH level, density, taste, moisture (10-15 minutes) and FDA for fluoride (1½-2 hours).

7. Ingredients pumped upstairs for assembly (rate: 4 gal/min., 60 lb./min.).

8. Station I: Tubes filled (80-84/min.); operator loads box of tubes into magazine that orients the tube.

9. Machine fills tube, crimps ends of tubes, then delivers tube to conveyor belt. (80-84/min.).

10. Conveyor belt (moves @ 30 ft./min.).

11. Station II: QC Operator picks up tube; makes sure crimp up to code & printing okay.

12. Operators insert tubes into boxes (83-84/min.) and place on conveyor; boxes slide off conveyor into Station III.

13. Tubes deposited onto belt (moves 30 ft./min.).

14. Station III: Shrink-wrapping Operator gathers 6 boxes by hand; heat chamber shrink-wraps 6 boxes together (13-14/min.).

15. 6-packs loaded by operator into master shipping cases (6 6-packs per case; 2 6-packs loaded per minute).

16. Station IV: Cases palletized Operator loads 100 master cases onto each pallet.

17. Pallets loaded onto conveyor; forklift takes downstairs.

18. Temporary storage downstairs (0-12 hrs.).

19. Product loaded into Tom's of Maine truck which delivers it to warehouse 1 mile away.

20. Product stored in warehouse (up to 3 months).

21. Orders filled — FIFO

22. Shipped to customers via common carrier.

= process step
= quality check
= storage or delay

carbonate, fluoride, carrageenan, glycerin, sodium lauryl sulfate, and natural flavors) are mixed at high speeds of up to 600 rpms for 2 1/2 to 3 hours (#5). This toothpaste mixture is called "slurry." After the mixing is complete the operators scrape the mixing blades to remove as much of the mixture as possible from them; this also facilitates cleaning and preparation of the equipment for the next batch.

Tom's of Maine has 10 varieties of toothpaste in all. In a 20-day month, 13 to 14 days are spent on making toothpaste. When filling 4-ounce tubes, four batches of one flavor are made in one 10-hour day, which yields approximately 46,000 4-oz. tubes. When filling 6-oz. tubes, six vats of the mixture are made, which results in 44,000 to 45,000 tubes in one 10-hour day. Tom's of Maine usually devotes two days per month to making spearmint toothpaste, which is its best selling flavor.

Tom's of Maine has an extensive preventive maintenance program which regularly checks parts and machines. These preventive checks are done during production. While a liquid product (such as deodorant or mouthwash) is being produced, maintenance checks are performed on the equipment used to produce "nonliquid" products, such as toothpaste. When nonliquid products are being processed, checks are performed on the equipment used to make liquid products. A supply of replacement parts is maintained at the factory.

Eleven people are involved in making a batch of toothpaste, including a foreman, a production supervisor, three mechanics (who work alternating shifts from 5 A.M. to 6 P.M., Monday through Friday), one quality assurance (QA) lab supervisor, one lab assistant, three batch mixers and one batch supervisor.

QC Tests on Slurry

After mixing is complete, each vat of slurry is tested for conformity to specs on pH level, density, taste, moisture and FDA requirements for fluoride level (#6). Two taste testers taste the slurry for acceptability. The reject rate on the slurry is below 0.01%. The rework rate on slurry is about 0.25%. QC tests on slurry generally take 10-15 minutes. Results of the chromatography tests for fluoride dispersion (samples of which are taken when the mixing is complete) are not available until after the toothpaste has been inserted into the tubes. This test takes about 1 1/2 to 2 hours. Tom's of Maine has only had to reject one batch in 14 years because fluoride was not dispersed evenly throughout the mixture. If a batch is rejected for any reason, the contents are discarded.

Pumping of Vat Contents to Production Line

After the QC tests on slurry are complete, the ingredients are pumped upstairs at a rate of 5 gallons or 60 lbs. per minute for continuation of the production process (#7). When all the slurry in the vat has been emptied, the empty vat is moved out of the way and a new vat of slurry that has been prepared off-line is rolled into place. This changeover takes about 5 minutes.

The empty printed tubes are loaded into a magazine that orients the tube.

Four Station Assembly Process

Upstairs in the plant the four assembly stations are set to receive the toothpaste slurry. The empty printed tubes, open at the bottom, have caps already screwed on. The aluminum tubes are lined with a plastic phenolic coating. Without this coating the calcium carbonate would react with the aluminum. Tom's of Maine uses aluminum tubes because they are recyclable. All hand operations in the production process serve a dual purpose: they perform a specific function (such as picking up a tube and placing it in the box) as well as serve as quality checks.

Thirteen people are involved in this serialized part of the production process including a foreman, a production supervisor, three mechanics, one operator at Station I, two operators at Station II, two operators at Station III, one "floater" (who keeps the floor clean, moves boxes, and generally troubleshoots), and one operator and one forklift operator at Station IV.

Stations I through III are synchronized by a conveyor. Operators' idle time is taken up by moving material from station to station, i.e., moving pallets on finished goods to the conveyor which takes them downstairs. The speed of the conveyors used is approximately 30 feet per minute.

Station I—At the first station, a machine operator loads a box of tubes into a magazine that orients the tube (#8). The empty boxes are collected, then sent back to the supplier for reuse. The machine fills each tube and passes each through a series of four crimping stations, which fold the end of the tube three times; the fourth station prints a batch code on the tube (#9). The machine then delivers the tube to a conveyor which moves at 30 feet per minute (#10). Tubes are filled at a rate of 80 to 84 per minute (6-oz. tubes at 80 to 81 tubes per minute; 4-oz. tubes at 83 to 84 per minute). The vice

A machine fills each tube and passes each through a series of four crimping stations, which fold the end of the tube three times.

president of manufacturing Gary Ritterhaus refers to these rates as the "quality sweet spots," the processing rates at which a less than 0.01% flaw rate is achieved.

Station II—Two operators pick up tubes off the conveyor and visually check that the crimp is up to code (it is not leaking) and that the printing is okay (#11). One operator loads a magazine with toothpaste boxes; two other operators then insert the tubes into the individual boxes (#12). These operators are rotated once every hour to minimize the adverse effects of repetition and "overuse" injuries. The filled boxes slide off the conveyor into Station III (#13). Air is blown toward the boxes as they slide so that if a box has not been filled, it is blown off the conveyor and onto the floor.

Station III—One operator gathers six toothpaste boxes by hand, stacked in 2 groups of 3 boxes (#14). The operator puts the six boxes under a plastic sheet and loosely wraps the boxes with the plastic. The six boxes are then shrink-wrapped together by a machine with a heat chamber. Six-packs are shrink-wrapped at the rate of 13 to 14 per minute.

Station IV—Another operator puts six 6-packs into a corrugated master case (#15), producing a little over 2 master cases per minute. The batch code is put on the case; then the case is loaded onto a pallet. An operator loads 100 master cases onto each pallet (#16). The master cases are "stretch-wrapped" onto the pallet. Pallets are then loaded onto a conveyor and taken downstairs (#17). From there, a forklift operator moves the pallets downstairs, where they are stored (#18) until the truck is loaded, either the evening of the day the tubes are filled or the next morning. The truck ships the pallets to the warehouse the next morning (#19).

Four people work in the warehouse. The product is stored in racks in the warehouse for up to 3 months (#20). Orders are filled on a "first-in, first-out" basis (#21). The warehouse is arranged randomly, each product marked with a product code. Shelf life for toothpaste is 12 to 18 months. Turnover at retail outlets typically is faster than this. The product is shipped to customers via common carrier (#22).

QC Charts and Checklists

Tom's of Maine uses a variety of QC charts and checklists during the toothpaste manufacturing process. These charts include a "Production Order" and "Toothpaste Manufacturing Checklist" used when the toothpaste ingredients are mixed; a "Toothpaste Checklist" used in preparing the four-station assembly process; a "Fill Report" which is a quality control check used after the tubes are filled; and a record of "Tube Defects Per Day" which records such problems as bad or loose caps, bumps in the tube lining, printing problems, defective liners, or metal shavings.

Forecasts and Production Scheduling

Forecasts for unit sales are generated by Sales & Marketing. Manufacturing looks at the 3-month history and considers the Sales & Marketing forecast in scheduling production. Because one marketing strategy Tom's of Maine uses is a revolving product "deal cycle" in which retailers can achieve discounts if they purchase products during specified periods, the company is able to anticipate increased demand and schedule production accordingly. Most retailers use the deal cycle to order product to take advantage of lower prices.

Quality Emphasis

Producing quality products is a major emphasis at Tom's of Maine. In a discussion of how Tom's of Maine defines "quality," operations manager Dave Pierce, vice president of manufacturing Gary Ritterhaus, and public relations associate Matthew Chappell cited the following quality concerns:

- Tom's of Maine products must have all natural ingredients that are environmentally safe. The products are packaged in ways which respect nature as well. Consumers are encouraged to recycle the product packaging. Tom's of Maine recycles all of the packaging in which its supplies are delivered.

- All products must be proven safe and effective for consumers. But Tom's of Maine's product ingredients are never tested on animals. To avoid animal testing, alternative tests are used. For instance, as an alternative to the traditional rabbit eye test for mucus membrane sensitivity, a vegetable protein is used which reacts in the same way as the rabbit's eyes.

- Consideration of customer needs and keeping customers happy is another quality concern at Tom's of Maine. "You don't have a job if you don't have a happy customer" is a common company refrain. Like all health- and environmentally-conscious consumers, users of Tom's of Maine's products are very astute. They demand more information about the products they purchase than the average customer. For this reason, Tom's of Maine includes three panels of product information on its toothpaste packages compared to one panel of information on the average toothpaste carton.

- Customer feedback is solicited on every package. Every letter or phone call Tom's of Maine receives—about 400 per month—is read (or received) and tabulated. Approximately 70 percent of the letters are positive. All letters are answered by a member of the Consumer Dialogue Team, which is comprised of 10 people from all departments in the company. The quality assurance and production departments receive copies of all complaints.
- Tom's of Maine also strives for a high quality of life for its employees. To this end, there are no weekend or night shifts at the plant and there is frequent rotation on production lines. Employees are members of various teams including the Consumer Dialog Team, Engineering Team, Production Team, Quality Assurance Team, and Scheduling Team. The teams, which are further broken down into task forces, are adaptations of quality circles. Tom's of Maine plans to have all self-managed teams in place in the near future.
- Service to community is an important quality consideration for Tom's of Maine. In keeping with Tom Chappell's theory of "Common Good Capitalism," employees are encouraged to devote 5 percent of their work time to doing volunteer work in the community. They are given time off—with pay—to perform this volunteer service.

What Does the Future Hold?

Tom Chappell says Tom's of Maine plans to grow at a rate of 15 percent annually by entering new markets and offering new products every year, while "keeping an eye on the inward or spiritual aspect of how we do things." (Spiritual growth is, unfortunately, not quantifiable.) In 1995 Tom's of Maine introduced several varieties of natural soaps, added two new toothpaste flavors (Wintermint and Cinnamint baking soda), and two new "gentle" deodorants for sensitive skin in Calendula and Woodspice fragrances.

For the near term Tom's of Maine plans to manage growth through continuing to improve productivity efficiencies with faster equipment instead of expanding the production facility or increasing the number of employees and work shifts. Tom's of Maine estimates that production can be doubled in the current facility by using faster machines. For the long term, if continued growth means expanding manufacturing facilities, Tom's of Maine will consider moving to another town, so its influence on Kennebunk will not "overwhelm" the community, according to Tom Chappell. Tom's of Maine also hopes to continue expansion into foreign markets. Currently they export their products to England, Canada, and Israel.

You might say that Tom's of Maine was slightly ahead of its time in 1970 in targeting the market for healthier, more environmentally-friendly products. In the 1990s more and more consumers are seeking products that are pure and natural, without chemical additives—a movement partially driven by fear of the environment.[2] Mass marketers such as Johnson and Johnson and Del Laboratories all recently have identified the natural products market as a growth area and are producing such products.

Tom Chappell believes that an emphasis on the science of plants will grow the business. To this end he has brought aboard a pharmacognosist (or specialist in plant chemistry and the medicinal value of plants) to ensure that Tom's of Maine's products remain not only natural but also provide the most efficacious use of natural ingredients for consumers.

Case Questions

1. Discuss how quality fits into Tom's of Maine's overall strategy and the relationship between strategy and success at Tom's of Maine.
2. Identify and discuss some of the tactics Tom's of Maine uses to fulfill its mission to provide natural and environmentally-safe products.
3. Discuss the eight dimensions of quality (performance, features, reliability, conformance, durability, serviceability, aesthetics, and perceived quality) with regard to Tom's of Maine tooth paste. Which dimensions do you think are most important to the success of this product and why?
4. What happens to toothpaste slurry? Detail its path step-by-step until the finished product is stored.
5. What happens to a 50-lb. bag of powdered raw materials? Detail the process steps from receipt of the materials until the empty bag is returned to the supplier.
6. Identify the costs Tom's of Maine incurs with regard to appraisal, prevention, and failure.
7. Identify the various quality control measures Tom's of Maine employs as related to raw material inspection, work-in-process inspection, and finished goods inspection.

Endnotes

1. Tom Chappell, *The Soul of a Business: Managing for Profit and the Common Good,* Bantam, October, 1993.
2. *Wall Street Journal*, Tuesday, May 11, 1993.

References

Tom's of Maine 1993 Annual Report, "The Private Good and The Common Good—Finding a Middle Way."

Mary Martin, "Toothpaste and Theology," *Boston Sunday Globe,* October 10, 1993.

Wall Street Journal, "Putting It Mildly, More Consumers Prefer Only Products That Are 'Pure,' 'Natural'." May 11, 1993 .

"Tom Chappell: Minister of Commerce," *Business Ethics,* Vol. 8, No. 1, January/February, 1994.

PROCESS FLOW

Design, Analysis, and Improvement

In Part III process flows are designed, analyzed, and improved. They are classified, charted, and quality control systems are devised for them. Managing technology and teamwork is crucial for OM success and a chapter is devoted to each topic. Capacity and facilities planning are integral parts of process flow design, analysis, and change for improvement.

Chapter 5 describes process configuration strategies. This refers to the variety of process methods used for doing work. The custom shop, the job shop, the flow shop, and flexible process systems are explained in detail so that a choice can be made. Processes also are differentiated with respect to assembly (synthetic), disassembly (analytic), and combinations. Knowing the type of process is necessary to manage it. Thus, work configurations and process types are classified in a useful way for operational purposes (design, analysis, and improvement).

Chapter 6 introduces flowcharting methods for process analysis and redesign. Process flowcharts and layout charts are crucial graphics for understanding the nature of the process. They organize the information to allow analysis and change. Three major areas for process improvement are reduction of set-up times, switching to better materials and improving maintenance procedures. Specific methods for process improvement are explained for all three areas.

Chapter 7 presents methods for quality control. It starts with the Olympic perspective, a mind-set required for quality achievement. Quality control methods are explained using the seven classical tools for achieving better quality. These include data check sheets, bar charts/ histograms, Pareto charts, cause and effect (Ishikawa) diagrams, statistical process control charts, and statistical run analysis. Acceptance sampling treats negotiations between suppliers and buyers using Type I and Type II errors as the basis.

In Chapter 8, management of OM technology is related to responsibility for the transformation process. As technology changes there is pressure to adopt the new technology while using the systems that characterize the old technology. This technology trap is discussed so that it can be avoided. Examples are given of process technology at different stages of development for glass and automotive windshield glass. This technology provides a unique model of constant change and adaptation by OM. Other technologies discussed include: packaging and delivery, testing, quality control, and design for manufacturing and assembly.

Chapter 9 details how planning teamwork and job design recognize the importance of people as part of the process. Effective training is dependent upon good job design. There is no point in training people to do badly designed jobs. Setting time standards for doing productive jobs is detailed. These are called job observation studies. Work sampling is described for improving the way teams function and job design. Managing the work system includes job evaluation and improvement, work simplification, and job enrichment.

In Chapter 10, capacity management is defined in various ways and related to peak and off-peak demand. Modular part design and group technology production methods are presented as methods for augmenting capacity. Bottlenecks limit capacity. They need to be identified and can be altered. The effects of delay on the capacity of the supply chain are illustrated by means of a supply chain game. The breakeven model for choosing among alternative capacity configurations is developed and linear programming is shown to be a preferred method for determination of how best to use and design capacities. The relevance of the learning curve model to capacity is explained.

In Chapter 11 the four major components of facilities planning are discussed. First is where to locate regionally; second is to find the specific structure and site; third is to choose the layout, and fourth is to select equipment. Service and manufacturing location decisions are discussed. The transportation model is used to minimize costs of shipping and production. It also allows consideration of marketing differentials. A scoring model is furnished that permits tangible and intangible factors to be considered for site selection. Design for manufacturing criteria are developed for equipment selection which also is related to layout interactions. The layout of the workplace is part of job design. Layout models are described which cover intangible factors as well as the flow costs that characterize layout load models.

Chapter 5

Process Configuration Strategies

After reading this chapter you should be able to...

1. Distinguish between processes and work configurations.
2. Explain the three major kinds of work configuration.
3. Discuss why maintenance of equipment is a process.
4. Explain how simulation can be used to improve a process.
5. Distinguish between synthetic and analytic processes.
6. Reveal why it is essential to chart every process step.
7. Compare the custom shop to the job shop.
8. Define cycle time and use it to design a flow shop.
9. Explain flexible process systems.
10. Relate all of the work configurations with respect to volume and variety.

Chapter Outline

5.0   WHAT IS A PROCESS?
Work Configurations Defined
Process Simulations
Classifying the Process
Charting the Process
Choosing the Process
Means and Ends

5.1   STRATEGIC ASPECTS OF PROCESSES

5.2   TYPES OF PROCESS FLOWS—SYNTHETIC PROCESSES

5.3   TYPES OF PROCESS FLOWS—ANALYTIC PROCESSES

5.4   SOURCING FOR SYNTHETIC PROCESSES

5.5   SOURCING FOR ANALYTIC PROCESSES

5.6   MARKETING DIFFERENCES
Marketing Products of Analytic Processes
Best Practice

5.7   MIXED ANALYTIC AND SYNTHETIC PROCESSES
The Opt System's V-A-T

5.8   WORK PROCESS CONFIGURATIONS

5.9   THE CUSTOM SHOP

5.10 THE JOB SHOP

5.11 THE FLOW SHOP
The Benefits and Constraints of Flow Shops

5.12 ECONOMIES OF SCALE—DECREASING THE COST PER UNIT

5.13 THE INTERMITTENT FLOW SHOP

5.14 FLEXIBLE PROCESS SYSTEMS

5.15 MOVING FROM ONE WORK CONFIGURATION STAGE TO ANOTHER
Expanded Matrix of Process Types

5.16 THE EXTENDED DIAGONAL
Adding the Custom Shop
The Role of FMS
Applied to Services

5.17 AGGREGATING DEMAND

Summary
Key Terms
Review Questions
Problem Section
Readings and References
Supplement 5: How Purchasing Agents Use Hedging

The Systems Viewpoint

Strategies are more likely to be successful if they have been formulated using the systems approach. An equivalent statement is that strategies are more likely to be successful if they have been formulated using all of the factors that are relevant to their success. The systems approach allows the process choice to be based upon what the competition is doing, the state of technology, what the customer will like, the costs of production, the control of quality, and the ability to schedule delivery. Note the systems implications of sourcing which are detailed in this chapter. First, however, focus on processes.

The process is composed of subprocesses. Each is like an input-output system that is connected to another input-output system. Each is a link of the means-end chain, and a process in its own right. The connected links form a chain of conversions. The chain is the total process.

Although the parts of the process can be studied individually, the systems approach ultimately always focuses on the whole. The systems approach often uncovers relationships in one part of the chain that affect relationships in other parts of the chain.

Before an appropriate process design can be fashioned, it is essential to have a common understanding of what a process is, and be able to define the term.

5.0
WHAT IS A PROCESS?

Process

A series of goal-oriented activities or steps; a detailed description of the input-output transformation sequence.

Process steps

Operations or stages within the manufacturing cycle required to transform components into intermediates or finished goods.

A **process** is a series of activities or steps which are goal-oriented. The APICS Dictionary defines **process steps** as: "The operations or stages within the manufacturing cycle required to transform components into intermediates or finished goods."[1]

There is no reason to limit this definition to manufacturing. It applies equally well to service functions. The steps of the process are specific and concrete. They tell "how to …" make things like air conditioners, or cars; or "how to …" fill things like cans with beans or tubes with toothpaste; or "how to …" provide medical or travel services.

Process steps often require the use of science and technology. Chemistry may be required to make the product. Instructions for what steps to take are varied; they may be recipes for making bread or blueprints for making the breadbox. Processes are named for what they make, like cheese, auto assembly, and research reports.

Work Configurations Defined

It is important to distinguish between processes which are the steps and activities required to make the product and work configurations that help accomplish production. There are many kinds of processes, but only a few basic types of work configuration. **The work configuration** is the physical set up used to make the product. It is the arrangement of people and equipment and the way materials and work in process flow from place to place in the plant, or in the office.

As noted, the type of activity flow used for the process is called the work configuration. There are six fundamental kinds of work configuration:

1. custom shops—for one of a kind items
2. job shops—for batch processing
3. flow shops—for serialized processing
4. flexible manufacturing systems
5. continuous flow processing
6. projects

Projects are a type of work configuration that is used for start-up purposes, such as for major new undertakings like building a bridge. The focus of this chapter is on processes for production systems that are up and running, so projects will not be treated here. Continuous flow processing will be discussed, but because it is very specialized and highly engineered, it will not receive the kind of attention the other five categories receive.

The kind of work configuration that is used is a function of volume and variety. Large volumes of identical items are not treated in the same way as small volumes of many different kinds of items. Work configurations constitute a main theme of this chapter.

The process steps must be fully specified. This means that the technology of the process is explicit and the work configuration for carrying out the process is explained in detail. This level of specification is absolutely essential for process design and control. No technicalities can be ignored.

The maintenance program for jet engines is a process. Inadvertently, an O-ring was left out of a jet engine on a 727 that plunged 20,000 feet before the pilot could shut down that engine and safely level out. Tragedy was averted and the same

Work configuration

Physical set up (activity flow) used to make the product; six types: custom shops, job shops, flow shops, flexible manufacturing systems, continuous flow processing, and projects.

Running a nuclear power plant requires a precise process. No technicalities can be ignored.
Source: Union Electric Company; Photo–Ken MacSwan

lesson was learned again: in a process, each part has a place and each step of the process must be taken correctly every time the process is used. In this case, the lives of many people depended upon rigorous adherence to process specifications.

Note that the concept of "inadvertent behavior" is totally unacceptable for proper process management. There can be no surprises. Accidents are inadmissible. Haphazard performance must be designed *out* of the process. Extensive use of check lists or other means of achieving conformance are mandatory. Further, the pilot's training included situations of this kind so that disaster could be averted. The maintenance process had failed, but the flying process was able to handle the contingency.

Process Simulations

Simulations

"Pretend" runs of what might occur using models of processes.

Simulations are "pretend" runs of what might occur using models of processes. For example, pilot training uses simulations of contingencies to eliminate the serious repercussions of unintentional errors. Process simulations are imitation processes that mimic the real-world process.

The board game Monopoly® is a simulation of real estate transactions. People buy and sell real estate. Some make money while others lose it. Except for the latter fact, Monopoly is not a very accurate simulation of real estate dealings. Another example of a simulation is "rotisserie baseball" where the performance of real players is used to simulate the behavior of fictitious teams on which the betting is fierce.[2]

Simulations are successfully used for training, testing and analyzing. Factory processes and production schedules can be created and improved. Under controlled conditions, contingencies artificially arise which allow individuals to practice remedial actions. This learning takes place without being involved in actual crises.

Computer simulations have been used to test the design of the Boeing 777. In this way, the plane has been flown before it is built. It is subjected to engine failures and turbulence beyond anything that will ever be experienced in actual flight. Simulation modeling also is used for production scheduling and inventory control. It is a powerful means of studying the performance of the system as a whole—which is in keeping with the systems approach. Simulation is one of the best methods available for designing good processes. It is a methodology which embodies the systems approach, and is especially effective where nothing should be left to chance.

The process for flying an airplane is detailed. The manuals for procedures are filled with information. All of the controls have to be coordinated. Each procedure for taking off and landing must be precisely stated and checked. Maintenance processing is intricate. Every wire, chip and instrument in the Boeing 777 must do its job correctly. Onboard instrumentation must call attention to problems and flaws. Designing the planes and then building them are two more processes. Four processes are mentioned in this paragraph, and all of them use simulation to improve the performance of that process.

Classifying the Process

The definition of a process must be able to encompass the great diversity of processes such as Disney theme parks, the Houston chain of restaurants, cable television services, power generation systems, automated teller machines (ATMs), newspapers produced and delivered every day, health care centers and academic centers that use pedagogical processes for educational purposes. With such enormous diversity, it is imperative that a strong classification system be developed for process types.

Charting the Process

It also is critical to be able to chart in total detail every step of the process. To get an idea of what is required, try to specify in detail the program that would enable a robot to dress itself. It is hard to separate the methods used from the technology required.

The program required to just tie shoelaces poses enough of a problem to make the point. There is difficulty in setting down the appropriate detail to enable machines to do things that people do by intuition.

Choosing the Process

It is necessary to be able to deal with the question of how to choose a process technology. In choosing the process, the work configuration tends to be chosen by default. Should it be one that excels with high volumes of a few varieties or one that excels with low volumes of a high number of varieties? Excels means that the process has some winning combination of higher quality, lower costs, and shorter delivery times than the processes used by competitors.

Consider again, what is a process? It is a detailed description of the input-output transformation sequence. Different kinds of equipment can be used to make the same things. The choice will depend upon what work configurations are best-suited for the expected volumes and varieties of product outputs.

Means and Ends

The study of OM requires understanding what is needed to achieve the desired ends: technology and training are essential at all levels within the organization, and in the suppliers' organizations to achieve the strategic objectives of the company. From a systems point of view, it is important to note that the ends of one part of the process are often the means of the next part of the process. A means-end model connects strategies with tactics. **Ends** are objectives; **means** are the process components designed to achieve them. Raw material transformations are the means to making components. Components are converted into parts. Parts are assembled into products. Products are grouped together into boxes. Cartons of boxes are delivered to carriers. Carriers move cartons to distributors and they in turn make deliveries to retailers who sell the products to customers. This supply chain is a specific example of a generic means-end chain.

The means-end supply chain does not stop with delivery. Customers use the products and begin the chain of payment. They return money to the retailer who pays the wholesaler who pays the producer or a distribution center if there are additional levels in the distribution hierarchy. The goal was to create the means for obtaining revenue. That objective started the entire systems process. Thus, when that customer's money enters the bank, there is a short-term conclusion to the process. In the long-term, other factors also must be considered to assure that customers remain loyal. The marketing process is a critical part of the means-end chain.

Ends

Objectives; part of a means-end model which connects strategies with tactics.

Means

Process components designed to achieve the ends (or objectives); part of a means-end model.

Boutwell Owens and Co., Inc.

A 'total package service' provider that takes packaging from the drawing board to the loading dock.

Boutwell Owens and Co., Inc. is a small family-owned manufacturer of paperboard cartons, inserts, "skin" packaging, and retail display cards in Fitchburg, Massachusetts. The company succeeds in a somewhat stagnant market because of its emphasis on quality and service. Boutwell sees itself not as a printer of product packages, but as a "total package service" provider that takes packaging from the "drawing board to the loading dock."

Boutwell Owens provides different levels of service to companies depending on their size. For small companies, they "get them into the marketplace with a sharp-looking package," said Ward McLaughlin, Boutwell's president. For large companies, Boutwell not only designs their packaging, but manages their inventories as well. Customers include such familiar names as Radio Shack, GTE, Polaroid, Milton Bradley and Disney.

Boutwell is successful because, according to vice president of operations William Murnane, "We're a job shop. We do whatever the customer needs done. We'll take on the jobs that no one else will do."

President Ward McLaughlin points out that "a company comes to us mainly because of our ability to…get them out of trouble, and when they find out that our quality and service stay equal to what we first initiated, they stay" with Boutwell.

The company runs three shifts to meet demand. Its work force has grown from 60 to 170 employees since 1985. Boutwell is proud of its ability to set up machines fast and meet tight deadlines. Fast turnover of orders is critical when dealing with large retailers. "The Wal-Marts, the Kmarts, the big retailers are pretty relentless," said Brian Jansson, vice president of finance. "Orders will be canceled automatically if you're late."

Source: "Building packages that move the product," by Bruce Phillips, (Massachusetts) *Fitchburg-Leominster Sentinel and Enterprise,* June 27, 1994.

5.1

STRATEGIC ASPECTS OF PROCESSES

All of the components of a means-end chain are driven by the strategic issues that underlie the purpose of the process. Another way of viewing the strategic drivers for the means-end chain is by identifying the process as an interconnected set of transformations. The sequences of linked, input-output systems form one or more chains which represent the process cycles for various products. Strategic decisions about processes must translate cost, quality and delivery requirements into specific methods used to realize the objectives. Thus, *process configurations bridge and link strategy with tactics.*

5.2

TYPES OF PROCESS FLOWS—
SYNTHETIC PROCESSES

There are many ways of categorizing process flows. One of the best ways is to note the difference between analytic and synthetic processes. Combinations of the two basic types of processes are common, but all processes tend to be more like one than the other.

It makes sense to broadly define analytic and synthetic processes at this time so that the comparison can be kept in mind. Synthetic processes combine things, usually putting many parts together to make one or a few products. Analytic processes take things apart creating many products from a few.

A **synthetic process** is a process in which a variety of components come together, ultimately to form a single product. Progressive assembly which follows the instructions of how to build sequentially the subassemblies and final assemblies is an example of the synthetic function. Instructions for construction emphasize the correct order for putting things together.

Any sequential process can be delineated by an assembly diagram such as the one drawn in Figure 5.1. This kind of schematic is used by process designers to make sure that all the nuts and bolts are assembled in the right order. The sequence of assembly goes from twigs to branches (which are subassemblies) to the trunk (which is the final assembly). If the trunk started to spread out again at the roots the process would be starting to be analytic. Processes that combine assembly (synthetic) and disassembly (analytic) are more common than pure processes of either kind.

Synthetic process

Process combining a variety of components to form a single product; employs synthesis to achieve a smooth process flow.

F I G U R E   5 . 1

AN ASSEMBLY CHART (MANY PARTS TO ONE)

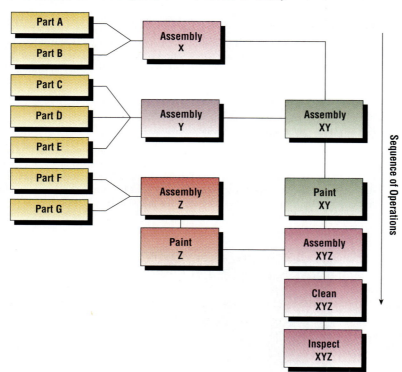

FIGURE 5.2

SYNTHETIC PROCESS — HOW AN AUTOMOBILE IS ASSEMBLED

❶ Stamping

The Stamping Department produces major body panels such as roofs, side panels, and fenders. Above, employees unload and stack stamped panels for delivery to the Body Assembly Department. (STEP 2)

❷ Body Assembly

The Body Assembly Department is probably the best place to see how the automated operations free the production team to perform other tasks which are less physically demanding. After assembly a completed white body is sent to the Paint Department. (STEP 3)

❸ Paint

Before primer and topcoat applications, various coatings are applied to each vehicle for corrosion protection, prevention of water leaks, wind noise, and paint chipping. Robots apply topcoat and perform some sealer and undercoating operations. (STEP 4)

❹ Trim & Final

Vehicle personality becomes evident here—"our car" starts to look like "your car." The Trim & Final Department uses innovative technology such as the tilt line, which allows employees to work comfortably to install underbody components such as fuel tanks and wire harnesses. Installation of front and rear windshields is an automated operation, as is engine and transmission installation.

❺ Quality Control

To ensure that all quality standards and engineering specifications are met, each employee inspects his/her own work and the Quality Control Department's employees give each vehicle a rigorous inspection. These activities occur throughout the manufacturing process and before shipping to customers.

Source: AutoAlliance International, Inc.

❻ Shipping

When a vehicle is ready for delivery, it is transferred to Mazda North America's (MANA) Shipping Yard where MANA coordinates delivery to both Ford and Mazda distributors.
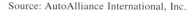

Synthetic processes employ synthesis which is defined as many inputs combining to form a whole. The purest example of synthesis is many to one, but in reality this is often tempered by a relatively synthetic process which combines the many to form a few.

Synthesis characterizes the automobile assembly plant where many components are brought together to form a few car models. Figure 5.2 illustrates an automobile assembly process, showing what goes into making an automobile and in what order these components are introduced to the assembly process.

Another typical synthetic process is that required to bring a number of different kinds of information together with the purpose of analyzing the data and then writing the final report. This is the main function of the Market Research Store, which should be as much a master of the synthetic process as the automobile assembly plant.

A prime objective for every kind of synthetic process is to have a smooth flow of the items being processed. As these items flow along they become more and more complete. The stage of completion could be specified by the percent of the total processing time that has been consumed up to that point.

For the auto assembly process a total processing time of 32 hours is not unusual; the engine might join the chassis at the point where more than 70 percent of total assembly time has passed. The first completion of data analysis for the market research report might occur when less than 30 percent of the total report writing process is completed.

To keep the process from being interrupted, it is necessary to have an ample supply of the right components (parts or information) on hand at the places where they are needed. They could be brought to where they are needed, as they are needed, which is called **just-in-time**. Alternatively, they could be brought to where they are needed long before they are needed to avoid any slip-ups in the timing of the deliveries. That kind of protection also is called **just-in-case**.

This issue of storage of components and timing of their delivery is directly related to the rate at which the process is working. Three important questions need to be answered:

1. How many units come off the line every hour? Say that there are ten. This means that every six minutes a unit is completed.
2. How long does it take to make a unit? The fact that a unit is complete every six minutes does not always mean that it took six minutes to complete it.
3. How many lines are operating and have to be fed with components or information?

Assume that the information assembly process to write a report is initially poorly balanced. This means that the process would be erratic, stopping and starting. Because it did not flow smoothly, there would be times when data was backed up waiting to be processed. At other times, people and computers would be sitting around idle waiting for data to work on.

The operations manager can smooth the flow by creating work stations at which specific tasks would be done. The tasks would follow the logical progression of the assembly chart. Some things cannot be done before others. The time to complete tasks at each work station would be made about equal, creating what is called a balanced line.

In this chapter, perfectly balanced lines are assumed. Later chapters treat the real-world complication of lines that are not perfectly balanced. This occurs when the amount of work at stations is unequal. Stations have different amounts

Just-in-time

Components (parts or information) are on hand and brought to where they are needed as they are needed.

Just-in-case

Components (parts or information) are on hand and brought to where they are needed before they are needed to avoid delivery slip-ups.

of idle time. Perfect balance also assumes that there is no rest time at any station. Perfect balance assumptions are difficult and sometimes impossible to achieve when people dominate the activities at work stations. Perfect balance can be approached using machine assistance for people at work stations. One can most easily visualize perfect line balance as resulting from a totally mechanized and robotic assembly system where the entire synthetic process is synchronized.

In this instance, assume that a paced conveyor belt is used which is stopped at each station for c minutes. The *cycle time* is c minutes which means that a finished unit comes off the production line every c minutes, the pace that is set for the conveyor belt.

5.3

TYPES OF PROCESS FLOWS— ANALYTIC PROCESSES

Analytic process

Progressive disassembly; reverse of synthetic process as it breaks up one thing into many things.

An **analytic process** is characterized by progressive disassembly. It is the reverse of the synthetic process as it breaks up one thing into many things. Going backwards on the assembly chart, the treeing process is reversed. It starts with the trunk and moves out to the branches and then to the twigs. A variety of products can be obtained from the "source." This type of process is illustrated in Figure 5.3.

The name "analytic" process is derived from the fact that analysis is involved. Analysis is defined as breaking up a whole into its parts to find out their nature. An analytic process breaks up the whole in order to use the parts for various reasons.

FIGURE 5.3

**ANALYTIC PROCESS
(ONE SOURCE TO MANY PRODUCTS)**

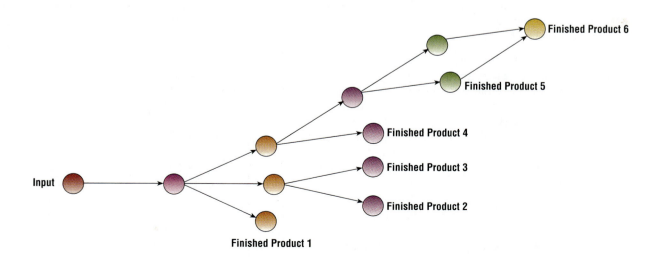

FIGURE 5.4

ANALYTIC PROCESS—CHOCOLATE PRODUCTS

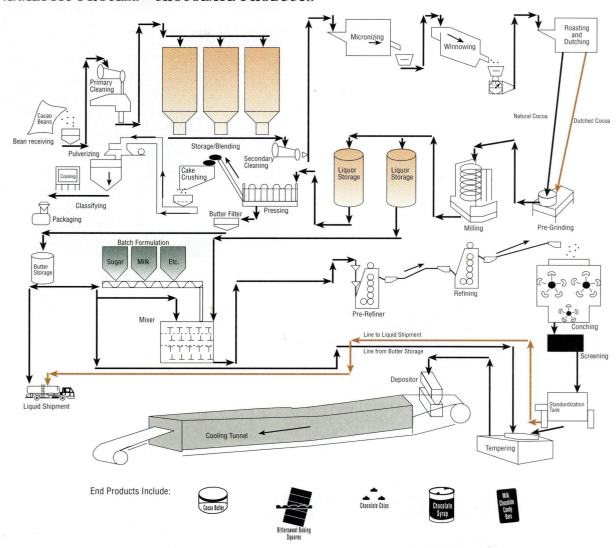

End Products Include:

Cocoa Butter

Bittersweet Baking Squares

Chocolate Chips

Chocolate Syrup

Milk Chocolate Candy Bars

- The basic process is composed of cleaning, grinding, conching, and mixing ingredients of various kinds in different amounts.
- Micronizing is a pre-roasting thermal (or heat) treatment, needed to obtain a clean separation from the husks (shells).
- Winnowing is shelling the beans.
- Cleaning at various stages is essential before grinding (milling) because various kinds of impurities which damage quality must be removed.
- Grinding (milling) precedes liquification.
- To produce a powdery product from high-fat cocoa mass, it must be partially defatted so cocoa butter is extracted by pressing which leaves a solid material known as cocoa press cake.
- Pulverizing cocoa press cake produces cocoa powder which is used in many ways by food industries and by final consumers.

- Cocoa butter is refined in several stages before conching.
- Dutching is the alkalization (adding alkalies) of powder and liquor to modify the flavor and color of chocolate, called dutched cocoa and chocolate.
- Conching is a high energy mixing to distribute cocoa butter and flavoring.
- Tempering is a heating and cooling process to set up the correct crystal structure in the fat.
- Differences in the quality of chocolate are due not only to beans and recipes used...they are dependent on the production process used.
- End products include cocoa powder, chocolate flavored syrups, chocolate chips, dark chocolate, milk chocolate candy bars, and cocoa butter.

Courtesy of Chocolate Manufacturer's Association.

Pork requires an analytic process
that transforms the commodity into
a variety of products.
Source: IBP, Inc.

The purest example of an analytic process would be where one thing is converted
into many derivatives without having to add anything else. Most analytic processes
have one main, basic ingredient which is converted into a number of derivative products often with the help of additional materials.

Chocolate requires an analytic process to convert the main raw material—cacao
beans cut from (9 inch) pods which grow year round on the cacao trees—into
many different products such as cocoa powder, unsweetened, bittersweet, semisweet,
sweet, milk, and white chocolate, in various forms such as chocolate bars, chips
and syrup. Other derivatives are chocolate chips, bakers' chocolate, and cocoa butter.
Sugar is an additive for many of the products. Milk chocolate requires the addition
of milk solids and sugar. Less expensive chocolates substitute fats and emulsifiers
for cocoa butter.

The analytic process also applies to information systems. Previously, it was
used to describe a synthetic process where data from a variety of sources were
combined to create a report. The analysis of data, used by laboratories, functions in
the opposite way. A blood sample is drawn and tested yielding data on blood sugar,
cholesterol, and red cell count, among other things. The final report decomposes
the blood sample into many end categories. These are used by doctors to evaluate
health and to diagnose symptoms of health problems.

5.4

SOURCING FOR SYNTHETIC PROCESSES

The flowcharts for synthetic processes show how various materials, components
and parts are fed into the main stream, where they are joined together to form the
product that marketing will sell. Purchasing the materials, components and parts (or
data) from different sources requires dealing with many different suppliers.

Typically, the operations manager is not responsible for the sourcing decisions. Because these decisions impact strongly on the performance of operations there is a need for coordination. The systems approach is essential to make certain that purchasing and OM are playing on the same team.

There are prices, qualities and delivery reliabilities to compare. These include on-time delivery of the correct amount with the agreed upon quality and cost. Acquisition strategies for synthetic processes are significantly different from those needed for analytic processes, though both require comprehension of the mechanisms that apply to their particular kind of supply chain.

Synthetic processes usually involve many parts and one or more suppliers for each of them. All of these parts cost the company consequential amounts of money. The cost of materials for synthetic processes has been a large percent of the cost of goods sold (COGS). It can be as high as 80 percent. It is, on average around 40 to 50 percent, and increasing for many firms.

Being the low cost producer is a major goal of most firms. Even Rolls Royce, with its very high price tags, has redesigned its operations to reduce costs significantly. Because of the impact of purchase prices on total costs, most synthetic process industries are very strict with their suppliers.

Increasingly, a rational cost structure for purchased products from all suppliers has become the objective of companies with synthetic processes. To achieve this, it is necessary to have suppliers that are the low-cost producers for all of the parts and components the company purchases. This creates the potential for cost advantages from savings that are made on all of the parts and components that are purchased. If the savings that can be made are not taken, they are called opportunity losses. Those firms that are on top of the situation, thereby avoiding opportunity losses and realizing the potential savings, achieve significant competitive advantages. Companies using synthetic processes are particularly vulnerable if they are ignorant of this fact.

The cost of materials component of the COGS is normally understated for synthetic processes. This means that the COGS is also understated. The costs are understated because the cost of materials purchased do not include all of the costs of quality.

In particular the costs of failures, as measured by losses of customer loyalty, are totally missed. Also the cost of rework due to suppliers' defectives is not usually associated with the cost of purchases. Missing these factors understates the costs. Further, the cost of designing and installing supplier certification programs has been growing as competitors pay increasing attention to the triad of critical dimensions: price, quality and speed of delivery. The cost of materials should reflect the cost of the supplier certification program.

Suppliers that do not deliver on time delay production and that, in turn, holds up deliveries to customers. Delivery on time is a critical quality dimension in its own right. Delivery delays can have a substantial impact on customer loyalty. With the above in mind, the choice and management of suppliers to insure that the highest delivery standards are set and met is becoming another competitive requirement for synthetic processes.

5.5

SOURCING FOR ANALYTIC PROCESSES

Typically, analytic processes begin with the purchase of the single commodity which the process transforms into many derivative products—such as cacao beans into chocolates, sweet crudes into various fuels, corn into breads, sugars, fuel additives, and grapes into wine and vinegar. For analytical processes, purchasing agents choose worldwide sources for the basic input component. The price, quality and delivery of the core purchasing ingredient are subject to the kinds of fluctuations that characterize that particular commodity market.

Commodities are considered materials that are generic. Within any one commodity class the materials are treated as being relatively indistinguishable from each other. "Relatively" is used because operations managers of analytic processes know that qualities of commodities vary which greatly affects the quality of their finished products.

The sourcing decisions have a major impact on how well the process functions. Because it is seldom OM that reaches these sourcing decisions, a systems approach is essential to coordinate the operations and procurement of materials. It also is important to note that the systems approach focuses on many different issues according to whether the materials will be subject to synthetic or analytic processing.

Analytic processes, by their definition, have a totally different supplier situation than that which applies to synthetic processes. There are many suppliers of the same few commodities. Frequently, price is the main factor that drives the acquisition of the input materials. Yet, as was stated above, it is always subject to quality considerations which OM and the agency doing the sourcing coordinate within the framework of the systems approach.

Prices are established by the specific commodity exchange that trades in that commodity. The Chicago Board of Trade and the New York Mercantile Exchange are well-known commodity trading exchanges in the United States for many of the raw material ingredients that require analytic processes. Most countries have equivalent exchanges.

Sourcing in the commodity markets requires a variety of special skills. Purchasing strategies play a major role in determining the average price paid for a commodity. The average is taken of the price paid across many transactions. This can include both buying and selling the commodity. Some contracts are made at present prices while others involve commodity prices at future dates. The operation of futures markets is discussed in courses and books on financial trading. There are different traditions and techniques required to successfully buy corn, coffee, salt, rice, gold, platinum, copper and other metals, cattle, hogs and other livestock, as well as fuels and oils from plants and flowers such as sunflowers and peanuts. There are different ways of bidding for commodities and establishing contracts in the futures markets.

Factors such as weather can affect crop prices for commodities such as oranges and grains. The prices of metals, including gold, copper and platinum, will vary according to local and world events. The average prices which are affected by such extraneous factors get translated into the eventual prices of the derivative products. A freeze in Florida drives the price of oranges up and a crisis in Russia causes the price of gold to jump. When many farmers have abundant crops which they bring to the market, the excess of bumper crops drops the prices.

Money is made and lost by managers of analytic processes who trade in their commodities as another necessary side to their business. The advantages that can be realized by astute maneuvering in the commodity markets is one of the features of dealing in analytic processes. It is hard to overlook the fact that some very large coffee companies have made more money trading in green coffee bean futures than in selling bags and cans of their brand of coffee. The same kind of parallel exists for other commodities where the trading required for purchasing provides a business opportunity in its own right.

5.6

MARKETING DIFFERENCES

For synthetic processes, it is important to be very focused and sharp about marketing and selling the basically one product to be sold. At the same time, one product permits concentration on the marketing plan. There is only one sales force, and it does not have to divide marketing resources among several different product lines. At the end of the supply chain marketing is specialized. Upstream, at the beginning of the chain, purchasing deals with an array of suppliers.

For analytic processes, there are often quite disparate end products. Each addresses a different market constituency. Each requires different marketing techniques. Gasoline for autos is an entirely different market from industrial fuels, and both contrast with kerosene for jets. A variety of marketing plans are needed and separate sales forces are required. The marketing budget must be allocated to diverse end products. Downstream, at the end of the supply chain, marketing is diversified. Upstream, where the supply chain begins, purchasing is focused.

Marketing Products of Analytic Processes

A commodity is often defined as a class of items which are indistinguishable from each other. There is reason to take exception to this characterization. The qualities of commodities are not homogeneous within any category. Differences are often overlooked or dismissed by the commodity purchasing process.

A major strategic advantage is available to those who capitalize on the differentiation of commodity qualities. They can buy on terms of quality as well as by price. The quality advantage can be utilized to advantage at the downstream end of the analytic process in each of the markets of the derivative products.

There are many types of wheat grains. Varieties of corn are so numerous that only experts can catalog them. Some soy bean products are cleaner than others. Some crops just barely qualify with respect to nutritional standards for cattle feeds; others are considered to be superior. The standards apply to quality requirements that range from minerals to vitamins.

Cranberries are graded by the cooperatives that sort and package them. Wet harvesting is less expensive and inferior to dry harvesting. Wet cranberries need to be dried and then sorted according to their color and bounciness, a surrogate for firmness. Within the grading categories for cranberries there exists a great deal of product quality variability.

Many commodities are shipped in bulk using railroad box cars, large trucks and the cargo holds of ships. The method of transportation and the care with which the shipment is arranged will affect the quality of the merchandise when it is received. The logistics of transport are very important in this regard for quality as well as with respect to the timeliness of delivery.

Educating the consumer about quality differentials for various commodities is not usual. Utilizing the differences for creating consumer preferences and price differentials has been done successfully for wine grapes. There is a complex mystique about vintages. Prices can range from under $5 to $500. To a somewhat lesser degree, the same marketing approach using quality differentiation has been applied effectively to coffee beans.

The choice of quality grades for commodity inputs to the analytic processes will have major impact on the quality of the derivative products. Instead of talking about GIGO (garbage-in, garbage-out), the time has come to talk about

QUIQUO, which emphasizes the more positive aspect of process management (quality-in, quality-out).

It is better to see the glass as being half filled than to see it as being half empty. Those who say that quality should play a larger part in determining purchasing decisions for analytic processes are on-target. Price alone cannot continue to be the key to sourcing and purchasing for analytic processes.

Chocolate fanciers know that all cacao beans are not alike. Most chocolate comes from forastero trees which grow in West Africa and Brazil. Scarcer and more flavorful beans come from criollo and trinitario trees which are found in Central America. Having the best beans is essential for preparing the best chocolates.

Best Practice

Best practice is a process standard which has strong roots in the systems approach. The idea is to identify those who perform the process in the best ways yet known. Striving to make one's own process the best in the "class" is like aiming to win the gold at Olympic events; the prior year's records of gold medal winners are the standards for best practice in that event. The procedure of comparing one's own process against the best is called **benchmarking**. It is used to compare various activities in OM, such as quality attainment, sales effectiveness and purchasing performance, and also to compare production processes and determine product excellence.

Best ingredients are a necessary but not sufficient condition for product excellence. Best process (BP) must be combined with best ingredients (BI) to bring out the best product (BPR). The factors in the equation are multiplicative:

$$(BP)(BI) = (BPR)$$

If either (BP) or (BI) = 0, meaning low quality, then (BPR) = 0. Thus, for example, in Figure 5.4, for the highest quality chocolate product, the conching step must knead the mixture until it is totally smooth. This process step can continue for as long as three or four days to achieve best process (BP) results. If this is combined with the best cacao beans (BI), then best product (BPR) can be attained.

There must be a three-way understanding of the system, calling for a meeting of the minds between marketing, purchasing and operations. The payoff for coordination of the functions is great. Purchasing and OM working together combine (BI) and (BP), while marketing exploits the advantage of superior quality (BPR).

The resulting differentiation raises commodity products above the customary concept of being indistinguishable and homogeneous and places them into special niches—like Godiva chocolates. This increases profit margins of the derivative products of analytic processes. It also provides an opportunity for developing competitive advantage in terms of the quality of commodities.

5.7

MIXED ANALYTIC AND SYNTHETIC PROCESSES

Most processes are combinations of synthetic and analytic processes. For example, maple syrup from the maple tree is a commodity that gets converted into maple sugar, pancake syrup, candies, and cakes. To produce each end product requires that other commodities be added to the mixture.

Best practice
Process standard with strong roots in the systems approach; striving to make one's own process the best.

Benchmarking
Procedure of comparing one's own process against the best.

TECHNOLOGY

National Semiconductor, a $2 billion electronic chip manufacturer, invested $77 million in creating a facility devoted to BiCMOS (bī-sē-mōs) technology. BiCMOS (which stands for Bipolar Complementary-Metal-Oxide-Semiconductor) is a semiconductor manufacturing technology used in cellular phones, laptop computers, and mobile communication networks.

BiCMOS Technology at National Semiconductor

BiCMOS technology requires Class 1 "clean-room" facilities, which are necessary to maintain the integrity of the tiny electronic circuits. During the manufacturing process, thousands of these tiny circuits are created, then interconnected in a thin chip of silicon the size of a human fingernail. A Class 1 clean-room is 10,000 times cleaner than a hospital surgery room.

The BiCMOS facility is an expansion of National Semiconductor's South Portland, Maine, location. The expansion will add 12,000 square feet of new manufacturing space to 444,000 square feet of existing space. The South Portland plant is one of the oldest semiconductor manufacturing facilities in the U.S., in continuous use since 1964. Existing facilities are to be converted from 5" to 6" wafer manufacturing lines when the expansion is complete. This conversion will enable increased product yield.

Applying BiCMOS technology in a manufacturing environment, National Semiconductor can provide customers with the break-through products needed in the expanding wireless communication marketplace.

The company will utilize BiCMOS technology to manufacture its SiRF (Silicon Radio Frequency) mixed-signal products. These include the DECT (Digital European Cordless Telecommunications) transceiver chip and the PLLatinum PLL (Phased Locked Loop) frequency synthesizer line for cellular phones and wireless communication systems.

According to Wayne Carlson, general manager of the South Portland facility, "BiCMOS will contribute to the future success of National Semiconductor well into the twenty-first century."

Source: "National Semiconductor plant expansion brings new technology to Maine," Business Wire, September 8, 1993.

Generally, one type of process predominates, that is, the process is primarily either synthetic or analytic. In the case of maple syrup, the process is predominately analytic even though additives are needed to produce the final set of products.

The assembly of computers is a synthetic process. Also, the computer chips inside the computers are constructed and assembled by a synthetic process. Yet, many of the computer chips—sourced from all over the world—are treated as if they are commodities. To counter the concept that all chips are alike, Intel spends freely to advertise its chips as being unique and special. Nevertheless, a computer is a mixture of A and V processes. So too are cars, VCRs, published books, clothing and cans of food.

The OPT System's V-A-T

OPT is a production process scheduling system. It will be encountered whenever serious production scheduling efforts are being used. OPT stands for *optimized production technology*. This system is highly sensitive to the existence of bottlenecks. It is often referred to as a production theory of constraints for finite scheduling—which means lot by lot scheduling.[3] The methods of OPT are discussed in the context of tactical scheduling activities.

The concepts of OPT were initiated by a professional consulting company set up by Eliyahu Goldratt who developed the OPT principles concerning synchronized production scheduling. The term appears to have been taken into common OM usage to represent finite scheduling principles.

OPT classifies processes by means of the acronym *VAT*. A *V-process* is analytic, starting with a single commodity at the bottom and branching out into refined products at the top. An *A-process* is synthetic, starting at the bottom with several inputs which are combined to yield a single marketable product. The combination is named by OPT as a *T-process*. The charts in this text have been guided by these conventions.

OPT focuses attention on the fact that each type of process (V and A and T) has unique production scheduling aspects. Also, each type of process has characteristics which work best with specific types of work process configurations, as described below. Proper matching of work process configurations with the synthetic components and the analytic components of process flows can provide significant advantages.

5.8

WORK PROCESS CONFIGURATIONS

There are fundamental differences in the way that both synthetic and analytic processes can be designed or configured. A primary factor for classification is the number of units produced at one time. That number is often referred to as the lot size, N. Lot size applies to discrete production runs as compared to continuous production output systems which include flow shops (assembly lines) and continuous (liquid) flow systems.

Another classification factor is the variety level. It is customary to find that as the lot size goes up the variety level diminishes. In part, this is a statement that reflects old technology where set-up time was inversely related to time and cost per part. New flexible process systems (FPS) technology has altered this formula.

5.9

THE CUSTOM SHOP

The **custom shop** is the work configuration in which products or services are "made-to-order." Custom work is like tailoring a suit of clothes to fit one specific individual. This kind of work process is usually associated with a lot size of one, $N = 1$.

Many custom shops, like machine shops and cabinetmakers, use synthetic processes, doing one custom job after another. As for analytic processes, laboratories might be viewed as custom shops where a specific ingredient (such as blood) is broken down into its components for detailed analysis.

What does the custom shop look like? Not too much space is required to house the limited materials needed for the jobs at hand. There is *general-purpose equipment (GPE)* needed by relatively few people with craft skills who turn out the product. Often, the workers are owners and entrepreneurs operating in a small workshop housed in a store or garage.

The information equivalent is a small office just large enough to allow for word processing equipment, some files and telephones. If a service is offered, such as a travel agency, there must be desks for clients and agents to sit at together. Behind the scenes, and occupying little space, are the telecommunications capabilities and computer reservation systems of the travel agency.

5.10

THE JOB SHOP

The job shop is not a custom shop or a flow shop which operates a serialized production line. The **job shop (JS)** is a work configuration for processing work in batches or lots. N, the lot size, can vary within these guidelines. The job shop produces N units at a time for shipment, for stock, or a combination.

Job shop (JS)

Work configuration for processing work in batches or lots.

There are over 100,000 job shops in the U.S.A., and a larger number in Europe. Job shops account for over 70 percent of all manufactured parts in the Western world. Both in the U.S. and in Europe, the typical lot size is 50 units.

The work configuration of the job shop has equipment at various locations and batches of work queue up waiting for processing at each location in accordance with what needs to be done. The equipment is of the general purpose type so it can be used to do many different jobs.

For most job shops, the variety of the work that can be accomplished is quite substantial. This is in keeping with the large array of process equipment, the broad range of skills of the machinists and the kind of demands that are the basis of the

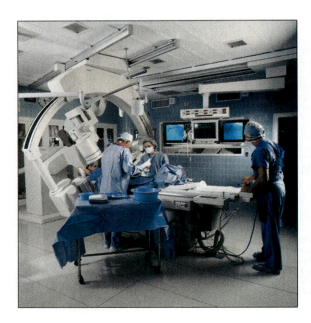

An operating room is a job shop in which the variety of work that can be accomplished is quite substantial. Source: Siemens Medical Systems, Inc.

job shop business. The demands received by job shops arise from industrial firms which place orders for the great variety of parts, often in substantial numbers, but not enough to warrant using the serialized production configuration of the flow shop.

The job shop is able to produce the entire line of products which any combination of general purpose machine tools can create. The capabilities of the job shop are extensive and it can turn out a lot of different items. For example, a drill press will make holes of different sizes in various materials. A lathe is used to turn the outside diameters of cylindrical parts.

Job shops have some special-purpose equipment (SPE) for larger batch sizes. Instead of using a turret lathe, an automatic screw machine can turn out many more cylindrical parts once it is set up. In general, for higher volumes of output, the job shop has intermediate-level equipment that falls between GPE and SPE.

Figure 5.5 illustrates an operations or routing sheet for a part manufactured in a job shop. The sequential steps for die-casting and machining are listed on the operations sheet. It furnishes information about where the equipment to be used is located.

F I G U R E   5 . 5

THE BILL OF MATERIALS AND THE OPERATIONS SHEET USED BY A JOB SHOP

Bill of Materials

Item ___Motor Switch MS60___   Sheet No. _1_ of _1_

Drawings _95.6—95.8_   Assembly _3 HP Motor X22_

Part No.	Part name	No./item	Material	Quantity/item	Cost/item	Operations
P14	Casing	1	Zn	0.2lbs	$5.15	Cast,trim
P21	Spring	2	Sprg. St.	—	1.64	Buy
•	•	•	•	•	•	•
•	•	•	•	•	•	•
•	•	•	•	•	•	•
P7	Clips	4	AL15	—	0.25	Buy

Operations Sheet

Part No. _____P14_____   Economic lot size ___750___

Part name____Casing____   Process time/lot ___6 hrs___

Blueprint No. _95.7_   Set-up time_____see below

Use for ___Motor Switch MS60___   Quantity per _____1_____

_____Motor Switch MS61_____   _____1_____

_____Subcontract Hamilton_____   ___500/order___

Material __Zn___   Supplier _Vaun___

% Scrap __10___   Weight __0.2lb/___   Cost _$1.65/___

Operation No.	Operation	Location/ Machine	Std. Time in minutes Operations	Set Up
1	Change die	F5	—	30
2	Injection molding	F5	3	—
3	Remove	F5	5	—
4	Deflash	F4	7	—
5	Tumble	F7	20	10
6	Plate	G3	10	12

For larger batch sizes, the job shop
has intermediate-level equipment.
Source: Ingersoll-Rand Company

The bill of materials, which often is found on blueprints, spells out instructions
about materials and tools. Thus, together, the routing sheet, the bill of materials and
the blueprints, provide all the information needed to do the job.

Information processing equipment is general purpose most of the time. Computers are wonderful general-purpose machines, which when loaded with software can
become highly specialized. In banks, dedicated, special-purpose equipment is required
to handle checks. The high volumes of transactions in check handling are treated
with both job shop and flow shop configurations, according to the processing transactions. Over time, banking equipment is moving towards total flow shop configurations. Many word processing applications are job shops.

What does the job shop look like? The space is much larger than for the custom
shop. It has to be big enough to contain the materials, general-purpose equipment
(GPE) and people with skills. Most job shops are running many jobs at the same
time, but there are considerable variations in size.

5.11

THE FLOW SHOP

Flow shops (FS) are serialized work configurations where one unit of work is at
various stages of completion at sequential work stations located along the production line. They are dedicated production facilities which, like auto assembly lines,
can be kept running day-after-day. Demand is sufficient to warrant a sequential
production process that moves work from one station to the next. Usually, a paced
conveyor carries the work from station to station, stopping for a carefully chosen
period of time at each station. The length of time that it is stopped is called the
cycle time. A great deal of effort goes into balancing the line. The cycle time includes
some time for rest, which reduces the output rate.

Flow shops (FS)

Serialized work configurations
where one work unit is
at various stages of completion at sequential work
stations located along the
production line.

In a typical flow shop configuration a conveyor is used to transfer work from station to station. This plant has a monorail system to transfer loads between production stations.
Source: Intel Corporation

The flow shop also is often described as being a serialized production process, which means sequentially in order, regular and continuous. To contrast the operating characteristics of the job shop with those of the flow shop see Figure 5.6.

Usually, the supply capabilities of the flow shop are equal to or greater than the average demand. For those intervals when demand exceeds supply, the flow shop might be replicated in parallel. This often is the case when the labor component of the flow shop is high and the capital investment component is low. Sometimes the rate of throughput can be increased by using more powerful technology.

As an alternative method of increasing the supply, the flow shop can be changed from a one-shift basis to a multi-shift basis. This can include overtime operations or a temporary shift to a two- or three-shift basis. When demand is seasonal, flow shops are configured to provide multi-shift operations to match peak load demands with a single-shift basis for off-peak times.

Flow shops are dedicated processes where materials and components from suppliers merge and join the production line in an uninterrupted sequence. It is more difficult to configure analytic processes, from start to finish, as a flow shop. Once the ingredients for the derivatives have been obtained, it is possible to run with the flow shop configuration.

A flow shop for chocolate making (Figure 5.4) can be initiated after the conching of the blended mixture of liquor and cocoa butter. At that point, a synthetic process can be used to make chocolate bars, chocolate kisses, and so forth. To feed the flow shop there always must be a supply of the chocolate blend.

Continuous processes

Work configuration where the product flows through pipes continuously as it is being treated.

Some analytic processes can be designed to permit the fanning out of the derivative products as part of a continuous flow process stream. These are called **continuous processes**, the work configuration where the product flows through pipes continuously as it is being treated and processed. Good examples of such continuous processes come from the chemical and petrochemical industries. They may be used for making plastics and synthetics such as nylon thread. Also, continuous process flow shops, designed to be high-volume and cost-effective systems, are typified by petroleum refining and beer making.

FIGURE 5.6

COMPARING JOB SHOPS AND FLOW SHOPS

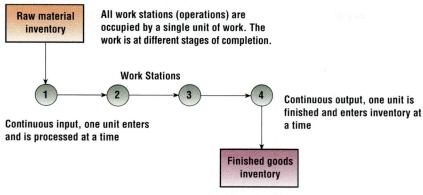

FLOW SHOP CONFIGURATION (Serial production)

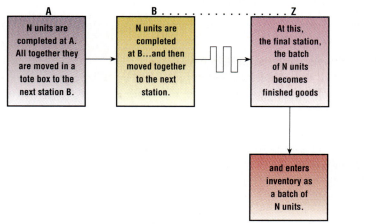

Entire batch of units is completed at one work station before moving to the next.
JOB SHOP CONFIGURATION (Batch production)

An excellent example of efforts to approximate a continuous flow for analytic processes is found in the area of distribution. Related to rationalization of the distribution process is the field called **logistics**. Logistics concerns the use of methods for obtaining effective procurement, storage, and movement of materials. Logistics processes operate the distribution systems which move finished goods downstream in the supply chain. Distribution of product is a fanning-out process. Shipments of goods go from the producers' sources to warehouses, and from there to regional distribution centers which move the goods to retailers. The retailers distribute the product to the ultimate consumers. Meanwhile, cash flow moves up the chain from the customers to the producer.

The supply chain epitomizes analytic processes. Great efforts are being made to keep product moving throughout the supply chain in quantities that are related to actual needs at various nodes in the network. Thus, value adding is maximized by

Logistics

Methods for effective procurement, storage and movement of materials.

the proper procurement schedule which minimizes storage at every stage. When material arrives as needed, the process is characterized as being just-in-time.

Henry Ford's sequenced assembly line received so much publicity that flow shops came to be identified with synthetic processes by terms such as "mass production." Analytic processes deserve as much attention. When there is sufficient demand, job shop assignments for batches of work that relate to analytical processes can be converted into the much more economic flow shops that disassemble the basic commodities into a set of final products. These often entail the transfer of liquids. They are continuous flow-through processes such as encountered in refining petroleum, but also used for milk and its by-products.

Combinations of analytical and synthetic flow shops abound in extracting and blending by vintners and beer makers. Telephone switching centers are able to continuously connect and disconnect lines with each other using process techniques that bring things together and disassemble them in great numbers and with incredible speed. The operations within computers that allow them to perform in the way that they do follow the same mixture of analytics and synthetics.

The Benefits and Constraints of Flow Shops

The advantages of converting both analytic and synthetic processes into flow shop form are three-fold:

1. the cost per unit can be significantly reduced
2. the qualities of the product can be raised to higher standards
3. the consistency of the qualities can be improved

Flow shops must be justified by volume. The fixed costs required to design and equip the process are high. The increased fixed costs needed to equip the system must be traded off against the decrease in variable costs associated with a less labor-intensive process. The marketing department and the sales force must be able to generate enough demand to allow capital expenditures to be amortized over a reasonable period of time.

5.12

ECONOMIES OF SCALE—DECREASING THE COST PER UNIT

Economies of scale

Reductions in costs per unit achieved as a result of producing larger volumes of work.

Economies of scale are reductions in costs per unit that are achieved as a result of producing larger volumes of work. Thus, the flow shop was first conceived by someone who gleaned that repetitive tasks could be done more efficiently if people specialized in particular aspects of the work process.

There had to be enough call for doing each of the repetitive jobs on a regular basis to allow for the development of specialized work stations where people did the same thing over and over again. Thus, person D is assigned the exclusive use of the hammer and person J is assigned to the exclusive use of the screwdriver. D becomes the hammer specialist and J, the screwdriving expert. Adam Smith called this method of organizing people working on the process, "the division of labor."

In present day terms, added efficiencies are gained from being able to train people to increase their competence as specialists. If there is no body of knowledge about

P E O P L E

Jeffrey Bleustein
. . .
Harley-Davidson, Inc.

Jeffrey Bleustein, who heads Harley-Davidson's motorcycle division, faces the challenge of expanding the business to keep up with the growing demand for Harleys. Harley-Davidson has been in an "allocation mode" with dealers since the mid-1980s, despite more than doubling its output of motorcycles since then.

Harley-Davidson's goal is to produce motorcycles at the rate of 100,000 units a year by the end of 1996. After a thorough study involving employees at all levels, the company decided it could achieve this output objective within its existing facilities by:

- continuing the conversion from a job shop to a flow environment
- providing product focused machining and assembly

- eliminating waste in all manufacturing processes through continuous improvement activities

The price tag on this effort was put at $80 million over a three-year period in addition to the normal capital expenditures of approximately $50 million per year.

"We are working hard and we are putting our money behind increasing output," Bleustein says. "We are very cautious in our approach to do this in a planned way and a careful way because we don't want to incur quality problems. There have been times in our history where, in increasing production in response to a rapidly growing demand, we let quality drop. We will not do that again."

In addition to increasing motorcycle output, Harley-Davidson also aims to lower costs on a per-motorcycle basis, improve productivity, continue to improve quality, and achieve a greater degree of flexibility in manufacturing processes to more closely meet customer desires.

Source: John Fauber, "Harley exec says success creates own problems," *The Milwaukee Journal*, June 5, 1994 and information from Jeffrey Bleustein.

how to use a hammer so that everyone is equally good whether they are trained or not, then specialization in the use of hammers is pointless. The same can be said if training would improve the way the job is done, but training is not used.

In many companies, there is a lack of understanding that training can improve any and every activity. In part, this is because there is a lack of know-how and, worse yet, no one even knows that the know-how exists. For whatever reason, if the body of knowledge for training is not available, then one of the main reasons for specialization is missing. Yet, there are additional reasons for flow shop type specialization.

Economies of scale can be gained from the use of special-purpose equipment. SPE often requires different work routines and leads to training by the supplier of the equipment. Even if that training is not offered, the worker learns to use the new equipment and a form of self-training occurs. The hammer specialist is given a pneumatic hammer and gradually develops new skills in hammering. Repetitive tasks can become boring, but alternatively, even without SPE, learning and self-training can reward the repetitive job assignment.

Further economies of scale become possible because of volume production. These include expanded levels of coordinated promotional, advertising, and marketing activities. High volume purchasing should earn well-deserved discounts. Work in process can be minimized in a well-designed flow shop process. In a related advantage, value adding can be maximized. There is focus of purpose because of concentration on doing the same things over and over. This can be turned to advantage.

When sufficient demand exists, the flow shop is the preferred process. It is the simplest and most efficient production configuration. Economies of scale make it worthwhile to try to aggregate enough volume to justify using it.

When sufficient demand does not exist, the use of the flow shop provides one of the best strategies for getting the demand level to be sufficient. The flow shop tends to support itself by lowering costs. This allows for decreases in price and/or increases in marketing support. The superior quality of flow shop product (if the flow process is properly designed) enhances customer preference and supports increased demand volume.

The negative press that the flow shop gets because repetitious work is boring and mind-numbing can be overcome with proper job design and the rotation of jobs. These subjects are addressed in discussions about human resources management.

The contrast between job shop and flow shop production is worth examining. Lacking specialization, working with batches is a less efficient production configuration. Being a greater producer of product variety is the direct cause of increased complexity. The traffic of batches of work moving around in tote boxes generates patterns of interference. Controlling traffic requires extra planning to avoid tie-ups and blockages. Job shop batches interact with each other, waiting for their turn, competing for facilities and floor space. The mix of jobs is seldom the same. The pattern is always changing. Repeated planning is required and costly. The correctness of each plan will vary. Sometimes the plan will be perfect and more often somewhat off-the-mark. Consequently, the precision, scope and timing of each day's job shop plan will be less on-target than for the flow shop.

Interesting alternatives are coming into being with the technological development of flexible systems. Their downside is expense and the need to pioneer in their use and development. Flexible manufacturing systems (FMS) and flexible process systems (FPS) are discussed under the generic terminology of FMS. It should be noted that flexibility is associated with the economies of scope, a uniquely different rule of economics than the economies of scale. **Economies of scope** are realized when set up cost reductions allow more frequent set ups and, thereby, smaller lot sizes can be run.

Economies of scope

Set-up cost reductions allow more frequent set ups and thereby, smaller lot sizes can be run.

5.13

THE INTERMITTENT FLOW SHOP

An alternative work configuration which has many flow shop benefits is called the intermittent flow shop. It can be used when demand volume is not sufficient to justify a full-scale flow shop. The design of an intermittent flow shop (IFS) allows it to be set up and run for an interval of time, then shut off.

IFSs are not as efficient as permanent flow shops because they need to be turned on and off as required. This means that the same complete engineering of the process will not pay. Also, to have a large flow shop investment not running for long periods of time is counterproductive. Nevertheless, the IFS can be viewed as a step in the transition of a company process as it moves from being a job shop to a flow shop.

IFSs are turned on when high output volumes are needed in short periods of time. They can be used to build up stock when the savings of flow shop speed counterbalance the cost of carrying stock. Sometimes a special order is received which requires larger volumes of product than are normally made. The intermittent flow shop is a possible solution.

The ability to use serialized production for work that formerly or also had been done in batches requires planning and investment. Once it is planned and the facilities are available, it gets called into being by need, but requires both set-up times and learning times to get the line running. As with pure flow shops, the intermittent form can be used for both synthetic and analytic processes. The criteria for establishing optimal runs apply.

A flow shop that can be shut down and dismantled does not have the same high level of special-purpose equipment as a full-time dedicated flow shop. It will not be as cost efficient, but it still makes use of the principles of flow shops.

These principles relate the total time required to make a complete unit of product, requiring i operations

$$\sum_{i=1}^{i} t_i$$

or steps, to n, which is the number of work stations (Figure 5.6), and C (the cycle time).

The relationship with no idle time at any station is specified by:

$$nC = \sum_{i=1}^{i=10} t_i$$

It is assumed that the job consists of ten steps. The sum of the times to do them is 30 minutes. The whole job, all ten steps, could be done at one station. Then $n = 1$ and $C = 30$. The cycle time is 30 minutes and, therefore, the production rate is 60/30 which is two units per hour.

Assume that the same job will be done with two stations. Then, $n = 2$, and $C = 15$. The cycle time has been cut in half. The production rate is doubled since $60/15 = 4$. On the ledger, the balance of costs must be tested. The cost of the extra station is to be weighed against the doubling of the output rate.

Determination of when to use the IFS form of flow shop is guided by the principle that flow shop production is more costly to set up than batch production, but thereafter per unit costs are lower. It is necessary to calculate the demand level at which the total costs are about the same for batch production or the intermittent flow shop. This is the *point of indifference* as to which work configuration is used. It also is referred to as the point where the two kinds of processes breakeven with respect to total costs.

5.14

FLEXIBLE PROCESS SYSTEMS

There is another alternative to the classic flow shop which is of relatively recent technological origin. It is the **flexible process system** (**FPS**). For manufacturing it is called the flexible manufacturing system (**FMS**). Applied to services it is called FPS or FOS for flexible office system.

Flexible process system (FPS)

Used like flow shops but produce variety due to computer programming and computer-driven equipment operating together to change set up in instants; include flexible manufacturing system (FMS) and flexible office system (FOS); and associated with economics of scope.

Flexible systems are able to be used like flow shops even though they produce variety. The reason is that computer programming and computer-driven equipment operate together to change set ups in instants. The economics of such systems facilitate increasing diversity, called *economies of scope*.

Each station is capable of doing a set of different things and switching from one member of the set to another in an instant. Table 5.1 is a simple illustration of the principle. Station 1 can shift instantly between the operations a, b, c, and d. Station 2 shifts between e, f, g, and h. Station 3 shifts between i, j, k, and l.

TABLE 5.1

FLEXIBLE PROCESS SYSTEM WITH THREE WORKSTATIONS AND FOUR PRODUCTS

Product	1	2	3	4
Station 1	a	b	c	d
Station 2	e	f	g	h
Station 3	i	j	k	l

One product that this FMS can make is aei. Other products are bfj, cgk and dhl. If only the columns are products, then four different products could be made with flow shop speeds and costs. If all possible combinations were products, then there would be $4 \times 4 \times 4 = 64$ different products that could be made with the work station advantages of flow shop processes.

If the three stations are connected and coordinated to work together, they constitute a cellular manufacturing system. This requires programming so that each station can do its work, and methods of conveyance can move the work from one station to another. The components of the cell are programmed to coordinate activities on a number of items having somewhat different designs. While they are different, as a general rule jobs of similar family characteristics will be assigned to particular manufacturing cells. In that way, variants of the programming can be economically developed. The efficient production of families of parts is a concept that is known as group technology. Consequently, this application of FMS is classified as **group technology cellular manufacturing systems**.

Group technology cellular manufacturing systems

Application of FMS; employs stations connected and coordinated to work together to efficiently produce families of parts.

High-technology systems are coming into use to convert paper trails into digital form. The processes that will foster and expedite paperless offices and check-free banking will be characterized by a great amount of flexibility. Freedom from paper requires radically new forms of processes.

There is ample evidence that the use of optical scanning methods and voice-activated request systems has decreased reliance on old-style record keeping. Credit card companies and many banks are employing new technologies to revamp the accustomed information systems approach for recording, storing and informing transactions.

Home shopping networks on TV are breaking new ground in information processing. Figure 5.7 depicts a home shopping process system where the customer communicates with the computer using a Touch-Tone phone to input the item wanted, quantity and credit card information. While the customer remains online, the computer completes the credit card check and determines item availability. If all checks out, the customer is given a confirmation number and told the amount that is being deducted from their account.

It is expected that with interactive cable, customers will be able to browse through catalogs and interact even more directly with the computer system. These are representative of the truly flexible service systems currently being developed.

F I G U R E   5 . 7

HOME SHOPPING ONLINE PROCESSING

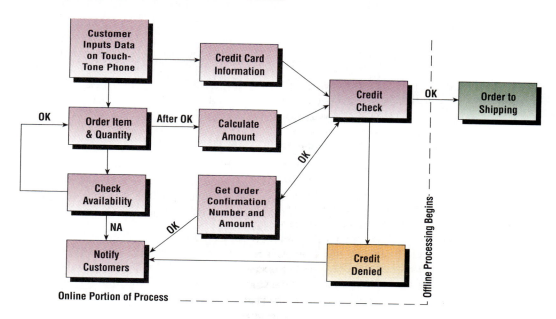

5.15

MOVING FROM ONE WORK CONFIGURATION
STAGE TO ANOTHER

Hayes and Wheelwright[4] developed the matrix shown in Figure 5.8. It captures an important aspect of the relationship between product life cycle stages and process life cycle stages.

The product life cycle stages are labeled product structure by the authors. This terminology is translatable to categories of volume and variety.

Process life cycle stages are labeled process structure by the authors. This terminology is equivalent to type of work configuration.

The relationship depicted by the matrix indicates that as products move from low volume and high variety (left) to high volume and commodity products with no variety (right), the process type moves from the job shop (top) to the continuous flow system (bottom). The diagonal, moving downward and from left to right, delineates the progression. Also, the authors distinguish between smaller job shops and larger ones by calling the latter batch work.

FIGURE 5.8

MATRIX OF: PRODUCT LIFE CYCLE STAGES (COLUMNS) VERSUS PROCESS LIFE CYCLE STAGES (ROWS)

Process structure Process life cycle stage	Product structure—Product life cycle stage			
	I Low volume-low standardization, one of a kind	II Multiple products low volume	III Few major products higher volume	IV High volume-high standardization, commodity products
JS	commercial printer			
Batch		heavy equipment		
FS			automobile assembly	
CONT				sugar refinery

Economies of Scale — — lower unit costs — —▶

Legend: JS = job shop, FS = flow shop, CONT = continuous flow

Reprinted with permission of the *Harvard Business Review*; An exhibit from "Link Manufacturing Process and Product Life Cycles: Focusing on the Process Gives a New Dimension to Strategy," by Robert H. Hayes & Steven C. Wheelwright, Jan.–Feb. 1979, pp. 133–140. Copyright © 1978 by the President and Fellows of Harvard College; all rights reserved.

This matrix takes an operations-oriented approach to defining product life cycles instead of the focus that marketing uses. It defines product life cycles in terms of increasing volume and decreasing variety. The power of this matrix is in the connection that it makes between the growth in demand volume and the nature of the process that can deliver enough product to balance supply and demand.

It is possible to use a flow shop for lower-volume and higher-variety products. This would be like moving heavy equipment in Figure 5.8 down one box and off the diagonal. Auto-assembly could be raised one row and made a batch process. Such off-the-diagonal positions are permissible and special. However, the following admonitions should be kept in mind:

1. If the actual company process falls above the diagonal, it is not well suited to handling the larger volume that the market demands, and it can provide more flexibility for variety than the market requires. Both of these deviations have costs which exert pressure on the company to move back to the diagonal to be on a par with its competitors.
2. If the actual company process falls below the diagonal, it is geared to handling higher volumes than the market demands, and is less able to cope with the flexibility needed for the variety levels that the market requires. Once again, both directions for the deviations have costs which exert pressure on the company to move back to the diagonal.

As the authors of the matrix state, "A company that allows itself to drift from the diagonal without understanding the likely implications of such a shift is asking for trouble."[5]

Expanded Matrix of Process Types

Six classes for work configurations were mentioned at the start of this chapter. Project work configurations were dropped from that list because they apply to start-ups and not online production systems. That leaves five to consider. Because the matrix in Figure 5.8 illustrates only four process types, a new matrix has been constructed to reflect the positioning of the five process types. Table 5.2 presents the classification in list form.

T A B L E   5 . 2

EXPANDED MATRIX OF PROCESS TYPES

Process Type	Variety	Volume
Customized	Highest Level	One at a Time
Job Shop	Medium Level	Medium Volume
Flow Shop (Assembly)	Low Level	High Level
Continuous Flow	Lowest Level	Highest Level
Flexible System	High Level	Few at a Time

In addition to Table 5.2, Figure 5.9 illustrates the expanded matrix. The directions for economies of scale are the same as in Figure 5.8. Flexible process systems have been added. Therefore, economies of scope are represented by the bottom (FMS) row of the matrix.

This makes clear how the FMS process category can act like a custom shop or a job shop. It has the economic advantages of the flow shop. It can emulate all of the different work configurations, except continuous process flow systems, as noted below.

F I G U R E   5 . 9

EXPANDED VERSION OF FIGURE 5.8

Expanded Version of Figure 5.8

	ONE	MED	HIGH	CONT
CS	X			
JS		X		
FS			X	
CONT				X
FMS	X	X	X	

Legend:
CS   = custom shop
JS   = job shop
FS   = flow shop
CONT = continuous flow
FMS   = flexible manufacturing system

The FMS-type process is used most effectively when it acts like a custom shop or a job shop, yet having the economic advantages of the flow shop. It also can emulate the job shop or the flow shop, but the economic advantages of being a flow shop in disguise are lessened as the lot size increases.

Because of the ability to handle the entire range of volumes and variety levels, the Xs in Figure 5.9 are entered into three stages of the product life cycle. The fourth stage, which represents continuous production of commodity products, is possible for FMS but not economical.

5.16

THE EXTENDED DIAGONAL

Adding the Custom Shop

The smooth progression along the diagonal is not disrupted by the addition of the customized shop. Customized process abilities are associated with the highest level of variety and the most unique products. Hence, the custom shop creates a new north-west corner for the matrix because it can produce higher variety at lower volumes more economically than the job shop.

This places the job shop second in position on the matrix where it fulfills the mission of producing lot sizes in batches of 50 units on average. As discussed previously, assembly line flow shops turn out only a few major product types. They produce a low variety of goods or services in quite high volumes on serialized production lines. Continuous flow systems are better served by dedicated equipment than by flexible systems.

The Role of FMS

The addition of flexible systems stops the diagonal progression and bends the variety level back to the highest level that can be obtained. The volume also circles back to one or a few at a time, like the custom shop. Because of the ability to achieve instantaneous changeovers from one job to another, the process acts as if it were a flow shop.

Flexible systems make a distinct break with the past. This is the first time that production technology permits hybrid results. The hybrid in this case combines the variety of the job shop with the economies of scale advantages of the flow shop. This new technology opens processing vistas not conceived of until recently. Flexible systems continue to gain in power as the underlying computer-driven technology evolves. Nevertheless, in spite of their promise and actual use in various factories, the overall developments remain tentative and experimental.

Applied to Services

It is also convenient to extend the basic matrix shown in Figure 5.8 (which applied to manufacturing) so that it can reflect the same set of considerations when applied to services. This results in Figure 5.10.

The systems view of relationships between service products and processes is similar to the relationships for manufacturing. The fact is that processes tend to develop from higher variety to lower variety as market success moves demand volume from low to high. Supply, in order to match demand, requires alterations of the process type.

MATRIX (FOR SERVICES) OF: PRODUCT LIFE CYCLE STAGES (COLUMNS) VERSUS PROCESS LIFE CYCLE STAGES (ROWS)

	No economies of scope			
JS	offices			
Batch		textbooks		
FS			fast foods	
CONT				newspapers

Economies of Scale — — lower unit costs — —▶

Legend: JS = job shop, FS = flow shop, CONT = continuous flow

Reprinted with permission of the *Harvard Business Review*; An exhibit from "Link Manufacturing Process and Product Life Cycles: Focusing on the Process Gives a New Dimension to Strategy," by Robert H. Hayes & Steven C. Wheelwright, Jan.–Feb. 1979, pp. 133–140. Copyright © 1978 by the President and Fellows of Harvard College; all rights reserved.

5.17

AGGREGATING DEMAND

Demand is affected by variety, price and the quality of the product. Using these controls, it is possible to aggregate demand. The accumulation of volume is done in order to move toward process configurations for higher-volume levels. One of the most used controls is lower price which usually increases demand in line with price-elasticity relationships. Variety is another means of increasing aggregate demand that often is used.

The elasticity relationship of product quality and price with demand volume is not to be forgotten. Customers cannot find substitutes for unique qualities of products, which makes those products relatively inelastic. Figure 5.11 relates supply capabilities, product variety, product quality and price to demand. They are all interrelated. Boxes 3 and 4 drive 1, but the work process configuration of 2 determines the economic feasibility of the system. Also, fundamental to the success of demand aggregation is product design.

FACTORS THAT AFFECT AGGREGATION OF DEMAND (ALL LINES FLOW TWO-WAYS)

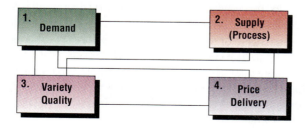

Flexible manufacturing systems are outstanding sources of control over the factors shown in the Figure 5.11. Because FMS process costs are at a minimum—paralleling flow shop economies of scale—it is possible to gain market share and aggregate volume by lowering prices. In spite of low prices, the profit margins remain intact and cash flow is not threatened. The same kind of reasoning applies to the advantages of high variety which FMS permit. Superior quality of FMS products is derived from programming the system, which calls upon the enhanced control features and excellent tolerances that are characteristic of FMS systems.

Thus, using an FMS permits a company to aggregate volume of new products which can then be assigned to conventional flow shop processes. In other words, flexible manufacturing systems can help facilitate the transformation of batch-type processes to dedicated flow shop facilities.

S U M M A R Y

One major differentiating characteristic of processes is whether they are synthetic or analytic. Synthetic processes assemble; analytic processes disassemble. It is noted that the OPT system first makes this structural distinction before trying to deal with synchronizing production scheduling.

Process configuration strategies are explained. There are basically six different ways to do work and five of these are appropriate for online production systems. The five types of work systems are: customized one-of-a-kind output, job shop batches, serialized output of the flow shop, small lots with high variety from flexible process systems, and continuous output flows from refinery-type systems. These work configurations are explained, including the cycle time of the flow shop.

▶ Key Terms

analytic process *176*	benchmarking *182*	best practice *182*
continuous processes *188*	custom shop *184*	economies of scale *190*
economies of scope *192*	ends *171*	flexible process system (FPS) *193*
flow shops (FS) *187*	group technology cellular	job shop (JS) *185*
	manufacturing systems *194*	
just-in-case *175*	just-in-time *175*	logistics *189*
means *171*	OPT *184*	process *168*
process steps *168*	simulations *170*	synthetic process *173*
work configuration *169*		

▶ Review Questions

1. What is meant by the statement that the management of supply is related to the management of demand?

2. It is said that the existence of the demand for a product leads to the creation of the industry that can supply products to satisfy that (need) demand. If need drives supply, then what accounts for the sales of VCRs, air conditioners and laptop computers? Consumers did not ask for these products until long after they were developed. Producers took the risk of tooling up for high volumes of production in order to obtain low enough prices so that consumers would buy these items.

3. Is it true that the drivers of demand are price, quality and variety? Are there other drivers?

4. What is a process?
   a. Does a process differ from a procedure? How?
   b. Does a process differ from a method? How?

   Hint: Use the dictionary to look up "procedure" and "method." Then, relate process to technology and note the relationship between process and product technology.

5. What is a custom shop and why is the custom shop a natural start for an entrepreneur?

6. What rules apply for managing a successful custom shop?

7. Au Bon Pain is a fast-food franchise which stresses high quality coffee, buns, sandwiches, and soups, among other things. Which are more applicable: the rules for running a good custom shop or those for a good job shop?

8. Describe a job shop. What rules apply for managing a successful job shop?

9. Would an entrepreneur prefer a job shop to a custom shop?

10. Contrast analytic processes with synthetic processes and give examples of both kinds of systems. Can simulation be used to study these processes?

11. Discuss the advantages and disadvantages of moving from one work configuration stage to another. Refer to Figures 5.8, 5.9 and 5.10. Consider moving down the diagonal to higher volumes as well as up the diagonal to higher variety. Discuss the use of simulation to evaluate using each work configuration.

12. What are economies of scope? Do they eliminate economies of scale or can these two kinds of process economies work together?

13. What kind of a business process is exemplified by a supermarket? Deal with the entire process of ordering, stocking the shelves, customers shopping, checkout counters, etc.

14. What kind of business process and work configuration is used by the post office? Discuss the process. Try to estimate how many transactions are handled per minute.

▶ Problem Section

1. Much used by the automobile industry for the assembly process is a paced conveyor to move the chassis from station to station. This is usually cited as an example of a pure synthetic process.
   a. Is this true?
   b. If the assembly process of an older auto assembly plant in New Jersey takes 32 hours, and there are 1000 work stations along the conveyor path, what is the length of time that the conveyor is stopped at each station?

2. A new auto assembly plant has been built in Kentucky. It has an assembly process that takes 22 hours, and there are 800 work stations along the conveyor path. What is the length of time that the conveyor is stopped at each station?

3. With reference to Problems 1 and 2 above, what is the cycle time for each auto assembly plant? Comment on the comparison of the New Jersey plant with the Kentucky plant.

4. With reference to Problems 1, 2 and 3, above, what is the hourly production output rate at each plant? Assuming two eight hour shifts, six days a week, compare the weekly output of the New Jersey plant and the Kentucky plant.

5. The average annual investment cost of a work station in New Jersey has been calculated to be $100,000. It has been calculated to be $150,000 in Kentucky. The hourly cost at a work station is $60 in New Jersey and $40 in Kentucky. How do the two plants compare with respect to the cost of labor for making a car?

6. An interesting comparison can be drawn between the Volvo method of building autos in Sweden and that of U.S. auto companies. Specifically, the Volvo Company has pioneered a team approach for assembling an auto. The Volvo team builds the car on a platform. Even the engine is put together by the same team and mounted on the chassis. This means that the workers come to the work instead of the work coming to the workers.
   a. If it takes Volvo workers 40 hours to assemble a car in this way, what is the cycle time?
   b. How many stations are there?
   c. Analyze the pros and cons of the Volvo assembly system.
   d. Does the Volvo system constitute a synthetic flow shop?
   e. Volvo does not use this team assembly method for its plant in the U.S. Why might that be the case?

7. The highest volume item in a job shop is made in lots of 1000 units once every week. As an alternative, it has been suggested that an intermittent flow shop (IFS) should be set up twice a year to satisfy the same demand. Use the following data to compare the costs of each production strategy.

   IFS set-up costs = $50,000 per set up

   Cost per unit with IFS = $5.00

   Time per unit with IFS = 0.1 minute per unit

   Regular batch set-up costs = $1,000 per set up

   Cost per unit with regular batch production = $8.00

   Time per unit with regular batch production = 1.0 minutes

   Carrying cost is one percent per month

8. Applying the information given in Problem 7 above, consider the following:

   Since an IFS has been suggested, why not analyze, at the same time, the use of a programmable FMS. This particular product lends itself to a family of parts which could be made with a group technology cellular manufacturing system. The investment for the manufacturing cell would be shared with other members of the family. An estimate has been made that its share of the annual investment would be $200,000 per year. The cost per unit would be $2.00 and the time per unit is 0.50 minutes.

   Do you recommend this option? Explain and discuss.

9. Define and compare the six basically different ways of doing work in terms of the seven factors listed below. Each kind of process system (i.e., the six types of work configuration) is best suited to a particular set of conditions. These conditions arise from factors such as the following:
   a. the character of the product line; how complex it is
   b. the equipment associated with the process
   c. the nature of the market
   d. the financial situation of the firm
   e. how many units are to be made or serviced
   f. how profitable the venture is
   g. what the competition is like

10. The approach to evaluating alternative work configurations can be modeled to provide more structure. To exemplify, a model is constructed below to evaluate under what conditions the intermittent flow shop (IFS) will be selected in preference to the batch production of a job shop (JS).

For the IFS, set-up, take-down and changeover costs have to be amortized over the lot size that is run. The IFS produces N units at a cost of c dollars per unit with a fixed set-up cost of S dollars.

Compare that cost with the cost of set up for the job shop production process, S'. The lot size is the same but the cost per unit, $c' > c$. The cost comparison is shown below with some hypothetical numbers for costs and a lot size of $N = 100$.

$$\text{IFS: } (cN + S)/N = (3 \times 100 + 500)/100 = \$8 \text{ per unit made}$$

$$\text{JS: } (c'N + S')/N = (4 \times 100 + 300)/100 = \$7 \text{ per unit made}$$

so choose to use the job shop to make the 100 units. Now, however, assume that the lot size is 500 units.

$$\text{IFS: } (cN + S)/N = (3 \times 500 + 500)/500 = \$4 \text{ per unit made}$$

$$\text{JS: } (c'N + S')/N = (4 \times 500 + 300)/500 = \$4.60/\text{per unit made}$$

so choose the IFS.

Now, assume that both set-up costs are double.
a. What is the choice when the lot size is $N = 100$?
b. What is the choice when the lot size is $N = 500$?

11. If the total time to make a computer mouse is 24 minutes and it is made at four stations, how long is the cycle time?

Hint: Use the equation

$$nC = \sum_{i=1}^{i} t_i$$

to determine the production output rate (productivity rate).

12. Create a board game to simulate the work flow through the four stations in Problem 11. Describe and discuss the procedures.

▶ **Readings and References**

William J. Abernathy, K. B. Clark, and A. M. Kantrow, *Industrial Renaissance*, NY: Basic Books, 1983.

Horace Lucien Arnold, and Fay Leone Faurote, *Ford Methods and the Ford Shops*, NY: The Engineering Magazine Company, 1915.

Eliyahu M. Goldratt, and J. Cox, *The Goal*, 2nd Rev. Ed., North River Press, 1992.

Eliyahu M. Goldratt, and R. Fox, *The Race*, North River Press, 1986.

Robert H. Hayes, and Steven C. Wheelwright, "Link Manufacturing Process and Product Life Cycles: Focusing on the Process Gives a New Dimension to Strategy," *Harvard Business Review*, January–February, 1979, pp. 133–140.

Terry Hill, *Manufacturing Strategy: Text and Cases*, 2nd Ed., Homewood, IL, Irwin, 1994.

R. Normann, *Service Management: Strategy and Leadership in Service Business*, 2nd Ed., NY, Wiley, 1991.

Thomas E. Vollmann, William L. Berry, and D. Clay Whybark, *Manufacturing, Planning and Control Systems*, 3rd Ed., Homewood, IL, Irwin, 1992.

S U P P L E M E N T   5

How Purchasing Agents Use Hedging

Process design should take into account purchasing strategies. When the costs of materials are volatile and fluctuate a great deal as a result of such things as weather (oranges) and gold (world tension), it is important to protect the production process from sudden exorbitant price increases in the basic materials that are used. That is where hedging comes into play.

Hedging is a purchasing strategy that often is utilized to protect against price fluctuations. Money is neither made nor lost because the buyer is locked into a fixed price as follows:

1. A purchase is made for delivery of the commodity in a future month—cost is x dollars.

2. When production needs material, a cash purchase is made for immediate delivery—cost is y dollars.

3. Simultaneously a sale is made for delivery in the same future month—revenue is y dollars.

If $x < y$, the price has risen just when OM needs to buy the materials. However, the buyer is protected with a net cost of x for materials. This is because of the profit $y - x$ made from buying and selling futures in Steps 1 and 3.

If $x > y$, the price has fallen. The OM buyer will not gain that advantage having net costs of x because of the loss $x - y$ from buying and selling futures in Steps 1 and 3.

With hedging, the buyer always nets out paying x. The company will neither make nor lose money.

The advantage of hedging is stability. It insulates the production system from wild fluctuations. However, some companies prefer to speculate in the commodity prices of the raw materials that are used. Often this is because these companies are major buyers in those markets and can influence the selling prices of the commodities.

Problems for Supplement 5

1. In July, the price of the material used by the process was $0.58 per pound. The purchasing agent expected the price to rise by December, and therefore bought a futures contract for 1000 pounds to be delivered in December. In September, the company needed this material because demand had exceeded forecasts. The price of the material had risen to $0.68 per pound.

    What course of action should the purchasing agent take? What amounts of money will be involved? Show the financial transactions that take place, including the sums of money for costs and any revenues, if there are any.

2. Now, treat an alternative scenario:

    In July, the price of the material used by the process was $0.58 per pound. The purchasing agent expected the price to rise by December, and therefore bought a futures contract for 1000 pounds to be delivered in December. In September, the company needed this material because demand had exceeded forecasts. The price of the material had fallen to $0.48 per pound.

    What course of action should the purchasing agent take? What amount of money will be involved? Show the financial transactions that take place, including the sums of money for costs and any revenues, if there are any.

Notes...

1. Jon F. Cox, John H. Blackstone, and Michael S. Spencer, *APICS Dictionary*, 7th Ed., APICS Educational and Research Foundation, 1992.

2. "In Virtual Mudville, the Outlook Is Joyless as Rotisseries Halt," *Wall Street Journal*, p. 1, August 11, 1994. This was the day before professional baseball went on strike.

3. Eliyahu M. Goldratt created OPT. His company, Creative Output in Milford, Connecticut, was the original consulting group. The theory is explained in novel form: *The Goal* by E. M. Goldratt, and J. Cox, 2nd Rev. Ed., North River Press, 1992. Also, see: *The Race*, E. M. Goldratt, and R. E. Fox, North River Press, 1986.

4. Robert H. Hayes, and Steven C. Wheelwright, "Link Manufacturing Process and Product Life Cycles: Focusing on the Process Gives a New Dimension to Strategy," *Harvard Business Review*, January–February, 1979, pp. 133–140. Specific reference is to Exhibit 1 on p. 135.

5. *Ibid.*, p. 135.

Chapter 6

Process Analysis and Process Redesign

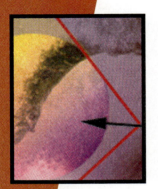

After reading this chapter you should be able to...

1. Explain why process analysis and redesign are valued OM capabilities.

2. Chart any process that needs to be studied using conventional symbols or those of your own choosing.

3. Explain what activities should be captured for process charts.

4. Delineate when more detailed symbols should be used on a process chart as compared to less detailed symbols.

5. Explain the use of process flow layout charts.

6. Discuss the choice of continual improvement of processes as compared to their redesign.

7. Apply process improvement concepts to information systems.

8. Describe the impact of set-up times and costs on processes.

9. Explain SMED and utilize SMED methods for reducing set-up times and costs.

10. Differentiate between internal and external set-up costs.

11. Discuss the benefits of a total productive maintenance (TPM) program.

12. Explain the steps TPM advocates, and under what circumstances.

13. Distinguish between preventive and remedial maintenance strategies.

14. Explain why new developments like computer-aided process planning are of growing importance.

15. Describe the use of Machine-Operator Charts.

Chapter Outline

6.0 PROCESS IMPROVEMENT AND ADAPTATION

6.1 PROCESS FLOWCHARTS

6.2 BACKGROUND FOR USING PROCESS ANALYSIS
Process Management Is a Line Function
Demand Is Greater than Supply
Seven Points that Deserve Consideration for Process Flow Design
The Purpose of Process Charts
The Use of Process Charts for Improvements
New Orders

6.3 PROCESS CHARTS WITH SEQUENCED ACTIVITY SYMBOLS
Computer-Aided Process Planning (CAPP)
Creating Process Charts
Reengineering Is Starting from Scratch in Redesigning Work Processes

6.4 PROCESS LAYOUT CHARTS

6.5 ANALYSIS OF OPERATIONS
Value Analysis for Purchased Parts or Services
Other Process Activities

6.6 INFORMATION PROCESSING

6.7 PROCESS CHANGEOVERS (SET UPS)
Single-Minute Exchange of Dies (SMED)

6.8 MACHINE-OPERATOR CHARTS: PROCESS SYNCHRONIZATION

6.9 TOTAL PRODUCTIVE MAINTENANCE (TPM)
The Preventive Maintenance Strategy
The Remedial Maintenance Strategy
Zero Breakdowns
TPM for the Flow Shop

Summary
Key Terms
Review Questions
Problem Section
Readings and References
*Supplement 6: A Total Productive Maintenance Model for Determining the Appropriate
        Level of Preventive Maintenance to be Used.*

The Systems Viewpoint

The systems approach is being used when process analysis for the job shop is focused on the total flow. This entails spanning the process from the first operation—including all the purchased supplies—to the last operation, which includes shipping, and contacting the customer to find out if the product was satisfactory and the delivery made on time. Because it is encompassing, the systems approach to improvement often entails altering the process technology internally and externally, at the suppliers' plant and at all points of distribution. Charting operations for process analysis and redesign should include all of the process transformations that affect customer satisfaction. Systems analysis and redesign always involve "the big picture," which is another name for the *systems level of process change*. Thus, total productive maintenance (TPM) is a systems approach to prevent failure and single-minute exchange of dies (SMED) is a systems approach to reduce set-up times.

6.0

PROCESS IMPROVEMENT AND ADAPTATION

Continuous improvement of processes is an excellent goal for, and a valued function of, OM. An existing process represents an investment in equipment and training which can be analyzed and redesigned to be as efficient as possible. Costs can be reduced without impairing (and perhaps increasing) reliability, which means only scheduled preventive maintenance. Output is often raised by productivity improvement studies (PIPS) without lowering quality, and sometimes raising it.

Citibank recognized the potential opportunities for continuously analyzing their customers' needs. Every branch bank is evaluated and adapted to the needs of its local customer markets. Changes in customer contact processes include tellers and platform officers and services. Paul Revere Insurance Company analyzed internal processes with emphasis on the fact that employees who work together are each other's customers. If every employee listens to the voice of their internal customers, the service experiences of external customers can be greatly enhanced. Airlines that empower their attendants to solve customers' problems and train them to do so are noticeably superior in customer ratings. This is hardly a surprise to those airlines that do it. In each case, it took conviction, analysis and forthright action to improve the process.

The methods for analyzing processes are worthy of study. They are effective and capable of making substantial savings and other improvements that permit the firm to experience greater profit margins, market share, and cash flow. The methods are aimed at achieving best practice in processes.

Continuous process improvement includes streamlining operations, introducing new equipment, improving the way the job is done and the training for it. Process innovations can occur through the work of R & D, or by means of copying the best.

Continuous improvement at
Chrysler led to the "doors-off"
system that helps technicians
install components more easily.
Source: Chrysler Corporation

The method of finding the best, called benchmarking, is based on comparative measures which identify leaders in specific kinds of processes. Benchmarking must identify external candidates for best processes, and then, by measuring, comparing, and analyzing, select those processes that will be used as a standard to be achieved. McNair and Leibfried stress continuous improvement using benchmarks as a guide, with the ultimate goal of being "better than the best."[1] They credit Xerox as having developed benchmarking in its present form, starting in the 1970s, and as a result of competitive pressure from Japanese copy manufacturers.

Benchmarked companies do not have to be in the same industry, at least for certain processes. Thus, Lands' End has been benchmarked by computer companies that are trying to become effective with mail order. Disney has been benchmarked by manufacturers that provide service to their customers. The Paul Revere Insurance Company has been benchmarked by banks for their team quality programs—modeled on the Olympics—which award bronze, silver and gold pins.

Good process innovations also can occur because of external forces which impose safety restrictions, pollution regulations and other governmental controls for the first time. Process adaptation also becomes essential when competitive process advantages occur which leave the others at a disadvantage.

6.1

PROCESS FLOWCHARTS

The use of process charts or process flowcharts is critical to understanding what is going on in the system, how the product is being made, or delivered as a service. **Process charts** are maps of the specific operations, transports, quality checks, storages and delays that are used by either the present or the proposed system. There is a solid benefit for every organization to chart the processes that it uses to obtain revenues. Wasteful steps that can be removed or remodeled can significantly decrease costs and cut time.

Process charts

Maps of specific operations, transports, quality checks, storages and delays used by present, or proposed, systems.

Figure 6.1 shows a process chart for manufacturing an electrical panel box. The process chart uses triangles for storage, circles for operations, and squares for inspections. These are some of the conventional symbols. Other symbols will be introduced later on. The person doing the process charting has discretion concerning what amount of detail is best for a particular purpose.

FIGURE   6 . 1

PROCESS CHART FOR MAKING ELECTRICAL PANEL BOX

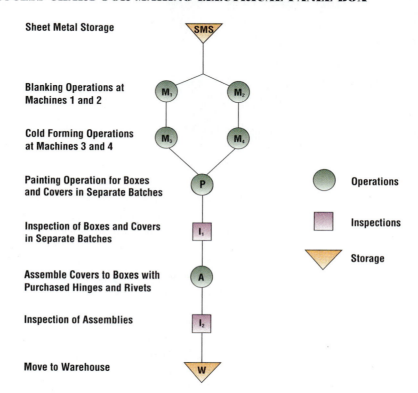

The chart shows that sheet metal is taken from storage for blanking operations on two different machines, which cut out the correct shapes. Two other machines do cold forming on the blanks for the boxes and covers in batches. Boxes and covers are painted and dried separately in batches. The chart does not point out that a label with the company's name, logo, and the model number of the panel box assembly, is applied to the cover. Then, covers and boxes are inspected and assembled together with purchased hinges and rivets. The assembled unit is inspected for appearance and cover operation. Satisfactory product is stored in the warehouse.

It is to be noted that much information is not included in this process chart. Times required for operations, lot sizes, and distances that batches have to be transported are all missing. Some process charts have all this information. Also, operations can be broken down into smaller segments. For example, cold forming takes place in three steps that are bundled together for the purposes of this chart. There are many options in drawing up process charts. Simple charts, such as the one in Figure 6.1, are meant to provide an overview of the process at a glance.

6.2

BACKGROUND FOR USING PROCESS ANALYSIS

Charting an existing process is a starting point for analysis of that process to see if it can be done in a better way. Process analysis also should be used for a process that has never been done before. **Process analysis** examines the purposes of the process and tries to find the best possible way to achieve those goals.

Process analysis is an important ongoing activity for job shops. When a job which has never been done before enters the shop, OM is responsible for setting up the process, often basing set up on similar jobs that have been completed. Therefore, OM's experience with prior processes is valuable. The existence of process charts that can be used as a basis for the new job is frequently accompanied by similarity of blueprints.

The new job offers an opportunity for modification and improvement of the old process. Using process analysis to adapt one system to fit another is a way of achieving improvement through redesign. Questions to be answered include which machines are to be run, when and where, by which operators, having what training? When orders are repeated, analysis and redesign of these processes makes sense.

For the flow shop, much time and attention needs to be paid to the design of the process sequence. It is more suitable to treat the flow shop as an engineering and capacity planning problem. Complete analysis is critical because the job runs continuously and even pennies that can be saved add up to big amounts. Most job shop processes do not have the volume to justify major pre-engineering with the use of simulation. The process charting described in this chapter applies to job shops.

Process charts for job shops trace the patterns of movement of all materials that come into and go out of the shop. From the receiving docks or the mail-in box to the shipping docks or the mail-out box, tracking by process charts should be done. Typically, the mix of orders in the job shop is so dynamic that it is necessary to improve old processes and invent new ones.

Process Management Is a Line Function

Process managers are responsible for the production line. That is why they are called line managers. They are *getting out the product*. Filling orders is their prime job, but they must take the time to plan the processes for the production line; proper planning saves time in the long run. With good systems in place, they can routinize the function of developing and improving process charts.

There are competitive advantages, including enhanced profitability, for those in OM who consistently avoid waste and disruption by improving process flows. A remarkable example of this is the hub and spoke process flow designed by the FedEx Company, which picks up packages at all locations and transports them by air to Memphis where they are sorted and transported by air to their final destinations. Using bar codes, the location of any package is quickly determined. The process flows are fast and secure. FedEx is a splendid model of what excellence in process design can achieve. **Time-based management** (TBM) of processes provides continuous improvement of delivery times which leads to important competitive advantages that need to be discussed.

In the same sense, there are great opportunities to gain increased profit margins from proper planning for job shop work. To achieve the gains entails allocating planning time for process management. This means assigning the task of creating new process charts to operations managers who have the appropriate training and skills.

Process analysis

Examines the purposes of the process and attempts to find the best possible way for goal achievement.

Time-based management (TBM)

One aspect of TBM provides continuous improvement of delivery times leading to competitive advantages.

Demand Is Greater than Supply

Successful job shops are generally backlogged. In other words, demand is greater than supply. Work must wait before it can be done, and there is pressure to make deliveries on time. This fact needs to be considered in reaching good process management decisions.

In a good job shop, there is concentrated effort on minimizing waste and disruption from machine breakdowns, worker errors, bad materials, and process distortions that cause defectives. Process managers jump to right the wrongs, which is called "putting out fires." The single, most important goal is keeping the orders moving out the door to customers. Therefore, it is an exciting challenge to prevent problems instead of correcting problems after they occur.

With a systems perspective at work, OM and marketing can determine which kind of job orders fit existing capabilities of the shop better than others. Process charts provide the kind of information that enables such determinations to be made.

Marketing can work at selling the optimal mix of orders for the existing machine capacities, worker skills and management competencies. Specific classes of new orders are complementary to the existing system while others do not fit as well. If the sales force is coordinated with OM, better focus for sales and marketing results in total systems improvement.

In a similar sense, some customers are prized for their steady, repeat business while others are occasional buyers who provide less profit for the company. OM and marketing need to be coordinated with respect to priorities for customers. Capacity changes can be made to fit the profile for profitable customers and eliminate resource drains for unprofitable customers.

Further coordination is required to recognize that certain new customers may have great potential for growing into the class of "most important" customers. This fact should be taken into account because when drawing up process charts, more attention is likely to be paid to streamlining the production process for prized customers. The effect of putting priorities on customers' orders is always encountered in materials dealing with scheduling.

Seven Points that Deserve Consideration for Process Flow Design

1. Alternative process flow designs are always possible. It is useful, therefore, to understand what reasons account for the process flow designs that are being used.
2. Scheduling decisions interact with the designs for process flows that have been developed. Customers are usually very sensitive to delivery lead times, ergo to process features.
3. Product quality is a direct function of process flow design.
4. Product cost (and productivity) are direct functions of the process flow design.
5. Customers prefer to choose products which maximize value, where:

$$\text{Value} = f \left\{ \frac{\text{Quality Worthiness}}{\text{Total Price Paid}} \right\}$$

6. Because process charting increases quality and decreases costs, it is a means of raising customer satisfaction.
7. Achieving the "best in class" value includes satisfactory delivery dates, and service excellence, both of which are components of quality worthiness. These are OM process management responsibilities which relate to process charts.

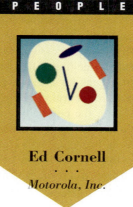

Ed Cornell
. . .
Motorola, Inc.

As a manufacturing manager in Motorola Inc.'s Cellular Products Division, Ed Cornell said his major responsibility is to provide customers with on-time delivery of the highest quality cellular infrastructure product. Motorola won the Malcom Baldrige National Quality Award in 1988, the first year the award was given. The company is known for its "Six Sigma Quality" initiative, which translates into a target of no more than 3.4 defects per million opportunities.

Cornell's group is responsible for the "back-end" areas of the Cellular Products division's factory. These areas include: base build, systems test, power amplifier, common kits, tanapas (or options), and associated engineering staff. One product the group manufactures are base stations, which Cornell describes as the "guts of the cellular network." Base stations are located in "cell sites." They receive and transmit the radio waves cellular phone send. The base station directs the call either to a "land line" phone line or relays the call to another cell site.

Employees in each area of the factory try to determine what they can do to ensure Six Sigma Quality. They track defects on a daily basis and "Pareto chart the problems," according to Cornell. Cross-functional teams (including representatives from manufacturing, engineering, finance, and purchasing) are encouraged to get together and work to solve quality problems that arise. "We get together each week to discuss those problems and figure out what our action plans are going to be to solve them," he said. Teams use process-mapping tools in analyzing problems and include all affected groups in formulating and implementing solutions. For example, teams will work with the purchasing group to ensure that any design changes are documented and communicated to suppliers. For new products, manufacturing works closely with the design group, so new products are easier to manufacture and more likely to be defect free.

Cornell, a five-year Motorolan, followed a somewhat nontraditional career path into manufacturing. With a bachelor's degree in finance from the University of Florida, he worked 11 years in his own business trading stock options and commodities at the Chicago Board of Trade. From there, Cornell landed a job at Motorola as a business analyst in the manufacturing group. While helping to coordinate the group's product-improvement efforts, the opportunity came along to be a manufacturing manager. Cornell applied for, and landed, the job!

Source: A conversation with Ed Cornell, Motorola, Inc.

The Purpose of Process Charts

Process charts make visible and accountable the appointment of people to machines, spaces and places. The charts make assignments of product orders to people, and of materials to machines and stock rooms. Only when charts exist can processing plans be evaluated and improved. Charts can assist intelligence and creativity but they are not a substitute for either of them.

Without the assistance of charts, assignments tend to be made in terms of managerial perceptions of existing situations and not in terms of planned-for conditions. The process flow designs are reactive to circumstances and not proactive to opportunities. Updating plans for processes grants a degree of control over circumstances that are dynamically changing. The charts are information support systems which provide a concrete means for analysis and improvement.

Orders that produce learning about a new process (routing, equipment, skills and training) can help the company develop competitive advantages. Embodying new technology in a process chart is equivalent to gaining competitive leverage.

Less dramatic, but of great importance to individual firms are discoveries that lead to the conversion of a batch-type process into an intermittent flow shop. Discoveries that permit a firm to set up jobs in minutes that used to take hours, and ways to bring new products to market in months instead of years are exciting opportunities for OM. All such major improvements stem from having a deep insight into the present processes which enable the invention of a much better process.

If it turns out that several managers have differing views as to which orders, customers and processes are important, the discrepancies should be addressed. There are methods for resolving the differences between multiple decision makers. One of the most used is a policy statement from the CEO or the president.

Distinguishing between orders, knowing which orders are more important than others, and why they are more important, is vital to achieving optimal process management. If, as is generally believed, important orders should go before less important orders, then agreement must exist as to what is important to the process managers and to other managers of the firm. Because they have different objectives, sales and process managers often do not see eye-to-eye on what is important. The systems viewpoint can help resolve this quandary.

A particular customer's order might be given top priority and assigned a preferred sequence by marketing. Process managers bristle at the notion that an order already in progress should be stopped and bumped. Major dislocations can arise that require two set ups for one order. Difficult clean ups can occur when a new order is pushed in front of other orders that are waiting. The sequence of set ups can increase costs greatly.

There usually will be agreement that orders that have large revenues are important. Often such orders will involve large unit volumes. Big orders are likely to be more important than orders that have small volumes. On the other hand, products that have high price tags and ample margins are likely to be more important than products that have low-profit value.

Total profit per order will not suffice as a measure of either order or customer importance. Major customers can place occasional orders for low-profit items. Most job shops associate the value of the customer with the importance of the order. The expected lifetime value of customers would be a fine measure of order importance, if it could be estimated. The investment in process analysis and improvement makes good sense when it can result in lower costs and better quality for valuable customers.

The Use of Process Charts for Improvements

Process charts can be used to achieve an optimal layout for the factory floor, warehouse or office. They permit measures of the effects of switching machines and people to alternative locations. A common objective is to minimize *total trans-*

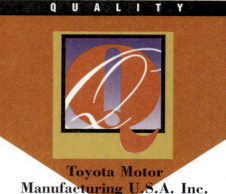

QUALITY

Toyota Motor Manufacturing U.S.A. Inc.

American workers in Kentucky adapt well to "kaizen," the Japanese practice that stresses continuous improvement.

When Toyota Motor Corporation announced plans for its first U.S. manufacturing facility in 1985, skeptics questioned how the company would adapt Toyota's production system to the American workforce. Today, Toyota Motor Manufacturing U.S.A. Inc. (TMM), located in Georgetown, Kentucky, is a thriving operation with 6,000 employees dedicated to teamwork and continuous improvement.

"Before I came here, I worried about operating this plant successfully because I didn't know of the American ways," said CEO Fujio Cho. Nevertheless, Cho formed an extremely effective work force, assimilating the Japanese work methods, by approaching the situation as a cultural learning experience. TMM's senior vice president Alex M. Warren Jr., said that: "One of the reasons we have

been as successful as we have to date, is [Cho's] willingness to do things the American way."

The TMM workers have adapted well to "kaizen," the name for the Japanese practice that stresses continuous improvement. Cho says the TMM team members are the "best work force I have experienced in my

25 years" of manufacturing. "Our team members come up with numerous bright ideas on their own. The employees are independent-minded, but they are also professional in their thinking. Of almost 40,000 ideas suggested by team members to boost production and cut costs in 1993, 38,000 were implemented."

Cho has set a high goal for TMM, which manufactures the Toyota Camry sedan and station wagon. "I hope this company becomes the number one auto plant in the U.S. in terms of productivity, quality, safety, and team members' morale," he said. "Even if it takes 10 years, I want to reach that goal of being the number one plant."

Source: Catherine Porter Prather, "Company of the Year: Remaining Positive and Leading Kentucky into the Future," *The Lane Report* (Lexington, Kentucky), January 1, 1994.

port—the sum of the number of units moved between adjacent workstations, multiplied by the transport distances—for frequently run jobs. If the weight or size is important, that can be factored in as total pounds or cubic feet moved. The right measure can determine when it would be sensible to alter the process by relocating workstations.

Shigeo Shingo, one of the most important founders of Japanese manufacturing methods, stated, "The prosperity of a resource-poor country like Japan rests solely on the high quality of its labor. Therefore, wasting labor on non-value-added work like transportation must be rejected. In fact, transportation should be seen as a grievous sin."[2]

In addition to transportation, there is a need to consider other important process functions and characteristics. These include: the work configuration and the technology of the process being used, the inspection methodology, quality procedures, and delays that occur as a result of the interaction between supply (process) and demand (orders).

Process charts bring up the possibility of replacing older, larger and slower machines with newer, smaller ones having faster processing rates. If the proper machines are chosen to be replaced, this could decrease the size of the waiting lines, also called queues or work in process (WIP). The WIP that forms between any two processing locations can be a profit drain. Decreasing queue length results in space saving which can be important for reasons other than just the freed-up space.

New Orders

Process flowcharts for job shops should be drawn up for new orders, parts that have not been made in the shop before. Job shops vary with respect to the percent of orders regularly repeated. Most experience a high level of new orders. For small new jobs the process charting is informal, even casual, whereas for large jobs greater care is taken in determining the routing and other particulars.

The custom shop does not experience the conflict of orders waiting for facilities and skills. There are only a few orders in the shop at any one time. Work is one-of-a-kind, so process repetitions are minimal. Therefore, process planning does not provide economic benefits. Further, reorders are nearly zero. This does not mean that it would not be helpful for a custom shop to develop process flowcharts. Such charts can help if the entrepreneurs of the custom shop wish to move toward a job shop, eventually processing large batches of work.

Many process flowcharts will be found on file in the job shop. This is a reflection of the fact that repeat orders are common, and the mainstay of the business consists of a catalog of parts that can be reordered. Most new work eventually joins the catalog of existing products.

New batch work interacts with the process flows of existing products. Scheduling conflicts between the newly received orders and older orders that already are on the shop floor must be avoided. Learning curves for new products include start-up problems with fixtures, jigs, software, and other tools that are used for the first time. Start-up problems are time drains. Pre-planning by use of process charts can help avoid blocking the use of equipment for which other jobs are waiting.

Overall, the use of process charts can improve operations in the job shop. They provide economic benefits by helping prioritize orders, eliminate conflicts, decrease waiting times, reduce traffic blockages, avoid bottlenecks, and minimize waste for repetitive activities.

6.3

PROCESS CHARTS WITH SEQUENCED ACTIVITY SYMBOLS

The six symbols explained in Table 6.1 are the American National Standard symbols for Process Charts (ANSI Y15.3M-1979) published by the American Society of Mechanical Engineers (ASME). They were reaffirmed in 1986 and withdrawn in 1994 by the ASME. Although first set over 50 years ago, these symbols are no longer regulated because of the growth in types of processes, the need to encourage flexibility, and the impact of computerization of process analysis and redesign.

TABLE 6.1

LEGEND FOR PROCESS CHART SYMBOLS

Symbol	Significance
⬤	An operation performed at a specific location
⬛	Inspection—a quantity check
◆	Inspection—a quality check
⬤	Transportation—the flow of materials, parts, components, among other things, from one point to another
▽	Uncontrolled storage—an item can be obtained without a withdrawal slip
▽	Controlled storage—a withdrawal slip is required to obtain an item

Figure 6.2 shows the "original" process chart for air conditioner chassis which move through a series of uncontrolled storages and transports until assembly. All of the symbols from Table 6.1 are displayed on each line of the chart. Only one symbol can be chosen for each line.

As shown by the process chart in Figure 6.2, the path used to transport air conditioner chassis from the press shop storage to the assembly area is long and circuitous. The distance that must be traveled between the press shop and the assembly building is about one football field in length. The chassis are moved to weekly storage, then to daily storage. From the daily storage they are moved to assembly every hour. This closely approximates just-in-time arrivals at assembly, but there are a lot of stops in between.

The process starts with uncontrolled storage and by means of transport moves to another storage, and so forth. A line connects the correct activity symbol for each row of the "original" process where "original" means that it is the existing process.

Various measures in the summary box describe the sequential list of activities along the connected pathway. There is the sum of the number of activities (eight). The total transport distance is 300 feet. In a manufacturing example, the length of time required for each unit of the order to go from the start of the process to its finish would be calculated and shown. This is the cycle time, C, which is an important process parameter.

FIGURE 6.2

"ORIGINAL" PROCESS CHART

Process Chart ___Original___ Analysis

Job Name ___Air Conditioner___   Date Charted ___8/24/95___
Part Name ___Chassis___   Part Number ___CH20___
Charted By ___PS___

							Total	Distance
Original	1	1	0	3	3	0	8	300′
Improved								
Difference								

Quantity	Distance	Process Symbols	Explanation
			Press shop storage
			Truck moves chassis*
			Weekly buffer
			Truck moves chassis**
			Daily bin
			Forklift truck***
			Assembled every hour
			Inspected every hour

*	4 times per week to the weekly buffer
**	once a day from weekly buffer to daily bin
***	hourly to assembly from the daily bin

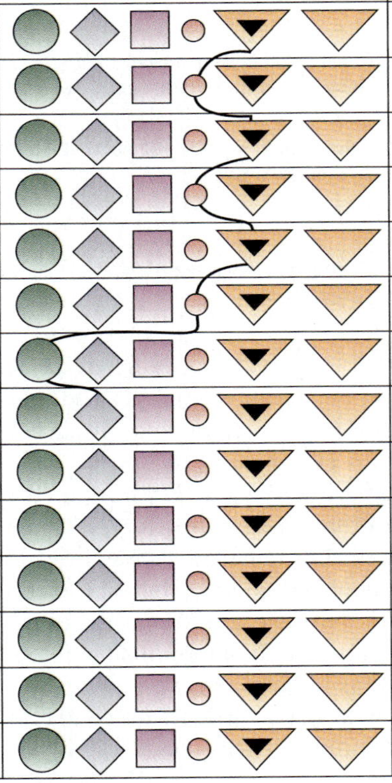

FIGURE 6.3

"PROPOSED" PROCESS CHART

Process Chart __Proposed__ Analysis

Job Name	__Air Conditioner__					Date Charted	__8/30/95__

Part Name __Chassis__   Part Number __CH20__

Charted By __PS__

								Total	Distance
Original	1	1	0	3	3	0	8	300'
Improved	1	1	0	1	1	0	4	80'
Difference	0	0	0	-2	-2	0	-4	-220'

Quantity	Distance	Process Symbols	Explanation
		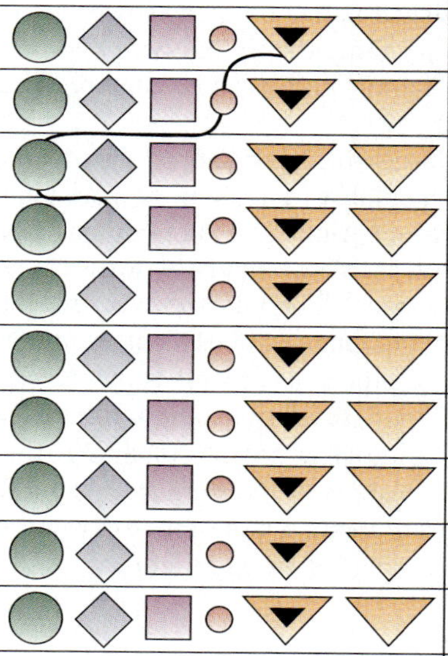	Press shop storage
	80 feet		Conveyor moves chassis*
			Assembled every hour
			Inspected every hour
*transfer rate to assembly is 1 chassis every 2 minutes			

The existing (original) process is analyzed and then, redesigned as an improved process. Figure 6.3 represents the "improved" process. Steps 2 through 6 are removed by eliminating both the weekly and the daily storage bank. Instead, a new step 2 is added, as shown in Figure 6.3.

This "proposed" process chart eliminates interim storage and moves the chassis directly to assembly. A heavy-duty, all-weather, covered conveyor is built between the two buildings to do this. A new opening is needed in the solid wall of the assembly building.

Press shop operations now can be synchronized to match assembly timing. The new distance that the conveyor will travel is about 80 feet. Transport distance is decreased by about 73 percent. The small trucks and forklift trucks are eliminated. Instead of eight activities there are only four. Space is freed up. With continuous improvement there may be even better numbers to report in the future.

Computer-Aided Process Planning (CAPP)

The ASME no longer sets the standard for the symbols. Process charting has become less regulated and more computerized. Many companies prefer to select symbols that suit their own purposes. There is increased use of process charts to get an overview which can be accomplished better by using fewer symbols. In Figure 6.1, the chart form is simpler, and only three of the six symbols in Table 6.1 are used to provide the kind of useful information which can be captured in computer form for computer-aided process planning.

For medium- and large-sized job shops, found in such industries as auto parts and paints, software is available to assist in process planning.[3] There also are factory simulations available from various software companies which examine the effects of using various process plans. These kinds of software will create process charts that permit analysis and redesign which take into account many processes moving simultaneously through the plant. Queues, bottlenecks and other interference can be altered by changing the timing of processes as well as by redesigning equipment assignments.

CAPP and factory floor simulations allow alternatives to be tested in pursuit of continual improvement, a goal of process planning. Benchmarking better processes, in other companies or divisions of the same company, should be brought into play if possible. Anheuser-Busch compares the process performance of all of its breweries on a regular basis. This is part of an ongoing process of continual improvement (CI).

Creating Process Charts

The various process charts should explicitly reflect everything that the process has to do with the cost of goods sold, cash flow, profit margins, productivity and value-adding transformations. This requires:[4]

1. Stating the exact materials to be used and in the precise amounts that will be consumed. Inventory records indicate purchasing sources. This step provides supply chain information which links the producer's process with its suppliers. New materials provide an opportunity to improve the process.
2. Noting the times required at every process step including storage and delays, transit, operations, quality control and inspections. Productivity measures are available from this step; bottlenecks can be identified and methods improved.
3. Indicating the explicit spatial movement patterns of each part as it is transported from place to place within the plant or the office. Plant layout deter-

mines the flow patterns. Traffic problems are related to layout. Transport time and cost are to be minimized subject to plant design constraints.

4. Specifying technological transformations that are the fundamental steps of the process. Materials are transformed, information is converted, people and cargo are transported. New technology offers an opportunity to improve the process.

5. Defining the conformance standards for acceptance or rejection of work in process and finished goods is part of process management. Inspection and quality control protocols are indicated on the process charts.

6. Being able to access process charts for each of the various part numbers and products that have been made in the shop. Frequently run products should be kept up-to-date and receive continuous improvement efforts. Less frequently run products are worth cataloging. A new order can be received that uses a process similar to one that was previously charted. This step advocates creating an encyclopedia of all the processes that have ever been done and that might need to be repeated. It reflects the extent of the product line. Process charts can be drawn up for products, parts and services that constitute potentials for future work. This can be an important systems link between operations and marketing.

7. Knowing how to assign or ship finished product that conforms to quality standards. This step provides supply chain data which links the producer's process with those of its customers.

8. Understanding that process flow charting is equally applicable for manufacturing and service applications. In the latter case, it can be used to systematize service processes. The symbols need to be changed to conform to the service. For example, computer programmers use their own special set of symbols to characterize information processes.

Reengineering Is Starting from Scratch in Redesigning Work Processes

Process redesign can take place in two ways. There is the method of continuous incremental improvements which the Japanese call "kaizen" and apply to quality enhancement.[5] An alternative is **reengineering**—starting with a clean sheet of paper to redesign the process.[6]

Reengineering
Starting from scratch in redesigning work processes.

Starting from scratch (zero-based planning) is not necessarily a requirement when new orders are received. Many organizations look for similarities with past orders for which process charts, tools and software, quality conformance standards, and experience with the process flows already exist.

There is a trade-off between zero-based planning and incremental improvement planning. Starting from scratch delays new order processing. It demands the use of the scarce time of process managers for planning instead of for turning out the product and shipping it to the customer. On the other hand, trying to adapt an old process for a new product usually creates suboptimal processes.

This is not always the case. If the degree of similarity between the old order and the new one is high, then there is an opportunity to use the existing process and improve upon it for the adaptation. Some analysis is needed before making that decision.

Although these are two different methods—continual improvement vs. starting from scratch—when properly done, they both result in a new process flowchart. The decision about which way is best depends upon management's willingness to use short-term instead of long-term goals.

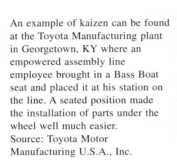

An example of kaizen can be found at the Toyota Manufacturing plant in Georgetown, KY where an empowered assembly line employee brought in a Bass Boat seat and placed it at his station on the line. A seated position made the installation of parts under the wheel well much easier.
Source: Toyota Motor Manufacturing U.S.A., Inc.

In some shops there is a knee-jerk reaction where orders for new products automatically raise the question, "What is most like this?" If there are strong similarities with prior work, the process charts from the close parallel are used to create the new process charts. Often, exact copies are made.

Why reinvent the wheel? This is a misleading rationale. The wheel is constantly being reinvented. Hard rubber tires replaced wooden wheels. R. W. Thomson invented the pneumatic tire. J. B. Dunlop developed the tire industry. Vulcanization (heating rubber with sulphur in a mold) gave rubber the required physical properties, but synthetic rubber tires were a later development. Radial-ply tires replaced the older bias-ply design with better wear and reduced blowouts. Invention of the wheel has not stopped, the design of new wheels is continuous. What did not need to be reinvented was the idea of a circular wheel used for cars, airplanes and trucks; what needed to be reinvented was how that wheel was to be made.

The idea of starting with a clean sheet of paper is appealing because it allows process managers to create improvements based on changes in technology. New people may have come on board with ideas for innovations that had occurred to no one previously. Different suppliers might contribute insights that had not been sought before.

Changes in materials can significantly alter processes. Some interesting substitutions have taken place in different industries. A quick sampling includes: steel in radial tires, vinyl in windshields, plastics in auto parts, aluminum for cans, paper for milk cartons and computer diskettes for paper.

Companies that are doing well have the time and money to take the longer view and go for major redesign. Companies under the gun do not feel that they have the necessary resources. Even if a firm must opt for the short-term approach, awareness of its options is better. After all, decisions based on instinct are better than decisions by default.

6.4

PROCESS LAYOUT CHARTS

The office or factory floor plan is often used to decide where to locate the computer, various machines, conveyors and storage. The floor plan shows the size and configuration of the work space and additional details such as the locations of doors, windows, elevators and closets. Three dimensional drawings are used when the height of ceilings is important.

"ORIGINAL" PROCESS FLOW LAYOUT

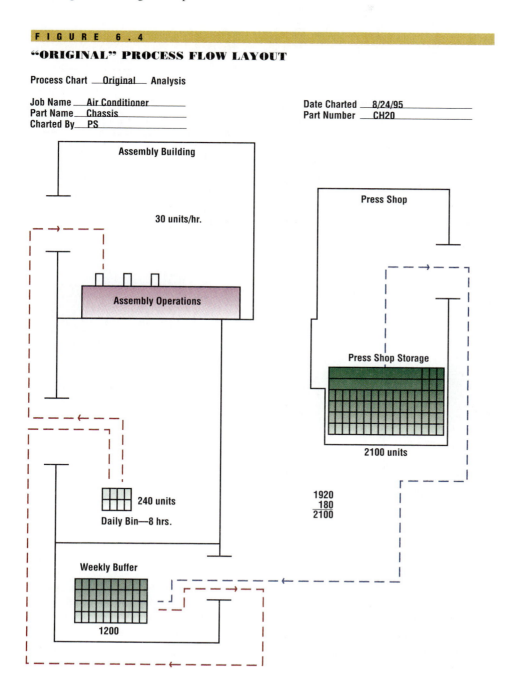

Process Chart __Original__ **Analysis**

Job Name __Air Conditioner__   **Date Charted** __8/24/95__
Part Name __Chassis__   **Part Number** __CH20__
Charted By __PS__

Assembly Building

30 units/hr.

Assembly Operations

Press Shop

Press Shop Storage

2100 units

240 units

Daily Bin—8 hrs.

1920
180
2100

Weekly Buffer

1200

"PROPOSED" PROCESS FLOW LAYOUT

Process Chart __Proposed__ Analysis

Job Name ___Air Conditioner___   Date Charted ___8/30/95___
Part Name___Chassis___   Part Number ___CH20___
Charted By___PS___

Among the reasons for detailing spatial dimensions are to:

1. calculate the distances that parts will be transported
2. choose the best means for transporting parts—by hand, truck or conveyor
3. be certain that there is clearance for large parts

4. understand the kind of space the process requires, including space for machines and people as well as tools and work-in-process storage space
5. understand the spatial aspects of what can be done to improve the process

Especially with regard to the largest repetitive batch jobs, it is usual to trace out the paths on the floor plan. It then is possible to improve the process layout. This ties the use of *process flow layout diagrams* together with the process charts for analysis already introduced in Figures 6.2 and 6.3.

Figure 6.4 is the physical description and layout of the "original" method for moving air conditioner chassis from press shop storage to the assembly building.

Figure 6.5 illustrates the new arrangement in the "proposed" process flow layout chart.

Compare Figures 6.4 and 6.5. The reduction in physical distance is immediately evident by comparing the process flow layout charts. This comparison illustrates how the layout diagrams can assist in making flow process improvements. The cost of the structural modifications to allow the conveyor's entry into the assembly building must be taken into account.

A wall has been broken through which yields direct conveyor access to the assembly function. The altered process flow is shown in Figure 6.5. As was known from the process chart analyses, transport distance is decreased markedly and space is freed up at the cost of structural change. The revised process is much closer to deliveries as needed or just-in-time systems.

6.5

ANALYSIS OF OPERATIONS

The process can be viewed as a total flow system. The activities that make up the flow are called *the operations*. Step by step, from start to finish, the term "operations" is generically employed to describe all operations on materials with machines in the factory and in the office.

The term operations often is applied to every kind of activity done in a process, but it is always applied to production transformations which are the circles in Figure 6.1. People who provide services to other people often are called "operators." Inspections for both quality and quantity are referred to as "operations." So is the action of storing raw materials and work in process in a warehouse or bin. Similarly, the term is used to describe the transportation of batches between workstations, cargo between marine terminals, and people between airports.

Operations can be decomposed into detailed elements such as searching for a part, reaching for it, grasping and turning it. This level of detail is generally reserved for discussion about time studies and human resources management. When changes are made to one or a few operations, then continuous improvement is the method being used. This is the operations level of process change as compared to the systems level.

Most process analyses start by looking at the operations level and noting that certain operations embedded in the processes are more likely than others to provide a basis for improvement. There also are typical transports and inspections that can be eliminated without losing delivery advantages and quality.

Improvements that are made to the individual operations of the flow process chart usually start with physical operations (the circles in Figure 6.1). Each of these operations can be analyzed separately for improvement. A new fixture might be designed which decreases cycle time or improves quality. A new cutting tool might be used to decrease machining time.

Value Analysis for Purchased Parts or Services

Value analysis (VA) is a systematic approach to investigating the functions served by purchased parts, their value to the final product, and whether they are being sourced at the best possible price and quality. It makes sense that high-dollar volume purchased parts receive attention, but often relatively unimportant parts shed light on great potentials for improvement. For value analysis, parts and services are equally applicable.

The function of a part was identified as fastening two parts together. A rivet was being used. VA asks its primary questions: is this rivet necessary, and what value does it add? After discussion, the answers were clear: the two parts being joined could be cast as one part; it would be stronger and less expensive to make. Thus, questioning the need for the inexpensive rivet led to an improved product at a lower cost.

Value analysis also queries if there are other sources for parts. It encourages reexamining make versus buy options. It examines the use of alternative materials that could provide equal or better properties at equal or lower costs. Too tight tolerances can be relaxed and process steps, such as an inspection, can be eliminated by making it right the first time. As part of the VA procedure, employees and suppliers are asked for their suggestions for improvement.

Value analysis has been used by many companies and governmental agencies to lower costs without impairing quality. A number of organizations stress the aspect of a methodical search for alternative materials to be used in place of the present materials. This is because technological change has been producing new materials in different industries. Awareness of applicability requires searching in places not normally accessed by designers and purchasing agents.

In the late 1940s, a large number of new materials began to appear such as special plastics, composites, ceramics and metal alloys. Value analysis was perfectly suited to the determination of whether the substitution of these new materials could lower costs and/or improve the performance of specific products.

There are many examples of the successful use of value analysis procedures. Performed under various names, VA is readily identified by the persistent use of a set of questions:

- What functions are served that justify the use of the present materials and what value do they have?
- What alternative materials might be used?
- What are the disadvantages of the alternative materials?
- What advantages (values) accrue by switching to an alternative?
- After reevaluation, are the present materials still the best ones to use?

When questions cannot be answered, appropriate assignments are made within the value analysis group or team. After researching the unanswered questions, the group meets again and goes through the same set of questions making comparisons between present and proposed materials. All qualities and costs are compared.

Value analysis is part of the systems approach, enabling cross-boundary queries in quest of superior materials and services. It is apparent that many factors play a part in material choices. To illustrate, the marketing advantages of dent-free plastic door panels played an important part in the design of materials for the Saturn. Crash tests for safety are an integral part of material choices for autos so the testing laboratories must be represented. Major savings in headlight design were made by Ford and in the front grille design by Buick. The U.S. Navy recorded many success stories in its earliest support for the use of VA on all kinds of Navy equipment.

Value analysis comparisons between present and proposed materials are facilitated by using process chart analyses. The analysis of materials often has side benefits of improving specific operations. A furniture manufacturer replaced some wood parts with plastic. This produced major cost savings and quality improvement in the woodworking process which traditionally does not consider plastic alternatives. Value analysis provided a nontraditional approach. In the same way, the elimination of paper trails in factories and offices, and the replacement with bar codes, optical scanners and computers has changed fundamental process procedures resulting in great improvements in quality, inventory and production scheduling.

Other Process Activities

A process chart can be used to compare the present method for a quantity check (a square in Figure 6.1) with a revised method of inspection. Inspection operations for quantity checks are often related to withdrawing materials from the storeroom, assuring that an order is properly filled, and paying on a piecework basis. The proposal might be to replace a person with a scanner. Alternatively, the proposal might be to postpone the quantity check until a later stage in the process.

An inspection that provides a quality check (a diamond in Figure 6.1) can be postponed or eliminated. The process chart provides both visual and quantitative data that summarizes the pros and cons of making the change. Recent best practice assigns responsibility for inspections to the worker making the part or providing the service. Quantity and quality checks can often be made together.

Improvements can be made to transport operations (a little circle in Figure 6.1). The recorded savings in Figure 6.3 were due to a creative idea about cutting unnecessary transportation. It is noteworthy that, in this case, the transport function was singled out of the overall process flow for attention.

Improvements can be made to storage operations. The air conditioner press shop to assembly example works well here too. Storage was eliminated in the proposed process flow layout charts (Figures 6.4 and 6.5). These were uncontrolled storages (a little dark triangle inside a bigger triangle—both with apex down—in Figure 6.1). This example also illustrates a case in which storage functions have been plucked out of the overall flow process to receive special attention.

With controlled storage (a triangle with apex down as shown in Figure 6.1), a withdrawal slip is required to obtain an item. The control feature is used to protect against pilferage and to insure that a known level of inventory exists. However, controlled storage is more expensive to administer than uncontrolled storage, where an item can be obtained without a stock request form. The loss of inventory control in the uncontrolled case can increase the out-of-stock, order and carrying costs. These additional costs must be compared with the savings obtained by removing administration of the storage.

6.6

INFORMATION PROCESSING

Figure 6.6 presents another application of a process flow layout diagram. It is the present floor plan schematic used for the information processing backroom of the Market Research Store. The pattern of movement that is shown is for the three most frequent kinds of market research reports the company sells. These are known as R_1, R_2, R_3.

There is much movement back and forth and there is crisscrossing of lines as sections of the research reports move out of initial process planning where work assignments are made. These are followed by data entry at dispersed terminals (*T*). Analysts are all grouped together. Sometimes additional data analysis is done at computer work stations (*W*). Approval must be obtained from various supervisors (*S*), who also are dispersed. Printers and copy machines (*C*) are scattered about the room. The assembly of final reports includes collating and binding. The manager oversees the entire process.

F I G U R E   6 . 6

"ORIGINAL" INFORMATION PROCESSING

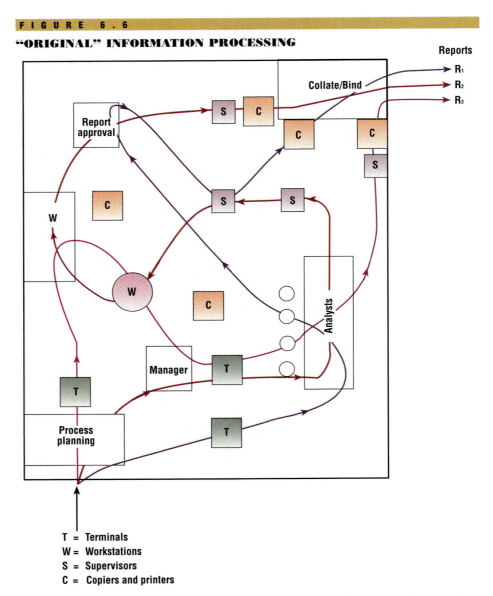

T = Terminals
W = Workstations
S = Supervisors
C = Copiers and printers

There are too many transfers of jobs resulting in more lines drawn on the floor plan than necessary. The present data transfer process appears to be chaotic and confused. Information should be handled once and not repeatedly. That is a standard axiom for information systems.

Figure 6.6 can have other jobs overlaid. Because the various orders will have different flow patterns, the number of lines will further increase. Simplification of the information transfer process for the Market Research Store can be achieved by standardizing some of the functions on the flow process layout diagram. Process improvement calls for reducing the complexity of the transaction paths; Figure 6.7 presents an abstract schematic diagram for a plan which appears to do just that.

FIGURE 6.7

PROPOSED PROCESS FLOW LAYOUT FOR THE MARKET RESEARCH STORE INFORMATION PROCESSING

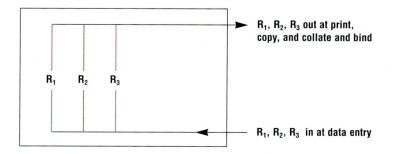

The amount of floor space is the same as in Figure 6.6, but the layout is different. In the proposed version, data entry, common to all reports, is done on a flow shop line as the first function in the shop. The data entry function will be engineered to be as efficient and cost effective as possible with data checking being computerized and automated. Analysis of each report is unique and done separately (as indicated in Figure 6.7 by the vertical lines marked R_1, R_2, R_3). At the final stages, all reports join a flow shop for printing, copying, collating and binding.

The frenetic process flow pattern has been simplified by removing the supervisors' desks and by making each individual responsible for checking their own work at each stage of the process. This organizational procedure increasingly is being used with success for improving the output quality of processes.

Companies everywhere have begun to employ this approach, disseminating responsibility and accountability to the people who are doing the work. Such empowerment removes bosses, chiefs, supervisors and inspectors from the approval process. This smooths process flows by removing interruptions. It allows problems to be spotted before they would otherwise be found and puts the responsibility for correction in the hands of those who can make the corrections.

The information processing backroom of the Market Research Store has installed flow shop procedures for functions that are identical, no matter what the specific nature of the report. By using universal bar codes and optical scanners, data entry might eventually be fed automatically into the correct data bank for analysts' use. Assembling reports, which takes place at the end of the process, also is controlled by bar codes. This is similar to auto assembly which uses bar codes to route parts so that all of the right components come together for a specific order.

Other ideas that could be tried for the modified process include the development of networking ability so that each person can transmit screens of data to every other person's screen in the room. This technology facilitates rapid sharing of information and decisions. Data entry can be accomplished by part-time employees working at home. Analysts also can work at home. Materials completed for the reports can be transferred from laptop and desktop PCs located anywhere to the "information factory" in the operations backroom of the Market Research Store. The backroom prints, assembles and distributes the reports. Major process and technology improvements of these sorts can give the MRS a succession of competitive advantages.

6.7

PROCESS CHANGEOVERS (SET UPS)

Changeovers

Needed steps to prepare equipment and people to do a new job; special form of operations.

Changeovers are the steps that need to be taken to prepare equipment and people to do a new job. The term **set up** is usually applied to the start-up of a new job, which means the cleanup from the old one. For process analysis, it is best to separate these two steps and deal with set up and **put-away** (cleanup) as two activities.

Set up

Applied to the start-up of a new job, which means the cleanup from the old one.

Of critical importance to job shop processes are the changeover costs and times. These are often treated as being synonymous so that mention of cost implies time and vice versa. Thus, the generic term, "set up," is often used to include time and cost and everything that has to be done to change the process from one product to another.

Put-away

In process analysis, the cleanup activity follows set up and production.

Set-up costs are often proportional to set-up times. However, the relationship breaks down when a lot of technology is devoted to allowing very rapid set ups. Then there is a cost of not using this technology for the purposes for which it was intended, namely, short runs of many designs that fit within the family of parts (technology group) that can be made on this equipment. Further, when the set up can take place offline, it can take longer and still cost much less than when it must interrupt the production process.

Machines have to undergo cleanup from prior jobs and then be reset. People have to shift jobs, often moving from one location to another. The learning curve comes into play every time operators bring a new order online, which also is part of the changeover process.

Changeovers are a special form of operations. They occur because shifting gears to convert the process to deal with new orders requires taking the gear box apart. Imagine the impact on driving if this were required to shift the gears of a car. Further, the analogy requires that many different cars are involved. Thus, a manual for changing over the gear boxes is needed for hundreds or even thousands of different car models. Each changeover is equivalent to a new order in the shop.

Leaving the analogy and returning to the job shop environment, envision this script: a "good" customer orders a few units of an item which takes a couple of minutes to make, but the set-up time is half a day. Here are numbers to consider:

Cost per unit　　=　　$0.80

Set-up cost　　　=　　$800

Order size　　　 =　　4 units

Lot size　　　　 =　　? units

To be realistic, assume that the set-up cost includes the cost of labor of two people working four hours each and the cost of lost production related to four hours of downtime of two machines.

This item costs less than a dollar if the set-up cost is ignored but if the set-up cost is amortized over the four items, they each cost $200.80. One additional assumption is that this item is reordered, from time to time, by various customers. What options are open to the process manager as he or she goes about preparing the production schedule?

An example of SMED set ups can be found in the activities of the pit crew which changes a racing car's tires without shutting down the engine.
Source: Pennzoil Products Company

The most apparent option is to run a larger number of items than four. If forty items are produced, then the cost per item is reduced to $20.80 which is still far from providing a reason to celebrate. Running a lot size of 400 provides a cost per unit of $2.80 which might be reasonable from the customer's point of view.

Single-Minute Exchange of Dies (SMED)

Another option is the one that Taiichi Ohno of Toyota Motor Manufacturing U.S.A. Inc. took. He asked for, and got, a revolutionary reduction of set-up times.

Before describing the character of that revolution, it is interesting to note the historical events that triggered the modification. Taiichi Ohno asked Shigeo Shingo, who was at that time with the JMA,[7] to cut set-up times for a particular operation from four hours to three minutes. Shingo, who was an advocate of cutting all set-up times to less than ten minutes, was shocked by Ohno's "utterly unreasonable directive"[8] into totally rethinking the anatomy of changeovers.

"In developing the idea of the *single-minute exchange of die* (SMED) setup, I blasted a tunnel through the mountain, so to speak, dramatically shortening the time it took to move between processes or operations. Cutting four-hour set-up changeovers to three minutes simply blasted out of existence the band-aid approach of economic lot sizes," Shingo wrote.[9]

The SMED system for cutting changeover times is based on the recognition that there are set-up steps which can be done without shutting the machine down. Examples of SMED set ups can be found in the activities of the pit crew which changes the racing car's tires without shutting the car off. Changeover time in the pit is part of the racing team's performance.

Watching set-up teams do their work, it will be seen that some steps could have been done before turning off the machine and starting the set up. Also, by redesigning set-up procedures, it is possible to increase the number of such steps.

With this in mind, here are four steps that lead to cost (and time) reductions for set ups:

1. Recognize those set-up tasks that can be done when the machine is running; tasks can be done before or after the machine is shut down. These are called external set-up elements. External set-up elements can be done while the machine is working on another order, or while it is running the new order.
2. Study those set-up tasks that can only be done when the machine is shut down. These are called internal set-up elements.
3. Redesign and convert as many internal set-up tasks as possible to be moved to the external category.
4. Every set-up operation's element should be scrupulously analyzed and examined in search of continuous improvement.

It used to be that set-up times were larger for machines that worked faster. The trade-off for paying more for setting up was the lower per unit costs of the outputs. While this relationship still generally holds, the application of SMED concepts reduces the downtime of high-volume equipment by doing much of the set up while the machine is working on another order or the new order.

Was Shingo's insight a special moment in history or inevitable? The same question can be asked about Henry Ford's development of sequenced assembly and Eli Whitney's invention of interchangeable parts. These are all major manufacturing inventions with lots of consequences for operations in general.

Perhaps the "time was right." Alternatively, maybe eventually the inevitable would have occurred. Such speculation does not diminish the contribution nor the credit that these inventors deserve.

Further, flexible manufacturing systems can provide rapid built-in changeovers for a specific menu of production items. The SMED concept is substantially assisted by the development of programmable set ups which are enabled by computer-driven mechanical equipment. Robots are a type of programmable set-up system that is gaining use globally.

Robots are ideal for rapid-changeover systems. The sound engineering concepts of SMED combine with the programmable capabilities of robotics to bring about a revolution in the engineering mind-set. The old engineering truism that changeover times and costs must increase with the use of volume-efficient equipment is disappearing.

The fact that SMED is having such a major impact on manufacturing, makes it easy to overlook the fact that SMED concepts are not limited to manufacturing. The ability to move from one order to another with minimum changeover delay is applicable to many service systems. The restaurant business is filled with changeover costs. Cooking is set-up intensive and dishwashing exemplifies the takedowns and cleanups.

In education, consider the set-up times required for each class. Every exam requires someone to write it. Many students must take it, at least enough to pay for the set-up costs. Ask a professor about the agony of writing a make-up exam for one or two students who failed to take the scheduled exam. Someone must grade that exam. In travel, think about the set-up times to get a cruise ship underway. Finally, consider the question: are there set-up times required to operate a computer? Because the set-up times can be substantial, every business that relies on computers can gain advantages by decreasing set-up times.

IBM's FlowMark —
Workflow Management Technology

As business processes become more complex, planning and managing all the activities and resources involved presents an enormous challenge for companies. In recognizing this challenge, the International Business Machines Corporation (IBM) developed FlowMark®, a workflow-management technology for businesses interested in improving their workflow processes. FlowMark is the workflow function of IBM WorkGroup, a family of integrated customizable, object-oriented software that provides workgroups with capabilities ranging from work and information technology to group communications. The workflow function operates on a business's local area network and provides the capabilities to model, document, test, and manage processes. It supports a variety of operating systems, including IBM's OS/2, Microsoft Windows, and AIX clients and servers.

The workflow function consists of three components: (1) *buildtime* software used to define how workflow applications will route data among users; (2) *runtime* software that follows workflow procedures through "to do" lists; and (3) a *server* that coordinates buildtime, runtime, and the database. To-do lists become online work lists and procedures become process models. In executing a process model, the application reminds users of activities that need to take place and links them automatically with the program the work requires.

"Workflow management streamlines critical business processes, providing employees and work groups with the proper information and work status to make the right decision quickly and effectively," said Richard Sullivan, IBM WorkGroup Marketing, Software Solutions Division.

An international venture, the workflow function was developed by the IBM Software Development Laboratories in Vienna, Austria and Boeblingen, Germany. In addition to English, FlowMark will be available in Japanese, Dutch, French, German, Italian, and Spanish.

Sources: "IBM Announces Object-Oriented Workflow Management System; Flowmark Designed to Streamline Business Processes," Business Wire, September 21, 1993, and information on IBM WorkGroup from Jennifer R. Surro of Brodeur & Partners, Inc., Waltham, MA.

In summary, set-up times are vital process parameters. The reduction of changeover times can dramatically improve the competitive advantages of a company. Consequently, this is one of the most promising methods for process improvement in the present operations environment. As set-up times are reduced to minutes and then seconds, the *economies of scope* come into being. This means that variety can be achieved without a penalty. With increased variety, marketing, operations and distribution have to coordinate their activities. A real systems effort is called for because increased variety adds more opportunity and complexity to challenge management.

6.8

MACHINE-OPERATOR CHARTS: PROCESS SYNCHRONIZATION

Machine-Operator Chart (MOC)

Provides a means for visualizing the way in which workers can tend machines over time.

Another process chart that deserves attention is the **Machine-Operator Chart (MOC).** The MOC provides a means for visualizing the way in which workers can tend machines over time. It encourages visual analysis of the interaction of operators with the machines they operate. It permits evaluation of the efficiency of the relationship. The MOC permits intelligent assignments of operators to machines, and allows for synchronization of operations (and operators) because time-matching is inherent in the charting.

The time that operators spend tending machines often includes set-up times which interrupt production. Product output is stopped while tool alignments are made and fixtures and jigs are adjusted. Replacing the paper for the fax or the inkjet cartridge for the printer are other illustrations of machine-operator interactions that stop production. The cleanup activity is such a major fact of life for many processes that there is irony in the fact that set up has come to mean the whole changeover process including cleanup.

Figure 6.8 shows a Machine-Operator Chart in which all of the make readys and put-aways are designed to be internal set-up elements. The machine and the operator are both idle 50 percent of the time. In sixteen time units the operator is idle 4 + 4 = 8, and the machine is idle 2 + 4 + 2 = 8. Operator and machine are badly coordinated.

FIGURE 6.8

MACHINE-OPERATOR CHART—ORIGINAL PROCESS

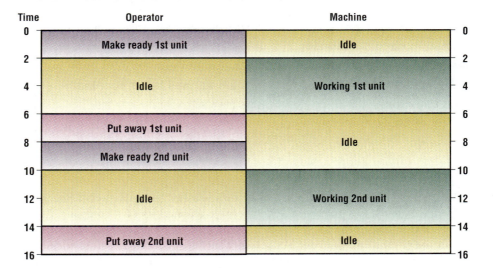

As an example, a part to be painted is mounted in the painting cabinet while the paint nozzles are shut off. The spray-paint nozzles are activated and then shut off so the part can be removed and a new one mounted.

In Figure 6.9, process designers have found a way to mount the next part while the prior one is being removed. The operator makes ready the next part to be painted while the prior part is being painted. The one exception is the first unit which is loaded before the start-up of the paint cabinet.

FIGURE 6 . 9

FIGURE 6 . 9

MACHINE-OPERATOR CHART—IMPROVED PROCESS

Time	Operator	Machine	Time
0	Make ready 1st unit	Idle	0
2	Idle		2
4	Make ready 2nd unit	Working 1st unit	4
6	Put away 1st unit	Idle	6
8	Idle		8
10	Make ready 3rd unit	Working 2nd unit	10
12	Put away 2nd unit	Idle	12
14	Idle		14
16	Make ready 4th unit	Working 3rd unit	16
18	Put away 3rd unit	Idle	18
20			20

FIGURE 6 . 10

MACHINE-OPERATOR CHART—IDEAL PROCESS

Time	Operator	Machine	Time
0	Make ready 1st unit	Idle	0
2	Make ready 2nd unit		2
4	Idle	Working 1st unit	4
6	Put away 1st unit		6
8	Make ready 3rd unit	Working 2nd unit	8
10	Put away 2nd unit		10
12	Make ready 4th unit	Working 3rd unit	12
14	Put away 3rd unit		14
16	Make ready 5th unit	Working 4th unit	16
18	Put away 4th unit		18
20	Make ready 6th unit	Working 5th unit	20
22			22

The operator has externalized the make ready element but is still constrained by the internal set-up characteristics of the put-aways, which means that the machine must be turned off to remove the part. This partial SMED-like improvement has decreased the operator's idle time from 50 percent to 33.33 percent. The same decrease (50 percent to 33.33 percent) applies to the machine's idle time. This ignores the two periods of idle time that are required for the first part to be painted. Continuing with the spray paint example, the part to be painted is loaded onto a fixture while another part is being painted. When the spray unit is shut off to remove the part, the fixture with the next part drops into place in the paint cabinet.

Figure 6.10 depicts the successful application of SMED. Both *make ready* and *put-away* have been converted into external set-up capabilities. Once the system is up-and-running, both the operator and the machine are 100 percent utilized.

A continuous circular conveyor enters the painting booth at the front. The part is spray painted as it moves along leaving the spray booth from the rear. The conveyor circles back to the station outside the paint booth where the operator loads parts onto the conveyor and removes painted parts from it.

6.9

TOTAL PRODUCTIVE MAINTENANCE (TPM)

Total productive maintenance (TPM)

Systematic program to prevent process breakdowns, failures and stoppages; strategies include preventive, remedial, and partial preventive maintenance (PPM).

Total productive maintenance (TPM) is a systematic program to prevent process breakdowns, failures and stoppages. Process disruptions can be caused by many factors such as poor materials, tool breakage, power failures, and absenteeism.

One important cause of disruption is machine breakdowns, including computer failures. Machine breakdowns can stop a process with large penalties to be paid. TPM is a systems approach to preventing such failures. TPM encompasses preventive maintenance strategies which service equipment at regularly scheduled intervals. Remedial maintenance (RM) is required to repair equipment when failure occurs. Partial preventive maintenance (PPM), discussed below, is incomplete servicing scheduled at regular intervals to decrease the probability of failure and the need for remedial maintenance.

Total productive maintenance is a program which addresses all factors that slow down or stop a productive process. If the computer goes down as the bank is about to transfer substantial funds overseas at the end of the day, the interest that would not be earned might well be in the millions. Because of the risk exposure, backup systems would be part of the failure prevention system. This is because computer failures can be costly. TPM programs are absolutely essential when process stoppage is life threatening. Backing up equipment and using redundant systems do not constitute preventive maintenance, but they are TPM strategies.

There are two questions that can help determine how large an expenditure for TPM is reasonable:

1. How vulnerable is the specific process to a breakdown that will slow or stop production output?
2. How costly is the production disruption that will occur?

Job shop machine breakdowns are serious, but they can be finessed in several ways. First, the equipment is general purpose and, therefore, similar machines are available on the same plant floor. Second, the order being processed at the time of breakdown is one of many orders that are being worked on in the shop, so the extent of the disruption is limited.

Nevertheless, a worst case scenario can occur where the order is already delayed, the customer is important and annoyed, and the breakdown occurs on the only machine that is well-suited for the particular order. An even worst case, discussed below, is stoppage of the flow shop.

When the costs of disruption are extreme, including life threatening, the vulnerability is often addressed by redundancy. That is, two pieces of equipment are used, one for the process and the other as backup. This is a TPM strategy calling for the redesign of the system. The cost of redundancy can be assessed as an alternative to the cost of failure. Used widely by banks, the design of "nonstop computers" (such as those of Tandem and Stratus) is based on duplication of functions, backup of components and redundancy of circuits. Commercial jets can fly on one engine, but two are provided.

It should be noted that **redesign** can mean changing the process methods, that is, the systems being used. It can mean changing the process technology, including substituting other equipment and/or by using redundant equipment for fail-safe performance.

Consider a ceramic fixture used to hold a unit of work in a heat-treating chamber. Suppose that this fixture shatters every now and then. Because the fixture cannot be replaced with something more durable, and because it is not too costly, the redundancy level that the process manager suggests is six fixtures.

Why did the process manager select six instead of two or three? The answer lies with other factors that determine redundancy levels, including the lead time (LT) or amount of time needed to secure a replacement. Lead time starts with the placement of an order and ends with the receipt of that order. The minimum order size might be six. In this case, assume that the LT is three to four weeks and the supplier is reluctant to produce only one fixture at a time since the set-up costs are too high to run one at a time.

For many machines, redesign is costly, redundancy is affordable, and the lead time is short. In these cases, the goals of TPM are achievable by keeping one or more backups in stock and keeping contact with the supplier to make certain that lead times and availability have not changed.

Redesign

Changing the process methods/systems being used; needed when cost of failure is high, the probability of breakdown is great, and it cannot be lowered by preventive maintenance or better practice.

Total productive maintenance (TPM) can set goals for zero machine and equipment breakdowns. Air traffic control equipment must be serviced with a zero-breakdown policy to ensure safe landings.
Source: Hughes Aircraft Company

Redesign is used when the cost of failure is high, the probability of a breakdown is too great, and it cannot be lowered by preventive maintenance or better practice. That is another example of total productive maintenance having a broader agenda than even the most dedicated preventive maintenance program.

It is useful to note that the cost of failure (c_f) and the probability of it occurring (p_f) lead to an expected cost of failure

$$\bar{c}_f = (p_f)(c_f)$$

Redesign is essential when this number is large. It can be a large number, even though p_f is small, if c_f is large. In this regard, it is often stated that $c_f = \infty$ when the cost of life is involved. The expected cost of failure is infinite in this case, and redesign is warranted. Similar concepts and terminology are used to determine optimal levels of preventive maintenance when the expected costs of failure are not excessive.

Total productive maintenance (TPM) is a complete plan to prevent failures of various kinds. It is an across-the-board attack on machine breakdowns. It includes shifting to new technology, which is more reliable. It employs statistical quality control methodology and, based on results, leads to the redesign of systems with planned redundancy.

When these are not required, remedial and preventive maintenance are the remaining options. There are many different levels of preventive precautions that can be taken. The alternative is to wait for a failure and then use remedial maintenance (RM) to repair the problem. The zero-level of preventive maintenance is RM.

TPM addresses two kinds of failures:

The *first* and worst kind of failure stops the process. Remedial maintenance (which is repair) is called for immediately. It is considered a critical failure.

The *second* kind of failure leads to degradation of the system's performance. Repairs are made, but this kind of problem is not critical and tends to reoccur. The total cost of remedial maintenance becomes a large enough figure to command attention.

The cost-effective scenario for each kind of failure is different. In both cases, however, it often turns out that some level of preventive maintenance is warranted.

There is willingness to spend a great deal to prevent a failure if the cost of that particular kind of failure is high. Analysis of alternative strategies for failure prevention include the degree to which failure probabilities can be reduced and the cost of achieving that reduction versus the penalty cost of that failure. The conditions for a relevant trade-off analysis exist.

Figure 6.11 illustrates the relationship between the probability of failure (p_f) and the cost of prevention (c_p). It is seen that as partial preventive maintenance (PPM) moves from remedial maintenance (no preventive maintenance) toward maximum TPM, the probability of failure decreases quickly at first, and then slows down, even though proportionately greater amounts of money are being spent.

A power failure in a plant or office constitutes a breakdown of the first kind. Without the backup of a redundant power supply, the ongoing systems will be forced to shut down. To avoid power failure, the processes of the power plant are examined in detail. The intent is to pinpoint the causes and the pre-conditions of power failures. The techniques that apply are identical to those used for prevention of defective products with TQM and when applying statistical methods of process analysis.

Every potential cause is studied. A maintenance agenda is set up to remedy problems and reduce the probability of failure. Aircraft maintenance programs have been worked out with care and provide an excellent model to copy.

The second type of failure is epitomized by light bulbs burning out. This kind of failure is seldom critical. There are usually several bulbs to light the space and as soon as one burn out occurs, remedial maintenance takes over.

A plan can be developed for decreasing the number of light bulbs that need to be replaced at one time, which costs a great deal because each replacement requires a new set up for correction of the fault. The use of one set up to replace all bulbs at the same time, whether or not they have burned out, is worthy of consideration.

One set up to replace all bulbs, burnt out or not, is a good example of the importance of set-up theory for process management. The example also under-scores the fact that TPM is part of process analysis. Should light bulbs be replaced as they burn out or should all bulbs be replaced at a proper interval?

It is generally less expensive to use preventive maintenance than corrective main-tenance. A trade-off model should be employed as discussed in the following sections.

FIGURE   6 . 1 1

PROBABILITY OF FAILURE AS A FUNCTION OF COST OF PREVENTION

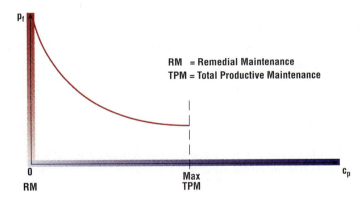

RM = Remedial Maintenance
TPM = Total Productive Maintenance

The Preventive Maintenance Strategy

There are 1000 light bulbs at risk for failure with rates described below. The TPM policy is to replace, all at once, the 1000 light bulbs 6.5 times a year (every eight weeks). Each replacement entails a cost of $5.00 per bulb and a set-up cost of $1,000. For this TPM policy, the total cost would be 6.5 (1,000 × $5 + $1,000) = $39,000.

Burned out bulbs are only replaced during the TPM sweep. This could mean that workplace light is not always sufficient. To avoid having insufficient light there might be a basis for having some extra bulbs installed. The cost of such redundancy is not included in this sample analysis, but it readily could be added.

The Remedial Maintenance Strategy

The alternative strategy is to replace each bulb as it burns out. That costs $10 per light bulb. The bulbs are rated as having an expected lifetime of 900 hours. The plant operates two eight hour shifts seven days per week for a total of 112 hours per week.

This means that on average, every eight weeks all of the light bulbs will be replaced. Thus, each light bulb must be replaced 6.5 times per year (8 × 6.5 = 52 weeks). On average, 6,500 bulbs burn out each year and that means the total cost of replacement on a one-by-one basis is $65,000. For this situation, since $65,000 − $39,000 = $26,000, the preventive maintenance strategy has a distinct advantage over remedial maintenance.

Zero Breakdowns

Just like quality programs designed to achieve zero defectives, TPM can set down goals for zero machine and equipment breakdowns. A breakdown, by definition, is a malfunction. The failure was supposed to be headed off by the preventive maintenance program. A policy of **zero breakdowns** specifies that everything possible will be done to eliminate malfunctions and failures. Air Force One, the President's plane, is serviced with a zero breakdown policy.

Zero breakdowns
Specifies everything possible will be done to eliminate malfunctions and failures.

When a failure occurs, it compromises the process and further analysis is warranted. This includes the possibility that shorter intervals between servicing could head off unscheduled repairs. The extra cost of replacing parts more frequently also can be examined.

Cost trade-off models are easy to construct if the cost of an unscheduled repair can be calculated. One factor which needs to be included is the cost of process disruption. Usually this is a large cost, which explains why aggressive total productive maintenance (TPM) programs are viewed with favor by companies that try them out.

TPM for the Flow Shop

For the flow shop process, the worst case scenario is when the breakdown stops the line. This occurs when a station along the path of the conveyor ceases being able to function. Unless the prior stations on the line are immediately notified to stop producing, WIP begins to accumulate forming queues in front of the nonfunctioning machine. To avoid severe congestion and eliminate safety hazards, it usually is necessary to shut down the prior workstations.

This scenario has many costs associated with it. There is the cost of repairing the equipment (remedial maintenance) at the station. The cost of lost production can be considerable. Another cost that often is overlooked is the cost of line start-up which might require relearning and the production of defectives.

Another kind of flow shop breakdown is when all the equipment stops because of a power failure. The same kind of effect is produced by a failure of the paced conveyor unless the work is suitable for hand carrying or moving by forklift between stations. In addition, some automobile companies empower workers to stop the line when serious problems develop with systems-wide implications. The kind of problems that are systems wide include passing defectives to downstream stations which wastes their work, complicates repair, and hides the defect.

The severe costs of flow shop stoppages generally guide the choice of an effective complete program of preventive maintenance. This is done after the shift ends or on a weekend when there will be no interference with the maintenance team. These maintenance procedures require similar considerations regarding what is done online and offline as those which applied to SMED changeover and set-up situations. The production line is gone over thoroughly, cleaned up, and prepared to run without further maintenance until the next scheduled overhaul. Aircraft receive major preventive maintenance at longer intervals than the more frequently scheduled minor preventive maintenance. The same applies to automobile servicing.

Equipment failure always is an important consideration for process designers. Serialized production systems that are prone to breakdowns cannot be tolerated. If redundancies are not feasible for protection, then it is essential to use TPM to bring the process as near to zero-breakdown levels as possible.

SUMMARY

How to analyze and redesign existing processes is the subject of this chapter. It provides information about how to create process charts and process flow layout charts. Process charts trace the present sequence of activity symbols. Process flow layout charts trace the paths of parts on the floor plan. Each can be analyzed leading to process improvements. Then each can show the proposed process side-by-side with the present one. The growing impact of computer-aided process planning is discussed. The use of reengineering, process planning with a clean sheet of paper, is contrasted with continuous improvement. An example of information process analysis is given.

The chapter also includes discussion of SMED (single-minute exchange of dies) methods for decreasing the costs and times required to change over from one process to another. This has become increasingly important as companies strive to offer their customers greater product variety. Machine-Operator Charts allow synchronization of operator set ups with machines. When set-up constraints are reduced, major improvements can be made in productivity. Total productive maintenance (TPM) is explained with respect to preventive maintenance strategies and remedial maintenance strategies. All of these topics add up to a strong competitive ability with respect to the analysis and redesign of processes.

▶ Key Terms

changeovers *230*	Machine-Operator Chart (MOC) *234*	process analysis *211*
process charts *209*	put-away *230*	redesign *237*
reengineering *221*	set up *230*	time-based management (TBM) *211*
total productive maintenance (TPM) *236*	value analysis (VA) *226*	zero breakdowns *240*

▶ Review Questions

1. What are the conventional symbols used for process charting? Are there any other activities that make up processes which are not covered by the conventional symbols?

2. What are the distinctions between process charts and process flow layout charts? When should one or the other be chosen?

3. How does process charting apply to job shops? How does it apply to intermittent (short-run) flow shops? How does it apply to long-term, continuous flow shops?

4. What is CAPP and why is it important?

5. How do process charts show the receipt of materials from suppliers as part of the process? How do they show shipments of finished goods to distributors?

6. Why should process charts be drawn up by process managers who are responsible for everything that happens on the production line?

7. Explain the following statement:

    It is unfortunate if process managers are so busy taking care of existing customers that they do not feel they have time to plan for new jobs. Time should be allocated to setting up the systems to routinize the function of developing and improving process charts.

8. Is it sensible and justified that "prized customers" get more time and attention for process charting than others?

9. What are the purposes served by Machine-Operator Charts?

10. How does process analysis aid in the discovery of new ideas?

11. What are the set-up times required to operate a computer? Describe them and address the issue of reducing them.

12. TPM programs are absolutely essential when process stoppage is life threatening. Give some examples and explain.

▶ **Problem Section**

Problems Concerned with Process Analysis and Charting

> *The idea is to create process flowcharts that are serious attempts to capture the actual process that is taking place. Some environments in which this can be done include service-oriented, fast-food chains. Note: If possible, visit a fast-food restaurant at off-peak hours and talk to the manager about this assignment. Explain that it requires your understanding the process and how it is managed.*

1. Analyze the way burgers are cooked and served at McDonalds. Then draw a process chart for the steps that are required. Draw a process flow layout chart for the McDonalds being studied.

2. Analyze the way french fries are made and served at Burger King. Then draw a process chart for the steps that are required. Draw a process flow layout chart for the Burger King being studied.

3. Analyze the way chili is made and served at Wendy's. Then, draw a process chart for the steps required to make and serve it. Draw a process flow layout chart for the Wendy's restaurant being studied.

   > *There are many processes with which most people have some experience, such as changing a tire (a service) or baking a cake (a product). These are presented below with some other processes that can be charted.*

4. Analyze the servicing process that takes place at a gasoline station. Then draw a process chart that compares self-service at the gasoline station with full-service for a fill-up and oil check. Draw the process flow layout chart for the station.

5. Analyze the service process of making travel arrangements. Then draw a process chart for a travel agent setting up the itineraries and reservations for a business trip to be taken by two company executives who are traveling by air from Chicago to Los Angeles. Draw a process flow layout chart for the agent and the agency.

6. Analyze the service process for putting a spare tire onto a car after having a flat. Then draw a process chart appropriate for changing the tire. Draw a process flow layout chart that shows where all the materials are kept and how they are moved around.

7. Analyze the process for baking a cake. Then, draw a process chart for baking a cake. Draw a process flow layout chart as well.

Problems Concerned with Creating a Production Process

> *Before it is possible to create a production process, it is necessary to know how the product, such as cheese or concrete, is made. Similarly, for services, it is necessary to know what steps are required to deliver the service, such as servicing copy machines or writing a research report. There are many sources of information about how to make and do things.[10]*

8. Research the process for making cheese. Create a process chart for producing wheels of cheese in volume as a business venture. What work configuration (job shop or flow shop) do you recommend?

9. Research the process for making concrete. Create a process chart for mixing cement to make concrete in volume as a business venture. What work configuration (job shop or flow shop) do you recommend?

10. Research how to make a jar of jam. Create a process chart for producing jars of jam in volume as a business venture. What work configuration (job shop or flow shop) do you recommend?

11. Research the process for making vinegar. Create a process chart for producing 16- and 32-oz. bottles of vinegar to be sold in supermarkets. What work configuration (job shop or flow shop) do you recommend?

12. Research how to service copy machines. Create a process chart for providing this service on a business-like basis. What work configuration do you recommend?

13. Analyze how to write a research report. Create a process chart for providing this service on a business-like basis.

14. With respect to Problem 13 above, the Market Research Store has considered using virtual offices, meaning that some percent of their report writers work at home on personal computers. How might this concept change the process chart? How might it change the process flow layout chart?

15. Draw the Machine-Operator Chart for the following situation:

   - there are two operators and one machine
   - make ready takes two time periods
   - machine processing time requires four periods
   - put-away takes two time periods
   - there has been no attempt to employ SMED concepts so both make ready and put-away have to be done as internal activities with the machine turned off

   Figure 6.8, entitled Machine-Operator Chart—Original Process, showed what occurs when this situation exists with one operator.
   a. What can be done with two operators?
   b. How does this compare with the 50 percent idle times associated with one operator in the original process?

16. Using the information in Problem 15 above, what can be done if SMED methods permit make ready for unit (n) while unit ($n - 1$) is working on the machine?

The Preventive Maintenance Strategy Problem

In the subway system of this city there are 100,000 light bulbs that are at risk for failure with rates described below. The complete preventive maintenance policy is to replace all of the bulbs at once, four times per year. These light bulbs are long-life light bulbs which cost more than regular ones.

Each of these bulbs has an expected average life of 3,000 hours and cost $7.00. The set-up cost for each subway station is $5,000 and there are 50 stations.

In the subway, each light bulb remains lit, day and night, all year long, which is a total of 8,760 hours. Burned out and broken bulbs are only replaced during the TPM sweep. This could mean that station platform light is not always sufficient. At an extra cost, remedial replacement could be used. The total sweep would still take place, replacing all light bulbs, whether or not they were newly replaced.

17. For this problem, ignore remedial replacement and its extra cost.
    a. Is a total replacement sweep four times per year likely to be sufficient? *Note:* The distribution is such that many light bulbs burn out long before the expected lifetime is reached and many last considerably longer than 3,000 hours.
    b. What is the total cost for following this TPM policy?

The Remedial Maintenance Strategy Problem

The alternative strategy (using the information in Problem 17 above) is to replace each bulb as it burns out. Replacement costs $5 per light bulb in addition to the $7 purchase cost per bulb. As noted in the previous problem, the bulbs have an expected lifetime of 3,000 hours. Each light bulb in the subway remains lit, day and night, all year long, a total of 8,760 hours. The expected number of times each bulb will burn out per year is 2.9. For ease of calculation, round this off to 3 times per year that each light bulb must be changed.

18. For this problem:
    a. What is the total cost of following the remedial maintenance policy?
    b. Which is preferred, TPM or RM?
    c. If the purchasing agent is able to buy the bulbs for $6.00, will it make a difference?

19. How does the service (set-up and changeover) procedure used for racing cars differ from that of regular auto servicing? Is there anything that can be learned from the difference? What does this query have to do with SMED?

20. Paint stores avoid having to carry thousands of different cans of paint for each of the colors customers want. The customer picks out the color that is wanted from color-coded chips or strips of paper. The code number is entered into the computer, which then generates the formula for that particular color. The machine that has the basic color ingredients is set to input six drops of this and four drops of that into a basic flat or semiglossy or glossy white. For example, to make a lemon white the recipe is two drops of *y*, one drop of *gy* and two drops of *oy*. Note that the colors are added to only three basic types of white paint.

Draw the process diagram for this method of producing variety needed by the customer. Compare it to the situation where the paint company is obliged to make all of the different colors in the factory and ship them to each retailer.

21. The cost of controlled storage (CST), where a withdrawal slip is required to obtain an item, is $70 per item so controlled per month. The cost of uncontrolled storage (UCST), where an item can be obtained without a stock request form, is $10 per item so stored per month. An uncontrolled item that costs $30 is known to suffer pilferage losses of three items a month. There are 100 of these items in storage. Which is preferred, CST or UCST?

▶ **Readings and References**

Michael Hammer, and James Champy, *Reengineering the Corporation*, NY, Harper Collins, 1993.

Michael Hammer, *Beyond Reengineering*, NY, Harper Collins, 1995.

M. Imai, *Kaizen: The Key to Japan's Competitive Success*, NY, Random House Business Division, 1986.

Junichi Ishiwata, *IE for the Shop Floor: Productivity Through Process Analysis*, CT, Productivity Press, 1991.

C. J. McNair and Kathleen H. J. Leibfried, *Benchmarking: A Tool for Continuous Improvement*, *Harper Business*, 1992.

Seiichi Nakajima, *TPM Development Program: Total Productive Maintenance*, CT, Productivity Press, 1989.

Taiichi Ohno, *Workplace Management*, CT, Productivity Press, 1988.

Alan Robinson, *Continuous Improvement in Operations: A Systematic Approach to Waste Reduction*, CT, Productivity Press, 1991.

Shigeo Shingo, *A Revolution in Manufacturing: The SMED System*, CT, Productivity Press, 1985.

Shigeo Shingo, *The Shingo Production Management System, Improving Process Functions*, Cambridge, MA, Productivity Press, 1992.

S U P P L E M E N T   6

A Total Productive Maintenance Model for Determining the Appropriate Level of Preventive Maintenance to be Used

The policies of TPM are employed to choose the optimal level of preventive maintenance to use, including the option of zero preventive maintenance. This is illustrated below.

A simple equation is written for the expected total cost of following three alternative strategies, called $x = 1, 2, 3$:

($x = 1$)   No preventive maintenance, called remedial maintenance (RM)

($x = 2$)   Partial preventive maintenance, called PPM

($x = 3$)   Complete preventive maintenance, called CPM

Using numbers based on the type of curve shown in Figure 6.11 allows a comparison between the expected total costs of the three maintenance strategies, $x = 1, 2, 3$.

$$p_f c_f + (1 - p_f)c_{nf} + c_p = \overline{TC}(x) \qquad \textbf{Equation 6.0}$$

p_f = the probability of failure associated with the cost of prevention, c_p, and time period t

c_p = the cost of prevention

c_f = the cost of failure

c_{nf} = the cost of no failure which is equal to zero

$\overline{TC}(x) =$ the expected total cost of maintenance strategy x

Table 6.2 provides relevant data and the expected total cost for each maintenance strategy.

TABLE 6.2

DATA NEEDED TO COMPARE MAINTENANCE STRATEGIES

x	Strategy	p_f	c_f	c_p	$\overline{TC}(x)$
1	RM	0.5	$10,000	—	$5,000
2	PPM	0.3	$10,000	$1,000	$4,000
3	TPM	0.1	$10,000	$2,000	$3,000

The TPM strategy is indicated because it has the lowest expected cost. Once again, it is useful to note that a superior TPM strategy (call it TPM+) might be developed which reduces the probability of failure to less than 0.1.

Problems for Supplement 6

1. Would a TPM+ strategy be warranted if by spending $2,500 for the cost of prevention, the probability of failure could be reduced to 0.08?

2. What level of probability reduction would make the TPM and TPM+ strategies equivalent? This value of p is called a breakeven point because, at least as far as the numbers are concerned, there is indifference between the choice of these two strategies.

Notes...

1. C. J. McNair, and Kathleen H. J. Leibfried, *Benchmarking: A Tool for Continuous Improvement*, *Harper Business*, 1992.

2. Shigeo Shingo, *The Shingo Production Management System, Improving Process Functions*, Cambridge, MA, Productivity Press, 1992.

3. *CAPP Report*, Waltham, MA, HMS Software Inc., June 1994.

4. Shingo lists four process functions that should be under constant surveillance for improvement. These are: processing, inspection, transportation, and delays.

5. M. Imai, *Kaizen: The Key to Japan's Competitive Success*, NY, Random House Business Division, 1986.

6. Michael Hammer and James Champy, *Reengineering the Corporation*, NY, Harper Collins, 1993.

   Michael Hammer, *Beyond Reengineering*, NY, Harper Collins, 1995.

7. JMA is the Japan Management Association.

8. Ibid, footnote 1, p. 150.

9. Ibid, pp. 147, 149.

10. For learning about processes that are actually used for making things, see the 27 volumes of the *Science and Invention Encyclopedia*, Westport, CT, H. S. Stuttman, Inc. Another good book to consult about real processes is the *Science Encyclopedia*, NY, Dorling Kindersley, Inc., 1993.

Chapter 7

Methods for Quality Control

After reading this chapter you should be able to...

1. Explain why quality is important to company success.
2. Determine how many pieces to make to fill an order when the probabilities of defectives are known.
3. Describe the seven classical process quality control methods.
4. Show how to use data check sheets to create bar charts, histograms and Pareto charts.
5. Discuss the use of cause and effect (Ishikawa/fishbone) diagrams.
6. Construct control charts for variables and attributes.
7. Explain how to interpret control chart signals for $\bar{x}$-charts, R-charts, p-charts, and c-charts, including upper and lower control limits (UCL and LCL).
8. Understand the meaning of "runs" on a control chart and know how to use them as early warning systems.
9. Discuss the use of inspection with and without control charts.
10. Describe acceptance sampling and explain when it is used.
11. Detail the basis of the negotiation between suppliers and buyers with respect to risks of accepting and/or rejecting lots.
12. Explain the use of the average outgoing quality limits.

Chapter Outline

7.0 THE ATTITUDE AND MIND-SET FOR QUALITY ACHIEVEMENT
     The Zero-Error Mind-Set
     The Olympic Perspective
     Compensating for Defectives
7.1 QUALITY CONTROL (QC) METHODOLOGY
7.2 INTRODUCTION TO SEVEN CLASSICAL PROCESS QUALITY CONTROL METHODS
7.3 METHOD 1—DATA CHECK SHEETS (DCSs)
7.4 METHOD 2—BAR CHARTS

7.5   METHOD 3—HISTOGRAMS
      Auto Finishes—Quality on the Line
7.6   METHOD 4—PARETO ANALYSIS
      The 20:80 Rule
      Pareto Procedures
7.7   METHOD 5—CAUSE AND EFFECT CHARTS (ISHIKAWA/FISHBONE)
      A Causal Analysis
      Scatter Diagrams
7.8   METHOD 6—CONTROL CHARTS FOR STATISTICAL PROCESS CONTROL
      Chance Causes Are Inherent Systems Causes
      Assignable Causes Are Identifiable Systems Causes
7.9   METHOD 6—$\bar{x}$-CONTROL CHART FUNDAMENTALS
      The Chocolate Truffle Factory (CTF)
      Upper and Lower Control Limits (UCL and LCL)
      How to Interpret Control Limits
7.10 METHOD 6—CONTROL CHARTS FOR RANGES—r-CHARTS
7.11 METHOD 6—CONTROL CHARTS FOR ATTRIBUTES—p-CHARTS
      c-CHARTS FOR NUMBER OF DEFECTS PER PART
7.12 METHOD 7—ANALYSIS OF STATISTICAL RUNS
      Start-ups in General
7.13 THE INSPECTION ALTERNATIVE
      100% Inspection, or Sampling?
      Technological and Organizational Alternatives
7.14 ACCEPTANCE SAMPLING (AS)
      Acceptance Sampling Terminology
      O.C. Curves and Proportional Sampling
      Negotiation between Supplier and Buyer
      Type I and Type II Errors
      The Cost of Inspection
      Average Outgoing Quality Limits (AOQL)
      Multiple-Sampling Plans
7.15 BINOMIAL DISTRIBUTION
7.16 POISSON DISTRIBUTION

Summary
Key Terms
Review Questions
Problem Section
Readings and References
Supplement 7: Hypergeometric O.C. Curves

The Systems Viewpoint

Quality achievement is a common goal which unites all of the components and constituencies of the system. All of the parts that go into an assembly are multiplicatively interrelated so that the probability of failure gets increasingly worse and the number of extra parts that must be made increases geometrically. This is an example of the effect of systems interdependencies. The mind-set for quality requires understanding the entire system. Typical of systems interdependencies is the fact that suppliers and buyers are always sharing risks. It is useful to be conscious of this when negotiating quality sampling plans. Also, it is unrealistic to set goals for only one cause of failure. In achieving effective team effort for the best quality, winners strive to be the best in the world, an Olympic perspective.

7.0
THE ATTITUDE AND MIND-SET FOR QUALITY ACHIEVEMENT

The most important goals for process analysis are higher quality, improved productivity, and lower costs. Quality is so important to the well-being of the firm that it is the key focus of this chapter. For quality methods to be successful, all participants in the organization must have their collective minds set on quality achievement. This is done by reducing the probability of failure (p_f) to meet a variety of standards for all components. If at the same time that p_f is being reduced, the standards are being raised, then the mind-set for being the best in quality achievement and competitive capability is functioning.

The Zero-Error Mind-Set

The appropriate mind-set is to abhor defectives and do as much as feasible to prevent them from occurring. When they occur, learn what caused them and correct the situation. Meanwhile, raise the goals and improve the standards. Tougher hurdles are part of the quality framework which seeks continuous improvement. *Do it right the first time* is the motto of the approach which champions zero errors.

Those who say that no defects are permissible are at odds with those who say that a few defects are to be expected. The latter group go on to say that one should learn from mistakes and take the necessary measures to prevent them from recurring. The safest position is to be flexible. Go for a "no defects" policy and when it fails to work switch to the "learn from mistakes" policy. After corrective action has been taken, switch back to the zero-errors goal.

The Olympic Perspective

The Olympic perspective calls for team play with everyone striving to achieve the company's personal best. To the managers of the firm, "going for the gold" is no less a meaningful objective than for Olympic teams. The plan calls for breaking process quality records with continued improvements.

Compensating for Defectives

In the job shop, the first items made, during and right after set up, are likely to be defectives. If the order size calls for 70 units and the expected percent defective to get the job right is $p = 5$ percent, then prepare to make 74 pieces.

$$Batch\ Size = \frac{Order\ Quantity}{(1 - p)} = \frac{70}{(1 - 0.05)} = 73.7$$

Expand this concept to an assembled unit (like a computer) that consists of n electronic components, each of which has a known probability p_i of being defective ($i = 1, 2, 3, ..., n$). Assuming that component failures are independent of each other, and that the components cannot be tested separately but only after assembly, then the number of electronic assemblies to make would be:

$$Batch\ Size = \frac{Order\ Quantity}{\prod_{i=1}^{n} (1 - p_i)}$$

Say the order quantity is 100 units and there are five components to be assembled, each of which has a ten percent probability of being defective. How many units should be made in the batch?

$$Batch\ Size = \frac{100}{(1 - 0.1)^5} = \frac{100}{0.59} = 169.5$$

With rounding, the decision is to make 170 units. The interesting effect of the multiplicative probabilities of five units is to increase the number of units that must be made to fill the order by 70 percent. If there had been only one component (with $p = 0.10$) that could lead to defective performance, then the correct number to make to fill the order would be $100/0.9 = 111$.

With such examples of systems interdependencies, the correct attitude for quality achievement requires wanting to understand the entire system. It is unrealistic to set goals for only one cause of failure. The effort must be made to perceive all of the causes of defectives in order to measure the interactive probability of failure p_f or the probability of proper performance which is $(1 - p_f)$.

7.1

QUALITY CONTROL (QC) METHODOLOGY

The body of knowledge that relates to quality attainment starts with the detection of problems. This is followed by diagnosis based on the analysis of the causes of the problems. Next comes prescription of corrective actions to treat the diagnosed causes. This is followed up through observations and evaluations to see how well the treatment works.

Figure 7.1 presents one version of the quality control model which is related to the fundamental cycle of scientific method.

CYCLE OF SCIENTIFIC METHOD

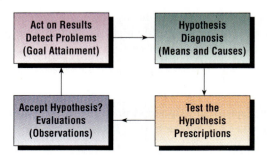

Deming applied the model of the scientific method to quality control. He used four steps (Plan-Do-Study-Act)[1], as shown in Figure 7.2. Shingo[2] felt that control of quality was missing from the Deming cycle—by interpretation, not Deming's intent. To avoid any confusion on this matter, the quality control function is shown in the *Act* box of Figure 7.2.

THE DEMING QUALITY CYCLE

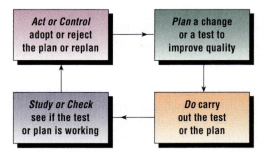

Often, the person doing the work does the checking and exercises the control. Sometimes checking and controlling are done by a member of the quality control department. There are many instances where both are used.

Quality originates with a thorough understanding of the process. If management does not have a complete understanding of all processes used by their businesses, the models for quality attainment in Figures 7.1 and 7.2 will not work. OM is involved in getting everyone in the company motivated to know all about the processes being used and changes that are being made to them.

7.2

INTRODUCTION TO SEVEN CLASSICAL PROCESS QUALITY CONTROL METHODS

Process quality control methods are used to analyze and improve process quality. There are seven time-honored and well established methods, as shown in Table 7.1. These are data check sheets, bar charts, histograms, Pareto analysis, cause and effect charts, statistical quality control charts, and run charts.

Process quality control methods

Seven methods used to analyze and improve process quality include: data check sheets, bar charts, histograms, Pareto analysis, cause and effect charts, statistical quality control charts, and run charts.

TABLE 7 . 1

SEVEN QC METHODS

1. *Data check sheets* organize the data. They are ledgers to count defectives by types. They can be spreadsheets which arrange data in matrix form when pairs of variables are being tracked. Sometimes, like spreadsheets they are called check sheets.

2. *Bar charts* represent data graphically. For example, the number of pieces in five lots can be graphically differentiated by bars of appropriate lengths. Often data is converted from check sheets to bar charts and histograms.

3. *Histograms* are frequency distributions. When data can be put into useful categories, the relative frequencies of occurrence of these categories can be informative.

4. *Pareto analysis* seeks to identify the most frequently occurring categories; for example, what defect is the most frequent cause of product rejections, next to most, etc. This facilitates identification of the major problems associated with process qualities. It rank orders the problems so that the most important causes can be addressed. It is a simple but powerful method that is the outgrowth of Methods 1, 2, and 3.

5. *Cause and effect charts* organize and depict the results of analyses concerning the determination of the causes of quality problems. The *fishbone (or Ishikawa) chart* is based on listing variables that are known to effect quality. *Scatter diagrams* help determine causality. Both are explained below.

6. *Statistical quality control (SQC)* charts—of which there are several main types—are used for detecting quality problems, much as early warning detection systems (EWDS). SQC charts also can help discern the causes of process quality problems.

7. *Run charts* can help spot the occurrence of an impending problem. Thus, run analysis is another EWDS. It also can help to diagnose the causes of problems.

Method 6 covers the preparation and employment of statistical quality control charts. It is the most elaborate and powerful approach of all the methods. Although Method 7 is commonly listed as a separate procedure, it is based on the use of the statistical control charts developed in Method 6.

The time period from which these methods are derived goes back to Shewhart.[3] Few modifications have been made, but all methods have been affected by computers and technology which can capture, analyze and chart necessary data.

The amount of computing required to track multiple measures of quality used to be prohibitive. Consequently, simple methods were developed during the period from 1930 through 1950. Data collection was straightforward. Uncomplicated charts were

W. Edwards Deming
. . .
The Father of Quality

Sources: John A. Byrne, "Remembering Deming, The Godfather of Quality," *Business Week,* Jan. 10, 1994, p. 44, and Peter Nulty, "The National Business Hall of Fame," *Fortune,* Apr. 4, 1994, p. 124.

In the early 1980s, corporate America began to search for ways to combat the intense competition from Japanese auto and electronics manufacturers. In their search they rediscovered W. Edwards Deming, a key figure in Japan's post-World War II economic recovery. A statistician with a Yale Ph.D. in mathematical physics, Deming first went to Japan as a U.S. War Department representative in 1947 to help rebuild the shattered country. He was invited back time after time to teach business executives and engineers his principles of quality.

Deming preached a systems approach to quality that emphasized identifying and eliminating defects throughout the production process rather than inspecting for defects at the end of the process. The methods he used, originated by Walter Shewhart and referred to as statistical process control (SPC), were widely accepted and applied by Japanese companies. In 1951 Japan established the prestigious Deming Prize, awarded annually to companies worldwide in recognition of outstanding quality.

A 1980 NBC television documentary entitled "If Japan Can…Why Can't We?" was largely responsible for American industry's rediscovery of Deming. The documentary recognized Deming's significant contribution to Japan's success in developing superior-quality products that were growing ever more popular with American consumers. Shortly afterward, Ford, General Motors and other U.S. companies sought Deming's help. From that point until his death in 1993, Deming worked ceaselessly to teach American companies how to achieve quality through continual improvement, teamwork, and customer satisfaction.

used to present results. These graphical techniques, described below, are so powerful that they still are being used with great success. They can be understood by workers who are only briefly trained in quality control methodology, yet another advantage.

Computer power has been used intelligently by OM to extend the reach of quality analysis to a wide range of measures, without complicating the analysis. The seven methods can be utilized across a great spectrum of quality dimensions without creating unnecessary analytic complications. There is every good reason to avoid unnecessarily complex analyses in favor of simple methods that are easy to use and straightforward to understand.

7.3

METHOD 1—DATA CHECK SHEETS (DCSs)

The first method to consider is the use of data check sheets. These are used for recording and keeping track of data regarding the frequency of events that are considered to be essential for some critical aspect of quality.

For a particular product there may be several data points required to keep track of different qualities that are being measured. It is useful to keep these in the same time frame and general format so that correlations might be developed between simultaneous events for the different qualities.

A data check sheet, such as shown in Figure 7.3, could record the frequency of power failures in a town on Long Island in New York State. The data check sheet shows when and how often the power failures occur. The DCSs is a collection of information that has been organized in various ways, one of which is usually chronologically.

FIGURE 7.3

DATA CHECK SHEET—TYPE OF POWER FAILURES

Date	A	B	C	D	E	F	Comments	Location
12/6/95	X						Crew delay	P21
1/2/96			X				6 hr. service	P3
2/15/96				X			3 hr. service	P22
3/7/96				X			Snowstorm	P21
4/29/96			X				3 hr. service	P3
5/15/96						X	Flooding	R40
6/7/96				X			Crew delay	P5
7/29/96		X					R40/no service	P21-R40
8/21/96			X				Road delays	B2
9/30/96			X				Crew delay	R40
11/2/96					X		Partial crew	P3
12/14/96		X					Flat tire	P5
Total	1	2	4	3	1	1		

The information about when power failures have occurred can be used in various ways. First, it provides a record of how frequently power failures arise. Second, it shows where the failures occur. Third, it can reveal how quickly power is restored. Fourth, it can show how long each failure has to wait before receiving attention, usually because work crews are busy on other power failures. None of the above measures explain the causes, however.

After the data check sheet is completed, problem identification follows. Then causal analysis can begin. The data check sheet can now be called the *data spreadsheet* to conform to the computer era. Data capture is more efficient now that it is operating with software on the computer. Organized collection of data is essential for good process management.

7.4
METHOD 2—BAR CHARTS

Bar chart

Graphical method for presenting statistics captured by data check sheets; data differentiated by bars.

An often-used graphical method for presenting statistics captured by data check sheets is a **bar chart**. It is simple to construct and understand. People who are turned off by tables of numerical data often are willing to study a bar chart. Figure 7.4 shows a bar chart for the recorded number of power failures which were listed on the data check sheet.

FIGURE   7 . 4

BAR CHART—TYPE OF POWER FAILURE

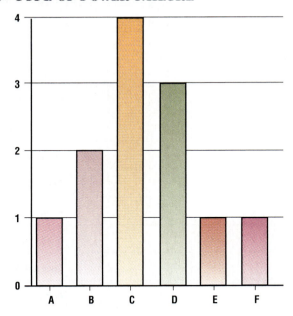

7.5
METHOD 3—HISTOGRAMS

Histograms

Bar charts measure frequencies with which each of all possible outcomes occurs.

Histograms are bar charts which measure the frequencies (f_i) of a set of k outcomes ($i = 1,2, \ldots, k$). Histograms show the frequency with which each of all possible outcomes occurs. When the sample size is n:

$$\sum_{i=1}^{n} f_i = n \quad (i = 1, \ldots, k)$$

Alternatively, the frequencies can be converted to probabilities by dividing them by n:

$$p_1 = \frac{f_1}{n}, p_2 = \frac{f_2}{n}, \ldots, p_k = \frac{f_k}{n}$$

and these sum to one:

$$\sum_{i=1}^{k} p_i = 1$$

Note that in Figure 7.5, the x-axis describes the number of paint defects ($i = 0, 1, 2, \ldots, > 8$) that have been found during an inspection of autos coming off the production line. The sample size is $\sum f_i = 100$ and f_i goes from zero to greater than eight. The y-axis records the observed frequency f_i for each number of paint defects.

FIGURE 7.5

HISTOGRAM OF PAINT DEFECTS

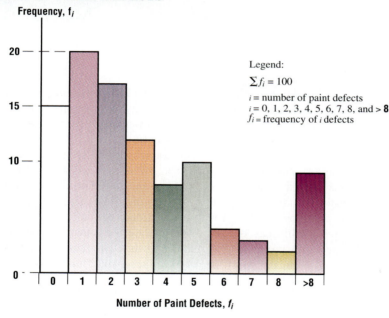

Legend:

$\sum f_i = 100$

i = number of paint defects
i = 0, 1, 2, 3, 4, 5, 6, 7, 8, and **> 8**
f_i = frequency of i defects

Auto Finishes—Quality on the Line

In this sample of $\sum f_i = 100$, the relative frequency for the number of times that no paint defects are found is 15 out of 100. This is quite low and needs an explanation. In many auto companies the standards for paint and finish are so high that few cars are found faultless. The definition of a defect is stringent. Figure 7.5 shows that one defect is found twenty percent of the time. More than one defect occurs 65 percent of the time.

Customers are very demanding about their car's finish. The paint standards are set very high by car manufacturers to reflect this demanding attitude on the part of new car buyers. The Big 3 auto makers of Detroit and the major auto makers in Europe benchmark against the paint jobs on Japanese cars which have set a new standard for best practice.

Perfect finishes in automobile painting take a great amount of technology and manufacturing care.
Source: Toyota Motor Manufacturing U.S.A. Inc.

Automobile painting is regarded as a particularly difficult production process. Perfect finishes take a great amount of technology and manufacturing care. Defects arise in ways that are hard to eliminate. The systems perspective, in this example, links OM, marketing and governmental concerns.

Quality control of paint finishes is a critical responsibility of OM. Marketing views auto finishes as a crucial sales dimension. The government is concerned about the pollution associated with the painting process and wants to reduce it through regulation.

As a result, U.S. car makers have been trying to develop new processes which decrease the pollution associated with conventional petroleum solvents used for the clearcoat, top-layer finishes. A new powder-paint process that Chrysler, Ford and GM are developing jointly promises more durable finishes with fewer defects and less pollution. "…The challenge is to efficiently produce paint finishes that rival those of Japanese carmakers."[4]

7.6

METHOD 4—PARETO ANALYSIS

Pareto analysis
Goal is to determine which problems occur most frequently and arranges them in rank order of relative frequency of occurrence.

The goal of **Pareto analysis** is to determine which problems occur the most frequently. Then the problems can be arranged in the rank order of their relative frequency of occurrence. Figure 7.6 shows a Pareto chart used to analyze a restaurant's most frequent quality problems.

Bad service is the number one complaint. Among all five complaints, it occurs 54 percent of the time. Cold food is the second most frequent criticism with 22 percent of the objections. The third grievance is about ambiance—"very noisy" (12 percent). The fourth criticism is that the restaurant is too expensive (8 percent). Excessively salty food is the fifth complaint (4 percent).

The rank order of the hierarchy of defect causes is evident in Figure 7.6. The probabilities reflect a high degree of skew (exaggeration) for the top two sources of complaints. This kind of skew is typical of Pareto frequencies that are found in actual situations. The first two causes of defects account for 76 percent of the complaints.

FIGURE 7 . 6

PARETO CHART PROBABILITIES RANK-ORDERED FOR VARIOUS TYPES OF DEFECTS EXPERIENCED

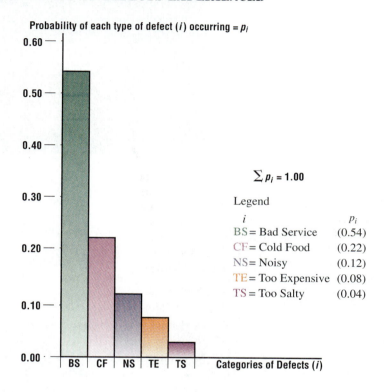

Probability of each type of defect (i) occurring $= p_i$

$\sum_i p_i = 1.00$

Legend

i		p_i
BS $=$	Bad Service	(0.54)
CF $=$	Cold Food	(0.22)
NS $=$	Noisy	(0.12)
TE $=$	Too Expensive	(0.08)
TS $=$	Too Salty	(0.04)

Categories of Defects (i)

The 20:80 Rule

Pareto analysis is designed to separate the most frequent quality problems and complaints from less frequent ones. Often, 20 percent of the total number of problems and complaints occur 80 percent of the time. This is called the **20:80 rule**. Some things occur far more frequently than others. In the restaurant case, 2 out of 5 quality problems account for 76 percent of the complaints. What is important about these numbers is the need to be sensitive to the real issues of "what counts." Pareto attempts to separate the critical minority from the inconsequential majority.

20:80 rule

20% of the total number of problems and complaints occur 80% of the time.

Pareto Procedures

The procedure to get Pareto-type information can take many forms. First, it helps to develop a comprehensive list of all the problems that can be expected to occur. To do this, many people should be consulted both inside and outside the company. Surveys, telephone interviews and focus groups can be used to find out what problems people say they have experienced.

The 20:80 rule applies to relative frequency. Most of the world's tulips are grown in Holland, a small percentage of the land.
Source: Breck's

Complaints are a valuable source of information. Accurate records can be maintained using data checklists; all the better if they are computerized. Organized efforts to secure serious complaint feedback is a marketing responsibility. Complaints are so vital to OMs that they should encourage marketing to keep Pareto data on complaints. They should monitor changes in complaints both by type and frequency. Clusters of complaints at certain times provide important diagnostic information for process managers.

This life insurance company charts customer complaints and has a system that provides comprehensive feedback from customers so that they can take care of potential problems before they become a source of customer complaints.
Source: Samsung Group

Gathered in whatever way, the data are there to reflect the frequency with which all of the problems on the list have been experienced. Problems mentioned in the text range from power failures to scratches on an auto's finish and complaints about restaurant service.

Some of the problems arise during inspection of the product while it is on or leaving the production line. These are internally measurable problems. Other problems must wait for customer experiences. They are learned about through complaints from customers who write, or call toll-free numbers set up by the company to provide information and receive complaints. Surveys also pick up complaints, or they become apparent when customers seek redress within the warranty period. Retailers, wholesalers and distributors know a lot and can communicate what they hear from their customers.

7.7

METHOD 5—CAUSE AND EFFECT CHARTS (ISHIKAWA/FISHBONE)

To determine the cause of a quality problem, start by listing everything in the process that might be responsible for the problem. A reasonable approach is to determine the factors that cause good (or bad) quality. These potential causes are grouped into categories and subcategories, as is done for the example below which lists many of the factors responsible for the quality of a cup of coffee.

After the list a chart will be drawn. This is shown in Figure 7.7. **Cause and effect charts** show at a glance what factors affect quality, and thereby what may be the causes of quality problems. These charts are organized with the quality goal at the right.

The processes used to grow coffee beans, harvest, roast, store and ship the coffee, would be included in this cause and effect diagram if there was control over these factors. This coffee making process begins with making a cup of coffee. All participating processes are part of the causal system.

Process charts play a crucial role in causal analysis. They map everything that is happening in the process and are often listed as an eighth process quality control method.

Cause and effect charts
Organize and depict the results of analyses concerning the determination of the causes of quality problems; include fishbone (or Ishikawa) charts and scatter diagrams.

A Causal Analysis

The quality of a cup of coffee is a business concern. Chains of coffee shops have opened hundreds of new outlets all over the U.S.A. A variety of business processes are involved in determining quality. Table 7.2 presents a list of causal factors that should be considered by all participants in the system that affects coffee quality.

After making the list, these variables should be charted using the fishbone construction method developed by Ishikawa.[5] There are nine main variables and different numbers of subcategory variables. The latter are shown in Figure 7.7 as fins radiating from the nine lines of the main variables, connected to what looks like the spine of a fish. The subcategories are listed in Table 7.2 under the nine main variables.

TABLE 7.2

VARIABLES FOR THE QUALITY OF A CUP OF COFFEE

Coffee Beans—Type Purchased
source location (Sumatra, Colombia, Jamaica, etc.)
grade of beans
size and age
how stored and packed (how long stored before grinding)

Roasting Process
kind of roaster
temperatures used
length of roasting

Grinding Process
type of grinder
condition of grinder
speed and feed used
fineness of grind setting
length of grinding
quantity of beans ground
stored for how long before use

Coffee Maker
drip type or other; exact specs for coffee maker
size and condition of machine (how many prior uses)
method of cleaning (vinegar, hot water, detergents)
how often cleaned
number of pounds of coffee used

Water
amount of water used
type of water used (chemical composition)
storage history before use
temperature water is heated to

Filter
type of filter used
    (material: paper (and type), gold mesh, other metals)
size of filter
how often used
how often replaced

Coffee Server
type of container
size of container
how often cleaned
how long is coffee stored

Coffee Cup
type of cup
size of cup
cleaned how (temperature of water, detergent)
cleaned how often

Spoon in Cup - Sweetener - Milk or Cream
type of spoon (material/how often cleaned and how)
type of sweetener used
type of milk, half and half or cream

FIGURE 7.7

ISHIKAWA/FISHBONE DIAGRAM

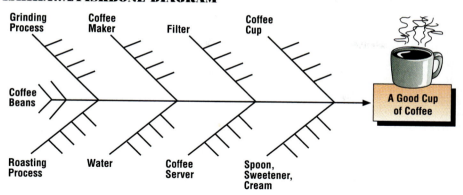

The diagram is conducive to systems-wide discussion, involving everyone, concerning the completeness of the diagram. Have all the important factors been listed? When the fishbone diagram is considered complete, then follows the determination of what variations should be tested. Market research is used to connect the process variations tested to consumer reactions.

Returning to the power failure situation, consider some of the many causes of power failures that exist. The household consumer of electricity is unlikely to know which cause is responsible for any failure.

The electric utility, on the other hand, can do the necessary detective work to determine what has caused each failure. Method 4 (using Pareto charts) can provide a useful approach to identify the most frequent types of power failures, probably by region and extent. Method 5 (using Ishikawa diagrams) now helps with the search for causes of these problems. To sum up, Ishikawa diagrams enumerate and organize the potential causes of various kinds of quality failures.

Scatter Diagrams

Scatter diagrams are composed of data points that are associated with each other, such as height and weight. The pairs of data are plotted as points on the chart. The idea is that if data are related—that is, if x is related to y—this fact will show up in the graph. The graph for height and weight could be expected to indicate that as height increases so does weight.

Scatter diagrams

Help determine causality; type of cause and effect chart composed of data points associated with each other (i.e., height and weight).

Because even strong relationships are far from exact, the plots of points will be scattered around an imaginary line that could be drawn to capture the relationship. If a relationship exists, it might be seen. It also can be measured by statistical means. Spurious relationships can show up. There are no guarantees with scatter diagram methods; they are, at best, a way to probe for relationships.

Let x be one of many factors that the fishbone diagram indicates as a possible cause of the quality problem, y. If that is true, then y is a function of x.

$$y = f(x)$$

A scatter diagram is created for various values of x and y that have been obtained by experimentation, observation, or survey. Two results are shown in Figures 7.8 and 7.9. First, consider Figure 7.8.

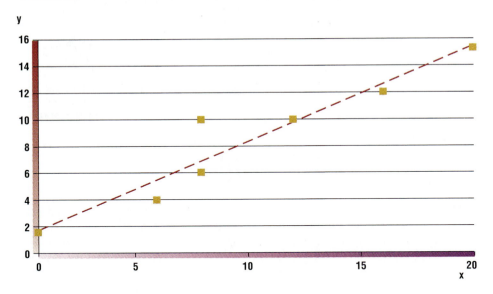

FIGURE 7.8

SCATTER DIAGRAM IMPLYING STRONG CORRELATION

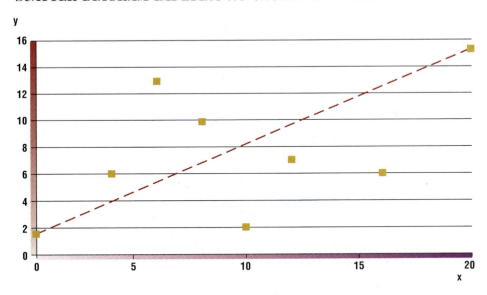

FIGURE 7.9

SCATTER DIAGRAM IMPLYING NO CAUSAL RELATIONSHIP

The various (x, y) points are tightly grouped around the line: although a relationship appears to exist, there is not sufficient information to assume that x and y are causally related. Further testing is in order and a logical framework for causality must be built. Still x and y appear to have something to do with each other instead of appearing to have no relationship.

In Figure 7.9 the various (x, y) points are far more widely scattered. This indicates that no strong relationship exists between x and y. The pattern is best described as a random assortment of points.

While the scatter diagram does not give a statistical measure of the correlation between x and y, it graphically suggests that such does exist. It is reasonable to use statistical methodology to confirm strong correlations so that prescriptions can be made to reduce and/or remove quality problems.

Revisiting the power failure case, for diagnostic and prescriptive purposes, the cause of each power failure should be tracked down and noted: lines are knocked down by cars and trucks as well as by wind storms; transformers fail more often under peak demands; and fires result from overloaded power generation equipment correlated with summer heat spells. Time of day, day of the week, date of the year may all be useful data for patterns to be found. The computer makes it easy to enter a lot of information and analyze the data for patterns.

Knowing the most frequently occurring types of failures might lead to solutions that could reduce their frequency. Overhead wires in certain locations could be buried at a cost that might be much lower than the expensive remedial actions taken after storm-caused failures. Power-grid sharing could be initiated before overloads occur. Strong correlations provide information that connects types of failures with the times and places that they occur, which may facilitate problem solving.

7.8
METHOD 6—CONTROL CHARTS FOR STATISTICAL PROCESS CONTROL

Output from a process is measured with respect to qualities that are considered to be important. For quality to be consistent it must come from a process that is stable, one with fixed parameters. This means that its average μ and its standard deviation σ are not shifting. To understand the concept of stability it is necessary to develop the statistical theory of stable distributions.

A *stable distribution* is described as being stationary. This means that its parameters are not moving. *Statistical process control (SPC)* can provide methods for determining if a process is stable. *Statistical quality control (SQC)* encompasses SPC as well as other statistical methods, including acceptance sampling, covered later in this chapter. In common usage, SPC and SQC both describe the application of control charts as a means to warn about unstable systems.

It is important to note that in real production situations, a lot of data are collected. First, enough data are obtained to create reliable control charts. Then, more data are collected to examine whether the process is stable. Evidence of temporary instability often is thrown out as being associated with start-up conditions. The process is reexamined to see if it has stabilized after the start-up. The initial sample size is often 25 to 30 sets of data, followed by another 25 to 30 sets of data. Once the process is considered stable, then regular observations are made at scheduled intervals.

If the process is not stable, and that often occurs when starting-up a new system, then Steps 1, 2, and 3 apply:

1. diagnose the causes of process instability
2. remove the causes of process instability
3. check results
4. go back through Steps 1–3 until stability is achieved

Adherence to a rigorous quality control program is not important only to a company's sales, market share, and customer loyalty. Quality errors resulting in the recall of products can have costly, even disastrous, consequences for a manufacturer. Just ask Copley Pharmaceutical Inc., a Canton, Massachusetts-based company that manufactures and sells a variety of prescription and over-the-counter drugs.

Copley recently incurred a $6 million expense in a voluntary recall of one of its asthma medications. Due to suspected microbial contamination of its 0.5 percent Albuterol Sulphate Inhalation Solution, Copley recalled 20 milliliter

Copley Pharmaceutical Inc.

Quality errors that result in the recall of products have costly, even disastrous, consequences.

vials produced in certain manufacturing lots. The solution is used by asthma sufferers and patients with other respiratory problems. Copley said sales of the product were "significant" to its

operations, accounting for about 14 to 16 percent of its total sales.

Copley was the first company to market with the product, notes Natwest's Jack Lamberton, an industry observer. "The first generic to get shelf space usually keeps it." However, the recall provided an opportunity for Copley's main competitor, Schering-Plough Corp. "Schering will fill the vacuum," Lamberton says. "It is unlikely Copley will ever get back the sales."

By Copley's conservative estimate, the 0.5 percent Albuterol will be back on the market within six months of the recall.

Source: "Copley says Albuterol recall to cost $6 million," *Reuter Business Report,* January 6, 1994.

Chance Causes Are Inherent Systems Causes

Variability exists in all systems. Two seemingly identical manufactured parts have their own signatures, like fingerprints, when measured to whatever degree of fineness is required. Systems with less variability may differ with respect to measurements in thousandths of an inch whereas systems with more variability may be noticeably different in terms of hundredths of an inch. The degree of fineness has to do with the kind of fit that is needed between parts, and this requirement is reflected by the tolerance limits the designers have set.

There are two basically different kinds of variability. The first kind is an inherent (intrinsic and innate) property of the system. It is the variability of a stable system that is functioning properly. That variability is said to be caused by chance causes, or simple causes. These causes cannot be removed from the system which is why they are called *inherent systems causes.*

A process that experiences only chance causes, no matter what its level of variability, is called a stable process. The variability that it experiences is called *random variation.*

Assignable Causes Are Identifiable Systems Causes

The second kind of variability has traceable and removable causes which are called assignable or special causes. Unlike chance causes, these causes are not background noise. They produce a distinct trademark which can be identified and traced to its origins. They are called assignable because these causes can be designated as to type and source, and removed.

The process manager identifies the sources of quality disturbances—a tool that has shifted, a gear tooth that is chipped, an operator who is missing the mark, an ingredient that is too acidic—and does what is required to correct the situation. Thus, tool, die and gear wear is correctable. Broken conveyor drives can be replaced. Vibrating loose parts can be tightened. Human errors in machine set ups can be remedied. These are examples of assignable causes that can be detected by the methods of statistical process control (SPC).

An alternative name that might be preferred to assignable causes is *identifiable systems causes*. Quality control charts are the means for spotting identifiable systems causes. Once identified, knowledge of the process leads to the cause which is then removed so that the process can return to its basic, inherent causes of variability. The ability to do this enhances quality attainment to such a degree that Method 6 is widely acknowledged as powerful and necessary.

A process that is operating without any assignable causes of variation is stable by definition, although σ can be large. Stability does not refer to the degree of variability of a process. Instability reflects the invasion of assignable causes of variation which create quality problems.

There are different kinds of control charts. First, there is classification by variables where numbered measurements are continuous, as with a ruler for measuring inches or with a scale for measuring pounds. The variables can have dimensions of weight, temperature, area, strength, variability, electrical resistance, loyalty, satisfaction, defects, typing errors, complaints, accidents, readership and TV viewer ratings. The fineness of the units of measure are related to the quality specifications for the outputs.

Second, there is classification by attributes. Output units are classified as accepted or rejected, usually, by some form of inspection device which only distinguishes between a "go" or a "no go" situation. Rejects result from measurements falling out of the range of acceptable values that are pre-set by design. Rejects are defective product units. Under the attribute system of classification, process quality is judged by the number of defective units discovered by the pass-fail test. It is charted as the percentage of defects discovered in each sample, and is called a *p*-chart. Also under the attribute system of measurement is the *c*-chart which counts the number of nonconformities of a part.

7.9

METHOD 6—$\bar{x}$-CONTROL CHART FUNDAMENTALS

The **$\bar{x}$-chart** uses classification by variables. A sample of product is taken at intervals. Each unit in the sample is measured along the appropriate scale for the qualities being analyzed. The **control chart** is a graphic means of plotting points which should fall within specific upper- and lower-bound limits if the process is stable.

$\bar{x}$-chart

Type of statistical quality control (SQC) chart; uses classification by variables to detect and discern causes of process quality problems.

Control chart

Graphic means of plotting points which should fall within specific upper- and lower-bound limits if the process is stable.

Table 7.3 illustrates a sequence of quality measures, called x_i (i = output number) where x_i can be measured on a continuous scale which permits classification by variables. The sample subgroup size, n, is 3 in Table 7.3, and the sample subgroups are named with Roman numerals I, II, and so on.

T A B L E   7 . 3

SUBGROUP SAMPLE SIZE IS 3 AND THE
INTERVAL BETWEEN SAMPLES IS 4

Subgroup Sample I	Interval Between Samples	Subgroup Sample II	Interval Between Samples
x_1 x_2 x_3	x_4 x_5 x_6 x_7	x_8 x_9 x_{10}	x_{11} x_{12} x_{13} x_{14}

There are two subgroup samples of 3. The output interval between them is 4. Subgroup sample sizes between 2 and 10 are common. The interval between subgroups varies widely, depending on the output rate and the stability of quality for the product.

The sample mean for each subgroup $\bar{x}_j$ is obtained.

$$\bar{x}_I = \frac{(x_1 + x_2 + x_3)}{3}, \qquad \bar{x}_{II} = \frac{(x_8 + x_9 + x_{10})}{3} \ \dots \ \text{etc.}$$

In general

$$\bar{x}_j = \frac{\sum x_i}{n}$$

The grand mean, $\bar{\bar{x}}$, is the average of the sample means, that is: where N is the number of sample subgroups being used.

$$\bar{\bar{x}}_j = (\bar{x}_I + \bar{x}_{II} + \dots + \bar{x}_N)/N$$

The Chocolate Truffle Factory (CTF)

Data have been collected by the OM department for the famous truffle bonbons made by the Chocolate Truffle Factory. These are marketed all over the world. Complaints have been received that the candies have been uneven in size and weight. This is an ideal opportunity to use an $\bar{x}$-chart to find out if the chocolate-making process is stable or erratic.

The standard weight set by CTF is 30 grams of the finest chocolate. A perfect pound box would contain 16 chocolates each weighing 28 grams. Company policy, based on principles of ethical practice, is to err on the side of giving more, rather than less, by targeting 2 extra grams per piece. However, extra chocolate is expensive. Worse yet, if weight and size are inconsistent, the larger pieces create dissatisfaction with the truffles that meet the standard. Also, consistency of size is used by customers to judge the quality of the product. Weight is easier to measure than size, and is considered to be a good surrogate for it.

The company wants to test for consistency of product over a production day. Samples were taken over the course of one day of production at 10:00 and 11:00 A.M., and at 1:00, 3:00 and 4:00 P.M. The hours chosen are just before the line is stopped for tea break and/or for cleanup. A sample subgroup size of four consecutive observations was chosen for each of the five samples. This yielded 20 measures and five sample subgroup means.

For instructive purposes, this sample is about the right size. For real results, all five days of production for a typical week should be examined. Except as an example, a one day sample is too small. There is a rule of thumb widely used:

> Practitioners consider 20 to 25 subgroup means as appropriate for a start-up or diagnostic study.

Table 7.4 offers five sample subgroup sets of data as the columns (j) with one for each time period. Measurements are in grams.

TABLE 7.4
WEIGHT OF CHOCOLATE TRUFFLES (IN GRAMS)

Subgroup j =	I	II	III	IV	V	
Time:	10 A.M.	11 A.M.	1 P.M.	3 P.M.	4 P.M.	SUM
x_i						
1	30.50	30.30	30.15	30.60	30.15	
2	29.75	31.00	29.50	32.00	30.25	
3	29.90	30.20	29.75	31.00	30.50	
4	30.25	30.50	30.00	30.00	29.70	
SUM	120.40	122.00	119.40	123.60	120.60	
$\bar{x}_j$	30.10	30.50	29.85	30.90	30.15	151.50
R_j	0.75	0.80	0.65	2.00	0.80	5.00

(Where $\bar{x}_j = \dfrac{\sum x_i}{n}$; $i = 1, 2, \dots 4$)

There is a range measure R_j in this chart which will be used later. It is the difference between the largest and the smallest number in the column.

The grand mean value is $(151.50)/5 = 30.30$ which indicates that, on average, chocolate truffles are 1 percent heavier than the standard. This average value seems well suited to the ethical considerations of the firm since it is above, but not too far above, the target value of 30.00 grams. CTF's policy is ethical but expensive. Customers get more than a pound of chocolates on a regular basis.

In Table 7.4 above, individual sampled values fall below the 30 gram target five times, even though in all cases they are well above 28. The lowest value sampled was 29.50. The highest value is 32. The three highest values all appear at the 3:00 P.M. point.

Upper and Lower Control Limits (UCL and LCL)

Control limits are thresholds designed statistically to signal that a process is not stable. Control limits will now be derived for the $\bar{x}$-chart. Before doing that, study Figure 7.10 to identify what these control limits look like. Note that the 3:00 P.M.

Control limits

Thresholds designed statistically to signal a process is not stable.

sample is identified on the chart as being nearly out of control. This $\bar{x}$-chart with UCL and LCL has no points outside the limits, and therefore no points are specifically out of control. Nevertheless, the 3:00 P.M. sample, when it is backed up by a lot more data, is the kind of visual warning that usually warrants attention.

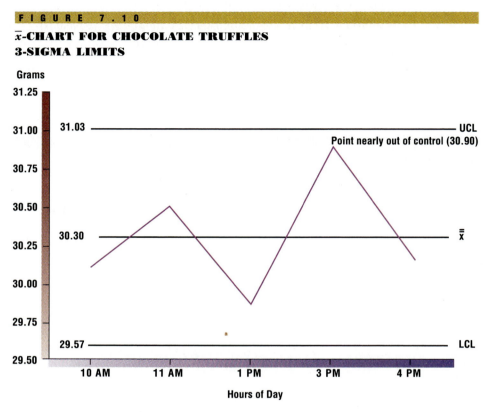

FIGURE 7.10

$\bar{x}$-CHART FOR CHOCOLATE TRUFFLES
3-SIGMA LIMITS

Perhaps, the 3:00 P.M. sample is taken during the afternoon tea break, during which the chocolate mixture becomes thicker and the molding machine tends to make heavier pieces. The cleanup comes after 3:00 P.M. Tracking and discovering such causes will lead to the elimination of the problems. Although the process is stable, it appears to have too much variability $(32.00 - 29.50)$. It may need to be redesigned. The chocolate moldfilling machines might need replacement or rebuilding. Workers using the filling machines might need additional training. From the above comment on variability, it becomes apparent that there is a need to study the range measure. That will be done shortly.

The control chart drawn in Figure 7.10 is based on plotting the sample means from Table 7.4. The point to emphasize is that the underlying distribution is that of the sample means. It is necessary to estimate the variability of this distribution of the sample means in order to construct control limits around the grand mean of a process. The upper control limit (UCL) and lower control limit (LCL), shown in Figure 7.10 above, could have been determined by calculating the standard deviation for the sample means $\bar{x}_j$, given in Table 7.4.

An easier calculation of variability uses the range, R. It is well suited for repetitive computations that control charts entail. Column values of R for CTF are given in Table 7.4. The range within each subgroup, is defined as:

$$R = x_{\text{MAX}} - x_{\text{MIN}}$$

The largest x value in the 10:00 A.M. sample subgroup is 30.50 and the smallest is 29.75, so $R = 0.75$. UCL and LCL will be computed shortly using the average value of R for the subgroup samples.

Statistical quality control works because the variation observed within each subgroup (R_{I}, R_{II}, ... etc.) can be statistically related to the variation between the subgroup means. Only if the process is stable will the two different measures of variability be the same.

The between group variability is related to the differences between the grand mean and successive sample means. Thus:

$$(\bar{x}_{\text{I}} - \bar{\bar{x}}), (\bar{x}_{\text{II}} - \bar{\bar{x}}), ..., (\bar{x}_N - \bar{\bar{x}})$$

whereas, the within group variability is related to the values (R_{I}, R_{II}, ... etc.). If the process is stable, the plotted points of the sample means—reflecting between group variability—will fall within control limits that are based on within group variability. If the process is unstable, some of the plotted points are likely to fall outside the control limits.

This reasoning applies to various kinds of control charts. Present discussion concerns the $\bar{x}$-chart. This is followed by application of control limits to the chart for ranges, called the R-chart, and then, the p-chart for attributes and the c-chart for defects per part.

FIGURE 7.11

DISTRIBUTION OF THE SAMPLE MEANS IS NARROWER THAN THE PARENT POPULATION

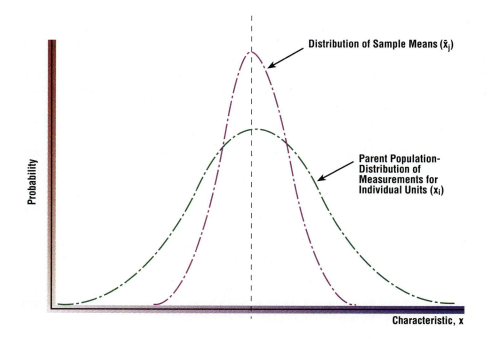

When the process is not stable because of the existence of assignable causes, between group variability exceeds within group variability. Note that the observations taken within the subgroup are always taken as close together as possible, whereas in the interval between subgroups, allow sufficient time for an assignable cause to enter the system. The subgroup size is kept small. That way, a high degree of homogeneity should exist for the observations within each subgroup. In a stable system, the same homogeneous condition will apply to all the samples, no matter when they were taken.

To construct the control chart:

1. Individual samples are taken from the parent population in groups of size n.
2. $\bar{x}$ and R measures are calculated for each sample subgroup.
3. The control limits UCL and LCL are computed.

The sample means $\bar{x}_I$, $\bar{x}_{II}$, etc., are distributed in accordance with the size of n. Figure 7.11 illustrates how the distribution of the sample means is narrower than that of the parent population. The distribution of the sample means is used for the control chart construction.

The relationship between the variability of a stable parent distribution and that of the distribution of the sample means is well-known as the central limit theorem. In the equation below, the left-hand side is

$$\sigma_{\bar{x}}^2$$

which is the variance of the distribution of the sample means. The right-hand side is the variance of the parent population distribution. The equation under the variance relationship is equivalent for the standard deviation parameter which is called the *standard error of the mean*.

$$\sigma_{\bar{x}}^2 = \sigma_{\bar{x}}^2 / n \quad \text{where } n = \text{subgroup size,}$$

$$\text{or} \quad \sigma_{\bar{x}} = \sigma_{\bar{x}} / \sqrt{n} \quad \text{in terms of standard deviations.}$$

If the subgroup size is equal to one, $n = 1$, then

$\sqrt{1} = 1$, and the two distributions are identical. A subgroup size of one defeats the purpose of measuring within subgroup variability, but it proves the logic of the equation. A subgroup size of $n = 2$ is the smallest feasible size for SPC.

If n is a very large number, $\sigma_{\bar{x}}$ gets small, eventually approaching zero. Then, the distribution of sample means takes on the appearance of a vertical line. This, too, is an extreme, mainly useful to test the sense of the equation.

Continuing with the fundamental equations that underlie control chart theory, it should be noted that:

$\sigma_{\bar{x}} = $ the standard deviation of the distribution of the sample means is measured as follows:

$$\sigma_{\bar{x}} = \sqrt{\frac{\sum_{j=1}^{N} (\bar{x}_j - \bar{\bar{x}})^2}{N}} \qquad \textbf{Equation 7.0}$$

$(j = 1,2,\dots N)$ where $N = $ the number of subgroup sample means that are calculated.

Rather than going through all of these calculations, R is the preferred measure. Meanwhile, to make certain that the contrast between the sample means parameter and that of the parent population is clear, review Equation 7.1.

$\sigma_x = $ the standard deviation of the parent population.

It is measured by:

$$\sigma_x = \sqrt{\frac{\sum\limits_{i=1}^{nN} (x_i - \bar{\bar{x}})^2}{nN}} \quad (i = 1, 2, ..., nN) \qquad \textbf{Equation 7.1}$$

where (nN) = the total number of observations made. It is the sample subgroup size (n) multiplied by the number of sample subgroups taken (N).

Finally, the grand mean used for the control chart can be measured in different ways, thus:

$$\bar{\bar{x}} = \frac{\sum\limits^{nN} x_i}{nN} = \frac{\sum\limits^{N} \bar{x}_j}{N} = \frac{606}{(4)\,(5)} = \frac{151.50}{5} = 30.30,$$

and the average range is:

$$\bar{R} = \sum \frac{R_j}{N} = \frac{5}{5} = 1.00$$

The data for $\bar{x}$ and $\bar{R}$ come from Table 7.4 above.

The R measure must now be converted to represent the sample standard deviation. The conversion factors relating the range to the sample standard deviation have been tabled and are widely available. The conversion factor for upper and lower 3-sigma limits is A_2. It is given below in Table 7.5.

T A B L E   7 . 5

CONVERSION FACTORS FOR $\bar{x}$- AND R-CHARTS

Subgroup Size (n)	A_2	D_3	D_4
2	1.88	0	3.27
3	1.02	0	2.57
4	0.73	0	2.28
5	0.58	0	2.11
6	0.48	0	2.00
7	0.42	0.08	1.92
8	0.37	0.14	1.86
9	0.34	0.18	1.82
10	0.31	0.22	1.78
12	0.27	0.28	1.72
15	0.22	0.35	1.65
20	0.18	0.41	1.59

A_2 is used for UCL and LCL for the $\bar{x}$-chart.
D_3 is used for the LCL for the R-chart.
D_4 is used for the UCL for the R-chart.

The meaning of A_2 multiplied by $\bar{R}$ is shown below:

$$A_2\bar{R} = k\sigma_{\bar{x}} = 3\sigma_{\bar{x}} = 3\frac{\sigma_x}{\sqrt{n}} \qquad \textbf{Equation 7.2}$$

The calculations for the upper control limit $UCL_{\bar{x}}$ and for the lower $LCL_{\bar{x}}$ using three standard deviations ($\pm 3\sigma$) are as follows:

$$\text{UCL}_{\bar{x}} = \bar{\bar{x}} + A_2\bar{R} \quad \text{and} \quad \text{LCL}_{\bar{x}} = \bar{\bar{x}} - A_2\bar{R}$$

The calculations for the Chocolate Truffle Factory are:

$$\text{UCL}_{\bar{x}} = 30.30 + (0.73)(1.00) = 31.03 \qquad \textbf{Equation 7.3}$$

$$\text{LCL}_{\bar{x}} = 30.30 - (0.73)(1.00) = 29.57 \qquad \textbf{Equation 7.4}$$

How to Interpret Control Limits

The limits used in Figure 7.10 resulted from the above calculations. The 3:00 P.M. point would have been outside the control limits if two standard deviations had been used instead of three. The 3:00 P.M. point was 30.90 and the $UCL_{\bar{x}}$ for two sigma would have been 30.79. Using Equation 7.2, $\sigma_{\bar{x}} = 0.243$, $2\sigma_{\bar{x}} = 0.487$, and from Equation 7.3, $UCL_{\bar{x}} = 30.30 + 0.49$. The number of standard deviations used will control the sensitivity of the warning system. Three-sigma charts remain silent when two-sigma charts sound alarms.

Understanding how to set the control limits is related to balancing the costs of false alarms against the dangers of missing real ones. One-sigma limits will sound more alarms than two sigma ones. The judicious choice of k for $\pm k\sigma$ is an operations management responsibility in conjunction with other participants to assess the two costs stated above. There must be an evaluation of what happens if the detection of a real problem is delayed. The systems-wide solution is the best one to use.

The probability of a subgroup mean falling above or below 3-sigma limits by chance rather than by cause is 0.0028, which is pretty small, but still real. For 2 sigma it would be 0.0456 which is not small. k also can be chosen between 2 and 3 sigma. In some systems, two successive points must go out of control for action to be taken. There are fire departments that require two fire alarms before responding because of the high frequency of false alarms. When a response is made to out-of-control signals, knowledge of the process is the crucial ingredient for remedying the problem.

7.10

METHOD 6—CONTROL CHARTS FOR RANGES—R-CHARTS

R-chart

Type of statistical quality control (SQC) chart; monitors the stability of the range (variability).

The $\bar{x}$-chart is accompanied by another kind of control chart for variables. Called an **R-chart**, it monitors the stability of the range. In comparison, the $\bar{x}$-chart checks the stationarity of the process mean, i.e., checks that the mean of the distribution does not shift about—while the R-chart checks that the spread of the distribution around the mean stays constant. In other words, R-charts control for shifts in the process standard deviation.

The $\bar{x}$- and R-charts are best used together because sometimes an assignable cause creates no shift in the process average, nor in the average value of the standard deviation (or $\bar{R}$), but a change occurs in the distribution of R values. For example, if the variability measured for each subgroup is either very large or very small, the average variability would be the same as if all subgroup R values were constant.

Remembering that the sampling procedure is aimed at discovering the true condition of the parent population, the use of $\bar{x}$- and R-charts provides a powerful combination of analytic tools.

THE *R*-CHART
3-SIGMA LIMITS

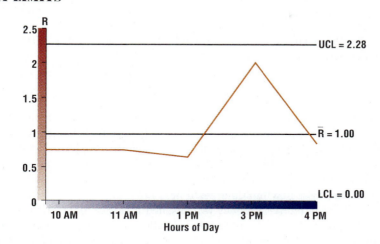

The equations for the control limits for the R-chart are:

$$\text{UCL}_R = D_4\bar{R} \text{ and } \text{LCL}_R = D_3\bar{R}$$

The Chocolate Truffle Factory data from Table 7.4 above can be used to illustrate the control chart procedures. Continuing to use $k = 3$ sigma limits and the conversion factors given in Table 7.5 above, the R-chart control parameters are obtained.

$$D_4 \; (for \; n = 4) = 2.28, \quad D_3 \; (for \; n = 4) = 0, \quad \text{and} \quad \bar{R} = \frac{5.00}{5} = 1.00$$

These are used to derive the control limits.

$$\text{UCL}_R = (2.28)\,(1.00) = 2.28 \text{ and } \text{LCL}_R = (0.00)(1.00) = 0.00$$

Figure 7.12 shows that the R-chart has no points outside the control limits. The same pattern prevails as with the $\bar{x}$-chart. The 3:00 P.M. point is close to the upper limit. This further confirms the fact that the process is stable and is unlikely to do much better. However, the sample is small and good chart use would throw out the 3:00 P.M. sample and take about 25 more samples before making any decision.

7.11

METHOD 6—CONTROL CHARTS FOR ATTRIBUTES—*p*-CHARTS

The ***p*-chart** shows the percent defective in successive samples. It has upper and lower control limits similar to those found in the $\bar{x}$- and R-charts. First, compare what is measured for the p-chart to what is measured for the $\bar{x}$-chart.

p-chart
Shows percent defective in successive samples.

Monitoring by variables is more expensive than by attributes. The inspection by variables requires reading a scale and recording many numbers correctly. Error prevention is critical. The numbers have to be arithmetically manipulated to derive their meanings.

In comparison, with the p-chart, a go/no-go gauge, a pass-fail exam, or a visual check—good or bad—can be used. All of these measurement methods are examples of the simplicity of monitoring by attributes.

Only one chart is needed for the attribute, p, which is the *percent defective*. The p-chart replaces the two charts $\bar{x}$ and R. Also, only the upper control limit really matters. The lower control limit is always better as it gets closer to zero.

The p is defined as follows:

$$p = \frac{\text{number rejected}}{\text{number inspected}}$$

Assume that the Chocolate Truffle Factory can buy a new molding machine which has less variability than the present one. CTF's operations managers in cooperation with the marketing managers and others have agreed to redefine acceptable standards so that profits can be raised while increasing customer satisfaction. Also, it was decided to use a p-chart as the simplest way to demonstrate the importance of the acquisition of the new molding machine.

Previously, a defective product was defined as a box of truffles that weighed less than a pound. What was in the box was allowed to vary, as long as the pound constraint was met. Now the idea is to control defectives on each piece's weight.

The new definition of a reject is as follows: a defective occurs when $x_i < 29.60$ grams or $x_i > 30.40$ grams. This standard set on pieces is far more stringent than the earlier one set on box weight. The new standards will always satisfy CTF's criteria to deliver nothing less than a pound box of chocolate.

It has been suggested that the data of Table 7.4 above be tested against the p-chart criteria. Rejects are marked with an R on the table. The number of rejects (NR) is counted, and p is calculated for each subgroup in Table 7.6.

TABLE 7.6

**WEIGHT OF CHOCOLATE TRUFFLES (IN GRAMS)
(R = REJECTS DUE TO EXCESSIVE WEIGHT)**

Subgroup j = Time:	I 10 A.M.	II 11 A.M.	III 1 P.M.	IV 3 P.M.	V 4 P.M.	SUM
x_i						
1	30.50R	30.30	30.15	30.60R	30.15	
2	29.75	31.00R	29.50R	32.00R	30.25	
3	29.90	30.20	29.75	31.00R	30.50R	
4	30.25	30.50R	30.00	30.00	29.70	
NR	1	2	1	3	1	8
p	0.25	0.50	0.25	0.75	0.25	

TABLE 7.7

SUMMARY OF THE FINDINGS (*NR* = NUMBER OF REJECTS)

Subgroup No.	*NR*	*n*	*p*
1	1	4	0.25
2	2	4	0.50
3	1	4	0.25
4	3	4	0.75
5	1	4	0.25
SUM	8	20	

The *p* values are extraordinarily high. CTF either has to go back to the old standard or get the new machine. This analysis will provide the motivation for changing machines. The *p*-chart parameters are calculated as follows: first, compute $\bar{p}$.

Average Percent Defective =

$$\bar{p} = \sum \frac{\text{number rejected}}{\text{number inspected}} = \frac{8}{20} = 0.40$$

Companies that subscribe to the concepts of TQM are committed to obtaining defective rates that are less than 1%, and this is 40%. It is evident that the new criteria and the present process are incompatible. Constructing the *p*-chart confirms this evaluation.

The control chart for single attributes, known as the *p*-chart (there is a *c*-chart for multiple defects described below), requires the calculation of the standard deviation (*s*). For the *p*-chart computations, the binomial description of variability is used:

$$s_p = \sqrt{\frac{\bar{p}(1 - \bar{p})}{n}} = \sqrt{\frac{(0.40)\,(0.60)}{4}} = \sqrt{.06} = 0.245$$

Shifting to more conservative 1.96 sigma limits means that 95% of all subgroups will have a *p* value that falls within the control limits[6] which are determined as follows:

$$\text{UCL}_p = \bar{p} + 1.96s_p = 0.40 + 1.96(0.245) = 0.40 + 0.48 = 0.88$$

and

$$\text{LCL}_p = \bar{p} - 1.96s_p = 0.40 - 0.48 = -0.08 < 0.000 = 0.00$$

The LCL cannot be less than zero, so it is set at zero.

Note that UCL_p and LCL_p are functions of s_p which can vary with the sample size *n*. Thus, if the 11:00 A.M. sample had nine measures, then

$$s_p = \sqrt{\frac{\bar{p}(1 - \bar{p})}{n}} = \sqrt{\frac{(0.40)\,(0.60)}{9}} = \sqrt{.027} = 0.163$$

has decreased, and this will reduce the spread between the limits, just for the 11:00 A.M. portion of the chart. The control limits can step up and down according to the sample size, which is another degree of flexibility associated with the *p*-chart.

The appropriate control chart for percent defectives for the Chocolate Truffle Factory is shown in Figure 7.13.

F I G U R E   7 . 1 3

THE *p*-CHART

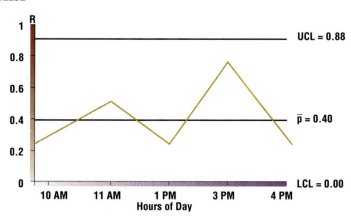

There are no points outside the UCL, although the 3:00 P.M. value of 0.75 is close to the upper limit's value of 0.88. The same reasoning applies, namely the process is stable, but the 3:00 P.M. sample should be eliminated from the calculations and 25 more samples of four should be taken.

c-Charts for Number of Defects per Part

***c*-chart**

Shows the number of defectives in successive samples.

The ***c*-chart** shows the number of defectives in successive samples. It has upper and lower control limits similar to those found in the $\bar{x}$-, R-, and p-charts.

The finish of an automobile or the glass for the windshield are typical of products that can have multiple defects. It is common to count the number of imperfections for paint finish and for glass bubbles and marks. The expected number of defects per part is $\bar{c}$. The statistical count c of the number of defects per part lends itself to control by attributes which is similar to p, the number of defectives in a sample.

It is reasonable to assume that c is Poisson distributed because the Poisson distribution describes relatively rare events. (It was originally developed from observations of the number of Prussian soldiers kicked by horses.) There is an insignificant likelihood that a defect will occur in any one place on the product unless there is an assignable cause. That is what the *c*-chart is intended to find out.

The standard deviation of the Poisson distribution is the square root of $\bar{c}$. Thus:

$$\text{UCL}_c = \bar{c} + k\sigma_c = \bar{c} + k\sqrt{\bar{c}}$$

and

$$\text{LCL}_c = \bar{c} - k\sigma_c = \bar{c} - k\sqrt{\bar{c}}$$

The number of defects are counted per unit sample and $\bar{c}$ is estimated as follows:

Sample No.	Number of Defects	Sample No.	Number of Defects
1	3	6	5
2	2	7	3
3	5	8	4
4	8	9	3
5	2	10	3

Legend:
Total Number of Defects = 38
Total Number of Samples = 10
Average Number of Defects per Sample, = 3.8

The control limits are calculated:

$$ULC_c = 3.8 + 2\sigma_c = 3.8 + 2(1.95) = 7.70$$

and

$$LCL_c = 3.8 - 2\sigma_c = 3.8 - 2(1.95) = -0.10$$

A negative LCL is interpreted as zero and the UCL is breached by the fourth sample. This violation seems to indicate that all is not well with this process. The control chart is shown below.

FIGURE 7.14

c-CHART FOR NUMBER OF DEFECTS PER PART

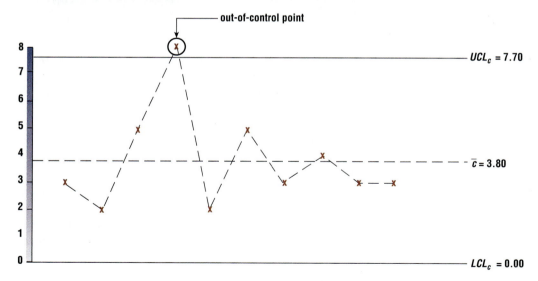

It is possible to throw out the fourth sample and recalculate the charts. In reality, larger samples would be taken to draw up the initial charts and then further samples would determine the stability of the process.

U.S. Postal Service Uses Optical Scanning to Speed Mail Sorting

n recent years, optical scanning has become commonplace in grocery stores, department stores, and other retail businesses. The use of optical scanning equipment, which reads bar codes affixed to products, improves the speed and accuracy of the customer check-out process. It also provides a continually updated record of sales and inventories. Many nonretail entities also use optical scanning to increase efficiency and reduce costs. The United States Postal Service, for example, uses scanning and other technology extensively in its operations.

Optical scanning greatly speeds up the mail sorting process. One person can manually sort about 500 pieces of mail an hour, but a scanning machine can sort 30,000 to 40,000 pieces an hour. In addition, the cost of sorting by hand is approximately $42 per 1,000 pieces, compared to about $4 per 1,000

pieces for automated sorting. The time and cost savings from optical scanning are enormous given the volume of mail handled each day—about one-half billion pieces.

An estimated 40 percent of the mail is bar coded when received by the Postal Service and can be scanned immediately. Another 40 percent bears typed or printed addresses that can be read by sophisticated computers, called optical character readers. After reading the address, this same

computer bar codes the mail, which can then be sorted by the scanning equipment. Mail with handwritten addresses or incomplete addresses goes through further electronic processing to supply the necessary bar-coding information. This processing is often done by private companies, such as DynCorp, operating at various sites throughout the country. This processing approach is viewed by postal officials as an interim step until technology is developed to the point that computers can read handwritten mail.

Source: "DynCorp Takes High-Tech to Mail-Carrying Business," *Central Penn Business Journal,* April 4, 1994.

7.12

METHOD 7—ANALYSIS OF STATISTICAL RUNS

Run

Count on a control chart, successive values that fall either above or below the mean line.

A **run** of numbers on a control chart is successive values that all fall either above or below the mean line.

Consider this scenario. For months, the control chart has indicated a stable system. However, the latest sample mean falls outside the control limits. The following conjectures should be made:

1. Has the process changed or is this a statistical fluke?
2. Was the out-of-control event preceded by a run of sample subgroup mean values?

> *A run is a group of consecutive sample means that all occur on one side of the grand mean.*

3. If there was a run of six or more, assume that something is changing in the process. Look for the problem. Sample the process again, as soon as possible. Make the interval between samples smaller than usual until the quandary is resolved.

Sometimes a separate run chart is kept to call attention to points that are falling on only one side of $\bar{\bar{x}}$. Separate charts are not required if the regular control charts are watched for runs.

The $\bar{x}$-chart in Figure 7.10 shows no runs. The R-chart in Figure 7.12 shows a run of 3 consecutive points below the grand mean. The p-chart in Figure 7.13 shows no runs. Charts often show short runs of two, three or four, because short runs are statistically likely.

Runs can appear on either side of the grand mean, $\bar{\bar{x}}$. The probability of two consecutive $\bar{x}$'s falling above $\bar{\bar{x}}$ is $1/2 \times 1/2 = 1/4$. The probability of two consecutive $\bar{x}$'s falling below $\bar{\bar{x}}$ is $1/2 \times 1/2 = 1/4$. The probability that there is a run of two, either above or below $\bar{\bar{x}}$ is $1/4 + 1/4 = 2/4 = 1/2$. The probability of a run of three above is $(1/2)^3 = 1/8$, and below it is the same. Above and below is $(1/8 + 1/8) = 2/8$. To generalize, the probability of a run of r above the grand mean is $(1/2)^r$, and the same below, so above and below is:

$$p(r) = (2)(1/2)^r = (1/2)^{r-1} = \frac{1}{2^{(r-1)}}$$

Then, the probability of a run of six ($r = 6$) is:

$$p(r = 6) = \frac{1}{2^{(6-1)}} = \frac{1}{2^5} = \frac{1}{32} = 0.03125$$

Note that a run of 6 has a probability of $1/32 = 0.03125$. This is about the same probability that a sample mean has of being above the UCL with 1.86σ limits. Thus, a run of six adds serious weight to the proposition that something has gone awry.

The run chart also shows whether the sample means keep moving monotonically, either up or down. Monotonically means always moving closer to the limit without reversing direction. The run of 3 in Figure 7.12 is not monotonic. A monotonic run signals even greater urgency to consider the likelihood that an assignable cause has changed the process's behavior. Statistical run measures neither take monotonicity nor the slope of the run into account. Steep slopes send more urgent signals of change.

Start-ups in General

Many, if not most, processes, when they are first implemented, moving from design to practice, are found to be out of control by SPC models. Process managers working with the SPC charts and 25 initial samples can remove assignable causes and stabilize the process. The continuity of the process that is required is characteristic of flow shop type configurations.

Nevertheless, the p-chart has been used with orders that are regularly repeated in the job shop.

7.13

THE INSPECTION ALTERNATIVE

Some firms have practiced inspecting out defectives instead of going through the laborious process of classifying problems, identifying causes and prescribing solutions. This approach to process management is considered bad practice. Inspection, in conjunction with SPC, plays an important role by providing sample means, allowing workers to check their own output quality, and inspecting the acceptability of the suppliers' shipments (called acceptance sampling).

> *Inspections and inspectors reduce defectives shipped. They do not reduce defectives made.*

100% Inspection, or Sampling?

An option for monitoring a process is using one hundred percent (100%) inspection of the output. There are times when 100% inspection is the best procedure. When human checking is involved, it may be neither cost-efficient nor operationally effective. Automated inspection capabilities are changing this.

The use of 100% inspection makes sense for parachutists and for products where failure is life-threatening. Inspections which are related to life-and-death dependencies deserve at least 100 and, perhaps, even 1000% inspection. For parachute inspections and surgical cases, a second opinion is a good idea.

NASA continuously monitors many quality factors at space launchings. Aircraft take-off rules are a series of checks that constitute 100% inspection of certain features of the plane. There are no serious suggestions about doing away with these forms of 100% inspection. There have been occasions when pilots forgot to do a checking procedure with dire consequences.

An economic reason to shun 100% inspection is that it is often labor intensive. This means it is likely to be more expensive than sampling the output. Sampling inspection is the other alternative. The few are held to represent the many. The body of statistical knowledge about sample sizes and conditions is sound and is treated shortly.

A common criticism of 100% inspection is that it is not fail-safe because it is human to err. Inspectors' boredom makes them very human in this regard. Sampling puts less of a burden on inspectors.

Inspection procedures performed by people are less reliable and have lower accuracy when the volume of throughput is high. Imagine the burden of trying to inspect every pill that goes into a pharmaceutical bottle. If the throughput rate is one unit per second and it takes two seconds to inspect a unit, then two inspectors are going to be needed and neither of them will have any time to go to the bathroom. With present scanner technology the dimensions of the problem are changed.

In such a situation sampling inspection is a logical alternative. If destructive testing is required (as with firecrackers), sampling is the only alternative. Otherwise, the firecracker manufacturer will have no product to sell. Destructive testing is used in a variety of applications such as for protective packaging and for flammability. The sampling alternative is a boon to those manufacturers.

Technological and Organizational Alternatives

The situation for inspection is changing in several ways, with important consequences for process management and quality control.

This robot has visual capabilities
and was developed to inspect and
monitor hazardous materials.
Source: Martin Marietta
Corporation

As automated inspection becomes more technologically feasible, less expensive and/or more reliable than sampling, the balance shifts to 100% technology-based inspection. This is a definite new trend which promises consumers a step function improvement in the quality of the products they buy.

Robot inspectors have vision capabilities that transcend that of humans, including infrared and ultraviolet sight. They have optical-scanning capabilities with speeds and memories that surpass humans. Additionally, they have the ability to handle and move both heavy and delicate items as well as dangerous chemicals and toxic substances.

There are other changes occurring with respect to who does inspections. In many firms, each worker is responsible for the quality control of what he or she has done. This is proving effective, but the job must be such that the operator is in a position to judge the quality, which is not always the case. For those situations, alternative quality controls are developed and supported by teamwork.

Teamwork originated with Japanese manufacturing practice, but has spread so widely that it cannot be associated with any one country's style. Rather, it should be identified with the type of organization that accepts empowerment of all employees. The employees are often called "associates."

Many organizations now champion the practice of having associates inspect their own work. No other inspectors are used. Quality auditors are gone. There may be a quality department, but it is there to back up the associates with consulting advice. This variant of traditional 100% inspection has been winning adherents in high places in big companies all over the world.

7.14

ACCEPTANCE SAMPLING (AS)

Acceptance sampling is the process of using samples at both the inputs and the outputs of the traditional input-output model. **Acceptance sampling** uses statistical sampling theory to determine if the suppliers' outputs shipped to the producer meet

Acceptance sampling

Uses statistical sampling theory in conjunction with SQC to check conformance to agreed upon standards; process stability is assumed.

standards. Then, in turn, AS is used to determine if the producer's outputs meet the producer's standards for the customers. Acceptance sampling checks conformance to agreed upon standards. It should be noted that SPC is used during the production process. Also, AS operates with the assumption that the production process is stable.

In the 1920s, at Western Electric Company, the manufacturing division of AT&T, statistical theory was used to develop sampling plans that could be employed as substitutes for 100% inspection. Usually, the purchased materials are shipped in lot sizes, at intervals over time. The buyer has set standards for the materials and inspects a sample to make certain that the shipment conforms. Sometimes the supplier delivers the sample for approval before sending the complete shipment.

Military purchasing agencies were one of the first organizations to require the use of acceptance sampling plans for purchasing. There were published military standards such as U.S. MIL STDS-105D, and more recently ANSI/ASQC Z 1.4-1981.

Similar standards applied in Canada and other countries as well. Sampling plans are commonly part of the buyer-supplier contract.

Acceptance sampling is particularly well suited to exported items. Before the products are shipped to the buyer, they are inspected at the exporting plant. Only if they pass are they shipped. The same reasoning makes sense for shipments that move large distances, even within the same country.

There are companies that specialize in acting as an inspection agent for the buyer. The inspectors sample the items in the suppliers' plants with their full cooperation. Suppliers that have a trust relationship with their customers sample their own output for acceptability to ship.

When done properly, acceptance sampling is always used in conjunction with statistical quality control (SQC). It is assumed that when the lot was made the process was stable with a known average for defectives ($\bar{p}$) which is used for either the average percent or average decimal fraction of defective parts. The sample is expected to have that same percent defectives.

Variability is expected in the sample. This means that there are more or less defectives in the sample than exactly the process average ($\bar{p}$). A sampling plan sets limits for the amount of variability that is acceptable. The reasons that the limits might be exceeded include:

1. The number of defectives found in the sample exceeded the limits even though the process was stable with $\bar{p}$. There is a limit as to how much variability will be allowed before this assumption (1) and the lot are rejected.
2. The statistical variability exceeded the limits because the process was out of control (not stable) during the time that the lot was being made.

If the sample has too many defectives, then the lot can be rejected and returned to the manufacturer or it can be *detailed*. This means using 100% inspection to remove defectives. If the inspection was done at the supplier's plant, then there is no need to pay to transport the defectives and then to return them. Also, there is advanced warning that the defective lot has been rejected. This can alter the production schedule and lead to expediting the next lot.

Acceptance Sampling Terminology

The basic elements of sampling plans are:

1. N = *the lot size*. This is the total number of items produced within a homogeneous-production run. Homogeneity means that the system remained unchanged. Often, conditions for a homogeneous run are satisfied only for the specific shipment. Large

shipments can entail several days of production. Each time the process is shutdown, or a new shift takes over, there is the potential for changes impacting the process. SQC can be used to check if the process remains stable.

Every time there is a change of conditions, such as a new start-up (new learning for the learning curve), the stability of the process has to be checked. Each time one worker relieves another, it should be assumed that a new lot has begun. There may be a new start-up after a coffee break or the lunch hour.

It is assumed that the sample is representative of the lot. It also is assumed that the average percent defectives ($\bar{p}$) produced by the process does not change from the beginning to the end of the run. If $\bar{p}$ does change, the sample should pick up that fact because there will be too many or too few defectives in the sample. The "too few" is not a cause for concern.

2. $n =$ *the sample size.* The items to be inspected should be a representative sample drawn at random from the lot. Inspectors know better than to let suppliers choose the sample.

3. $c =$ *the acceptance number.*

4. The sampling criterion is defined:

n items are drawn from a lot of size N,

k items are found to be defective,

if $k > c$, reject the entire lot,

if $k = c$, accept the lot,

if $k < c$, accept the lot.

Example: 5 items ($n = 5$) are drawn from a lot size of 20 ($N = 20$). The acceptance number is set at $c = 2$. Then if 3 items are found to be defective ($k = 3$), reject the entire lot because $k > c$. If 2 or less items are found to be defective, $k \leq c$, accept the lot. It makes sense to remove the defectives from the lot. This means that $n - 2$ units are returned to the lot.

If the acceptance number is $c = 0$, one defective rejects the lot.

O.C. Curves and Proportional Sampling

Operating characteristic curves (O.C. curves) show what happens to the probability of accepting a lot (P_A) as the actual percent defective in the lot (p) goes from zero to one. Each O.C. curve is a unique sampling plan. The construction of O.C. curves is the subject of the material that follows. However, to become familiar with reading O.C. curves is the first requisite.

A good way to start is to examine why the use of proportional sampling does not result in the same O.C. curve for any lot size. Proportional sampling is equivalent to taking a fixed percent (PRC) of any size lot (N) for the sample (n). Thus:

$$\text{PRC} = 100 \left(\frac{n}{N}\right)$$

and

$$n = \frac{\text{PRC} (N)}{100}$$

Set PRC at (say) 20%. Then if the lot size is $N = 100$, the sample size $n = 20$. If $N = 50$, $n = 10$; and if $N = 20$, $n = 4$. Figure 7.15 shows the three O.C. curves that apply where $c = 0$.

Operating characteristic curves (O.C. curves)

A unique sampling plan which shows what happens to the probability of accepting a lot as the actual percent defective in the lot goes from 0 to 1.

O.C. CURVES FOR THREE PROPORTIONAL SAMPLING PLANS

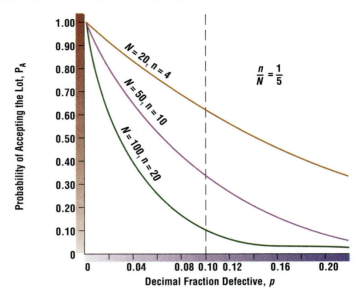

The three O.C. curves show that proportional sampling plans will not deliver equal risk. The $n = 4$ plan is least discriminating. Its probability of accepting lots with 10 percent defectives is about 60 percent. The $n = 10$ plan has a P_A around 35 percent and the $N = 20$ plan has a P_A about 10 percent. Each plan produces significantly different P_A's.

O.C. CURVES WITH $c = 0$ AND $c = 1$

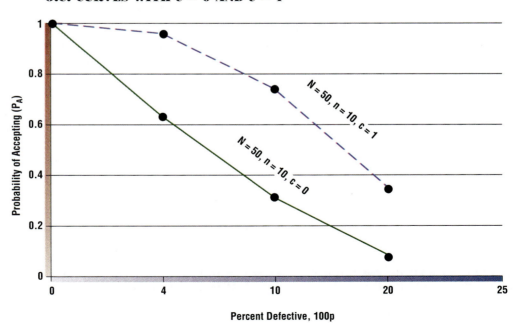

It has been established that O.C. curves with the same *n/N* ratio are not the same curves. The acceptance number *c* has a big effect which can be seen by studying the O.C. curves in Figure 7.16.

When *c* is increased from 0 to 1 for $n = 10$ and $N = 50$, there is a marked decrease in the discriminating ability of the sampling plan. This is because the probability of accepting a lot with a given value of *p* rises significantly when the acceptance number, *c*, increases from zero to one. Consequently, *c* provides control over the robustness of the sampling test. $c = 0$ is always a tougher plan than $c > 0$.

The other controlling parameter is *n*. Return to Figure 7.15 and notice that as *n* goes from 20 to 10 to 4, the probability of acceptance increases. The larger the value of *n*, the more discriminating is the sampling plan. The sample size *n* drives the value of P_A, whereas the effect of *N* is not great, except when it is quite small. The curves in Figure 7.15 above would not differ much from those that would apply there if all of the sampling plans had the same $N = 50$.

Designing a sampling plan requires determining what the values of P_A should be for different levels of *p*. This is to provide the kind of protection that both the supplier and the buyer agree upon. Then, the O.C. curve can be constructed by choosing appropriate values for *n* and *c*. Note that an agreement is required. The design is not the unilateral decision of the buyer alone. The situation calls for compromise and negotiation between the supplier and the buyer. Figure 7.17 shows what is involved.

F I G U R E   7 . 1 7

BUYER'S AND SUPPLIER'S RISK ON O.C. CURVE

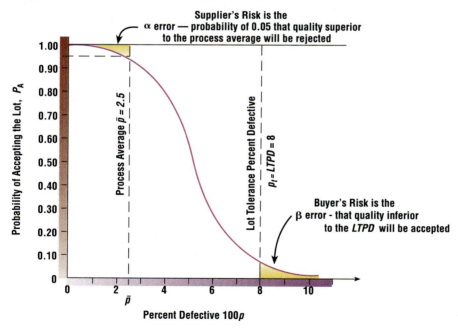

Negotiation between Supplier and Buyer

Two shaded areas are marked alpha & beta (α and β) in Figure 7.17. The α area is called the supplier's risk. α gives the probability that acceptable lots will be rejected by the sampling plan which the O.C. curve represents. Acceptable lots are those which have a fraction defective *p* that is equal to or less than the process average $\bar{p}$, which operates as the limit for α.

The $\overline{p}$ threshold for α is also called the average quality limit (AQL). It makes sense that the supplier does not want lots rejected that are right on the mark or better (lower) than the process average. A sensible buyer-supplier relationship will be founded on the buyer knowing the process average.

The beta area represents the probability that undesirable levels of defectives will be accepted by the O.C. curve of the sampling plan. Beta is called the buyer's risk. The limiting value for beta, defined by the buyer, is called the lot tolerance percent defective, LTPD or p_t. If decimal fraction defective is used instead of percent defective, the limit is called lot tolerance fraction defective (LTFD). It is also known as the reject quality limit (RQL).

LTPD (or LTFD or RQL) sets the limit that describes the poorest quality the buyer is willing to tolerate in a lot. The buyer wants to reject all lots with values of p that fall to the right of this limit on the x-axis of Figure 7.17. The use of sampling methods makes this impossible. There is a probability that the sample will falsely indicate that the percent defective is less than the limiting percent defective p_t. This is the probability of accepting quality that is inferior to the LTPD. Therefore, the buyer compromises by saying that no more than beta percent of the time should the quality level p_t be accepted by the sampling procedure.

Type I and Type II Errors

Decision theory has contrasted the two kinds of errors with which the alpha and beta risks of error correspond. These are Type I and Type II decision errors. The α error falsely rejects good lots, equivalent to making the mistake of finding an innocent person guilty.

Alpha is known as a Type I error and identified as an error of omission.

The β error is the probability of incorrectly accepting bad lots, equivalent to making the mistake of finding a guilty person innocent.

Beta is known as a Type II error and identified as an error of commission.

The Cost of Inspection

Given the process average, $\overline{p}$, α, and the consumer's specification of p_t, and β, then a sampling plan can be found that minimizes the cost of inspection. This cost includes the charge for detailing, which entails inspecting 100 percent of the items in all rejected lots to remove defective pieces.

Such a sampling plan imputes a dollar value to the probability (α) of falsely rejecting lots of acceptable quality. The supplier can decrease the α-type risk by improving quality and thereby reducing the process average. Real improvement to $\overline{p}$ might be costly. It is hoped that the buyer would be willing to accept part of this increased cost in the form of higher prices. In one way or another, the supplier and the buyer will share the cost of inspection. Therefore, it is logical to agree that the rational procedure to be followed is to minimize inspection costs as they negotiate about the values for LTPD and β.

The sampling plan will deliver the specified α and β risks for given levels of $\overline{p}$ and p_t. If the process average, $\overline{p}$, is the true state of affairs, then α percent of the time, $N - n$ pieces will be detailed. Then, the average number of pieces inspected per lot will be $n + (N - n)(\alpha)$. This number is called the average sample number (ASN).

Average Outgoing Quality Limits (AOQL)

An important acceptance sampling criterion is average outgoing quality (AOQ). AOQ measures the average percentage of defective items the supplier will ship to the buyer, under different conditions of actual percent defectives in the lot.

In each lot, the sample of n units will be inspected. The remaining $(N - n)$ units will only be inspected if the lot is rejected. The expected number of defectives in the unsampled portion of the lot is $\bar{p}(N - n)$. Out of every 100 samples, P_A will be the percent of samples passed without further inspection. The defective units in the samples passed will not be replaced (detailed). On the other hand, $1 - P_A$ of the 100 samples will be rejected and fully detailed identifying all defectives. Thus, $P_A\bar{p}(N - n)$ is the expected number of defectives that will be passed without having been identified for every N units in a lot.

The average outgoing quality (AOQ) changes value as p goes from zero to one. Thus:

$$\text{AOQ} = P_A p (N - n)/N \qquad \text{Equation 7.5}$$

Note that P_A changes with p in accordance with the specifics of the sampling plan, which is based on the values of n and c that are chosen. Thus, AOQ is a function of all the elements of a sampling plan. It is, therefore, a useful means of evaluating a sampling plan. Specifically, for each value of p, an AOQ measure is derived. It is the expected value of the fraction defectives that would be passed without detection if the process was operating at value p. The AOQ measure reaches a maximum level—called the average outgoing quality limit (AOQL)—for some particular value of p. This is shown in Figure 7.18.

F I G U R E   7 . 1 8

AVERAGE OUTGOING QUALITY AS A FUNCTION OF THE PROCESS FRACTION DEFECTIVE, p

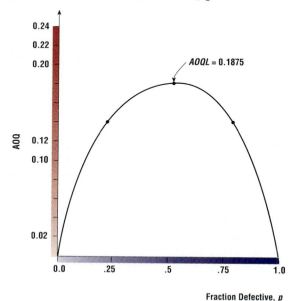

Fraction Defective, *p*

Table 7.8 presents the AOQ computations which are based on Equation 7.5 above. The AOQL value occurs when $p = 0.5$. This particular sampling plan (P_A as a function of p) is developed by the hypergeometric method described (see Supplement 7).

TABLE 7.8

COMPUTING AVERAGE OUTGOING QUALITY LIMIT

p	P_A	$(N - n)/N$	AOQ
0	1	3/4	0
1/4	3/4	3/4	9/64
1/2	1/2	3/4	12/64 ← AOQL
3/4	1/4	3/4	9/64
1	0	3/4	0

The maximum value of AOQ, the AOQL has been calculated as 12/64. This average outgoing quality limit (AOQL) describes the worst case of fraction defectives that can be shipped, if it is assumed that the defectives of rejected lots are replaced with acceptable product and that all lots are thoroughly mixed so that shipments have homogeneous quality.

Multiple-Sampling Plans

Multiple-sampling plans are utilized when the cost of inspection required by the single-sampling plan is too great. Single-sampling plans require that the decision to accept or reject the lot must be made on the basis of the first sample drawn. With double sampling, a second sample is drawn if needed. Double sampling is used when it can lower inspection costs.

The double-sampling plan requires two acceptance numbers c_1 and c_2 such that $c_2 > c_1$. Then, if the observed number of defectives in the first sample of size n_1 is k_1:

1. Accept the lot if $k_1 \leq c_1$.
2. Reject the lot if $k_1 > c_2$.
3. If $c_1 < k_1 \leq c_2$, then draw an additional sample of size n_2. The total sample is now of size $n_1 + n_2$.

The total number of defectives from the double sample is now $k_1 + k_2$. Then:

4. Accept the lot if $(k_1 + k_2) \leq c_2$.
5. Reject the lot if $(k_1 + k_2) > c_2$.

Double sampling saves money by hedging. The first sample, which is small, is tested by a strict criterion of acceptability. Only if it fails are the additional samples taken. Double-sampling costs have to be balanced against the costs of the single large sample.

Multiple sampling follows the same procedures as double sampling where more than two samples can be taken, if indicated by the rule: $c_j < k_j \leq c_{(j+1)}$. As with double sampling, the sample sizes are specified with upper and lower acceptance numbers which get larger for additional samples. When the cumulative number of defects exceeds the highest acceptance number c_n, the lot is rejected. Thus, reject the lot if:

$$(k_1 + k_2 + \ldots + k_j + \ldots + k_n) > c_n$$

In the same sense, the costs of multiple-sampling plans can be compared to single or double sampling. Sequential sampling may be cost effective where many samples of large size have to be taken regularly. The sequential sampling procedure is to take successive samples with rules concerning when to continue and when to stop. In effect, it determines the spacing shown in Table 7.3 above.

Tables exist for single- and double-sampling plans which make the choice of a plan and the design of that plan much simpler.[7]

7.15

BINOMIAL DISTRIBUTION

For the binomial distribution to apply, the sample size should be small as compared to the lot size. The criterion is given after Equation 7.6, which is the binomial distribution.

$$P_A = \text{Prob } (j;n,p) = \frac{n!}{j!(n-j)!}(p^j)(q^{n-j}) \qquad \textbf{Equation 7.6}$$

n = sample size; p = the decimal fraction defective and $q = 1 - p$.

P_A is the probability of acceptance as previously used in the chapter. The lot size N is assumed to be infinite so it does not appear in Equation 7.6. The assumption of large enough N is considered to be met when $n/N \leq 0.05$.

When $j = 0$, the probabilities are calculated for the sampling plan with $c = 0$. Just to be sure that the factorial is known, $n! = n(n-1)(n-2)...1$ and $0! = 1! = 1$.

The O.C. curve for $n = 100$, $c = 0$, is calculated from the computing Equation 7.7 for sample size $n = 100$, and the acceptance number $c = 0$. With this sample size, it is expected that the lot is 2,000 or larger. Also, this sample size has been chosen to enable a later comparison with the use of the Poisson distribution for creating O.C. curves and because most tables for the binomial distribution do not go up to $n = 100$.[8]

$$\text{Prob } (0; 100, p) = \frac{100!}{0!(100-0)!}(p^0)(q^{100}) = q^{100} = (1-p)^{100}$$

$$\textbf{Equation 7.7}$$

Note that 100! is divided by 100! which equals one and $p^0 = 1$. Table 7.9 is derived by using this binomial computing formula with the beginning of the entire range of possible values of fraction defectives.

T A B L E   7 . 9

BINOMIAL DERIVATION OF O.C. CURVE WITH $j = 0$

Fraction Defective (p)	$1 - p$	$P_A = (1-p)^{100}$
0.00	1.00	1.000
0.01	0.99	0.366
0.02	0.98	0.133
0.04	0.96	0.017
0.05	0.95	0.006
.	.	.
.	.	.
1.00	0.00	0.000

BINOMIAL O.C. CURVE–c = 0

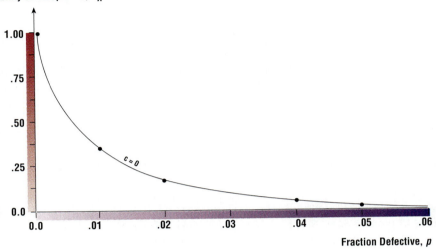

Probability of Acceptance, P_A

$c = 0$

Fraction Defective, p

Figure 7.19 shows the plot of this O.C. curve.

For the acceptance number $c = 1$, it is necessary to derive the binomial for $j = 0$ and $j = 1$. Add these results together, thus: Prob(0; 100, p) + Prob(1; 100, p) will be used to derive the O.C. curve for $n = 100$ and $c = 1$. The computations follow in Equation 7.8:

$$\text{Prob } (1; 100, p) = \frac{100!}{1!(100 - 1)!}(p^1)\,(q^{99}) = 100p\,(1 - p)^{99} \quad \textbf{Equation 7.8}$$

In this case 100! is divided by 99! which leaves 100 as a remainder. The result can be used for computing Table 7.10 with the beginning of the entire range of possible values of fraction defectives, p.

BINOMIAL DERIVATION OF O.C. CURVE WITH $j = 1$

Fraction Defective (p)	$q = 1 - p$	$P_A = 100p\,(1 - p)^{99}$
0.00	1.00	0.000
0.01	0.99	0.370
0.02	0.98	0.271
0.04	0.96	0.070
0.05	0.95	0.031
.	.	.
.	.	.
1.00	0.00	0.000

Calculation of $P_A = 100p(1 - p)^{99}$ when $q = 0.99, 0.98$, and so on is done readily with a math calculator or the computer. However, for those who want a computing method without electronic aids, use logarithms and antilogs remembering that decimals have negative characteristics.

To calculate the $c = 1$ plan, add the P_A values for each row of Tables 7.9 and 7.10 above. Thus,

TABLE   7 . 1 1

BINOMIAL DERIVATION OF O.C. CURVE WITH $c = 1$

Fraction Defective (p)	$P_A (c = 0)$ $P_A (j = 0)$	$P_A (j = 1)$	$P_A (c = 1)$ Sum $P_A (j = 0$ and $j = 1)$
0.00	1.000	0.000	1.000
0.01	0.366	0.370	0.736
0.02	0.133	0.271	0.404
0.04	0.017	0.070	0.087
0.05	0.006	0.031	0.037
.	.	.	.
.	.	.	.
1.00	0.000	0.000	0.000

Figure 7.20 shows the plot of this O.C. curve and that of Figure 7.19 so that a comparison can be made.

FIGURE   7 . 2 0

BINOMIAL O.C. CURVE–$c = 1$

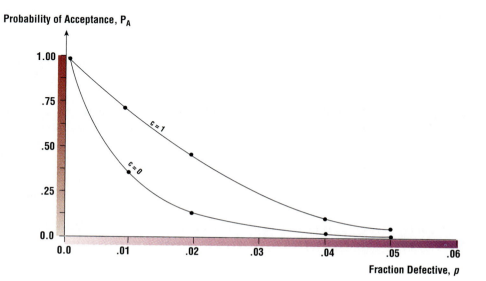

As expected, the O.C. curve for $c = 0$ is far more stringent than for $c = 1$.

7.16

POISSON DISTRIBUTION

An even easier method for creating O.C. curves uses the Poisson distribution[9] to approximate the binomial distribution when p is small (viz., when $p < 0.05$). The mean of this distribution is $m = np$ with n the sample size and p the fraction defective. A sample size of $n \geq 20$ is customary. Like the binomial, it assumes sampling with replacement which means that N is large enough to be treated as not affected by sampling.

The fundamental equation for the Poisson distribution is:

$$P_A = \text{Prob}(j; m) = \frac{m^j e^{-m}}{j!}$$ **Equation 7.9**

Let $j = 0$, which is equivalent to $c = 0$, and let $n = 100$. Then, because $m = np$, the mean $m = 100p$. Thus, Prob $(0; m) = e^{-100p}$. It is simplest however to use the mean value in the center column of Table 7.12 to calculate the values of the probability of acceptance in Table 7.12.

TABLE 7.12

POISSON DERIVATION OF O.C. CURVE WITH $j = 0$

Fraction Defective (p)	$100p = m$	$P_A = e^{-m}$
0.00	0	1.000
0.01	1	0.368
0.02	2	0.135
0.04	4	0.018
0.05	5	0.007
.	.	.
.	.	.
1.00	100	0.000

Observe that this result is essentially the same as that obtained by using the binomial distribution. As with the binomial, a sampling plan where $c = 1$ requires computations of P_A for $j = 0$, and $j = 1$.

The computing equation is:

$$P_A = \text{Prob}(1; m) = \frac{m^1 e^{-m}}{1!} = me^{-m}$$ **Equation 7.10**

Again, it is easier to use m in the middle column of Table 7.13 to calculate the amount that $j = 1$ increases the probability of acceptance.

TABLE 7.13

POISSON CALCULATIONS FOR O.C. CURVE WITH $j = 1$

Fraction Defective (p)	$100p = m$	$P_A = me^{-m}$
0.00	0	1.000
0.01	1	0.368
0.02	2	0.271
0.04	4	0.073
0.05	5	0.034
.	.	.
.	.	.
1.00	100	0.000

Summing the third column contributions to P_A row by row from Tables 7.12 and 7.13 above, Table 7.14 is derived.

TABLE 7.14

POISSON DISTRIBUTION FOR O.C. CURVE WITH $c = 1$

Fraction Defective (p)	P_A ($c = 0$) ($j = 0$)	($j = 1$)	P_A ($c = 1$) Sum P_A ($j = 0$ and $j = 1$)
0.00	1.000	0.000	1.000
0.01	0.368	0.368	0.736
0.02	0.135	0.271	0.406
0.04	0.018	0.073	0.091
0.05	0.007	0.034	0.041
.	.	.	.
.	.	.	.
1.00	0.000	0.000	0.000

The Poisson approximation of the binomial is good for small values of p. Compare the binomial tables with the Poisson tables through the range of $p = 0.00$ to 0.05. These O.C. curves are sufficiently similar that either approach can be used. However, at 0.05 the error rises above 10 percent. Previously, it was stated that the Poisson should be used with $p < 0.05$, or 5 percent. Because the binomial is always more accurate, if there is doubt about using the Poisson, the binomial is not a difficult alternative.

It will be instructive to compare one larger value (for $c = 0$ and $p = 0.10$). The binomial result is:

$$P_A = (1 - 0.10)^{100} = 0.000027$$

The Poisson result is:

$$P_A = e^{- 100(0.10)} = 0.000045$$

The Poisson approximation yields an error of 67 percent when $p = 0.10$, which is substantial.

S U M M A R Y

The methodology of quality management is extensive. This chapter presents the major approaches used including data check sheets, histograms, scatter diagrams, Pareto charts, Ishikawa/fishbone diagrams. Then the most important of all, statistical control charts. ($\bar{x}$-, R-, p- and c-charts) are explained—how they work and why they work. Average outgoing quality limits and multiple sampling are part of the treatment of acceptance sampling methods used for buyer-supplier quality relations.

▶ Key Terms

acceptance sampling   *283*
cause and effect charts   *261*
histograms   *256*
Pareto analysis   *258*
run   *280*
$\bar{x}$-chart   *267*

bar chart   *256*
control chart   *267*
operating characteristic curves (O.C. curves)   *285*
process quality control methods   *253*
scatter diagrams   *263*

c-chart   *278*
control limits   *269*
p-chart   *275*
R-chart   *274*
20:80 rule   *259*

▶ Review Questions

1. Develop a data check sheet for:
   a. basketball scores
   b. top ten players for runs batted in (RBIs)
   c. some football statistics (provide some options)

2. Develop a data check sheet for set-up times for:
   a. service time in a sit-down restaurant
   b. service time in a fast-food restaurant

   Note: First define service time explicitly.

3. Draw a fishbone diagram for the causes of poor reports.

4. An airline has requested a consultant to analyze its telephone reservation system. As the consultant, draw a fishbone diagram for the causes of poor telephone service.

5. Prepare a cause and effect analysis for the qualities of a fine cup of tea.

6. How do run charts give early warning signals that a stable system may be going out of control?

7. What is the purpose of control charts?

8. What would a go/no-go gauge which has tolerances of ± 0.04 look like for 2.54 centimeter nails? Draw it.

9. Explain why service systems are said to have higher levels of variability from chance causes than manufacturing systems.

10. A job shop that never produces lot sizes greater than 20 units cannot employ the $\bar{x}$- and R-charts of SQC. Why is this so? What kind of quality management can be used?

11. Stylish Packaging is a flow shop with a high-volume, serial-production line making cardboard boxes. An important quality is impact resistance, which requires destructive testing. Can SQC methods be used with destructive testing? What quality control method(s) are recommended? Discuss.

12. Can the consumer tell when an organization employs SQC? Can the consumer tell when it does not employ SQC?

13. Differentiate between assignable and chance causes of variation. What other names are used for each type of cause? Give some examples of each kind of variation. How can one tell when an assignable cause of variation has arisen in a system?

14. What is a stable system? Why is the condition of stability relevant to the activities of P/OM?

15. Why is the use of statistical quality control by drug and food manufacturers imperative? Would you recommend 2-sigma or 3-sigma control limits for food and drug applications of SQC? Explain.

16. One of the major reasons statistical quality control is such a powerful method is its use of sequenced inspection. Explain why this is so.

17. How should the subgroup size and the interval between samples be chosen? Is the answer applicable to both high-volume, short-cycle time products (interval between start and finish) and low-volume, long-cycle time products?

18. Differentiate between the construction of the $\bar{x}$-, R-, p- and c-charts. Also distinguish between the applications of these charts. Are the charts applicable to both high-volume, short-cycle time products (interval between start and finish) and low-volume, long-cycle time products?

19. What is meant by "between-group" and "within-group" variability? Explain how these two types of variability constitute the fundamental basis of SQC.

20. Use the concepts of "within-group" variability and "between-group" variability to answer the following questions:
    a. Are the designer's specifications realistic?
    b. Is the production process stable?
    c. Is the process able to deliver the required conformance specifications?

▶ **Problem Section**

1. Draw a scatter diagram for the following data where it is suggested that df is a function of T.

Sample No.	T	df
1	120	80
2	110	75
3	105	65
4	112	73
5	118	78

If T is the temperature of the plating tank and df is the measure of defective parts per thousand, after plating, does there seem to be a special relationship (correlation) between T and df?

2. After set up in the job shop, the first items made are likely to be defectives. If the order size calls for 125 units, and the expected percent defective for start-up is 7 percent, how many units should be made?

3. A subassembly of electronic components, called M1, consists of 5 parts that can fail. Three parts have failure probabilities of 0.03. The other two parts have failure probabilities of 0.02. Each M1 can only be tested after assembly into the parent VCR. It takes a week to get M1 units (the lead time is one week), and the company has orders for the next five days of 32, 44, 36, 54, and 41. How many M1 units should be on hand right now so that all orders can be filled?

4. Use SPC with the table of data below to advise this airline about its on-time arrival and departure performance. These are service qualities highly valued by their customers. The average total number of flights flown by this airline is 660 per day. This represents three weeks of data. It is suggested that a late flight be considered as a defective. Draw up a p-chart and analyze the results. Discuss the approach.

THE NUMBER OF LATE FLIGHTS (NLF) EACH DAY

Day	NLF	Day	NLF	Day	NLF	
1	31	8	43	15	22	
2	56	9	39	16	34	
3	65	10	41	17	29	
4	49	11	37	18	31	
5	52	12	48	19	35	
6	38	13	45	20	44	
7	47	14	33	21	37	

5. New data have been collected for the CTF. It is given in the table below. Start your analysis by creating an $\bar{x}$-chart and then create the R-chart. Provide interpretation of the results.

WEIGHT OF CHOCOLATE TRUFFLES (IN GRAMS)

Subgroup j = Time: x_i	I 10 A.M.	II 11 A.M.	III 1 P.M.	IV 3 P.M.	V 4 P.M.	SUM
1	30.00	30.25	29.75	29.90	30.05	
2	30.50	31.05	29.80	29.00	29.60	
3	29.95	30.00	30.05	29.95	29.90	
4	30.60	29.70	29.80	29.65	29.85	

6. For the data in Table 7.4 discussed earlier, throw out the 3:00 P.M. data and substitute instead the following four values of x_i:
   30.15, 29.95, 29.80, and 30.05.
   Recalculate the parameters for the $\bar{x}$-chart and draw the revised chart. Discuss the results.

7. Recalculate the R-chart parameters without 3:00 P.M. data and using the new data supplied in Problem 6 above for Table 7.4. Discuss the results.

8. What is the probability of a run of eight points all above the mean? What is the probability of a run of ten points above or below the mean?

9. For the data tabulated below, draw the p-chart. Compare these results with those based on Table 7.7.

Subgroup No.	NR	n	p
1	1	9	–
2	2	9	–
3	1	9	–
4	3	16	–
5	1	9	–
	SUM	SUM	

10. A food processor specified that the contents of a jar of salsa should weigh 14 ± 0.10 ounces net. A statistical quality control operation is set up and the following data are obtained for one week:

Sample No.				
	1	2	3	4
1	14.10	14.06	14.25	14.06
2	13.90	13.85	13.80	14.00
3	14.40	14.30	14.10	14.20
4	13.95	14.10	14.00	14.15
5	14.05	13.90	13.95	14.60

a. Construct an $\bar{x}$-chart based on these 5 samples.
b. Construct an R-chart based on these 5 samples.
c. What points, if any, have gone out of control?
d. Discuss the results.

11. Use a data check sheet to track the Dow Jones average, regularly reported on the financial pages of most newspapers. Record the Dow Jones closing index value on a data check sheet every day for one week. Do an Ishikawa analysis, trying to develop hypotheses concerning what causes the Dow Jones index to move the way it does. Draw scatter diagrams to see if the hypothesized causal factors are related to the Dow Jones.

12. A method for determining the number of subgroups that will be sampled each day will be developed. Apply it to the situation where the total number of units produced each day is 1,000. The subgroup sample size is 30 units and the interval between subgroups is 200 units.

a. How many subgroups will be sampled each day?
b. How many units will be sampled each day?
c. What should be done if the sampling method damages one out of three units?

Method: Dividing the production rate per day by the subgroup sample size plus the interval between samples (given in units) determines the number of samples to be taken per day. If the subgroup sample size is $n = 3$ units, and the interval between sample subgroups is $t = 4$ units, and the total number of units produced each day is $P = 210$, then the calculation to determine the number of subgroups sampled each day, called NS, is as follows:

$$NS = P/(n + t) = 210/(3 + 4) = 30$$

The total number of units that are sampled and tested for quality is $n(NS) = 90$. This means that 90 units would be withdrawn from the production line and tested. If the quality-testing procedure damages the product in any way, the sample interval would be increased and the sample size might be reduced to two.

▶ **Readings and References**

G. Bounds, L. Yorks, M. Adams, and G. Ranney, *Beyond Total Quality Management*, McGraw-Hill Book Co., Inc., 1994.

P. E. Crosby, *Quality Is Free*, NY, McGraw-Hill Book Co., Inc., 1979.

P. E. Crosby, *Quality Is Still Free*, 1st Ed., NY, McGraw-Hill Book Co., Inc., 1995.

D. J. Crowden, *Statistical Methods in Quality Control*, Englewood Cliffs, NJ, Prentice-Hall, Inc., 1957.

W. E. Deming, *Out of the Crisis*, Boston, MA: MIT—Center for Advanced Engineering Studies, 1986.

Harold F. Dodge, and Harry G. Romig, *Sampling Inspection Tables, Single and Double Sampling*, 2nd Ed., NY, Wiley, 1959.

Eugene L. Grant, and Richard S. Leavenworth, *Statistical Quality Control*, 6th Ed., NY, McGraw-Hill Book Co., Inc., 1988.

Richard Tabor Greene, *Global Quality: A Synthesis of the World's Best Management Methods,* Burr Ridge, IL, Irwin, 1993.

A. V. Feigenbaum, *Total Quality Control*, 3rd. Ed., NY, McGraw-Hill Book Co. Inc., 1991.

K. Ishikawa, *What is Total Quality Control? The Japanese Way*, Englewood Cliffs, NJ: Prentice-Hall, Inc., 1985.

J. M. Juran, *Quality Control Handbook*, 4nd Ed., NY, McGraw-Hill Book Co., Inc., 1988.

S. B. Littauer, "Technological Stability in Industrial Operations," *Transactions of the New York Academy of Sciences*, Series II, Vol. 13, No. 2, (Dec. 1950), pp. 67–72.

Jeremy Main, *Quality Wars: The Triumphs and Defeats of American Business*, A Juran Institute Report, The Free Press, NY, 1994.

P. R. Scholtes, et. al., *The Team Handbook*, Joiner Associates, (608-238-8134), Madison, WI, 1989.

W. A. Shewhart, *Economic Control of Quality of Manufactured Product*, Princeton, NJ, D. Van Nostrand Co., Inc., 1931.

Shigeo Shingo, *Zero Quality Control*, CT, Productivity Press, 1986.

H. Wadsworth, K. Stephens, and A. Godfrey, *Modern Methods for Quality Control and Improvement*, NY, Wiley, 1986.

Hypergeometric O.C. Curves

With this supplement, three methods of designing O.C. curves for single sampling will have been developed. Why bother with this third method when the binomial and the Poisson have been shown to be close enough under the demonstrated conditions? The reason for choosing one method instead of another is based on the relationship of the sample size n to the lot size N. In the case of the hypergeometric, the small size of N makes it necessary to do sampling without replacement.

This means that if there is a lot size of $N = 4$, and it is known that there is one defective in that lot, then for the first sample there is a one in four chance of drawing the defective part. If the defective is not found in this first sample, then the second sample has a one out of three chance of being the defective. The probability goes from 25 to 33.3 to 50 percent. The hypergeometric distribution takes this effect into account, whereas the binomial and the Poisson do not.

Stating this in yet another way, the hypergeometric distribution is used when the lot size N is small and successive sampling affects the results. Each next sample reduces the number of unsampled units remaining in the lot. Thus, the sample is first drawn from N units, then from $N - 1$, $N - 2$, ... units, etc.

If items are replaced to keep N of constant size, there is a significant chance that the same item would be sampled more than once. Also, with destructive testing, sampled items cannot be replaced because they have to be destroyed to be tested.

The hypergeometric distribution given in Equation 7.11 applies to a sampling plan with acceptance number, $c = 0$. (Repeated note: the factorial of any number is written $m!$ and is calculated by $(m)(m - 1)(m - 2)...(1)$.)

$$P_A = \frac{(N - x)!(N - n)!}{(N - x - n)!N!} \qquad \textbf{Equation 7.11}$$

$x \quad =$ the possible number of defectives in the lot and
$x/N = p$ is the possible fraction defective of the lot.

As x is varied from 0 to N, the appropriate values of P_A are determined. This allows a plot of P_A vs. p (or $100\,p$, the percent defective). Thus, in the case where: $N = 4$, $n = 1$, and $c = 0$, Equation 7.12 for P_A is:

$$P_A = \frac{(4 - x)!(3)!}{(3 - x)!4!} = \frac{(4 - x)!}{4(3 - x)!} = \frac{4 - x}{4} = 1 - p \qquad \textbf{Equation 7.12}$$

where p is the symbol for fraction defective. The last term of Equation 7.12 is $(1 - p)$, and it is obtained by substituting $x = Np = 4p$ into the next to the last term of the equation.

Varying p to derive P_A results in Table 7.15. This is the same distribution that was used to derive the values of AOQ and AOQL in this chapter.

TABLE 7.15

HYPERGEOMETRIC DISTRIBUTION

x	$\dfrac{x}{N} = p$	P_A
0	0.00	1.00
1	0.25	0.75
2	0.50	0.50
3	0.75	0.25
4	1.00	0.00

The hypergeometric formula when $c > 0$ has the same additive requirements for the probability of acceptance with $j = 0$ and $j = 1$ that were shown for the binomial and Poisson distributions.

Equation 7.13 gives the general hypergeometric equation.

$$P_j = \frac{C_{n-j}^{N-x}\, C_j^{x}}{C_n^{N}} = \frac{(N-x)!\,x!\,n!\,(N-n)!}{(n-j)!\,(N-x-n+j)!\,j!\,(x-j)!\,N!} \qquad \text{Equation 7.13}$$

The numerical example first derives the hypergeometric O.C. curve for $c = 0$ with $j = 0$, and then, using addition, for $j = 1$ and finally for $c = 1$. Also, for this example, the lot size is $N = 50$ and the sample size is $n = 10$. All calculations are based on Equation 7.13 above.

TABLE 7.16

HYPERGEOMETRIC FOR O.C. CURVES WITH $j = 0$ AND $c = 0$

x	$p = x/N$	P_A	
0	0.00	$\dfrac{50!\,10!\,40!}{10!\,40!\,50!}$	$= 1.000$
2	0.04	$\dfrac{48!\,2!\,10!\,40!}{10!\,38!\,2!\,50!}$	$= 0.637$
5	0.10	$\dfrac{45!\,5!\,10!\,40!}{10!\,35!\,5!\,50!}$	$= 0.311$
10	0.20	$\dfrac{40!\,10!\,10!\,40!}{10!\,30!\,10!\,50!}$	$= 0.083$

The next step is to determine the additions to the probability of acceptance when $j = 1$.

TABLE 7.17

HYPERGEOMETRIC FOR O.C. CURVES WITH $j = 1$

x	$p = x/N$	P_A+
0	0.00	There are no defectives in the lot, therefore the additional probability of acceptance with the relaxed acceptance criterion of $j = 1$ is zero. P_A+ signifies that this is measuring the additional probability of acceptance.
2	0.04	$\dfrac{48!2!10!40!}{9!39!50!} = 0.326$
5	0.10	$\dfrac{45!5!10!40!}{9!36!4!50!} = 0.432$
10	0.20	$\dfrac{40!10!10!40!}{9!31!9!50!} = 0.268$

Table 7.18 adds the respective rows of Tables 7.16 and 7.17 above.

TABLE 7.18

THE $c = 1$ HYPERGEOMETRIC DISTRIBUTION

x	p	$c = 0$ $j = 0$	$j = 1$	P_A $c = 1$
0	0.00	1.000	0.000	1.000
2	0.04	0.637	0.326	0.963
5	0.10	0.311	0.432	0.743
10	0.20	0.083	0.268	0.351

This $c = 1$ plan is a nonrigorous acceptance plan because it has such high probabilities of accepting 10 percent and even 20 percent defectives.

Problems for Supplement 7

1. Develop the hypergeometric for $N = 5$, $n = 2$, and $c = 0$. Write out the table for P_A. Draw the O.C. curve.

2. Develop the hypergeometric for $N = 5$, $n = 2$, and $c = 1$. Write out the table for P_A. Draw the O.C. curve.

3. Develop the hypergeometric for $N = 50$, $n = 2$, and $c = 0$. Write out the table for P_A. Draw the O.C. curve. Compare this to the hypergeometric distribution that was developed in the supplement for $N = 50$, $n = 10$, and $c = 0$.

4. Develop the hypergeometric for $N = 50$, $n = 2$, and $c = 1$. Write out the table for P_A. Draw the O.C. curve. Compare this to the hypergeometric distribution that was developed in the supplement for $N = 50$, $n = 10$ and $c = 1$.

5. The situations in Problems 3 and 4 meet the binomial requirement that $n/N < 0.05$. Use the binomial distribution to create the table for P_A. Draw the O.C. curve. Compare the results of the hypergeometric in this case with those of the binomial distribution.

6. Comment on the use of the Poisson distribution for the situations given in Problems 3 and 4 above.

Notes...

1. W. E. Deming, *The New Economics, for Industry, Government, Education*, Cambridge, MA, MIT Center for Advanced Engineering Studies, 1993, p. 135. Deming used *Study*, not *Check*.

2. Shigeo Shingo, *Zero Quality Control*, CT, Productivity Press, 1986, p. 32.

3. Walter A. Shewhart, *Economic Control of Quality of Manufactured Product*, Princeton, NJ, D. Van Nostrand, 1931. Walter Shewhart's seminal work dates back to the 1930s.

4. "They use a less environmentally friendly system," Tom Meschievitz, director of Paint Engineering, General Motors–North American Operations. "Big 3 in low-pollution paint venture," *The Boston Globe*, May 26, 1994.

5. K. Ishikawa, *What is Total Quality Control? The Japanese Way*, Englewood Cliffs, NJ, Prentice-Hall, 1987.

6. It is useful to note the effect of $\pm k\sigma$ on the probability ($P_{\bar{x}}$) that a sample mean falls within the control limits.

k	$P_{\bar{x}}$	k	$P_{\bar{x}}$
1.00	68.26%	2.00	95.44%
1.64+	90.00%	3.00	99.73%
1.96	95.00%	4.00	99.99%

7. Harold F. Dodge, and Harry G. Romig, *Sampling Inspection Tables, Single and Double Sampling*, NY, Wiley, 2nd ed., 1959.

8. A classic collection of mathematical tables is in *The Handbook of Mathematical Functions with Formulas, Graphs, and Mathematical Tables*, U.S. Department of Commerce, National Bureau of Standards, Applied Mathematics Series 55, June, 1964.

9. The Poisson Table in the Appendix can be used to determine the probabilities of c or less defects where c and j are equivalent.

Chapter 8

Management of OM Technology

After reading this chapter you should be able to...

1. Explain the role of technological transformation as it relates to OM.
2. Discuss the specifics of technology management relating it to the management of skills.
3. Specify a ratio percent that reflects the mixture of skills and technology used over time.
4. Explain what happens when new technology is applied to old systems, and why this is called the technology trap.
5. Explain technology timing.
6. Discuss the evolution of the technology of an industry such as automotive windshield glass.
7. Deal with advanced technology for a complex industrial product such as automotive windshield glass.
8. Explain the interaction of process technology with the supply chain to deliver the product.
9. Differentiate between the management of technology (MOT) for the original product market and the replacement product market.
10. Explain OM's role in the management of the technology of packaging and delivery systems.
11. Discuss the reason for including the technology of testing all aspects of product qualities as part of OM's responsibility.
12. Explain how the management of testing technology relates to OM's awareness of legal statutes and governmental rules.
13. Reveal why design for manufacturing (DFM) and design for assembly (DFA) are vitally important concepts.
14. Explain why design for manufacturing and design for assembly are treated as a management of technology challenge.
15. Discuss procedures for evaluating and selecting designs based on DFM and DFA criteria.
16. Illustrate OM's technology management responsibilities for changeover capabilities.
17. Describe the new technologies that OM will be managing in the future including flexible systems, expert systems, cyborgs and robots, and miniaturization.

Chapter Outline

8.0 TECHNOLOGY COMPONENT OF TRANSFORMATION

8.1 MANAGEMENT OF THE TECHNOLOGY COMPONENT

8.2 THE TECHNOLOGY TRAP

8.3 TECHNOLOGY TIMING

8.4 THE ART AND SCIENCE OF TECHNOLOGY

8.5 THE DEVELOPMENT OF GLASS PROCESSING TECHNOLOGIES
Technological Innovations—Product and Process
Supply Chain Technology
Replacement vs. Original Market Technology

8.6 THE TECHNOLOGY OF PACKAGING AND DELIVERY
Systems Analysis

8.7 THE TECHNOLOGY OF TESTING
Testing by Simulation

8.8 QUALITY CONTROL TECHNOLOGY

8.9 TECHNOLOGY OF DESIGN FOR MANUFACTURING (DFM) AND
DESIGN FOR ASSEMBLY (DFA)
Changeover Technologies

8.10 THE NEW TECHNOLOGIES—CYBERNETICS
Flexibility—FMS, FOS, FSS, and FPS
Expert Systems
Cyborgs and Robots
The Technology of Computer-User Interfaces
Other New Technology Developments—Materials and Miniaturization

Summary
Key Terms
Review Questions
Problem Section
Readings and References
Supplement 8: The Poka-yoke System

The Systems Viewpoint

The management of technology is a shared responsibility with powerful leverage for success and failure. Correct timing for the acquisition of new technology is more successful when it is a team effort because it involves many forms of expertise. These consist of estimates of when competitors will change technology; OM evaluations of the effect of technology changes on productivity, unit costs and quality; judgments about the effects of the OM changes on customers' satisfaction; forecasts about how technology is changing and its rate of change; financial advice concerning the best timing of major expenditures; and management consideration of the availability of management as a resource that can deal with the shift in technology. The measuring concepts that go along with the systems approach to managing technology timing include payback period, breakeven time, and cost recovery period. The text explains why the management of the technology for packaging and delivery of products underscores the need for the systems approach. The same can be said about the technology of testing.

8.0

TECHNOLOGY COMPONENT
OF TRANSFORMATION

Technology

Consists of equipment employed in a manner to bring about specific transformations.

Technology plays a commanding role in accomplishing the input-output transformation that is OM's main concern. From the OM perspective, **technology** consists of equipment employed in a manner—defined by scientific knowledge about the conditions necessary—to bring about specific transformations. The equipment includes computers, machinery, tools and communication systems. Combined with people's skills, materials are transformed and services are provided. Technology components of transformation also can be categorized by types of processes such as physically transforming glass, rubber or steel, assembling parts to make toasters, airplanes or computer chips, transferring information, and providing services.

8.1

MANAGEMENT OF THE TECHNOLOGY
COMPONENT

Management of technology (MOT)

Component of the input-output systems transformation functioning at the societal, company, and international levels.

Management of the technology component of input-output systems transformations is often called **management of technology (MOT)** when it is taught as a subject in business schools. The management of technology functions at three distinct levels.

1. At the societal level, management of technology for products and processes includes governmental regulations controlling technology for health and safety purposes. Also governmental, patent jurisprudence operates to stimulate invention of new technological transformation systems by protecting inventors against competition for lengthy periods of time. Illegal copying of an invention and profiting thereby is called *infringement*. The legal definition of infringement and the courts' punishment for this crime are continually undergoing revision, with resulting stricter laws and harsher punishments for infringement.

2. At the company level, management of technology is focused on satisfying customers and gaining competitive advantage. Some firms invest in the development of new technology to gain an advantage which could be patent protection. Another option is to copy the new technology of competitors. This is not acceptable when it leads to successful charges of infringement. To avoid litigation, many firms pay competitors for the use of their new technology by licensing the rights.

3. Management of technology at the international level, involves all of the areas of Levels 1 and 2 above. Each country has different rules and policies. There are many international agreements which are constantly being altered by new trade agreements. A challenge to those managing technology is to keep up with the international scope of changes regarding the safety of the workplace and the protection of customers. Applicability of patents, the courts' rulings on infringement and the regulation of license agreements are international subjects which require systems cooperation between the legal and operations management participants in the firm. To have international P/OM capability requires extensive knowledge and constant updating. Companies cannot do business on a global scale without this expertise.

8.2

THE TECHNOLOGY TRAP

Technological changes occur in clusters or droves. The *discontinuity* of major waves of new technology *can create a trap*. Consider this headline: "Companies have spent vast sums on technology. Now they have to figure out what to do with it all."[1] This *Wall Street Journal* feature refers to information technology (IT). The problem applies to other kinds of new technology, but almost everything new about technology in the period 1990–2020 will be related to IT in one form or another.

The trap is wanting to be current, but not knowing how to use the new technology to reshape old systems. The key to the problem is recognizing that information systems have been in place for decades. Traditional ways of doing things are hard to change, thus new technology has been made subservient to the old systems. In other words, old systems are frozen into place and the new technology has not been able to thaw out the old procedures in order to establish new ones.

The tendency to use new technology to mimic old systems is likely to be detrimental. Companies that have started out with a reengineering point of view and a reinvent business processes mind-set have succeeded in gaining technological advantages. Dr. Zuboff of the Harvard Business School has shown that those companies that have rethought the systems they use for doing work have profited from introducing computer technology. Those that computerized without altering the existing systems did not do well.[2] To be successful, companies must not only continually change their equipment, but also their processes, and most importantly, the mind-set of their management and employees.

8.3

TECHNOLOGY TIMING

Management of technology involves knowing when to shift from one form of technology to another. Should the old computer technology be replaced by the latest available, or should the organization wait for future developments? What does waiting cost? It entails continuing use of an older technology that might be slower or more expensive. Newer technology might produce a higher quality product. On the other hand, new technology often entails costs of learning and *shakedown risks* of unexpected bugs and glitches. Shakedown and break-in are both concerned with start-up problems.

The timing of shifts in technology involves assessment of the costs and savings made by changing now as compared to the costs and savings from deferring the acquisition. These should include the advantage, particularly with industrial customers, of having the reputation of being a technology leader instead of a follower.

If the benefits of being a leader in technological development are not significant, then copying technology without infringing across the global boundaries that apply can prove advantageous. Being second can be a successful management of technology strategy. Postponement and deferral of new technology development is a sound MOT strategy when:

1. development costs are high
2. shakedown problems can alienate customers
3. infringement cannot be substantiated
4. the benefits of new technology are likely to require more years of development

Payback period

Estimated number of years required to produce enough profits or savings to offset expenditures completely. Also called breakeven time and cost recovery period.

Usually, investments are made in developing new technology when the payback period is relatively short compared to the expected lifetime that the development will be generating profits and savings. **Payback period** is the estimated number of years that will be required to produce enough profits or savings to offset the expenditures completely. Some companies prefer to talk about breakeven time, essentially the same concept at work. Cost recovery period also is used. It is the time required to accumulate profits which fully offset the expenses entailed in replacing old technology with new technology. All of these measure the time required to pay off alternative investments. It is a favored way to evaluate choices in technologies.

Different criteria will apply according to the specifics of the situation. However, as a rule of thumb, in stable markets that are estimated to last about ten years, a three-year payback period was often considered to be the allowable upper limit. In the days of rapidly evolving technology, much shorter time periods have to be set.

Net Present Value (NPV)

Financial management tool which discounts future earnings by prevailing expected interest rates.

An alternative for evaluating technology changes is **Net Present Value (NPV)** which discounts future earnings according to the expected interest rates that will prevail. With a six percent interest rate, earnings at the end of one year would be divided by 1.06. Earnings that apply only to the second year would be divided by $(1.06)^2$. Earnings that apply only to the nth year would be divided by $(1.06)^n$. The calculations are explained in the supplement in the appendix.

NPV is a financial management tool. The fact that OM and finance must use it together to manage technological decision making is illustrative of the need for using the systems approach for MOT.

The OM decision to "make" or "buy" components is often related to technological expertise and to the cost of leading or following in technology. *Buy* avoids having to invest in and learn the new technology that *make* entails. The decision to buy uses someone else's expertise, which makes sense during the present times of great technological volatility. It made less sense during past periods of technological stability, which have been called "the quiet times."

To keep abreast of technological developments, it may be essential to make certain components. Making also can provide opportunities to achieve best quality and gain competitive advantages. Evaluation of the make or buy decision is a constant challenge to OM.

Many companies have appointed people to the job of technology assessment and management. This job requires being able to evaluate the present status of technology and future state probabilities. It is part of the job to coordinate OM, finance and marketing to determine as a team how to manage technological timing. Among the technology timing issues to be considered are:

1. Assessing technology plans of competitors and the importance of matching them. Can customers tell the difference?
2. Knowing the cost and/or benefits of the present technological level of the firm.
3. Understanding the costs of transition and the payback period.
4. Knowing when to move up the technological ladder. Periods of volatility are to be avoided whereas periods of relative stability are preferred.
5. Involving technological forecasting, which requires knowing that part of technology development is in the laboratory and the other part is on the factory floor or out in the field with the service crews.

8.4

THE ART AND SCIENCE OF TECHNOLOGY

The nature of work blends people and machinery together to form the technology of processes. What often is missed is the interplay of equipment capabilities (industrial science) and human skills (industrial arts). Both play a role in determining how specific forms of technology create processes for making and doing things. Technology is a synthesis of these two factors. Managing technology requires managing the combination of people skills with computers, robotics and other equipment.

Before 1950, definitions of technology placed more stress on the "arts" part of the interactions that occur between people's skills using tools and machines than on the materials and machines themselves. This makes sense because skills with machines used to constitute a larger part of the technologies responsible for all kinds of products. Such skills included the ability to measure and test using calipers, micrometers and all kinds of laboratory equipment, including microscopes, to assess the chemistry of products. At one time, rulers and scales were high-powered technological developments.

In their early interactions, machines did less and people did more. Go back to the seventeenth century, before the torrent of inventions associated with the Industrial Revolution. Look at how the percent contribution made by skills (as compared to the percent contribution made by machines) goes down over time. The equation for the percent contribution of skills over time $P(S)_t$ might be written:

$$P(S)_t = \frac{skills}{skills + technology}(100)$$
Equation 8.0

Alternatively, an equation for the percent contribution of technology $P(T)_t$ can be written:

$$P(T)_t = \frac{technology}{skills + technology}(100)$$
Equation 8.1

Figure 8.1, using Equation 8.0 above, shows the decline in the percent participation of human skills—as a generalized shape for an average process—over a period of four hundred years.

FIGURE 8.1

PARTICIPATION OF SKILLS AS A PERCENT OF TOTAL LABOR PLUS MACHINE INPUTS—ESTIMATES FROM 1600 TO 2000

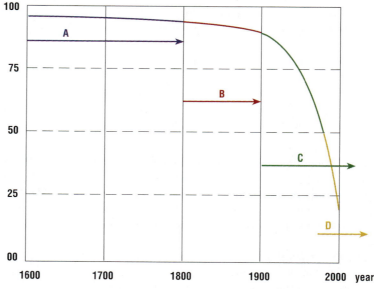

A   water, wind and animal power 1600—1800
B   steam power 1800—1900
C   mechanical systems and electrical power 1900—???
D   electronic digital control of power 1980—???

As time goes on, engineering and science have made people increasingly peripheral to operations. The art skills are built into the equipment. The further back one goes in time, the more reliance on people's skillful contributions, the smaller the contributions of machine technology, and the more that materials used were closer to basic commodities and less specialized. Now, there is more value added by advanced technology and less by people's skills.

The decrease is very slow for many years and then speeds up. Recent usage accelerates the decrease of the arts and skills portion relative to technology and science. The mixture within technology has shifted. In the first half of the twentieth century, machines developed capabilities for moving and lifting power. They were acclaimed for their sheer strength and endurance, not for their smartness and adeptness which had to be derived from the human operator.

Starting with the 1950s, machine intelligence began to be developed. At first, the speed of doing simple calculations and the size and persistence of memory seemed to spell out the basic advantages of machine calculations. It was firmly believed that machine intelligence had serious limitations which would never approach human capabilities. These restrictions are falling by the wayside. Computers working together with machines now can perform an array of functions that surpass what humans alone can do in all forms of arts and skills.

Geometrically increasing numbers of applications can be cited where industry now places greater reliance on machines than people for physical skills, performance learning abilities, and fast decision making with large databases. The technological replacements apply to making things and providing services. They apply to the skills and equipment required to inspect and test results. Here too, human efforts are being replaced by technology.

Tool and die masters working materials by hand are replaced by people with creative knowledge and special training. Here, a technician produces tooling by adjusting the model on her screen and sending a code to the milling machine.
Source: Cooper Industries, Inc.

TECHNOLOGY

Neural Network Programming at Mellon Bank Corp. and LBS Capital Management, Inc.

Neural network programming is being used in both the consumer lending and mortgage banking divisions at Mellon Bank Corp. Neural networks, a form of artificial intelligence, differ from more common methods of software programming in part because the neural systems are able to replicate more complex concepts and learn from their mistakes. In neural networks, layers of simulated neurons copy the human ability to recognize patterns.

Mellon Bank Corp. uses a neural network software product called Colleague™, an automated underwriting and decision-support system. Developed by San Diego-based HNC Software Inc., Colleague increases Mellon Bank Corp.'s speed and efficiency in evaluating loan applications. The neural network system interfaces with the bank's existing application processing system to quantify risk and decrease response time. HNC both developed the system and trained Mellon Bank Corp. employees to use it.

Mellon Bank Corp. is not alone among financial service companies in applying neural networks to programming for financial operations. LBS Capital Management Inc. of Safety Harbor, Florida, applies neural network programming to asset management operations.

LBS, a company striving "to transform money management from an art into a science," combines an expert system called Omni with neural networking in making investment decisions. Omni is responsible for recognizing market trends and making allocation decisions between cash and stocks.

The neural network part of the system makes decisions regarding which stocks to purchase, quantities to purchase, and optimum purchase times. The system "will outperform traditional number-crunching computers" because it adapts and learns new rules daily. Unlike human managers, the neural system is unemotional, unwaveringly objective, and allows continuous monitoring of 2,000 stocks.

Sources: Patty Tascarella, "Mellon technology assists loan service," *Pittsburgh Business Times and Journal*, September 5, 1994, and Rex Henderson, "Smart Machines," *Tampa Tribune*, October 17, 1993.

On the other hand, the expanding role of technology has increased the amount of knowledge work required by industry. Such creative knowledge work, essential for planning, designing and programming, is predominantly done by people with special training who are aided by computers. It has been said that the OM work-force has been replacing blue collars with scholars.

Thus, the increasing role of technology has changed the way operations must be managed in a number of ways. OM people must understand more about the technology of the process and the computers that are used to control the process. Skilled machinists, tool and die makers, and artisans with tools are being replaced by CAD-CAE-CAM (computer-aided design, computer-aided engineering and computer-aided manufacturing) experts using electronic data interchange (EDI) via satellites and the Internet. Knowledge work has grown with greater technology and managing such creative efforts has become part of the OM responsibility.

It is useful to assess the management of technology in terms of the degree of participation and the coordination of the components, which are the:

1. arts and skills of people in making things and providing services
2. industrial science capabilities of machines in making things and providing services
3. applied science properties of materials
4. increasing intelligence of machines
5. important role of art in managing technology at the strategic level
6. increasing role of science as a support for art in managing technology at the strategic level

With respect to Points 5 and 6 above, it should be noted that the ratio of art to science in managing technology at the strategic level is very high. Planning and programming are essentially arts reminiscent of the skills of artisans in the 1500s. Science is gradually bringing greater support to strategic levels much as it did to tactical skill levels in the late 1800s.

8.5

THE DEVELOPMENT OF GLASS PROCESSING TECHNOLOGIES

With respect to Points 3 and 4 above, there are many different materials that could be used to illustrate the shift from art-based skills to technology-based processes; steel, rubber, paper, glass are among them. Each provides wonderful examples of processes that increasingly make use of intelligent machines and a wealth of scientific knowledge about the properties of materials.

OM manages technology of products that are so taken for granted that few stop to wonder how this is done in a factory. What skills and technologies must be managed to make windshields? Everyone is familiar with windshields, having spent enough time sitting behind them. The manufacture of windshields will be addressed in this section with perspectives about glass in general, the history of change in technology and amazing challenges that OM has faced up to in this multibillion dollar industry.

It is simple to illustrate the shift from art to advanced technology using various glass-making categories. Glass processes run the gamut in the degree to which they require art and science in varying amounts. Masterful human skills are required for glass blowing, the art of shaping a mass of molten glass into various shapes by blowing a controlled stream of air through a tube into the mass. Glass containers have been found that date to the second millennium B.C.

Glass blowing is an example of one of the oldest industrial arts. The same low-technology equipment in the hands of a skilled artisan produces art work, whereas in the hands of a novice it produces poor work. Glass blowing for valuable vases and other glass art is done in custom shops all over the world. Glass blowers can do things with glass that no machines can even approximate; various works of art in glass, such as produced by Steuben of Corning Incorporated, command prices that attest to this. The photograph on the next page shows a glass blower at work.

Having looked at the highest percent participation of art-skill level (near 100 percent) for glass blowing, next move to flat glass which has occupied a range of positions over time. Prior to 1959, flat glass had to be cast, rolled and polished through a labor-intensive process with serious quality problems.

Then the British firm of Pilkington introduced an entirely new technology called the *float process*. It had taken them seven years of experimentation to make the continuous ribbon of glass (melted sand) floating on a bath of molten tin acceptably flat. Float glass does not have the imperfections that rolling glass produces. Not only does it have higher quality, but it is easier and less expensive to make.

When the method was perfected, Pilkington started to license it, which it still does today. The result was that OM in the glass industry experienced a sudden revolution in technology. There was an abrupt and severe decline in the skill to technology ratio for the new process. Conversely, the technology component of the new process was much higher than in any other glass-making operation. Replacing casting with the float-glass process had to be carefully managed as the float method became the primary method worldwide for making flat glass. At this time it is used in first, second and third, world countries.

Flat-glass sales generate more revenue than any other glass product. This product is demanded worldwide for window panes in houses. The revolution in flat-glass production fell on OM's shoulders because the conversion had to be made from outmoded technology to the new technology without disrupting the supply of product to customers. An entirely new process had to be learned and brought online. The success of the industry to adapt is a tribute to the operations managers of the flat-glass industry.

Glass blowing is an art which utilizes simple equipment and basic materials in combination with great operator skill.
Source: Steuben

Technological Innovations—Product and Process

The glass business has been an innovative industry. Improved products and processes have continually appeared. One reason for this success is that the customers of glass companies are the auto, construction, and telecommunication industries (fiber optics). Taken together, this constitutes a multibillion dollar market which aside from economic business cycles is growing on a global scale.

New product adaptations flourish when raw materials are readily available and inexpensive. Global companies aggressively compete for market share by utilizing new technologies. In many instances, the new products require increasing the technological composition of the system to be managed by OM.

The story about managing technology in the glass industry continues with the recognition that float glass is the starting component for automotive glass. Automotive glass requires the use of much higher levels of technology than household glass. Safety glass is mandated for automobile windshields in the U.S. and many other countries because it will not shatter under impact.

Safety glass is a sandwich of two pieces of float glass (called lites) that are annealed by heating and cooling for strength. Placed between the inboard and outboard lites is a polyvinyl butyryl laminate. The sandwich is bonded without adhesives using temperature and pressure.

The process for making windshield glass requires a mixture of high levels of labor skills with corresponding levels of technology or very advanced automated systems of technology working without human operators. The number of steps that must be taken and the need for skills and technology are evident from the process flowchart for making windshields, shown in Figure 8.2, which presents an illustration of the generic windshield-making process.

Many scientific breakthroughs in glass-manufacturing processes allow companies like Guardian Industries, Libbey-Owens-Ford (LOF) Co., and PPG Industries, to equip cars with the large curved automotive glass windshields that distinguish modern-day autos from older ones. It should be noted that windshields can include special edge markings around the entire perimeter. These borders or frits are used as a theft-prevention system. Radio antennas and heating grids are embedded (respectively) in front and rear window car glass.

As the amount of glass increases, the heating and cooling of the car demands more energy. To counteract that costly effect, solar glass is installed which controls all three parts of the solar spectrum—visible, ultraviolet (UV) and infrared (IR) light. Solar glass uses a multilayer metal/metal oxide coating on the inboard surface of the outboard glass in the laminated safety glass windshield structure. Silver coating and other layers reflect solar energy and control glare. This replaces tinting which is done after manufacturing.

Windshield production necessitates a highly-mechanized, mass-production flow shop process when the demand volume is sufficiently great. This kind of volume characterizes windshield production by the supplier for the original use by the auto manufacturer. Less highly-mechanized shops that permit smaller batches with quick model changeovers are used for the replacement market. Even so, replacement volumes are large enough to warrant relatively continuous production processes.

PROCESS FLOWCHART FOR MAKING WINDSHIELDS

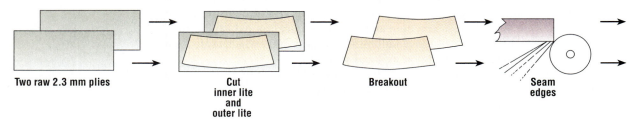

Two raw 2.3 mm plies   Cut inner lite and outer lite   Breakout   Seam edges

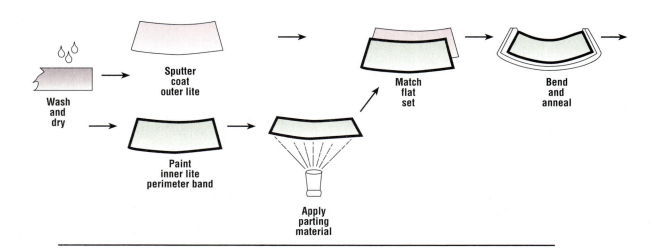

Wash and dry   Sputter coat outer lite   Paint inner lite perimeter band   Apply parting material   Match flat set   Bend and anneal

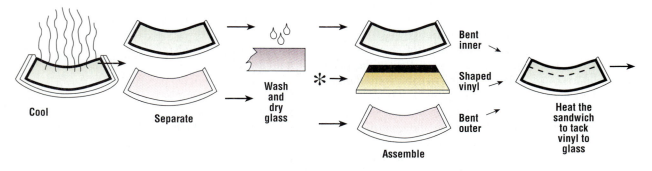

Cool   Separate   Wash and dry glass   Bent inner   Shaped vinyl   Bent outer   Assemble   Heat the sandwich to tack vinyl to glass

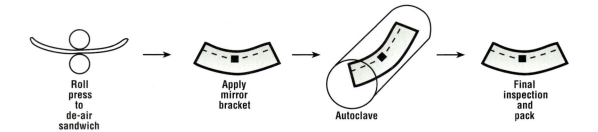

Roll press to de-air sandwich   Apply mirror bracket   Autoclave   Final inspection and pack

FIGURE 8.3

HIGH PRODUCTION AUTOMOTIVE IN-LINE WINDSHIELD FORMING AND ANNEALING SYSTEM

PRODUCTION PROCESS FOR CHRYSLER AUTOMOTIVE WINDSHIELDS

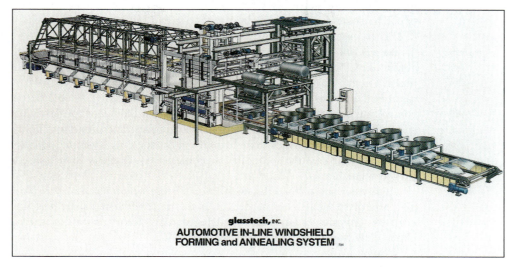

glasstech, INC.
AUTOMOTIVE IN-LINE WINDSHIELD
FORMING and ANNEALING SYSTEM TM

The glasstech portion of the production process is indicated by a red circle around the bullet. The other bullets are non-glasstech parts of Chrysler's process for automotive windshield glass.

- Rectangular flat glass brackets, purchased from an outside supplier, are automatically loaded onto two parallel pre-processing lines.
- The glass brackets are cut into templates on these two lines using CNC (Computer Numerical Control) machines which will become the inner and outer lite of the finished windshield. Each line processes an inner lite, an outer lite, an inner lite, an outer lite, etc., in a continuous series.
- After cutting, the templates are transferred to CNC grinders where a Type 1 ground edge (SAE J 673 b Standard) is applied around the entire periphery.
- Cut and ground templates are automatically indexed into a print room where a black enamel frit is silk screened onto the glass obscuring the border. This black border protects ultraviolet degradation of the urethane used to install the windshield into the vehicle. The trademark also is silk screened onto the glass at this time.
- There also may be an additional silk screen of the inner template with a grid of lines of silver enamel frit at the lower area of the windshield. This feature, know as EWD (Electric Wiper Defrost), heats the area of the windshield where the windshield wipers would park, helping prevent ice buildup. The electric grid on the windshields is visible from the outside of the vehicle and works on a principle similar to that of heated back lites.
- The silk-screened templates are automatically loaded onto the Glasstech, Inc. Automotive In-Line Windshield Forming and Annealing System (see Figure 8.3). This furnace simultaneously forms the templates in each line to the required design surface shape at a cycle time of 15 seconds.
- Glass templates are positioned automatically onto the load area of the air-flotation furnace. The lites are indexed through heating zones of the gas-fired furnace until they reach the desired forming temperatures.
- Heated glass templates exit the air flotation zones of the furnace and travel beneath a patented friction-free air suspension conveyor. A patented computer-controlled positioner maintains accuracy to 0.5mm and deposits both inner- and outer-glass templates onto a full-faced mold. Aerodynamic shapes that require inverse and compound curvature are accurately formed to tolerances of ±0.75mm to design surface.
- While the glass is sagging on the mold, it is conveyed to the second forming station where the lites are pressed and then fully formed by a vacuum.
- The formed lites are then placed on an annealing ring where they are cooled to develop acceptable stresses.
- The unload conveyor's mechanical set of fingers gently lifts the glass to the cooling conveyor. This conveyor can be tilted down to reduce the unload height to 36 inches from the factory floor.
- The entire Glasstech forming system occupies a space of 13.5 by 112 feet. Rapid part-to-part changeovers are possible because of self-aligning tooling. By forming parts on the same tooling in each line, any inner glass will mate with any outer glass.
- The formed lites are unloaded onto racks where they are washed and separated. A plastic interlayer is placed between the inner- and outer-formed glass.
- The glass-plastic-glass "sandwich" is put into an autoclave where heat and pressure bond the separate components into a single, strong, transparent windshield.
- Electric leads are soldered onto the windshield to complete construction of the EWD feature. The electric leads connect the EWD to the wiring harness to allow energizing of the silver grid area of the windshield.
- The production capacity is 240 windshields per hour.
- Distribution is handled just-in-time from the glass to the assembly plants. MOPAR distributes replacement parts to the aftermarket.

It should be apparent that production of automotive glass is not a static process. There is a lot of change going on as a result of research and development (R & D) that is invested in automotive windshield processes. Solar glass is a good example, but there are many others as well. In most companies, R & D is a separate department which works closely with OM to convert research ideas from theory to practice. The systems approach is required to blend the needs and interests of OM and R & D in pursuit of company objectives.

As a result of much invention and the use of computers, Glasstech, Inc., has been able to develop a process for producing automotive windshields that operates at the highest level of technology. In this case, the percent of art skills required for running the process was zero. It is 100 percent technology. People do participate in controlling the process, and they are more critical than ever for designing the windshield, planning its production, setting up the process and programming the Glasstech "automotive in-line windshield forming and annealing system." People are essential for deriving creative solutions for problems associated with ever larger and more-curved windshields.

Figure 8.3 shows the Glasstech, Inc., system which has a major role in the overall automotive windshield process. It is the most advanced system of its kind and is used by Chrysler to produce windshields for factory installation of automotive glass (called originals instead of replacements).

Comparison between the glass-blowing photograph and Figures 8.2 and 8.3 illustrates how differently the same basic material can be treated. Managing glass blowers creating designs for Steuben is a different kind of OM challenge than making float glass at LOF, which, in turn, is a world apart from creating windshields for Chrysler using the advanced technology of Glasstech, Inc.

For all kinds of materials including metals, plastics, ceramics, glass, paper, and wood, there are advances in technology that require new processes to take advantage of the enhanced properties of the materials. For the most part, these processes utilize computers to design the product, and often to run the process. People are needed to plan the process, program the machines, and deal with inevitable contingencies.

The connection between the design of the product and the process is increasingly of major technological consequence. In Section 8.9 below, that connection is explored with respect to designing for manufacturing (DFM) and designing for assembly (DFA). It is worth noting (Figure 8.3 above) how large a role assembly plays in the production of automobile windshields.

The linking of design with manufacture, assembly, and distribution depends upon the most advanced forms of new computer technology, such as CAD-CAE-CAM. Yet, at the same time, design is not an automated function. Designers are knowledge workers whose art is dependent upon innate skills.

Supply Chain Technology

Supply chains epitomize technological linkages which are managed by operations personnel. The design, production and distribution of windshields is a supply-chain system that is sensitive to technology at every one of its stages. Suppliers' technologies must be congruent with the producer's technologies which, in turn, must be complementary to the customers' technological needs. OM is either responsible for coordination of this system or an integral part of the team.

Every time a new car model is launched, windshields, rear windows and side glass must be designed to fit the model and to meet designers' specifications. That glass product must be fabricated in volume that is commensurate with the production schedules of the auto assembly plant. The glass must be on hand to be assembled with the rest of the car. Simultaneously, the same glass parts need to be stocked as replacement products. The following three "ifs" are relevant:

1. *If* the windshield fabricating process is a flow shop, then continuous delivery of flat glass from suppliers must be achieved and maintained. This condition is dependent on the suppliers using appropriate technology to achieve the needed volumes.
2. *If* the customer is an auto assembly plant, then the technology of the windshield producer must be capable of achieving a production schedule that matches the requirements of the assembly plant. Shipments will be made regularly in large quantities.
3. *If* the customer is a distributor for replacement glass, then the technology upon which production scheduling is based must be able to produce a variety of models quickly to meet retail demand.

In fact, Points 2 and 3 are often both true at the same time. The technology of the windshield producer must be able to combine the requirements of both kinds of customers. In general, technologies along the entire supply chain must be compatible and coordinated with respect to both quality and quantity to insure smooth running of the system. Flows need to be balanced, matching supply and demand.

Auto factories (the customers) require enough inventory of windshields for all models scheduled to be made within the next production lead-time period. Lead time, in this case, is the elapsed time between ordering and receiving the windshields that are needed.

While just-in-time deliveries are usually the goal, just-in-case stocks are kept on hand to provide buffer protection against the contingencies of the real world. As the key to success or failure, the technologies of the producers of windshields and their suppliers of float glass must interact properly with the technologies of their customers, who manufacture automobiles.

P/OM accountability starts with coordinating the design phase with fabrication and assembly (manufacture). Supply chain responsibility continues with setting up the inventory, delivery and stocking policies. The entire system is technology driven.

It is not surprising that technology for the supply chain that deals with automotive replacement glass is different from the technology that is the foundation for the original glass market. Replacement glass is often produced as an added amount of product with the original production run. It is then sent in small batches to the distributors dealing with the service outlets that include glass retailers, collision and body shops and dealer service stations, a unique supply-chain system.

The demand for replacement auto glass in the U.S.A. is large. Windshield replacement alone is in excess of 8 million units per year. This demand is generated by thousands of different kinds of vehicles on the road. "There are over 15 thousand automotive glass SKUs…in North America, and this number is growing exponentially."[3]

Any one of them can experience glass breakage, which means that OM must adjust production scheduling to handle smaller batches of many more varieties. This necessitates other kinds of OM rules for fabrication, inventory scheduling and warehousing based on alternative technologies.

P E O P L E

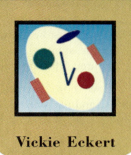

Vickie Eckert
. . .
Megatest Corporation

Vickie Eckert, vice president of Business Process Improvement for Megatest Corporation, was elected to both the 1994 and 1995 Malcolm Baldrige Board of Examiners. She was one of approximately 200 quality experts chosen to sit on the Baldrige Board each year, of the 600 to 700 people who apply. Eckert spearheaded Megatest's bid for the Baldrige Award in 1993.

"When we submitted our application, we'd been involved with our quality system for a couple of years," said Eckert. "We knew we'd been making progress but didn't have anything to measure ourselves against." Eckert said the Baldrige Examiners' feedback "told us where we were doing very well and areas that we might want to look toward improving for the future. We got our feedback at the end of 1993 and used it to develop the '94 and '95 quality program. The Baldrige Award criteria was an excellent source for allowing us to look at the total company and not just isolated segments, or individual successes."

As a Baldrige Examiner, Eckert attended a three-day training class in Gaithersburg, Maryland. She said it takes each examiner about 30 to 40 hours to review a Baldrige Award application and comment on it. "There's no pay involved; you do it for the love of it," Eckert said. "It's a great learning experience because you have the opportunity to see how other companies have interpreted the guidelines and how they integrated quality into their culture and their environment."

Eckert's current focus at the San Jose, California-based Megatest is "taking our quality system to what we call The Next Level…training and education for the whole work force,"

she said. The new program emphasizes not only quality tools "but additional skills such as project management, basic math, English as a second language. We are looking at our employees as a total package and what competencies are required for their—and our—future."

Megatest Corporation designs, manufactures, and markets automatic test equipment (ATE) for the semiconductor industry. This equipment is directly involved with semiconductor manufacturing quality control processes. Eckert explains, "We provide the test equipment used to ensure that high quality semiconductor devices get to the market. The quality chain is only as strong as its weakest link—we have to deliver quality products!"

Source: A conversation with Vickie Eckert, Megatest Corporation.

An important part of the technology of supply chains is the means by which the various players (suppliers, producers and customers) are able to know where things are located. This includes the trucks, trains and planes used to carry the goods. Customers want to know when they can expect delivery of the various products they have ordered from suppliers.

The Association of American Railroads (AAR) has an information system that keeps track of where all railroad boxcars, tank cars and hopper cars are in the country. Using this information, individual companies monitor their rolling stock.

DuPont has a group called Cartrak that uses computers to regulate the movement of its railcars. They note cars that are not moving, taking actions to expedite their return for reuse.

UPS and FedEx use the combination of optical scanners, network telecommunications and computer-database management to track package movements for their customers. Bar-code readers continuously send information to the tracking computer. Similar bar-coded information technology is used within factories to track the status of products and components.

Everyone that is part of the supply chain has something they are shipping or something they are receiving, or both. That model applies to adjacent stations within factories, banks, investment brokers, hotels and restaurants. It is worth noting that the supply chain concept applies equally well to external and internal suppliers, producers and customers, the systems point of view always encompasses the trio.

Replacement vs. Original Market Technology

The automotive replacement glass market is a system which entails dealing with greater variety than the automotive original glass market. Systems with greater variety generally also must cope with more variability, risk and uncertainty than systems with lower levels of variety. The technology with which OM must deal is large batch and intermittent flow shop configurations.

To make the analogy more broadly applicable, consider the similarity that exists with replacement glass and the replacement tire market. There are so many tire sizes, grades and styles for cars, trucks and vans, that the variety of models is the greatest source of problems. Like auto replacement glass, replacement tires have to be distributed through many different dealerships.

One company which has won acclaim for its success in the difficult tire replacement market is Kelly-Springfield. Their success has been based on their ability to:

1. Forecast well the need for each variety of product in the different market locations.
2. Produce the right amounts of each kind of tire to meet demand when it arises instead of by excessive stock levels.
3. Distribute tires so that the right models are in the correct places to fill customer demands on an as needed basis. It should be noted that "fill or kill" tends to characterize this market where other brands are available as substitutes.

Good forecasting models and understanding of the markets are both critical ingredients of success. Next, strong quantitative scheduling models are used which take into account the fact that tire-replacement producers have to cope with a great diversity of molds and many changeovers. No matter how continuous the engineers and process managers have been able to make the flow of materials, they still are dealing with the management of dissimilarity.

The management of technology for the production of variety must be able to deal with dissimilarity. Product multiplicity has more code numbers for stocking and inventory. It has more settings for machinery and greater control must be exercised over the paths and storage of batches. It is common terminology to say that the entropy is higher for systems with product multiplicity and dissimilarity. The term entropy is a measure of disorder from physics which has been absorbed as a systems term to describe complexity. Entropy increases with greater uncertainty and increased disorder.

This results in higher unit costs for replacement products, even though replacement technology is more likely to use general purpose equipment (GPE) than special purpose equipment (SPE). The classic trade-off applies. General-purpose technology is less expensive than special-purpose technology and it can be used to produce greater variety. The unit cost differentials can be considerable.

8.6
THE TECHNOLOGY OF PACKAGING AND DELIVERY

Figure 8.3 above does not show the technology for packaging and delivering glass windshields, which can shatter and chip if not properly handled. Transport, handling and packaging technologies differ according to whether the market that is being served is for original transparencies (high-volume shipments) or for individual windshields that are shipped from the distribution center when requested by the auto service for specific customers.

Most products need some kind of protection during shipping. The production process can have near-zero defectives as a result of excellent quality management. Unfortunately, the damage done during delivery of the product can zoom the real defective rate—as seen by the customer—to outrageous levels. The manufacturer may not know about this situation for a long time. Complaints are not made on products not purchased. Feedback from warehouses, truckers, and retailers can take a long time. Awareness of such problems dawns slowly.

A consumer package goods company experienced a downturn in sales of its bar soap in a specific sales region. As part of the investigation for the causes of the problem, it was found that the product on the shelves of the supermarkets had damaged boxes. This condition was attributed to a change in handling procedures in the distribution center which had started using a forklift truck to move the product. The forklift truck operator had previously worked in a scrap metal business. The problem was remedied by training.

A firm that manufactures dinnerware of medium-grade porcelain china discovered that full sets of dishes and cups were being delivered with three out of every ten pieces broken. They also found the company was infrequently called by customers to ask for replacements. More often, customers accepted the loss because it was not fine china, but they never ordered from this company again. Further, when the company was called to ask for replacements, the person representing the company had no interest at all in tracing the shipment to learn what caused the damage.

OM was on top of the issue of production defectives. The production process was in control and excellent with low-defective rates for this china product. The shipment-defective rates completely altered the "delivered rate of defectives." From the systems point of view the defective rate was unacceptable.

The quality of the delivered product is part of TQM accountability. The packaging technology and handling methods are part of that accountability. The chinaware company must question what caused the breakage and what should be done to prevent it from happening again.

Systems Analysis

P/OM is responsible for delivering a quality product under all circumstances. P/OM must undertake a systems analysis of the delivery process where packaging is the technology. The systems analysis entails getting answers to three lines of inquiry.

1. What product-protection parameters have been specified for the packaging? What has the designer done to achieve these specifications? Typical questions would be:
   - From what height can the package be dropped without breakage?
   - Has the package been designed to withstand travel-specific stresses and strains (air, boxcar, truck, etc.)?
   - How much vibration can the packaged product tolerate?
   - What compression force will the package support?
   - What happens if liquids are spilled on the packaging?
   - Does humidity affect the performance of the package in terms of any of the above quality dimensions?
   - What are the temperature extremes the package can withstand?
   - Finally, has the package design been adequately tested?
2. Did the actual handling of the package which resulted in broken product fall outside of the designer's specifications?
   - If it did, in what ways did the actual treatment exceed the limits for which this product was protected?
   - If it did not, can the packaging system be redesigned?
3. Has something permanently changed in the delivery system's way of handling the package?
   - If so, can the packaging technology be upgraded, or can a new delivery system be found?
   - If a new delivery system is feasible, will the present packaging technology suffice?
   - If nothing has permanently changed in the delivery system, is the present package design satisfactory?

The systems approach focuses on getting to the root of the problem. Was the package designed correctly? Was the package tested correctly? If it was designed correctly, then what caused the breakage? If it was not designed correctly, how should the package be designed and tested?

The china had been packed in a fiberboard box which was placed inside of a corrugated box. There were no styrofoam inserts which strengthen the box and protect its contents (called unitized packaging). Plastic "popcorn" was used.

Packages can be made of cardboard, plastics, wood, and combinations of materials. Inserts and fill offer even more options. There are too many kinds of packaging materials and systems of packaging to try to do more than present the basic scope of this subject.

8.7

THE TECHNOLOGY OF TESTING

Each kind of packing material has its own unique characteristics. Various tests have been developed to determine strengths and weaknesses. Corrugated boxes are marked with a figure to indicate the number of impact pounds the box can withstand before bursting. For example, "100 test" indicates less strength than "200 test" or "300 test."

Packaging experts have expressed concern that too few and too simple tests should not be interpreted as more than vaguely indicative of package integrity. The bursting-strength test reflects just one type of resistance on which a package must be rated. Even this one rating for impact is suspect for a variety of reasons. Humidity can affect the strength of cardboard and will not be reflected by the test measure. What passes a bursting-strength test in Maine may fall short in Florida. In short, correct measurement of package strength is dependent on a thorough understanding of packaging technology.

In general, testing technology is conditional upon a complete comprehension of the process being tested. The ramifications of testing technology apply to a great range of applications which include designing products to have minimum variability in manufacturing.

Testing by Simulation

Empirical testing

Experimental testing which can simulate worst cases of the kind of handling the package design can be expected to undergo; examines alternatives.

An important testing alternative is **empirical testing**, simulating the worst cases of the kind of handling the package design can be expected to undergo. With this kind of experimental testing it also is possible to examine alternative package designs.

It is best to combine structural knowledge of packing materials—package engineering—with testing by simulation which is a pragmatic procedure. Simulation for package testing means dropping, kicking, gouging, vibrating, pouring liquids onto, heating up and cooling down, among other things. It is not easy to simulate all of these conditions. Still, it may be easier to use simulation methodology than engineering knowledge of the complex processes.

Simulation testing technology plays a most important part in designing windows for homes and office buildings. High floors of the latter do experience severe winds. The testing method should consider both broken panes and blown-in window frames. In addition, it must consider leakage. How high a wind should these windows protect against? Hurricanes are classified as winds above 75 miles per hour and window designers need to provide protection for occasions above that level in hurricane zones.

Simulation testing technology plays an important role in the design of windows and doors for homes and office buildings.
Source: Campbell Mithun Esty/Andersen Windows

OM working together with product designers and R & D can test the output of the process and reconfigure the way the product is designed and made. Testing technology is part of the OM kit of tools. Excessive auto recalls can be avoided by pre-testing for contingencies. The silent interior of the car is a function of the way car doors are sealed. Models can be tested before the production system is finalized. Wind tunnels which are used for testing airplane performance also are used for testing the performance of windows.

Legal statutes and governmental rules are the concern of OM. Building to specifications requires knowing what the changing specifications are and what must be done to meet expectations. Knowing the law becomes part of the system. Fulfilling the law often is a matter of process, and testing is needed to determine if the requirements are being met.

Thus, after Hurricane Andrew the building codes for southern Florida, devastated by that storm, were changed so that roofs would not blow off and windows implode. The California highway system adopted more stringent building codes following the collapse of sections of highway during the 1994 earthquake.

8.8

QUALITY CONTROL TECHNOLOGY

One of the most important aspects of total quality management (TQM) is inspection and testing technology. Such quality testing technology is at the heart of quality control. The technology of testing has been advancing at a rapid rate. As in other aspects of technological development, increasing amounts of the art of quality achievement and control are being built into the machines using science.

Inspection is epitomized by visual abilities although there are many other senses for gathering information. Using vision, inspectors can check quality (is the finish on the car acceptable?) or quantity (how many parts were finished, or what is the number of air bubbles in a piece of glass?). The quantity of product is readily determined by optical scanners which can count bar codes with more reliability than human inspectors.

Optical scanners at checkout counters reading the *universal product codes (UPC)* which are printed on all packages reflect applications of bar code technology. Now commonplace in supermarkets, they are being used in factories and banks to determine the status of transactions. They are used increasingly because their capabilities are improving. They have a great deal of information redundancy which means that even damaged packages can be read as well as bar codes on round and odd-shaped surfaces.

The multiple functions of adding up the bill and deducting the item from inventory are readily programmed into the supermarket's computer system, with which the optical scanner communicates. The same technology working with a filling machine can determine how many pills to put into a bottle or when the fill level is reached for a bottle of juice.

For quality control applications, scanners must be able to perceive and differentiate along each of the critical quality dimensions. The criteria for product acceptability can be programmed precisely by computer software and translated into the language of the scanner so the inspection process can distinguish between acceptable and defective products. Inspection technology continues to develop along both quantitative and qualitative lines.

For perfumes, a good nose is a requirement. Are there machines that have a better analytic smelling capability? To judge wines, taste, vision and smell are essential. Are machines in the wings to take over these tasks? For now, these sensory skills are best supported by machines; not replaced by them. At the same time, there has been continuous development of machine inspection abilities. These include sensing systems for temperature, pressure, and weight. Visual sensors deal well with shape and size.

Sensory capabilities used to be only people skills; now technology has been developed to replace people in some areas. The complete replacement of human inspectors by machine inspectors is still far from being accomplished. Nevertheless, OM is experiencing a major "change" in inspection technology that is targeted right at the heart of operations.

New technology permits automation of the statistical quality control process. SQC monitoring equipment inputs data directly into a database. The statistical characteristics of the processes being monitored are continuously calculated. Appropriate quality control charts can be drawn automatically and submitted on a regular basis to QC inspectors for their interpretation. Symptoms and signs of systems being out of control can be incorporated in the programs.

Computer software is programmed to increase the frequency with which samples are taken as well as their sizes when the defective rate increases. When runs are detected and/or out of control points are noted, the size and frequency of samples are adjusted. Under specific circumstances, the appropriate employee is notified about the situation so that interventions by workers can occur. The application of these methods to employee drug testing is an interesting extension of the same kinds of methodology and technology.

Testing procedures are the technology that quality control relies upon. Optical scanning is among machine sensory systems that are more sensitive than human ones, and are preferred. Computer programming of the SQC function is one of the most important technological developments at this time.

8.9

TECHNOLOGY OF DESIGN FOR MANUFACTURING (DFM) AND DESIGN FOR ASSEMBLY (DFA)

Design for manufacturing (DFM)

Starts with the goal of feasible fabrication; compliments DFA; objective of minimum variability.

Design for assembly (DFA)

Starts with the goal of feasible assembly; compliments DFM; objective of minimizing the assembled system's variability and ease of assembly.

It used to be that designers only considered the needs of end users when designing the product. OM was not consulted about how best to fabricate and assemble the product. Later, designs that were not feasible to manufacture would have to be redesigned, which compromised other aspects of the product. The manufacturing process was jerry-rigged as a measure of last resort to make the product.

Design for manufacturing (DFM) and **design for assembly (DFA)** start with the goal of feasible fabrication and assembly. They then move on to higher-order quality objectives as discussed below. It is now acknowledged to be good practice to have the design team include OM consultation from the start of the project to achieve feasibility and sensibility.

Because fabrication is distinctly different from assembly, it is useful to maintain separate agendas for DFM and DFA. On the other hand, often, but not always, the manufacture of parts leads to their assembly, so that both functions are complimentary. They should be dovetailed and designed to work well together. When a producer combines components from different suppliers, DFA must be coordinated by the assembler. Windshield glass epitomizes DFM and DFA combinations.

Even when it may not seem as if assembly is part of the product manufacturing cycle, it should be considered relevant. Pouring vinegar into a bottle is a kind of assembly of the product with its package. Packaging engineers and process managers for soft drink and beer companies view their technologies as an advanced form of serialized flow shop production. Packaging is often the ultimate stage of the manufacturing process, as in canning foods and boxing toothpaste. Consequently, the product and packaging should be designed and planned together.

The second order objective of DFM and DFA is to design parts that can be made with minimum variability. This may require testing the design by running *pilot studies* to produce sample product using the kind of equipment that will be employed in manufacture. Pilot studies are reduced-scale simulations of the full-scale production process; they make it possible to approximate the variability of dimensions during manufacturing. This is an example of the use of simulation to achieve *design modification*.

The objective is to pursue an aggressive strategy for designing a product which can be made by a process such that the manufactured product has minimal variation.

Figure 8.4 depicts this idea as a design feedback system:

FIGURE 8.4

DESIGN-TEST-REDESIGN TOLERANCE MATCHING

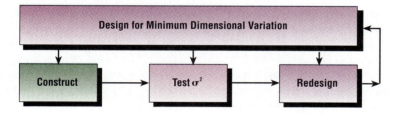

Note: Test σ^2 is testing dimensional variation. Redesign starts the cycle all over again. Thus, redesign and readjust are followed by reconstruct, retest, and redesign until they get it right.

This process is continued until the criteria for nearly optimal (rather than simply satisfactory) manufacturing and assembly conditions have been met. Nearly optimal means:

- the best design that could be found is adopted
- the number of design-redesign feedback cycles is large enough so that the rate of improvement from continuing to cycle is considered negligible

Table 8.1 shows how the variances (σ_i^2) of three designs can differ with respect to component assemblies. Each component is associated with a quality dimension (called Q_i). The third column, labeled TM ($\sigma_1^2 + \sigma_2^2$), is a measure of how well the two parts fit together, called **tolerance matching**. It is often the case that this number is the sum of the variances of the two parts that fit together.

Tolerance matching

Measure of how well two parts fit together.

TABLE 8 . 1

VALUES OF $\sigma_i{}^2$

	$Q_1 (\sigma_1^2)$	$Q_2(\sigma_2^2)$	$TM (\sigma_1^2 + \sigma_2^2)$
Design 1	6	8	14
Design 2	5	3	8
Design 3	7	5	12

The fabrication variances of Design 2 are uniformly lower than those of Designs 1 and 3 for both quality dimensions. The column headed by $TM (\sigma_1^2 + \sigma_2^2)$ for tolerance matching is taken as the sum of the variances of the two parts which are assembled into each other like a key fitting into a cylindrical lock.

It is fundamental to statistical theory that the variance of two well-behaved distributions that are taken together is the sum of the individual variances of each distribution. Poisson and normal distributions are considered well-behaved; both are unimodal and have stable statistical characteristics.

From the manufacturing point of view (DFM), the goal is to minimize the variability of each important dimension. From the assembly point of view, the goal is minimizing the assembled system's variability, TM ($\sigma_1^2 + \sigma_2^2$).

In searching for the optimal, product design and process design technology are being modified using a feedback system similar to that of quality control. The comparable cycle of improvement is the one used in quality control which is plan-do-study-act or plan-do-check-act.

The difference from the quality control application in this case is that design of the product has not yet been completed and accepted. Further, the process for making the product has not yet been finalized, and the product has not been released for manufacture. There is time to experiment with pilot plant-type production.

The idea is similar to the package testing example previously suggested. The design is pilot tested in a manufacturing setting or assembly environment that comes as close as possible to the full-scale real-world situation. Simulation of manufacture and/or assembly is the appropriate way to describe these experiments.

It may be best to start with two designs, both of which are considered feasible alternatives. The two variants are fabricated with a process that captures as much of the actual production system process as possible, with as large a sample size as possible.

The guiding principle of diminishing returns is given above, but for those who are more comfortable with a number, 50 is used as an acceptable minimum and several hundred is a good target, if it can be achieved economically. The realities of the situation will determine what can and cannot be done in assembling a sample of sufficient size to use for DFM and DFA comparisons. Further, if it is feasible to produce a large sample, and if it is reasonable to employ statistical design methods, such as the analysis of variance, the power of the DFM/DFA study is enhanced.

The system must be stable by the criteria of SQC. The best quick check on this is to plot the values of the sample in a histogram to see if a unimodal distribution emerges. Bimodality is a clear indication that the system has to be stabilized. Other techniques should be used if possible to assure stability or to help stabilize the system.

Given a stable system, the comparison between the variances of the two designs can be made. The design alternative with the lowest variance may give direction to the search for another design alternative. It is best that a group consisting of designers, process managers, marketing managers, R & D scientists and development engineers interpret the data and plan for the next set of designs to be produced.

Elkem Metals Co.
· · ·
Expert system minimizes impurities in ferroalloy refining process.

Elkem Metals Co., a manufacturer of silicon and ferroalloys, installed an expert system at its Alloy, West Virginia, facility to control the plant's refining process. Design and installation of the expert system came with a price tag of $250,000. The system increased the plant's output by 15 percent in three years.

The expert system was needed to create the uniformity lacking in Elkem's products. Variations occurred because some of the people involved in the process had developed their own recipes for the company's ferroalloy products. With the expert system, the computer controls not only the amount of each element to be combined, but also the amount of time the process will take. The system controls the temperature as well, which is important to minimizing impurities in the refining process. "The expert system…

basically took all the guesswork out of the refining process," said plant manager, George E. Tabit.

In designing the system, Oxko Corp. of Annapolis, Maryland, started with interviewing the Alloy plant's best workers. This technique, known as "knowledge engineering," is basic to the development of all expert systems. The knowledge of the process culled from the workers is

then replicated in a computer program.

Steven W. Oxman, president of Oxco, does not believe that expert system technology in any way replaces machines with people. "Once we invented the saw, should we still continue to cut wood by chopping on it?" said Mr. Oxman. With an expert system, we allow "our thinking to get to the next level of use."

Elkem Metals is owned by Elkem AS of Oslo, Norway. Elkem AS is one of the world's largest manufacturers of ferroalloys.

Source: Consella A. Lee, "Expert system blends knowledge, precision to increase productivity," (Baltimore, Maryland) *Sun*, December 2, 1993.

Changeover Technologies

The technology used for set ups and model changeovers is a major process driver. That technology determines the flexibility of the line. If the technology permits, a great number of models can be run in small batches with the cost efficiency of the flow shop, but the variety output of a job shop. DFM can play a crucial role. It allows the choice of a design with excellent set-up characteristics instead of accepting a changeover situation by default.

Changeover technologies have been substantially decreasing the time it takes to shift what is being produced from one model to another. The engineering of changeovers, almost always dependent on the creation of new technology, has undergone a major revolution as envisioned by Shingo when he introduced his single-minute exchange of dies (SMED) philosophy. The new set-up methods often increase productivity and improve quality at the same time.

8.10

THE NEW TECHNOLOGIES—CYBERNETICS

For OM to effectively manage technology, it is necessary to distinguish between old technologies and new ones which promise to alter the landscape significantly. Some of the new technologies have been the subject of study for a long time, but recent developments indicate they needed twenty or thirty years to reach the point of practical application. Some new technologies have not yet reached that point.

Cybernetics

Study of control systems; employs similarities between human brains and electronic systems.

Cybernetics[4] is the study of similarities between human brains and electronic systems, including sensory devices, computers and robotic analogs for doing work. It compares the way electronic signals are generated by information in the computer system with the human use of light and sound, among other things. In the human, signals are transmitted between adjacent nerve cells called neurons, across a boundary called a synapse. The state of the synapse determines whether or not a signal can pass across the boundary. This description has many similarities to electronic systems.

The resemblance between human systems and those of computers provides opportunities to learn by making comparisons. The outgrowths are impacting technology, including the development of:

1. new machines that permit *flexible manufacturing* and *flexible processes*
2. *adaptive systems* that can emulate and surpass many human abilities
3. *neural networks* for creating *artificial intelligence* as well as learning organizations and *expert systems*. (Neural networks are computer programs modeled like synaptic neuron networks in human beings.)

Flexibility—FMS, FOS, FSS, and FPS

Flexible manufacturing systems (FMS)

Individual machines and/or groups of machines (called cells) programmed to produce a menu of different products.

Flexible manufacturing systems (FMS) are individual machines and/or groups of machines (called *cells*) which can be programmed to produce a menu of different products. The one thing the products on the menu have in common is the fact that the machine or cell can make all of them when supplied with appropriate tooling.

Flexible office systems (FOS)

Outgrowth of the new technological capabilities applied to office systems.

Flexible service systems (FSS)

Outgrowth of the new technological capabilities applied to service systems.

The changeover times for these computer-controlled systems are negligible, meaning that very small batches (typical of the job shop) can be produced with the economic advantages of serialized flow shops. Theoretically, the batch size can be one. **Flexible office systems (FOS)** and **flexible service systems (FSS)** are also an outgrowth of the new technological capabilities. **Flexible process systems (FPS)** is a generic name for the same technological capability when it is applied to any work situation that utilizes the flexibility concept in a manufacturing or nonmanufacturing environment.

Flexible process systems (FPS)

Generic name for the same technological capability when it is applied to any work situation utilizing the flexibility concept in a manufacturing or nonmanufacturing environment.

Flexible warehouses are a good example. Areas of the warehouse are not dedicated to particular classes of items. Instead, products can be stored wherever space allows. The computer remembers where the items are by storing the product SKU code, the space code and the number of units at that location.

With technological flexibility, questions arise concerning how many different products can be produced on the same flexible system. How big is the menu? That seems to depend on the culture of the FPS users. For FMS, American companies use fewer menu options than Japanese firms, which use fewer options than German companies. The numbers are not precise, but they are sufficiently different to override statistical differences. Table 8.2 provides some approximate numbers.[5]

TABLE   8 . 2

COMPARISON BY COUNTRIES
NUMBER OF VARIETIES PRODUCED BY THE FMS

Germany	80
Japan	50
U.S.A.	7

Adaptive capabilities bring flexibility to automated systems. The latter are often called fixed-automation to indicate that they are economic to use if the demand exists for high-output volumes. They can deliver high quality, but have no capability to produce anything other than the single product that they were designed to produce. They are in trouble when demand falls off for the high-volume product they produce.

Fixed-automation does not even permit changeovers at high cost because they have been dedicated to one thing, such as a V8 engine. Ford Motor Company had built such a plant, but could not use it when the petroleum crisis in the 1970s killed the possibility of selling V8 engines. On the other hand, the windshield system of Glasstech, Inc. represents adaptive automation at its best.

Expert Systems

Expert systems are *computerized sequences* or strings of intelligent inquiries and answers that are copies of what the "experts" do to solve problems or to accomplish specific kinds of tasks. These smart procedures are developed by copying the way the experts go about solving the problems. The problems are complex so it is necessary to have a means of catching all the details. It is not always clear why experts do what they do. The simplest procedure is to ask.

Ultimately, all of the expert responses will have been obtained and emulated. For complex problems, it can take a long time to piece together the necessary conditions to be able to emulate the expert. The expert system may be dedicated to few or many variants. Different external factors can be encountered. Once completed, this expert program must be tested and debugged.

Consider an insurance expert in marine underwriting. This is a specialized insurance policy writer for ships who knows what questions to ask in order to determine the appropriate premium to charge. The expert system is built to conform to the questions and decisions made by this expert, incorporating experiences with many kinds of ships of different ages, and other relevant data.

Figure 8.5 illustrates decision-tree representation of the marine underwriting expert system. Figure 8.5 is a simplified version of the questions that must be asked. After these questions have been answered, the policy situation is identified as one out of 15,552 different possibilities. Real marine underwriters have written policies for each of these situations and determine the appropriate premium in each case. The expert system is therefore able to identify the type of policy and premium that applies.

Expert systems
Computerized sequences for smart procedures developed by copying the way experts solve the problems.

FIGURE 8 . 5

EXPERT SYSTEM DECISION TREE FOR A MARINE UNDERWRITER

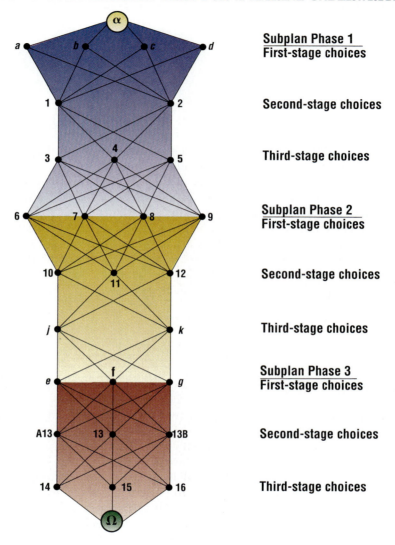

Subplan Phase 1
First-stage choices

Second-stage choices

Third-stage choices

Subplan Phase 2
First-stage choices

Second-stage choices

Third-stage choices

Subplan Phase 3
First-stage choices

Second-stage choices

Third-stage choices

The list of questions that goes with the decision tree:

0. α—Start!
1. What type of policy is wanted—a, b, c or d?
2. Inboard or outboard engine—1 or 2?
3. What type of ship is it—3, 4, or 5?
4. Where is the ship berthed—6, 7, 8 or 9?
5. What kind of maintenance policy—10, 11 or 12?
6. Fire protection, built-in or portable—j or k?
7. Type of policy allowed—e, f or g?
8. Optionals: radar, loran, etc.—A13, 13 or 13B?
9. Crew training—14, 15 or 16?
10. Ω Choose policy and determine premium!

This requires obtaining all the necessary information an expert marine under-writer should have to draw up an appropriate maritime insurance policy for vessels of given sizes and characteristics. The variety (in details) of ships to be insured is enormous. Classes of situations must therefore be created to establish coverage and rates to be charged.

Questions are followed by appropriate answers. Once the expert program is written, all of the vessel's specifications can be input to the computer. The program will follow the inquiry to response strings. If successful, it will fill out the policy form and determine the coverage options and premiums to be charged.

These decision trees branch over and over, with many conditional paths leading to highly-complex networks. If two more three-part questions were added to the simple tree shown in Figure 8.5 above, it would describe 139,968 unique policy situations.

Adding to the complexity are the various paths of best practice for the many kinds of situations that can arise. The networks are converted into computer programs that copy the best practice of experts. They have the advantage of being able to copy more than one expert and then combine all the best results. Expert systems can combine experts of one kind with experts of another kind. Also, an expert system's capabilities can learn and improve with experience.

It is noted that when an expert is confronted with a choice previously made, say A, sometimes the expert opinion has changed, or there can be recognition that this time A is accompanied by M whereas the previous time there was no M to consider. The expert now chooses B. The database must be amended and M may have to be added as another consideration.

Why are expert systems used? There are many reasons, including the fact that experts are in short supply. Also, expert systems can combine the expertise of more than one person which results in a meta-expert system. Expert systems can be used to reduce bottlenecks when demand is greater than the supply. Applications to complex manufacturing processes are warranted where there is a shortage of experts. The falli-bility of humans, even experts, can be alleviated. An expert system can provide a solution which can be reviewed by a human expert in much shorter time than that expert would need to derive the solution.

Imagine an expert system which provides an airline pilot with a recommenda-tion about whether or not to land in bad weather. It could save lives. Expert systems can be used to monitor in moments, complex multidimensional quality control systems readings that would take many hours and/or days for a human judge to evaluate.

Expert systems can: guide ships through channels that are difficult to navigate; monitor and shut down NASA space launches when everything does not check out; provide safety checks (and shutdowns) for nuclear power plants. Expert systems can replace outmoded and archaic organizations which rely on too many people deal-ing with too much information that needs to be rapidly integrated.

Air traffic control (ATC) is an excellent candidate for expert systems. ATC is a network of people responsible for the position of planes in their sectors which then get handed over to people in adjacent sectors. Expert systems provide instantaneous and seamless communications, just what is needed to assist the human controllers who at present are too fragmented.

Expert systems can read many controls at the same time and integrate the systems data. The applications of these ideas promise changes in the management of tech-nology that are going to alter substantially the way things are done.

Artificial intelligence (AI)

Systems which have the capability of sensing and learning; analogous to the brain and its functioning.

Neural networks

Mechanisms that are analogous to human nervous systems designed to enable computers to emulate functions associated with human intelligence.

Discussion about expert systems should draw the distinction between them and **artificial intelligence (AI)**. Expert systems are based on copying the procedures of the experts. AI is based on copying the nervous system of humans to develop mechanisms which have the capability of sensing and learning; there are strong analogies to the structure of the brain and the way the brain works.

Mechanisms that are analogous to human nervous systems are called **neural networks**. Neural networks are being designed to enable computers to read handwriting and to emulate the kinds of functions associated with intelligence in human beings. A machine with artificial intelligence should seem intelligent to an intelligent human being.

Part of the mystique of AI is defining a test that will conclusively indicate an intelligent machine. The search for the test is philosophical and the field has attracted many who are more interested in the means to the end than the end. Still, though there are claims of success, artificial intelligence remains a research area. There will be increasing applications of intelligent machines in the smart factory and office.

Cyborgs and Robots

Cyborg

Technological pairing of people and machines to achieve results neither could otherwise obtain.

Cyborg is a coined word which combines the cyb of *cyb*ernetics with orgs from *org*anisms. Cybernetic organisms are a form of technology which pairs people and machines to achieve results neither could otherwise obtain. Someone wearing a pair of reading glasses is a cyborg. If the glasses changed depending on whether the person was looking near or far, the more supportive would be the interaction and the more interesting the cyborg. Strong cyborg interactions are reflected by highly adaptive behavior on the part of both the person and the machine.

Heroic measures to keep a person alive combine machine technology with the human body as a cybernetic organism. Anyone with a pacemaker is a grateful cyborg. The same principles apply to scuba diving. Air tanks strapped to the diver's back and a breathing device with a regulator which adjusts to depths are crucial to life in an underwater environment.

Jacques Cousteau advanced the technology which permits people to stay underwater for long periods of time. "Cousteau was engaged in an extended stay program in Conshelf III, an 18 foot diameter sphere in which his team of six aquanauts spent 27 days at 328 feet (100m), followed by a three-day decompression."[6] Although the predecessors of humanity were likely to have been water breathers, it is not a matter of skill to stay underwater for any length of time. Technology alone makes it possible.

Cousteau and others have pursued the idea that people could be given breathing implants similar to the gills of a fish (tissue capable of absorbing oxygen from water and releasing carbon dioxide from the blood stream). If that ever is done, a better example of a cyborg will be hard to find. Some practical reasons for wanting to be able to live underwater include the ability to mine for minerals and extract oil on the sea floor. Such cyborgs would provide a technological breakthrough.

It is easy to extrapolate to sustaining life in the space lab, on space trips and in communities on the moon and Mars. Some practical reasons for wanting to be able to live in outerspace include the ability to use gravity-free factories to produce products to much finer tolerances (i.e., ball bearings). Certain chemicals are dangerous to handle in the Earth's atmosphere which would be stable on the moon. Various medical conditions would be advantaged by low-gravity treatment centers. Process management will never be the same, given the significant technological changes that will occur.

At the other end of the spectrum are the robots that do many jobs in the factory. The painting robots and the welding robots take careful programming to enable them to do the job right for each of the various car models that come down the line. They can changeover from the positions required for one model to that of another in nanoseconds. Robots are excellent changeover devices. Consequently, when model flexibility is desired, the use of robots is generally warranted.

Yet to come are the home cleaning robots and gardener robots who would combine the motor skills of a human with the strength of a machine and the intelligence that a programmer can instill. Already in existence is an egg-moving robot which can handle the egg with such daintiness of movement that the shell remains intact.

The Technology of Computer-User Interfaces

Programming is another illustration of the way technology necessitates bringing together machines interacting with operator skills. Instead of using their hands to set the machines and make them do their work, people program the machines or the computers which interface with the machines. Because programming is done by people to tell machines what to do, somehow they must both speak the same language.

Thus, the technology of computers is hidden from users. Programmers instruct machines using special languages. When a user clicks a mouse, hundreds or even thousands of lines of code can be the internal response of the system which is translated into the configuration on the screen.

Even computer-literate users do not understand "machine language." Instead, their literacy is bounded by understanding the software they use. There are enough problems dealing with relatively user-friendly software such as WordPerfect, Word, Quattro Pro, and Lotus.

Machines that are run by programmed instructions can, in turn, respond with instructions for the people that run them. For both inputs and outputs, the form communication can take varies. It can be typed, handwritten or vocal communication.

Programming technology is required for the regulation of factories and offices. It also is critical to make correct machine settings, for control over set ups and changeovers, to schedule production, to maintain inventory control, and to schedule maintenance activities. The programming languages needed for many of these activities are not user friendly.

Individuals have begun to take for granted this entry of programming technology to communicate with their systems. They have become accustomed to user-friendly desktop publishing, graphics, word processing, spreadsheet calculations, database management, indexing, outlining, and spell checking. Scientists can work at home on simulations, scientific model building and analysis, and myriad other applications of methods so advanced they were once considered feasible only on mainframes.

This form of technology, if it is to be effectively used, requires higher levels of training than the kinds of technology that preceded it. Going and gone are the skills of tool and die masters who worked the materials by hand. They are replaced by the considerable knowledge required of those who must write machine instructions for CAD/CAM software for designers. The programs for computer-aided design must be congruent with the programs for computer-aided manufacture because CAM programs convert CAD instructions into directions for factory machinery.

There is every reason to encourage the development of new arts and skills of people in the use of powerful office network systems. This includes the need for knowing how to use hardware and software for telecommunications, office networks

(LAN and WAN), sharing of laser printers and other equipment. LAN is local area networks and WAN is wide area networks. These acronyms describe the scope of interconnected computers which share databases and equipment.

Other New Technology Developments—Materials and Miniaturization

When reference is made to new technology, this usually is taken to mean that an industrial science improvement has occurred in equipment and machinery. For example, Intel and Motorola have continuously increased the power of their computer chips. Usually, new technology replaces existing technology which results in old technology. Gone are the ancient bulky and weighty calculating machines which were slow and costly.

The new technology also relates to stronger and better materials that have been developed. The weight of a typical VCR has been cut by more than 80 percent since the first unit was introduced by Sony. Automotive designers have had to reduce the weight of a car in order to satisfy government emission requirements. To do this, glass, plastics, aluminum and laminates have been used extensively. The sweeping windshields of cars have saved much weight, added beauty and provided this chapter with an industrial focus that reflects OM challenges and excitement.

Miniaturization has been pioneered by Sony for its own product line. To make small products requires processes that can fabricate and assemble miniaturized parts. This has led the company to manufacture its own equipment. As a result, Sony has developed entire manufacturing systems that are required for fabrication and assembly of products with very small components.

Improvements in management technology qualify as new industrial arts.[7] Advances in management that improve quality and productivity are legitimate candidates. Accordingly, TQM is replete with technological inventions. MRP, JIT and SMED are other aspects of new and improved technologies. Another facet of new technology is rapid project development which permits new products to be brought to market in half the conventional time. Flexible manufacturing equipment has been discussed in a prior section. It constitutes new technology which creates a unique class of work configuration which, if properly used, can combine the economies of scale of the flow shop with batch shop economies of scope.

S U M M A R Y

This chapter begins with an explanation of why OM must understand how to manage technology. The reason is that technology is one of two crucial components of the input-output transformation model. The two components are people with production skills and technology with engineering capabilities to improve upon what people can do. Managing the combination is an OM responsibility.

There is a technology trap. By default, new technology is often applied to the management of the old system with unfavorable results. Technological upgrading requires excellence in timing. This is illustrated by using the automotive glass industry and, in particular, windshield manufacturing. Levels of technological sophistication are explored for the replacement vs. the original market requirements and the supply-chain technology appropriate to each. The technology of packaging and delivery is included, as is the technology of testing and quality control technology. The technologies of design for manufacturing (DFM) and design for assembly (DFA) are explored, including an example of how to evaluate and select the preferred design.

Discussion of managing technology could not be complete without mention of changeover technologies and description of the technologies including cybernetics, flexible processing systems, expert systems, cyborgs and robots. The supplement treats Poka-yoke, the Japanese name for technology to prevent defectives.

▶ Key Terms

artificial intelligence (AI) *336*	cybernetics *332*	cyborg *336*
design for assembly (DFA) *328*	design for manufacturing (DFM) *328*	empirical testing *326*
expert systems *333*	flexible manufacturing systems (FMS) *332*	flexible office systems (FOS) *332*
flexible process systems (FPS) *332*	flexible service systems (FSS) *332*	management of technology (MOT) *308*
net present value (NPV) *310*	neural networks *336*	payback period *310*
technology *308*	tolerance matching *329*	

▶ Review Questions

1. What are the effects of rapid technological change?

2. How is technology a component of the I/O transformation?

3. Explain the management of technology in three ways: societal, company, and international levels.

4. What is the technology trap?

5. What are the basic elements of good packaging and how do they relate to the management of technology?

6. What are shakedown risks?

7. How does packaging relate to the supply chain and what is supply-chain technology?

8. What is the technology status of a glass blower?

9. When is the advantage of having a reputation for being a technology leader likely to pay off?

10. What is good technology timing?

11. Why is OM concerned about the licensing of technology? Relate the answer to an example.

12. How was flat glass made before the float-glass process was developed?

13. Explain the process for making windshields by referring to the process flowchart.

14. Relate OM's management of technology to R & D.

15. Compare the effects of replacement and original market demands on OM and MOT.

16. What is entropy and how is it relevant to MOT?

17. What should be done about delivery processes that increase the defective rate substantially?

18. Discuss OM's responsibilities with respect to testing technology.

19. What steps need to be taken to achieve DFM and DFA?

20. What are pilot studies?

21. Describe some new technologies.

22. How do expert systems work? Does OM have any use for them?

23. What methods are available for assessing the technology plans of competitors?

24. How can it be determined if customers are aware of and sensitive to the use of different technologies by competitors?

25. What is meant by knowing the cost and/or benefits of the present technological level of the firm?

26. How can the costs of transition from one technology level to another be estimated and used to determine the payback period?

27. What is the difference between payback period, breakeven time and the cost recovery period?

28. Explain why it is considered better to change technologies when the development cycles enter a period of relative stability instead of during periods of volatility.

▶ **Problem Section**

1. Bursting strength and product support (within the container) are listed for two different package designs. The container will be used to ship printers:

	Bursting Strength	**Product Support**
Box A	100 test	8
Box B	200 test	6

Assume that the computed value of the package design V is modeled as follows: $V = (BS)(PS)^2$ and the biggest number is the best. What does the scoring model value of each package indicate?

2. Compare technologies as follows: turn on various computers and note the time it takes each to boot up. The distribution of times will represent a range of set-up times for the computer. How variable are these set-up times? Explain the meaning of this exercise to the management of technology by OM.

3. Is it reasonable to compare word processors (as technology) by the following method: test how long it takes each word processor—on the same computer, to do the same task—such as repositioning from the first page to the fiftieth page of a 100 page document.

4. Use a multiplicative scoring model ($V_{D_i} = Q_1^2 Q_2^3 TM$) to evaluate alternative designs with respect to excellence for DFM and DFA. The matrix below presents the variances for two components and their tolerance match. In addition, an importance weight is given to each component's quality-fit dimension and to the tolerance match between the components. The smaller the weight the more important it is, which is in keeping with the preference for small variances.

T A B L E   8 . 3

VALUES OF Σ_j^2 FOR EACH DESIGN

	$w_1 = 2$ $Q_1\,(\sigma_1^2)$	$w_2 = 3$ $Q_2\,(\sigma_2^2)$	$w_{TM} = 1$ $TM\,(\sigma_1^2 + \sigma_2^2)$
Design 1	6	8	14
Design 2	5	6	11
Design 3	7	5	12

How does this situation differ from that in Table 8.1 shown earlier?

5. Calculate the payback period for a technology expense of one million dollars if the new technology immediately starts to earn revenues of $100,000 per month. Operating costs plus depreciated fixed costs are estimated to be $30,000 per month. How does this payback period differ from the cost recovery period?

6. Create an expert system for determining an optimal diet for anyone interested. Build a decision tree such as the one shown in Figure 8.5 above.

7. Show how Figure 8.5, which designs an expert system for marine underwriting, leads to 15,552 uniquely different situations.

8. With respect to the material in Problem 7, explain the statement that two more three-part questions would lead to 139,968 uniquely different situations.

9. With respect to the material in Problem 8, how many uniquely different situations would three more three-part questions yield?

10. Why is air traffic control (ATC) an excellent candidate for an expert system?

11. What is Glasstech's output rate and cycle time?

▶ Readings and References

W. Ross Ashby, An Introduction to Cybernetics, NY, Wiley, 1956.

Stafford Beer, *Decision and Control: The Meaning of Operational Research and Management Cybernetics*, NY, Wiley, 1994.

M. P. Hessel, M. Mooney, and M. Zeleny, "Integrated Process Management: A Management Technology for the New Competitive Era," *Global Competitiveness: Getting the U.S. Back on Track*, Ed. M. K. Starr, Norton for the American Assembly, 1988.

Douglas Hofstadter, and the Fluid Analogies Research Group, *Computer Models of the Fundamental Mechanisms of Thought*, NY, Basic Books, 1995.

Billy Mac Jones, *Magic with Sand: A History of AFG Industries, Inc.*, Wichita State University Business Heritage Series, Center for Entrepreneurship, College of Business Administration, WSU, Wichita, KS, 1984.

Michael E. McGrath, *Product Strategy for High-Technology Companies: How to Achieve Growth, Competitive Advantage, and Increased Profits*, Burr Ridge, IL, Irwin, 1995.

Michael G. Pollack, "The AGR Channel-to-Market," *US Glass*, Feb. 1994, p. 59. (AGR is auto replacement glass.)

Edward B. Roberts, Ed., *Generating Technological Innovation*, Sloan Management Review, Executive Bookshelf, NY, Oxford University Press, 1987.

Shigeo Shingo, *Zero Quality Control: Source Inspection and the Poka-Yoke System*, Norwalk, CT, Productivity Press, 1986.

F. V. Tooley, Ed., *Handbook of Glass Manufacture* (2 volumes) NY, Ashlee Publishing Co., Inc., 1985.

U.S. Congress Joint Economic Committee, "The U.S. Trade Position in High Technology: 1980–1986."

Shoshana Zuboff, *In the Age of the Smart Machine, The Future of Work and Power*, NY, Basic Books, 1988.

The Poka-yoke system

Poka-yoke is the Japanese name for technology specifically designed to prevent defectives from occurring. It is proactive technology used to speed up quality control. This is in contrast to SPC which is reactive, allowing the control charts to indicate that something has gone awry. Poka-yoke aims at prohibiting the possibility of certain specific kinds of problems.

Shigeo Shingo developed the Poka-yoke system for problem-prevention. These Japanese words mean "mistake-proofing" which can be considered in the same vein as weatherproofing. In the case of weatherproofing, steps are taken to prevent wind, rain, cold and/or heat from entering the building.

A Poka-yoke system mechanically or electronically carries out 100% early-warning inspection. When and if abnormalities arise with the process, this information is immediately fed back to people and/or machines that are trained and/or programmed to take immediate corrective action. The instantaneous feedback function is crucial because it allows corrections to be made on the spot.

Mistake-proofing technology follows mistake-eliminating methodology. This means that steps are taken to prevent problems from arising in the first place. Then, if problems do occur, the system is designed to catch them on the spot. This rapid feedback eliminates defectives and prevents them from *corrupting* other dependent parts of the system. Using SPC, it is possible to generate large amounts of defective product before correcting the situation and eliminating the problem.

Timing is of the essence. Poka-yoke systems are designed to prevent defectives, so this approach approximates the goal of zero defect programs.

One option is for machines to be designed to shut down automatically in the presence of a problem. Sensors can often determine machine abnormalities faster than product variability. Technology allows machines that are acting erratically to shut themselves down. Many kinds of operations can be brought to a halt before defect product is produced in quantity. A sensor determines that temperature is too high. NASA space launches are replete with Poka-yoke technology to prevent liftoff from occurring by reading equipment rather than product performance.

Put in terms of SQC, the goal of Poka-yoke is to take action before assignable causes are detected by a control chart. This requires inventiveness to configure a "mistake-proofing" system. In a case where the defectives could not be prevented, technology was used to mark the location of defects so they could be repaired quickly. This prevented having to shut down the machine until regular maintenance time.

Poka-yoke includes various warning systems like lights that flash and buzzers that sound when abnormalities or imbalances are detected. Figure 8.6 illustrates a Poka-yoke system application used in a Japanese company.[8]

Mechanical and electronic devices are not the only source of sensors. People are often able to sense process abnormalities. Training and technology combined with a sixth sense can provide early warnings that something is not right.

Noting the condition of machines is not part of normal quality control practice. Technology awareness makes mistake prevention an area of opportunity for OM and process managers. Sensing devices, switches and motion detectors are among the arsenal of mechanisms that are mentioned by Shingo in his book which cites scores of examples of Poka-yoke.[9]

To determine how much it is worth spending on technology to prevent defects, answer the following questions and then combine them in a cost evaluation model as shown below.

F I G U R E   8 . 6

A POKA-YOKE DEVICE TO INSURE THE ATTACHMENT OF LABELS

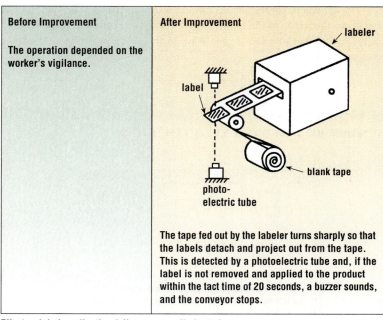

Before Improvement

The operation depended on the worker's vigilance.

After Improvement

labeler

label

photo-
electric tube

blank tape

The tape fed out by the labeler turns sharply so that the labels detach and project out from the tape. This is detected by a photoelectric tube and, if the label is not removed and applied to the product within the tact time of 20 seconds, a buzzer sounds, and the conveyor stops.

Effect: label application failures were eliminated.
Cost: ¥ 15,000 ($75)

Source: From *Zero Quality Control: Source Inspection in the Poka-yoke System* by Shigeo Shingo. English Translation © 1986 by Productivity Press, Inc., P.O. Box 13390, Portland, OR 97213-0390, (800) 394-6868. Reprinted by permission.

1. What is the probability of the fault x occurring?
   Say that $p_x = 0.01$ which means there is one defective unit per 100 units produced.

2. How many units are produced per day?
   Say that $V = 4000$ per day.

3. How much does it cost when *x* occurs? This cost should include the cost of repair and the affect of the fault on other units as well as process equipment.
   Say that $c_x = \$100$ per occurrence of x.

4. How much will it cost to use technology to prevent the fault from occurring?

   Say that the technology costs $50,000 and it is company policy to amortize the expense of such technology within one year or 250 working days. Therefore, this cost is:

$$C(T_x) = \frac{\$50,000}{250} = \$200 \; per \; day$$

Then, the cost of prevention and the cost of defectives can be contrasted. The cost of defectives is $C(D_x) = (p_x)(c_x)V$, and this is to be compared to the cost of technology, $C(T_x)$. Thus, $C(D_x) = (p_x)(c_x)V = (0.01)(100)4000 = \$4,000$ per day. Because the cost of defectives is 20 times larger than the cost of technology to prevent those defectives, the investment for Poka-yoke type technology is a clear winner.

Problems for Supplement 8

A curved piece is inserted into a fixture which holds the piece while it is being stamped. Defectives are made whenever the piece is inserted upside down. When correctly inserted, the piece curves to the left as in Figure 8.7. When incorrectly inserted, it curves to the right. Training is ineffective because the people who do this job are from a constantly shifting group of part-time workers.

FIGURE   8 . 7

PIECE P MUST BE INSERTED INTO THE DIE D

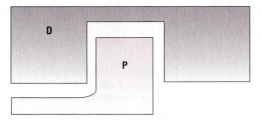

The piece P must be inserted into the die D with the curve to the left as shown. A defective is produced if the curve is to the right.

1. What technology can be used to prevent this defective from occurring?
2. How much can be spent on this technology, given the following information?

$$p_x = 0.10$$
$$V = 5000 \text{ per day}$$
$$c_x = \$1.00 \text{ per occurrence of x}$$

Company policy amortizes the costs of preventive technology over one year or 250 working days.

Is it likely that the company will invest in the Poka-yoke?

Notes...

1. From the Technology Report section of the *Wall Street Journal*, June 24, 1994.

2. Shoshana Zuboff, *In the Age of the Smart Machine, the Future of Work and Power*, NY, Basic Books, 1988.

3. Michael G. Pollack, "The AGR Channel-to-Market," *US Glass*, Feb. 1994, p. 59. (AGR is auto replacement glass.)

4. Stafford Beer, *Decision and Control: The Meaning of Operational Research and Management Cybernetics*, NY, Wiley, 1994.

5. Studies from the Columbia Business School Center for the Study of Operations found results similar to those reported by Jay Jakumar of the Harvard Business School in *Technology Review*, July, 1985, pp. 78–79.

6. *The New Illustrated Science and Invention Encyclopedia*, H. S. Stuttman, Inc., CT, Westport, 1987, p. 2972.

7. M. P. Hessel, M. Mooney, and M. Zeleny, "Integrated Process Management: A Management Technology for the New Competitive Era," *Global Competitiveness: Getting the U.S. Back on Track*, Ed. M. K. Starr, published by Norton for the American Assembly, 1988, pp. 121–158.

8. This example is taken from the book cited in the next footnote.

9. Shigeo Shingo, *Zero Quality Control: Source Inspection and the Poka-yoke System*, Norwalk, CT, Productivity Press, 1986. See in particular, Chapters 6 and 7, pp. 99–261.

Chapter 9

Planning Teamwork and Job Design

After reading this chapter you should be able to…

1. Explain why, in times of technological volatility, training people in teamwork is increasingly important for successful OM.
2. Describe the older training methods.
3. Describe the newer training methods.
4. Discuss the teamwork problems of performance evaluation.
5. Explain why activity-based cost systems are increasingly vital for proper job design and OM decision making.
6. Describe how to do a job observation (or time study) to determine standard output rates and costs.
7. Explain leveling and normal time.
8. Describe the method for determining adequate sample size for the time study.
9. Explain how work sampling is used to improve job design.
10. Describe the method for determining adequate sample size for the work-sampling study.
11. Explain how job observation can improve job design.
12. Detail the nature and applicability of synthetic time standards.
13. Describe job design, improvement and enrichment.
14. Discuss the issue of wage determination.

Chapter Outline

9.0   TRAINING AND TEAMWORK

9.1   TRAINING AND TECHNOLOGY

9.2   OPERATIONS MANAGERS AND WORKERS

9.3   PRESENT TRAINING METHODS

9.4   NEW TRAINING METHODS
        Action Learning
        Team Learning
        Computer-Based Training (CBT)
        Distance Learning

9.5   PERFORMANCE EVALUATION (PE)

9.6   USING PRODUCTION AND COST STANDARDS
        Activity-Based Cost (ABC) Systems and Job Design

9.7   HOW TIME STANDARDS ARE SET—JOB OBSERVATION
        Cycle-Time Reduction
        Computers Substitute for Stopwatches
        Shortest Common-Cycle Time
        Productivity Standards
        Training for Job Observation
        Setting Wages Based on Normal Time
        What Is Normal Time?
        Learning Curve Effects
        Job Observation Corrections
        Expected Productivity and Cost of the Well-Designed Job
        Job Observation Sample Size

9.8   WORK SAMPLING AND JOB DESIGN
        Work Sample Size

9.9   THE STUDY OF TIME AND MOTION—SYNTHETIC TIME STANDARDS
        Evaluation of Applicability

9.10 MANAGING THE WORK SYSTEM
        Job Design, Evaluation, and Improvement
        Wage Determination
        Work Simplification and Job Design
        Job Design and Enrichment
        The Motivation Factor

Summary
Key Terms
Review Questions
Problem Section
Readings and References
Supplement 9: Ergonomics or Human Factors

The Systems Viewpoint

P/OM, with its process orientation, manages technology and human resources. The latter is greatly facilitated by strong systems identification with the department of Human Resources Management (HRM). P/OM depends on the assistance of HRM for hiring and training that enhances teamwork for the mutual pursuit of company goals. One of the difficulties of making the systems approach work is that people from different departments have different perspectives and orientations. It takes dedicated effort to get finance, marketing, OM and other functions to work as a team.

When people recognize the strengths that come from their individual differences, teamwork is more effective. Staying within the bounds of each department is like asking the same person who packed your parachute to inspect it. Creative insights have a higher probability of occurring when there are more independent minds cooperating. HRM and P/OM work well together when they share knowledge of the company strategy and commitment to the systems approach, which fosters coordination between the functional departments. Training for teamwork is a systems-wide concern. It is the foundation for quality of products as well as internal processes. For the systems approach to work, job design must encourage and facilitate teamwork. Generalists who share knowledge about each other's areas of expertise should be rewarded as a team.

9.0

TRAINING AND TEAMWORK

These are times of technological volatility. The result is that personnel are dealing with rapid shifts in what they are doing and how they are doing it. Operations managers must teach themselves first, and then others, how to adapt to extreme and sudden changes. Training in new process technologies is enhanced by teamwork which starts with a cooperative relationship between OM and the Human Resources Management (HRM) Department.

In many companies, most of the people hired are interviewed by HRM. In general, the staffing of jobs in the company is influenced by the HRM Department. HRM personnel play a major role in matching people and jobs. Many companies now screen for team players. These are people who enjoy cooperating with fellow workers instead of competing with them. Japanese companies in the U.S. regularly use this approach to hire workers. They have endeavored to steer clear of unions. However, when this has not been possible, efforts have been made for management to team up with the unions, as is the case with Toyota and GM's joint venture NUMMI.

HRM, in addition to playing a key role in hiring, is also the department that organizes training programs throughout the company. Training can foster the teamwork that P/OM's success requires. In addition, P/OM works closely with HRM to inform those who train and hire about requirements as operational needs are modified by competitive and technological changes. Training and teamwork are foundations for quality products and processes. Teamwork is a function of morale, which HRM influences through administration of payment and pension plans and in a variety of other ways as well. Close coordination of OM with HRM helps OM keep in touch with employee morale.

9.1

TRAINING AND TECHNOLOGY

All OM systems are dependent upon the participation of people. The human factor always plays a part and teamwork counts. Even for entirely automated systems, people make plans, build plants, maintain and change them. Generally, semi-automated combinations of people and machines provide greatest flexibility. Adaptive systems can be guided by managers and engineers to achieve flexible automation. The basic equation has been used for many years:

Workers + Machines = Process Capabilities

What is lacking in the above equation is the fact that people need training and teamwork to use technology effectively. Training provides skills, and skills relate to quality and productivity. The systems approach provides recognition of the fact that machines are too narrow, confined and isolated in the picture of what makes a process work. Technology is a better word because it includes product design, materials, process methods and flows, information systems and software. Therefore, the preferred equation is:

Teamwork + Training + Technology = Process Capabilities

For each dollar spent on technology, companies can spend two or three dollars training people to use that technology in a skilled and efficient manner. The goal is to determine the correct blend of technology and training. The best guidelines for deciding this are as follows:

- Try to keep it simple.
- Avoid going beyond the limits of technological expertise.
- Count on people for quality, productivity and good ideas.
- Finally, when technology is upgraded, provide sufficient training to make it work.

It has been widely written that GM expected Toyota to employ robotics and complex equipment at NUMMI (New United Manufacturing Motors Inc.), a joint venture between GM and Toyota. Instead, intense training was used with simple technology to produce a high quality car in the old GM Freemont, California plant that had been rated substandard.

This is a real example of choosing workers skillful with general-purpose equipment over automated operations using robotics and other special-purpose equipment. Thus, the decision to invest in elaborate machines has to be regarded with suspicion and carefully compared with the decision to train workers.

An interactive, multimedia, shop-floor training program is used to access information about a sophisticated gas turbine.
Source: Cooper Industries, Inc.

Finally, some things can only be done by one or the other. Technology and training can be traded off but not completely. A glass blower's job description is heavy on training and light on technology. The auto windshield maker tips the scales in the other direction. The chef of a great Italian restaurant eschews a microwave while Burger King revels in its continuous-chain infrared broiler which cooks the burger just the right amount of time for the carefully controlled thickness of the patties.

9.2

OPERATIONS MANAGERS AND WORKERS

The daunting problem that P/OM has faced since it began to be studied by organizational theorists and industrial engineers has been how managers should relate to, work with, and deal with employees. Production and operations managers need to know how to foster competence, how to obtain commitment, how to develop and maintain motivation, and how to secure the loyalty of people who work in the system. All of this can be summed up as OM seeking to develop successful teamwork.

The past decade has seen such turbulence in downsizing of companies that the concept of loyalty to the firm by both workers on the floor and management is stretched thin. There also are numerous examples of adversarial relationships between management and workers, especially when unions are involved.

The manufacturing workforce in the U.S. is shrinking in size. But, at the same time, manufacturing employees play a more important part than ever in achieving competitive performance. The service workforce continues to expand, but low productivity is distressing. Good relationships between managers and workers characterize globally successful competitors. In most prosperous domestic firms, management and workers are a team. Some firms use participative management, where workers help management reach decisions and share decision-making responsibilities.

Closer harmony with unions is evident in such plants as NUMMI and Saturn. On the floor, workers are making important creative contributions. The HRM component is increasingly recognized as a part of successful competitive strategies, with emphasis on selection, training, motivation, and commitment to excellence. HRM also must forge new relationships with unions that bring out the best of P/OM practice.

The U.S. steel industry has returned to profitability after years of serious "hard times."[1] Two factors that are usually cited are:

1. The change of technology to continuous casting, processing the metal while red hot into final products, constitutes a flow shop. Traditional methods can be categorized as labor-intensive, job shop operations. First the metal was poured into molds; then it was cooled; next the mold was stripped; then the metal was reheated for processing into the final product. The change saved energy and labor costs.
2. The changing relationships of the companies and the union which were traditionally highly adversarial became cooperative. Since the "hard times," the United Steel Workers and the companies have agreed on flexibility in assignments and cooperation allowing for job rotation and job cross-training. Unions have been given more say, often with a seat on the board.

In the steel industry, unions and management have learned to work together, using technology and training to survive and prosper. Similar accommodations have been made in aerospace, machine tools and the chemical industries. The new connections between technology and human resources management are still being forged and tested in the automotive industries.

9.3

PRESENT TRAINING METHODS

Traditional training methods that are widely in use include on-the-job training, classroom instruction, audiovisual methods, workshops and seminars, case methods, games and simulations, role-playing, and self-instruction. There are no mysteries about any of these traditional training methods. They have long been in place and everyone who has held a job and gone to school has been exposed to all of them.

Training program effectiveness is determined by four basic elements:

1. objectives of training (what is needed)
2. methods and design of the training programs
3. delivery of the training program
4. measurement of training program effectiveness

Figure 9.1 shows present methods for achieving traditional objectives.

RELATING TRAINING OBJECTIVES AND METHODS

Objectives	Methods
Information	Lectures
Skills	Demonstration, Practice Group Exercises
Attitude	Experiential Learning
Individual Behavior	Practice, Feedback and Experience
Organizational Behavior	Team-building, Management Reinforcement

For any of the methods, support and reinforcement of training in the actual work environment is critical if behavior change objectives are to be met. Unfortunately, reinforcement is often bypassed for practical reasons like lack of time and money.

Current training trends show utilization of both traditional and innovative methods for a variety of applications, including organizational development, management and supervisory development, and technical skill development.

9.4

NEW TRAINING METHODS

The repertoire of training techniques has been expanding to include more experimentation and technology. It now includes cross-training, called action learning, team learning, computer-based training (CBT), and distance learning, which uses interactive video and multimedia systems. Leading companies worldwide expect OM involvement in the development of advanced training methods. The list of companies to which this applies includes Citicorp, GM, HP, Honda, IBM, Motorola, and Toyota.

Action Learning

Action learning
Trading jobs for a period of time; encourages cooperation and team building.

Action learning, trading jobs for a period of time, has been very effective. It encourages systems thinking by allowing people to stand in each other's shoes and play each other's roles. Action learning encourages team building and leads to the cooperative environment needed for team learning.

Team Learning

Team learning
Emphasizes interdependencies that exist between people working toward a common goal; also called cooperative learning; requires application of the systems approach.

Team learning adds to personal learning by emphasizing the interdependencies that exist between people working toward a common goal. It also is called cooperative learning. An example is teaching the baseball team to play together as a coordinated group instead of teaching each player how to be individually better at their assigned positions.

Classroom instruction is a widely
used traditional training method.
Source: ConAgra, Inc.

Team learning follows **team building** which involves training groups of people to have a common focus. Often referred to as obtaining a personal commitment to the company "family," this approach is intended to do away with adversarial relations. The belief is that when internal competition is replaced with cooperation, then learning can begin.[2]

Although team building methods remain experimental, they are being widely used. "Determined to achieve a 'cultural revolution,' Oldsmobile is putting—forcing, really—its independent-minded dealers through the team- and trust-building training used by another General Motors division, Saturn...."[3]

Team learning requires applying the systems approach. In every organization there are people who, like a pitcher, catcher, or batter, can make greater contributions to winnings by being team players than by being individual stars. For business, team learning targets cooperation instead of what has come to be known as "personal best."

Honda of America has utilized team learning for its P/OM managers to enable them to shift easily across a broad spectrum of jobs. At Honda, the job description at any moment in time is related to what needs to be done. The Honda P/OM "associates" are not interchangeable, but they can play all the positions as needed. Hewlett-Packard employs team learning to achieve a higher level of performance than otherwise could be realized.

Team building

Training groups of people to have a common focus; replaces internal competition with cooperation.

Team learning targets cooperation rather than what has come to be known as "personal best." Source: McDonnell Douglas Corporation

Computer-Based Training (CBT)

Scenarios and simulations can be run on the computer which imitate real life. The advantages of such instruction are that time can be compressed. Many things can be made to happen which would take years to occur in the real world. Extremes in demand and breakdowns in supply can be experienced by the student, enabling him or her to see how well a course of action works.

CBT is like training with a flight simulator. Such horrible events as losing an engine while encountering hurricane-force winds can be dealt with over and over again using a simulator before they are experienced while flying a real plane. Computer-based training offers the advantages of repeating a situation as needed, and avoiding the risks of real losses if the wrong things are done. Such failures are simply part of the computer record.

There are several disadvantages. One is the cost of creating computer programs that capture all of the necessary elements of reality. Another disadvantage is that it may not be possible to make a computer behave like the real world. The full system of interconnected factors may not even be known to the simulation-model builder and programmer. Therefore, CBT is best used to deal with situations where training on the computer can be said to parallel reality. A reality check is always necessary. It must be explicit and convincing to the trainers and trainees.

Distance Learning

Distance-learning training is delivered through the use of teleconferencing or videoconferencing, often via satellites, to reach trainees in multiple physical locations simultaneously. Networked computer-based training simulation methods are widely employed and there are many other forms that it takes.

Computer-based training like this
interactive video training program
prepares workers before they begin
an assignment.
Source: ARCO Photo Collection

One of these is interactive video, which represents a range of options using TV
and two-way communication. At its simplest, the camera is located in only one
location, but the ability to communicate is two-way. Often, the camera follows the
teacher, charts and diagrams. Students can be located miles away, as is the case
with the British Open University, which has successfully educated many thousands
of students, every year, from all over the U.K.

Within companies, there can be cameras at two locations so that the operations
managers can converse and see each other. This also is called videoconferencing. It
can be used in two ways: teams can train each other in respective skills, and they
can transfer knowledge in a far more effective way than by telephone alone.

9.5

PERFORMANCE EVALUATION (PE)

Evaluation of individual performance (PE) is commonly used but can be counterproductive because the people who are measured distrust it. There is resentment
based on the feeling that PE is not accurate, yet workers are permanently categorized by it. It should be noted that management performance evaluation may be less
tangible than worker performance evaluation. In many organizations both tangible
and intangible factors are used.

As a counterpoint to negative reactions about PE, many companies claim that
its use is essential to remove poor performers as quickly as possible, to provide incentives for excellent performance and to provide feedback to people so they can improve
their performance. Also, PE is a part of management practice all over the world.

Distance learning enables
employees at Ford Motor Company
to increase their knowledge by
taking university courses right at
their workstations via closed-
circuit television.
Source: Ford Motor Company

Average performance

Employees are grouped above
and below an average
performance measurement.

Therefore, it is necessary to understand that a person's real contribution to the group may not be captured by the PE measures that are used. Most performance comparisons are made by finding **average performance** and then grouping those above and below the average. By definition, about half of the people measured will be below the average. This immediately punishes part of the group, as always happens when the average is used, even if everyone in the below average group outperforms everyone else in the world.

Since the dimensions of the measures may miss some important contributions that individuals make to the system's performance, one must conclude that there are dangers in this approach. The measurement of team or group performance is not subject to most of these criticisms. Because performance measures are widely used for reasons that are logical, it may be possible to temper them with group performance measures.

Evaluation of worker performance begins with the measurement of output. Then, a standard is required to compare with the actual output. Process managers can define expected outputs for machines, but find definition elusive for people.[4]

Why is the question of performance evaluation so important? The reason is that the human resources component of the total cost of goods and service is sizable. It is larger in the service sector than in manufacturing, and it has been growing at a fast rate. The supply chain, of which manufacturing is just a part, has a great deal of direct labor in it. Warehousing, delivery and retailing tend to be labor-intensive services in the supply chain that the producer cannot do without.

It reaches into the suppliers organizations as well. Suppliers have varying levels of labor in their products and then these too must be delivered to the manufacturers. Human resources plays a significant role in management and staff functions. Many of these costs show up as overhead costs instead of labor. In short, there are many places where people play a part and the human resources costs are hidden.

Some industries are more labor-intensive than others. The level of direct impact on factory costs is usually related to the efforts made to measure and control labor costs. In the banking industry, check processing remains labor intensive and performance evaluation is an ongoing activity. Major efforts are made to measure and control labor costs.

The problem of measuring on-the-line labor costs will continue to interest many companies. Measuring workers' performance remains critical wherever management wants to understand the company's costs and people play more than a trivial part in the cost structure of the system. The service sector can expect continuous growth in the performance evaluation area and in cost control of its labor component.

9.6

USING PRODUCTION AND COST STANDARDS

Production standards are criteria specifying the amount of work to be accomplished in a given period of time. Standards are statements of expected output from which costs can be estimated. Production standards are widely used by manufacturing and service industries in this way. Standards are needed to enable management to compute the real costs of production. Measurement of costs provides estimates of the cost of goods and services.

Production standards

Criteria specifying the amount of work to be accomplished in a given period of time; used to compute real costs of production.

How standards are set will be discussed later. How standards are used will be examined now. Japanese companies use standards as *targets to beat*. U.S. companies use standards as *goals to meet*. The meeting goal can put a lid on accomplishments, whereas the beating goal motivates continuous improvement.

Below are a few P/OM situations that require good production standards.

1. How much to charge for a new product or service is dependent on what the real costs are. These costs are to be recovered, then a profit is added on top of them.
2. Fixed and variable costs are used for breakeven analysis. This provides a capacity goal for the company. It cannot make money until it generates the breakeven volume. Direct labor is part of the variable cost component. Indirect labor, administration, and R & D are part of the fixed cost component.
3. Plant selection needs estimates of labor costs for different areas of the country.
4. Production scheduling allocates jobs, thereby assigning labor costs. Inventories of work-in-process and finished goods are storing labor costs on the shelves and in storage bins.
5. Determining an optimal product mix is dependent on being able to estimate profit per piece—based on cost per piece—for each product that could be in the mix.
6. The selection of technology in trade-off for labor requires knowing what the labor costs are now and what they are likely to be in the future. If labor costs are rising and the costs of technology are decreasing, when will it be reasonable to shift from labor to technology?
7. Companies that bid for jobs with municipalities, or bid as suppliers to companies, must know their labor costs. Otherwise they could lose money by winning contracts.
8. Capacity planning requires estimates of variable costs for different work configurations.

Especially with respect to services, the variability of labor expenses adds more uncertainty than most other costs to the organization's competitiveness. It is simple to predetermine the costs of materials, energy, space, insurance, and equipment.

The quandary of estimating how much labor will be used to set up a job, and then produce a batch of given size is complicated by behavioral factors. In this regard, the service estimate is more uncertain than the manufacturing estimate.

Labor estimation for the flow shop is easier than for the job and batch shop. This is another reason the flow shop is a preferred work configuration. Reduction of variability is a favorite reason P/OM tends to choose machines in place of people.

Even with the flow shop in place, there has been growth in the wage rates of indirect labor such as administrative, sales, promotional, clerical and research personnel. It was anticipated that the growing number of computers would decrease such overhead charges. The reverse is true as, on average, they have continued to increase. It is, therefore, important to examine the basis for assigning indirect and overhead costs.

Activity-Based Cost (ABC) Systems and Job Design

Activity-based cost (ABC) systems

Actual resource expenses are subtracted from revenues to yield the short-term contribution margin (STCM); general expenses are subtracted separately from STCM to yield operating profit.

The method of assigning overhead costs to products based on direct labor dollars associated with specific job designs has been traditional. Many times the mix of resources required to make a product is badly reflected by volume-driven measures such as direct labor. In such cases, **activity-based cost (ABC) systems** should replace the traditional cost-accounting methods. Under such circumstances, actual expenses (AE) for resources including charges for materials, energy and direct, short-term labor and overtime are subtracted from revenues (R) to yield the short-term contribution margin (STCM). Thus:

$$STCM = R - AE$$

The other portion of operating expenses are assigned to resources that were acquired prior to actual use. These general expenses (*GE*) cover resources committed for purchasing, machine time, engineering and administration. Permanent direct labor is a part of these expenses which are divided into used and unused portions (*UGE* and *UNGE*). How much of purchasing's cost should be added to this particular job? How much inventory remained unused? These costs are subtracted separately from the short-term contribution margin to yield operating profit (*OP*). Thus:

$$STCM - UGE - UNGE = OP$$

To illustrate, assume that all costs are multiplied by 10^4 and apply to one job and one week:

Materials of $5 + Energy of $3 + Direct Labor of $4 = AE of $12

STCM = Revenue of $18 − AE of $12 = $6

The overhead to be allocated is measured by the company for one week and applied in normal fashion to this job. However, it will be broken into two parts as follows:

Assume that:

GE = $3 with UGE = $1 and UNGE = $2

Then,

STCM of $6 − UGE of $1 − UNGE of $2 = OP of $3

The approach makes evident that overhead charges of $2 are labor costs that would have occurred whether the job was done or not.

It should be emphasized that ABC systems make clear the labor costs that would not have been incurred if the job had not been done. ABC also makes evident the labor costs that would have occurred anyway, and divides these into used and unused portions. This facilitates evaluation of job and process design.

Traditional costing systems lump and therefore hide the charges for resources that have been paid for whether or not they are utilized. The equation used by the leading proponents and creators of ABC to explain the subject[5] is (in units or dollars):

$$\text{Activity Available} = \text{Activity Usage} + \text{Unused Capacity}$$
$$\text{GE} = \text{UGE} + \text{UNGE}$$

It may be easier to understand what is being suggested by converting the above equation into a cost equation as follows:

Costs Assignable for Products Made (Activity Usage, UGE)

= Costs of Resources Available (Activity Available, GE)

− Costs of Resources Unused (Unused Capacity, UNGE)

The costs of resources that are used are considered to be real costs that can and should be assigned to costs of goods sold. The costs of unused resources are cause for concern. Too much capacity may have been put into place. If that is the case, and there is repeated evidence of unused capacity, it should be reduced or alternative uses for that capacity should be found longer term. Unused capacity of all kinds, including labor and technology, should be tracked. ABC is the accounting system that accomplishes that goal.

This subject is usually explored in accounting texts. More recently, P/OM has included this accounting issue, generally when discussing the increasing irrelevance of using labor charges as the guide for allocation of overhead charges, or when discussing capacity planning. These are the two places where a discussion of ABC seems most appropriate.

Without ABC, the decision to buy new technology can be negatively impacted. Traditional Net Present Value (NPV) accounting penalizes long-term planning where large income streams will not begin until far in the future. Meanwhile, labor-intensive situations that could be relieved by technology are penalized with large overhead charges. Job design is negatively impacted. As a result, the company tends to avoid these kinds of activities and loses out on potential opportunities to be a leader in the area.

An illustration of how the problem of assigning overhead costs to output has relevance for P/OM follows. In a lock manufacturing company, incorrect allocation of overhead caused the spring-making department to be closed. Thereafter, springs were purchased, which led to an increase in real costs and a loss of control over quality. The traditional accounting system had done a real disservice to P/OM. This was not a surprise to the process managers who had resisted the decision to shut down springs. The process design was flawed.

ABC can be credited with raising awareness of misleading information that traditional accounting methods can provide. The approach ties in with results that were obtained employing linear programming to optimize the use of limited resources years before ABC was developed.

When linear programming (LP) is used to determine an optimal product mix, it is frequently the case that full capacity utilization results in lower profits than properly configured partial capacity utilization. LP depends on good measures of labor productivity, availability of capacity, utilized capacity and unused capacity. The labor productivity is a function of the design of jobs.

PEOPLE

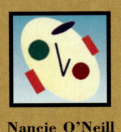

Nancie O'Neill
• • •
Westin Hotels and Resorts

As manager of Training, Development, and Performance Systems for Westin Hotels and Resorts, Nancie O'Neill works closely with the company's operations. "The human resources department is here to serve the operations of the business. The company is designed to manage hotels, and as part of that operation, human resources must respond to what the business needs, not just what we think we need," she said.

O'Neill works out of company headquarters in Seattle, Washington. Her responsibilities are to consult with the more than 75 Westin properties with regard to training activities, the companywide performance review system, and associate recognition programs. O'Neill was instrumental in developing "Westin University℠," a training and development program for corporate associates based on the company's fourteen core competencies.

"Westin University℠ is the comprehensive corporate effort to develop associates in the corporate facility as well as the corporate associates at the properties," O'Neill explains. "We are trying to move towards being more of a learning organization and so have started

mandating some of the classes," she said. For 1994–1995 required classes were Introduction to Quality (also known as "the Westin continuous improvement process"), Meeting Internal Customer Needs, Diversity, and Sexual Harassment.

In implementing continuous improvement, each associate learns how to use a five-step problem solving model. O'Neill said the five-step process is designed to help Westin be a "service quality organization focused on preventing problems" rather than a "customer-service based organization which deals with problems as they come up." The five steps include (1) identify the improvement opportunity, (2) gather data using the various tools and methods of total quality, (3) analyze the data, (4) identify solutions, and (5) implement the solution. O'Neill said step five is

"where we use the 'plan-do-check-act'—the continuous improvement process—we plan the change, we document it, we try it out, we check for revisions, and we standardize it." According to O'Neill, although not all problems are analyzed with this five-step process, it can be applied to a wide range of problems, from those "involving rags used in rooms to glitches in the reservation systems."

She notes that in the world of continuous improvement and downsizing, less emphasis is placed on job descriptions than in the past. "Job descriptions are an anchor to what someone's responsibilities are, but we have to look more specifically at the objectives within that job description that need to get done…we don't expect our associates to say 'I'm sorry but that's not in my job description'—that's a very bureaucratic approach to running a business," O'Neill said.

Source: A conversation with Nancie O'Neill, Westin Hotels and Resorts.

9.7

HOW TIME STANDARDS ARE SET— JOB OBSERVATION

Get a group of people together and compare their golf or tennis scores. Big differences exist. Check how many push-ups they can do and how fast they can input data on the computer terminal. Variability in the time it takes different people to do the same job description can be reduced by training, but never eliminated. In many cases, companies use so little training that the distribution of times is larger than one might guess would be the case.

Some people might be good golfers, but poor typists. Variances in "time to accomplish" exist among people within jobs and different jobs require different skills. To find a common ground for setting standards and evaluating the efforts and outputs of workers, it is necessary to build a sound structure and begin the analysis on a simple level.

Diagnostic production studies, where the worker is observed constantly over a long period, go back in time to the early 1900s. The approach can be compared to 100% inspection. Both the job and the worker are examined in great detail. This method has shortcomings so it is seldom used. Its primary weaknesses are high costs, unreliable results, worker confusion and hostility. However, when diagnostic studies lead to improvement of job design they are focusing on the right target.

One hundred percent observations gave way to sampling procedures which were identified by the name *time and motion studies*. Time study and work-sampling methods also were derived from statistical developments that date back to the 1920s. It is now politically correct for P/OM to eliminate references to time and motion studies, stopwatches and time studies. That does not mean that the functions are eliminated, just the references. The functions are being done, often in other ways as described ahead. However, the distinction should be made at this point between the study of how long it takes to do specific operations which is a **time study**, and the study of the motions of which the job is comprised, a **time and motion study**.

Instead of tracking the worker continuously, time and motion studies were based on obtaining a sample of observations sufficient to answer such questions as how long it takes to do each of the component tasks. The main or core process steps are called *basic elements* that compose the job. Support steps are called *extraneous elements*. From this information, the expected daily output of workers on the line can be determined.

Time study

Study of how long it takes to do specific operations.

Time and motion study

Study of the motions of which the job is comprised; based on sample observations.

Cycle-Time Reduction

The job observation analyst studies the overall job with the purpose of breaking the job down into basic and extraneous elements. These basic elements, when added together, form the *job cycle*. This job cycle will be the shortest cycle of basic elements required to make the product. It is the repetitive core of the job. Short cycles constitute the main portion of the job. Having a short-cycle time is equivalent to gaining high output rates. Essentially the same considerations apply to process studies leading to cycle-time analysis and reduction.

In many instances, the spotlight shifts from the individual to groups of individuals or to the combination of people and technology. The fundamental idea is to time all elements that make up the cycle. This is another aspect of process analysis with emphasis on the basic elements and on the times required to perform them. These are often referred to as diagnostic observations instead of as time studies a term that carries some negative connotations.

Diagnostic observational methods are not well suited for long-cycle jobs which do not have short-cycle repetitive basic elements following each other in sequence. Work sampling, used for long-cycle studies, is discussed ahead. The strength of job observational methods lies in their ability to deal with repetitious basic elements.

Job observational methods can be used to set standards or to study how work is done with the purpose of improving the job. The use of time studies for analyzing the performance of individuals has declined as technology has come to play a larger role and labor a smaller one. The decrease in piece-work pay, or hourly pay, in favor of pay by the day, or salaried pay, also has lessened the need for standards that state the expected number of pieces per hour.

It also has decreased as emphasis is placed on the contribution of group performance to the reduction of cycle time. It has been said, if there is only so much time and money to be spent on organizing and improving the process, then process cycle-time reduction and benchmarking are areas of study that are likely to have greater leverage than classical kinds of time or job studies.

Computers Substitute for Stopwatches

The reasons for job observation include:

1. improving the job
2. simplifying work
3. determining expected output rates
4. determining each individual's real output rate

All four are applicable to service and manufacturing. The fourth reason is often prompted by need to determine if individuals are performing well, and for payment purposes. Check processing in the bank has a room filled with people doing the same things. Job performance evaluation is ongoing for each one of them. These same labor-intensive conditions that applied to most manufacturing operations circa 1940, still are used in many banks, insurance companies, and travel agencies in various places worldwide.

Although the names "time studies" and "time and motion studies" are out, old-fashioned and unloved, similar procedures are still widely used. They are called by other names including process analysis and cycle-reduction studies. They are done in other ways. The difference may be that *cycle-time analysis* places more stress on the group. Some say there is less focus on minutia and unnecessary detail, they say that detail has been replaced by the bigger picture.

While there is truth to this statement, nevertheless, standards continue to be set and used everywhere. They are still based on logic, measurement and analysis. The new methods relying on computers and programming may be even more detailed than the old-fashioned time study.

Figure 9.2 is the *job observation study sheet* used by the Formatted Floppydisk Corporation (FFDC). It is being used to study the packaging of ten 3.5″ double-sided, high-density floppydisks.

FIGURE 9.2

FFDC'S JOB OBSERVATION SHEET

Operator____LM____
Operation____R96____
Machine____F9____
Materials____DK____
Speed_____✓____
Feed_____✓____

Date_____10/22/99_____
Location_____16_____
Time Begin__10:30_____
Time End____10:38_____
Observer____PS_____

ELEMENT	LEFT HAND	RIGHT HAND	EXTRANEOUS
1	Place 10 diskettes in box	Hold and position box, move to get cellophane wrapper	
2	Place box in wrapper, drop in chute Pick up 10 diskettes	Machine seal wrapper Pick up box	
A			Replace tote box every 100 diskette boxes
B			Get 500 diskettes every 50 boxes
C			Get 50 boxes every 50 boxes
D			Get cellophane wrapper every 500 boxes

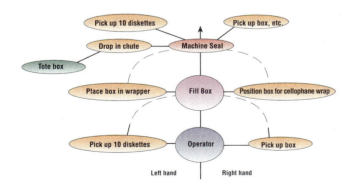

Job Layout Diagram

Observations (all times in hundredths of a minute)
Elements

Cycle	1	△	2	△	A	△	B	△	C	△	D	△
1	14	14	20	6								
2	31	11	36	5								
3	48	12	53	5								
4	65	12	70	5								
5	83	13	86	3								
6	97	11	104	7								
7	117	13	122	5								
8	134	12	139	5								
9	151	12	156	5								
10	167	11	171	4								
11	184	13	207	4	203	19						
12	218	11	222	4								
13	233	11	237	4								
14	249	12	256	7								
15	269	13	289	6			283	14				
16	301	12	306	5								
17	318	12	338	6					332	14		
18	350	12	355	5								
19	368	13	376	8								
20	389	13	394	5								
21	406	12	410	4								
22	423	13	429	6							448	19
23	461	13	467	6								
24	478	11	482	4								
25	493	11	497	4								
26	507	10	510	3								
27	520	10	526	6								
28	537	11	542	5								
29	554	12	560	6								
30	573	13	579	6								
Total		359		154		19		14		14		19

Two basic and four extraneous elements comprise the packing process. The two basic elements set the tempo for the complete short-cycle job. The four extraneous elements support the two main components of the packing process.

Stopwatches, once the mark of a time study observer, now are rarely seen in the U.S.A. This is because technology has developed machines that can be set to pace the worker; computers (PCs and notebooks, as well as computers adapted to timing) are used to time workers with a lot more accuracy than stopwatches offered. Computer databases have been developed with the aid of the computers mentioned above that are timing workers; and computer databases contain synthetic time standards.

In a real sense, stopwatches with external observers metamorphosized into internalized computer timers that could observe a great deal more without being as intrusive. When computers time workers, they are capturing the moments when the worker feeds the next unit (say a check to the optical reader). They can record every time a worker types the letter "q" or stops feeding data to the machine for any interval greater than 5 seconds. Computers using satellite hookups determine when truck drivers stop and how many miles they log between stops.

Computers determine and calculate the intervals between successive steps in the process just as if a stopwatch were at work in the hands of the old-fashioned time study observer. Surveillance has not disappeared. It has become more obscure and hidden. Labor, and especially unions, did not like the stopwatch. The surveyor is now an unseen machine and, in many organizations, employees do not know what about their performance is being scanned and tabulated.

Company computers count the number of entry data errors made as a check on clerical quality. Computers observe how long mail-order or airline reservation people spend on the phone. These are just a few more examples of a growing use of time studies without human observers, often done under the name of diagnostic observations.

Because of the parallel between computer and stopwatch as simple timers, it remains useful to consider how stopwatch methods are applied to time both the basic and extraneous elements of the cycle. Various kinds of stopwatches and stopwatch methods were available for different applications as well as to satisfy personal preferences. The major distinction was between *continuous* and *snap-back use of stopwatches.*

The snap-back approach was used to time each work element. When an element was completed, the watch hand returned to zero to begin timing the next element. Continuous readings are cumulative and require subtraction to determine element times. The computer notebook can be programmed to record in either mode. Subsequent calculations follow the same routines as with the stopwatch, but faster.

"Continuous" is the method shown in Figure 9.2 above as can be verified by studying the columns headed 1, 2, A, B, C, and D. The snap-back system saved having to do repeated subtractions, but it was more demanding on the job observation analyst and needed a larger sample for the same precision that could be obtained from the continuous method. The computer as an observation tool can do a 100 percent study without error or intrusion.

The stopwatch was attached to a time study clipboard, which held the job observation or time study sheet. FFDC's completed time sheet for 30 cycles is shown in Figure 9.2. There is partial information about each basic element and how long it takes. More complete information is available on the left-hand, right-hand operation charts. In addition, there is information about the operation, the operator, the observer, the date, the equipment being used, the set up, materials used, and the speed and feed rates of the equipment.

Figure 9.2 also provides a left-hand, right-hand activity analysis sheet which gives the basic elements and their sequence. It explains the extraneous elements and gives the point at which they enter the work sequence. Job observation by computer can pick up the tasks done by each hand, if that seems useful. FFDC's job layout diagram for diskette packing is shown.

Enough data must be taken to capture all conditions under which the particular job might be done. This could include different plant settings where the domestic plant is older than the one abroad. Standards may be different in each location. Special methods, materials, locations, and other factors can produce different results. Special factors should be minimized.

Job-observation studies should not be conducted for poorly designed jobs. They should be made only when it is agreed that the best design has been developed. Design the job before setting any standards. Because of variability, it makes sense to check on a study by replicating the results a sufficient number of times. When job-observation results are contested, the variability of study conditions and/or their variance with actual conditions are frequently introduced to win the argument.

The job design (tasks, activities, operations) in Figure 9.2 have been broken down into six elements, but only two elements dominate the process. In this simple case, activities of an operator's left and right hands are described for each basic element. The other elements are long cycle ones used to support the short-element cycles. The short-cycle job is filling tubes of toothpaste. The long-cycle jobs are to refill the bins with tubes, caps and boxes.

Basic element 1 dictates that the left hand should be placing 10 floppydisks, which are loose but grouped in tens, into a box while the right hand holds and positions the box so the floppydisks can be inserted. The left- and right-hand operations are coordinated through job design and training.

Element 2 consists of the left hand holding the box for the right hand, which seals the box with a cellophane wrapper. Then the left hand releases the box over a conveyor which carries it away to a tote box at the end of the conveyor. Each box gently slides into the tote box, which can hold up to 105 of the diskette boxes.

Enter in the first column the time it takes to complete element 1 in the first cycle. The time it takes to complete element 2 in the first cycle is recorded next. The stopwatch is continuous, so the first value is 14 and the second is 20, meaning that element 2 required $20 - 14 = 6$ hundredths of a minute.

Next, cycle 2 is observed with values of 31 and 36 which entails $31 - 20 = 11$ and $36 - 31 = 5$, all in hundredths of a minute. The time unit is chosen because it fits the job and can be read on the stopwatch. The stopwatch is a decimal-minute type which reads directly in hundredths of a minute. The translation of these terms into computer capabilities is straightforward. Specific actions trigger the computer to note the time just as if a stopwatch had been built into the system.

The sample size is 30 cycles, all of which are listed on the sheet. Thirty is chosen as a convenient start-up number that fits the size of the single sheet on the clipboard. Later, it will be determined if a larger sample is needed. The computer has no such restraints and limitations.

When the observations have been completed, the computations start. The sum of the times of the two core elements constitutes the job-cycle time. Cycle 1 required 0.20 min. Element 1 needed 0.14 min. and element 2 consumed the remaining 0.06 min. The second cycle ends at 0.36 min. This means that the second cycle consumed $0.36 - 0.20 = 0.16$ min. In this second cycle, element 1 needed 0.11 min. and element 2 used the remaining 0.05 min. Cycles 3 to 30 are obtained in the same way.

The extraneous elements listed on the time sheet are operations which must be done once for every n times that the core cycle is repeated. They support the basic elements of the core process. The right- and left-hand activity chart lists the extraneous element A as the requirement that after 100 diskette boxes have been packed and shunted along the conveyor, the tote box they fall into at the end of the conveyor must be taken away and replaced with another one. This occurs once for every 1000 floppydisks that are boxed and removed.

Extraneous element B requires the worker to interrupt the basic work cycle once for every 50 diskette boxes completed. This means stopping after 500 floppydisks have been packed to get a new supply of floppydisks. *Better job design* would allow A and B to occur at the same time, i.e., resupply floppydisks in lots of 1000. Apply the same thinking to extraneous element C, which resupplies the diskette boxes in 50 unit lots. This could be increased to 100. Combining A, B and C might be feasible. Extraneous element D obtains cellophane wrappers for 500 boxes.

Extraneous elements break into the short-cycle system regularly. Therefore, when they are included, the total job-cycle time is of much longer duration than the basic elements-cycle time. They disrupt the process and set back the learning curve. It is not unusual for a worker to slow down for a while on the basic elements after an extraneous element occurs. Also, the worker might meet a friend while going to the store room and decide to chat a bit. That would not occur during the job observation study, but would occur in reality. The company would realize less production than the job-observation study indicated.

Shortest Common-Cycle Time

For extraneous element D, the worker replaces the used up role of cellophane with a new one in the cellophane wrapping machine. This takes place every 500 diskette boxes, or once every 5000 floppydisks. Since D has the longest extraneous cycle it constitutes the *shortest common-cycle time* necessary to encompass all element cycles that are part of the process.

This is something like finding the lowest common denominator to determine the value of a series of fractions. In this case, it means that 500 cycles of the basic elements are required to include one cycle of the longest extraneous element. A totally proper job observation study would include at least two occurrences of the longest extraneous cycle (D).

To observe 500 cycles imposes such a great observational burden that it may be decided to ignore perfection and include only one of these longest-term cycles. If the time to perform element D is variable, then it could be simulated several times to get a reasonable average value. The same applies to other extraneous elements which have low frequencies of occurrence in the sample. The effects of extraneous elements on the basic element times is another matter which can be simulated if the standard times (the ultimate goal of the study) appear to be sensitive to such effects.

Almost always, even the shortest-cycle, repetitive jobs include long-cycle, extraneous elements. The job observer must be on top of all of the longer-cycle elements which are to be included in the study. The same requirement concerning extraneous cycles applies to computers programmed to make job observations.

The layout chart could have added distances traveled to the picture on the time sheet. How far does the operator have to go to get floppydisks, boxes and cellophane? The more information the better when trying to reproduce the job observation study results at a later date. It may not be possible to recover finer points about the extraneous elements unless the necessary documentation exists somewhere. The computer (as observer) must be programmed to capture the way in which the basic and extraneous elements interact with the layout.

Productivity Standards

Productivity or production standards are measures of the expected output rate. They can be developed for short-cycle jobs using time study data. How much effort should be spent on developing standards depends to a great extent on the kind of work configuration being used. Enough product must be run in the job shop to warrant developing productivity standards. The flow shop qualifies, but in many cases the standards will have been predetermined by engineering design.

The ten-diskette box packer is part of a flow shop assembly operation for FFDC. Several workers do the same operations at the same time. FFDC workers are cross-trained which means they are able to shift operations. Some make floppy-disks others format them, label them, and still others pack them. Furthermore, there are different kinds of floppydisks and each variant has its own characteristics and output standards.

The productivity standard is used to calculate expected output. It is adjusted average performance with an allowance for rest and delay. The productivity standard enables the process manager to determine how many workers are required to meet demand. Labor costs can be estimated for the total job.

Table 9.1 summarizes and operates on the information collected for FFDC. Basic and extraneous elements 1, 2, A, B, C, and D are listed. Total times, as summed on the time study sheet of Figure 9.2 above, are carried into the first row of Table 9.1. The total time for element 1 is 359; for element 2 it is 154. Then, for the extraneous elements A, B, C, and D, total times are 19, 14, 14, and 19, respectively.

TABLE 9.1

FFDC'S–JOB OBSERVATION SUMMARY:
10 HD 3.5" DISKETTE BOX CELLOPHANE WRAPPED

Element	1	2	A	B	C	D	Total
TTO	359	154	19	14	14	19	
NO	30	30	1	1	1	1	
ECPO	1	1	1/100	1/50	1/50	1/500	
AT or ST	11.97	5.13	0.19	0.28	0.28	0.04	
LF or AL	0.95	1.00	1.00	1.00	1.00	1.00	
NT or AT	11.37	5.13	0.19	0.28	0.28	0.04	
CFr&d	110%	110%	110%	110%	110%	110%	
ST	12.51	5.64	0.21	0.31	0.31	0.04	19.02

TTO = Total time observed
NO = Number of observations
ECPO = Expected cycles per observation
AT or ST = Average time or selected time (in 0.01 minutes)

LF or AL = Leveling factors or allowances
NT or AT = Normal time or adjusted time
CFr&d = Correction for rest and delay (r&d)
ST = Standard time (in 0.01 minutes)

The third row of the summary sheet lists the number of occurrences that can be expected for each observation of the basic elements. This is 1 for the basic elements of the core cycle and the appropriate fraction for the extraneous elements.

The first row is divided by the second row yielding the average time for each element. This quotient is multiplied by the third row, yielding an average time per core cycle for each element. The results are shown in the fourth row as average time (also called—*selected time*) per core cycle. To illustrate, average time for element 1:

$$359/30 = 11.96666; 11.97 \times 1 = 11.97$$

This is in hundredths of a minute or 0.1197 minutes. The average time for element 2 = 0.0513 minutes, and so forth.

Training for Job Observation

Leveling factors

Used to convert average time to adjusted or normal time.

What if the observed worker is believed to be faster than average? It is necessary to adjust such performance to what is considered normal. **Leveling factors** are used to convert average time to adjusted or normal time. When the computer is used to record times, it captures all workers and the leveling problem is eliminated.

Process managers used to say that the toughest job in the time study procedure was picking an allowance or leveling factor. It is historically interesting that training for leveling was widely practiced. One of the most popular training methods used "movies" made by motion picture cameras. A movie was made of a normal worker doing the job. The picture was shown using projectors with variable speed controls. By adjusting the speed controls it was possible to make a normal worker seem to work faster or slower. This illustrates another interaction between time studies and time and motion studies.

The important point to emphasize is that all of the time study raters within the company had to agree with each other on what constitutes normal performance speeds. This high degree of conformity meant there would be consistency of ratings for all employees in the same company.

Setting Wages Based on Normal Time

Normal time

Adjusted average time.

Normal time is adjusted average time. Pay scales can be set so that normal time workers are being paid wages similar to those being paid by other companies in the region for the same kind of jobs. This removes pay scale inequities between employees in the same company and among workers in other companies. Such parity is particularly desirable for workers within the same industry. Synthetic time standards provide benchmarks for stabilizing what is normal time in similar jobs across companies and industries.

What Is Normal Time?

Before making a job observation study, workers should be told the full story. The purpose of the study is to develop standards which permit the company to anticipate costs and meet demand requirements. These standards are not meant to exceed what is reasonable. They are intended to prevent overworking operators. Good time standards remove unreasonable expectations which entrap both management and employees.

An average worker should be selected for the study. The observer tries to make sure that the subject is operating under standard conditions, using routinized methods. When the subject of the study seems to work at a normal rate, the leveling factor is determined. For each element, the average time (row 4) and the leveling factor (row 5) are multiplied, which yields the normal time (row 6).

Assume that the observed worker is faster than normal. The observer estimates that this worker is producing 10% more output than an average worker. The observer assigns the worker a leveling factor of 110 percent. The normal time will be greater than the average time, meaning that the normal (slower) worker can be expected to take longer than the observed study time.

Examine element 1 in Table 9.1 above. The operator is working at 95 percent of normal in the judgment of the observer. Multiplication produces a normal time of 0.1137 minutes, which is less than the average time of 0.1197 minutes. Less time to complete an element means that the normal operator works faster and will be able to produce more units than the observed worker.

When the leveling factor is 100 percent, as it is for the remaining five elements in Table 9.1, average time and normal time are equal. This leaves the remaining numbers in rows 4 and 6 equal. Many studies require a leveling factor that is not equal to 1. This is because operator skills differ in a variety of ways that are dependent on their motor skills, body size and experiences.

Learning Curve Effects

The observer must allow time for learning to take place so that work time stabilizes. Repetition improves performance. Start-up is getting back in practice. Learning increases capacity to produce, raising productivity and lowering defectives produced. This subject is found in discussions of capacity management.

Even with the most familiar, repetitive work, start-up is slower. With repetition, as the job is being relearned, the operator is said to start moving up the learning curve. The effect is substantial when the job is done for the first time. Nevertheless, learning takes place every time that the operator comes on to start a new shift.

Computer monitoring picks up the learning curve phenomenon. It provides a great deal of useful information about the learning curves of workers. It is valuable to study these records to determine how to help people move up their respective learning curves faster.

Job Observation Corrections

Leveling estimates introduce one kind of measurement correction. Time study errors also can occur when the stopwatch is misread or the reading is entered incorrectly. SQC concepts apply. Look at the distribution of element time readings. Extreme numbers which do not seem to belong to the distribution are outlying values that can be thrown out of the sample. Increased sampling may be warranted.

Subtraction errors can occur when the stopwatch provides continuous readings. They can best be corrected by using several analysts. It is more difficult for one person to redo the calculations. Snapback readings are more demanding mechanically than continuous readings. Errors can be made in hesitating to snap back or forgetting to do so. Two observers, working together, can be used when breaking in a new observer.

There is yet another correction to be made. Called the *rest and delay correction factor (r&d)*, it is shown in row 7 of Table 9.1. Some say that r&d is the "mother" of all corrections. The estimation of appropriate r&d factors is required for computerized observation systems.

Generally, the number chosen for r&d results from negotiations with unions or with workers and their representative groups. The agreed upon numerical factor is often affected by who is the best negotiator, union or management. After all the finessing with exact observational methods and leveling, it comes as a shock that assigned values for r&d commonly range from 5 to 15 percent, depending upon the character of the job, the degree of personal needs, union-management negotiations, and so on.

India's Programming Industry

Many major U.S. companies are availing themselves of the global workforce in fulfilling their software programming needs. Companies like Dun & Bradstreet, Digital Equipment, Motorola, IBM, and Intel are looking to India for computer programming talent.

A significant industry for India, its software companies exported over $125 million in programming services during a one-year period from 1992 to 1993. U.S. companies generally hire Indian programmers for on-site, short-term assignments and at lower salaries than U.S. programmers command. This wage disparity prompted the U.S. government to propose a change in the visa requirement, which would involve a longer application process for Indian nationals.

Indian officials claim that only about 2,000 visas were obtained by Indian software programmers in 1993 and that "most of these people had skills rare in the U.S." U.S. companies maintain that use of Indian pro-

grammers not only cuts costs, but also complements the programming work done in the U.S. They say that most of the "cutting-edge innovation and strategic design" work still is done at home, while the Indian programmers make the "Lego blocks" or components of standard programming.

Not only are U.S. companies hiring Indian talent for software programming, they also are entering into joint ventures with Indian software firms. The Dun & Bradstreet Corporation, a marketer of information, software, and services for business decision making, has entered into a venture with Satyam Software, one of India's fastest-

growing software companies.

"The joint venture will help us address several new markets, and more rapidly introduce new products and services throughout the world," says Kumon Mahadera, D&B's senior vice president, Asian Business Development. "This is another manifestation of D&B's global strategy of actively participating in all the major markets, and tapping into the world's talent pool to find the best people regardless of location."

Sources: "Indian software firms join to push for easing of restrictions," (New York) *Journal of Commerce*, April 11, 1994; "Dun & Bradstreet and Satyam Computer Services join forces for new enterprise," Business Wire, January 27, 1994; and Maggie Farley, "Cheaper software-writing talent draws companies to Indian city," *Boston Globe*, July 5, 1993.

The point to keep in mind is that without the r&d factor, workers might never have a moment to stop. In this case, the left and right hands would constantly be moving packing the floppydisks. There would be no time to rest and no time to use the rest room. The best way to set the r&d factor is to enlarge the extraneous elements to include a given amount of time to use the rest facilities. A proper scenario should be developed and examined with an appropriate simulation.

Normal time is multiplied by the rest and delay factor, which is always more than 100 percent, yielding standard time. Thus, an r&d correction of 110 percent is to be applied to all of the elements as shown in Table 9.1. If some elements are particularly onerous, they might be given a larger r&d buffer. Also, it is reasonable, and simpler to calculate r&d assigned only to the basic elements.

Total standard time is the basis for calculating production output standards. The sum of the standard times for all the elements is the total standard time of the operation. For the FFDC example, total standard time is 0.1902 minutes.

Expected Productivity and Cost of the Well-Designed Job

Before measuring productivity and getting a cost for the job, it is essential to evaluate the job design and improve it as much as possible. One method that is used is called work simplification. It removes unnecessary steps and streamlines work flows based on common sense considerations.

Assuming that this has been done, how many boxes of 10 floppy disks will the normal FFDC worker produce per hour?

$$\text{POS} = 60 \frac{\text{min.}}{\text{hour}} \div 0.1902 \frac{\text{min.}}{\text{box}} = 315.46 \frac{\text{boxes}}{\text{hour}}$$

where,

$$\text{POS} = \text{productivity output standard}$$

(It is to be noted how the dimensions result in minutes cancelling minutes and yielding boxes per hour.)

If 60 minutes per hour had not been used, then 1 divided by 0.1902 = 5.26 boxes per minute. Multiplying 5.2576 by 60 yields 315.46 boxes per hour. For simplicity, round this output rate to 315 boxes per hour and call it the expected output rate for the job.

Assume that the wage rate for this job classification is $2 per hour. Then the labor cost for packing each box of floppys would be $2/315 = $0.00635. The expected cost of the operation is a little more than a half-cent apiece.

The major weaknesses of this form of study are: defining a normal worker, application of the leveling factor, choosing values for the rest and delay correction, and overlooking influential extraneous elements. Defenders of time study methods claim that satisfactory solutions have been found for all of these problems.

Job observation is a skill, not a science. The use of the stopwatch and leveling both require training and practice. Time study practitioners seldom break down elements into less than 0.04 minutes because smaller intervals are difficult to observe. Another fact worth considering is that the worker's performance may not be stable in the Shewhart sense, especially because of "learning." The time study should not begin until the worker's learning ceases and performance has stabilized.

There remains the issue of determining an adequate sample size for the job observation or time study. How many cycles should be observed? Using automated (100 percent) computer observation eliminates this issue.

Job Observation Sample Size

Especially for service operations where no computer monitoring system is feasible, observations of performance may need to be made on a sampling basis. This is done to understand the nature of the job, to improve it, to establish pay scales and to set productivity expectation levels. Under such circumstances, there is a need to set appropriate sample sizes.

Job observation studies are based on sampling n basic cycles of a job. Accordingly, a statistical issue arises about how large a sample (n_j) should be taken for each element, called j. Various formulations can be used depending upon the assumptions made.

Equation 9.0 controls statistical measurement error in terms of sample size for each element j. It answers the question: how large a sample (n_j) is needed (how many cycles should be observed) to obtain 95% confidence that the true element time (t_j) lies within the range:

$$\bar{t}_j \pm 0.05\ \bar{t}_j \qquad\qquad \textbf{Equation 9.0}$$

which is ±5 percent of the observed average time? This is the value 0.05 which appears in the denominator of Equation 9.1.

$$n_j = \left(\frac{1.96 \cdot s_j}{0.05 \cdot \bar{t}_j}\right)^2 \qquad\qquad \textbf{Equation 9.1}$$

As for the 1.96 which appears in the numerator of the equation, note that $\pm 1.96\sigma$ creates two tails of a normal distribution each of which has an area of 0.025. This provides the confidence interval within which the true element lies 95% of the time.

Assume that $i = 5$ cycles have been observed and that these are the first five cycles in Table 9.1. The average times

$$\bar{t}_j = \frac{\sum_{i=1}^{i=n} t_{ij}}{n} = \bar{t}_j = \frac{\sum_{i=1}^{i=5} t_{ij}}{5}$$

of the first and second elements will be found to be:

$$\bar{t}_1 = 12.4 \quad \text{and} \quad \bar{t}_2 = 4.8$$

These are now used to compute the sample standard deviation as shown in Equation 9.2.

$$s_j = \sqrt{\frac{\sum_{i=1}^{i=n}(t_{ij} - \bar{t}_j)^2}{(n-1)}} \qquad\qquad \textbf{Equation 9.2}$$

Obtain s_j for the first five cycles of elements 1 and 2, as follows:

$$s_1 = \sqrt{\frac{5.20}{(n-1)}} = \sqrt{\frac{5.20}{4}} = \sqrt{1.3} = 1.140$$

$$s_2 = \sqrt{\frac{4.80}{(n-1)}} = \sqrt{\frac{4.80}{4}} = \sqrt{1.2} = 1.095$$

Now, put these values of $\bar{t}_j$ and s_j, into Equation 9.1. This creates Equations 9.3 and 9.4

For element 1:

$$n_1' = \left(\frac{(1.96) \cdot (1.140)}{(0.05) \cdot (12.4)}\right)^2 = (3.60)^2 = 12.96 \approx 13 \qquad \textbf{Equation 9.3}$$

For element 2:

$$n_2' = \left(\frac{(1.96) \cdot (1.095)}{(0.05) \cdot (4.8)}\right)^2 = (8.94)^2 = 79.9 \approx 80 \qquad \textbf{Equation 9.4}$$

Whereas $n = 5$ is the number of cycles observed up to this point, n' (as used in Equations 9.3 and 9.4) is the number of cycles that should be observed. Specifically n' is the required number of cycles to be observed so that there can be 95 percent confidence that the true element time lies within the range $\bar{t}_j \pm 0.05\ \bar{t}_j$, which is ± 5 percent of the observed average time.

The required sample sizes are observations of 13 basic cycles for element 1 and 80 basic cycles for element 2. One element dominates the sample size decision. It is the element that requires the largest value of n'. The largest n', derived in Equation 9.4, is 80. Element 1, with an indicated study sample size of 13, is dominated by element 2 with $n' = 80$. The present sample size is 5. Therefore, 75 additional observations should be made to satisfy the statistical criterion just developed.

Extraneous elements should have no effect on the sample size for core elements. In this case, $s_j = 0$ for all of them since only one value was observed. If a larger sample indicated high variability, then it would be fitting to study extraneous elements apart from the overall core study. It would be unusual to base the study sample size on extraneous elements.

Since the largest value for n' which was 80 is much larger than the actual n which is 5, additional observations should be taken. The procedure for sample size evaluation is repeated until the largest value of n'_j is equal to or less than the actual number of observations made, that is, max $n'_j \le n$.

Good procedure, given the sample of 5 and the indicated need for 75 more observations to get to 80, is to take another 5 or 10 readings and then test again. Problem 2 in the Problem Section requests computations for the first 10 cycles.

These job observation procedures are suitable for short-cycle operations. They provide reasonable cost and output information. For long-cycle, nonrepetitive operations, work sampling is used. Synthetic time standards are used to make estimates for jobs that are being designed and do not yet exist. Both methods will be explained briefly.

9.8

WORK SAMPLING AND JOB DESIGN

Work sampling is applicable to both manufacturing and services. It is singularly well suited to services which do not lend themselves to time studies of the type previously discussed.

Services are less repetitive. They have longer cycles and they are less routinized. They can be categorized in broad terms, as in the illustrative case below.

Workers often cannot account for the way in which they spend their time. This applies, in particular, to creative and knowledge workers, office and research personnel, and middle and top managers. Their work does not tend to be repetitive like manufacturing and assembly where time studies can define jobs and rates of output.

Recognition of the opportunity to design jobs that foster teamwork has led P/OM to employ **work sampling**, also called *operations sampling*. This method can determine with reasonable accuracy, the percentage of time workers are engaged in different tasks before and after job design.

In the 1930s, the English statistician, L. H. C. Tippett, reported on his experiments with work sampling in Great Britain's textile factories.[6] Unlike most procedures for sampling the quality of materials, observations of workers' activities

Work sampling

Also called operations sampling; method to determine with reasonable accuracy, the percentage of time workers engage in different tasks before and after job design.

are best made randomly over time. As with any sampling system, observations should be of sufficient number to paint an accurate picture of what the workers are doing.

Work sampling methods have to take into account the fact that employees who know they are going to be observed, may act differently than if they do not think they are being observed. This means that the best sample is one which has not been anticipated by either the observer or the worker.

Work sampling often is used to determine what fraction of the time workers are idle and how they spend their time when engaged. For office workers, a general classification might include phoning, filing, typing at the computer terminal, and talking with others. It may be difficult to use the computer to track these kinds of jobs. An observer randomly sampled what was happening in the office to develop the data shown in Table 9.2.

TABLE 9.2

WORK SAMPLING STUDY IN OFFICE

	Number of Observations	Fraction of Total (p_i)
Phoning	100	0.20
Filing	50	0.10
Computer	305	0.61
Talking	45	0.09
Total	500	1.00

When a sufficient sample has been taken, ratios can be formed as descriptive measures of what goes on in the system. In Table 9.2, 500 observations of what four people are doing at randomly chosen moments have been made over 5 days. This represents 125 observations of each person. There are 25 observations per day, about three per hour. Assuming that each observation captures what is going on in a given minute, then this is about a 5 percent sample per day.

From Table 9.2, computer use (0.61) is six times as likely as talking (0.09) and filing (0.10) and three times as likely as phoning (0.20). Perhaps it seems to the manager that there is too much phoning. Greater controls over personal calls may be in order. For effective job design, it might be wise to expand the study to do a *content analysis* of transactions that are taking place over the phone.

P/OM's use of work sampling is not designed to catch workers offguard. It is intended to map out the way they spend their time leading to teamwork improvements. Most people are not aware of their allocation of time to activities. This can help employees utilize their time more fully. Therefore, instead of just observing whether workers are idle or busy, the expanded analysis notes specific allocations of time to functions. This type of study is more appropriately called *operations sampling* than work sampling.

Work Sample Size

How large a sample is needed? This is the same question asked previously with respect to time studies. The method employed is similar. The equation for *n* is:

$$n_i = \left(\frac{1.96}{0.30}\right)^2 \left(\frac{1 - p_i}{p_i}\right)$$

Equation 9.5

n_i = The number of observations to be taken to provide a sufficient sample for the *ith* activity.

p_i = The *ith* operation's activity level observed as a fraction of total observations that an operation occurs. When using this formula, the dominating activity for determining n will be the one with the smallest p.

k = The number of normal standard deviations required to give a specified probability for the confidence interval. In Equation 9.5, $k = 1.96$ so the probability is 95% that the true value of p_i falls in the range $p_i \pm (0.30)p_i$.

0.30 = The accuracy range specified by management such that the true value of p_i falls within the range $p_i \pm (0.30)p_i$.

The values of 1.96 and 0.30 were chosen to make the illustration easy to follow. They can be altered to suit the situation.

The smallest $p_i = 0.09$. This means that $p_i \pm (0.30)p_i$ spans the range from 0.063 to 0.117, which seems a reasonable range for the fraction associated with talking to people in the office.

Using Equation 9.5:

$$n = \left(\frac{1.96}{0.30}\right)^2 \left(\frac{1 - 0.09}{0.09}\right) = (42.6844)\,(10.1111) = 431.59$$

Equation 9.6

The actual sample taken was 500 observations, therefore no more sampling is required.

9.9

THE STUDY OF TIME AND MOTION— SYNTHETIC TIME STANDARDS

A solution to problems associated with setting standards for new products, or where people are opposed to being studied on the job, was provided by **synthetic** or **predetermined time standards**. These standards are created from a subset of elements common to all jobs. The nature of any job can be described by the way in which the common alphabet of work elements is arranged.

Frank Gilbreth was the management pioneer who created the first such alphabet of job elements or modules more than 85 years ago. He called these modules *therbligs* and named 17 of them.[7] It takes too much detail to enumerate all of the symbols, but a representative group of four tells the story.

1. *Grasp* begins when a hand touches an object. It consists of gaining control of an object, and ends when control is gained.
2. *Position* begins when a hand causes the part to line up or locate, and ends when the part changes position.
3. *Assemble* begins when the hand causes parts to begin to go together; it consists of actual assembly of parts, and ends when the parts go together.

Synthetic or predetermined time standards

Created from a subset of elements common to all jobs.

4. *Hold* begins when movement of the part or object, which the hand has under control, ceases; it consists of holding an object in a fixed position and location, and ends with any movement.

Once the standard work elements were identified, it became feasible to study all kinds of jobs, real and imagined. The predetermined times were based on the study of thousands of different operations in which each of these elements appeared. Motion pictures were made of many different kinds of jobs. These were analyzed to determine the appropriate statistical distribution of element times. Expected standard times were obtained from these distributions.

Tables of standard times for a representative set of work elements ("reach", "move", "grasp", and "position") are shown in Table 9.3. They are part of the total system of synthetic standards known as MTM-1, the methods-time-measurement system. In addition to the four work elements listed above, the total set of predetermined standards includes: move, turn, apply pressure, release, disengage, eye travel and eye focus, body, leg and foot motions, and simultaneous coordination of all of these. Using these, (as with Table 9.3) a standard time for any usual job can be derived.

The time measurement units (TMU) are given in terms of 0.00001 hour. This means that 1 hour = 100,000 TMU. The procedure for using the synthetic time standards is:

1. Describe the job design in detail and identify the work elements.
2. Determine the appropriate times for each work element.
3. Add the element times together. This requires that individual work elements be independent of each other. If not, interactions need to be taken into account. The sum must reflect the true total time for the job.

The MTM Association for Standards and Research has developed computer systems for combining the synthetic elements and deriving work standards. Two of these systems are called 2M and 4M representing increasing levels of computer application. "4M can function as a stand-alone system, or be integrated with CAPP, CAD/CAM or MRP systems."[8] MTM-UAS (Universal Analyzing System) is a collaborative effort with German, Swiss and Austrian MTM Association groups specifically designed to create standards for batch production.

Table 9.4 illustrates a 2M application which can be run on PCs or mainframe computers. The specific operations to "form grooves" is shown as a computer print-out. Time is in fractions of minutes, a software option, but the basic time unit remains the standard TMU.

Evaluation of Applicability

Some advantages of synthetic time standards, as compared to those derived from conventional study methods are:

1. The leveling factor problem is eliminated. Synthetic time standards average out rating differences across many operators.
2. Synthetic production standards are founded upon element times derived from very large samples of observations. This provides increased reliability of the derived standard time.
3. The speed of preparing cost estimates, as well as their reliability for new jobs is improved. Production schedules can be quickly determined and modified.

TABLE 9.3

TIME VALUES FOR VARIOUS CLASSIFICATIONS OF MOTIONS

TABLE I — REACH — R

Distance Moved Inches	Time TMU				Hand In Motion		CASE AND DESCRIPTION
	A	B	C or D	E	A	B	
3/4 or less	2.0	2.0	2.0	2.0	1.6	1.6	**A** Reach to object in fixed location, or to object in other hand or on which other hand rests.
1	2.5	2.5	3.6	2.4	2.3	2.3	
2	4.0	4.0	5.9	3.8	3.5	2.7	
3	5.3	5.3	7.3	5.3	4.5	3.6	**B** Reach to single object in location which may vary slightly from cycle to cycle.
4	6.1	6.4	8.4	6.8	4.9	4.3	
5	6.5	7.8	9.4	7.4	5.3	5.0	
6	7.0	8.6	10.1	8.0	5.7	5.7	
7	7.4	9.3	10.8	8.7	6.1	6.5	**C** Reach to object jumbled with other objects in a group so that search and select occur.
8	7.9	10.1	11.5	9.3	6.5	7.2	
9	8.3	10.8	12.2	9.9	6.9	7.9	
10	8.7	11.5	12.9	10.5	7.3	8.6	
12	9.6	12.9	14.2	11.8	8.1	10.1	
14	10.5	14.4	15.6	13.0	8.9	11.5	**D** Reach to a very small object or where accurate grasp is required.
16	11.4	15.8	17.0	14.2	9.7	12.9	
18	12.3	17.2	18.4	15.5	10.5	14.4	
20	13.1	18.6	19.8	16.7	11.3	15.8	
22	14.0	20.1	21.2	18.0	12.1	17.3	**E** Reach to indefinite location to get hand in position for body balance or next motion or out of way.
24	14.9	21.5	22.5	19.2	12.9	18.8	
26	15.8	22.9	23.9	20.4	13.7	20.2	
28	16.7	24.4	25.3	21.7	14.5	21.7	
30	17.5	25.8	26.7	22.9	15.3	23.2	
Additional	0.4	0.7	0.7	0.6			TMU per inch over 30 inches

TABLE V — POSITION* — P

CLASS OF FIT		Symmetry	Easy To Handle	Difficult To Handle
1—Loose	No pressure required	S	5.6	11.2
		SS	9.1	14.7
		NS	10.4	16.0
2—Close	Light pressure required	S	16.2	21.8
		SS	19.7	25.3
		NS	21.0	26.6
3—Exact	Heavy pressure required.	S	43.0	48.6
		SS	46.5	52.1
		NS	47.8	53.4

SUPPLEMENTARY RULE FOR SURFACE ALIGNMENT	
P1SE per alignment: $>1/16 \leq 1/4''$	P2SE per alignment: $\leq 1/16''$

*Distance moved to engage—1" or less.

TABLE II — MOVE — M

Distance Moved Inches	Time TMU			Hand In Motion B	Wt. Allowance			CASE AND DESCRIPTION
	A	B	C		Wt. (lb.) Up to	Dynamic Factor	Static Constant TMU	
3/4 or less	2.0	2.0	2.0	1.7				
1	2.5	2.9	3.4	2.3	2.5	1.00	0	
2	3.6	4.6	5.2	2.9				**A** Move object to other hand or against stop.
3	4.9	5.7	6.7	3.6	7.5	1.06	2.2	
4	6.1	6.9	8.0	4.3				
5	7.3	8.0	9.2	5.0	12.5	1.11	3.9	
6	8.1	8.9	10.3	5.7				
7	8.9	9.7	11.1	6.5	17.5	1.17	5.6	
8	9.7	10.6	11.8	7.2				**B** Move object to approximate or indefinite location.
9	10.5	11.5	12.7	7.9	22.5	1.22	7.4	
10	11.3	12.2	13.5	8.6				
12	12.9	13.4	15.2	10.0	27.5	1.28	9.1	
14	14.4	14.6	16.9	11.4				
16	16.0	15.8	18.7	12.8	32.5	1.33	10.8	
18	17.6	17.0	20.4	14.2				
20	19.2	18.2	22.1	15.6	37.5	1.39	12.5	
22	20.8	19.4	23.8	17.0				
24	22.4	20.6	25.5	18.4	42.5	1.44	14.3	**C** Move object to exact location.
26	24.0	21.8	27.3	19.8				
28	25.5	23.1	29.0	21.2	47.5	1.50	16.0	
30	27.1	24.3	30.7	22.7				
Additional	0.8	0.6	0.85		TMU per inch over 30 inches			

TABLE IV — GRASP — G

TYPE OF GRASP	Case	Time TMU	DESCRIPTION	
PICK-UP	1A	2.0	Any size object by itself, easily grasped	
	1B	3.5	Object very small or lying close against a flat surface	
	1C1	7.3	Diameter larger than 1/2"	Interference with Grasp
	1C2	8.7	Diameter 1/4" to 1/2"	on bottom and one side of
	1C3	10.8	Diameter less than 1/4"	nearly cylindrical object.
REGRASP	2	5.6	Change grasp without relinquishing control	
TRANSFER	3	5.6	Control transferred from one hand to the other.	
SELECT	4A	7.3	Larger than 1" x 1" x 1"	Object jumbled with other
	4B	9.1	1/4" x 1/4" x 1/8" to 1" x 1" x 1"	objects so that search
	4C	12.9	Smaller than 1/4" x 1/4" x 1/8"	and select occur.
CONTACT	5	0	Contact, Sliding, or Hook Grasp.	

EFFECTIVE NET WEIGHT			
Effective Net Weight (ENW)	No. of Hands	Spatial	Sliding
	1	W	$W \times F_c$
	2	W/2	$W/2 \times F_c$

W = Weight in pounds
F_c = Coefficient of Friction

TABLE 9.4

COMPUTER PRINTOUT FOR 2M ELEMENT ANALYSIS

2M DATA SYSTEM   R5715B   **2M ELEMENT ANALYSIS**   **Requested by MMM**   **Time 13.46.46**   **Date 04/22/94**

Element: MARK-T3 MARK ROLL BY FORMING GROOVES WITH MARKING TEMPLATE   Learning Level . . . .:100

Scope
Starts: Reach to marking template.
Includes: Position template, inspect roll, aside template, straighten roll, position template,
           inspect, place template on roll, straighten template, mark roll, and aside template.
Ends: Aside template and release.

UPDATED   TOTAL Mins. –Manual: .3360
                                       –Process

Line	Sub Element		Frequency	Process Mins	Manual Mins
010	HB3	Get template, position to roll, and aside after inspection	1.0000		.0450
020	VA	Inspect roll	1.0000		.0090
030	AA1	Get roll and straighten	10.0000		.1200
040	HB2	Get template, place to roll, and aside after use	1.0000		.0360
050	VA	Inspect	1.0000		.0090
060	PB2	Position template on roll	1.0000		.0180
070	AA3	Get ends of template and straighten	2.0000		.0600
080	AA1	Move hands to center and press	1.0000		.0120
090	AB2	Position template and press	1.0000		.0270

TOTAL Mins: .3360

Synthetic time standards are in a state of flux. Their technology is changing as they convert from tables and charts toward total computerization. Using an international perspective, they provide a common base for work being done at multiple global locations. They are likely to be adopted by companies in parts of the world where labor costs are low.

Transnational and multinational corporations will continue to shift labor-intensive work to areas of the world where there are low costs. Companies will outsource those parts which require labor-intensive inputs. Control over costs requires standards; synthetic time standards will find increasing use in this regard.

The approach used by synthetic time standards is, in modified form, applicable to the standardization of programming work. Thus, in the U.S., systems for developing standards that apply to the cost of writing lines of code have been studied for many years by a variety of companies. The synthetic time standards also have applicability to information processing and other services which conform to the manufacturing orientation.

Services continue to be labor intensive. Service operations in airports, hospitals, banks and schools might be studied and systematized by the application of the synthetic approach. Service operations have many characteristics that lend themselves to

QUALITY

Martin Marietta Corp. instituted TQM in 1985. The TQM program formalized the defense contractor's commitment to attaining quality ahead of cost and schedule concerns. By the end of 1990 the program saved the Martin Marietta Electronics & Missiles Group more than $250 million through cost avoidance.

Employee involvement was among the TQM program's five major initiatives. (The other initiatives include management commitment, customer requirements, continuous process improvement, and supplier partnerships.)

Performance Measurement Teams, organized around specific projects and empowered to take immediate action to resolve issues, were established in 1988. The teams include employees from every "discipline" which might affect a

TQM at Martin Marietta Corp.

Performance Measurement Teams empowered to resolve issues.

project. Disciplines include assemblers, inspectors, expediters, engineers and facilities planners. In weekly meetings, teams discuss their performance, identify ways to improve, and then develop action plans for key issues. In 1993 the Group had 368 teams in which more than 4,000 employees participated.

The continuous process improvement initiative embraces concurrent engineering, a method used to ensure that quality is designed into a product from inception. Concurrent engineering involves staff from diverse areas of the Group in the initial design of a product (such as engineering, design, mechanical, industrial, testing, and product assurance).

The Group's field achievements are proof of the quality program's success. The Patriot air defense missile was successful in all of its test flights, the vertical launch system successful in all test firings, and the night navigation targeting system achieved an extraordinary 97 percent overall mission-effectiveness rate during initial flight operations.

Source: (Pittsfield, Massachusetts) *Berkshire Eagle,* March 1, 1993.

better organization and control using predetermined time standards. The potential application for service operations is significant. Thus, 2M with "MTM-HC is a data system developed from MTM-UAS to provide time standards for tasks in a health care environment."[9]

9.10

MANAGING THE WORK SYSTEM

What is the status of the people part of OM systems? Are people working smarter, not harder? Is there a common goal to produce a quality product? Are managers viewed as consultants helping workers instead of as judges grading students?

Job Design, Evaluation, and Improvement

The answer seems to depend on where and when one looks. Many companies have adopted new approaches in dealing with employees. When times get rough, some of them revert back to the old ways. Other firms seem to have permanently embraced the integrated empowering approach to employee relations. There is evidence that work system improvements can pay off handsomely. Companies worldwide which follow job design improvement principles which enhance teamwork and workers' sense of security outperform competitors.

There is benefit to be gained by viewing the entire set of jobs in a company as a system involving all employees, workers and managers. Striving together as a team, their outputs are the revenue earners. Job improvement will result in a blend of effects, including lower costs, increased output rates, better deliveries, greater variety, and above all else, better quality.

Job evaluation is a term often used to describe the analysis of a job and its improvement. It is more valuable to consider job evaluation as *system evaluation*, in which sets of interrelated jobs are analyzed and improved as an entire network of linked efforts. Traditional job evaluation looked upon each job as a separate and independent part of the workplace. This view is not as powerful as the system view.

Wage Determination

Wage determination for OM employees is based upon the:

1. difficulty of jobs performed
2. amount of training required
3. amount of skill required
4. value of these jobs to the company

There are problems defining an unambiguous basis for establishing wages. For example, the difficulty of a job and its value to the company may not be correlated. Skill required and value to the company may even be inversely related. There is a logical flaw in the old approach to wage determination that cannot be remedied by reasoning. This is apparent in reading about legal efforts to prove wage discrimination—say, between women and men.

The traditional approaches to wage determination are still practiced by many organizations, but they are changing. There is no redeeming benefit to studying the old ways. The newer approaches are to set acceptable costs for labor and target those costs as goals. In addition to such target costing and the use of activity-based costing (ABC), the wage levels are set by market forces for certain skills and experience.

Note that ABC-type thinking results in the attempt to determine what each job contributes to the profitability of the firm. Efforts to define the value of jobs to the company pay off not only in helping develop rational pay scales, but by putting emphasis on the design of jobs to maximize contribution.

At the experimental end of the spectrum, groups of workers participate in deciding who gets raises and who does not get them. This is *wage self-determination* done by the group for themselves and the people with whom they work. There are many other kinds of arrangements for self-participation in setting wage rates and making joint decisions that improve *conflict resolution* and employee morale.

Old ways of determining wages are being replaced by a systems approach in world-class organizations. The systems approach is based on a *common pay rate* for team members working together. To the extent that the group is successful in generating profit for the company, they are rewarded as a group by bonuses.

In many companies now there is *cross-training*, so that many people do different jobs at different times. The old system would have trouble sorting out what each person should be getting paid. Accounting is far easier with the new approach. Everyone in the group receives the same pay, and all work for the same system. Some companies call the bonus system *gainsharing*. An increasing number of unions are allowing cross-training. There is more emphasis on training, including the use of quality circles, quality of work life groups, and participation groups.

Work Simplification and Job Design

Work simplification should be used first to develop as good a job as possible. This takes into account the systems view of each job as a service to the next worker in the sequence. When this thinking has been applied, the application of time studies and job observations, including realistic productivity measures, can lead to significant improvements.

The success of *suggestion systems* in companies where employees are not driven by work standards and where there is little fear of losing one's job, indicates the extent to which improvement can be continuous. Good suggestions are rare in companies where the suggestion might result in someone being fired because a job no longer is necessary. Continuous job design improvement programs work when there is worker security; they work best when there is worker empowerment in a teamwork environment.

Because continuous improvement hinges on people involvement and communications, employees' ideas are displayed in the break room at this credit company.
Source: Dana Corporation

In recent years, there has been de-emphasis on work simplification in many industries. This is because job shop lots are decreasing in size. Line changeovers to new and modified outputs are becoming more frequent. Computer-integrated flexible manufacturing systems and adaptive automation require intense preplanning that obviates the need for ongoing improvement of operations. Also, strong methodologies of operations research and management science, such as simulation, are increasingly being used to improve the performance of systems. When these methods compete for investment study dollars with the older methods, the new methods win. Individual job design improvement is disappearing as systems design improvement redefines the nature of all jobs as doing whatever it takes to foster teamwork and achieve objectives.

Job Design and Enrichment

As technology changes, operators tend to communicate with the computers that run the machines rather than with the machines themselves. This changes the pattern of communication in the factory and in the office. There is more time and opportunity for people to speak to each other. L- and U-shaped workstations have become commonplace in systems of interrelated jobs designed to facilitate communication between workers. Groups of people get together more often to discuss the system of work and how it can be improved. When the ideas of people are taken seriously, and when people know what happens to their work down the line, each job is enriched.

There is ample research evidence that whether the work is in Mexico, China or the U.S., some workers prefer simple, highly repetitive tasks, and others do not. Some like to switch roles. Others prefer to stay at the same tasks. When groups of employees are offered job enrichment opportunities, the group splits into those that want them and those that reject them. Consequently, there is an opportunity for management to explore preferences with each worker. Work assignments can be made according to worker preference in many work systems, because both types of work exist.

The flow shop can be particularly tedious for workers who like variety. As more stations are added, productivity rises. However, cycle time gets shorter, placing an increasing burden on station workers, who have shorter jobs to do with greater frequency. The quality of work can deteriorate under such circumstances. Worker dissatisfaction can increase. In the 1970s, the Vega plant of General Motors at Lordstown, Ohio, suffered a long and difficult strike shortly after it was built because of worker dissatisfaction with the flow-shop pace.

During the 1970s, organizations started to redesign their production lines so workers would be required to do many different jobs. Volvo of Sweden began experimenting with teams that built an entire car. Other organizations decreased the level of specialization in their line-balanced systems. This increased the size of each workstation and the work content per employee.

A group of GM employees sponsored by the United Auto Workers were sent to Volvo in Sweden for six months. Less than half preferred the Volvo system. GM has tried to make its flow-shop design less tedious. By the late 1980s, General Motors set up lines where teams worked at different stations and the car was transferred by conveyor from one station to another. Again, the work content per employee is larger.

Overall, with cross-training and employee involvement in quality, the work content per employee continues to increase. Most of the complaints center around the paced-conveyor flow shop as well as its intermittent forms instead of the job shop, but there are tedious aspects of the job shop as well. The project is usually immune to such complaints. If there is any criticism on the part of project workers, it is the continual crises that afflict their nonroutinized workdays. There may be complaints for FPS, because human support teams can perform dull tasks to service their computer masters.

Cross-training for job rotation (where workers exchange jobs) has become increasingly acceptable. Job enlargement (with its larger work content) also has taken hold. Greater worker participation in decision making has been endorsed by both public and private organizations. The ultimate criterion is where the objectives of the organization can be met with less specialization and with less authority over workers.

The Motivation Factor

A remarkable case history was documented in the 1930s by a study group from Harvard at the Hawthorne Works of the Western Electric Company in Chicago. The study concerned levels of illumination and their affect on productivity. It was discovered that whether the illumination was raised or lowered, productivity was improved. The key finding (known as the Hawthorne effect)[10] was that employees responded positively to management's interest and attention. This response level overrode the functional affects of the illumination level. Such complex behavior differentiates people and machines.

Critics of the Hawthorne study claim that the result was a self-fulfilling prophecy. The workers gave the researchers what they wanted. Even if this is true, it still appears reasonable to expect positive reactions from workers to care and attention.

Motivation can be both positive and negative. The latter is associated with poor employee morale. When discussing incentives and motivation, the major difficulty is the measurement problem. Nevertheless, accepting the lack of precision involved, incentives are a real causal factor that can affect worker behavior. Incentives include wages, job title, office size and floor, organizational importance, ability to participate in decisions, vacations, leisure time, and variety of tasks assigned. For the most part, these categories represent intangible qualities that escape definition and measurement.

In summary, P/OM is a people-oriented function. P/OM deals with workers and suppliers. To make the process work, P/OM must be able to understand the changing role of people in the process. This has become even more true as OM moves onto an international stage. People throughout the global network are indispensable.

SUMMARY

Chapter 9 begins by making clear that planning teamwork is the mutual goal of OM and HRM. What constitutes the traditional and then newer methods of training is discussed. The reason that performance evaluation is a controversial topic is succinctly stated. The chapter develops the ideas behind designing jobs for production and cost standards. This is followed by the study of time standards comparing traditional time study methods with a parallel discussion of how computers are used to determine how many pieces (or service steps) to expect per hour, the production standard. Next comes the study of time and motion for operations. This includes job design improvement and the use of synthetic time standards to set production standards. Work sampling is covered. The chapter concludes with discussion of how to manage the design of larger work systems, including job improvement, evaluation, enrichment, and wage determination.

▶ Key Terms

action learning *352*

ergonomics *389*

production standards *357*

team learning *352*

work sampling *373*

activity-based cost (ABC) systems *358*

leveling factors *368*

synthetic or predetermined time standards *375*

time and motion study *361*

average performance *356*

normal time *368*

team building *353*

time study *361*

▶ Review Questions

1. Why is training for teamwork a concern of P/OM?

2. What is action-learning training?

3. What is computer-based training?

4. Explain why good estimates of labor costs are needed to determine how much to charge for a new product or service.

5. Explain why good estimates of labor costs are needed to determine fixed and variable costs to be used for breakeven analysis.

6. Explain why good estimates of labor costs are needed to determine plant location and selection.

7. Explain why good estimates of labor costs are needed to develop excellent production schedules.

8. Explain why good estimates of labor costs are needed to determine an optimal product mix.

9. Explain why good estimates of labor costs are needed to determine when a technology upgrade should be made.

10. Explain why good estimates of labor costs are needed to make intelligent bids for jobs with federal or local municipalities, or with companies.

11. Frederick W. Taylor thought the relationship between the well-designed job, the best worker, and the reasonable wage scale could be determined by logical analysis and intelligent experimentation. The concept of the well-designed (specialized and efficient) job has been attacked. The same is true of the notion of best workers and worst work-

ers. A guaranteed wage and pay not related to individual productivity have been suggested as the only reasonable wage scale. Discuss from the systems point of view.

12. What is meant by the statement: "Observations of productivity can be used for purposes of estimation and bidding, but they are unacceptable as a means of setting work standards for the amount of output each employee is expected to produce"?

13. Why is leveling no longer a major concern for job observation?

14. For the work sampling study in the text, what should be done about the manager's belief that there is too much phoning?

15. For the work sampling study in the text, the manager believes too much time is spent filing. What can be done about that?

16. What is meant by the statement: "The design of industrial jobs is being replaced by systems design of interrelated jobs"?

▶ Problem Section

1. Show all computations in tabled form needed to determine the largest value of n' for the first 5 cycles of elements 1 and 2 in the table of Figure 9.2 above.

2. Show all computations in tabled form needed to determine the largest value of n' for the first 10 cycles of elements 1 and 2 in the table of Figure 9.2.

3. Package Design International (PDI) is a medium-sized company. Typically, because of overload, its executives work 10 to 12 hour days. The president calls for reduction of the average period from order-entry date to delivery date.

   The OM department suggests building a supply chain simulator. The simulation would capture all factors that affect order to delivery times. This would be a training and problem-solving device costing an estimated $25,000. The operations manager estimates it will take 40 hours for each executive to learn how to use this simulator. There are five executives that would be expected to work with the simulator.

   OM has told the president that the simulator would result in $250,000 of additional profit for PDI in the next year.

   Should the contract for the simulator be given?

4. Finding the *shortest common-cycle time* for time studies is like finding the lowest-common denominator for describing a series of fractions.

Cycle Number	Core Elements		Extraneous Elements			
	1	2	A	B	C	D
1	5	4	—	—	—	—
2	3	2	—	—	—	—
3	4	3	—	—	—	—
.	.					
.	.					
100	3	4	20	20	25	30

What is the shortest common-cycle time for the time sheet shown here?

5. As output volume (*OV*) rises, so do wages (*W*). This is shown by a plus sign at the end of the arrow connecting *OV* and *W*. As *W* increases, so does worker motivation (*M*). This is shown by a plus sign at the end of the arrow connecting *W* and *M*. Continuing

in this way, increasing motivation conditions worker performance to raise output volume. Figure 9.3 illustrates this feedback relationship. However, other factors might be considered in this systems dynamics diagram.[11] Thus, additional inputs (arrows) can be added to the figure to indicate that other factors also affect output volume, wages, and worker motivation, such as quality (Q). Do this for quality (Q).

FIGURE 9 . 3

SYSTEMS DYNAMICS DIAGRAM FOR OUTPUT VOLUME (OV), WAGES (W), AND MOTIVATION (M)

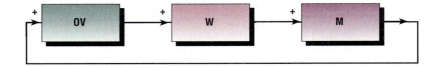

6. It has been decided to study the activities of the Market Research Store using work sampling. First, however, it has been concluded that any office can be used as a pilot operation where the methods can be developed and tested. Using Table 9.2 discussed earlier as a guide, choose any office and drop in five to ten times, noting each time what every person who works in that office is doing. Then, develop the kind of information that is shown in Table 9.2. Even though the sample is small, analyze this data. How big a sample will be needed?

7. For the MRS study described in Problem 6 above, what categories of activities might apply? How often should observations be made? Would you recommend using random sampling or taking observations at set times that are known to the employees of the MRS?

8. A Spanish company packs green olives that have been soaked in brine into one pound jars. The olives are first divided into one-pound-plus batches using a digital scale. On average, there are 32 olives to the pound. The olives are inserted in a jar and brine is added. Then the jar is sealed with a twist cap and put onto a conveyor. The left-hand, right-hand chart is drawn below.

	Left Hand	Right Hand
Element 1	Get olives. Put on scale. Measure 1 lb. +.	Get jar.
Element 2	Put olives in jar.	Clamp jar. Get brine.
Element 3	Get screw cap.	Add brine.
Element 4	Screw on cap.	Unclamp jar.
Element 5	Move toward olives.	Put jar on conveyor.

a. Sketch the process flow and layout.

c. Develop a sequence of work elements.

d. Prepare a job observation or time sheet.

TABLE 9.5

OLIVE PACKING TIMES (IN HUNDREDTHS OF A MINUTE)

		Element				
		1	2	3	4	5
Cycle	1	15	8	7	10	10
	2	13	8	8	11	10
	3	15	7	8	12	10
	4	17	9	9	11	11
	5	12	9	6	10	10
	6	14	7	7	10	9
	7	14	8	6	9	9
	8	13	8	6	9	10
	9	15	7	8	9	11
	10	16	9	7	10	9
Total time		144	80	72	101	99
No. of observations		10	10	10	10	10
Expected cycles		1	1	1	1	1
Average time		14.4	8.0	7.2	10.1	9.9
Allowance		1	1	1	1	1
Normal time		14.4	8.0	7.2	10.1	9.9
R & D Correction		1	1	1	1	1
Standard time		14.4	8.0	7.2	10.1	9.9

9. Use the information in the problem above to determine the standard time for the job.

10. Can predetermined time standards be used for the Spanish olive packing company? Develop the appropriate report.

11. A new set of data have been collected by FFDC using operations sampling instead of time study-job observation.

work Study

Activity	Description	Number of Times Observed
1	Place 10 diskettes in box.	212
2	Seal box, etc.	90
A	Put away 100 boxes.	10
B	Get 500 diskettes.	8
C	Get 50 boxes.	7
D	Get cellophane every 500 boxes.	12

339

How do these results compare with those derived by the time study in Figure 9.2 shown earlier?

12. With respect to job observation sample size and Equations 9.3 and 9.4 above, do the calculations for the following data and determine whether it is OK to stop taking additional sample observations.

Cycle Number	t_1	t_2
1	13	5.0
2	13	5.0
3	13	5.0
4	13	5.0
5	13	5.0
6	13	5.0
7	13	5.0
8	13	5.0
9	13	5.0
10	13	5.0
Totals	130	50.0

$$\bar{t}_1 = 130/10 = 13.0$$
$$\bar{t}_2 = 50/10 = 5.00$$

$$\text{Data and Computations for } t_j = \sum_{j=1}^{10} \frac{t_j}{n}$$

How could this result have occurred?

13. Use activity-based costing to show how much overhead is not applicable to Job X. Assume that all costs are multiplied by 10^4 and apply only to Job X for a one week period.

$$Materials\ Costs\ \ = \$\ 7$$
$$Energy\ Costs\ \ = \$\ 3$$
$$Direct\ Labor\ Costs = \$\ 5$$
$$Revenue\ \ \ \ \ = \$22$$

What is short-term contribution margin?

14. Using the information in Problem 13 above, continue with the following additional data:

The overhead to be allocated is measured for one week.

$$GE = \$4, \text{ with } UGE = \$2, \text{ and } UNGE = \$2.$$

What is the operating profit?

15. Using the information in Problems 13 and 14 above, what is the unused portion of overhead that should not be associated with the profitability of the job?

16. Using the information in Problems 13, 14 and 15 above, what is the used portion of overhead expenses that should be associated with the profitability of the job?

▶ **Readings and References**

Dennis R. Briscoe, *International Human Resource Management*, Englewood Cliffs, NJ, Prentice-Hall, 1995.

Sam Certo, *Human Relations Today: Concepts and Skills*, Burr Ridge, IL, Irwin, 1995.

Robin Cooper, and Robert S. Kaplan, "Activity-Based Systems: Measuring the Costs of Resource Usage," *Accounting Horizons*, September 1992.

F. B. Gilbreth, *Primer of Scientific Management,* NY, D. Van Nostrand, 1914.

J. W. Forrester, *Industrial Dynamics*, NY, Wiley, 1961.

Louis W. Joy III, and Jo A. Joy, *Frontline Teamwork*, Burr Ridge, IL, Irwin, 1993.

Jon R. Katzenbach, and Douglas K. Smith, *The Wisdom of Teams: Creating the High-Performance Organization*, Harper Business, 1994.

Brian H. Maskell, *Performance Measurement for World Class Manufacturing*, Cambridge, MA, Productivity Press, 1991.

F. J. Roethlisberger, and W. J. Dickson, *Management and the Worker*, Cambridge, MA, Harvard University Press, 1939.

P. R. Scholtes, et al., *The Team Handbook*, Joiner Associates, (608-238-8134), Madison, WI, 1989.

Oded Shenkar, Ed., *Global Perspectives of Human Resource Management,* Englewood Cliffs, NJ, Prentice-Hall, 1995.

A. R. Tilley, *The Measure of Man and Woman: Human Factors in Design*, Henry Dreyfuss Associates, NY, from The Whitney Library of Design, imprint of Watson-Guptill Publications, 1993.

L. H. C. Tippett, "Statistical Methods in Textile Research," *Journal of the Textile Institute Transactions*, Vol. 26, February, 1935.

SUPPLEMENT 9

Ergonomics or Human Factors

The traditional part of managing human resources from the point of view of OM is taking care of production employees. What is learned about doing that well can be extended to taking care of customers, a very valuable human resource to be managed. OM can help care for people making deliveries and those who work in the suppliers' organizations. OM may even be in a position to help the company provide for the well-being of anyone coming in contact with the company. OM knowledge about ergonomics could spare the company a lawsuit brought by someone walking on the sidewalk who trips and breaks a leg.

Ergonomics, originally a British term, is the systematic study of how people physically interact with the working environment, as well as their equipment, facilities and products. Ergonomic design of the system is the goal. Equivalent terms include *human factors, human engineering,* or *biomechanics.*

The prime rule of ergonomics is *safety first* for workers at their jobs and consumers using products produced by the company. In the U.S., workplace safety is monitored by OSHA, the Occupational Safety and Health Administration created in 1970. There also is a Consumer Product Safety Commission to deal with the safety of products as diverse as children's toys and automobiles.

Acceptable levels of safety are difficult to specify. The ideal situation is "perfect" safety, but like zero defects, perfect safety is an unobtainable state. So-called "safety factors" are designed into bridges, ships, and planes. This is done to raise safety to as near-perfect levels as possible. Such safety levels are called *fail-safe systems*. This is supposed to mean that the system is immune to failure. Interpret this as meaning the probabilities of such occurrences are so small they can be ignored. According to Borel, these might be events associated with probabilities in the neighborhood of one in a million (0.000001).[12]

Ergonomics

Systematic study of how people physically interact with the working environment, as well as their equipment, facilities and products; also called human factors, human engineering, or biomechanics.

The design of facilities and the working environment is a form of industrial architecture. In many ways, it represents the same kinds of approaches and the same set of human factors objectives as those of industrial designers. Industrial design is the profession that assumes responsibility for the architecture of products, machines, and living spaces, among other things. Industrial designers are responsible for achieving safety, comfort, and aesthetically pleasing design results.

As a concrete example of the effort required to do this, look at Figure 9.4 which pictures the "constant factors in vehicle seating." This is the work of Henry Dreyfuss Associates in Anthropometry which deals with the measuring of men, women and children.[13]

Note the detail with respect to how people of different sizes are positioned and how the standard displays are tilted to avoid reflection. There is discussion in the text about forces required for hand and foot controls. The diagram is packed with information for anyone involved in race cars, sport cars, autos, trucks, among others. Such detailed diagrams are derived from equally detailed studies. They enable proper design of the workplace, in this case for truckers, taxi and race car drivers. There is similar information available about interactions with the environment, including the way in which people work together.

FIGURE 9.4

CONSTANT FACTORS IN VEHICLE SEATING

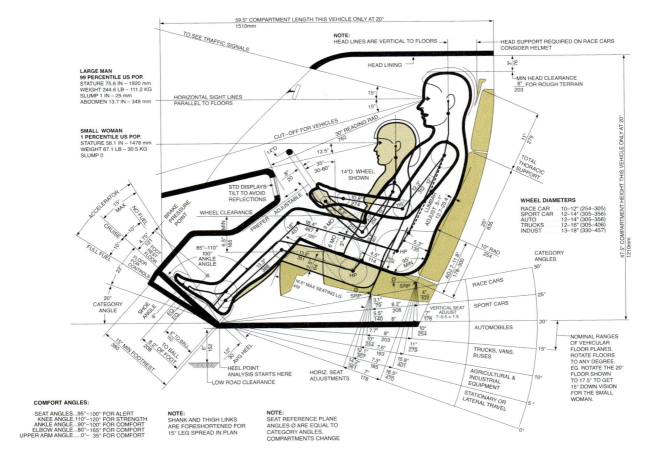

Office ergonomics includes proper positioning of the hands and wrists on the keyboard.
Source: Bayer Corporation

The human factors area treats both the physiological and psychological characteristics of people. It attempts to provide high levels of safety and comfort. It is concerned with the appearances of things, and the way they affect efficiency. All the senses and interrelationships of the senses to both motor and mental responses are part of the fabric of these factors, which describe the interactions of the worker and the workplace, the plant and the office as well as the customer and the product. The systems point of view is essential to capture the interactions between the workforce and the workplace.

Not just sight, illumination, and color concern the designer; hearing, noise, taste, smells, temperature changes, and body orientation are other factors that condition the attitudes, performance, safety and comfort of the system's workforce.

What is not immediately apparent is the flaw in logic of an individual using himself or herself as a physical model of what other people need or want. Self is not a sufficient design guide for the consumer product design or for employee workplace considerations. The requirements of a representative distribution of people can be determined. A design based on statistical knowledge regarding these dimensions would satisfy the minimum requirements for a majority of potential users.

Individual preferences vary with respect to the amount of light that is both comfortable and satisfactory for accomplishing a given job. Many studies have been conducted to determine superior designs for controls such as dials, gauges, and tracking devices on airplanes, cars, and industrial machines.

The sense of hearing has been studied with equal fervor. Many production operations are quiet, but others produce extremely high-noise levels with affects on workers and potential damage to hearing.

Problems for Supplement 9

1. What is the derivation of the word ergonomics?

2. It has been said that OSHA has more claims for repetitive stress or overuse syndrome (such as carpal tunnel) than any other occupational illness or injury. How can the design of the environment be put to work to alleviate this serious situation?

3. Because repetitive stress syndrome is the result of continuous repetition of certain motions, what sort of statistical studies might be undertaken to improve the design of the job?

4. What is meant by a fail-safe system?

5. Certain workers have a history of numerous accidents, whereas most workers have few accidents. What can be done for those who are accident prone?

6. Industrial designers have an orientation to the systems approach which is critical for their success. Why is this so?

Notes...

1. "Why American Steel Is Big Again," *New York Times*, July 21, 1994, p. D6.

2. Jon R. Katzenbach, and Douglas K. Smith, *The Wisdom of Teams: Creating the High-Performance Organization*, Harper Business, 1994.

3. "Team Spirit is New Message at Olds," *New York Times*, June 23, 1994, p. D1, by-line James Bennet.

4. Brian H. Maskell, *Performance Measurement for World Class Manufacturing*, Cambridge, MA, Productivity Press, 1991.

5. Robin Cooper and Robert S. Kaplan, "Activity-Based Systems: Measuring the Costs of Resource Usage," *Accounting Horizons*, September, 1992.

6. L. H. C. Tippett, "Statistical Methods in Textile Research," *Journal of the Textile Institute Transactions*, Vol. 26, February, 1935, pp. 51–55.

7. F. B. Gilbreth, *Primer of Scientific Management*, NY, D. Van Nostrand, 1914.

8. *The 4M System*, publication of the MTM Association for Standards and Research, 1991, p. 3.

9. *2M—The Macro Matic Data Handler*, published by the MTM Association for Standards and Research, 1990, p. 6.

10. F. J. Roethlisberger, and W. J. Dickson, *Management and the Worker*, Cambridge, MA, Harvard University Press, 1939.

11. J. W. Forrester, *Industrial Dynamics*, NY, Wiley, 1961. This book develops the methods of systems dynamics.

12. Emile Borel, "Valeur Pratique et Philosophie des Probabilités," *Traite du Cacul des Probabilités et de ses Applications*, Tome IV, Fascicule 3, Ed. Emile Borel (Paris: Gautier-Villars, 1950).

13. A. R. Tilley, *The Measure of Man and Woman: Human Factors in Design*, Henry Dreyfuss Associates, NY, from The Whitney Library of Design, imprint of Watson-Guptill Publications, 1993, page 66.

Chapter 10

Capacity Management

After reading this chapter you should be able to...

1. Define capacity and explain how it is measured.
2. Discuss how efficiency and utilization relate to actual measured capacity.
3. Explain why the goal of 100 percent utilization is an outdated objective that can be counterproductive.
4. Discuss the problem of planning capacity to deal with peak and off-peak demand.
5. Explain how bottlenecks affect capacity and discuss what to do about them.
6. Describe the effects of delay on the capacity to make and deliver products and services throughout the supply chain.
7. Set up the beer game to help team members visualize the effects of delay on out-of-stocks and over-stocks.
8. Explain the application of breakeven as a percent of maximum capacity to determine work configuration preferences.
9. Describe why breakeven analysis is called a quintessential systems model.
10. Explain how linear programming relates capacity and optimal product mix determination.
11. Describe how linear programming depends upon systems-wide inputs.
12. Explain how modular production and group technology are approaches that increase realizable capacity.
13. Describe what the learning curve reflects and explain how learning increases actual capacity.

Chapter Outline

10.0 DEFINITIONS OF CAPACITY
Standard Hours
Maximum Rated Capacities
Difficult to Store Service Capacity
Backordering Augments Service Capacity

10.1 PEAK AND OFF-PEAK DEMAND
Qualitative Aspects of Capacity
Maximum Capacity—100% Rating
Capricious Demand
Capacity Requirements Planning (CRP)

10.2 UPSCALING AND DOWNSIZING

10.3 BOTTLENECKS AND CAPACITY

10.4 DELAY DETERIORATES PERFORMANCE IN THE SUPPLY CHAIN
Contingency Planning for Capacity Crises
A Supply Chain Game—Better Beer Company
Cost Drivers for Capacity Planning
Forecasting and Capacity Planning

10.5 CAPACITY AND BREAKEVEN COSTS AND REVENUES
Interfunctional Breakeven Capacity Planning

10.6 BREAKEVEN CHART CONSTRUCTION
Three Breakeven Lines
Profit, Loss and Breakeven

10.7 ANALYSIS OF THE LINEAR BREAKEVEN MODEL
Linear Breakeven Equations
Breakeven Capacity Related to Work Configurations
Capacity and Margin Contribution

10.8 CAPACITY AND LINEAR PROGRAMMING

10.9 MODULAR PRODUCTION (MP) AND GROUP TECHNOLOGY (GT)
Group Technology (GT) Families

10.10 LEARNING INCREASES CAPACITY
The Learning Curve
Service Learning

10.11 CAPACITY AND THE SYSTEMS APPROACH

Summary
Key Terms
Review Questions
Problem Section
Readings and References
Supplement 10: The Breakeven Decision Model

The Systems Viewpoint

Capacity management relates to how the existing system is used. The systems viewpoint broadens the scope of inquiry to include questions about the existing arrangement and whether alternative configurations might not provide superior alternatives. The discussion of options cannot be pursued without consultations between marketing, finance, R & D and OM. Capacity is always limited by the supply chain bottleneck. The viewpoint must be systems-wide to scan across facilities with excess capacity in order to spot the overloaded resources that are struggling to keep up with the rest of the system. Delays are another source of capacity problems which are often caused by factors that are outside the organization's boundaries. Thus, suppliers must be viewed as part of the system to understand swings in demand caused by delays that create above peak and below normal requirements throughout the entire supply chain. Anyone who has ever played the beer model is permanently converted to systems thinking. The breakeven model (BEM) deserves an award for being the quintessential systems model of the organization. In spite of the fact that a breakeven model relates every function, it has been taught for years as capital budgeting and not much more. Linear programming is another model that demands systems thinking even though it can be taught as a means of modeling resource management, or simply as "the blending problem." LP is a way of looking at capacity management through the eyes of all participants in the big system. Modularity of parts (design) and group technology (for production of families of parts) are powerful systems concepts that increase an organization's capacity to competitive proportions. The learning curve is another example of a systems concept that relates the learning individual and the learning organization to higher capacity achievement without spending. In this case, "capacity is free."

10.0

DEFINITIONS OF CAPACITY

Actual Capacity
Greatest output rate achieved with the existing configuration of resources and the accepted product or service mix plans.

Actual capacity is the greatest output rate that can be achieved with the existing configuration of resources and the accepted product or service mix plans. Altering the product or service mix can (and usually will) change actual realizable capacity. Modifying the existing configuration of resources, equipment and people in the work force, alters real capacity. The systems point of view includes cash as part of the resources because cash can be converted into new machines which alter real capacity. The systems viewpoint also includes good ideas which can increase capacity with minimum expenditures.

The classical formula for actual measured capacity is:

$$C = T \times E \times U$$ Equation 10.0

where: C = actual measured capacity (in standard hours)
T = real time available
E = efficiency
U = utilization

Standard Hours

T is determined by calculating the amount of *time* that is available when fully utilizing the resources that are in place to make the product. Doubling the number of machines doubles the amount of available time, T.

E is the *efficiency* with which time T can be utilized to make different kinds of product. Then, $T \times E$ is equivalent to standard hours available to make the products.

U is how much of the available capacity can be (or is) *utilized*. Lack of orders or breakdowns of machines diminishes U. When T and E and U are multiplied, the product is C, the actual capacity that is being (or has been) utilized.

Table 10.1 is used to illustrate the calculations.

TABLE 10.1

CAPACITY UTILIZED (C)

| Product | Standard Hours of Output | | | | |
	Mo	Tu	We	Th	Fr
T101	25	20	65	18	30
T102	65	10	25	40	0
MW11	40	90	50	70	80
TOTAL	130	120	140	128	110

Assume that this plant is rated with a *maximum capacity* of 150 standard hours of output. It never achieves the maximum, but comes closest to doing so on Wednesday. There are a variety of reasons why the plant does not achieve the maximum. Two factors are embodied in E and U, explained below. Another reason could be that the product mix has caused bottlenecks and flow disruptions.

E, the efficiency, is a proportional factor needed to convert to standard time. Machines or people that work slower have lower efficiency than those that have a higher productive output. Often, the best in the class is given an efficiency of one. It is to be expected that variations in efficiency will occur. Sometimes the source of the variation can be traced. If it is significant variation, then it should be studied and corrected. Unexplained fluctuations in capacity are unpleasant and unprofitable.

If a job is being done at 90 percent of the standard time because the supplier delivered defective product, remedial action must be taken with supplies on hand and the problem must be corrected with future deliveries.

For example, if toothpaste tube caps have a burr on the thread, then a tentative solution for the toothpaste producer is to assign a worker to file off the burrs. The supplier of caps is notified and it is expected that the problem will be rectified in future shipments. Meanwhile, actual toothpaste line capacity is reduced by 10 percent.

U, the utilization, is applied as a proportional correction to standard time when there are machine breakdowns and/or people are absent. Even when everything is running as planned, the value of U is usually less than 100 percent. If the system is being run faster by turning up the speed of the machine, the value of U can exceed 100 percent. There are pros and cons related to running above maximum rated capacity. How long the maximum capacity is exceeded also counts.

U is a measure to be wary of when it becomes an objective of management to keep it as close to 100 percent as possible. There are sound economic reasons not to produce when the production quota has been met and the safety stock is sufficient. Shutting down the machine for two hours of an eight hour day means that the U measure goes to 75 percent. Actual measured capacity is going to be reduced by a fourth.

Management must establish the fact that getting the job done early can be a good thing worthy of reward. Normal inclination is to avoid criticism for not fully utilizing the resource. A special effort is required to break through this illusion. The costs of making goods which are not yet sold must be analyzed. How long will they remain as inventory? Are there fluctuations in demand that they will buffer or is the buffer in place already? The costs of arbitrary utilization of capacity should be recognized for what they are.

In the example below, U is assumed to be 0.963, which is likely to be viewed as a more reasonable utilization factor than 0.750. P/OM might not want a permanent situation where the utilization factor was much below 0.900. The desired numbers will be dependent on the situation. For example, in service organizations, a much lower number might be considered desirable in order to keep queues short. Cyclical industries expect to cycle between utilization factors in the 0.700 range and then up to more than 100 percent capacity. Overall, many companies prefer to have capacity in reserve and expect to operate effectively below the misleading ideal of 100 percent utilization.

For the data in Table 10.1:

$$C = T \times E \times U \text{ for Monday might be:}$$
$$C = 150 \times 0.9 \times 0.963 = 130 \text{ actual standard hours}$$

The ratio of actual standard hours to maximum standard hours would be:

$$R = 130/150 = 0.87 \text{ or } 87 \text{ percent}$$

Maximum Rated Capacities

Capacity
Maximum containment and/or maximum sustained output.

Capacity can be defined as maximum containment. Containment capacity takes many forms (i.e., the auditorium has a seating capacity of 360, or the gas tank will hold 16 gallons).

It also can be defined as maximum sustained output. Both meanings are regularly used by P/OM. The capacity to contain or store inventory is the first of the two definitions of capacity.

The capacity to make inventory or provide service is the second definition of capacity. Production and/or operations capacity describes how many units can be served or made per unit of time. For services, a bank might compare the maximum number of people the bank teller can process per hour with the maximum number of people the ATM can process per hour. This is a service capacity comparison.

For manufacturing, compare the maximum number of hot dogs Oscar Meyer can make per hour with the maximum number of hot dogs Hebrew National can make per hour. This comparison is one that both companies would like to make in order to compare their PMC's (productivities at maximum capacity). PMC makes an excellent measure for benchmarking. Benchmarking is a systematic comparison of fundamental measures with those of competitors and the best departments of the company itself.

Consider using maximum containment as a benchmarking measure. Generally, the larger the storage facility, the more material sitting around without having value added, and the poorer the performance measure. However, if a company built substantial petroleum storage facilities early in 1973, it would have been in the catbird seat when the oil embargo in the fall of 1973 created severe petroleum shortages. The harsh effects of the oil crisis lasted through the end of March 1974.

FIGURE 10.1

SUPPLY AND DEMAND RELATIONSHIPS WITH CAPACITY

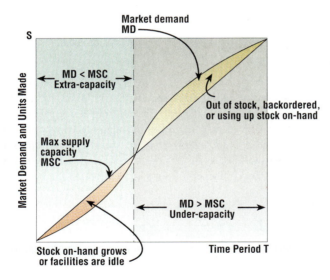

The utilization of containment capacity is another important factor for P/OM to consider. Having extra capacity would not have been a boon to Mobil Oil unless the tanks were filled with crudes and fuels. On the other hand, high utilization of containment capacity runs counter to the desire for low inventory levels, just-in-time deliveries, and constant value adding.

The two kinds of capacity situations have a trade-off relationship. Extra-capacity and under-capacity are shown in Figure 10.1. When market demand falls below maximum supply capacity (MD < MSC), the production system can feed storage and build up inventories. When market demand exceeds maximum supply capacity (MD > MSC), the inventory can be used to help meet that demand.

In Figure 10.1 the diagonal line is the maximum capacity to supply product. It is a rate of output which at the end of period T has the capability of producing S units. Realistically, the curved line is sometimes below the diagonal and sometimes above it. For convenience, the market demand rate over the entire period T accumulates total demand of S.

Difficult to Store Service Capacity

The concept developed above and shown in Figure 10.1 does not apply to services as readily as to manufacturing. This is because most services cannot be stored. The extra-capacity on the left side of Figure 10.1 is wasted idle time for service personnel. For goods it can represent building stock on-hand that can be drawn down when the under-capacity, right side of Figure 10.1 occurs.

Backordering Augments Service Capacity

If the customer is willing, backorders can be used to satisfy demand when there is no inventory available. This applies to services which cannot be stored. It applies as well to the manufacturer who is out of stock. The overloaded system does not have to turn away orders if the customer agrees to wait until other customers' jobs are finished.

10.1

PEAK AND OFF-PEAK DEMAND

There is another question about planning capacity. Should the maximum output capacity be great enough to handle peak load? This would be equivalent to providing enough electrical generating power to supply all needs for air-conditioning demand on extremely hot business days. Alternatively, should the telephone companies have enough capacity to take care of all phone calls without any delays for Valentine's Day and Mother's Day? These holidays are known as the heaviest traffic days for phone companies.

The model for peak vs. non-peak capacity is:

$$\text{Buy peak capacity when: } C_p < C_{np}$$

C_p is the investment for peak capacity less that required to meet average demand, depreciated over the life of the system. Say it takes an extra $5,000,000 to increase capacity from average to peak. This amount is spread out over 10 years, or $500,000 per year.

C_{np} is the cost of not having capacity to meet demands that are greater than the average demand. It is determined by the costs of such events as brownouts, power failures, and loss of goodwill in the community. There are some lost revenues as well. Say that the total cost averages $30,000 per incident.

Figure 10.2 provides an illustration of the difference between peak and off-peak average demand, as well as the level of normal supply which is above off-peak average in this case.

FIGURE 10.2

PEAK AND OFF-PEAK AVERAGE SUPPLY AND DEMAND

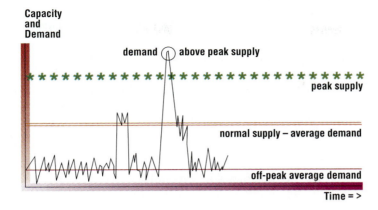

To illustrate with a case, assume that the probability that actual demand exceeds average demand is estimated to be 5% of the year or roughly 18 summer days. Total cost associated with C_{np} is:

$$365 \times 0.05 \times \$30,000 = \$547,500$$

This is more than the investment to prevent any brownouts, so the best advice, based on these figures, is to buy the equipment to generate sufficient power to meet peak demands.

Qualitative Aspects of Capacity

> *"As is our confidence, so is our capacity."* —William Hazlitt[1]

If our confidence is high, so is our capacity. But this is a different definition of the word. Usually, when capacity is treated in a P/OM context, the quantitative point of view prevails. There are, however, qualitative aspects that are important to cite.

It is not unusual to hear "he or she has the capacity to be a fine ballplayer, chef, manager, … etc." This invokes mental ability or physical skills. Organizations have capacities to deal with problems and opportunities. It is like an inventory of capabilities. Oftentimes, only when tested does the capacity to out-perform the normal emerge.

In that sense, Motorola has the capacity to meet quality standards that are far superior to most other companies. Robert Galvin, former chairman of Motorola challenged the company to develop peak capacity for quality production.

The company had been embarrassed by the Japanese company Quasar's purchase of the TV division which had a history of severe defectives reported between 150 and 180 defects per 100 sets. Within three years traditional Japanese quality had asserted itself. The defect rate had dropped to 3 or 4 per 100 sets. Service calls dropped from $22 million to less than $4 million. In-plant repair staff went from 120 to 15. Quality people liked to tell the story. Galvin did not like to hear the story being told, and threw down the gauntlet.

Motorola has since become a world leader in the production of quality products. Motorola set for itself an incredible quality goal of 3.4 defects per million parts. This is associated with six-sigma limits as compared to three-sigma limits in quality control. The company has reported being very close to achieving the goal. Motorola was the first company to win the Baldrige Award for quality ability. The turn-around is impressive.

Skillful management can increase the level of capacity that is achieved. Managing installed physical capacity properly means obtaining the maximum available capacity. This goal is particularly relevant when dealing with management of peak demand. If the maximum capacity of the process is less than the peak demand, knowing how to assign priorities will influence real capacity levels achieved.

Maximum Capacity — 100% Rating

Capacity as measured by maximum output volume per unit time, or throughput rate, comes closest to capturing the P/OM concept. That does not make it easy to measure. It is possible to produce at more than 100% of capacity for a period of time. Maximum capacity depends on who is doing the work and what is being made or serviced. Although 100% and maximum capacity are illusive concepts, they are useful standards to go by as long as the users are aware of their arbitrary nature.

When people are part of the process, some of the qualitative aspects of capacity cannot be ignored. Some people work faster than others. Most people fluctuate in their rates of output. There are learning curves at work. People can do some jobs better than others and, in this regard, all people are not alike. Ambiguity abounds in this arena.

Machines can work at different speeds. Driving an automobile illustrates that. Different velocities are best for optimal fuel consumption, minimum tire wear, and life of the car. Machines can be set for certain speeds which engineers would say are equivalent to 100% of their capacities. Ask the engineer what will happen if the machine is set to work faster. A typical answer will be that the machine will wear out sooner.

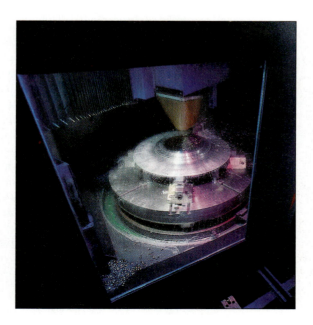

For heavy and difficult work especially in environments that are very hot, noisy, and even life threatening, machines are faster than people.
Source: Cooper Industries, Inc.

Figure 10.3 shows two patterns that are typical of wearout life and failure as a function of accumulated usage and age. Pattern A illustrates the output lumens of a lightbulb which is diminishing before final failure. Pattern B shows a Weibull-type distribution which reflects the high initial mortality of certain products (such as lightbulbs). Then there is a period of low product mortality until it reaches the expected time of failure. Such time-of-failure distributions must be understood if capacity is to be managed properly.

FIGURE   10.3

**WEAROUT AND FAILURE PATTERNS AS A FUNCTION
OF ACCUMULATED USAGE AND AGE—
THE FAILURE DISTRIBUTION IS THE WEIBULL DISTRIBUTION**

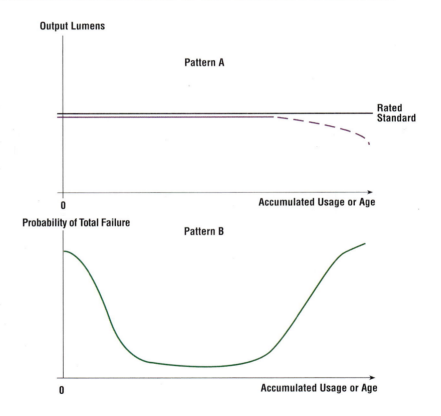

Setting the standard for the maximum capacity creates an interesting comparison between what people and machines can do. For highly repetitive jobs, most machines can be set to work faster than people. For heavy and difficult work, especially in environments that are very hot, noisy and even life threatening, machines win without question. Thus, robots are the best choice for nuclear experimentation. Also affecting the determination of capacity is the fact that machines can break down and people can be absent. Maximum capacity is usually determined with all the systems on go.

On the other hand, automated voice mail systems, such as those operated by financial service companies and phone companies, are tedious. They can require listening to endless options. It is not unusual to have to listen to seven or eight alternatives before getting to the one that applies. It is so much faster for the customer to speak to an operator that many people who have Touch-Tone phones prefer to act as though they have a rotary phone in order to get a human operator.

The capacity of such systems may be difficult to evaluate because people manipulate the system to their own advantage. In the same regard, the bank teller can move certain requests faster than the ATM and vice versa. Further, especially with services, capacity measures are a function of the demand on the systems. Excess demand tends to deteriorate the performance of service systems. Requirements for special services which are common in many systems such as health care and education also make it difficult to establish maximum capacity figures.

Capricious Demand

Capacity planning is one of the most important business activities. It is filled with opportunity to manage to advantage and fraught with difficulties. This is especially true if demand is capricious and tough to forecast correctly. Capacity planning is done to reach optimal supply decisions which, it is hoped, will match future demand patterns. This means that capacity planning does not have to be a single frozen value but can be a dynamic trajectory with fluctuations and oscillations.

Capacity Requirements Planning (CRP)

There are two aspects to CRP. The strategic issues related to long-term planning including breakeven points and optimal allocation of resources are being treated in this chapter. The shorter-term, tactical issues are operating issues which properly belong to a discussion of material requirements planning.

The ability to alter capacity in steps or stages, as shown in Figure 10.4, is valuable for strategic planning.

FIGURE   1 0 . 4

DYNAMIC ADJUSTMENT OF MAXIMUM CAPACITY

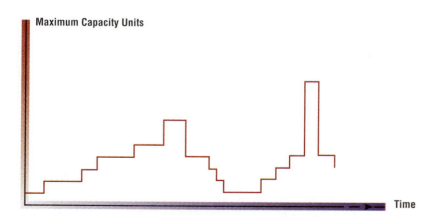

It takes planning to keep as flexible as possible with respect to capacity. Combinations of negotiations (as in a power grid) and technology can be brought to bear on a temporary basis. Consider the discussion of rentals which follows.

Part-time employees are a familiar means for making capacity adjustments, up or down. They do not require either severance pay or fringe benefits. Part-time employees relieve the firm of many other obligations associated with full-time employees. Rentals are an equivalent method for making capacity adjustments with equipment and space without taking on long-term commitments that lock the company into higher capacity than it is likely to need over the long-term.

Coproducers and copackers are firms used to increase supply by subcontracting with them on a short-, medium- or long-term basis. For example, a company that sells margarine might make enough to supply 60 percent of its market. The rest of its market demand is supplied by the output of a coproducer which has excess capacity. The product of the coproducer is packed in the subcontracting company's familiar package.

10.2

UPSCALING AND DOWNSIZING

Capacity planning is often constrained to an optimal size by the engineering requirements of the process. Many chemical processes must be designed and built to a specific size in order to operate properly. Glass windshields require a given number of steps that take exact amounts of time.

Serialized flow-shop processes also are constrained by engineering factors, although to a lesser extent than continuous chemical flow process. Job-shop and batch processes are far more flexible with regard to optimal production volume sizes.

To increase maximum capacities, processes can sometimes be operated above their rated capacities by using overtime and faster conveyor speeds. There are quality repercussions and the equipment is seldom designed for sustained overloads.

To meet greater demand, it is usually advisable to subcontract with coproducers until such time that additional demand warrants building another plant. Such processes do not have the flexibility of batch shops which can be expanded and contracted within reasonable limits.

A coil of flat rolled steel can be annealed in less than 20 minutes on a continuous anneal line, as compared with several days for batch annealing.
Source: LTV Steel, Cleveland Works

As the labor component of the process changes, the capacity adjustment reflects this fact, allowing an increase or decrease in the workforce. Because many services are notable for their labor percentage of the cost of goods sold, the capacity adjustment in services often involves people. The increase in service jobs continues to dominate the growth of jobs for the workforce in the U.S.

Capacity planning involves not only machinery and labor, but also management. When demand slackens, the notion of reducing managerial capacity emerges. This is particularly the case where management size has grown because of the bureaucratic tendency to hire assistants for assistants. The Parkinsonian rule that "work expands to fill the time available for its completion"[2] can be carried one step further.

In many organizations "work expands until there is no time available for its completion." This means that there are never enough people; everyone is busy. Capacity is increased until there are so many levels in the organizational hierarchy that communications almost never make it all the way up. The management at each level becomes self-protective and an outside consulting firm is often hired to downsize the company. This usually means slicing out the bulge of middle management.

For growth operations, excluding continuous processes such as the chemical plants described above, there is a need to adjust capacity to match increases in demand. Typically adjustments lead or lag, depending on the industry. Figure 10.5 shows an industry capacity-adding pattern that tries to stay ahead of the curve. This is the situation for electric utilities.

F I G U R E   1 0 . 5

GROWTH ADJUSTMENTS FOR MAXIMUM CAPACITY

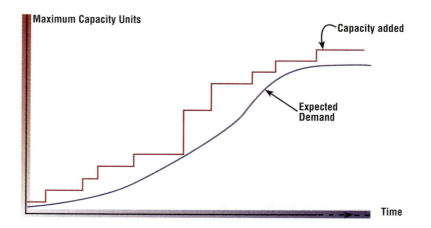

Ideally, the adjustments can be accomplished in an ascending staircase of steps. The timing and the size of the increments need to be carefully considered. The issue of whether there are natural increments of added capacity needs to be addressed. These natural increments are like buckets of supply that have minimum cost per unit of added capacity. If not, it may be necessary to reengineer the system by scrapping the present one and introducing a new one.

10.3

BOTTLENECKS AND CAPACITY

Another major issue in determining maximum capacity is the effect of bottlenecks which limit throughput rates. Real capacity can be much lower than the apparent capacity of rapidly performing machines. The system's point of view requires that the subject of bottlenecks be thoroughly addressed when dealing with the design and measurement of capacity.

Toward this end, the effects of finite scheduling, and synchronized manufacturing should be considered when dealing with capacity. A review of the V-A-T categorization developed by OPT for classifying processes is applicable. V-processes are analytic, starting with a single commodity at the bottom and branching out into refined products at the top. A-processes are synthetic, starting at the bottom with several inputs which are combined to yield a single marketable product.

OPT calls combinations of A and V the T-process. While bottlenecks can and do appear in A- and and V-processes, they are most likely to occur in the T-process because of the conflicting patterns of synthesis and analysis; also known as aggregation and disaggregation, assembly and disassembly, and combining and partitioning. Also, essential is line balancing the flow shop, an OM topic that is discussed in material on serialized production systems.

When determining or designing capacity, it is essential to consider at least two kinds of bottlenecks. The first is a fixed bottleneck in an unbalanced line of a flow shop. Recall that the flow shop is an A-type system.

Figure 10.6 illustrates a fixed bottleneck in such a system.

F I G U R E   1 0 . 6

**HOW A FIXED BOTTLENECK DECREASES CAPACITY
(UPSTREAM STARTS AT A AND DOWNSTREAM CONCLUDES AT E)**

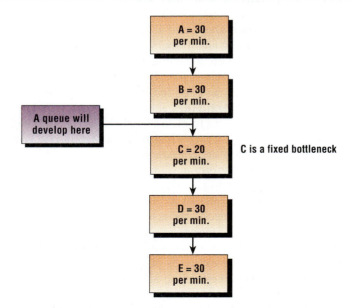

Station C will dominate the throughput rate until it is replaced
with a new facility that has a matched production rate of 30 per min.

Note that Station C has a maximum throughput rate of 20 per minute. All the other stations are rated 30 per minute. Station C will continue to dominate the throughput rate until it is replaced with a new facility that has a matched production rate of 30 per minute.

The line is unbalanced by Station C. A queue, also called a waiting line, will develop upstream from Station C. Flow starts upstream and moves downstream as shown. As the queue grows, upstream stations will cut their output rate to 20 per minute.

They will do so because they see that the queue is growing and may even be impinging on their own workspace. It is also possible that Station C will refuse to accept any more deliveries. The stations downstream from C have no options. They receive 20 units per minute. Maximum capacity is determined by the bottleneck. OPT principles call for feeding the bottleneck for the reason that C sets the production pace. If it fails to have work, the production rate falls below 20 per minute.

The second type of bottleneck can float from station to station in a job or batch shop identified with V-type systems. These temporary bottlenecks are associated with orders that have large lot sizes. The large lot ties up each of the machines on the order's routing schedule. Floating bottlenecks can make the determination of maximum capacity extremely difficult.

One approach is to select some typical order schedules from past records. By simulating the performance of the batch shop under the condition of efficient scheduling, it is possible to determine the maximum capacity of the system.

It also is possible to note where the bottlenecks are most commonly found. By altering the number and types of general-purpose equipment (GPEs), bottlenecks can be shifted and capacity increased.

A few additional points:

1. In some batch shops, the degree of repetition is sufficiently high to allow the scheduling for an average or typical day of orders (called an average order-schedule day) to be used as the basis for maximum capacity determination. Better scheduling practices might push the upper limit higher.

2. The V.A.T. classification associated with OPT reflects different possible definitions of maximum capacity. The V and T classes are producing mixed models and some standard such as standard hours is required. In addition to A and V there is the T-type classification, a combination of A and V.

3. Finite scheduling is what it sounds like. Specific lot sizes are finite, bounded and restricted. They are not unbounded, as is the potential output from a continuous production system or a flow shop. The lots are scheduled on a finite set of machines in such a way as to maximize throughput and speed deliveries.

4. Synchronized manufacturing occurs when finite scheduling succeeds in fitting jobs to machines so that flows are continuous, stations are balanced and bottlenecks are fed without queues forming. This is another definition of maximum capacity for a given configuration of equipment.

5. Because equipment and jobs interact to create multiple bottlenecks as well as temporary and floating bottlenecks, the achievement of maximum capacity is a function of the jobs the sales department obtains as well as the facilities P/OM has selected. The mix is dynamic and can change rapidly. Measures of maximum capacity may have to be estimates of averages.

6. Permanent bottlenecks, when they can be identified, are true determinants of capacity. They are easiest to spot and deal with in the flow shop, but can exist in the batch system. A good illustration of this is when the tool crib has a long queue in front of it. Every worker must go to the tool crib to requisition tools for the next job. The bottleneck at the tool crib lengthens the set-up and changeover times.

QUALITY

Grand Island Contract Carriers . . .

To improve the quality of its flatbed-trailer trucking service, Grand Island Contract Carriers (GICC), a Nebraska trucking firm, installed a satellite-based computer tracking system. "It's about utilizing the equipment to its full maximum capacity," said Richard Cordray, safety director for the company.

In the satellite-based tracking system, a signal from a satellite dish attached to each truck is transmitted to a computer center in San Diego, then to GICC every five minutes. The system allows real-time tracking and communications between GICC's dispatchers and the drivers. It also permits constant monitoring of the trucks' locations. A computer screen at the company's office maps the locations of the trucks. The map can either show the entire U.S. or only a certain area, including both city

Trucking firm operates at maximum capacity with installation of satellite tracking system

streets and the nationwide highway system.

"We used to have trucks passing by each other," said Cordray. For instance, one truck would be delivering a load with another on its way to the same location to pick up a

load. "The program helps eliminate waste and allows quicker response time," Cordray said. It also enables the office staff to respond more quickly to customer inquiries.

In addition to tracking, the computer system also affords communication with the drivers at any time. The drivers like being able to communicate directly with the office without having to stop or drive out of the way to find a telephone. The system also helps locate trucks "in the event of accidents, breakdowns or thefts," Cordray said. The equipment can be retrieved and put back on the road more quickly.

Source: Pat Dinslage, "High-tech truck tracking," *Grand Island Independent*, October 23, 1994.

7. Ingenuity is required to deal with measuring actual capacity utilization and defining a meaningful rating of real maximum capacity. Further resourcefulness is needed to spot and remove bottlenecks to increase both measures of capacity.

Queues are the result of bottlenecks caused by unbalanced capacities. Discussion of queues is appropriate in a chapter on capacity, but the equations are a bit tangential to the essence of strategic planning for capacity.

10.4

DELAY DETERIORATES PERFORMANCE IN THE SUPPLY CHAIN

Any one of the components in the supply chain can be a bottleneck. The capacity of a supplier can be a bottleneck for instance. Such a bottleneck will cause the producer to cut production runs. Simultaneously, the producer will notify all other suppliers to ship only a proportion of the regular order. Meanwhile, the manufacturer's warehouse will receive reduced shipments of the finished goods.

The distributor also will be shortchanged. Because the distributor does not have a sufficient inventory to meet all of the demands, retailers' orders will be cut. Finally, customers in the stores will be informed that they have to wait until more goods arrive. This situation, if not quickly remedied, can get progressively worse. Competitors, if they can, will surely take advantage of the failure to "hear the customer's voice."

In the above scenario, every supply chain player is suffering because of the problems of one supplier. The systems approach deals with such interconnectedness of the supply chain. It focuses on finding ways to reveal what kinds of problems each participant can visit on the others, and on how to remedy failures to meet demand with quality products.

F I G U R E   1 0 . 7

THE SUPPLY CHAIN GAME

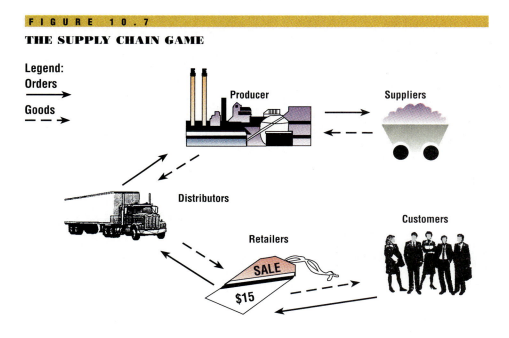

Lead time delays and poor forecasting cause oscillations
leading to shortages and overstocking

Contingency Planning for Capacity Crises

The producer may look around for a new supplier, or at least one that can make up the shortfall. A potential backup, even though more costly while the emergency exists, should have been known on a contingency basis. Before the supplier failed to deliver, were there any pre-warnings? Did the supplier provide all necessary information about the pending problem? Could the supplier and the producer have worked together to eliminate or reduce the severity of the problem? After-the-fact is too late. Such contingency planning for capacity crises deserves to be done beforehand.

At the same time, the other suppliers will be looking for new customers. Their contingency planning should have taken into account the fact that, through no fault of their own, this crisis has arisen for each of them. When these suppliers find other customers, they may have reduced ability to supply the producer. The distributor may start stocking a competing brand with serious long-term consequences for the producer's market share.

The retailers may find that the new competing brand is a hit with their customers. The loss of loyal customers is a serious blow stemming from this supply problem. One supplier's failure can destabilize an entire system, causing reduction in profit and losses in competitive leverage that will be hard to regain. Parts of the process linkages are shown in the systems dynamics chart of Figure 10.7 above.

A Supply Chain Game—Better Beer Company

The effect of timing and delays on the performance of a linked system with respect to the adequacy (or inadequacy) of existing capacity is incredibly important to understand.

Jay Forrester used simulation to demonstrate many of these effects in 1961 with his seminal work on systems dynamics.[3] This was at the Sloan School at M.I.T. where the "beer game" was developed in the 1960s. It is a supply chain simulation game[4] that is still entirely applicable and used for instructive purposes.

The "beer game" is quite involved and takes a long time to play. Figures 10.8 and 10.9 replace the game with some charts of simulations that reflect the effects of delays on decisions made by producers, wholesalers, retailers, suppliers and customers. The simulations are called the Better Beer Company. The product that is at the heart of the simulation is called Woodstock.

FIGURE 10.8

SUPPLY CHAIN SIMULATION
DISTRIBUTOR–RETAIL LINK

Week	Begin SOH	Supply	Net SOH	Demand	End SOH	Order Quantity	Delivery Week
1	12	4	16	-4	12	4	5
2	12	4	16	-8	8	8	6
3	8	4	12	-8	4	16	7
4	4	4	8	-8	0	16	8
5	0	4	4	-7	-3	12	9
6	-3	8	5	-6	-1	8	10
7	-1	16	15	-5	10	4	11
8	10	16	26	-4	12	4	12
9	12	12	24	-4	20	0	13
10	20	8	28	-4	24	0	14
11	24	4	28	-4	24	0	15
12	24	4	28	-4	24	0	16
13	24	0	24	-4	20	0	17
14	20	0	20	-6	14	4	18
15	14	0	14	-8	6	8	19
16	6	0	6	-8	-2	10	20
17	-2	0	-2	-4	-6	4	21
18	-6	4	-2	-4	-6	4	22
19	-6	8	2	-4	-2	4	23
20	-2	10	8	-4	4	4	24

Begin Week 1 with 12 cases of stock on-hand (SOH). Supply at the start of the week is 4 cases so Net SOH is 16 cases. As expected, the demand is for 4 cases so end of Week 1 SOH is 12 cases. The retailer orders 4 cases and since lead time is

3 weeks, the 4 cases will be delivered at the beginning of Week 5. Continue to read Figure 10.8 in this fashion. Note that in Weeks 10, 11, and 12, the End SOH rose to 24 cases, double the normal amount.

In Figure 10.9, the next link in the supply chain connects the producer and the distributor.

F I G U R E   1 0 . 9

SUPPLY CHAIN SIMULATION
PRODUCER-DISTRIBUTOR LINK

Week	Begin SOH	Supply	Net SOH	Demand	End SOH	Order Quantity	Delivery Week
1	64	16	80	-16	64	16	6
2	64	16	80	-32	48	32	7
3	48	16	64	-32	32	64	8
4	32	16	48	-32	16	64	9
5	16	16	32	-28	4	48	10
6	4	16	20	-24	-4	32	11
7	-4	32	28	-20	8	16	12
8	8	64	72	-16	58	16	13
9	56	64	120	-16	104	0	14
10	104	48	152	-16	136	0	15
11	136	32	168	-16	152	0	16
12	152	16	168	-16	132	0	17
13	152	16	168	-16	152	0	18
14	152	0	152	-24	128	16	19
15	128	0	128	-32	96	32	20
16	96	0	96	-32	64	48	21
17	64	0	64	-64	0	80	22
18	80	0	80	-94	-16	128	23
19	-16	16	0	-64	-64	80	24
20	-64	32	-32	32	-64	16	25

Figure 10.9

To read the chart, follow the same procedure as with Figure 10.8. Thus, Week 1 begins with 64 cases of beer at the distributor and End SOH is 64. Lead time is four weeks, so the order placed at the end of Week 1 is received at the beginning of Week 6.

These charts show how the retailer, spotting what seems like a major increase in demand, orders substantially more stock to avoid outages. The retailer's order for more product takes three weeks lead time between placing the order and receiving the shipment. If the increase was just a random pulse in demand, then other retailers might counterbalance the effect by ordering less, because, by chance, they had less than normal demand. The distributor would hardly notice such effects which tend to cancel out each other. The retailer would gradually correct for the extra stock on-hand that resulted.

On the other hand, if there is a common cause for many retailers to experience increased demand, they would all increase their order sizes, perhaps even overreacting with an extra large order to avoid going out-of-stock. The escalation in order sizes from many retailers would be a red flag to the distributor, who would greatly increase orders placed with the producer.

The four week lead time that the producer takes to fill the distributor's orders adds to the delay and increases the chances of large oscillations occurring for all supply chain participants. The producer experiencing such a surge in demand would be best advised to try to meet it with overtime.

Figures 10.10 and 10.11 show the stock on-hand (SOH) and orders placed by the retailer and distributor. It is evident that large oscillations are costing all participants a great deal. This is in spite of the fact that a review of the orders made by both the retailer and the distributor leads to the conclusion that the ordering policies followed were sensible.

SUPPLY CHAIN SIMULATION RETAILER'S STOCK ON-HAND OSCILLATES

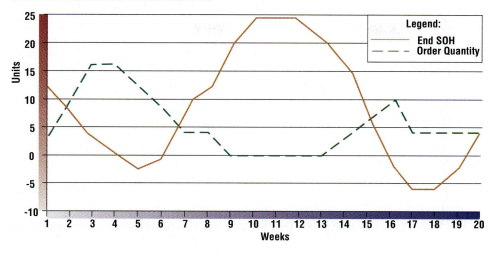

Figure 10.12 compares the end stock on-hand results for the retailer and the distributor. The effect had seemed enormous to the retailer. However, when the comparison is made with the distributor, the retailer's swings were gentle. The effect is going to be even worse at the producer's level.

If the increased demand seems to be sustained over a reasonable period of time, the producer might invest in more capacity (equipment and people) for what seems like a reasonable change in demand levels. Patient evaluation of the permanency of common causes is easy to counsel, but hard to realize in the face of possible stock outages and unhappy customers, retailers and distributors. The producer could not resist and invested in more capacity.

SUPPLY CHAIN SIMULATION DISTRIBUTOR'S STOCK ON-HAND OSCILLATES

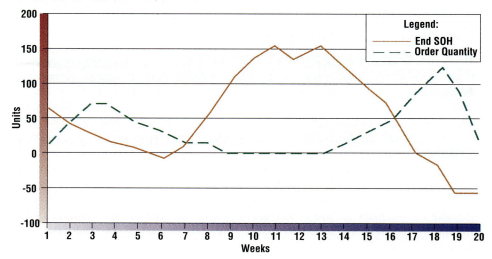

FIGURE 10.12

SUPPLY CHAIN SIMULATION
COMPARISON OF DISTRIBUTOR AND RETAILER SOH CURVES

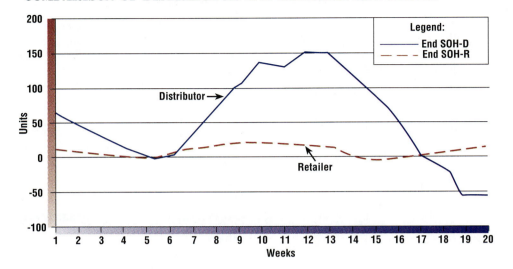

The final episode in this scenario is that the cause of the increased demand at the retail level was temporary. It was a common cause, such as might be created by a sensational TV program or a published newspaper article about the product. In reality, the glamour and appeal were gradually forgotten and the special selling effects eventually disappeared. Remember, this is a game and the scenario plus what follows are part of role playing, not reality.

The producer now has more capacity than is likely to be used for a long time, if ever. The producer drops the price to be able to sell everything that its new capacity enables it to make. The distributor orders more to take advantage of the probable increase in demand that follows a drop in price. The retailer also is working on lower margins and prefers to push the competitive products.

The product can be beer or soft drinks, cosmetics or food. Supply chains also can be extended to industrial products for original equipment manufacturers (OEM). The basic idea is valid as long as a relevant supply chain can be identified.

Cost Drivers for Capacity Planning

Activity-based cost (ABC) systems can always help in capacity decisions because they separate costs into those that are associated with resource capacities used and those that are unused. The producer, by monitoring production performance in terms of what is really contributing to the costs of output, would have identified the cost drivers, the costs of the main resource capacities being used. They literally drive costs and profitability. Traditional accounting systems lump many of these costs together and call them *overhead* or *burden.* ABC assigns overhead as correctly as possible, true to the spirit of the purpose. The purpose is to make each product or service reflect its true cost to the company so that it is possible to determine its true contribution to company profits.

In this way, specific costs are obtained for all equipment and facilities being used to make the products of the product line. Each product can be differentiated by the amount of resources that it actually uses. This leads to knowing the real profitability of each product in the line.

Before the TV event that caused sales to jump, Woodstock beer was being charged with the same percent of overhead costs for sales as all other beers made by the company. This charge under conventional accounting hid the fact that Woodstock did not use sales force time to the same extent as the other beers. In fact, sales did almost nothing to sell it.

Sales force and selling expenses are important to this product category. The activity-based cost driver might well be sales force hours required. Various products in this line could be differentiated by the sales force cost driver. Knowing the real costs and profitability of Woodstock might make a big difference in the decision to add new capacity.

Woodstock literally sold itself to a special market segment which found the name appealing. It was unlikely that segment could continue to grow. Even permanent growth was not likely. It was believed that the brand franchise would continue to get smaller as the Woodstock generation disappeared. This logic should drive the capacity decisions for Woodstock. The advantage of ABC when making capacity decisions is that it reveals fundamental information about costs and profit margins.

ABC is used in other ways for capacity-related decisions. It can provide process cost information to product designers to help them control manufacturing costs. Such cost information is related to volume and capacity. If the cost drivers are related to capacity, as well they might be, with respect to particular machines and often to the cost of set ups, then designers for DFM and capacity decision makers need ABC.

Forecasting and Capacity Planning

When forecasting can be done with a reasonable degree of accuracy, reactive capacity decisions can be made. If the summer that is about to start is expected to average ten degrees hotter than usual, and the air-conditioner manufacturer has a chart like the one shown in Figure 10.13, then production schedules can be set with some confidence.

FIGURE 10.13

**AIR-CONDITIONER SALES VS. AVERAGE
WEEKLY SUMMER TEMPERATURES**

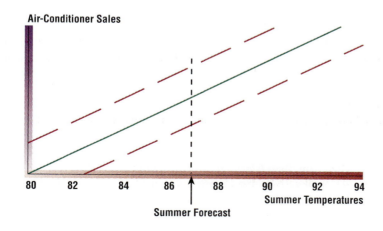

Figure 10.13 plots AC (air-conditioner) sales vs. average weekly summer temperatures. A band is drawn above and below the line. Within that band 90 percent of results have occurred. The band defines a confidence range. The forecast of weather has been correct within ± 3 degrees. Putting it all together, there is a reliable forecast and a strong correlation between that forecast and sales.

Does the process possess the capacity to supply the demand? If it has more capacity than needed, the ABC issue arises of adjusting downward the resource availability by selling it off or by finding alternative products to utilize the capacity. If the process has insufficient capacity to supply the demand, then either sales will be lost, or more capacity will be developed by using extra shifts and overtime, new equipment, and coproducers.

10.5
CAPACITY AND BREAKEVEN COSTS AND REVENUES

The breakeven model may be the most significant basis for making intelligent capacity decisions. To use breakeven, it is important to review direct and indirect costs.

The breakeven model uses the same costs and revenues as the input-output model which has been developed. This is not a review of costs, however, because the breakeven discussion of costs requires points be made in addition to those which applied to the efficiency of the input-output transformation process.

Variable costs per unit = vc are the inputs that tend to be fully chargeable and directly attributable to the product in the ABC accounting sense. They also are called direct costs because they are paid out on a per-unit of output basis. Direct costs for labor and materials are typical. Variable costs can often be decreased by learning to do the job better.

Material costs can sometimes be reduced using value analysis to improve material usage or find better materials at lower costs. Changing order sizes can increase discounts and grouping purchases can result in lower transportation costs for full car loads. Lower cost suppliers directly decrease the variable cost of inventory being held and stored. OM has many avenues to pursue as it strives to continuously improve quality while reducing variable costs.

There are some classification problems for variable costs. Differentiating between direct and indirect labor costs provides an example. Office work is indirect because such costs usually cannot be attributed to a particular unit of output on a cost per piece basis. If a direct link can be made so that overhead costs can be assigned directly, then such costs should not be equally shared in the overhead pool with other products.

Total Variable Costs = TVC = (vc)V are the result of multiplying the variable costs per unit (vc) by the number of units or Volume (V). These are also total direct costs.

Fixed Costs = FC have to be paid, whether one unit is made or thousands. For this reason, administrative and supervisory costs are usually treated as fixed charges. Together with purchasing and sales, they are all bundled together as overhead costs. Salaries and bonuses paid to the CEO and to every other manager are overhead because they cannot be assigned to specific products and services.

It is possible, with activity-based costing, to more carefully assign fixed charges to each specific product. This is important because breakeven analysis is done for each product and not for a product mix. Incorrect assignment of overhead to the fixed costs will affect the breakeven value. It is generally believed that the basic propor-

Variable costs per unit = vc

Inputs that tend to be fully chargeable and directly attributable to the product; also called direct costs.

Total Variable Costs = TVC = (vc)V

Result of multiplying the variable costs per unit (vc) by the number of units or Volume (V); also direct costs.

Fixed Costs = FC

Overhead costs paid regardless of production quantity; nonattributable to specific products or services.

Bellcore TEC ushers in new era of telephony

Does the telecommunications industry have the capacity to handle increasing demand for area codes and telephone numbers? Bellcore TEC, the New Jersey baby bell which administers the North American Numbering Plan for Canada, the U.S., Bermuda, and 15 Caribbean countries, assures us that it does. In fact, the new era of "telephony" (as the industry calls itself) arrived as of January 1995.

The proliferation of products using telecommunications technology (such as cellular telephones, computer modems, fax machines, pagers, and PBX switchboards) propelled the need for more telephone numbers and area codes. The previous numbering plan, which served from 1947 to the end of 1994, set up the original three-digit area code system which allowed for 160 possible combinations. Those area codes contained either a "0" or "1" as the middle digit, which identified the number as an area code. "The

original 160 codes were supposed to last to the end of the century, and they very nearly did," said Ron Conners, Bellcore's director for North American Numbering Plan Administration.

The new rules do away with the "0" or "1" middle-digit requirement. This allows for up to 640 new area code combinations, which means that 6 billion new numbers across North America will now be available. "It's impossible to predict the future," said Conners, "but we're confident these 640 new codes will last us through the year 2025. ..."

The new numbering system also requires new switching technology to read the new combinations and the conversion of telephone lines from analog to digital. Digital technology will actually serve to cut down on the need for new phone numbers because it permits the simultaneous transmission of faxes, data, video, and voices—all on one line.

Sources: "Area codes changing; Bellcore seminar explains it all," Business Wire, May 11, 1994; Eliot Kleinberg, "Number crunch: Phone companies are out of area code-combinations," *Palm Beach Post*, September 20, 1993; and Susan E. Kinsman, "In-state toll calls soon to require more dialing," *Hartford Courant*, August 29, 1994.

- -

tions are evident and that roughly the right numbers are used. ABC can help confirm this and/or correct deviations from the fundamental rule which states, if they do not vary with volume, then they are fixed.

ABC may also help move some charges from the fixed to the variable category. This is possible when the cost per unit made or serviced can be identified by the accounting system and moved to the variable category.

Total Costs = TC = FC + TVC are the sum of the Fixed and the Total Variable Costs.

Total Revenue = TR = (p)V is the volume V multiplied by the price per unit p. When goods and services are sold in the marketplace, they generate revenue. The breakeven model treats revenue as a linear function similar to variable costs. Each

Total Costs = TC = FC + TVC
Sum of the Fixed and Total Variable Costs.

Total Revenue = TR = (p)V
Volume (V) multiplied by price per unit p.

Marketing at this natural gas company manages price volatility by hedging a portion of the company's gas production through futures contracts on the New York Mercantile Exchange.
Source: FINA, Inc.

unit sold generates the same amount of revenue equal to the price (p). Cumulative or Total Revenue (TR) is equal to the number of units sold to date (V) times the price (p), i.e., $TR = (p)V$. Total Revenue equals zero when $V = 0$ because no units are sold. An assumption of breakeven analysis is that all units made are sold.

The breakeven analysis question is: how many units need to be made and sold in order to recover costs and breakeven? The results change over time as factors vary. Fixed costs often grow, prices change and variable costs per unit decrease with proper OM attention.

Marketing is responsible for setting prices to create the required demand volume while still generating a fair profit. The way the demand volume changes as a function of the price set is described by the "price elasticity" of the product or service. Price elasticity is affected by customer quality expectations and the degree to which competitive products are substitutable. Advertising and promotion strategies are costs usually assigned to overhead. All these market drivers are crucial to the P/OM domain because they determine output volumes, and thereby affect capacity plans.

Interfunctional Breakeven Capacity Planning

As was the case with input-output process analysis, for breakeven, too, every functional manager should be involved. There is something for everyone to be concerned about with the breakeven model. Without interfunctional planning, all relevant information cannot be secured. Successful coordination of process activities is not likely to be achieved without the systems approach.

How much capacity as well as what kind of capacity need to be addressed. What work configuration is to be used? Where should the capacity be located? Price will play a part in the revenue line. Technology used affects fixed and variable costs. Method of depreciation is another accounting issue that interacts with the finance-P/OM decision concerning technological investments to be made. The P/OM tie in with marketing and finance is inescapable.

The domains of cost concerns are:

1. P/OM is accountable for variable cost per unit, a function of the technology employed. Production volume and technology used are strongly correlated. P/OM should participate in decisions about both. These responsibilities determine $TVC = (vc)V$.

2. P/OM and finance are mutually involved in major determinants of fixed cost, FC. At the root of this relationship are the technology options and their affect on work configurations that can be used. Work configurations affect the variable costs and quality that can be achieved.

3. P/OM and marketing have to work out the price per unit (p) that customers are to be charged, and the demand volume (V) that should result. These decisions lead to estimates of total revenue (TR), total profit (PR), and margin contribution which is $(p - vc)V$.

10.6

BREAKEVEN CHART CONSTRUCTION

Breakeven charts were developed in the 1930s by Walter Rautenstrauch, an industrial engineer and professor at Columbia University.[5] Breakeven analysis (BEA) has become one of the most fundamental models of P/OM, providing important information for capacity decisions. Figure 10.14 illustrates the basic chart for one product and a given period of time.

Breakeven analysis can be done in either mathematical terms or in graphical form. To understand the breakeven approach to capacity planning it is easier to start with the graphical approach. Then the equivalent mathematical formulation will be developed.

FIGURE 10.14

LINEAR BREAKEVEN CHART

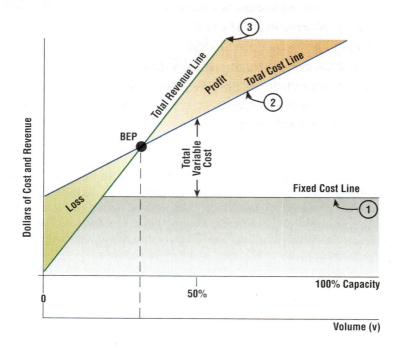

The ordinate (*y*-axis) is dimensioned in dollars of fixed cost for line (1), total costs for line (2) and dollars of revenue for line (3). The abscissa (*x*-axis) of the chart represents supply volume in units or the percent of capacity utilized for a given period of time. The *x*-axis terminates at 100 percent of capacity and whatever volume of units are equivalent for the given period of time.

Three Breakeven Lines

Line 1's fixed costs (*FC*) are derived in terms of a specific period of time such as a week, month, or year, which is consistent with the amount of capacity expressed by the x-axis. Fixed charges do not change as a function of increased volume or increased capacity utilization. They do change as a function of the time period. Depreciation is greater for three months than for one month. Fixed costs are relatively linear.

Line 2 in Figure 10.14 is the total cost line where total cost ($TC = FC + TVC$) is the sum of fixed and total variable costs. *TC* is a function that increases linearly as production volume gets larger. Total variable costs do not begin at the zero level. They are added to fixed costs, which are usually assumed to be the same at all volumes, even at zero production level. Vertical distances in the triangular area lying between the fixed cost line and the total cost line measure the variable costs (*vc*'s).

If overtime is needed, then the total cost line starts to curve up faster. This is illustrated in Figure 10.15 below which serves to show that breakeven models can be designed for nonlinear systems.

Depreciation (normally a fixed cost) should be added to the total variable cost, $TVC = (vc)V$ if it can be calculated as a function of machine utilization and wear. Reduction in the machine's life is directly attributable to having made the part. Taxes that are levied on the basis of units produced or revenue obtained also would be appropriate to add to the variable cost.

Line 3 in Figure 10.14 is the total revenue line, $TR = (p)V$. It is a linear function that increases with greater production volume when operating at a low enough share of the total market that free competition can adequately describe the situation. Companies with large market shares often experience diminishing effectiveness with advertising and promotions.

The saturation effect may lead to price discounts and decelerating revenues with increasing volume. This effect is illustrated in Figure 10.15 below which serves to show that breakeven models can be designed for nonlinear systems.

Profit, Loss and Breakeven

In Figure 10.14, both profit and loss are shown. The lower-shaded area between the total cost line (2) and the total revenue line (3) represents loss to the company. Loss occurs to the left of the breakeven point (BEP). The upper-shaded area between the same lines represents profit to the company. Profit occurs to the right of the breakeven point. Because it is a linear system, maximum loss begins at the left side of the diagram. With growth of output volume, the loss decreases until it reaches zero at the BEP. Then profit starts and it increases linearly throughout the range of positive profits until maximum profit is achieved at 100% of capacity.

Figure 10.15 shows a nonlinear breakeven chart where there are two breakeven points. Operating at capacities that fall between the BEPs is profitable. Using capacities to the right or left of the BEPs results in a loss. It also is worth noting that there is an indicated capacity which will produce the maximum profit. The graphic model can be used to determine this optimal capacity if the real shapes of the cost and revenue curves can be determined.

FIGURE 10.15

NONLINEAR BREAKEVEN CHART

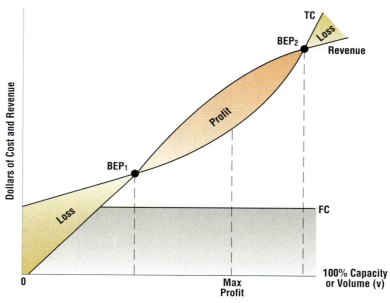

The definition of the BEP is that volume (or percent of capacity) at which there is no profit and no loss. Literally, it is the point at which the total costs of doing business exactly balance the revenues, leading to the name "breakeven." This is:

$$TR - TC = 0 = \text{profit}$$

The amount of loss is decreasing and cost recovery is improving as volume increases up to the breakeven point. At breakeven, cost recovery is completed. The BEP occurs at a specific volume of production (in units) or a given percent utilization of plant capacity. The values of BEP are critical for the diagnosis of healthy production systems. The concept is equally applicable to manufacturing and services.

10.7

ANALYSIS OF THE LINEAR BREAKEVEN MODEL

Figure 10.14 has been redrawn in Figure 10.16 so that the *y*-axis becomes profit and loss. This is accomplished by rotating the total cost line to be parallel with the *x*-axis.

F I G U R E   1 0 . 1 6

PROFIT AND LOSS BREAKEVEN CHART

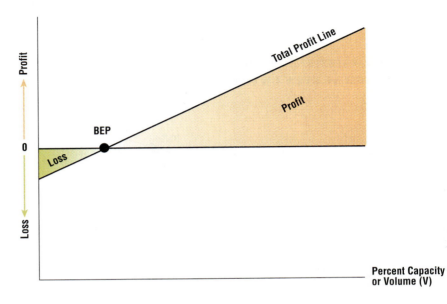

Slope

Reflects the rate at which profit increases as additional units of output (percent of capacity) are sold.

The **slope** of the line in Figure 10.16 reflects the rate at which profit increases as additional units of capacity are sold. Above breakeven, it means making profit faster but below breakeven it means creating losses faster. Because companies expect to operate above breakeven, the larger slope is preferred.

F I G U R E   1 0 . 1 7

PROFIT AND LOSS AS A FUNCTION OF VOLUME OF PRODUCTION (OR PERCENT OF CAPACITY) FOR ALTERNATIVE WORK CONFIGURATIONS, A AND B

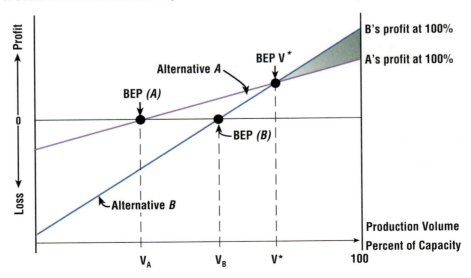

Next, consider the position of the BEP. If it moves to the right because of changes in the costs and revenues, then the organization must operate at a higher level of capacity before it starts to make a profit. Conversely, by reducing the breakeven point, profit can be made at lower volumes. Because companies want to operate above breakeven, lower BEPs are preferred.

Figure 10.17 shows two profit-loss lines marked A and B. Each is associated with a specific process alternative.

Alternative A has a lower BEP than alternative B. This makes A more desirable than B with respect to the position of the BEP. On the other hand, B generates more profit once the point V^* has been reached.

At the demand volume represented by point V^*, both alternatives yield equal profit. V^* is the point of equivalent profit for alternatives A and B. If greater demand than V^* can be generated, alternative B is preferred. If V^* cannot be achieved, then choose alternative A. Figure 10.18 represents the four different situations that can arise in a two-by-two matrix.

FIGURE 1 0 . 1 8

FOUR OPTIONS FOR CHOOSING AMONG ALTERNATIVES CONSIDERING THE BEP AND THE PROFIT RATE

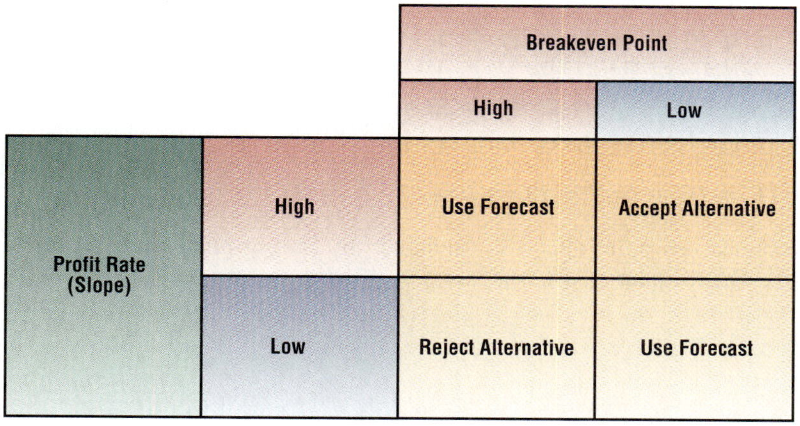

		Breakeven Point	
		High	Low
Profit Rate (Slope)	High	Use Forecast	Accept Alternative
	Low	Reject Alternative	Use Forecast

Two of the cells lead to clear-cut decisions. An alternative that has a lower BEP and higher profit-rate slope is the winner. An alternative that has a higher BEP and a lower profit-rate slope is rejected. An alternative that has a lower BEP and a higher profit-rate slope is accepted. The two other cells are ambiguous situations requiring a forecast of the expected demand volume.

If the volume forecast is above the highest BEP (V^* in Figure 10.17), then choose the alternative with the highest profit rate. If the volume forecast is below the lowest BEP (V_A in Figure 10.17), then choose the alternative that produces the least loss because it has the smallest slope.

Other combinations can occur. The forecast can fall between the two BEPs. Then, the procedure is to determine the expected profit by combining the profit and loss function with the probability distribution of demand using a decision matrix. The approach is shown in a supplement to this chapter.

Linear Breakeven Equations

The math equivalent for the graphic breakeven chart is useful. The symbols of the equations have been presented previously. The equations can be written in terms of the output volume (V) or percent of capacity $(100)V/V_{max}$.

The percent of capacity measure is used with the flow shop because it is designed with a known maximum capacity rating. Capital-intensive industries also relate their performance to the percent of capacity used. On the other hand, labor-intensive industries (such as check processing) are more likely to use measures of production volume than percent of capacity.

TR	=	total revenue per time period T
p	=	price per unit
V	=	number of units made and sold in time period T
FC	=	fixed costs per period T
vc	=	variable costs per unit of production
TVC	=	total variable costs = $(vc)V$
TC	=	total costs per period $T = FC + TVC$
TPR	=	total profit per period T

Total revenue for period T is:

$$TR = (p)V$$

Total cost for period T is:

$$TC = FC + (vc)V$$

Total profit for the interval T is (in terms of volume):

$$TPR = TR - TC = (p)V - FC - (vc)V = (p - vc)V - FC \qquad \text{Equation 10.1}$$

The breakeven point (V_{BEP}) is calculated by setting profit equal to zero, thus: $TPR = 0$. Then, Equation 10.1 becomes:

$$(p - vc)(V_{BEP}) = FC$$

and

$$V_{BEP} = \frac{FC}{(p - vc)} \qquad \text{Equation 10.2}$$

Breakeven Capacity Related to Work Configurations

The fixed costs and capacities of flow shops are much higher than the fixed costs and capacities of job shops. General-purpose equipment (GPE) used in the job shop is designed for high variety in low-capacity volumes. Special-purpose equipment (SPE) characteristic of the flow shop is designed for low variety in high-capacity volumes. Even the movement of materials between equipment is mechanized for the flow shop, whereas it is done mostly by hand in the job shop.

In trade for the higher fixed costs of the flow shop, P/OM expects to obtain lower variable costs. On the other hand, the manager of the job shop does whatever is possible to keep high-variable costs under control. If there is sufficient stable demand to warrant investment in flow-shop capacity, and if technology exists to deliver low-variable costs at the required quality levels, P/OM will opt to invest in flow-shop capacity.

Dialogue should be initiated by P/OM with marketing to investigate the possibility of accumulating sufficient stable demand if it does not presently exist. Modular production abilities allow P/OM to assist in achieving the demand levels that are appropriate for flow-shop capacities.

Flexible-purpose equipment (FPE) is the third category of process capacity to consider. The fixed costs of *flexible manufacturing systems* (FMS), *flexible processing systems* (FPS), and *flexible office systems* (FOS) are even higher than those of the flow shop. Since flexible, or adaptive, automation can be operated remotely 24 hours a day, 7 days a week, with time out for maintenance, the high fixed costs can be amortized across a broad product line.

It is necessary to have planned ahead for the proper use of FPSs. The high fixed costs require that multishift production be used to increase the volumetric capacity of the system. This allows the BEP as a percent of capacity to occur at a reasonable point, whereas single-shift underutilization of the equipment can prove costly and ineffective.

With limited use of flexible technology, meaning that it plays a small role in plant processes, then the problem of underutilization is minimal. However, when the overall technology of the process is FPE, what is required is careful pre-design of the use of the potential capacity to produce a broad mix of products that have strong market appeal. Large investments in flexibility demand great capacity planning.

Capacity and Margin Contribution

Breakeven analysis provides important parameters for evaluating capacity decisions. The critical rate of return on capacity is based on the slope of the profit line. The equally critical BEP is determined by the ratio of fixed costs to the per unit margin contribution which is equal to $(p - vc)$.
That is:

$$(p-vc)\ V_{BEP}=FC$$

which means that total margin contribution at breakeven completely recovers the fixed costs and no more. In general, total margin contribution is equal to $(p - vc)V_{actual}$.

Japanese industry specialized in flow-shop capacity following the MacArthur initiative to rebuild the Japanese economy after World War II. This concentration on flow shops challenged the Japanese export industries to amass the necessary demand levels to support the cost advantages gained from great production capacities. They met the challenge, at first by specializing in products with low prices and high reliability. Later, having penetrated the markets, they went upscale, still mastering new approaches to high-capacity production.

10.8

CAPACITY AND LINEAR PROGRAMMING

Job-shop capacity can be mismanaged. The mixture of jobs on different machines can block access by other jobs which would enhance margin contribution. **Linear programming (LP)** is a modeling technique used by OM to achieve optimal use of

Linear programming (LP)
Modeling technique used by OM to achieve optimal use of resources.

resources. LP shows that it is often not optimal to allow low margin products to block capacity that could be used by higher margin products. Sometimes, the optimal product mix does not fully utilize capacity because of this effect.

It is useful to demonstrate the point LP makes about unused capacity. Table 10.2 gives the production capacities of two departments (D1 and D2) which can make different amounts of two products, V and W. Also shown is the profit per unit for each product.

TABLE 10.2

PRODUCTION CAPACITIES OF TWO DEPARTMENTS (IN UNITS)

Product	$V_{capacity}$	$W_{capacity}$
Department 1 - Forge	10.0 per day	20.0 per day
Department 2 - Heat Treat	12.0 per day	12.5 per day
Profit	$2 per unit	$3 per unit

Table 10.3 shows various combinations of V and W that can be made under the constraints of Table 10.2. It also shows the total profit for the day based on the computing formula $2V + 3W$. Unused departmental capacity is called **slack**. The computation of slack is given after Table 10.3.

Slack

Unused departmental capacity.

TABLE 10.3

TRIAL AND ERROR COMPUTATION FOR SEVEN LP SOLUTIONS

Plan	V	W	D1 Slack	D2 Slack	Profit
1	5	5	25.00%	18.33%	$25.00
2	6	5	15.00%	10.00%	$27.00
3	7	5	5.00%	1.67%	$29.00
4	7.83	4.35	0.00%	0.00%	$32.20
5	5	7	15.00%	0.58%	$31.00
6	0	12.5	37.50%	0.00%	$37.50
7	8	4	0.00%	1.33%	$28.00

Slack is calculated—for each department:

$$\text{capacity used} = V/V_{capacity} + W/W_{capacity}$$

and

$$\text{slack} = \text{capacity unused} = 1 - V/V_{capacity} - W/W_{capacity}$$

To illustrate: for Plan 1

Department 1 capacity used: $5/10 + 5/20 = 3/4$

Department 1 slack: $1 - 3/4 = 1/4$

Department 2 capacity used: $5/12 + 5/12.5 = 0.81667$

Department 2 slack: $1 - 0.81\ 2/3 = 0.18333$

Linear programming finds the optimal solution without employing trial and error, the method used to create Table 10.3. The trial and error method was not aimless or random, however; it was based on learning in which way improvement lies and then moving in that direction.

The important conclusion is that maximum profit is achieved with some unused capacity for Department 1. This is Plan 6 which makes zero units of *V* and puts all of its capacity into the higher margin product, *W*. All of D2's capacity is used up, which means that nothing else can be made.

It is possible to prove with logic that this is the maximum profit solution, however, the LP method does this automatically by efficiently finding the point of maximum profit.

Plan 4 has used up all of the capacities of both D1 and D2, but its profit is 32.20, 5.30 less than Plan 6 profit. This is 16% less profit and all capacity has been depleted. With Plan 6, 3.75% of D1's capacity is still available to be employed in other ways or sold off if it cannot be used. Thus, LP ties in with ABC accounting, and with capacity planning overall.

10.9

MODULAR PRODUCTION (MP) AND GROUP TECHNOLOGY (GT)

Goods and services with large markets lend themselves to flow-shop production. Those with small markets are not good candidates. Flow-shop capacities are too large for jobs that should be run or done in finite lots. With creativity, there are times when a conversion can be made from job-shop operations to intermittent and even permanent flow-shop status. The transition can take place gradually. Both MP and GT are examples of design for capacity (DFC).

It is easier to design high-capacity processes when starting from scratch than when attempting to change the system. Reengineering is the approach to use when wanting to bring about dramatic, instead of continuous gradual change of an existing system. On the other hand, when starting up a new system, without the clear goal of achieving a flow shop, it is more likely that a job shop will result. There are at least three good reasons for this:

1. There is a strong job-shop tradition in the United States and Europe. It is based upon technology which preceded computer-controlled equipment and upon entrepreneurial endeavors of small- and medium-sized businesses. It also is based upon the kind of markets that characterized the 1950s and 1960s. They were smaller than national and international markets are today, and not large enough to support high-volume processes.
2. It is easier to settle for a job shop that offers good profits than to persevere for a flow shop that has potential for even more profits, but with much more set-up work and risk.
3. Some goods and many services cannot be produced by a flow shop. Artists' work belongs in the custom shop. Personal services require attention to details that, at least on the face of it, defy flow-shop capacities.

In spite of these beliefs, opportunities often exist which permit aggregation of demand to flow-shop proportions even though the markets of the products are relatively small. Modular product and process design presents one such opportunity.

This means specialization in the production of particular parts or in specific sequences of activities for services and programming strings of code, which can then be included as components of more than one product or service. The roots of such efforts exist in the well-developed concept of standard parts, such as screw threads and lightbulbs.

The reason for wanting to achieve such commonality is that one part, operation, or activity, if used in several products or services, can accumulate sufficient demand volume to warrant investing in flow-shop capacities.

The **principle of modularity** is: *to design, develop, and produce the minimum number of parts (or operations) that can be combined in the maximum number of ways to offer the greatest number of products or services.*

Table 10.4 illustrates the modularity of N parts named (PA_i), $i = 1, 2, ..., N$, used in M products named (PR_j), $j = 1, 2, ..., M$.

TABLE 10.4

SPECIFICATION OF PARTS USED IN PRODUCTS

Variety of Parts (PA_i)			Variety of Products (PR_j)					
	PR_1	PR_2	PR_3	PR_4	...	PR_j	...	PR_M
PA_1	1	0	1	1	...	0	...	0
PA_2	0	1	1	2	...	0	...	0
PA_3	0	0	0	0	...	1	...	0
PA_4	0	1	1	0	...	0	...	0
.	.	.	.	.	...	.	...	.
.	.	.	.	.	...	.	...	.
PA_i	0	0	1	0	...	2	...	0
.	.	.	.	.	...	.	...	.
.	.	.	.	.	...	.	...	.
PA_N	0	0	0	1	...	1	...	1

Legend:

PA_i denotes the part identified by the stock number i.

PR_j denotes the name or stockkeeping number of the product.

All j varieties are listed in the finished goods catalog.

As shown in the j-column of the matrix, the product j assembly requires one unit of part 3, two units of part i, and one unit of part N. The sequence of assembly is not indicated.

PA_N denotes the last part listed in Table 10.4.

PR_M denotes the last product name listed in Table 10.4. It is a product using one unit of part PA_N.

With N different kinds of parts, a total of M different product configurations can be offered to customers. Some of these products require several units of a single part. For instance, PR_4 requires two units of PA_2, and PR_j requires two units of PA_i.

The maximum possible variety, which is the maximum value of M that can be obtained with N different parts, may be very large. This is especially so when there are different possible combinations of the parts and varying numbers of them in combination. The general objective is to have M as large as possible and N as small as possible.

This is equivalent to saying that the general objective is to have as many products as possible made with the smallest possible number of parts. A useful measure of the effective degree of modular design might be the ratio of the number of products (columns) that can be generated from a given number of parts (rows); that is, the objective: *MAX M/N*.

Many of the possible combinations cannot exist and would have no appeal as a product choice for customers. It is up to the team to find the right parts for modular specialization. Computers epitomize the modular concept. Customers can specify many different computers which are made of components that mix and match with ease. Companies specialize in a relatively few computer parts that make a relatively large number of computer products. The parts are made by flow shops with enormous capacities. This accounts for the ever lower price of computers.

Modularity applies to services and projects. When certain activities are used repeatedly in some service function, the steps can be standardized and the process to deal with them can be serialized for the flow shop or even automation. The ATMs used by banks provide a fine illustration of a service that has been designed to provide high-capacity, lower-cost equivalents of teller service.

Long ago it was recognized that parts of computer programs appear over and over. They were standardized and are regularly inserted as modules into computer programs by information systems programmers. Rewriting lines of code is like reinventing the wheel. Modular constructed housing can produce many versions of homes using the same basic parts. Furniture designers are masters of modular sections. Project managers have developed modular techniques to cut the costs of portions of what otherwise is a unique venture. Building developers thrive on separating what is unique about each structure from what is a repetitive activity.

Modular constructed housing can produce many versions of homes using the same basic parts. Source: Acorn Structures Inc.

Scott Haag
· · ·
Cinergy Corp.

As manager of Performance Services for Power Operations at Cinergy Corp., Scott Haag is involved with short-term capacity planning. A public utilities company created in 1994 with the merging of Indianapolis-based PSI Resources Inc. and Cincinnati Gas & Electric Co., Cinergy serves southwestern Ohio, central Indiana, and northern Kentucky. In this part of the country peak capacity for electricity occurs in the "summertime when everybody's running their air conditioners… usually between the hour of 4:00 and 5:00 o'clock in the afternoon," said Haag, a 1977 mechanical engineering graduate from Purdue University.

To ensure that generating units are available to meet peak demand for energy, Cinergy conducts routine maintenance on the units during off-peak periods. "Especially during the spring for the summer, we will have scheduled outages for those generating units to really maintain them, get them into shape," Haag said, so that in the summer they will be "online, reliable, and available to meet the peak capacity."

In forecasting how large the peak demand will be and when it will occur, Cinergy monitors the weather which involves temperature, relative humidity, and illumination—the measurement for brightness of the sun. The day of the week also is an important factor in forecasting demand for electricity. "The week days have significantly higher loads than weekends," Haag said, because of business operations Monday through Friday.

To make projections on the demand for electricity "we have computer programs that can work up to 30 days in advance," Haag said. This "allows us to determine which units to bring online. We want to bring on the most economical units first. If that happens to be a steam unit, it could take up to 24 hours to bring online, so we have to prepare for that process well in advance." In predicting demand for next-day usage, he said he tries to "make sure we have at least 6% additional capacity online to cover contingencies," such as generating unit outages or extreme temperatures.

In the event demand for electricity exceeds Cinergy's capacity to produce it, "the system dispatchers will call around to other utilities in the area to purchase power," Haag said. These purchases can be prescheduled "a week to a day in advance or they can be scheduled on the hour," according to Haag.

Source: A conversation with Scott Haag, Cinergy Corp.

Successful modular design increases demand, promoting increased capacity, allowing flow-shop economies to be utilized. The kind of matrix shown in Table 10.4 above must have been in the minds of the inventors of Lego and the earlier Erector sets. Modularity helps to assemble the flow shop by creating production systems that specialize in parts instead of products. Process planning that assists in the part orientation calls upon group technology.

Group Technology (GT) Families

Group technology deals with processes that specialize in families of similar parts and/or activities. It is technologically feasible to develop efficient flow shops that can make gears of similar design, but different size. The same applies to cams, springs, and so on. The changeover routines to go from one size gear to another are fast and inexpensive.

GT parts have similar design, and essentially the same production operations can be used to make them. There is a functional similarity to FMS but the variety level is far simpler. The technology required assures minimum transition costs as the production line shifts from one variant to another.

A company using group technology to improve the profit margin of its line of products or services might become so efficient in this specialized family of operations that it could gradually shift its emphasis from products to parts. Eventually, it would become a parts subcontractor to industry and supply institutions with the output of its most efficient operations. In Best Practice cases, divisions of companies achieve this status.

10.10

LEARNING INCREASES CAPACITY

Economic theory indicates that product can be delivered at lower unit costs when the output volume increases. This is well-known as "economies of scale." The Boston Consulting Group (BCG) confirmed this effect with many empirical studies of a wide variety of industries.[6] Figure 10.19 shows some general shapes for this function where the decrease varies between 20 and 30 percent with each doubling of volume.

FIGURE   1 0 . 1 9

AVERAGE UNIT COSTS DECREASE WITH
INCREASING EXPERIENCE

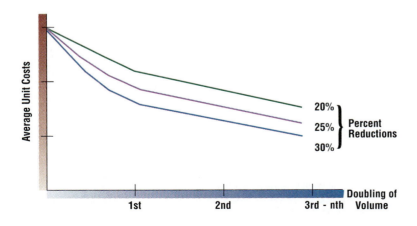

BCG calls this the **experience curve**. Generally, the rate of decrease in per unit costs is between 20 percent and 30 percent for each doubling of volume. Volume is a surrogate for experience and consequent learning. After doubling of volume has occurred several times, most of the payoff from learning will have been obtained. Reductions in average unit costs have to get smaller and smaller. Thus, price reduction potentials disappear as the market matures.

As the output volume gets larger, many factors operate to make the process more efficient. Workers learn their jobs better. It takes less time to do the jobs. More work gets turned out in a period of time, one of the main factors for the decrease in costs. This is equivalent to increasing C, the actual measured capacity. As a reminder:

$$C = T \times E \times U$$

where T = time available, E = efficiency, and U = utilization.

There are two ways of accounting for the increase in C. T becomes a larger number of available hours; since the same amount of work takes less time, more work can be done. Another way is that E becomes greater, even greater than 100%. It is just a formula which can be treated in either way as long as the effect is to increase actual capacity.

The Learning Curve

The **learning curve** is a model for individuals or groups. It describes how repeated practice decreases the time required to do the job. Thus, the learning curve captures the benefit of repetition and experience in quantitative terms for an employee or groups doing the same job routinely.

Two related learning curves are shown in Figures 10.20 and 10.21. They are explained below with their respective quantitative models.

FIGURE 10.20

LEARNING CURVE CUMULATIVE TIME

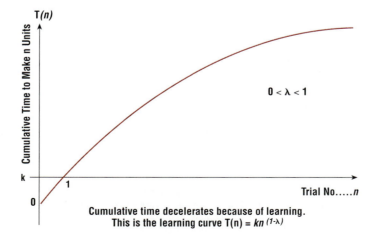

Cumulative time decelerates because of learning.
This is the learning curve $T(n) = kn^{(1-\lambda)}$

F I G U R E   1 0 . 2 1

LEARNING CURVE AVERAGE TIME

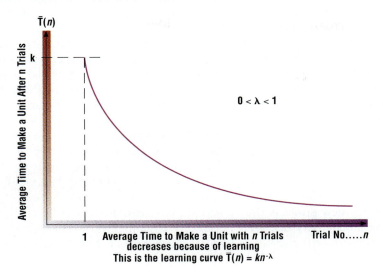

Average Time to Make a Unit with *n* Trials decreases because of learning
This is the learning curve $\bar{T}(n) = kn^{-\lambda}$

The equation: $T(n) = kn^{(1-\lambda)}$ is read as follows:

$T(n)$ = cumulative time to make *n* consecutive units

k = time required to make the first unit, $n = 1$

n = the number of units to be made

λ = the learning coefficient, $0 < \lambda < 1$

When $\lambda = 0$, the cumulative time $T(n) = kn$. Thus, $T(n)$ increases linearly (kn), and there is no learning since each unit requires the same amount of time, k.

When $\lambda = 1$, $T(n) = k$, which means that the job has been learned so well that after the first time, the rest of the units are made (literally) in no time at all. $\lambda = 1$, is, therefore, generally unrealistic and a goal never to be achieved.

As $\lambda \rightarrow 1$, successive units take less time. The curve drawn in Figure 10.20 reflects learning decelerating; however, it never approaches a limit as an asymptote. The value of λ affects the speed. The larger the lambda, the greater the deceleration.

Another useful form of the learning model is shown in Figure 10.21 where the average time to make a unit is calculated by dividing the cumulative time $T(n)$ by n.

$$\bar{T}(n) = \frac{T(n)}{n} = kn^{-\lambda}$$

k, n and λ are the same as before.

When $\lambda = 0$, $\bar{T}(n) = k$, which is the condition for no learning taking place. Every time the job is done it takes the same k time units. When $\lambda = 1$, an unrealistic limit is reached whereby average time k/n is geometrically reduced with each trial. In between are values of λ which yield reasonable decreases in the average time to do the job.

In reality, with continued practice, learning reaches a *plateau* where the output rate is relatively constant. When the employee takes a break there may be some relearning needed, but it is minimal. The effect becomes greater when the time between stopping and starting is longer. Therefore, for lunch breaks it might not be noticeable. After vacation, it is apt to be strongest.

The learning curve has been around a long time.[7] It has not proved directly useful, but it has provided conceptual insight into the problems associated with people doing small lots of different kinds of work. More time is spent learning and getting up-to-speed than producing effectively. Learning time can diminish productive capacity in the job shop, but good process management can alleviate the damage.

The learning effect for workers in flow shops is pronounced because of the high degree of repetition. There is a reverse effect with productive output declining because of monotony. The learning curve effect on production line start-up times, and the reverse effect, should be considered when designs for manufacturing (DFM) are prepared. New flexible technology provides opportunity for better productivity and quality for small lots.

Costing new products should take learning into account by pricing jobs at the average productivity rate for a reasonably large number *n*. Time observations for work standards do not begin until after workers' times stabilize at some plateau level.

Service Learning

In labor-intensive situations, learning curve concepts can be put to good use. How many repetitions does it take to learn a job? How do people differ in catching on? Does a master teacher help? How do the trainees of different trainers perform? What aspects of training make a difference in how quickly and how well experience is converted into improved performance? Who says there is only one plateau?

Learning improves productivity and excellence in service industries.
Source: Ritz-Carlton Hotel Company

There are better learning theories that allow for multiple plateaus or a succession of plateaus,[8] where learning starts again after a stable interval of continued practice on a lower plateau.

The learning curve's relationship to providing quality services efficiently should be highlighted. Productivity of service jobs has been a recognized problem which is often associated with the lower salaries that characterize services. In the same sense, there is an opportunity to systematize the learning function for the attainment of excellence in service quality.

The idea would be to raise the quality standards after a job has successfully plateaued at a lesser quality standard. Thus, learning curve plateaus could relate productivity with levels of quality improvement. They also apply to the time-based management system of stair-step reductions in process times which was effectively employed by Japanese automotive manufacturers to lower costs. These kinds of efforts must be fully supported by employees or else they are doomed to failure. On the other hand, there would be great support for learning to improve excellence if it was accompanied by financial bonuses and other kinds of recognition. These kinds of ideas about learning and work standards are particularly applicable to services where they have been consistently underutilized in spite of the voice of the customer.

10.11

CAPACITY AND THE SYSTEMS APPROACH

Capacity decisions are among the most crucial P/OM responsibilities. They set in motion scenarios for start-up and growth which are difficult to alter, unless they have been planned to be dynamic. Following the curve means forecasting diligently and always playing catch-up. This is characteristic of the way some companies proceed to invest with the introduction of new products. The policy to lead the curve requires that OM works with marketing to create specific demand levels. The capacity to satisfy that demand has been designed and built ahead of time. This is characteristic of companies that prefer to accept risk in order to be the market share leader.

Unused capacity is to be avoided, but some of it may be inevitable. It is not possible to forecast perfectly, or feasible to plan perfectly. Teamwork is no guarantee of being able to create exactly the demand for which capacity has been prepared. The idea is to perform as well as possible while continuously improving.

Capacity can be installed all at once, or it can be gradually increased. The incremental approach entails low risk, but it can foul up meeting market demands, produce higher unit costs, and worst of all, lead to poor and inconsistent quality of product. The capacity decision must have financial support, market rationalization, and OM knowledge of how to convert the strategic plan into production-transformation capacity to be installed.

The resulting too much or too little of this or that leads to downsizing, which is demoralizing and costly. Alternatively, the last minute realization of the need for greater capacity is fraught with difficulties and buys the additional volume at a higher price than would have been incurred from original planning stages. Strategic plans shared by finance, marketing and OM are the foundation for better capacity decisions, which with team learning can get better all the time.

S U M M A R Y

This chapter begins with the definition of capacity. The determination of capacity is one of the most important OM responsibilities. Capacity to meet peak and off-peak loads is discussed. The storage of capacity is an important topic. Services are especially affected by the difficulties of storing service capacity. It is shown how backordering augments service and manufacturing capacity.

Capacity requirements planning (CRP) is introduced. Adding and subtracting capacity through upscaling and downsizing are treated. Bottlenecks are the constraints on supply capacity. Understanding how to manage bottlenecks is an OM necessity. In the same way, delay creates oscillations of supply, which deteriorate productive performance in the supply chain. This leads to discussion of contingency planning for capacity crises and a supply chain game or simulation called the Better Beer Company.

In turn, cost drivers and forecasting for capacity planning are introduced, leading to the use of the breakeven model for capacity planning. It is shown to be a powerful planning tool which epitomizes the systems viewpoint by requiring inter-functional cooperation. Breakeven capacity is related to work configurations, margin contribution and/or profitability.

Another significant interfunctional capacity planning model which is introduced is linear programming, known as LP. Modular production (MP) and group technology (GT) are useful concepts for adding capacity creatively through design concepts that could be called design for capacity (DFC).

The chapter concludes with a discussion of how learning increases capacity. The learning curve is introduced and applied to service learning. The relationship of capacity planning to systems thinking is reiterated. The chapter supplement combines forecasting, decision making and breakeven analysis.

▶ Key Terms

actual capacity *396*	capacity *398*	experience curve *432*
Fixed Costs = FC *416*	learning curve *432*	linear programming *425*
principle of modularity *428*	slack *426*	slope *422*
Total Costs = TC = FC + TVC *417*	Total Revenue = TR = (p)V *417*	Total Variable Costs = TVC = (vc)V *416*
variable costs per unit = vc *416*		

▶ Review Questions

1. Define capacity.

2. Explain what is meant by the statement that maximum containment is a type of capacity which can have special meaning for services.

3. What motivated a company to build excessive petroleum storage facilities early in 1973? If events had worked out differently, would the storage facilities have been excessive?

4. Explain why maximum output is only one measure of capacity.

5. What is the relationship of maximum output to peak demand?

6. Give an example of a situation where what some might call excessive inventories turned out to be a blessing in disguise.

7. Give an example of constraints in capacity proving to be a blessing in disguise.

8. Should an airline have enough operators and trunk lines to prevent any calling customers from having to wait when they are responding to a one-day offer of half-fare travel rates?

9. List 6 variable costs.

10. List 6 fixed costs.

11. To what extent are accounting data likely to be available in manufacturing organizations with respect to fixed and variable costs that are used in breakeven analysis?

12. To what extent are accounting data likely to be available in service organizations with respect to fixed and variable costs that are used in breakeven analysis?

13. To what extent are accounting data likely to be available in government organizations with respect to fixed and variable costs that are used in breakeven analysis?

14. What do FC and $TVC = vc(V)$ and $TC = FC + TVC$ have to do with the employment of breakeven analysis?

15. What is overhead cost or burden?

16. How should overhead be treated in breakeven analysis (BEA)?

17. Relate activity-based cost (ABC) systems to BEA.

18. Regarding airline capacity, air shuttles provide continuous, as-filled capacity, instead of scheduled flights. How does this concept increase capacity by innovative use of equipment?

19. What is capacity requirements planning (CRP)?

20. How do bottlenecks affect capacity and what can be done about them?

21. What is modular production (MP) and how is it a form of design for capacity (DFC)?

22. What is group technology (GT) and how is it a form of design for capacity (DFC)?

23. What method is appropriate for determining the optimal product mix? What data are needed to accomplish this goal?

24. The capacity planning problem requires good estimates of the variable costs for different work configurations and different volumes of supply. Explain.

25. What form of capacity control is downsizing? What are its goals and disadvantages?

26. Delay in the supply chain wreaks havoc. Why is this true?

27. Discuss the following observation: the breakeven chart can represent only one product at a time, whereas most companies produce a product mix consisting of a number of different items or services which must share resources, including capital and management time.

28. Describe situations in which nonlinear analysis might be required for the breakeven chart.

▶ Problem Section

1. The peak load for Swimsuits Inc. (SI), a manufacturer of bathing suits, occurs in March and April when the cutters and sewers occupy every square foot in the plant. This determines the maximum output capacity of the company which is 250 suits per hour. SI works a 10-hour day, six days a week. Adding shift time is considered impossible. Therefore, it has been suggested that floor space be expanded by renting an adjoining loft for $1,200 per week. This would increase the maximum output capacity by 12 percent. The sales manager states that SI could sell 18,000 bathing suits per week during the 10-week season, at a profit of $1.00 per suit based on materials and labor. During the summer, SI's plant is less than 50 percent utilized.

Explain why the loft should or should not be rented. Are there any other capacity-based observations to be considered?

2. The new manager of the electric utility notes that recent summers have been warmer than average. Consequently, for next year she revises upward the probability that actual demand will exceed average demand; estimated to occur 8 percent of the year or roughly 29 summer days. She estimates that each power incident can cost far less than her predecessor had spent. Using different management techniques to deal with brownouts, she hopes to reduce the cost per incident to $10,000. Because the previous manager had not invested in additional capacity to cover maximum peak-load requirements, that option is still available. She faces the prospect that it will cost an extra $3,000,000 to increase capacity from average to peak. This amount is spread out over 10 years, or $300,000 per year.

   Should peak-load capacity be installed?

3. The classical formula for actual measured capacity is:

$$C = T \times E \times U$$

   where:

   C = actual measured capacity (standard hours)

   T = time available (actual hours)

   E = efficiency

   U = utilization

   If T = 240 standard hours, how can the actual capacity be more than that, say 280 standard hours?

4. In Alba, Italy, many famous restaurants rely on having enough truffles for the culinary delights which customers travel miles to enjoy. While customers pay a hefty price for the truffle dishes, the export market for truffles has much larger margins. Truffles are found by using pigs who can smell the strong aroma, then digging under the ground to get them. Weather conditions have caused the truffle hunters to stay at home, so a truffle shortage is arising at the distributors' level. There are several regional distributors to whom the truffle hunters sell their findings. What are the constraints on capacity? *Hint:* It would help to draw the supply chain which characterizes this product. Make certain to indicate that the supply chain starts with a scarce resource that has to be found in the woods. It cannot be farmed. Also, at the end of the supply chain there are two competing sources of demand, one domestic and the other international. Note how currency fluctuations might affect the shipments of truffles at home and abroad.

5. Company capacity is one million units per year for a new food product. The sales force can sell five million units over a five-year period, but only 100,000 in the first year. The breakeven volume is V_{BEP} = 50,000 units with p = 3, vc = 2, and FC = 50,000. What is the situation? What recommendations do you support?

6. The Global Company has engaged a management consultant to analyze and improve its operations. His major recommendation is to "conveyorize" the production floor. This would represent a sizable investment for Global. In order to determine whether or not the idea is feasible, a breakeven analysis will be utilized.

   The situation is as follows: the cost of the conveyor will be $200,000 to be depreciated on a straight-line basis over ten years. The conveyor will reduce operating costs by $0.25 per unit. Each unit sells for $2.00. The sales manager estimates that, on the basis of previous years, Global can expect a sales volume of 100,000 units. This represents 100 percent utilization of capacity. Present yearly contribution to fixed costs is $100,000. Present vc rate is $0.50 per unit.

Should Global install this conveyor?

7. Zeta Corporation is considering the advantages of automating a part of its production line. The company's financial statement follows.

Zeta Corporation

Total sales	$40,000,000	
Direct labor	$12,000,000	
Indirect labor	2,000,000	
Direct materials	8,000,000	
Depreciation	1,000,000	
Taxes	500,000	
Insurance	400,000	
Sales costs	1,500,000	
Total expenses		$25,400,000
Net profit		$14,600,000

The report is based on the production and sale of 100,000 units. The operations manager believes that an additional investment of $5 million can reduce the variable costs by 30 percent. The same output quantity and qualities would be maintained. Using five-year straight-line depreciation of $1,000,000 per year, construct a breakeven chart. Based on the breakeven analysis, should Zeta introduce the automation?

8. Referring to the linear programming problem discussed in the text of this chapter, show the calculations required to determine D1 and D2 slack for Plans 2 through 7 which appear in Table 10.3.

9. A product is presently moved on pallets by a forklift truck. Should a conveyor belt be installed? The capacity would remain the same, but the cost structure differs. The company uses a one-year time period for comparison.

	Option 1 Forklift Truck	Option 2 Install Conveyor
V_{max}	18,000 units per year	18,000 units per year
FC	$10,000 per year	$12,000 per year
vc	$0.50 per unit	$0.45 per unit
p	$2.00 per unit	$2.00 per unit

10. Machine 1 is being challenged by Machine 2 which has greater capacity, lower variable costs and higher fixed costs because it requires a much larger investment. Should the organization buy Machine 1 or Machine 2? The sales department states that

demand is larger than supply and that, if they had them, they could deliver 6,000 units per quarter. The organization uses a 3-month time period ($T = 3$) for its breakeven analysis. It could also be $T = 1$ quarter.

	Option 1 Machine 1	Option 2 Machine 2
V_{max}	5,000 units per quarter	10,000 units per quarter
FC	$2500.00 per quarter	$6500.00 per quarter
vc	$0.50 per unit	$0.10 per unit
p	$2.00 per unit	$2.00 per unit

11. The learning equation

$$T(n) = kn^{(1-\lambda)}$$

is said to apply to a particular job with the following values:

k = 3 minutes to make the first unit, $n = 1$

n = 25 units which is the number of units to be made

λ = 0.5 = the learning coefficient

Plot $T(n)$ which equals the cumulative time to make n consecutive units, from 1 to 25.

12. Using the information in Problem 11 above, plot

$$\bar{T}(n) = \frac{T(n)}{n} = kn^{(1-\lambda)}$$

from $n = 1$ to $n = 25$. k, and λ are the same as in Problem 11.

▶   **Readings and References**

N. Baloff, "Estimating the Parameters of the Startup Model—An Empirical Approach," *The Journal of Industrial Engineering*, 18, 4 (April 1967), pp. 248–253.

J. H. Blackstone, *Capacity Management*, APICS South-western Series in Production and Operations Management, 1989.

Boston Consulting Group, Inc., *Perspectives on Experience*, a collection of articles published by BCG, 1972.

Martin Christopher, *Logistics and Supply Chain Management*, Burr Ridge, IL, Irwin, 1994.

J. W. Forrester, *Industrial Dynamics,* Wiley, 1961.

Christopher Gopal and Gerry Cahill, *Logistics in Manufacturing*, Burr Ridge, IL, Irwin, 1992.

George Leonard, *Mastery*, NY, Plume, Penguin Group, 1992.

C. Northcote Parkinson, *Parkinson's Law*, 1962.

W. Rautenstrauch, and R. Villers, *The Economics of Industrial Management*, NY, Funk & Wagnalls Co., 1949.

Peter M. Senge, *The Fifth Discipline*, Doubleday/Currency, 1990.

The Breakeven Decision Model

In this supplement, breakeven analysis and decision making are interrelated. When demand is uncertain it must be forecast for the purpose of using breakeven analysis to do capacity planning. Capacity planning, given uncertain demand, will require a forecast. The issue being treated in this supplement is how to put BEA and capacity planning together.

It is easier to describe this combined model using an example which could be from either a manufacturing or a service industry. The choice was made to use an unknown industry called ANY Industry (ANYI).

The company is comparing Present Technology with Advanced Technology. The problem involves capacity planning, cost analysis, and demand analysis. The latter is based on prices charged and qualities perceived by the company's customers. The problem is that there is no clear cut answer about what to do. One choice has a better BEP; the other choice has a better rate of return slope and demand is uncertain. This is a challenging P/OM problem.

ANYI's P/OM department has drawn up breakeven charts for the alternatives. (Requested in the problem section for this supplement.) They have calculated the breakeven point and the rate of return of profit for both the Present Technology and the Advanced Technology.

Table 10.5 presents the data to calculate the BEP and the profit rate. The time period is assumed to be one year.

T A B L E   1 0 . 5

ANYI'S BEA DATA

	Present Technology	**Advanced Technology**
Fixed costs*	$1,000,000.00	$1,200,000.00
Per unit variable costs ($)	50.00	25.00
Per unit revenue ($)	200.00	200.00
Annual capacity	10,000 units	10,000 units

*Fixed costs are depreciated on an annual basis.

The breakeven point is 6,667 units or 66.7 percent of total capacity for the Present Technology. The breakeven point for the Advanced Technology is 6,857 units or 68.6 percent of full-capacity utilization.

These may seem to be relatively high breakeven points, but they are not unusual for many industries. While it would be nice to have breakeven percentages ranging from 30 to 45 percent, capital intensive industries, such as farm equipment and auto production have high BEPs. In this case, the Advanced Technology option is worse than the Present Technology option. At such high BEPs, there is an immediate and strong preference for the lower BEP. ANY Industry prefers a lower BEP, but this preference gets stronger when the choice is between already high values.

However, before any decision can be made, it is necessary to examine the relative profitabilities of the two plans at greater than breakeven percentages of capacity. The easiest comparison is at 100 percent of capacity.

Profit at 100 percent of capacity:

Present Technology	Advanced Technology
+ $500,000	+ $550,000

There is a 10 percent advantage for the Advanced Technology.

It also will be useful to examine the relative losses of the two plans at less than breakeven percentages of capacity. The easiest comparison is at 50 percent of capacity, well below breakeven.

Profit at 50 percent of capacity:

Present Technology	Advanced Technology
− $250,000	−$325,000

There is a 30 percent advantage for the Present Technology.

In summary, the Present Technology has the BEP advantage. It also has an advantage when demand falls below the BEP. The Advanced Technology has the advantage above the BEP.

Which should be chosen, the lower BEP with its lower rate of profit and lower rate of loss, or the higher BEP with its higher rate of return and higher rate of loss? This question cannot be answered by ANYI without a believable forecast and use of the decision model.

ANYI's forecast of annual customer demand is given below in Table 10.6. How it was derived and its believability should be stated clearly in an actual situation.

TABLE 10.6

FORECAST FOR ANYI

Demand Level D	Probability of Demand Level D
40% of capacity	0.05
50% of capacity	0.10
60% of capacity	0.15
70% of capacity	0.20
80% of capacity	0.25
90% of capacity	0.20
100% of capacity	0.05
	1.00

A specific amount of profit is associated with each level of demand. The total amount of profit (TPR) can be read from the breakeven charts or calculated directly using the equations below. $TPR = (p - vc)V - FC$ which becomes:

For Present Technology: $TPR = 150V - 1,000,000$

For Advanced Technology: $TPR = 175V - 1,200,000$

For Present Technology, at 0 percent demand, there is a loss of $1 million. At 100 percent capacity, 10,000 units, there is a profit of $500,000.

Relate the probability distribution to the profit at each Demand Level D. Multiply profit at D by the probability of D. Add up these products for all demand levels to obtain the average (or expected) profit for the given strategy. As derived in Table 10.7, ANYI's expected profit for Present Technology is:

$$\text{Expected Profit} = \sum_{V=0}^{V=100\%}(P_V)(TPR_V) = \$95,000$$

TABLE 10.7

ANYI'S EXPECTED PROFIT FOR PRESENT TECHNOLOGY

(1) Demand (D)	(2) TPR_V Profit @ D	(3) P_V Probability of D	Columns $(P_V)(TPR_V)$ 2 × 3
0%	− 1,000	0	0
10	− 850	0	0
20	− 700	0	0
30	− 550	0	0
40	− 400	0.05	− 20
50	− 250	0.10	− 25
60	− 100	0.15	− 15
70	+ 50	0.20	+ 10
80	+ 200	0.25	+ 50
90	+ 350	0.20	+ 70
100	+ 500	0.05	+ 25
		1.00	+ 95

Expected Profit = $95,000 (Profits in the Table in 000s)

$$\text{Expected Profit} = \sum_{V=0}^{V=100\%}(P_V)(TPR_V) = \$95,000$$

The BEP for Present Technology is 66.7 percent. Multiplying columns (1) and (3) of Table 10.7 determines the expected demand, which is 73 percent:

$$0.05(40\%) + 0.10(50\%) + 0.15(60\%) + 0.20(70\%) + 0.25(80\%)$$
$$+ 0.20(90\%) + 0.05(100\%) = 73 \text{ percent}$$

The same kind of calculations are now done for Advanced Technology. The expected demand is unchanged because the same demand distribution applies. Determine the profit at 73 percent capacity utilization. It is $77,500. That is the benefit of knowing the expected demand. Expected profit is calculated the long way in Table 10.8 where profits are in 000s.

TABLE 10.8

ANYI'S EXPECTED PROFIT FOR ADVANCED TECHNOLOGY

Demand (D)	Profit @ D TPR_V	Probability of D P_V	2 × 3 $(P_V)(TPR_V)$
0%	− 1,200	0	0
10	− 1,025	0	0
20	− 850	0	0
30	− 675	0	0
40	− 500	0.05	− 25
50	− 325	0.10	− 32.5
60	− 150	0.15	− 22.5
70	+ 25	0.20	+ 5
80	+ 200	0.25	+ 50
90	+ 375	0.20	+ 75
100	+ 550	0.05	+ 27.5
		1.00	+ 77.5

Expected Profit = $77,500

$$\text{Expected Profit} = \sum_{V=0}^{V=100\%}(P_V)(TPR_V) = \$77,500$$

Following this analysis, it is recommended that ANY Industry retain its Present Technology. The Advanced Technology will lower expected profits.

Management should note a few cautions, however. First, the competition may be shifting in the direction of the new technology which could be evolving. It might be advisable for the company to invest minimally at this time in order to learn the new technology and be better able to assess it the next time around. At a later point, both the fixed costs and variable operating costs of the Advanced Technology may decrease. ANYI also may be able to increase capacity, presently limited to 10,000 units per year. Quality issues and long-term TQM perspectives also should be considered.

If the Advanced Technology in any way affects the probability distribution, that must be taken into account. Perhaps the new equipment will permit faster delivery and/or more reliable tolerances or delivery schedules. The decision regarding technological choices might be sensitive to shifts in the distribution of the probability of demand.

With the addition of forecasting, the breakeven model has become a unifying systems-wide decision model. All of the functions should be playing a part in preparing the forecast and estimating the costs and revenues.

The decision model with forecasting overrides the problem of what to do when one strategy has a better BEP, but a poorer marginal rate of return, than another strategy. The forecast changes the approach from what is traditional breakeven analysis, where the BEP number is derived and all decision makers have in mind an estimate of the likelihood that the company will operate above or below that point. Without this estimate, the breakeven point is meaningless. The decision model has merged the BEP and the marginal rate of return into a single problem.

Problems for Supplement 10

1. Draw the breakeven chart for ANYI's Present Technology. Indicate the breakeven point and the rate of return of profit.

2. Draw the breakeven chart for ANYI's Advanced Technology alternative.

3. Table 10.9 presents a shift in the forecast.

TABLE 10.9

ANYI'S EXPECTED PROFIT FOR ADVANCED TECHNOLOGY WITH REVISED DEMAND FORECAST (PROFIT IN 000S)

Demand (D)	Profit @ D	Probability of D	2 × 3	1 × 3
0%	− 1,200	0		
10	− 1,025	0		
20	− 850	0		
30	− 675	0		
40	− 500	0		
50	− 325	0.10		
60	− 150	0.15		
70	+ 25	0.20		
80	+ 200	0.25		
90	+ 375	0.20		
100	+ 550	0.10		
		1.00		

Determine the effect of the revised forecast.

What is the effect of the change in the forecast and how will it affect the recommendation?

Notes...

1. This 1823 quotation can be found in *Characteristics*, page 89. William Hazlitt, *Characteristics*, publisher unknown, London, England, 1823.

2. C. Northcote Parkinson, *Parkinson's Law*, 1962.

3. The Sprague Electric Company played an important role in this research which was published as follows: J. W. Forrester, *Industrial Dynamics*, Wiley, 1961.

4. Peter M. Senge, *The Fifth Discipline*, Doubleday/Currency, 1990, pp. 27–54.

5. W. Rautenstrauch and R. Villers, *The Economics of Industrial Management*, NY, Funk & Wagnalls Co., 1949.

6. Boston Consulting Group, Inc., *Perspectives on Experience*, a collection of articles published by BCG, 1972. An article by Bruce D. Hendersen describes the experience curve.

7. There was real interest during the 1960s and a belief that productivity could be improved substantially by observing the learning curve phenomenon. See for example: Nicholas Baloff, "Estimating the Parameters of the Startup Model—An Empirical Approach," *The Journal of Industrial Engineering*, 18, 4 (April 1967), pp. 248–253.

8. The text reference is Mastery of Self, a book about multiple plateaus for learning tennis and other sports. George Leonard, *Mastery*, NY, Plume, Penguin Group, 1992.

Chapter 11

Facilities Planning

After reading this chapter you should be able to...

1. Explain the four distinct parts of facilities planning.
2. Discuss who is responsible for doing facilities planning.
3. Describe the nature of facilities planning models.
4. Explain why facilities planning requires the systems approach.
5. Describe the application of the transportation model.
6. Apply the transportation model to solve one aspect of the location decision problem.
7. Determine the relative advantages of rent, buy or build.
8. Describe the use of scoring models for facility selection.
9. Deal with the effect of multiple decision makers on the location, site, and building decisions.
10. Describe the factors that should be taken into account when equipment is being selected.
11. Explain why facilities planning is said to be interactive meaning that consideration goes back and forth between facility selection, equipment selection and facility layout.
12. Describe what facility layout entails.
13. Explain how job design and workplace layout interact.
14. Evaluate the use of quantitative layout rules (algorithms).
15. Discuss the use of heuristics to improve layouts of plants and offices.

Chapter Outline

11.0   FACILITIES PLANNING
          Who Does Facilities Planning?

11.1   MODELS FOR FACILITY DECISIONS

11.2   LOCATION DECISIONS—QUALITATIVE FACTORS
          Location to Enhance Service Contact
          Just-In-Time Orientation
          Location Factors

11.3 LOCATION DECISIONS USING THE TRANSPORTATION MODEL
        Suppliers to Factory—Shipping Costs
        Factory to Customers—Shipping Costs
        Obtaining Total Transport Costs
        Obtaining Minimum Total Transport Costs
        Testing Unit Changes
        Northwest Corner (NWC) Method
        Allocation Rule: $M + N - 1$
        Plant Cost Differentials
        Maximizing Profit Including Market Differentials
        Flexibility of the Transportation Model
        Limitations of the Transportation Model

11.4 STRUCTURE AND SITE SELECTION
        Work Configurations Affect Structure Selection
        Facility Factors
        Rent, Buy, or Build—Cost Determinants

11.5 FACILITY SELECTION USING SCORING MODELS
        Intangible Factor Costs
        Developing the Scoring Model
        Weighting Scores
        Multiple Decision Makers
        Solving the Scoring Model

11.6 EQUIPMENT SELECTION
        Design for Manufacturing—Variance and Volume
        Equipment Selection and Layout Interactions
        Life Cycle Stages

11.7 LAYOUT OF THE WORKPLACE AND JOB DESIGN
        Opportunity Costs for Layout Improvement
        Sensitivity Analysis

11.8 LAYOUT TYPES

11.9 LAYOUT MODELS
        Layout Criteria
        Floor Plan Models
        Layout Load Models
        Flow Costs
        Heuristics to Improve Layout

Summary
Key Terms
Review Questions
Problem Section
Readings and References

The Systems Viewpoint

Facilities planning requires expertise from so many areas that it is folly to think that one person alone can do it successfully. International facilities planning takes global team efforts with country specialists and broad-visioned generalists working as a team. When generalists and specialists have to work together, pooling their viewpoints and knowledge, it is a certain sign that a systems approach is required. When locations and specific structures must be considered together, the joint decisions are best approached using the systems approach.

When the real problem is too big to grasp because of the many possible options, the systems approach is indicated. Dividing the problem into regional subproblems to make it workable can jeopardize the quality of the solution yielding a suboptimal (less than best) solution. Location, site and building decisions are major ones. Strategies that yield suboptimal results can damage competitive capabilities. Layout and job design are partners in the flow shop. In the job shop, production scheduling, shop-floor layout and job design are interdependent and best approached as a system. In all such aspects, facilities planning is a multidimensional systems problem.

11.0

FACILITIES PLANNING

Facilities are the plant and the office within which OM does its work. In addition to the buildings and the spaces that are built, bought or rented, facilities also include equipment used in the plant and the office. There are four main components of **facilities planning** and they strongly interact with each other. These four are:

Facilities planning

Composed of the location, structure and specific site selection, equipment choice, and layout of the plant and office within which OM does its work.

1. *Location of facilities*—Where, in the geographic sense, should the various operations be located? Location can apply to only one facility but more generally includes a number of locations for doing different things. Thus, "where is it best to…" fabricate and assemble each product, locate the service center, situate the sales offices, and position the administration?

   Many location factors have to be dealt with in a qualitative fashion. Common sense prevails. Qualitative factors such as being near supplies of raw materials and/or sources of skilled labor vie with being close to customers. Seldom can all desires be satisfied. The options may be to choose one over the other or to compromise with an in-between location. There are similar problems for distributors who want to be located close by all kinds of transportation facilities (road, rail, airports, and marine docks).

Other location factors are better dealt with in a quantitative fashion because common sense does not work. The reason is that some real problems are too big to grasp all of the possible options. Because of this, optimal location assignments are often counterintuitive. For example, if a transportation analysis is made manageable by dividing the problem into regional subproblems, the solution may be seriously suboptimal (far from being the best that is possible). The systems approach which endeavors to include all relevant factors is needed in this case.

At the same time, there are strategic issues concerning centralization or decentralization of facilities. Should there be a factory in each country? Would a central warehouse be more desirable than regional warehouses or can they both be used to an advantage? Such decisions might be best reached with a combination of qualitative and quantitative considerations.

2. *Structure and specific site selection*—In what kind of facility should the process be located? How should the building or space by chosen? Is it best for the company to build the facility, buy or rent it? Choice of a specific building is often decided after the location is chosen. However, there are circumstances where the location, site and structure should be considered together. This makes structure and site decisions complex problems best resolved by using the systems approach.

3. *Equipment choice*—What kind of process technology is to be used? This decision often dominates structure and site options. In turn, there may be environmental factors that limit the choice of locations. Equipment choice can include transport systems and available routes. Interesting systems problems arise which combine all three aspects of facilities planning.

Planning the layout of a customer-service center facility must address seating arrangements, location of phone and telecommunications lines, space for computer equipment, and other information resources.
Source: Maytag Corporation

P E O P L E

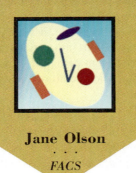

Jane Olson
. . .
FACS

When FACS (Financial and Credit Services), a division of Federated Department Stores, Inc., outgrew its Mason, Ohio facility, Jane Olson coordinated the company's effort to relocate operations to a new custom-built facility in the same office park. Olson, group manager, administrative services for FACS (which provides proprietary credit services to Federated's retail stores), explains the decision to stay in the same location: "We wanted to provide as little disruption for our associates as possible....If we moved somewhere different, we might have lost some of our valued associates."

Olson emphasizes that the move was a team effort, with her role as coordinator. "My main function was to represent the FACS point of view, to shepherd the planning of the space for the FACS departments, and to coordinate the physical move from the old building to the new building." The team consisted of a project manager from Federated Department Stores; representatives from Duke Construction, the company that built the new facility; representatives from FSG (Federated Systems Group), a division of Federated concerned with telecommunications and data for all Federated operations; and the Federated architect who planned the space.

"About ten months before the move we started having weekly phone conferences with the FSG people about the whole data and telecommunications side of the move," Olson said. That was important because: "We're a business that can't ever afford to be closed....we're always open when our stores are open," Olson said. During construction the team met weekly for progress reports and to raise issues that had come up since the previous meeting.

In planning the space, a major systems consideration was the adjacency of departments to each other. "We wanted to have certain departments in close proximity to each other to leverage the use of workstations...because some people can work in more than one department," Olson said. Also if one department is using fewer workstations while another is "flexing up," the new hires can move into workstations not in use. During the holiday season, FACS hires 400 to 600 extra people, "so seating is a challenge," said Olson.

Source: A conversation with Jane Olson, FACS.

4. *Layout of the facility*—Where should machines and people be placed in the plant or office? There is interaction of equipment choice with layout, structure, site and location. The size of the facility is determined by both present needs and future projections to allow for growth. Thus, facilities planning is systems planning. Once the location, site and structure are determined, layout details can be decided.

Thus, what path do conveyors, AGVs, and other transport systems take? AGVs are automatic guidance vehicles which are computer controlled along prescribed transport paths. Other transport systems include manual systems (like wheelbarrows), forklift trucks, among others. Layout is an interior design problem which strongly interacts with Components 2 and 3 above.

Usually, location is addressed first because its consequences may dominate other decisions. This is particularly true when transportation costs are a large part of the cost of goods sold. Location also may be crucial because of labor being available only in certain places. Closeness to market can affect demand. Closeness to suppliers can affect prices and the ability to deliver just-in-time. Tax considerations might play a part.

Once the location is known, a search for appropriate building structures can be done. However, when several locations are considered suitable and when buildings have been identified that qualify then, the various packages of locations and structures have to be considered together.

Who Does Facilities Planning?

The treatment of all four of the facilities planning issues is classic P/OM. Their importance is not in question, but the way that these four issues are done and the role of P/OM in the planning process has been changing. In the global world of international production systems, international markets and rapid technological transfers, facilities planning requires a team effort. It is no longer the sole responsibility of P/OM.

Team effort is required to deal with the issues properly. They deserve consideration by the entire strategic planning team of which P/OM is one important member. Besides the fact that all functional areas should be participants in location decisions, there are additional considerations.

Governmental regulations—from international to national to municipalities—have to be addressed. Legal issues need to be resolved. Lawyers from many countries often are consulted. There are negotiations with communities regarding such matters as financial incentives and tax breaks.

To rent, buy or build requires specialists who know domestic and international real-estate markets. Trade rules and tariffs are special for trading blocks and their partners. They are constantly changing for GATT, NAFTA, EU and other trade groups.

Currency conditions and banking restrictions must be interpreted by specialists. The list of special complications to consider is long and particular to each situation.

11.1

MODELS FOR FACILITY DECISIONS

P/OM's role and its contributions to facilities management are primarily of two kinds. First, P/OM's experience, with what succeeds and what fails for facilities planning, adds valuable insights to the planning team. Second, P/OM knows how to use facilities planning **models** developed by operations researchers, management scientists, and location analysts.

Models

Developed by operations researchers, management scientists, and location analysts for P/OM's use in facilities planning; types include: location decision, transportation, scoring, center of gravity, plant layout, among others.

Location decision models can use measures of costs and preferences to reach location decisions. Transportation models use costs and scoring models use preferences. Other models—such as breakeven analysis—can help to select a facility and to choose equipment. Plant, equipment and tooling decisions need close consultation between engineering, P/OM, and finance. These issues need to be guided and coordinated with top management.

Supermarket and department store locations are often chosen to be at the center of dense population zones. A center of gravity model can be used to find the population center, or the sales volume center—instead of the center of mass—which an engineer finds for structures or ships. Columbus, Ohio is a popular distribution center because a circle around it within a 500-mile radius encloses a large percentage of U.S. retail sales.

Plant layout models for flow shops require detailed engineering in cooperation with P/OM process requirements. The technical aspects of the process are vitally important in determining layout for a flow shop. Layout for the job shop may be one of the few areas where P/OM is expected to do the job on its own. In actuality, good layout decisions may be based more on models that assist visualization than on models that measure layout-flow parameters.

Nevertheless, there is a need to know that there are layout models for minimizing material transport distances. This is the measured path that work in process must travel in the plant. These models are interesting for special applications, as well as for their quantitative properties. Although their use tends to be limited, they are discussed in this chapter.

How models are used is what counts. Some applications can be criticized as overriding complex issues with simple metrics. The study of P/OM provides the basis for exercising good judgment about choosing models and methods. Layout goals now encourage communication and teamwork instead of imitation of machine performance which lends itself to precise measurement.

11.2

LOCATION DECISIONS— QUALITATIVE FACTORS

Best location

Related to the function of the facility and the characteristics of its products and services.

Best location is related to the function of the facility and the characteristics of its products and services. Also, location decisions always are made relative to what other alternatives exist—just as building decisions are made relative to what else is available. When the building-location decision is taken jointly, it is possible to decide to wait for another alternative to appear. This enlarges the scope of the problem and provides another example of the need for a systems-oriented perspective to resolve some location issues.

Location to Enhance Service Contact

Service industries locate close to their customers to achieve the kind of contact that characterizes good service. Bank tellers and ATMs (Automatic Teller Machines) are such contact points. No one wants to travel many miles to make deposits or withdrawals. The closer bank will get the business.

Branch banks, gas stations, fast-food outlets and public phones are scattered all around town, because distance traveled is one of the main choice criteria used by customers. Shopping malls are located so that many people find it convenient to drive to them. For retail business, best location is decided by the ability to generate high customer contact frequency.

An interesting exception to the advantage of proximity for contact are the services rendered to vacationers. Traveling many miles for sun and surf, or for snow and skiing, the service starts with the airline providing transportation. Then the hotel or resort offers food, shelter, sports and entertainment.

Facilities planning and management are crucial to success in the hotel and resort business. Location may be at the top of the list. Services, in general, are strongly affected by location, structure, site, equipment and layout because they all participate in making contact with customers successful.

Government institutions locate services close to the citizens who need them. Municipal governments provide police and fire protection to those who live within the municipality and pay the taxes. Effective federal service requires regional offices.

Just-In-Time Orientation

Extractors like to be close to their raw materials. On average, gold mining reduces a ton of ore to 4.5 grams of gold. The reduction process has to be near to the mines. Fabricators like to be close to their raw materials and customers. A balanced choice has to be made which will depend on the specific product. Assembly plants like their component suppliers to be close. They encourage suppliers to deliver on a just-in-time basis. The advantage to the supplier is that enough volume of business must be given by the producer to justify the location. Services usually profit by being close to the customer. Creative solutions are a goal.

The decision to be a nearby just-in-time supplier requires a mutual interchange of trust and loyalty. Just-in-time suppliers benefit from the proximity by being a single source (or one of very few sources) to their customer. Communication can be face-to-face and frequent. Customers benefit from reduction of inventory. Suppliers benefit from stability and continuity.

Location Factors

Six factors that can affect location decisions are:

1. *Process inputs*—Closeness to sources is often important. The transportation costs of bringing materials and components to the process from distant locations can harm profit margins.

Facilities planning and management are crucial to success in the hotel and resort business. Source: Club Med Sales, Inc.

2. *Process outputs*—Being close to customers can provide competitive advantages. Among these are lower transport costs for shipping finished goods, the ability to satisfy customers needs and respond rapidly to competitive pricing—which can provide market advantages.

3. *Process requirements*—There can be needs for special resources that are not available in all locations (i.e., water, energy, and labor skills).

4. *Personal preferences*—Location decision makers, including top management, have biases for being in certain locations which can override economic advantages of other alternatives.

5. *Tax, tariff, trade, and legal factors*—Trade agreements and country laws are increasingly important in reaching global location decisions.

6. *Site and plant availabilities*—The interaction between the location and the available facilities makes location and structure-site decisions interdependent.

Shipping costs are the primary concern for the first factor, however, there also can be price advantages for raw materials in certain regions and countries among others. For the second factor, being close to the customer has delivery and design advantages. For the third factor, alternatives exist such as processing bulk materials at the mining site to reduce their mass, and then further refining at a location close to the customer. Thus, process and transportation costs interact.

The fourth factor is individual, intangible and often encountered. For the fifth factor, taxes and tariffs can add costs to either the production or the marketing elements. Legal factors are difficult to estimate and can be substantial. The sixth factor can require the decision to defer making a decision.

11.3

LOCATION DECISIONS USING THE TRANSPORTATION MODEL

Transportation costs are a primary concern for a new start-up company or division. This also applies to an existing company which intends to relocate. Finally, it should be common practice to reevaluate the present location of an on-going business so that the impact of changing conditions and new opportunities is not overlooked. Relocating and moving costs can be added to the transportation model (TM).

When shipping costs are critical for the location decision, the TM can determine minimum cost or maximum profit solutions which specify optimal shipping patterns between many locations. Transportation costs include the combined costs of moving raw materials to the plant and of transporting finished goods from the plant to one or more warehouses.

It is easier to explain the TM with the following numerical example than with abstract math equations. A doll manufacturer has decided to build a factory in the center of the U.S.A. Two cities have been chosen as candidates. These are St. Louis, Missouri and Columbus, Ohio. Several sites in the two regions have been identified. Real-estate costs are about equal in both.

Suppliers to Factory—Shipping Costs

The average cost of shipping the components that the company uses to the Columbus, Ohio location is $6 per production unit. Shipping costs average only $3 per unit to St. Louis, Missouri. In TM terminology, shippers (suppliers in this case) are called sources or origins. Those receiving shipments (producers in this case) are called sinks or destinations.

Factory to Customers—Shipping Costs

The average cost of shipping from the Columbus, Ohio location to the distributor's warehouse is $2 per unit. The average cost of shipping from St. Louis, Missouri to the distributor's warehouse is $4 per unit. The same terminology applies. The shipper is the producer (source or origin) and the receivers are the distributors or customers (destinations or sinks). The origins and destinations are shown in Figure 11.1.

FIGURE 11.1

PLANT LOCATION DECISION WITH ONE ORIGIN (SUPPLIER) AND ONE DESTINATION (MARKET)

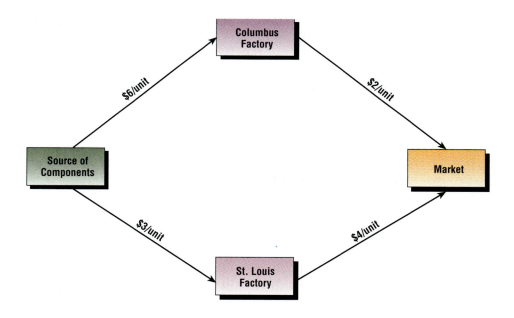

Total transportation costs to and from the Columbus, Ohio plant are $6 + $2 = $8 per unit; for St. Louis, Missouri they are $3 + $4 = $7. Other things being equal, the company should choose St. Louis, Missouri. However, the real world is not as simple as this.

The complication appears as soon as there are a number of origins competing for shipments to a number of destinations. Another complication is when the supply chain involves what is called transshipment where destinations become origins. The Columbus factory was a destination for components and an origin for shipments of finished goods to the market. Where the supply chain has many linkages the transportation model is structured to reflect the transshipment alternatives.

Two plants and two markets are shown with one source of components in Figure 11.2.

FIGURE   11.2

PLANT LOCATION DECISION WITH ONE ORIGIN (SUPPLIER), TWO FACTORIES, AND TWO DESTINATIONS

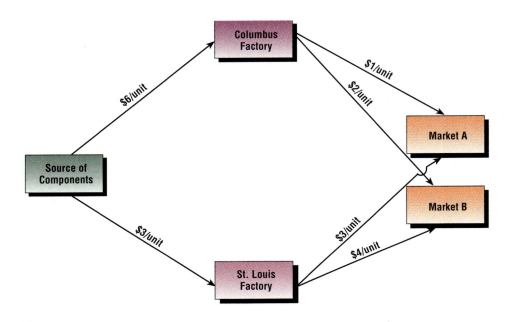

With these multiple facilities the TM comes closer to capturing the character of real distribution problems. The issue is: which plants should ship how much product to which distributors? If it turns out that the solution indicates that one plant should ship no units at all, that is equivalent to eliminating that plant and selecting the other one.

The data for the problem is only partially represented by the costs of transportation that are shown on the arrows in Figure 11.2 above. In addition, supply and demand figures are needed. The supply is what the plants can produce at maximum capacity. The demand is what the markets want to buy and consume. The minimum cost allocation can be determined by the transportation model which can be solved by using linear programming or various network methods. The method shown below is trial and error but there are other systematic ways, called algorithms, for solving the transportation problem. Table 11.1 presents the data.

TABLE 11.1

DATA FOR TRANSPORTATION ANALYSIS

Plant i	Supplier Transport Costs	To Market j Transport Costs A	B	Supply Quantities
P_1	$6/unit	$1/unit	$2/unit	90 units/day
P_2	$3/unit	$3/unit	$4/unit	90 units/day
Market Demand (units/day)		40	40	

The matrix entries are the costs of transporting finished goods from Factory i to Market j. Factories are distinguished by rows, $i = 1, 2$. Markets are represented by columns, $j = A, B$. In this case, $i = 1$ stands for the Columbus, Ohio plant location and $i = 2$ represents St. Louis, Missouri.

As for supply and demand, each market requires 40 units per day to be shipped from either P_1 or P_2, or a combination. Each plant can be designed to have a maximum productive capacity of 90 units per day—which in total is larger than the sum of the demands of both markets. This means that one plant *could* supply all demands. What is best—Columbus, St. Louis or both?

Total daily supply potential of 180 units per day exceeds total daily demand of 80 units per day by 100 units. To correct the imbalance between supply and demand, a slack, or dummy Market (M_D) is created to absorb 100 units per day. The dummy market does not exist. A plant that is assigned the job of supplying only the dummy market would be eliminated.

A possible pattern of shipments is shown in Table 11.2.

TABLE 11.2

TRIAL AND ERROR PATTERN OF SHIPMENTS

Plants i	Markets j M_A	M_B	M_D	Supply
P_1			90	90
P_2	40	40	10	90
Demand	40	40	100	180

If this was the minimum cost solution, then P_2 would be the best plant location. It supplies the real Markets M_A and M_B. It still has slack of 10 which is assigned to the dummy Market M_D so P_2 will work at 8/9 of capacity, supplying only real demand. Further, the Plant P_1 will be eliminated because it has been assigned the task of supplying *only* the dummy market.

Is this shipping pattern the best solution—resulting in minimum total transportation costs? It is now necessary to test this pattern to find out if there is any better arrangement.

The matrix of total per unit transportation costs is shown in Table 11.3. See Table 11.1 above for the basic data.

TABLE 11.3

TOTAL PER UNIT TRANSPORTATION COSTS

Plants i	Markets j M_A	M_B	M_D
P_1	6 + 1 = $7	6 + 2 = $8	0
P_2	3 + 3 = $6	3 + 4 = $7	0

Supplier per unit transportation costs have been added to the finished goods per unit transportation costs. Shipments to the dummy market cost $0 because it does not exist.

Obtaining Total Transport Costs

Combine the data in Tables 11.2 and 11.3 above multiplying costs by quantities to derive Table 11.4. This shows a total cost of $520 for the specific shipping pattern of Table 11.2 which uses only Plant 2.

TABLE 11.4

TOTAL COST TRANSPORT PATTERN USING ONLY PLANT 2

Plants i	Markets j M_A	M_B	M_D	Total Cost
P_1			90 × 0 = 0	0
P_2	40 × 6 = 240	40 × 7 = 280	10 × 0 = 0	$520
				$520

Next, it is appropriate to develop the method for finding the optimal (minimum cost) solution.

Obtaining Minimum Total Transport Costs

To obtain minimum total transport costs, there is a three step procedure:

1. Start with a feasible solution. It should balance supply and demand. It also should leave some cells unassigned as was done in Table 11.2. A method will be given shortly for getting started in a systematic way. It is called the Northwest Corner (NWC) method.

2. Select the lowest-cost, nonassigned cell. It is $7 at $P_1 M_A$. Determine the change in cost if one unit is moved to that cell. Make sure that the supply and demand totals are unchanged. This requires subtracting one unit and adding one unit at appropriate places in the matrix. If no saving is obtained (costs increase or stay the same) go to the next lower cost cell and repeat Step 2. Ultimately, every nonassigned cell will be tested.
3. If a decrease in total cost is obtained, then ship as many units as possible to the location where 1 unit produced a decrease.

Testing Unit Changes

What happens to the total cost of $520 if one unit is shipped from P_1 to M_A? Rearrange the total shipping schedule as shown in Table 11.5.

T A B L E 1 1 . 5

TOTAL TRANSPORTATION COST—ONE UNIT AT $P_1 M_A$

Plants i	Markets j M_A	M_B	M_D	Total Cost
P_1	$1 \times 7 = 7$		$89 \times 0 = 0$	7
P_2	$39 \times 6 = 234$	$40 \times 7 = 280$	$11 \times 0 = 0$	$514
				$521

The total cost of this shipping arrangement is $521 per day. Because 1 unit shipped from P_1 to M_A produces a greater total cost, more than 1 unit shipped in this way will be even worse.

Can the total cost be lowered by shipping 1 unit from P_1 to M_B? Review Table 11.6.

T A B L E 1 1 . 6

TOTAL TRANSPORTATION COST—ONE UNIT AT $P_1 M_B$

Plants i	Markets j M_A	M_B	M_D	Total Cost
P_1		$1 \times 8 = 8$	$89 \times 0 = 0$	8
P_2	$40 \times 6 = 240$	$39 \times 7 = 273$	$11 \times 0 = 0$	$513
				$521

The total cost result is again $521 per day. No way to decrease total cost has been found. No other possibilities exist for alternate shipping routes. It can be concluded that the first solution shown in Table 11.2 above is optimal, providing minimum total cost.

The location chosen is St. Louis, Missouri. The exact location of the plant has not been decided. Also, it has not been determined whether the plant will be rented or built. These issues are not addressed by this model. The plant selection was made on the basis of minimum transport costs.

Another factory is now added to demonstrate some other points. The supply and demand quantities, and the per unit costs are given in Table 11.7.

TABLE 11.7

SUPPLY, DEMAND, AND PER UNIT COSTS FOR THREE PLANTS AND THREE MARKETS

	Markets *j*			
Plants *i*	M_A	M_B	M_D	**Supply**
P_1	$ 7	$ 8	$ 0	50 units/day
P_2	$ 6	$ 7	$ 0	90 units/day
P_3	$ 8	$10	$ 0	90 units/day
Demand	40	40	150	230 units/day

Northwest Corner (NWC) Method

As before, the assumption is that transportation costs dominate the plant location decision. The initial assignment pattern shown in Table 11.8 is derived by the Northwest Corner method described below. Supply and demand are balanced with a dummy market, M_D.

TABLE 11.8

NORTHWEST CORNER (NWC)—INITIAL FEASIBLE ASSIGNMENT

	Markets *j*			
Plants *i*	M_A	M_B	M_D	**Supply**
P_1	40	10		50 units/day
P_2		30	60	90 units/day
P_3			90	90 units/day
Demand	40	40	150	230 units/day

Northwest Corner (NWC) method

A supply and demand matrix used to determine transportation costs which dominate the plant location decision; used for an initial feasible solution.

To use the **Northwest Corner (NWC) method**, begin in the upper (northern) left-hand (western) corner of the matrix. Allocate as many units as possible to P_1M_A—this is 40 units. More than 40 units exceeds demand. Assign as many units as possible without violating whichever constraint dominates—because it is smaller. The row constraint is 50 units; the column constraint dominates with 40 units.

P_1 still has 10 units of unassigned supply. Move east to allocate those 10 units at P_1M_B. All of P_1's supply is now allocated. M_B still requires 30 units to sum to the demand row value of 40. Moving south, these units are assigned from Plant 2. Movement from the NWC is always either to the right or down.

P_2 still has 60 unallocated units remaining. These are assigned to the dummy market, M_D. They will not be made or shipped. To complete the matrix, P_3's supply of 90 units must be allocated. Place them in the P_3M_D cell. Supply and demand quantities tally now.

The NWC method always can be made to satisfy the requirement for an initial feasible solution. The procedure could start at any corner, but the NWC method is conventionally accepted as the way to start. Whatever method is used to obtain an initial feasible solution, it must produce $(M + N - 1)$ assignments for a matrix with N rows and M columns. The condition that there will be $(M + N - 1)$ assignments applies to the initial feasible solution, all intermediate solutions, and the final and optimal solution.

If upon testing the shipment of one unit into an empty cell, the result is an improvement, then all the units that can be moved are moved into the new cell. No change is made if there is no improvement. The NWC provides just enough shipments so that every shipment has at least one other shipment in its row or column. That is necessary for being able to move units to empty cells without changing the row and column totals.

Allocation Rule: $M + N - 1$

The number of shipments used should never exceed:

$$(M + N - 1)$$

where:

$$M = \text{the number of destinations or markets}$$

and:

$$N = \text{the number of origins or plants}$$

In Table 11.8, there are $3 + 3 - 1 = 5$, which is the number of shipments derived by means of the NWC rules. A better solution can never be obtained with more than five shipments. Usually, the solution with more cells assigned would be worse.

There are $M \times N$ cells $(3 \times 3 = 9)$ with $M + N - 1 = 5$ assignments and therefore, $(M \times N) - M - N + 1 = 4$ boxes that are empty with no assignments.

The logic of the $M + N - 1$ assignments also is related to the mathematical nature of the system of equations that would be used to solve this problem. Thus, the problem is solvable using linear programming where each row and column are expressed as a separate inequation called a constraint. (An equation has an $=$ sign. An inequation has either $\leq$ or $\geq$ sign.)

There are supply and demand constraints. Because total supply and total demand sum to the same amount (using the dummy markets or, if needed, dummy plants) the number of independent constraints is one less than the sum of the number of rows and columns.

The exception to the rule which results in less than $(M + N - 1)$ assignments is known as *mathematical degeneracy*. The condition is purely mathematical and always can be resolved by adding a small extra amount to the appropriate row or column total. There should not be more than $(M + N - 1)$ assignments.

Plant Cost Differentials

The plants (being in different regions of the country or different countries in the world) might produce units at different costs. Some explanations for the differences might be:

1. different type processes and cultural factors
2. tax, tariff, and government regulation differentials
3. different labor and material costs

Also, for the global setting, transportation costs between plants and markets might play a role in location decisions.

The per unit production cost differentials reflect any factors that relate to the source. They are simple to factor into the matrix of transport costs. To every element in a row, the cost is added that characterizes the source of that row. Table 11.9 reflects the addition of the special production costs (shown in parentheses next to the plant) to the transportation costs of Table 11.7 above. Plant 2 has a per unit cost disadvantage of $16 as compared to Plant 3, and $8 as compared to Plant 1.

TABLE 11.9
PER UNIT TRANSPORTATION AND SPECIAL COSTS

Plants i	Markets j			Supply (units/day)
	M_A	M_B	M_D	
P_1 ($20)	$27	$28	$0	50
P_2 ($28)	$34	$35	$0	90
P_3 ($12)	$20	$22	$0	90
Demand	40	40	150	230 units/day

Maximizing Profit Including Market Differentials

Instead of using a cost matrix and minimizing total costs, the procedures can be reversed to permit maximization of a profit matrix. The power of including market-by-market revenues is shown below. For each cell, calculate the per unit revenue less the per unit shipping costs and other special per unit costs. This provides cell-by-cell data on per unit profit. Table 11.10 shows how this information is derived where per unit revenues are: $50 for Market A, $60 for Market B and the dummy market M_D has zero profit.

TABLE 11.10
PER UNIT PROFITS TAKING INTO ACCOUNT PER UNIT TRANSPORTATION AND SPECIAL COSTS

Plants i	Markets j			Supply (units/day)
	M_A	M_B	M_D	
P_1 ($20)	$50 − 27 = $23	$60 − 28 = $32	$0	50
P_2 ($28)	$50 − 34 = $16	$60 − 35 = $25	$0	90
P_3 ($12)	$50 − 20 = $30	$60 − 22 = $38	$0	90
Demand	40	40	150	230 units/day

The four step procedure for profit maximization is:

1. Start with a feasible solution which balances supply and demand and follows the $(M + N - 1)$ rule. The NWC method yields the same result as Table 11.8 above. Using the per unit profit data from Table 11.10 above, calculate the total profit for the first feasible (NWC) solution:
$$TPR_1 = 40 \times 23 + 10 \times 32 + 30 \times 25 = 920 + 320 + 750 = \$1,990$$

2. Select the highest-profit, nonassigned cell. It is $38 at P_3M_B. Determine the change in profit if one unit is moved to that cell. Make sure that the supply and demand total constraints are satisfied. This requires subtracting one unit and adding one unit at appropriate places in the matrix. If no profit increase is obtained go to the next highest per unit profit cell and repeat Step 2. Ultimately, every nonassigned cell will be tested. Table 11.11 shows the unit shift. It is followed by the next calculation of total profit, TPR_2.

TABLE 11.11
UNIT SHIFT FOR THE PROFIT-BASED TM

Plants i	Unit Shift				Total Profit TPR_2		
	M_A	M_B	M_D		M_A	M_B	M_D
P_1	40	10	0		40	10	0
P_2	0	30	60	$\rightarrow$	0	29	61
P_3	0	0	90		0	1	89

The profit change based on shifting one unit (indicated by Δ) is calculated:
$$TPR\Delta = 40 \times 23 + 10 \times 32 + 29 \times 25 + 1 \times 38$$
$$= 920 + 320 + 725 + 38 = \$2,003$$

Profit has increased by $13 for the single unit change, so move as many units as possible to P_3M_B, as explained in Step 3.

3. When an increase in total profit is obtained, ship as many units as possible to the location where a single unit produced a profit increase. The maximum amount that can be shifted for this example is 30 units, as shown in Table 11.12.

TABLE 11.12
UNIT CHANGE TO TOTAL CHANGE

Plants i	Unit Shift				Total Profit TPR_2		
	M_A	M_B	M_D		M_A	M_B	M_D
P_1	40	10	0		40	10	0
P_2	0	29	61	$\rightarrow$	0	0	90
P_3	0	1	89		0	30	60

$$TPR_2 = 40 \times 23 + 10 \times 32 + 30 \times 38 = 920 + 320 + 1,140 = \$2,380$$

The increase in profit over TPR_1 is 13×30 units $= \$390$.

4. Continue in this way, evaluating changes until profit cannot be increased any further. No further changes should be made because all evaluations of unit changes are negative and would decrease profits if used. The final solution is called TPR_4 because it is the fourth iteration result.

TABLE 11.13

PER UNIT PROFIT CHANGES FOR ALL NONASSIGNED CELLS OF THE FOURTH MATRIX

| Plants i | Fourth Matrix | | | | Evaluation of Unit Changes | | |
	M_A	M_B	M_D		M_A	M_B	M_D
P_1	0	0	50		−7	−6	•
P_2	0	0	90	→	−14	−13	•
P_3	40	40	10		•	•	•

Legend:
• = assigned shipments

$$TPR_4 = 40 \times 30 + 40 \times 38 = 1,200 + 1,520 = \$2,720.$$

The increase in profit of TPR_4 over TPR_1 is $730 which is a 37 percent improvement in profit.

Flexibility of the Transportation Model

The TM can include much more than just transportation costs. It can reflect price differences in markets (regional, domestic and international). Production efficiency differentials can be shown, as well as special location-based costs such as taxes, tariffs and wage scales. The costs of governmental rules concerning pollution control, employee health and safety, and waste disposal vary widely around the world.

Imbalance in supply (S) and demand (D) can be handled by creating dummies to absorb whichever is greater. If $S > D$ (as was the case for all of the previous examples), create market dummies that can absorb part of the oversupply. If $S < D$ (which is the case when peak demands are not being met), then create dummy plants to absorb all of the demand. Markets that are supplied by dummy plants are not being supplied at all. This raises the question of increasing capacity in optimal locations in order to satisfy demand most effectively.

The TM is a systems organizer for more than just the distribution cost data. Market and process differentials can be represented so their spokespeople also should be represented when the model is used. The TM as a profit maximizer is a good systems tool, whereas, as a cost minimizer it reflects P/OM's tactical interests.

Limitations of the Transportation Model

The limitations of the transportation method also should be understood by management. The cost equations and constraints on supply and demand are all linear. This means that the model does not allow discounts. It does not permit cost savings that learning curves and the experience horizon promise when larger volumes are handled.

QUALITY

Disney's America

To maintain quality image, Disney retreats on historical theme park concept

Public relations with a community or geographic area is important for companies to consider in choosing locations for new facilities. This is the lesson The Walt Disney Co. learned after selecting and purchasing a site in Prince William County, Virginia, to build Disney's America. The project met with fierce opposition from historians and other local observers who viewed it as a corruption of "hallowed historical ground." Manassas National Battlefield Park is about five miles east of the site Disney chose for the park.

Another perceived public relations faux pas in the eyes of the local opposition was Disney's buying the 3,000 acre tract near Haymarket, Virginia, in secret—a tactic the company normally uses in purchasing land for new facilities. (If the sellers of the land know Disney is interested in buying, they would be sure to raise their prices.)

Disney's America began as Disney Chairman Michael Eisner's personal vision—"a way to make American history come alive" for visitors. "I'm shocked because I thought we were doing good," Eisner told a group of reporters and editors from the *Washington Post*.

Disney crusaded for months to locate the new park on the land it had purchased. However, the project is now on a "back burner," according to Michael Johnson, manager of communications for Disney Development. In reevaluating the situation, Disney realized that the public in Virginia objected to the land being "Disneyfied" in the sense of Disney World in Orlando, Florida, and Disneyland in Anaheim, California. But Disney wishes to provide not just entertaining but hands-on, educational and intellectual vacation experiences for both children and adults. Before it again pursues the Disney's America project, the company will work to raise public awareness of the new concept.

Sources: William F. Powers, "Eisner Says Disney Won't Back Down," *Washington Post*, June 14, 1994; Richard Turner, "Disney to Move Planned Theme Park To 'Less Controversial' Site in Virginia," and "Disney Hopes Retreat Is Better Part of Public Relations," *Wall Street Journal*, September 29 and 30, 1994; and a conversation with Michael Johnson, Disney Development Co.

The model does not reflect other factors that influence location decisions such as the kind of community, weather, crime, and quality of schools, among others. Intangible factors are bypassed by this quantitative method. When intangibles are critical, then transportation cost differentials of the various alternatives can be combined with other costs and used with the intangible factors in a dimensional scoring model analysis.

11.4

STRUCTURE AND SITE SELECTION

It would be highly unusual to choose a structure without having carefully considered location preferences. It is quite usual to choose a location and then search for a specific structure and site. Often, the list consists of combinations. Various locations, each with attractive sites and/or structures, are cataloged. Location and structure-site decisions eventually are considered simultaneously.

Thus, North Carolina and Tennessee could be chosen as tax-advantage states in the U.S.A. If market differentials exist, they also could play a part in the selection among regions of the country. After choosing location, the search for sites and structures becomes more specific. In comparing alternatives, all relevant elements of location, site and structure should be included. If relocation is involved, the decision may be to select a new site or stay with the present one, and redesign and rebuild the facility.

Work Configurations Affect Structure Selection

Flow-shop structures permit serialized, sequential assembly with materials being received and introduced to the line as close as possible to the point of use. Access by suppliers at many points along the walls of the building is needed. Gravity-feed conveyors can be used instead of mechanized conveyors when the structure is multistoried. There are many similar issues that arise which relate the building type with the work configuration. Part of the design of a good flow shop is the design of the structure.

Job shops do not entail large investments in the design of the process—as does the flow shop—or an extensive FPS system. Therefore, the kind of structure that will be adequate to house job-shop processes is less restricted than for flow shops. More real-estate choices are suitable. Rentals are more likely to be feasible.

Service industries are often associated with particular kinds and shapes of structures. Airports, hospitals, theaters and educational institutions typify the site-structure demands for service specifics. Technological information and knowledge of real-process details are required to reach good decisions.

A fixed-position layout is used for items that are too large to move on a production line.
Source: McDonnell Douglas Corporation

Facility Factors

Companies that build their own facility to match work configuration requirements make fewer concessions. Continuous-process industries—like petrochemicals—have to build to process specifications. Even for the job shop, special requirements for space and strong floor supports—say for a large mixing vat—can influence the choice of structure. When renting, building, or buying, expert help from real-estate specialists, architects and building engineers should be obtained to insure proper evaluation of an existing facility or to plan a new structure.

Among the facility elements to be considered are:

• Is there enough floor space?
• Are the aisles wide enough?
• How many stories are desirable?
• Is the ceiling high enough?
• Are skylights in the roof useful?
• Roof shapes permit a degree of control over illumination, temperature, and ventilation. What are the maintenance requirements?

For new construction, in addition to costs, speed counts. Building codes may not be acceptable. Industrial parks may be appealing. Special-purpose facilities usually have lower resale value than general-purpose facilities. Good resale value can be critical allowing a company flexibility to relocate when conditions change.

Company services should be listed. Capacities of parking lots, cafeterias, medical emergency facilities, male and female restrooms—in the right proportions—must be supplied. Adequate fire and police protection must be defined. Rail sidings, road access and ship-docking facilities should be specified in the detailed facility-factor analysis.

External and internal appearance are factors. An increasing number of companies are using the factory as a showroom. Some service industries use elegant offices to impress their clients. Others use simplicity to emphasize frugality and utilitarian policies. Some consider appearance to be a frill. Others take appearance seriously and illuminate their building at night. Japanese management stresses cleanliness as a requirement for maintaining employees' pride in their company. When Sanyo acquired dilapidated facilities they painted the walls and polished the floors.

Rent, Buy, or Build—Cost Determinants

The costs of land, construction, rental rates, and existing structures are all numbers that can be obtained and compared with a suitable model. That model must be able to reflect the net present value of different payment schedules over time. An appropriate discount rate must be determined in conjunction with the financial officers of the firm.

Cost-benefit analysis is a model analogous to a ledger with dollar costs listed on one side and dollar equivalents for the benefits on the other side. The lists are summed. Net benefit is the difference. With intangibles such as community support, workers' loyalty and skill, union-management relations, and flexibility to relocate, it is difficult to get dollar figures. Thus, estimates of qualities need to be made and used in a scoring model.

Location, structure and site come as a package of tangible and intangible conditions that have costs—some of which are difficult to determine but which are worth listing.

Cost-benefit analysis
Model analogous to a ledger with dollar costs and dollar equivalents for the benefits listed which are summed and net benefit is the difference.

1. The opportunity cost of not relocating.
2. The cost of location and relocation studies. The collection of relevant data is facilitated through the cooperation of regional Chambers of Commerce.
3. The cost of moving may have to include temporary production stoppage costs. Inventory buildups may be able to help manufacturers offset the effects of stopping production. Inventory cannot help service stoppage.
4. The cost of land—often an investment. Renting, buying, or building have different tax consequences which can play a part in reaching decisions.
5. The costs of changing lead times for incoming materials and outgoing products as a result of different locations.
6. Power and water costs differ markedly according to location.
7. Value-added taxes are used in many European countries. VAT is proportional to the value of manufactured goods.
8. Insurance rules and costs are location sensitive.
9. Labor scarcities can develop which carry intangible costs.
10. Union-management cooperation is an intangible cost factor.
11. The intangible cost of community discord can be significant.
12. Legal fees and other costs of specialists and consultants are location sensitive, especially for small- and medium-sized organizations.
13. Workmen's compensation payments and unemployment insurance costs differ by location.
14. The costs of waste disposal, pollution and smoke control, noise abatement, and other nuisance-prevention regulations, differ by locations.
15. Compliance with environmental protection rules differ by location.
16. The costs of damage caused by natural phenomena are affected by location which determines the probabilities of hurricanes and earthquakes, as well as floods and lightning. Insurance damage rates have risen markedly and do not cover production stoppages.
17. Costs of reducing disaster probabilities—such as using raised construction to reduce flood damage risk.
18. Normal weather conditions produce costs associated with location. Heating, air conditioning, snow removal, frozen pipes, and more absentee days because of the common cold.
19. Facilities deteriorate faster in extreme cold or heat.

How can the many tangible and intangible costs be brought together in a unified way? Scoring models provide a satisfying means for organizing and combining estimates and hard numbers. This topic is covered below.

11.5

FACILITY SELECTION USING SCORING MODELS

If all of the facility costs were readily measurable, then an equation could be written for total costs—of the general form:

$$\text{Total Costs} = \sum x_i \quad (i = A \text{ through } Z)$$

where *A* through *Z* includes costs associated with the location, structure and site. Unfortunately, many factors cannot be reduced to dollars and cents. Part of the equation can be in dollars and the rest has to be cast as an estimation of costs. The total cost equation only can be symbolic:

$$\text{Total Costs} = f(\text{Tangible Costs} + \text{Intangible Costs})$$

The objective is to choose the location/structure/site option or alternative that minimizes Total Costs.

The **scoring model** permits the simultaneous evaluation of tangible and intangible costs. The method allows intangibles to be dealt with in a quantitative fashion with as many factors considered as seems necessary. However, the most practical idea is to concentrate on the important factors.

The comparison among strategic location alternatives is equivalent to measuring the ratios of the cost estimates for the different intangible and tangible cost factors. This is a way of measuring relative advantage. Each ratio portrays opportunities to improve. The numerators and denominators are equivalent to the utilities of different options.

Intangible Factor Costs

The ratios measure the degree to which one option is better than, or preferred to, another. The dimension of community attitude (CA) is a good instance of an intangible factor. CA cannot be measured in a precise and nonambiguous way—as for example, dollars of rent. Therefore, an estimate is made by someone who knows the community attitude aspect of the location decision. In effect, the estimate states a preference for one community's attitude as compared to the other.

Assume that Orlando, Florida has been given the top rating for CA. Because this problem is being stated in terms of costs, the best rating is "one" which is the smallest positive integer. Scoring models can be created with the goal of profit maximization or cost minimization. All terms must be consistent with whichever is used. For profits, the preference is large numbers. This will not apply to weights which are used in this scoring model. Large weights always are more important than small weights.

It is possible to score for any number of comparisons, but to keep things simple for the present explanation, just two cities will be used. The model can be extended to make more comparisons without any changes in method. The value of this ratio will be indicated by V_{CA}.

Orlando, Florida is being compared with Edison, New Jersey, by a high-tech manufacturer presently located in Buffalo, New York. The old Buffalo facility is to be closed. There is general agreement to move south, but healthy debate about how far south to move. Edison, New Jersey has been rated a 2 for supportive community attitude.

There are a number of ways to interpret the ratio:

$$V_{CA} = \frac{\text{Orlando}}{\text{Edison}} = \frac{1}{2}$$

First, the rater's preference for Orlando, with respect to community attitude, is two times greater than for Edison. The other side of the coin is that the rater's preference for Edison is half that of Orlando. Getting something that is less preferred is

Scoring model

Permits simultaneous evaluation of tangible and intangible costs; means for organizing and combining estimates and hard numbers.

construed as a cost. The ratio of the cost estimate for the best alternative to the other alternative with respect to community attitude is obtained by dividing 1 by 2. Because costs are best when they are small, 1 is better than 2.

Second, the cost of choosing Edison instead of Orlando with respect to CA is $2/1 = 2$. This could be interpreted as $2 or $2,000 or $2,000,000.

Third, the scoring model deals with the importance of CA relative to all the other factors by weighting the value of the ratio, thus:

$$(V_{CA})^{W_{CA}}$$

This enables the model to consider simultaneously the costs of both tangible and intangible factors. The preference measures and actual cost dimensions are weighted to express their importance to the total evaluation of the location alternative. To mix tangible costs with intangible costs, weights are used to express the relative importance of each. Then, all of the costs are combined. This treats measured values as though they also are preferences.

In fact, because apples and oranges are being compared, the scoring model can only work by combining the dimensionless ratios into a dimensionless value, V. This is a multiobjective model. Some of the objectives are in conflict. There is the objective to move to a location which has the best CA. This location might not be the one with the lowest taxes. A list of six objectives—three tangible and three intangible factors—that apply to this sample location, site, and structure problem follows.

Tangible Costs—Measured in Dollars:

1. Building Costs—based on annual (straight line) depreciation
　　　　　　　　　—fully equipped with general-purpose equipment
2. Taxes per year
3. Energy costs per year

Another tangible factor is labor. Companies choose locations based on cheaper labor markets (including low-cost labor countries). Material costs also can differ according to location. To keep this scoring model manageable, labor and material costs have been treated as equivalent at both locations.

Intangible Costs—Measured in Preferences:

4. Ease of relocating if and when such becomes necessary
5. Product quality as a function of skills and morale
6. Community attitude

If one location is best for all six objectives, and the location planners agree that the list of objectives is correct and complete, then there is no need for the scoring model. The choice is obvious. If, however, as is more usual, one location is best for some of the six objectives, and the other location is best for the remaining objectives, then the use of the scoring model is not only warranted but needed. Six different dimensions, representing conflicting objectives, must be combined to provide a reasonable basis for evaluation.

Developing the Scoring Model

The tangible objectives are directly measurable. The other three objectives are intangible and require an estimate of preference. To systematize the scoring model, each objective can be assigned a scale position between 1 (best) and 10 (worst). Assume that the proposals for the Orlando and Edison locations have been evaluated as shown in Table 11.14.

TABLE 11.14

FACTORS FOR COMPARISON OF ALTERNATIVE LOCATIONS

Objectives (1)	Orlando	Edison	Weight (4)
Tangible Costs ($)			
Building Costs (2)	$500,000	$300,000	4
Taxes	$ 50,000	$ 20,000	4
Energy Costs	$ 20,000	$ 30,000	4
Labor Costs	- - - - - -equivalent - - - - - -		
Material Costs	- - - - - -equivalent - - - - - -		
Total Tangible Costs	$570,000	$350,000	4
Intangible Costs (3)			
Relocation Flexibility	1	6	3
Workforce Quality	2	3	5
Community Attitude	1	2	1
Pollution Regulations	- - - - - -equivalent - - - - - -		
Room for Expansion	- - - - - -equivalent - - - - - -		

Notes:
(1) Cost minimization objectives for all factors; 1 best.
(2) Based on annual (straight line) depreciation including general purpose equipment costs, also depreciated.
(3) Measured in preferences with the scale of 1 to 10, where the lowest cost is the most preferred. The value 10 would be optimal if the table had been constructed for profits.
(4) The larger the weight, the more important the factor is considered to be relative to the other factors. The weights have been chosen to go from 1 to 5.

The first three tangible factors are dollars which can be added together since all costs are based on a one-year period. This yields $570,000 and $350,000 for the Orlando and Edison locations, respectively. Two additional dollar costs are not included because they are equivalent and in ratio cancel each other out.

Discounting methods can be used to compare net present values for time streams of money. Total Tangible Costs can be used as a single factor to represent all of the cost factors. It has the same weight of 4 as the other dollar factors. Companies will have different measures for the relative importance or utility of dollars. Organizations with large asset bases will consider small amounts of money to have low weights. Conversely, small- and even medium-sized companies usually treat cash-flow expenses as having large weights. In this case, management has evaluated dollars with an importance weight of 4 which is high.

This makes sense because to this company the purchase of a new building is a substantial investment. Edison has a cash outlay advantage over Orlando that is noteworthy at the outset.

Dollar expenditures are more important than any other factor—except Workforce Quality which carries a weight of 5. The factor called Workforce Quality is considered to be a surrogate for product quality that results from a skilled workforce with high morale and pride of workmanship.

Consider the intangibles. Orlando leads in all cases. For Workforce Quality and Community Attitude, both Orlando and Edison have fine ratings although Orlando has the edge. When it comes to Relocation Flexibility, Edison does not seem to satisfy the company's expectations. As for Pollution Regulations and Room for Expansion the two locations are equivalent and need not be considered because they would cancel out in ratio.

Weighting Scores

The fourth column of Table 11.14 above, captioned "Weight" presents weighting factors, or index numbers which have been scaled from 1 to 5. Choosing the weights is a subjective task which may not be easy to accomplish. The weights need to be chosen to capture the relative importance of each objective.

According to the weights assigned in Table 11.14, Workforce Quality (at 5) already has been discussed. Community Attitude (at 1) is least important. Relocation Flexibility (at 3) is somewhat less important than costs (at 4).

The arrangement of weighting values could change if the company's capitalization were altered or if the location objectives were modified. It is useful to keep in mind that the numbers used for the scoring model were derived by a team of company managers after discussion and analysis of the situation.

The weighting is arbitrary, but not random. The same can be said about the preferences for the intangible factors. Say that the management team—composed of five executives including the CEO—agreed to accept the numbers in Table 11.14. Hopefully, such agreement means that there is consensus about both preferences and importance of those preferences. It could be that four out of five want to please the CEO.

If factors such as community attitude, product quality, and flexibility could be associated with dollar values, there would not be a dimensional problem to resolve. Everything would be measured in dollars, and there would be no need for weights.

Multiple Decision Makers

Each manager on the site-selection team will provide individual ratings (numbers) for all of the factors. If averages are used to combine the preference scores and the weights, the effect would be to decrease the differentials between locations. Averaging the scores of multiple decision makers reduces variability. This is because individual extremes tend to cancel each other. If the P/OM rated Relocation Flexibility for Edison as nil and gave it a 10 while the CFO considered it fair and rated it 4, then the average would be 7. If averaging is used, the results can be compared with the results for the individuals. The managers can identify problems and areas to be further studied by comparing their preference scores and weights.

This is similar to the Delphi method (considered earlier) which identifies differences to encourage convergencies of informed opinion. Managers with dissimilar preferences and weights can discuss why they selected the ratings they did.

Scoring models organize a lot of information which is relevant for location decisions. Managers can study what is known and what is not; what is agreed upon, and what is not; what is important and what is not; whether consensus exists about what is important; what appears to require additional research, etc. Ultimately, the decision has to be made about accepting or rejecting the solution indicated by the scoring model.

Pooling and polling opinions of many people in the company about location decisions increases involvement and builds pride in the company. It leads to interfunctional communication about the facility decision. If the company has agreed to move people who want to be relocated, the approach tends to motivate more people to participate in the move.

Information about the choices should be made available throughout the company. If feasible, cost factors should be shared. When the factor lists are made up, *broad participation leads to better idea generation*. Notions appear that otherwise might be overlooked and the process moves faster. The decision to relocate is better shared than sprung as a surprise.

Solving the Scoring Model

The scoring model takes ratios like the one previously developed for Community Attitude:

$$V_{CA} = \frac{\text{Orlando}}{\text{Edison}} = \frac{1}{2}$$

and raises them to a power which is the importance weight of the factor. Thus:

$$(V_{CA})^{W_{CA}}$$

These are then multiplied together to derive V, which is the combined score.

$$V = (V_S)^{W_S}(V_{RF})^{W_{RF}}(V_{WQ})^{W_{WQ}}(V_{CA})^{W_{CA}}$$

where:

$\$$	=	Building Costs and other dollar expenditures
RF	=	Relocation Flexibility
WQ	=	Workforce Quality
CA	=	Community Attitude

Each ratio $(V_j)^{W_j}$ has Orlando in the numerator and Edison in the denominator. Therefore, if $V = 1$, then both locations have the same cost score. If $V < 1$, choose Orlando because it has a lower cost score than Edison. If $V > 1$, choose Edison because Orlando has a higher cost score than Edison.

This is the symbolic representation of the scoring model. In words, it is a multiplicative model which derives the products of the preference (and cost) ratios raised to powers which represent the importance of each factor.

$$V = \frac{\text{preference for location } a}{\text{preference for location } b}$$

V is a **pure number**, meaning that it has no dimensions. For example, if small size, little weight and low cost are the desirable components of a product design, then a and b can be compared, as follows:

Pure number, A number that has no dimensions.

$$V = \frac{\text{preference for product design } a}{\text{preference for product design } b}$$

$$= \frac{\text{dollars } a}{\text{dollars } b} \times \frac{\text{lbs. } a}{\text{lbs. } b} \times \frac{\text{centimeters } a}{\text{centimeters } b}$$

$$= \text{ a pure number (dimensionless) which means that}$$
$$\text{dollars, lbs., and centimeters cross out}$$

Using the numbers given in Table 11.14:

$$V = \frac{\text{preference for location } a}{\text{preference for location } b}$$

$$= \left(\frac{\$570,000}{\$350,000}\right)^4 \times \left(\frac{1}{6}\right)^3 \times \left(\frac{2}{3}\right)^5 \times \left(\frac{1}{2}\right)^1 = 0.002 = \frac{1}{500}$$

With this result, Orlando will be chosen because the ratio is less than 1. The combined costs of Edison (in the denominator) are greater than the combined costs of Orlando (in the numerator).

The scoring model is useful for a wide range of applications in addition to the location problem. It is appropriate for product-, process- and service-design decisions, equipment selection, warehouse-location plans, etc. The scoring model's multiplication method of evaluating alternatives by means of weighting factors is reasonable to handle multidimensional problems and is frequently used.[1]

11.6

EQUIPMENT SELECTION

The building blocks for resolving the equipment selection problem have already been put into place. The first issue that has to be decided is: what work configuration is going to be used? This is a capacity planning problem. It must be resolved before any equipment decisions can be made. General-purpose equipment for the job shop and batch production come in many varieties and can be purchased new or used. Continuous production processes and flow-shop systems must be designed using careful engineering analysis. The analysis along the lines of Points 1 through 4 follows:

1. For choosing between specific machines, conveyors, forklift trucks and other pieces of equipment, the net present value of the investments and operating costs should be determined. The reason that traditional financially discounted cash-flow analyses should be used is that it permits purchase price to be combined with the time stream of operating costs.
2. The classical breakeven model can provide critical data about equipment selection. This provides a benchmark regarding the volume of work that must be done on the alternatives before they start making a contribution to profit.
3. The scoring model, described for site selection, also is useful for evaluating some of the intangible factors that new equipment decisions involve. These include: difficulty learning to use new equipment, assessing the competitive benefits of new equipment, improvements in quality and productivity that could be expected, and the variety of jobs that can be done on the new equipment.
4. Set-up times of equipment alternatives must be compared. Using activity-based costing it is possible to estimate downtime expenses as well as to determine the effect of set-up costs and time on optimal lot sizes.

New equipment decisions are a P/OM responsibility. Still, the financial involvements require teamwork with financial control and accounting managers. Cooperation with engineers and R & D is warranted to make certain that all technical issues are properly addressed.

Design for Manufacturing—Variance and Volume

Without proper tools and equipment, even a skilled and motivated worker is hard pressed to do more than a mediocre job. Poor tools compromise quality. TQM requires that all the homework should be done concerning equipment performance, costs and quality, before reaching equipment-selection decisions. The estimates of performance should not be naive. They must be used to explain expected variation in the qualities of the product.

Part of the work entailed in design for manufacturing (DFM) is equipment performance evaluation. First, the productivity of equipment must be related to the volume of work that has to be done. This means comparing the production rates of each alternative. Second, different degrees of variability are associated with the equipment alternatives. This relates quality of product to equipment selection.

Like fingerprints, each piece of equipment produces unique variations. Original equipment manufacturers (OEM) can categorize the tolerance capabilities of the machines that they sell. This helps to select equipment but tests must be conducted to determine the output variability of specific products made on the machines. The observed level of variability must be compatible with the designer's tolerances. It also must meet the DFM expectations of the process managers. There often is a great deal of difference between various manufacturers' equipment which otherwise appears to be of the same class. Equipment choice that ignores this information is selling process management short.

Technical information and knowledge of real process details are needed to design service structures like airport terminals.
Source: HNTB Corporation

Equipment Selection and Layout Interactions

Equipment selection precedes facility layout. The two interact so strongly that it is logical to have an idea about layout requirements before facility choice. Size, weight, height and number of pieces of equipment have their footprints (space needed) and signatures (vibration, noise among others).

Movement of materials is another equipment factor. Elevated conveyors can carry materials near the ceiling whereas sufficient aisle clearance must be available for AGV and trucks. Equipment dictates facility layout design. It may influence facility location—when the ideal building exists somewhere.

P/OM's have a good idea of the equipment that will be required at the chosen facility, and therefore, of the general adequacy of a particular structure. The process of deciding what to do is interactive, going back and forth between facility selection, equipment selection and facility layout. It is not unusual to find that equipment selection must accommodate an existing facility.

Life Cycle Stages

The effect of life cycle stages (start-up, growth, maturity, and withdrawal) on equipment selection is significant. Equipment selection is dominated by volume considerations. Volume increases are what life cycle stages are all about. Therefore, it is to be expected that equipment changes will be made over time in keeping with the life cycle stages.

The dynamics are hard to manage unless they have been consciously planned. It might seem reasonable to start off with machines that are economic to use at low volumes. Then, as growth in demand occurs, growth in supply would follow. However, if the risk is low that such growth will occur, then there is reason to start with equipment that has greater capacity than will be needed at first. The period of underutilization can be a short time in the life of a machine. As the equipment approaches full utilization, the costs per unit tend to decrease. With machines that are economic at low volumes, success brings volume increases. This leads to over-time which inflates per unit costs.

Planning equipment selection should include timing of transitions in supply capability. Planning also must address the work configuration that will be attained eventually. The flow shop remains a worthy goal. When and if that state of affairs is achieved, low volume general-purpose equipment will not be called upon unless, the planning process includes the launching of new products at the proper points in time.

Replacement for wear is part of the planning. There is gradual deterioration in the performance of equipment which should be factored into the life cycle considerations. Maintenance can change the rate of deterioration and must be factored into the plan. While some equipment is nearly maintenance-free, other equipment requires costly programs with skilled labor and downtime expenses. Appropriate equipment for start-up may not be able to be fully depreciated at replacement time. There is a market for used general-purpose equipment, which may be a consideration that can add flexibility in the early stages of product life. Planning is required to have used-up, and phased-out, low-volume equipment, or to have found new uses for it.

Citicorp moved the research and development (R & D) operations for its Global Finance Division into a 5-story, 115,000-square-foot building in Westlake, Texas, in February 1994. The facility in the Solana office park, about 10 miles north of Ft. Worth, also serves as an education center for 27 Citicorp technology centers worldwide.

Citicorp located the facility in this area due in part to the availability of a trained workforce. IBM, co-owner of the 900-acre Solana office park, had just reduced its work force by 800; there were also technology professionals recently laid off from the defunct Superconducting Super Collider project. The Solana unit had 40 employees when Citicorp moved into the facility and in 1996, will have approximately 150 employees.

Establishment of the Solana technology unit is part of Citicorp's effort to remain a technology leader.

Citicorp Technical Center

The facility is being used to develop the technology and software Citicorp needs to shift its technological emphasis from large mainframe computer systems to networks of individual workstations. Its emphasis in software programming is on object-oriented technology, which allows programmers to borrow components from existing programs to develop new applications.

"There's a tremendous number of competitive threats, and technology is the way to maintain the forefront in quality, speed and timeliness," said Mary Cirillo, senior vice president of the Global Finance Operations Technology Group. "We're working on rapid deployment of new technology in diverse markets," she said.

Banking analyst Paul Mackey of Dean Witter Reynolds considers the move a "smart" one for Citicorp: "The Achilles' heels of banks is operational capabilities. Where they can get a competitive advantage is when they can operate more efficiently."

Source: "Citicorp Unveils Plan to Develop Software at Solana: Banking on Technology," *Fort Worth Star-Telegram*, May 14, 1994.

11.7

LAYOUT OF THE WORKPLACE AND JOB DESIGN

It is hard to believe that until the 1980s, workers in American and European auto-assembly plants had to bend down to install the new tires on the wheels, and then tighten the tire bolts. The auto-assembly conveyors ran flat out near the ground. The layout made the job not much different than changing a flat tire on the road. Visitors to Japanese auto plants saw that the conveyor lifted the car up at the point of tire attachment. This carried the car above the worker and allowed putting the tire on at shoulder level. Workers could stand tall and attach the tires to the wheels without bending. Thereafter, conveyors that lifted the car were installed in world-class auto-assembly plants.

This simple anecdote illustrates how *good layout interacts with good job design*. It also shows how **layout** of the workplace is a three-dimensional situation which lends itself to creative solutions for alleviating back problems. It is noted that job design in the flow shop is partnered with layout.

Productivity increases and quality improves when the layout and job are properly designed in the job shop. Travel time and distance traveled are reduced. Paths are kept clear. Queues at workstations are reduced. Production scheduling, shop-floor layout and job design are connected in a systems sense. The benefits of improved quality derived from better layout must be continuously monitored because process changes are constantly occurring.

Layout

A three-dimensional situation which lends itself to creative solutions for alleviating problems; a good layout interacts with a good job design; leads to productivity increases and quality improvements.

Opportunity Costs for Layout Improvement

It is evident that proper layout design can provide quality improvements (*QI*), productivity improvements (*PI*), and health benefits (*HB*) that result in higher profits. The evaluation of jobs and workplaces within the process could lead to layout improvements.

The paradigm that is used for such evaluations is like a balance scale or a seesaw. All of the costs of doing *X* are put on one side of the ledger. All of the costs of not doing *X* are put on the other side of the ledger. The side with the least amount wins the contest. This describes the way that a trade-off model works.

Improving layout and job design involves opportunity costs trade-off analysis. **Opportunity costs** are the costs of doing less than best. By doing something which is not the best, the cost to be paid is the difference in net benefits. This follows the same kind of reasoning that underlies the improvement method of the transportation model.

In this case, there are opportunity costs for not having the best possible layout design. The equation for this trade-off model states that layout design improvement should be made if:

Opportunity costs

Costs of doing less than best; cost to be paid is the difference in net benefits.

$$CPI \quad < OC(QI + PI + HB), \text{ where}$$

$$CPI \quad = \text{Cost of layout design plan improvement}$$

$$OC \quad = \text{Opportunity costs incurred for not having used the best possible layout with respect to QI, PI and HB}$$

$$OC(QI) = \text{Opportunity costs for quality improvements}$$

OC(QI) could have been obtained if layout design improvement had been made. They also are the opportunity costs incurred by not rectifying quality deficiencies.

Improved conveyor layout would yield a better car which is then translated into:

- larger market share
- greater revenues
- fewer warranty claims
- less service calls at company expense
- higher prices that could be charged
- smaller company discounts on ticket prices
- better worker attitudes and morale
- less dealer discontent
- more effective advertising campaigns

$OC(PI)$ = Opportunity costs for improved productivity

$OC(PI)$ could have been obtained if layout design improvement had been made. They also are the opportunity costs incurred by not rectifying factors that deteriorate productivity. Not having to bend down and work in a crouched position would enable workers to work faster as well as better. The cost of installing four tires to each car could be reduced. It is possible that one or more workers could be freed up to do other jobs in the time now allotted for the repetitive job at the production rate. Job observations and time studies can be used to get good estimates for the opportunity costs for productivity. Unlike $OC(QI)$, the costs $OC(PI)$ and $OC(HB)$, discussed below, are often measurable.

$OC(HB)$ = Opportunity costs for health benefit savings

Now overlooked, $OC(HB)$ could have been obtained if layout design improvement had been made. They also are the opportunity costs incurred by not rectifying layout factors that diminish health benefits.

The costs associated with bending and crouching include back problems for workers that result in medical claims, higher health insurance, absenteeism and lost time on the job. Computer operators have been experiencing a problem with their hands and wrists that is called carpal tunnel syndrome. Redesign of the computer keyboard, pads for the wrists to rest on, and other redesigns of the wrist support system have improved the situation, but the constancy of this repetitive work has led to the conclusion that computer typing jobs are best redesigned to include a regular rest period every hour. Job design, work schedules and layout must be treated together as an integrated system to deal with this issue.

Sensitivity Analysis

It is noteworthy that many of these opportunity costs are not easy to measure. Nevertheless, they are real and can be important. They do not disappear because they are considered difficult to measure with accuracy.

Estimates can be made more easily for a range running from a high to a low figure. By using high and low estimates it is possible to test the sensitivity of the equation to the range of estimation. Also, it is possible to find out what values of $OC(QI)$, $OC(PI)$ and $OC(HB)$ can lead to equality, thus:

$$CPI = OC(QI + PI + HB)$$

is a quality breakeven value for layout improvement. These steps are associated with the procedure called **sensitivity analysis**.

If the inequation $CPI < OC(QI + PI + HB)$ holds, then the present design can be improved and should be altered. However, there can be circumstances where the inequation indicates that no change should be made, but a change should be made anyway. The reason is that the opportunity costs, say for health benefits, have been underestimated. Safety comes under health benefits and when threats to safety exist, the opportunity costs of HB can be regarded as infinite. This means that CPI is always less than the sum of the opportunity costs. Then, continuous layout design and workplace improvement are warranted.

In the same sense, if managerial intuition rejects the model's solution, then deeper analysis of the quality opportunity costs is likely to show that they have been underestimated. The job and workplace design may be creating a situation that has

Sensitivity analysis
Procedures using high and low estimates to test the equation's sensitivity to the range of estimation.

not yet surfaced in traceable costs. Deterioration of worker morale is often detectable before it impacts quality on a measurable level. P/OM's hunches about negative interactions between layout and process design are worthy of respect.

11.8
LAYOUT TYPES

There are at least five basic types of layouts that will be found in plants and offices. These are:

1. *Job shop process layouts*—which means that similar types of equipment or jobs are grouped together. The lathes are in one place and the presses in another. For services, filing is in one room and copy machines are in another. Inspectors are in one place, designers are in another. Job-shop process layouts should facilitate processing many different types of work in relatively small lots. Mobility of equipment enables OM to set up intermittent flow shops when that configuration is attainable. Space must be allocated for work completed at one station and waiting for access to another station. Equipment mobility and layout flexibility allow this configuration to be rearranged to suit the high variety of order mixtures that can occur.

2. *Product-oriented layout*—typical of the flow shop. Equipment and transport systems are arranged to make the product as efficiently as possible. The layout is designed to prohibit disruptions to the flow. The product-oriented layout is most often associated with assembly lines.

3. *Cellular layout*—used with a group of machines that work with each other to produce a dedicated family of parts, as in group technology. The layout is engineered to facilitate efficient transfer of parts between machines in the cell. Tooling and transfer are part of the computer programming system which is capable of nearly instantaneous changeovers enabling small runs of a limited number of parts.

4. *Group technology layout*—used to efficiently produce a family of parts but there is no emphasis on computer controls and machine cells as in Type 3 above. The layout is focused on taking advantage of the similarity of design of the parts.

5. *Combinations of product and process orientation*—very common. Some of the products in the job shop achieve demand volumes that allow them to be run for long periods of time as intermittent flow shops. Modular product-designed parts often have the high volumes that allow cellular manufacture or group technology layouts with, in some cases, serialized flow shop advantages. At the same time, other jobs in the shop remain at low volumes that are only suited to the process layout of job shops. The combination produces a mixed-layout orientation which also is called a hybrid layout.

The complaint department of a large organization with a number of different products has the job-shop process layout. Within that group, there are certain subgroups that handle high frequency requests with flow shop dedication. Sixty percent of the complaints can be treated in the highly repetitive fashion. The remaining 40 percent require special treatment for customer satisfaction. The analogy can be extended to families of complaint types and the use of group technology process layout.

Conceived by employee team members from operations, engineering, and gas acquisitions, this gas processing plant was built on a limited budget using surplus compressors and equipment. Source: Phillips Petroleum Company

Volvos are manufactured in Sweden on fixed platforms. The work does not move. Workers move to it and around it. Ship builders and home builders move around the work. Modular housing exemplifies hybrid layout where the parts are made in a factory and then brought to the site where they are assembled, similar to Volvos.

With fixed platforms, workers carry or drive their tools to where they are needed. Commercial airplanes are moved along a production line, albeit slowly. Commercial airline layout is a different form of hybrid; one that combines fixed and moving positions. The fixed-position layout is necessary for power plants, refineries and locomotives. It is much more controversial (i.e., Volvo), so that deserves an extra word.

Volvo uses the fixed platform because Swedish workers find it more interesting to participate in building the whole car than doing a repetitive step along an assembly line. Worker motivation is improved. In Sweden, worker motivation is important because unemployment benefits are a large percentage of earned wages and readily available. Bored workers are likely to prefer the government benefits to the company wages.

The United Auto Workers (UAW) arranged an interesting experiment in which a group of GM workers were given the opportunity to work with Volvo for six months. Most of these workers said they preferred to work on the GM flow shop assembly layout in spite of, and because of, its repetitive nature. The Volvo layout required knowing and doing too many things. Some preferred the Volvo system and remained with Volvo.

Group technology (GT) for families of parts is a layout that many companies, such as Caterpillar, John Deere and Cummins Engine, have found rewarding. GT cells are usually part of hybrid layouts. The group technology concept uses a product layout with the capability of producing an entire family of parts. Special equipment is required which has rapid changeover capabilities to make parts that are identical except for size and/or other dimensional characteristics. Crankshafts, motors, and pumps are typical parts for GT. The production cells use flexible machine-tool organization. Group technology layout is usually a self-contained part of a process-layout job shop.

11.9

LAYOUT MODELS

The plant layout problem can be tackled with different kinds of models. There is good engineering knowledge for laying out the flow-shop product layout. There is less solid knowledge about excellence for job-process layouts.

Four points to bear in mind:

1. Job shops and the batch production environment are subject to major changes in the product mix. What is optimal for one set of orders may be downright poor for the jobs in the shop one month later. Therefore, layouts which are likely to be good for the expected range of order types are better than those which are excellent for one type and not good for others.
2. The degree to which a few order types dominate the job shop will modify the statement in Point 1 and allow some product layout to be mixed in with the process layout. It will be found that the per unit costs of operating product layouts will be significantly less than the costs associated with process layouts. The comparison is even more extreme if GT cells can be set up.
3. Flexibility is desirable, but on the other hand, it is expensive and disruptive to keep moving equipment around the plant. A balance has to be found between these two objectives. If the character of the batch work changes a great deal over time, it is best to go for the most general form of process layout. If layout is to be changed from time to time, it is essential to set up the layout system with this purpose in mind. Modular office layouts which are well-planned have remarkable flexibility for making quick changes without paralyzing the workforce.
4. Use quantitative models with caution. Creative thinking and common sense pay off in achieving good layouts. Precise measurements count when trying to squeeze a machine into a tight spot. More questionable is the use of elaborate mathematical models to minimize total distance traveled or handling costs.

Layout Criteria

What criteria determine a satisfactory layout? Some measures of layout effectiveness are:

1. *Capacity*—output or throughput. Goal: Maximum.
2. *Balance*—the degree to which the output rates of consecutive operations are balanced. This is particularly applicable for flow shops and projects. Job shops should be evaluated on the same basis. Goal: Perfect balance.
3. Amount of *investment and operating costs.* Goal: Minimum.
4. *Flexibility* to change layouts. Goal: Maximum.
5. Amount of *work in process* (WIP). Goal: Minimum.
6. *Distance* that parts travel—which is something like passenger miles traveled. Goal: Minimum.
7. *Storage for WIP and handling equipment* to move it from one part of the facility to another. Goal: Minimum.

Floor Plan Models

All seven of these criteria can be evaluated crudely yet with reasonably correct perceptions by using **floor plan models**. The floor plan drawings are graphic methods of trial and error which are used by interior decorators. They use little paper cutouts that represent the furniture. A doll house is the three dimensional deluxe version.

Plant layout models can be two or three dimensional. Often two-dimensional floor plans, with cutouts made from templates, are used to represent the various pieces of equipment. When conveyors are employed, overhead space requirements may be important, and three-dimensional "doll house" models are preferred.

These techniques are useful for an incremental approach to a satisfactory layout. They do not bring any quantitative power to bear. While an optimal layout based on the numbers may be far from satisfactory, the quantitative information might assist creativity.

Floor plan model

Graphic two- or three-dimensional method of trial and error used to evaluate the seven layout criteria of capacity, balance, investment and operating costs, flexibility, work in process (WIP), distance parts move, and storage for WIP.

Layout Load Models

A relatively simple quantitative approach examines alternative layout plans in terms of the frequency with which certain paths are used. Usually, the highest frequency paths are assigned the shortest plant floor path distances to travel. The objective is to minimize the total unit distances traveled.

The physical space is divided into locations. This is shown in Figure 11. 3 where the locations are designated as *A, B, C, D*, and *E*.

FIGURE 11.3

LAYOUT OF PLANT OR OFFICE FLOOR (WITH AREAS A, B, C, D, AND E AND WORK CENTERS 1 THROUGH 5)

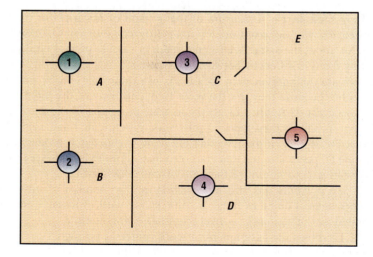

The average distance that must be traveled between areas *A, B, C, D*, and *E* are shown in Table 11.15. These distances (in feet) are designated as d_{IJ}. They are physical measurements related to the floor plan shown in Figure 11.3 above. The matrix is read from row *I* to column *J*.

**DISTANCES MEASURED BETWEEN LOCATIONS
A, B, C, D, AND E (d_{ij})
FROM ROW I TO COLUMN J**

I \ J	A	B	C	D	E
A	0	10	20	32	40
B	10	0	16	18	20
C	20	16	0	12	15
D	32	18	12	0	10
E	40	20	15	10	0

This matrix is symmetric so the distance from A to D is 32 which is the same as the distance from D to A. Often symmetry does not hold because of one-way passages and other simple things like the difference between up and down.

Layout planning involves locating equipment and people to do certain kinds of jobs which are designated as **work centers** at specific locations on the plant floor. Work centers are equipped to do specific kinds of jobs—like the press shop, copy room, hamburger grill and darkroom. Processing orders on-hand causes materials to move between the work centers and therefore, between the physical locations. The materials are at different stages of work in process. Further, they represent various orders and job types. Also, order-batch sizes vary.

What is needed next is a measure of the amount of materials or the number of trips between the various work centers that different jobs entail. If order-process batches are of about the same size, then a random sample of jobs could be taken to estimate how many trips occur between each center.

If batches are not similar, then the idea is to capture those jobs which are most important either because they are the most frequent, involve the largest number of units, are the most difficult and costly to transport, or are the most profitable. Matrices appropriate to these special jobs should be created and studied. Perhaps no more that 20 percent of the jobs that are done in this batch production environment consume 80 percent of the transport resources. For the jobs that are most important, the objective is to minimize the total distance traveled.

Table 11.16 presents the information relating the number of units that move between work centers on an average day in a stable work-flow environment. These data are called wc_{ij}. The matrix is read from row i to column j.

Table 11.17 shows the assignment of work centers 1 through 5 to the floor plan given in Figure 11.3. This pairs up the locations A through E with the work centers 1 through 5. The assignments are: *A1, B2, C3, D4, E5.*

Work centers

Locating specific locations on the plant floor to do specific kinds of jobs.

TABLE 11.16

TABLE 11.16

NUMBER OF UNITS FLOWING BETWEEN WORK CENTERS 1, 2, 3, 4, AND 5 (wc_{ij}) FROM ROW i TO COLUMN j

i \ j	1	2	3	4	5
1	x	100	60	80	20
2	40	x	50	10	90
3	80	90	x	60	30
4	120	10	40	x	70
5	110	5	5	30	x

TABLE 11.17

LAYOUT ASSIGNMENTS

I_i \ J_j	A	B	C	D	E
1	x				
2		x			
3			x		
4				x	
5					x

Multiplying the two matrices in Tables 11.15 and 11.16, is equivalent to multiplying $(d_{IJ})(wc_{ij})$ which yields unit-distances traveled. The equation for this volume of flow can be written:

$$f_{IJ, ij} = (d_{IJ}) (w_{ij})$$

Table 11.18 shows the result.

TABLE 11.18

TOTAL DAILY WORK FLOWS

I_i \ J_j	A1	B2	C3	D4	E5
A1	0	1000	1200	2560	800
B2	400	0	800	180	1800
C3	1600	1440	0	720	450
D4	3840	180	480	0	700
E5	4400	100	75	300	0

Number of Units Moving Between Work Centers Multiplied by the Distance Traveled Between Work Centers

$$f_{IJ,ij} = (d_{IJ})(w_{ij})$$

The total number of unit-feet traveled in this layout is 23,025. There are better layouts; a reassignment is suggested. Calculations show that an improvement results when *A1, B4, C5, D3,* and *E2* are used. These numbers will be found in Table 11.19 where the reassignment of work centers to locations is reflected by the column headings and row names.

TABLE 11.19

TOTAL DAILY WORK FLOWS—REASSIGNMENT MATRIX

I_i \ J_j	A1	B4	C5	D3	E2
A1	0	800	400	1920	4000
B4	1200	0	1120	720	200
C5	2200	480	0	60	75
D3	2560	1080	360	0	900
E2	1600	200	1350	500	0

Number of Units Moving Between Work Centers Multiplied by the Distance Traveled Between Work Centers

$$f_{IJ,ij} = (d_{IJ})(w_{ij})$$

The total number of unit-feet traveled per day in this layout is 21,725.

Unit-feet traveled has been decreased by 1300 which is almost 6 percent. It is likely that additional decreases can be achieved by studying the matrices and trying to eliminate the assignment of heavy-unit volumes to large distances.

Only two configurations have been tried out of 120 possibilities. The original could have been selected randomly. A starting point is needed to achieve improvements. The starting point concept bears similarities to the NWC method for the TM.

In fact, the original was chosen because it could be improved with a sensible heuristic (rule of thumb). This explains the reassignment. There remain 118 other possible assignments of work centers to locations. This is because there are 5! ways $(5 \times 4 \times 3 \times 2 \times 1 = 120)$ to assign five work centers to five locations.

The improved version is based on a reasonable heuristic, explained below. This means that it should be better than an alternative picked at random. Still other layouts exist which conform with the general idea of the heuristic. More calculation is warranted.

How was Table 11.19 constructed? The table is derived by literally rearranging the rows and columns from 1, 2, 3, 4, and 5 in Table 11.16 above to 1, 4, 5, 3, and 2 as shown in Table 11.20. The most practical approach is to read off each entry in Table 11.16 and enter it in the appropriate cell in Table 11.20. Thus, 80 should appear at the intersection of $i = 1, j = 4$ in both tables.

TABLE 11.20

NUMBER OF UNITS FLOWING BETWEEN WORK CENTERS 1, 2, 3, 4, AND 5 (wc_{ij}) FROM ROW i TO COLUMN j

i \ j	1	2	3	4	5
1	x	80	20	60	100
4	120	x	70	40	10
5	110	30	x	5	5
3	80	60	30	x	90
2	40	10	90	50	x

Then, multiply Table 11.20 by the numbers in Table 11.15 to derive Table 11.19.

Flow Costs

Say that the cost of a unit moving one foot is the same between all locations and work centers. Then, the costs of transporting units will not lead to a new matrix for Total Daily Work Flow Costs, as described by the equation below.

$$f_{IJ,ij} = (d_{IJ}) \, (w_{ij}) \, (c_{IJ,\,ij})$$

If the costs of moving units are different then it is necessary to multiply the matrix of Total Daily Work Flows so that the appropriate costs per unit and per foot are expressed. Note that these transport costs need not be symmetric. The cost of gravity-flow down is usually much less than the cost of carrying units up to higher floors.

If the cost is $1 per unit per foot traveled, then the matrices in Tables 11.18 and 11.19 also are representative of Total Daily Work Flow Costs. These costs would be $23,025 for the first assignment tested and $21,725 for the second and improved assignment.

Heuristics to Improve Layout

Two **heuristics (rules of thumb)** come to mind when searching for rules by which to improve plant or office layout.

1. Assign the centers with large work flow rates between them to locations as close as possible.
2. Assign the centers with small work flow rates between them to locations as distant as possible.

In Table 11.16 above, look at those pairs of work centers which have the largest work flow rates between them. These are in descending order:

Work Center 4 to Work Center 1 has a w_{ij} of 120 units.
Work Center 5 to Work Center 1 has a w_{ij} of 110 units.
Work Center 1 to Work Center 2 has a w_{ij} of 100 units.

Assign them to locations that are as close as possible. Note the distances between A and B (10 feet) and C and D (12 feet).

Using the data in Table 11.15 above, the biggest distances are between locations A and D, and A and E.

This means that the original layout of $A1$, $B2$, $C3$, $D4$, and $E5$, assigned 120 units to the second biggest distance (32 between A and D) in the matrix of Table 11.16. Further, the original layout assigned Work Center 1 to location A and Work Center 5 to location E which means that 110 units moved the largest distance between locations—40 between A and E.

To follow the heuristic "assign the largest number of units to the shortest distance" would mean assigning Work Center 1 to A and Work Center 4 to B. That was done in the second matrix which reassigned Work Center 4 to location B. Because Work Center 1 was already assigned to A, the next best distance from A is 20 at C. Therefore, Work Center 5 was assigned to C in the second matrix.

The second part of the heuristic calls for assigning the centers with the smallest work flow rates between them to the locations that are as distant as possible.

The two smallest flow rates between centers—in the matrix of Table 11.16—are both 5 as follows:

Work Center 5 to Work Center 2 has a w_{ij} of 5 units.
Work Center 5 to Work Center 3 has a w_{ij} of 5 units.

Work Centers 1, 4, and 5 have been assigned to locations A, B, and C. Therefore, 5 is assigned to C. The shortest distances from C are to D and E—12 and 15 respectively. Assign Work Center 3 to location D and Work Center 2 to E. This is the second matrix (called the reassignment matrix) that was developed.

The heuristic did its job, but it is not clear whether further improvement is still possible. Trial and error with the improved matrix can be used to test further shifts.

S U M M A R Y

The subject of facility management comprises choosing the location, determining the structure and the site, selecting equipment and preparing the plant or office layout. There are numerous qualitative factors to take into account. Then, the quantitative modeling can begin.

The transportation model is used to minimize shipping costs or maximize profits where production location and market differentials exist. Scoring models are discussed which can be applied for site or equipment selection. The layout of the workplace and job design interact strongly which is analyzed. Floor models that minimize distance traveled are explained.

After quantitative modeling it is necessary to return to the qualitative factors to evaluate the relevance and quality of the solutions offered by the models. Throughout this facilities planning chapter, the systems approach is explained because teamwork is desired in making location, site and equipment decisions.

▶ Key Terms

best location *452*	cost-benefit analysis *467*	facilities planning *448*
floor plan model *483*	heuristics (rules of thumb) *488*	layout *478*
models *451*	Northwest Corner (NWC) method *460*	opportunity costs *478*
pure number *473*	scoring model *469*	sensitivity analysis *479*
work centers *484*		

▶ Review Questions

1. What is meant by: "The basis for location decisions differ according to which parts of the supply chain are involved?"

2. How many assignments of four work centers to floor locations can be made with a four-story building? (No splitting of work centers is allowed.)

3. With four work centers there are how many possible layouts?

4. Finding the best locations for police and fire stations is a pressing urban problem. Discuss the nature of this problem and what variables are likely to be important. How does this compare to the facility layout problem?

5. Is it true that plant selection should in part be based on estimates of labor costs for different areas of the country?

6. For the scoring model, what happens when a weight of zero is chosen for one of the factors?

7. What enables the horizontal and vertical "stepping-stone" pattern to work for shifting assignments when using the transportation model?

8. The industries listed below tend to form high-density clusters in specific geographic areas. Try to find the rational explanation for the specific clusters?
   a. Financial Services
   b. Automobiles
   c. Stockyards
   d. Steel
   e. Textiles

f. Semiconductors
g. Aerospace
h. Resort Hotels
i. Motion Pictures
j. Publishing
k. Cigarettes
l. Petroleum
m. Credit Card Processing

9. Gasoline station location often is based on traffic density studies. What would be the criteria for a good location in terms of traffic density and patterns?

10. Recent emphasis on variety has led to flexible product layouts which can be quickly converted to produce mixed models. How can this be done using a product layout rather than the more typical process layout?

▶ Problem Section

1. Use the trial and error heuristic to determine the best layout for the following data:

		Locations	
Distances in feet	**A**	**B**	**C**
A	0	4000	2000
B	4000	0	3000
C	2000	3000	0

Daily flow costs per unit distances from and to work centers

	1	2	3
1	x	$50	$70
2	$20	x	$40
3	$30	$80	x

2. Use the scoring model to resolve the following equipment selection problem. A choice is to be made between two alternative telecommunications systems. The following data apply:

	System 1	System 2	Weights
Investment Cost	$5,050,000	$5,060,000	3
Operating Costs (Annual)	$400,000	$500,000	3
Downtime	3	2	7
Ease of Repair	2	3	5
Ease of Learning	4	5	2
Flexibility	4	2	4

Characteristics are scaled so that large numbers are less desirable than small numbers. Large weights are more important than small ones and the scale runs from 1 to 7.

3. A matrix of supply and demand is given below:

						Supply
						20
						30
						50
Demand	10	10	10	20	50	100

   Use the Northwest Corner method to assign shipments for the first feasible solution.
   A two-fold problem is encountered.
   a. What are these conditions called?
   b. What can be done about them?

4. The American Company has two factories, A and B, located in Wilmington, Delaware and San Francisco, California. Each has a production capacity of 550 units per week. American's markets are centered in Los Angeles, CA, Chicago, IL, and New York City, NY. The demands of these markets are for 150, 350, and 400 units, respectively, in the coming week. A matrix of shipping distances is prepared. Determine the shipping schedule which minimizes total shipping distance.

   The estimates of shipping distances (in miles) are shown in the matrix below along with supply and demand.

	Los Angeles	Chicago	New York City	Supply
Wilmington	3,000	1,000	300	550
San Francisco	400	2,000	3,000	550
Demand	150	350	400	

5. The NWC solution is given below for a new plan that has been proposed for the three-plant, two-market problem.

NORTHWEST CORNER—INITIAL FEASIBLE ASSIGNMENT

Plants i	Markets j			Supply
	M_A	M_B	M_D	
P_1	40	20		60 units/day
P_2		40	30	70 units/day
P_3			60	60 units/day
Demand	40	60	90	

The profits per unit shipped are:

Plants *i*	Markets *j*		
	M_A	M_B	M_D
P_1	$23	$32	$0
P_2	$16	$25	$0
P_3	$30	$38	$0

What is the total profit associated with the NWC assignment?

6. Using the information in Problem 5 above, ship as many units as possible into P_3M_B. How many units will that be? Draw the revised TM. Compare the profit of this shipping plan with that of the NWC assignment.

7. Using the information in Problem 5 above, ship as many units as possible into P_1M_D. How many units will that be? Draw the revised TM. What is the profit of that shipping plan? Compare the profit of this shipping plan with that of the NWC assignment.

8. Using the information in Problem 5 above, ship as many units as possible into P_2M_A. How many units will that be? Draw the revised TM. What is the profit of that shipping plan? Compare the profit of this shipping plan with that of the NWC assignment.

9. Using the information in Problem 5 above, ship as many units as possible into P_3M_A. The stepping pattern is more difficult but still uses horizontal and vertical moves from one shipping assignment to another. How many units will be shipped to P_3M_A? Draw the revised TM. What is the profit of that shipping plan? Compare the profit of this shipping plan with that of the NWC assignment.

10. The cost of improving plant layout is $100,000 to be depreciated over a five-year period. The estimated annual improvements in profit are: $6,000 from better quality, $4,000 from higher productivity, and $11,000 from health benefits. Is the layout improvement recommended?

11. The cost of improving office layout is $27,000 to be treated as a one-time expense. The office manager does not have any way of estimating improved profits resulting from quality changes but says that the new layout will save time and miles of walking for the 40 office employees. She estimates that the new layout will cut the amount of walking for the average employee by 100 miles per year. How much will the layout improvement be worth?

12. Using the layout load model determine a good arrangement for work centers at locations for the numbers given in the matrices below.

DISTANCES MEASURED BETWEEN LOCATIONS A, B, C, D, AND E (d_{IJ}) FROM ROW I TO COLUMN J

I \ J	A	B	C	D	E
A	0	10	20	32	40
B	10	0	16	18	20
C	20	16	0	12	15
D	32	18	12	0	10
E	40	20	15	10	0

NUMBER OF UNITS FLOWING BETWEEN WORK CENTERS 1, 2, 3, 4, AND 5 (wc_{ij}) FROM ROW i TO COLUMN j

i \ j	1	2	3	4	5
1	x	50	60	80	20
2	40	x	50	10	90
3	80	90	x	60	30
4	50	10	40	x	70
5	60	5	5	30	x

▶ Readings and References

P. W. Bridgman, *Dimensional Analysis*, New Haven, CT, Yale University Press, 1922; Paperback Edition, 1963.

John M. Burnham, *Integrative Facilities Management.* Burr Ridge, IL, Irwin, 1994.

R. L. Davies, and D. Rogers, *Store Location and Assessment Research*, NY, Wiley, 1984.

Christopher Gopal, and Harold Cypress, *Integrated Distribution Management*, Burr Ridge, IL, Irwin, 1993.

John R. Hauser, *Applying Marketing Management: Four PC Simulations*, Danvers, MA, The Scientific Press, boyd and fraser publishing company, (text with diskette for various kinds of scoring models), 1986.

Kevin Howard, "Postponement of Packaging and Product Differentiation for Lower Logistics Costs," *Journal of Electronics Manufacturing*, Vol. 4, 1994, pp. 65–69.

Henry J. Johansson, Patrick McHugh, A. John Pendlebury, William A. Wheeler III, *Business Process Reengineering*, NY, John Wiley, 1993, pp. 180, 186.

G. J. Karaska, and D. F. Bramhall, *Location Analysis for Manufacturing: A Selection of Readings*, Cambridge, MA, M.I.T. Press, 1969.

James R. Koelsch, "Work Smart," *Manufacturing Engineering*, November 1994, pp. 65-67.

Gary L. Lilien, *Marketing Management: Analytic Exercises for Spreadsheets*, Danvers, MA, The Scientific Press, boyd and fraser publishing company, (text with diskette for various kinds of scoring models), 1993.

James M. Moore, *Plant Layout and Design*, NY, McGraw-Hill, 1962.

A. G. Munton, N. Forster, Y. Altman, and L. Greenbury, *Job Relocation*, NY, Wiley, 1993.

Jean V. Owen, "Making Virtual Manufacturing Real," *Manufacturing Engineering*, November 1994, pp. 33-37.

Ruddell Reed Jr., *Plant Layout*, Homewood, IL, Irwin, 1961.

P. Vincke, *Multicriteria Decision-Aid*, NY, Wiley, 1992.

Notes...

1. These references are presented for historical information. The Readings and References section above presents present-day sources. L. Ivan Epstein, "A Proposed Measure for Determining the Value of a Design," *Operations Research,* Vol. 5, No. 2, April, 1957, pp. 297–299. This method is used to evaluate alternative aircraft designs by a major aircraft manufacturer. Other applications are described by C. Radhakrishna Rao, *Advanced Statistical Methods in Biometric Research*, NY, Wiley, 1952, p. 103. Also, see Walter R. Stahl, "Similarity and Dimensional Methods in Biology," *Science*, Vol. 137, No. 20, July 1962, pp. 205–212. The basic explanation was stated by Paul W. Bridgman, *Dimensional Analysis,* New Haven, CT, Yale University Press, 1922; Paperback Edition, 1963.

Industrial Perspective on Service

ROSENBLUTH INTERNATIONAL

Rosenbluth International is a privately held, worldwide travel management company headquartered in Philadelphia. Specializing in corporate travel, Rosenbluth's products include a variety of corporate, vacation, and meeting travel services. Rosenbluth International has annual sales in excess of $2.5 billion. With more than 825 U.S. locations, and locations in more than 30 countries, Rosenbluth has approximately 3,000 employees.

The company was established in Philadelphia in 1892 by Marcus Rosenbluth as a steamship ticket office. Rosenbluth built his business by providing exceptional service to customers and exploiting an historic change—*the European immigration to America*. Rosenbluth's first clients were immigrants who wanted to bring their relatives to America from Europe. They entrusted Marcus Rosenbluth with their savings until $50 could be sent overseas for the transatlantic passage to New York and a train seat to Philadelphia.

Over one hundred years later, President and Chief Executive Officer Hal Rosenbluth continues to provide superior service to customers and to exploit change in his growing company. Hal Rosenbluth began working in the family business the day after he graduated from the University of Miami in 1974. In the late 1970s, when the company had sales of $20 to $30 million annually, Rosenbluth identified *airline deregulation* as another monumental change to be exploited in the travel industry. Prior to deregulation, the processing of airline reservations was fairly simple. A travel agent's tasks consisted of asking clients if they wished to fly first class, coach, or economy; finding the most convenient arrival and departure times; and securing seats on those flights. With deregulation, new airlines, new routes, myriad new fares, and constant changes in all of these became a huge source of frustration for travelers and posed an enormous challenge to the travel industry.

Rosenbluth capitalized on the confusion created by deregulation by providing *guaranteed lowest priced air fares* to customers, a much needed, distinctive service. The company purchased computers and leased access to the airlines' new computerized reservation systems (CRSs) for direct access to flight information. Because the airlines' information technology had certain biases (their flights were listed first and so were easier to book), Rosenbluth set out to develop its own independent reservation software with a "bias" toward the lowest possible airfares.

Corporate Culture

Employees: Hal Rosenbluth believes that employees serve customers better in a caring, nurturing atmosphere where co-workers are encouraged to work as a team. All Rosenbluth employees, who are referred to as "associates" and "leaders," receive a company orientation for two days. Reservation agents, who are referred to as travel services associates (TSAs), additionally receive formal reservations training for two to eight weeks. This training consists of role-playing, computerized tutorials, and mentoring. TSAs are taught to extend "elegant service" to clients, which begins with the consistent greeting, "Welcome to Rosenbluth International; this is *Crystal Larson*, how may I help you?" Elegant service continues throughout the client contact with TSAs using language such as "Certainly" (rather than "Sure") and "May I place you on hold?" (rather than "Can you hold?").

Once on the job, training continues. All TSAs are "cross-trained" in at least 3 areas. This helps them understand the entire reservation process better and provides the opportunity to spread tasks out among a larger group of associates if necessary. Rosenbluth associates are guided by a formal total quality management program to constantly review the company's processes and identify areas of possible improvement. Ad hoc teams are formed to implement continual improvement actions.

Client Relationships: Rosenbluth opened its corporate travel division in 1964. In 1984 the DuPont Company was the first company to consolidate its travel program with one travel agency. DuPont selected Rosenbluth and remains one of its largest clients. Every corporate client's relationship with Rosenbluth is different. Sales associates nationwide work to bring in new business for the company. Clients work with the sales associates before signing on to achieve mutual business plans which ensure that all client requests are addressed. Some client offices have on-site reservation centers.

Supplier Relationships: Rosenbluth sees developing close relationships with suppliers to be a win-win situation. For example, Marriott trains and empowers Rosenbluth associates to solve problems in the same way as Marriott employees.

Linton—Centralized Support Facility

Many of Rosenbluth's centralized support operations are conducted at its Linton, North Dakota, facility. In 1988 Hal Rosenbluth set up a temporary data processing office at a vacant John Deere warehouse in Linton as a philanthropic gesture to help the farmers whose crops were being destroyed by drought. Because the Linton associates learned so quickly and so well and possessed the cooperative spirit that Rosenbluth looks for, Rosenbluth made the office permanent. Linton is a branch office and "business unit" for reservations. (Rosenbluth is organized into approximately 45 business units domestically; most of the business units are comprised of more than one branch location.) Today the Linton facility employs over 150 associates and has annual sales of approximately $15 million. Employee turnover at the Linton facility is roughly 2%.

The following discussion focuses on the operations in Linton in examining the systems involved in Rosenbluth's computerized reservation process, customer service, and accounting operations for processing airline tickets. In addition to reservation operations, Linton also provides 75% of customer service for all post-reservation operations for Rosenbluth and operations for processing all airline tickets issued by the company.

(Reservation operations and customer service are referred to as "front room" operations in a travel agency. Processing of airline tickets and client reporting activities are referred to as "backroom" operations.)

Computerized Reservation Process

Approximately 38 travel services associates (TSAs) work in reservation operations at Linton. TSAs handle 30 to 40 calls per day. Reservation operations at the Linton facility include the following areas:

(1) *Corporate Reservations:* Approximately twenty travel services associates (TSAs) handle corporate clients for Linton business unit reservations (60% of the time) and overflow from other business units (40%).

(2) *International Reservations:* Branch offices take international reservations and e-mail them to Linton for processing by one of four TSAs.

(3) *Internal reservations:* Three TSAs handle travel for Rosenbluth employees who are traveling on business.

(4) *Hotel Desk:* Four TSAs handle high volume hotel reservations; requests for these are taken by the branch offices, then sent to Linton for processing. The Hotel Desk is usually used for smaller hotels and clients who request that reservations be made directly with the hotel rather than through the airline reservation systems.

(5) *Data Entry:* The data entry team of four TSAs builds customer profiles based on the profiles submitted by the client on paper.

(6) *Satellite Ticket Printing (STP ticketing):* 380 clients have satellite ticket printers in their own offices; processing these tickets is handled at Linton.

(7) *Leisure Travel:* Three TSAs in Linton provide vacation planning services to employees of corporate clients. Reservations for leisure travel involving airline tickets are handled at the National Leisure Center in Pittsburgh. Land-only reservations are made at Linton.

Since gross margins in the travel industry are less than 3 to 4%, success depends on the efficiency of the reservation process. This process converts a client's request for travel services into the tickets, itineraries, and other travel documents sent to clients. Altogether Rosenbluth's business units issue between 50,000 to 60,000 tickets a week; in the event of fare wars or carrier bankruptcies, this could go up to 100,000 to 120,000 per week. Rosenbluth has approximately 2,100 TSAs working at all business units on a typical day. TSAs handle 2 1/2 to 3 phone calls for every transaction completed. An average reservation takes approximately 13 to 15 minutes from the time the call is answered to finalizing the "after-call work" in the documentation of the record. Total phone time in a recent month for all of Rosenbluth's reservation centers was nearly 2 million minutes.

About 95% of corporate reservations are taken by telephone. In addition to telephone reservations, Rosenbluth also accepts reservations in person at a branch location; by e-mail; by voice mail; and by fax. All reservations pass through the same automated

EXHIBIT 1

RESERVATION FLOW CHART

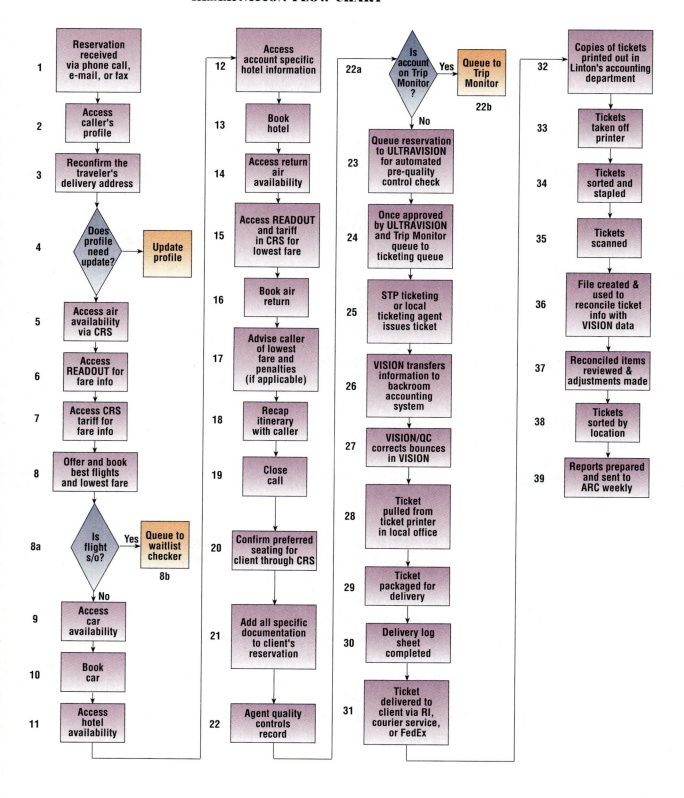

1. Reservation received via phone call, e-mail, or fax
2. Access caller's profile
3. Reconfirm the traveler's delivery address
4. Does profile need update? → Update profile
5. Access air availability via CRS
6. Access READOUT for fare info
7. Access CRS tariff for fare info
8. Offer and book best flights and lowest fare
8a. Is flight s/o? → Yes → Queue to waitlist checker (8b) / No
9. Access car availability
10. Book car
11. Access hotel availability

12. Access account specific hotel information
13. Book hotel
14. Access return air availability
15. Access READOUT and tariff in CRS for lowest fare
16. Book air return
17. Advise caller of lowest fare and penalties (if applicable)
18. Recap itinerary with caller
19. Close call
20. Confirm preferred seating for client through CRS
21. Add all specific documentation to client's reservation
22. Agent quality controls record

22a. Is account on Trip Monitor? → Yes → Queue to Trip Monitor (22b) / No
23. Queue reservation to ULTRAVISION for automated pre-quality control check
24. Once approved by ULTRAVISION and Trip Monitor queue to ticketing queue
25. STP ticketing or local ticketing agent issues ticket
26. VISION transfers information to backroom accounting system
27. VISION/QC corrects bounces in VISION
28. Ticket pulled from ticket printer in local office
29. Ticket packaged for delivery
30. Delivery log sheet completed
31. Ticket delivered to client via RI, courier service, or FedEx

32. Copies of tickets printed out in Linton's accounting department
33. Tickets taken off printer
34. Tickets sorted and stapled
35. Tickets scanned
36. File created & used to reconcile ticket info with VISION data
37. Reconciled items reviewed & adjustments made
38. Tickets sorted by location
39. Reports prepared and sent to ARC weekly

quality control checks as traditional phone reservations. (Rosenbluth Vacations, the business unit which offers vacations to consumers, accepts reservations through online services, such as Prodigy® and CompuServe®, and is testing interactive TV and Travel Channel applications.)

The computerized reservation process for reservations received by phone, e-mail, or fax is shown in the reservation flow chart in Exhibit 1. This process is used by TSAs at all locations, which are networked with the computer hardware housed in Philadelphia. This OM process will be discussed using the steps which are listed in the order of the script on the TSA's computer screen. (Numbers in parentheses refer to the numbered items on the Exhibit 1.) The following discussion assumes that an airline reservation is received via telephone call.

Upon answering the call (#1), the TSA follows a script displayed in a Windows format on the terminal which prompts him or her through the reservation. The TSA accesses the caller's profile ((#2), and reconfirms the traveler's delivery address (#3), if it has changed, the profile needs to be updated; the TSA makes changes as necessary (#4). The TSA then asks the client for information about desired travel destinations and times. The TSA then accesses airline availability using the airline-owned CRS (computer reservation system) (#5) and calls up Rosenbluth's proprietary READOUT® database which lists airfares in ascending order of price in one quadrant of the work screen (#6).

The TSA accesses tariff (fare plus tax) information through the computerized reservation system (#7). The TSA scans all the information on the screen and tells the client the best times and lowest fares available for the desired outgoing flights (#8). If the desired flights are sold out (#8a), the associate will send the information to an automated waitlist checker (#8b), so if a seat on the flight opens up, the information is relayed to a queue in the associate's computer terminal. If needed, car and hotel availabilities are accessed and bookings completed (#9–13).

Next, READOUT and tariff information are examined for return flights (#14 & #15). The associate books the return flight (#16), advises the caller of the lowest fares (#17), recaps the itinerary for the caller (#18), then closes the call (#19). "Talk time" on the call, from answer to close is typically 2½ to 3½ minutes.

Post-call, the associate confirms the client's preferred seating through the CRS or the telephone when necessary (#20). The software automatically appends the reservation with client-related information such as corporate ID numbers, frequent flyer numbers, meal requests and seat assignment information (#21).

Reservation teams maintain daily service statistics which are updated on a real-time basis. The goal is to answer 90% of calls within 10 seconds and the remaining 10% within 20 seconds. For example, the following statistics might be gathered on a typical day:

	Number of Calls	Average Speed of Answer (seconds)	Number of Abandons	Total Service Factor
Corporate/Domestic	357	6	12	94%
International/VIP Departments	69	5	1	96%
Group Department	20	2	0	100%

"Abandons" occur if the caller hangs up before the call is answered. Average abandon time is less than one minute. A promise to clients regarding how fast calls are answered is part of Rosenbluth's "Total Service Factor." The Total Service Factor is the total percentage of calls answered within a specified number of seconds. This percentage averages the number of calls answered, the average speed of answer, the number of calls abandoned and their abandoned time. With many clients Rosenbluth International contracts for a specific service level. Both staffing and network design/monitoring are many times dependent on this service level.

If the TSA has noted any errors on the screen (e.g., the phone number for a hotel is incorrect), he or she gives the changes to a data entry associate.

The TSA then runs the reservation record through one or two automated quality control systems (#22 & #23): the ULTRAVISION® system seeks to eliminate errors in the reservation; its fare checking module ensures the lowest fare was identified and booked. In addition, Rosenbluth's patented Trip Monitor™ system is a value-added service available to customers. Trip Monitor repeatedly rechecks the reservation for a lower fare through the departure date (#22a & #22b).

After the reservation record goes through quality control, the ticket, itinerary, and any other travel documents such as boarding passes are printed out at the local business unit or at a Satellite Ticket Printer (#24, #25, & #28). If printed at the business unit, an associate packages the ticket for delivery (#29) and then completes a delivery log sheet (#30). Tickets are delivered to the client via Rosenbluth International couriers, an independent courier service, or FedEx (#31).

If TSAs have downtime, they:

(1) work on records in the queue,
(2) work in the total client satisfaction center if cross-trained to do so, or
(3) update their knowledge of the travel industry with computer-based learning modules or articles from periodicals.

Backroom Operations

As the tickets are printed, the reservation record is simultaneously transferred to the backroom system Rosenbluth calls VISION® (#26). The quality control editor program within the VISION system automatically checks the reservation template to make sure that information in fields such as customer name, ticket numbers, form of payment, credit card numbers, agent numbers, and customer numbers are correct. VISION flags errors such as missing data or incorrectly formatted data. VISION also checks the "dollar savings comparison fields," the accuracy of the internal Rosenbluth number, and different fare remarks and reason codes. The system "bounces" the reservations with errors (less than 1% of total ticket volume) to a QC associate's computer screen (#27); the QC associate fixes problems 100 percent of the time. Reports of errors go back to business units requesting them. Eight associates work in the VISION quality control area.

Two copies of all airline tickets issued by Rosenbluth are printed out in the backroom operations area at Linton (#32). The first copy is for the agency's records, the second copy for the Airline Reporting Corporation (ARC) in Louisville, Kentucky. If the ticket was charged to a credit card, there is a charge form as well. The agency copy is stored for 2 years in the "ticket library" in Linton; international tickets are stored at overseas business units. Currently 5 to 6 million ticket copies are stored in Linton.

After the ticket is taken off the printer (#33), all copies are sorted and stapled (#34); then the ticket is scanned (#35). (For Worldspan and SABRE, the tickets are compared to a written report rather than scanned.) A file is created and used to automatically reconcile the ticket information with the data in VISION (#35). Reconciled items are reviewed and adjustments made (#37). Tickets are then sorted by ticketing location (#38).

Linton backroom associates prepare reports for each ticketing location for ARC weekly (#39). Every Thursday over 1,000 reports are sent via FedEx to ARC. Jointly owned by the airlines, ARC is responsible for reconciling sales of tickets each week for all participating airlines. Accredited travel agencies must report ticket sales for the previous week to ARC and also send them credit card charge forms for tickets purchased. ARC divides the money received among the airlines and pays the travel agencies their commission on ticket sales, which is typically 10%.

If the agency's reported information does not reconcile with the airlines' information, the airline notifies the agency (in the form of a debit or credit memo) and differences are examined. Less than one percent of the tickets need to be reconciled for differences. Situations that would cause these differences include the agency charging too much or too little for a ticket, taking an incorrect commission on a ticket, or using an expired or invalid fare.

If a customer is due a refund, the ticket is sent back to the business unit which booked the reservation; the business unit calculates the amount due to the client, then sends the information to Linton. Linton verifies the information and includes it with the weekly report to ARC.

In the case of a lost or missing ticket, the traveler who lost the ticket contacts the TSA or business unit that took the reservation. This information is relayed to the Linton backroom operations area. Rosenbluth notifies the airline in writing of the lost ticket. If the ticket is not used in 60 to 90 days, the airline will reimburse the travel agency (less a $25 to $50 administrative fee).

Another function handled by Linton backroom associates is commission tracking which ensures that Rosenbluth is paid the proper commissions from its nonairline suppliers. Associates follow up on commissions due to make sure they are collected. They send the suppliers notices which enumerate the amount due when commissions have not been paid. Large suppliers, including Marriott, Holiday Inn and Choice Hotels send a magnetic tape of all transactions, so the process in these cases is automated.

Total Client Satisfaction Center (TCSC)

If problems arise after the trip or additional information is needed, the client calls the reservation center where he or she made the reservation. If the issue requires additional research, it is forwarded to the TCSC for resolution via an electronically filed Detailed Information Request (DIR) script. In Linton, 28 TCSC associates take DIRs from TSAs who submit them on behalf of clients with problems, complaints, or questions. Linton TCSC associates typically handle 150 to 175 DIRs per day and may receive 180 to 200 calls per day from TSAs who handle clients' inquiries. TCSC associates respond to all inquiries within 24 hours. Front-line empowerment of associates provides immediate resolution to many problems. TCSC associates not only address and resolve issues involving clients, but also provide customer service assistance for other associates and suppliers. If a client has changes or problems during a trip, the client can call the business unit or Rosenbluth International's En Route Service.

Total Client Satisfaction is measured by a complaint ratio (reported complaints to transactions) which is typically .05%; periodic client surveys; and other measurements.

The top two requests are for copies of tickets or invoices that the client misplaced or lost. Many client calls deal with fare inquiries; this is not surprising since the airlines issue 50,000 to 60,000 fare changes per day. Other common inquiries deal with requests to check on frequent flyer mile accounts and hotel no-shows.

Bottlenecks

Bottlenecks in the computerized reservation process are caused by volume fluctuations and failure of the technical systems. One common potential bottleneck is overflow of incoming calls. This problem was solved in 1994 with the installation of a state-of-the art telecommunications network, the only one of its kind in the travel industry. With Rosenbluth's network operations, overflow can be handled by any office on the network. Call volumes at all reservation centers are monitored by the Network Operations Center (NOC), Rosenbluth's telecommunications control facility located in Philadelphia. If the NOC detects high call volumes at a facility, it can automatically reroute the overflow to where the available associates are. Routing for a particular client is worked out when the account is set up. During the winter of 1994, the NOC moved over 80,000 calls in January and February.

In the summer of 1994, Rosenbluth opened a new facility in Fargo, North Dakota, largely devoted to handling excess capacity during peak travel seasons. Known as an "IntelliCenter," this 200+ agent office is electronically networked with every Rosenbluth location around the world. IntelliCenter TSAs, like those in all Rosenbluth offices, have instant access to each caller's travel data, including company policy, reservation records, and personal preferences through Precision and VISION. Future opportunities for the IntelliCenter include multiple shifts and potential around-the-clock service.

Core Measurements

Rosenbluth collects data for various core measurements or "Business Unit Quality Indicators" in the areas of staffing (e.g., associate satisfaction, client retention), telephone statistics (e.g., average speed of answer) and service levels (e.g., air dollar savings). One of the most important of these measures to Rosenbluth is

EXHIBIT 2

**CORPORATE CORE MEASUREMENT:
ERRORS TO TRANSACTIONS INDICATOR**

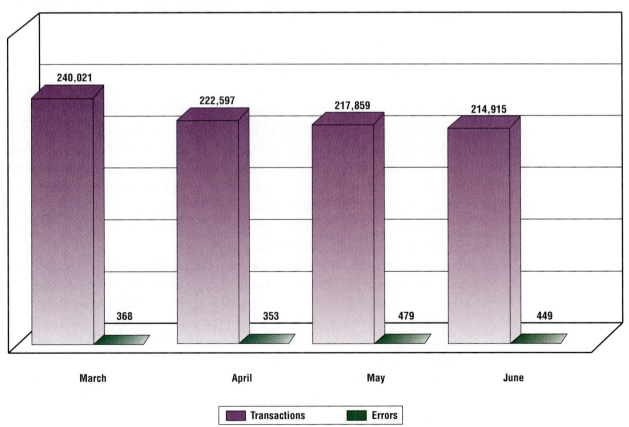

	March	April	May	June
Transactions	240,021	222,597	217,859	214,915
Errors	368	353	479	449

errors to transactions, an example of which is shown in Exhibit 2. The average ratio of errors to transactions for this period is 0.18%. Some of the other most important measures include associate satisfaction, service issues, transaction levels, and productivity.

Future of Rosenbluth and the Travel Industry

Rosenbluth International faces a series of trends which will affect the way it operates in the future. These trends include economic realities, technology, globalization of business, and reengineering.

Trend number one is the economic climate. The whole travel industry is changing as players look for ways to decrease costs while increasing revenues. (For example, Delta Airlines announced a cost-reduction program in April 1994 that has a goal of lowering its operating costs by approximately $2 million by June 1997.) The price of an average airline ticket decreased about $100 from 1993 to 1994. In their struggle to maintain profitability, the major airlines imposed a travel agency commission cap of $25 for any one-way domestic ticket and $50 for any round-trip domestic ticket in February 1995. This represented a significant savings for airlines, because travel agents sell about 80% of airline tickets and agent commissions account for 12% of airlines' operating expenses. At the same time, customers are demanding the lowest unit costs possible, while expecting a high level of service. In direct response to the airline commission caps, Rosenbluth introduced a 1-800 service business travelers can use to gain instant access to TSAs to book international and domestic reservations. This service was launched due to Rosenbluth's expectation that as a result of the commission caps, airline reservation centers and other travel agencies will be inundated with phone calls, and so will not be able to maintain high service levels. The new Rosenbluth service targets business travelers who do not have time to wait when trying to book travel arrangements. Hal Rosenbluth believes that the economic climate will work to improve his company's position in the travel market. Because of increasing travel costs, potential clients will recognize the significant benefits of applying the value-added services and information Rosenbluth can provide. This also will leverage Rosenbluth's position with its suppliers.

A second trend is the proliferation of technology. New technology will allow customers to purchase an airline reservation at an automated teller machine or through a personal computer. Technology also will allow for a "ticketless environment." Hal Rosenbluth predicts that at first travelers will still want documents to prove they have reservations; but 10 years from now, documents will exist only 5% of the time. Technology will affect travel agency operations as well. Colleen McGuffin, general manager at Linton, predicts that dealing with ARC will be "paperless" by 1997. Further, technology will allow for reservationists to "telecommute" to work; that is, to perform their jobs from a terminal in their homes. Hal Rosenbluth foresees that the environment won't be "agentless," but there will be less agent involvement.

The globalization of business is the third trend affecting the travel market. With its roots in international travel, more than 30 international locations and growing, and a vast "global distribution network" already in place, Rosenbluth International is well poised to serve customers around the globe.

Finally, Rosenbluth International also is exploiting the trend of reengineering operations. Rosenbluth and DuPont already have begun to map out an "ideal state" for business travelers, where there are no tickets or boarding passes, no check-ins, no car rental forms, no cash advances, and no expense reports. As reported in the September 12, 1994 *Business Travel News*, the two companies aren't quite sure of all the steps they will take to get there, but they are committed to the journey. The companies don't want to make miniature changes, but want to revolutionize what's going on. They are looking at smart-card technology on which all travel information can be stored. "With one swipe—flash—you're on the plane, in your hotel room."

Case Questions

1. Every Rosenbluth ticket jacket says: "We promise to apply the lowest applicable airfare for routing flown based upon time parameters requested by the traveler or traveler's corporate travel policy." Describe the systems Rosenbluth has in place to fulfill this promise to customers.

2. Describe the way in which Rosenbluth uses technology to solve potential bottleneck problems with incoming calls.

3. Explain the differences between ULTRAVISION and Trip Monitor. Why does Rosenbluth International charge an additional fee to clients who use Trip Monitor?

4. How does Rosenbluth measure productivity?

5. Discuss the potential for the use of EDI in this system. What advantages would it offer over voice or FAX communications?

6. Discuss ways to continue to reduce errors in ticketing.

7. Using Exhibit 1 as a guide, discuss ways to reduce the steps in the reservation process and thus increase customer service. How and when can errors be reduced in the process?

8. Discuss ways in which Rosenbluth can reduce the internal paperwork and copies its systems generate.

9. Given Hal Rosenbluth's vision for revolution, brainstorm ways to reengineer operations in light of future changes projected in the travel industry.

10. Based on predictions for future trends, what actions should Rosenbluth take?

References

Felicity Barringer, "From Milking Cows to Manning Computers," *New York Times*, August 11, 1989.

Charles Boisseau (*Houston Chronicle*), "Airlines experimenting with ticketless systems," *The Cincinnati Enquirer*, September 18, 1994.

Mary Brisson, "Rosenbluth Starts in on Res Reengineering," *Business Travel News*, November 22, 1993.

Donna J. Israel, "Technology Today: The Future Is Now," *Business Travel Manage-ment*, Vol. 6, No. 6, June 1994.

Jennifer McNamara, "Starting from Square One: Times Call for Reengineering, Megas Discover," *Business Travel News*, September 12, 1994.

"Many Happy Returns," *Inc. Magazine*, October 1990.

"Rosenbluth International Honored for Excellence in Customer Service," *CIO:* The Magazine for Information Executives, special issue, August, 1993.

Hal Rosenbluth, "Tales of a Nonconformist Company," *Harvard Business Review*, July-August 1991.

Hal Rosenbluth, "What's Ahead In Back-Office Automation," *Business Travel News*, June 3, 1991.

Julie Schmit, "Delta trims travel agent commissions," *USA Today*, October 29, 1994.

Thomas G. Watts, "New Visions on the Old Frontier," *Dallas Morning News*, April 28, 1991.

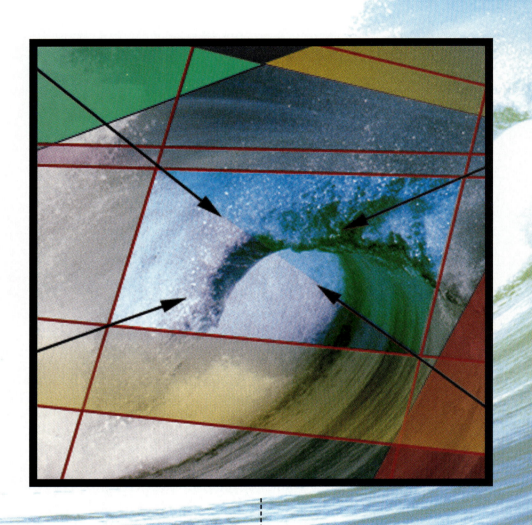

OPERATING THE SYSTEM

Part IV

Part IV provides the particulars required to make the system run, producing goods and services that customers want. Part IV is concerned with obtaining quality outputs, on time, with great efficiency so costs are competitive.

Tactics, not strategies, are at issue. Strategies already have been developed and implemented. They are responsible for setting the stage. This means the scenarios are drawn up which materials management and production scheduling must now carry out in the most productive way possible. Tactics are the details (not included in strategic planning) which fulfill the objectives of the strategies.

Chapter 12 explores materials management, which is responsible for purchasing from external suppliers. Some materials are more critical than others, which should affect how they are treated. Materials also need to be differentiated with respect to dollar volume. When dollar volume is greater there is more opportunity to save money by carefully managing ordering policy. The information system supporting purchasing, receiving, inspection, and storage is explained. Correct selection and certification of suppliers is detailed. The systems perspective is essential for successful materials management in the job shop and flow shop environments.

Chapter 13 moves the focus inside the plant or service facility. There it deals with workforce planning and the control of inventories based on forecasts of demand. The policy of a constant workforce for level production is contrasted with a chasing policy using hiring, layoffs, overtime and subcontracting in an effort to match demand forecasts. Combination policies also are explored using various methods for aggregate scheduling that are developed. Aggregate scheduling applies to batch production and mixed-model flow shop environments.

Chapter 14 explains how to manage inventories of materials that have continuous demands typical of flow shop production. They also are materials that are independent of other end items. The economic order quantity model for such materials is developed, then the economic lot size model is designed for intermittent flow shop production. Using these concepts, perpetual and periodic inventory systems are explained including when each is preferred. The quantity discount model provides further insight for inventory management.

Chapter 15 explains how to manage inventories of materials that have sporadic demands. Often their use is dependent on one or more job shop end-items which necessitates good organization of information. The appropriate information system facilitates material requirements planning which is known as MRP. The methods, strengths, and weaknesses of MRP are evaluated. The chapter also examines capacity requirements planning known as CRP, and distribution requirements planning known as DRP.

Chapter 16 introduces production scheduling of loading and sequencing for the job shop. Loading consists of assigning jobs that are on-hand to departments or work centers. It applies to both services and manufacturing. Load charting is detailed and three loading methods are explained. These are based on the assignment method, the transportation method and a heuristic method. Sequencing is the next logical procedure because it completes the assignment by specifying the order in which jobs are to be done. Various sequencing policies are described.

Chapter 17 explains what cycle-time management is and why it has become so important to manufacturing and service organizations. Line balancing for cycle-time control is applied to intermittent and continuous flow shops. Differences are drawn between deterministic, stochastic and heuristic line balancing situations. Cycle-time management is shown to be one aspect of time-based management (TBM). Queuing models are introduced in this context of balanced work flows when arrivals and servicing times are random variables. A supplement provides simulation of queuing models.

Chapter 12

Materials Management

After reading this chapter you should be able to...

1. Explain what materials management (MM) is and how it connects many functions.
2. Discuss the systems perspective applied to materials management.
3. Explain the "big picture" of materials as part of the cost of goods sold (COGS).
4. Discuss differences in materials management that apply to each kind of work configuration.
5. Explain the materials management information system.
6. Lower the risks that buyers take.
7. Describe "turnover" and "days of inventory" and explain why they are critical measures of systems performance.
8. Explain what is needed for successful centralization of the purchasing function.
9. Discuss the receiving, inspection and storage functions.
10. Discuss why bids are often required before purchasing is allowed.
11. Explain the management of spare parts.
12. Describe the use of value analysis for materials management.
13. Explain the management of critical parts.
14. Describe why parts with high-dollar volume deserve special attention.
15. Explain how to treat parts that have lower-dollar volume.
16. Describe why certification of suppliers is important to MM.

Chapter Outline

12.0   WHAT IS MATERIALS MANAGEMENT?
       The Systems Perspective for Materials Management
       Taxonomy of Materials
       Cost Leverage of Savings in Materials
       The Penalties of Errors in Ordering Materials

12.1   DIFFERENCES IN MATERIALS MANAGEMENT
       BY WORK CONFIGURATIONS

12.2   THE MATERIALS MANAGEMENT INFORMATION SYSTEM
       Global Information System—Centralized or Decentralized
       The Systems Communication Flows

12.3   THE PURCHASING FUNCTION
       The 21st Century Learning Organization
       Buyers' Risks
       Turnover and Days of Inventory—Crucial Measures of Performance
       Turnover Examples
       Purchasing Agents
       The Ethics of Purchasing

12.4   RECEIVING, INSPECTION, AND STORAGE

12.5   REQUIRING BIDS BEFORE PURCHASE

12.6   MATERIALS MANAGEMENT OF CRITICAL PARTS
       The Complex Machine—How Many Parts to Carry
       The Static Inventory Problem

12.7   VALUE ANALYSIS (ALTERNATIVE MATERIALS ANALYSIS)
       Methods Analysis

12.8   ABC CLASSIFICATION—THE SYSTEMS CONTEXT
       Material Criticality
       Material Dollar Volume

12.9   CERTIFICATION OF SUPPLIERS

Summary
Key Terms
Review Questions
Problem Section
Readings and References

The Systems Viewpoint

Materials management is a system of broad-based planning and control over one of the most important factors in the cost of goods sold (COGS). Two trends have been pervasive. The first is the marked decline of the direct labor component of the COGS. The second is the marked rise in the direct and indirect (overhead) cost of materials. Because materials costs are now critical to profitability, most organizations have created a position of responsibility to oversee the many parts of the system that have to be integrated for materials management. The integrator that keeps the materials management system together is the information system. It coordinates functions and permits purchasing to lower the buyer's risks. The two systems parameters— "turnover" and "days of inventory"—are developed. They are critically important measures of systems performance. Certification procedures demand the use of the systems approach.

12.0

WHAT IS MATERIALS MANAGEMENT?

Materials management (MM)

Organizing and coordinating all management functions responsible for every aspect of materials movements and transformations.

Materials management system (MMS)

System used in materials management which is triggered by demands.

Materials management (MM) involves organizing and coordinating all management functions that are responsible for every aspect of materials movements and transformations—called the **materials management system (MMS)**. This system is triggered by demands (including those forecasted) which deplete stocks causing inventory management to request replenishment through purchasing agents or direct contact with suppliers or vendors.

The distinction between vendors and suppliers is a matter of local usage. One, the other, or often both, are used by various companies and/or industries in different regions of the U.S. and the world. Digital Equipment Corporation uses "vendors" to sell its products. Ford Motor Company employs "certification" of suppliers to assure quality and reliability.

The Systems Perspective for Materials Management

Materials management as defined above is an interlinked system as shown in Figure 12.1. Note that the demand for stock is based on production needs and forecasts.

MM connects the external sourcing of supplies to the internal scheduling of product to be delivered to the customer. Mismanage any part of the interlinked system and the adage that "a chain is as strong as its weakest link" applies.

FIGURE 12.1

SYSTEMS PERSPECTIVE OF MATERIALS MANAGEMENT FOR INCOMING STOCK REQUIREMENTS

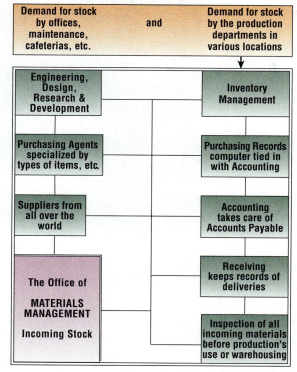

The **internal MM system** requires control of production materials which are process flows. Further, there is control of work in process on the plant floor and finished goods in the warehouse. Some companies consider external control over shipments of finished goods to distributors and customers as part of MM. Other companies limit MM to the use of incoming stock.

The materials management system must be synchronized and coordinated to be effective. The key to synchronization is knowing "when" materials are needed "where." Coordination results in meeting the deadlines. It is worth noting in Figure 12.1 above how many functions must be coordinated. Each deals with different aspects of materials management as part of the internal supply chain of the company. The incoming logistics system of the company is the outgoing logistics system of the supplier. **Logistics** are the strategies and tactics of materials and product distribution.

Taxonomy of Materials

Every business needs to acquire materials to function. The process of acquisition is known as buying or purchasing.

There are three main classes of materials that have to be purchased and managed. First, there are raw materials (RM). These are generally extracted from the ground, then refined, but still the basic ingredients. Examples include: mined metals such

Internal MM system

Requires control of production materials which are process flows.

Logistics

Strategies and tactics of materials and product distribution.

A materials management system enables a company to produce and deliver products on time to its customers.
Source: AlliedSignal

as copper, gold, and platinum, chemicals such as sodium and potassium salts, manganese and phosphates, grains such as wheat and rye, beans such as coffee, natural gas, and petroleum.

Raw materials have value added by operations. Value adding occurs when purchased components are further transformed by the company's production process. Thus, refining, processing, packaging, and shipping when done by the organization are value-adding and profit-making processes.

All buyers of raw materials specify their required quality standards. Grains can be too dirty. All soy is not alike. Coffee prices vary with the perceived quality of the taste of the beans. Raw materials are often bulky and require special spacious storage bins. Companies prefer to locate their refining operations near the source of raw materials so they do not have to transport tons of materials from which pounds or even ounces are eventually derived for use.

Organizations that are in the business of supplying raw materials at the very start of the upstream acquisition process are themselves dependent on purchasing. A quick summary includes the equipment to dig in the mines or harvest the crop. The deposits in which the mines are located must be acquired. The land to be planted and farmed must be procured. Mining requires tools and lubricants, and farming demands seeds and fertilizer. This leads to the somewhat trite statement that "every organization has a supplier." Both miners and farmers have offices to run to keep the records for themselves, their owners and the tax office. Everyone needs to buy paper and clips. Few are without computers. Miners, refiners, farm workers, and office personnel all need the basic necessitates for proper hygiene in the office or plant.

Second, components (C) and subassemblies (SA) are purchased materials that have greater value added than the raw materials. They are, in fact, composed of raw materials that already have experienced value adding. C and SA are characterized

Firms that supply raw materials
are themselves dependent on
purchasing materials.
Source: FMC

by some degree of fabrication, assembly and manufacture. They are assembled into
higher-order products by combining them with each other and with other parts
made by the producer.

This results in work in process (WIP) which can be stored or shipped as
finished goods (FG). WIP has more value added than the purchased subassemblies.
There is a progression of value adding that starts at raw materials and moves up to
finished goods.

Clearly defined quality standards for the components and subassemblies must
be set by the materials managers for the producer. As some materials are transformed
from raw materials to finished goods, they become less bulky. This characterizes
the analytic processes which start with tons of raw materials and reduce them to
smaller amounts of work in process based on the thought that less bulky still
require storage space.

The opposite effect occurs for many manufactured products which employ synthetic
processes. These assemble components into bigger and heavier subassemblies and
eventually finished products like farm tractors, diesel locomotives, automobiles, and
commercial airliners result. There is clearly an advantage in having such large and
heavy finished goods near to their marketplace so they do not have to be transported great distances to the customers.

Materials managers choose suppliers based on their locations. There has to be a
balance between where to buy bulky raw materials and where to assemble big products. Decisions about where to buy materials and where to locate assembly for
shipment to the marketplace are based on analyzing the costs of transport, handling
and storage. Being well-informed always represents an advantage because the alternatives being considered are better ones.

To illustrate, some organizations purchase all or part of the finished goods that
they sell as if they made it. This may be because the organization cannot meet demand,
and therefore employs the services of coproducers to augment their own output.

The finished goods are either shipped directly to customers or put into finished goods inventory. There have to be tight quality controls on the standards and the meeting of those standards.

Coproduced (often called copacked) products have much of their value added by the supplier. The purchaser's profits are related to marketing, selling and shipping the product. There are instances of where firms buy finished goods for sale in certain countries while making them in others. Analysis is based on running the numbers for alternative taxes and tariffs in the countries where finished goods are purchased and sold.

Cost Leverage of Savings in Materials

Purchases from suppliers represent a large percentage of costs that must be recovered before the organization can start to make a profit. Looking at this issue in another way, if either direct labor or materials costs could be reduced by the same percentage, the average organization would save twice or three times as much by choosing the materials costs reduction. The numbers differ by industry ranging as shown in Table 12.1.

TABLE 12.1

COMPARISON OF LABOR AND MATERIALS COSTS AS AN AVERAGE PERCENT OF THE COST OF GOODS SOLD FOR AN AVERAGE ORGANIZATION

	Percent Range	Average %
Direct Labor Costs	05–15 percent	10
Direct Materials Costs	20–40 percent	30

Say that the expenditure of $5,000 on an operations improvement study could reduce labor costs by 10 percent from $150,000 to $135,000. There is a net saving of $10,000. Using the percentages in Table 12.1, in the average organization, materials costs would be three times larger, or $450,000. If the $5,000 study could reduce materials costs by 10 percent, that would produce a saving of $45,000, or a net saving of $40,000. Spending to reduce materials costs provides savings that are four times greater than spending to reduce labor costs. Thus, investments in improving materials costs instead of labor costs are preferred because they have better leverage.

When only manufacturing industries are considered, the direct materials costs in Table 12.1 above range between 40 and 80 percent. The average percent for direct materials would be 60 percent. Thus, an expenditure of $5,000 on an improvement study to reduce material costs by ten percent would produce savings six times greater than the $15,000 savings associated with reduction of direct labor costs by 10 percent. The net saving would be $90,000 − $5,000 = $85,000. Spending to reduce material costs provides savings that are 8.5 times greater than spending to reduce labor costs. Thus, manufacturers should prefer investments that reduce material costs rather than labor costs because they have even better leverage than the average organizations described in Table 12.1.

These examples lead to conjecture about the persisting pressure for downsizing and continued adversarial relationships between unions and management. Until 1960, manufacturing was labor intensive. Efforts to control labor costs were rewarding.

A news service uses information as a material.
Source: Digital Equipment Corporation

Now, many services are labor intensive. In some service industries, Table 12.1 is reversed with materials being the smaller percentage of the cost of services rendered. Efforts are being made to control service labor costs. There also are numerous examples of the unionization of service industries.

Information is one of the "materials" used by services. When properly managed, information can reduce the labor costs of services. A well-run travel agency uses the computer to provide quality service at low labor costs. Mail-order companies can trade-off manual labor costs for computer input-output costs. The latter wins handily.

It should be noted that when purchasing pursues a high quality, low-cost buying program, the pressure to control prices is passed to suppliers' firms which are upstream from the producer in the supply chain. The expected pass-along reaction of suppliers will be to put their own suppliers, who are further upstream, under pressure to reduce prices. Thus, at each level suppliers may first try to reduce costs by downsizing to control labor costs. Ultimately, they will turn to their own suppliers, searching for additional cost advantages.

It is to be expected that analysis will continue to indicate that for all suppliers, their materials costs have greater leverage than their labor costs. This may not always hold true, but it is worth noting that replacement of labor with technology continues unabatedly throughout most of industry.

It is not certain that these relationships will be found throughout every supply chain. Still, it is noteworthy that at the raw materials level, increasingly technology and energy pay a much larger part of the bill than labor. It also is likely that when averaging across the whole supply chain, the major leverage factor is materials.

The Penalties of Errors in Ordering Materials

Order errors include wrong specifications, incorrect quantities, and overpriced materials. Orders placed for materials should be as correct in every aspect as possible. Mistakes are more costly in some situations than in others, but in all cases, errors are to be avoided.

Worst cases are likely to be for start-ups, and for the mature flow shop. In both of these cases, the process may have to be stopped until the materials problems are corrected. The least worst case is likely to be for the job shop where other jobs can be done while the materials errors are corrected.

Flexible and programmable systems also have some built in advantages of being able to shift from one product to another, with minimum penalty, when materials problems are encountered. Projects, such as bringing out a new product (start-ups), or building an office tower, can be flexible or vulnerable depending upon the stage of the system.

Specifications of materials must be defined exactly. They must be priced right, delivered on time, and inspected for faults. Information about materials must be managed with skill. Stock outages can delay or halt production of goods and services. The need for on-time quality materials is as crucial for services as for manufacturing. Consider the factory that runs out of boxes to package products, the printer without paper, or the fast-food restaurant that runs out of french fries.

The costs of mistakes can be exorbitant. In an actual case, an Exxon refinery needed a replacement gasket costing $85. Several gaskets were listed as being in stock, but when examined, they turned out to be the wrong size. The cost to the refinery was substantial because it had to cut production and send a chartered plane 700 miles to get a proper gasket. The actual cost of this inspection error was not announced but it was costly. The penalty for mistakes depends on the kind of product, the life cycle stage of the product, and the process configuration that is being used.

12.1

DIFFERENCES IN MATERIALS MANAGEMENT BY WORK CONFIGURATIONS

The management of materials requires special rules and decision models for different markets, life cycle stages and production configurations. That is, certain kinds of problems are posed by the materials requirements of flow shops that are not the same as for job shop and batchwork systems. Differences also apply to projects and flexible processing systems.

Flow shops require a continuous supply of a specific set of unchanging materials. The line must be shutdown when stock outages occur. Alternatively, carrying large amounts of inventory can be unnecessarily costly, requiring more storage space than is justified. Obsolescence, deterioration, and pilferage (usually in that order) are the curse of carrying large amounts of inventories.

Trade-off models

Needed to evaluate and balance the disadvantages against the benefits.

Trade-off models are needed to evaluate and balance the disadvantages of large amounts of stock on-hand (SOH) against the benefits of large order quantities placed long in advance of delivery so that substantial quantity discounts can be obtained. The trade-off model also must consider the possibility that too little inventory on-hand followed by supplier strikes, delivery slowdowns, and acts of nature, can lead to production shutdowns caused by a lack of supplies.

Knowing when to order and how much to order is the management imperative of the supply chain. Therefore, an order policy that specifies "when" and "how much" to order for each item, independent of what is happening to other items, is a sensible approach to use for items that have smooth and continuous demands. The flow shop is the work configuration which satisfies this demand pattern.

Wendell M. Kelly and Dennis J. Garrett

• • •

QualitiCare, Inc.

In researching business opportunities in health care and life sciences for the state of Maryland and the Baltimore business community, Wendell Kelly and Dennis Garrett identified an open niche in the distribution end of the supply chain for medical and surgical supplies. The two men exploited this opportunity themselves with the founding of QualitiCare, Inc. They began with a multimillion dollar commitment from the University of Maryland Medical System and Johns Hopkins University and have since expanded their customer base to include 45 medical institutions.

A computer system controls all activities in QualitiCare's 80,000-square foot Baltimore warehouse. For example, when products are received, the computer system tells the technicians where to place the products; fast moving items are placed close to the shipping area and slow movers are placed on higher shelves and to the rear of the warehouse. The system assists technicians in "picking orders" by telling them where products are located. It also generates automatic pick sequences which indicate the most efficient way to pick each order.

The company provides three categories of delivery service including scheduled distribution, just-in-time delivery and "stockless" programs. Scheduled customers may place their orders once or twice a week and receive supplies on a set day. Just-in-time customers place their orders between 10:00 A.M. and 2:00 P.M. and receive supplies no later than 4:00 P.M. the same afternoon. Customers with stockless programs keep no inventory on hand. QualitiCare delivers these supplies directly to the nurses' station of the department that ordered them.

QualitiCare's computer system uses electronic data interchange (EDI) with some customers and manufacturing partners. EDI allows customers to place their orders online from their computer to QualitiCare's computer. All order and usage information is automatically transmitted to the manufacturer. This builds what Mr. Garrett calls "the cycle of continuous replenishment." He explains that "if the manufacturers are able to see your usage on a regular and consistent basis, they can automatically ship things to you once you reach a preset reorder point."

Garrett attributes QualitiCare's success to the partners' ability to understand the changing dynamics of the industry and to bring value-added services to the customer. One value-added service they provide to hospitals is "a materials management software package that instantly allows them to be online to transmit data to us electronically and also to manage their own in-house inventory," said Mr. Garrett.

Sources: Patricia Meisol, "Medical supply firm off to a quick start," (Baltimore, Maryland) *Sun*, October 13, 1993, and a conversation with Dennis Garrett.

The flow shop consumes supplies constantly. Its demand for supplies can be considered to be smooth and regular. This kind of a pattern applies also to the intermittent flow shop, and to stable, big-batch order environments. Suppliers actively compete to supply the materials requirements of high-volume demand systems.

Job shops have a varied and changing set of materials requirements. These may include certain materials, the uses of which are repeated with some degree of regularity and others that are unique or infrequent orders. Frequency of usage is reflected in the dollar demand which is a good measure of item importance.

Generally, job shops involve smaller order quantities of far more kinds of materials than flow shops. When discounts can be obtained, they are lower. Being able to track stock levels for many items is important. It requires an information management capability. One widely-used information planning system—for job shops—is called material requirements planning (MRP). The job shop work configuration can be viewed as having a continuing stream of minor projects with short start-ups, which lack the serious consideration given to major projects with complex start-ups.

The job shop needs to have planning done for materials requirements. Such planning involves having a good information system to track the components used for each job in the shop. The time when those components are required must be specified and acted upon with sufficient lead time which is the replenishment time for an order. Lead-time data is part of the information system that materials management must track. Lead-time management is required to make certain that expected lead times are met and when possible to find ways to decrease the replenishment period.

Many suppliers are involved in job shop management. Changes from one to another are not unusual. A single decision that deviates from optimal materials planning cannot do too much harm. The cumulative effects of repeated instances of bad quality and/or late deliveries can impose significant penalties such as the loss of good customers.

Small penalties accrue to flexible manufacturing systems (FMS), especially if a large variety of products on the menu permit flexible shifting to other work orders while errors in materials are corrected. Minor penalties also characterize service-oriented job shops where flexible processing routines enable switching to alternative tasks while corrective procedures are undertaken with suppliers.

Projects such as building ships or theme parks have stages when certain materials are used that will not be used again. Because the activity stages are nonrepetitive, the majority of orders are preplanned and placed once. Nevertheless, materials requirements can be planned and information systems can be used to track the timing of materials needed for specific project activities.

Order policies concerning when to order and how much apply only to those project materials that are used consistently and uniformly throughout the project. Such ordering policies are appropriate when demands are independent of the specific project activities. This is the case for lubricants for machines, food for workers, paper for the office and lightbulbs. In contrast, if materials requirements are properly tracked, the right amount of cement can be ordered to be delivered on time for a specific construction job of known dimensions.

Often, much work on project materials is subcontracted to suppliers who as specialists are familiar with particular stages of materials requirements for the particular project. Cost of materials are often less crucial than on-time delivery. Sometimes, as with start-ups for new products, the materials required are unique. There is little experience in producing them. Discounts for quantity purchases are rare, whereas special set-up charges are not unusual. Careful follow-up with suppliers is essential when experience with their practices is limited.

When suppliers have minimum familiarity with the products they are supplying, this can lead to production difficulties. To illustrate, the supplier of the stage sets for a Broadway play, may have never before dealt with some of the special effects that are to be used. Without follow-up the director and cast may be in for some surprises when the dress rehearsals begin. On another tack, the unseasoned entrepreneur discovers a succession of bugs which prevent smooth start-up and constancy of output. Professional training in OM would have led to checklists and other procedures aimed at avoiding problems.

There are many tales or "war stories" that can be told about mistakes in materials management. The requirements are all the more difficult because of the fact that while purchasing and process management are part of the P/OM team, they are concentrating on different aspects of the system. Purchasing as a centralized or decentralized buying function is serving different product masters, various departments, diverse divisions and even plants, not all in the same country.

When there is antagonism between the buyers and those whom they buy for, many problems can surface. To prevent this from happening, close communication should be established. Working with purchasing, a good relationship should be established with suppliers. There is increasing belief in a materials system with fewer suppliers and a greater trust relationship between all parties.

Expectations of buyers for the performance of their suppliers change when there is continuity of their supply-and-demand relationship. Buyers expect that improvements will occur in quality, price, and speed of delivery. Some buyers may expect that shipping patterns will move closer to just-in-time (meaning delivered as needed and often abbreviated JIT). The suppliers look for continuity and loyalty from their buyers. A good relationship is derived from mutual benefits perceived by both parties.

12.2

THE MATERIALS MANAGEMENT INFORMATION SYSTEM

The **materials management information system (MMIS)** provides online information about stock levels for all materials and parts (RM, C, SA, WIP, and FG). Stock on order, lead times, supplier information, costs of materials and discount schedules also are part of the MMIS.

An **intelligent information system** is desired rather than a system that acts like a ledger providing only the information that is specifically requested. An intelligent MMIS can provide unsolicited data that is timely for decision making. It can request action (i.e., noting that a withdrawal from stock should trigger a buy decision). That order may be automatically created and sent to the supplier's computer. The idea of a smart information system is not just user friendly. It is a system that can prompt decisions, detect errors and provide a rapid-access, multifunctional systems database.

An intelligent information system is designed with parameters that capture the management issues of *what to buy, when to buy, from whom to buy, how much to buy, how much to pay*. This MMIS is constantly working overtime. The same material management system may have to be able to consider when to make the parts instead of buying them, and when to employ coproducers and copackers.

Global Information System—Centralized or Decentralized

Companies that can compete globally have been moving toward an organizational integration of the information required for materials control. They have been doing this on a global basis. In the past, a variety of materials management activities existed

Materials management information system (MMIS)

Provides online information about stock levels for all materials and parts; other data as well.

Intelligent information system

Provides unsolicited data that is timely for decision making; prompts decisions, detects errors, provides a rapid-access, multifunctional systems database.

as individual operations, each attended by managers who seldom communicated with each other. Eventually, in the search for greater effectiveness, a single, central materials control department has appeared in numerous organizations.

Centralization in purchasing may not be best for every organization, but it is more likely to be the best approach when the information systems infrastructure supports a global network. If the worldwide database is not properly organized, then decentralization is the only organizational format that can cope with the numerous tasks that procurement entails.

Many, if not most, organizations have a vice president in charge of materials control. This senior position may be essential to permit the centralization of the global materials management tasks. The responsibilities vested in a materials control department include at least four subfunctions, each of which has major international implications.

1. Purchasing from suppliers all over the world and having close relations with them. Regulating shipments as to carriers and delivery dates.
2. Maintaining stocks of inventory at various places worldwide, while knowing how much inventory is in place at each location (i.e., inventory stock control, on a global basis). Status in the warehouse includes control over the withdrawal procedures such as "first-in, first-out" or "last-in, first-out." **Expediting** means following up on orders and doing whatever can be done to move materials faster.
3. Inspecting incoming materials using various statistical methods of quality control for acceptance sampling. These inspections are taking place worldwide and must be coordinated by an information system that tracks location and status.
4. Running an international accounts payable department that can take advantage of terms for paying, such as 30 days before interest begins to be charged. Especially, in inflationary economies, such as Brazil's, the interest rate terms can turn out to be as important as the buying price. The Brazilian annual inflation rate was officially put at 6,000 percent, unofficially at 12,000 percent. Ford Motor Company's decision to set up a plant in Sao Paulo undoubtedly would have been reversed if such extreme inflation could have been forecast. Currency ratios also must be considered.

Expediting
Following up on orders and doing whatever can be done to move materials faster.

Each materials management activity interfaces with the others. They are the mutual responsibility of many departments within the organization. Coordination of activities is needed which is why information management is stressed.

The Systems Communication Flows

Many forms of communication unite the various materials management groups within the larger organization. Some companies use paperless systems—at least in parts of the organization. Tandem developed a paperless factory in Texas. IBM in Santa Palomba, Italy, uses electronic data interchange (EDI). The system links other European installations through an online, real-time information system aimed at providing assembly instructions for order fulfillment. Hewlett-Packard in California uses bar-code control of assemblies. Chrysler connects Detroit with its suppliers everywhere using EDI.

Still, most companies have paper-rich trails which are coordinated with computer records. The objectives for paperless organizations remain worthy goals that are gradually being approached. Satellites enable e-mail transmissions to abound raising communication levels by exponential amounts. EDI using computer-aided design permits The Limited to transmit clothing manufacturing patterns and materials require-ments to suppliers in Hong Kong and Indonesia moments after finalizing patterns in Columbus, Ohio and Waltham, Massachusetts.

Electronic information mixes with paper for invoicing accounts payable and for tracking stock withdrawals and stock on-hand. Verbal communications and TV pictures augment the multimedia system. The communications are especially helpful when problems arise due to contingencies, which in the course of otherwise normal mate-rials management, can be counted on to occur. Many times, contingencies lead the materials control department to seek help with problems from its own research and development (R & D) department as well as the supplier's R & D department. Such interchanges can take place on a global basis.

12.3
THE PURCHASING FUNCTION

Purchasing agents (PAs) and their buying organizations have the traditional role of bringing into the organization the needed supplies. While this is very important, the role is changing, becoming more integrated within the organization, and bring-ing in more important information, as well.

Purchasing agents (PAs)
Bring into the organization the needed supplies.

The 21st Century Learning Organization

The purchasing department in the twenty-first century is an information gathering agency. Totally on the up-and-up, by being everywhere and able to listen, it is able to learn about new technologies being used by suppliers and the organizations they supply worldwide. It is on top of new materials, new suppliers, new distribution channels, new prices and new processes that produce quality levels previously not attainable. It has a global reach via satellites and telecommunications capabilities that constantly expand horizons.

This purchasing department is responsive to the marketing strategies of the various suppliers from whom it obtains the required materials. The old and waning role of purchasing is to push suppliers on price. Shopping around for the best prices is no longer done. It is far from "best practice." The price-tag approach has been thrown out in favor of a long-term relationship with a few trusted suppliers.

The importance of the buying function depends upon the extent to which the company requires outside suppliers. The determination of what to make and what to buy is an OM decision, however, purchasing department information can be crucial. It is evident that the decision often depends on the terms to buy, including price, quality, delivery, and innovations, among others, which purchasing learns about and communicates to the OM team.

Certainly, if the production department cannot make the product, then the importance of the buying function increases. Few mail-order companies and/or depart-ment stores, produce any of the materials that they offer for sale. Supermarkets use the capacity of copackers to offer products with their name. Purchasing provides the leverage for such products.

Buyers' Risks

Mail-order and department store PAs generate such large volumes that they are of equally great importance to the mail-order company for which they work—and to the producers from whom they buy. Purchasing agents for flow shops also deal in such large volumes that making them happy is considered part of the supplier's business strategy. In many circumstances, purchasing managers are responsible for the success of their companies. In keeping with this responsibility is the salary levels that such buyers receive.

Purchasing for mail-order and retail eventually drives the logistics of the supply chain. This OM sales-service area is involved in a situation where the ability to shift rapidly to new product designs and suppliers may be essential for success.

Fashion is fickle. Women's clothing styles change in complex ways. Manufacturers have been caught with large unsold inventories as a result of committing to a style that didn't gain acceptance. When turnover is poor for retail or mail-order, the goods are marked down and sold quickly at large discounts that provide no profit, and often at a loss. Books and toys that do not sell are marked down and remaindered.

Turnover and Days of Inventory—Crucial Measures of Performance

Days of inventory (DOI)

Measure equivalent to dividing total stock on-hand by the average demand per day; expected length of time before running out of stock if no further stock is added; inventory backlog to be worked off.

The performance of the purchasing agent in the supply chain is related to the speed (or velocity) with which goods sell. There are various ways of measuring that rate. Velocity is measured as units sold—moved to the customer—per period of time. **Days of inventory (DOI)** is a measure that is regularly reported by many firms. DOI is equivalent to dividing total stock on-hand by the average demand per day. It is the expected length of time before running out of stock if no further stock is added. DOI changes when new stock is added and when expected demand speeds up or slows down.

Automobile companies, furniture manufacturers, and makers of baby strollers reflect the broad spectrum of companies that customarily report DOI. There is a comfortable level in each case, such that one month's supply of autos would usually be considered too little—six or more months of supply would be considered excessive. Ten months might be interpreted as a signal of dangerous accumulation of inventories. The computation of days of inventory is shown below, with a numerical example, where *AVG INV* is average inventory:

$$\text{DAYS OF INVENTORY} = (\text{AVG INV})/(\text{DAILY COGS})$$
$$= 365 \times (\text{AVG INV})/(\text{ANNUAL COGS})$$
$$= 365 \times (5)/(50) = 36.5 \text{ Days}$$

where days of inventory equals:

1. The average inventory (*AVG INV*) valued in millions of dollars is $5; it is identical with average dollar inventory (*AVG $ INV*).
2. *AVG INV* is divided by the daily cost of goods sold (*DAILY COGS*). *DAILY COGS* is derived from *ANNUAL COGS* by dividing the latter by 365 days.
3. $5MM divided by $50MM multiplied by 365 days = 36.5 days. This is equivalent to (*AVG $ INV*) being divided by the daily cost of goods sold.

$$\frac{AVG\ \$\ INV}{COGS\ PER\ DAY} = \frac{(\$)}{(\$)\ /day} = days\ of\ inventory$$

The dimensions of this equation are derived to show the way in which "days of inventory" result from the calculations.

Another measure that is used which captures the essence of supply chain objectives regarding throughput and value added is turnover. **Turnover** is measured as the number of inventory turns (which is the number of times that the warehouse is emptied and refilled per year). Inventory turnovers can be measured with different time periods than a year such as turns per month.

The same concept of turnover can be applied to retail stores, department stores, mail-order firms, manufacturers and distributors moving goods off the factory floor to the customer. The computation of inventory turnover is shown below, with a numerical example:

$$\text{INVENTORY TURNS} = \text{(ANNUAL NET SALES)/(AVG INV)}$$
$$= 60 \text{ MILLION }/6 \text{ MILLION}$$
$$= 10 \text{ TURNS PER YEAR}$$

The numbers that are used for this example are related to sales dollars rather than cost dollars. The latter were used to calculate days of inventory. Numbers, in this case, are based on a sales price of $1.20 per dollar cost of goods sold. Average inventory is estimated in terms of sales value rather than COGS.

Note the relationship between 10 inventory turns and 36.5 days of inventory. Thirty six and one-half days is the interval between turns. Both measures tell different but related stories about the same system. Inventory turns give a sense of the velocity with which inventory moves out of the warehouse. Days of Inventory states the inventory backlog to be worked off.

Using these measures, companies can determine how well they are doing in moving materials throughout the supply chain. If a soft drink company fails to make as many units of beverage as had been forecast, then days of inventory of sugar may grow too large and new orders will not be placed. For toys, books, VCR tapes, and autos, excessive days of inventory are costly.

Turnover Examples

The Limited aims to achieve better than 100 turns per year in their warehouses. Wal-Mart targets turns as great or even greater. Thomasville, a furniture company, has taken the traditional two to four inventory turns for suites of furniture into ten or better per year.

Good turnover brings large rewards and a reputation for purchasing success. On the other hand, there are great risks of failure. Poor measures mark the marginal competitor. Buyers and purchasing agents have to accept the risk of making errors such as overestimating demand or paying too high a price. Penalties of being wrong are high.

Purchasing Agents

Purchasing records provide a history of what has been done in the past. Documenting history is useful including what the costs were, who the major suppliers were, what discounts were obtained, what quality levels achieved, and delivery periods for specific items. Without documentation, the supplier history of a company can be lost.

Turnover

Measured as the number of inventory turns (number of times warehouse is emptied and refilled per time period); good turnover brings large rewards and a reputation for purchasing and sales success.

The Limited aims to achieve better than 100 turns a year in their warehouses.
Source: The Limited, Inc.

The skills and experience of purchasing agents are not readily transferable between different industries. They may vary between companies even in the same industry. Differentiation exists by types of materials, buying and shipping terms, and supplier purchasing traditions.

It is not possible to discuss all of the intricate relationships that have been developed by buyers and suppliers in order to achieve maximum satisfaction for both parties. A number of important procedures are covered, but new ones are being developed all of the time which take advantage of changes in the information, storage and transportation technologies.

The purchasing function is responsible for bringing the exact materials that production needs in time. Purchasing is the interface between P/OM and suppliers. P/OM may have some decisive requirements and some exceptional advice for suppliers. It is not surprising, then, that this part of the supply chain is tightly coordinated with P/OM. Whatever the organization's structure, purchasing must be part of the OM team.

When the purchasing process is technical, the purchasing agent may be an engineer or a person who has worked with the production department. Purchasing is usually responsible for the following functions which might be called the purchasing mission:

1. Ordering what is needed in the right quantities. Then meeting all quality standards at the best possible prices—always achieving delivery reliability. This mission must be coordinated with P/OM and marketing with respect to what is going to be needed when. Purchasing must be assisted by OM in predicting the amount of scrap. By increasing order sizes to compensate, costly orders can be avoided for a small number of units needed to complete orders.

2. Receiving inventories is part of the materials management function. Usually "receiving" is the responsibility of purchasing. How else can it be determined that the deliveries are on time and that P/OM will have what it needs as it schedules production runs? Will it be just-in-time, or will extra stock be carried just-in-case?

3. Inspecting the incoming goods to make sure that their qualities meet specifications.
4. Purchasing is the specialist in knowing which suppliers to use. Keeping in touch with change is necessary.
5. Purchasing also is the materials management function to turn to when Engineering Design Changes (EDCs) occur which demand changes in the specifications of purchased materials. What happens to the stock on-hand that is outmoded? How fast can the new specifications be made and shipped? If the company is constantly changing designs (which many are doing) then knowing which suppliers can cope with shifting demands is important.
6. Stability of supply relationships has value.
7. Purchasing must be adept at coordinating the materials that are needed for start-ups. The management dynamics that are associated with start-ups are entirely different from those which operate successfully for mature products.

Coordinating the goals of materials management with those of process management is one of P/OM's greatest responsibilities. Scenarios applicable to Functions 1 and 4 are common where a variety of suppliers are dealt with and about one out of ten has serious failures. Consequently, there is regular shifting of suppliers over time requiring P/OM adjustments.

A scenario applicable to Function 2 has the receiving dock forgetting to log in a shipment which results in a "false" crisis when that item seems to have run out of stock. In some instances, the mistake is not traced and even though there is the necessary stock on-hand, it is lost in the warehouse.

The following story for Function 3 occurs repeatedly. Inspectors fail to check all of the quality standards and put defective received goods into inventory. The production line may be forced to shutdown.

The situation that is involved with Function 5 is extremely serious in companies where technological change is moving at a rapid rate. Product malfunctions in the field lead to parts being redesigned and EDCs issued, sometimes in bunches, in the hope that the problems will be corrected. Even with mature aircraft such as the Boeing 737, design changes in the reverse thruster were called for after it had been flying for many years.

The Ethics of Purchasing

Purchasing agents make buying decisions that involve enormous amounts of money. As labor costs have contracted and materials costs have increased, the power to buy is vested in few hands. All over the world, there are cases of supplier companies attempting to influence purchasing decisions. In some places in the world this is not considered illegal or unethical.

In the U.S. it is not ethical or legal. The bidding section below mentions the way in which companies use bids to provide the appearance of purchasing decisions that are not influenced by gifts of any kind. The only sure way to avoid problems is to make certain that purchasing agents are trustworthy and moral.

It is understandable that purchasing agents achieve special arrangements with suppliers that they trust. After years of dealing with a supplier, a friendly relationship can develop which is an entirely ethical alliance. Both buyer and supplier value the long-term stability and goodwill of their relationship.

Eastman Chemical Company

Rated "number one" supplier by exceeding customers' expectations.

A 1993 Malcolm Baldrige Award winner, Eastman Chemical Company's vision is "to be the world's preferred chemical company" by exceeding customers' expectations. The company has fulfilled this vision. For four years running, more than 70 percent of Eastman's worldwide customers have rated the company as their number one supplier. Founded in 1920, the $4 billion chemical company manufactures and markets more than 400 different chemicals, fibers, and plastics for 7,000 customers around the world.

For seven consecutive years Eastman has earned an "outstanding" rating from its customers with regard to product quality, product uniformity, supplier integrity, correct delivery, and reliability. Eastman's "no-fault return policy" on its plastics products is believed to be the only one of its kind in the chemical industry. Based on the findings of Eastman's extensive customer surveys, the policy states that a customer may return any plastics product for any reason for a full refund.

By dialing 1-800-EASTMAN, customers can contact anyone in the company—including the company president—24 hours a day, seven days a week. The hotline receives more than 3,000 calls daily. All calls are answered by the second ring.

Online technical databases also are accessible to customers around the clock.

All employees are trained to gather complaint information and enter it into a companywide database. Customer advocates follow up and resolve the complaints. A Customer Interface Core Competency Team monitors and detects changes in customer satisfaction measures, determines root causes, and develops ways to improve.

Eastman is the tenth largest chemical company in the United States and 34th in the world. With approximately 18,000 employees, the company operates manufacturing facilities at its headquarters in Kingsport, Tennessee, and in New York, South Carolina, Texas, Canada, and the United Kingdom.

Source: Malcolm Baldrige National Quality Award *Profiles of Winners*, 1988–1993.

In the business environment of the U.S., personal relationships are not considered to be a reasonable basis for enterprise decisions. They exist nevertheless in less blatant form than in other cultures. For example, in Latin America and the Middle East, personal friendships are held to be business assets that reduce risk and have monetary value.

Part of this cultural difference can be traced to the importance placed upon legal contracts in the U.S. that does not exist elsewhere. Japan has far fewer lawyers than the U.S. The Japanese do not value legal contracts in the same way as in the U.S. With the growth of global business, such factors play a major role in determining management's success in handling the affairs of subsidiaries outside the U.S.[1] This emphasizes another international aspect of materials management and OM.

The design of a storage facility depends on what type of supplies will be unloaded.
Source: Digital Equipment Corporation

12.4

RECEIVING, INSPECTION, AND STORAGE

An important part of the materials management job is receiving shipments from suppliers. There is a need for a receiving-unloading facility designed to take the supplies out of the shippers' conveyance. After unloading the supplies, there is usually a storage area to put them. The design of this facility differs depending upon: what is to be unloaded (type of supplies), what the supplies are unloaded from (trucks, freight cars, hopper cars, ships, planes, etc.), and where they are to be unloaded.

The receiving facility is often called the receiving dock and there is another location for shipping called a shipping dock. In many instances these are the same place. In some instances, in the morning they are receiving docks and in the afternoon they are shipping docks.

In most situations, they are completely separate facilities. Wal-Mart uses **cross-docking** to transfer goods from incoming trucks at the receiving dock to outgoing trucks at the shipping docks. This means that a large percentage of goods never enter the warehouse but cross from one dock to the other. Such cross-docking has been credited with saving substantial amounts of money and time. It is often cited as an example of how P/OM's creativity improved the logistics of distribution operations.

It is not unusual for the freight cars or hopper cars to be used as storage facilities with materials being unloaded as needed. Instead of moving chemicals and plastics from the hopper car to the warehouse, the factory draws directly on the reserves in the hopper cars. The DuPont Corporation has successfully cut down on the number of hopper cars with inventory that are sitting around on sidings by coordinating customers' needs with shipping schedules.

Supplies must be inspected to control the quality and quantity of what was ordered. Is the shipment exactly correct and has it been received undamaged? Specific quality checks are made using acceptance sampling methods. Acceptable materials are moved to the storage facility—often the company warehouse.

Cross-docking

Transfer of goods from incoming trucks at the receiving dock to outgoing trucks at the shipping docks; goods never enter the warehouse; saves time and money.

12.5

REQUIRING BIDS BEFORE PURCHASE

Bidding

Buyer requests competing companies to specify how much they will charge for their product; can involve costs, quality, and delivery among others.

Bidding is a process by which the buyer requests competing companies to specify how much they will charge for their product. Competitive bids can involve more than price. Sometimes, purchasing requests suppliers to submit competitive bids for both cost and delivery time.

In some industries and government systems, purchasing is required to use the bidding process. It should be noted that there are always two points of view with respect to bidding—the buyer's and the seller's. Materials management is interested in the buyer's point of view of bidding which is lowest costs, best quality, and fastest delivery, among others.

With discretionary bidding, purchasing agents might decide to use it to prevent charges of favoritism. It controls expenditures when there can be significant variation in supplier charges. Bid requests state specifically all of the conditions that must be met and ask for details of what the supplier intends including prices, delivery dates and quality specifications and assurances.

Bidding can be a costly process for the MM buyer. This is especially the case when there are many criteria upon which competing suppliers will be rated. Cost also rises when there are many firms that are bidders. On the other hand, for bidding to work, there must be at least two suppliers willing to bid for the job.

Bidding is commonly required of government agencies. The IRS reviewed bids before choosing laptop computers. The Armed Services utilize bids for military acquisitions. Bids are familiar in situations where industrial firms have no prior supplier arrangements and in which costly purchases (including engineering and construction jobs) are to be made. The Federal government almost always awards contracts to the lowest bidder. Other concerns that are taken into account by private industry are seldom allowable with government awards.

Bids can be requested where the price is fixed and the creativity and quality of the solution is at stake. Advertising agencies bid for accounts which have a set budget. Alternative bids are based on campaign creativity. The same applies to a P/OM request for proposals from a consulting organization where the budget allocation is fixed. Competing bids for a computerized materials management system is common.

With so much bidding action, it is not surprising that bidding models have been developed. The models can assist both the buyer who makes the requests for proposals (RFPs) and the sellers who offer the bids. Bidding models indicate that as the number of bidders competing increases, the size of the winning bid decreases. With more bidders there is more variability.

Perhaps this is a good reason for the buyer to include many bidders. On the other hand, each extra bid adds to the ordering costs. There also is the fear that a low bid will be made by an organization that is less likely to produce quality work. Steps must be taken to insure that quality is not compromised by price. Also, too many bidders may drive the expected profit so low that qualified suppliers refuse to join the bidding, leaving the field open to the less qualified.

A company may not choose to buy from the organization presenting the lowest bid. Price is almost never the only factor that needs to be taken into consideration when awarding a contract. Among other things, it is essential to consider quality and guarantees of quality, the experience of the supplier, the uncertainties of delivery, and the kind of long-term, supplier-producer relationship that is likely to develop.

Bidding is applicable for project procurement policies and for start-up purchase arrangements for the flow shop. It is less relevant for the job shop, although it may make sense when costly components are involved. Bidding is a useful protection when there is suspicion that special purchasing deals are being made between suppliers and company personnel.

12.6
MATERIALS MANAGEMENT OF CRITICAL PARTS

For projects and the flow shop, certain parts can fail which will shutdown the line or seriously delay project completion. These are called critical parts. An entire refinery can be shutdown. The cost of lost production may well run into millions of dollars.

How many spares of the various parts that are judged to be critical parts should be kept in stock? How likely is it that a spare part kept in stock for an emergency will ever be called upon to be used? Often, severe technical problems are involved in purchasing critical parts for the maintenance function of complex technological systems.

This is a problem for the materials management-OM team which is familiar with the specific production equipment. It is able to evaluate the failure characteristics of the technology. Policies for stocking spare critical parts also are a function of the type of maintenance that is used. If preventive maintenance calls for replacing critical parts once every year, then that decision establishes the base requirements for stocking the part. When reliability is important, a technical basis must be used for purchasing spare parts.

For an important class of maintenance inventories, spare critical parts can be obtained inexpensively only at the start-up of the facility that will be needing the spare critical parts. If it turns out, later on, that an insufficient supply of these critical parts was acquired, the cost of obtaining additional spares is much higher.

A *failure model* can be constructed to determine how many spare parts should be kept in stock. This is best done for a specific case. It is then possible to generalize from that example. Start with engineering data or an accurate estimate. Assume that the failure data indicate that a particular critical part has a probability of i failures (p_i) over the lifetime of a flexible manufacturing machine, called the complex machine.

There is a cost c for each spare part purchased at the time that the complex machine is acquired. When a spare part must be purchased at a later time, due to an insufficient supply purchased at the start-up, the cost is estimated to be c'.

c' can be much larger than c because it includes the cost of doing whatever has to be done to fill the production void caused by failure of the part. Thus, c' might reflect the cost of buying finished parts from a coproducer. There also is a large cost per replacement part charged by the original equipment manufacturer (OEM). OEMs charge high prices for spare parts not acquired originally, especially when they are not carried in stock. Then, a special set up is required.

Replacement part manufacturers have different work configurations than original equipment manufacturers. The latter are seldom set up to make replacement parts economically. This characterizes the aftermarket for tires, auto parts and windshield glass. In this example, it is assumed that the spare part cannot be obtained from a n aftermarket replacement source.

The Complex Machine—How Many Parts to Carry

For the complex machine, let $p_i \geq 0$ for $i = 0, 1, 2, 3$. If $p_0 = 0$, then there must be at least 1 failure and no more than 3 failures over the lifetime of the machine.

Also, assume that the probability of failure is distributed as follows:

$$i = 0 \; failure; \; p_0 = 0+$$
$$i = 1 \; failure; \; p_1 = \tfrac{1}{2}$$
$$i = 2 \; failures; \; p_2 = \tfrac{1}{3}$$
$$i = 3 \; failures; \; p_3 = \tfrac{1}{6}$$

The first line: $i = 0$ failure; $p_0 = 0+$ needs interpretation. The value $0+$ is meant to imply that while the probability of zero failures is rated as 0.00 over the life of the complex machine, nevertheless, there is a small possibility ($+$) that it might occur. For the basic calculation this minimal probability can be ignored, but the data are included in the decision matrix shown in Table 12.2 below.

The sum of these probabilities equals 1:

$$\sum_{i=0}^{i=3} p_i = 1$$

Let $c = \$5$ and $c' = \$400$. The question is: how many spare parts (k) should be ordered at the time of the original purchase? The spare part is inexpensive if purchased initially. However, later replacement of the failed part costs 80 times the original price because of lost production time and large set-up costs. The decision matrix for spare-part strategies is given in Table 12.2.

TABLE 12.2

FAILURE COST MATRIX FOR THE COMPLEX MACHINE

	Number of Failures i Occurring During Machine's Lifetime				
	0	1	2	3	
p_i	$0+$	$\tfrac{1}{2}$	$\tfrac{1}{3}$	$\tfrac{1}{6}$	Expected Cost
k					
0	0	400	800	1200	$666.67
1	5	5	5 + 400	5 + 800	$271.67
2	10	10	10	10 + 400	$ 76.67
3	15	15	15	15	$ 15.00…MIN

Note: the initial number of spares = k.

The minimum expected cost is obtained by ordering 3 spares—a result that could have been anticipated by the size of the numbers for c and c'. At least 1 failure will occur in line with the given probability distribution. As is shown in the Problem Section at the end of this chapter, the result could be quite different if there was a substantial probability for zero failures.

The steps required to obtain the expected values in Table 12.2 are as follows. The outcome entries in the matrix are computed by two different relationships. First, when the number of failures equals or is less than the number of parts originally ordered with the machine, the cost is simply kc.

Second, when the number of failures is greater than the number of parts originally ordered, the cost is $kc + (i - k)c'$. For example, if three failures occur ($i = 3$) and only two parts were originally ordered ($k = 2$), then the cost is:

$$(2 \times 5) + (3 - 2)\, 400 = 410$$

After the matrix of total costs is completed, the expected values are obtained in the usual fashion.

$$\text{Expected cost for } k = 0:\ 400(\tfrac{1}{2}) + 800(\tfrac{1}{3}) + 1200(\tfrac{1}{6}) = 666.67$$
$$\text{Expected cost for } k = 1:\ \ \ \ 5(\tfrac{1}{2}) + 405(\tfrac{1}{3}) + \ \ 805(\tfrac{1}{6}) = 271.67$$
$$\text{Expected cost for } k = 2:\ \ 10(\tfrac{1}{2}) + \ \ 10(\tfrac{1}{3}) + \ \ 410(\tfrac{1}{6}) = 76.67$$
$$\text{Expected cost for } k = 3:\ \ 15(\tfrac{1}{2}) + \ \ 15(\tfrac{1}{3}) + \ \ \ \ 15(\tfrac{1}{6}) = 15.00$$

For realism, assume that all numbers are in thousands. Having the lowest expected cost of \$15,000, three spares should be ordered with the complex machine. This is \$61,670 cheaper than ordering two, and more than a quarter of a million dollars cheaper than ordering only one spare part with the complex machine.

When the three parts arrive with the machine, they should be inspected carefully to make certain that they are properly made according to specifications. They must be able to do the job, if and when they are called upon.

The problem could be complicated and made more realistic by: adding a charge for carrying a part in stock, changing the probabilities of failure after a failure occurs, allowing more than one spare to be reordered after failure, etc. All such issues, and others as well, can be treated in a more realistic albeit complicated model.

The Static Inventory Problem

The decision-matrix model effectively represents the **static inventory problem**. In materials management terms, this is deciding how much to buy when there is only a one time purchase. The model is not representative of flow shop problems but it is fully applicable to the job shop and project manager.

Variability of demand (in this case, the spare parts failure distribution) is only one way in which uncertainty about the order size can arise. Other causes are defectives (requiring additional parts to be made to fill the order), spoilage, and pilferage. All such factors can be accounted for with probability estimates and the decision-matrix methodology. A, B, and C are identical.

Static inventory problem
Deciding how much to buy when there is only a one time purchase.

Materials innovations, like fire-proofing products, can improve the quality and safety of a structure. Source: © W. R. Grace & Co., Conn.

12.7

VALUE ANALYSIS (ALTERNATIVE MATERIALS ANALYSIS)

Value analysis (VA)

Analysis of the value of alternative materials; includes consideration of improvements, opportunities and materials innovations.

Value analysis has been a part of materials management since the 1940s. The idea behind **value analysis (VA)** is that:

1. Materials are always being *improved* because of constant technological developments.
2. Unless a watch is kept on the relevance of materials, changes to the company's products and *opportunities* to shift to new and improved materials will be missed.
3. Efforts should be made continuously to improve the material qualities of the product, and to decrease product costs through *materials innovations*.

Value analysis is applied to all materials used by the process and for the product. VA works for materials in all domains including hospitals, hotels, restaurants, offices, manufacturers, and airlines. Airlines are beginning to save money by substituting voucher numbers for paper tickets.

Methods Analysis

Methods analysis

Systematic examination of all operations in any process in search of a better way of doing things; used to work smarter and not harder.

Value analysis and methods analysis are often seen as close relatives. **Methods analysis** is the systematic examination of all operations in any process in search of a better way of doing things. It encourages alternative processing by combining operations, eliminating unnecessary steps, and making work easier. Method analysts are widely employed to help OM work smarter and not harder.

To distinguish value analysis from methods analysis, the latter is primarily concerned with process improvement and secondarily with materials. Specifically, value analysis is the "analysis of the value of alternative materials." Inevitably, value and methods analyses must share common ground. Starting with different perspectives, they converge on the same kind of problems, providing similar kinds of problem resolution.

To reiterate, growth of interest in value analysis increases whenever materials technology seems to start a new wave of rapid and dramatic changes for particular industries. Such waves occur from time to time in metals, plastics, ceramics and most recently in glass (i.e., expansive solar glass windshields).

On a related front, materials shortages raise interest in value analysis aspects of materials management. The petroleum shortage of the 1970s led to major P/OM distortions. Companies were faced with the need to make swift alterations in the composition of their products. Goods and services were affected. Other temporary shortages that occurred during the last 20 years include yellow fats, paper products, lumber, copper, and water.

Material shortages hit particularly hard at the flow shop which is least flexible in adapting to changes in materials. Increasing numbers of organizations try to prepare for shortage situations by doing contingency planning. Sometimes, this even includes R & D efforts to prepare alternative formulations.

In addition to shortages, government agencies increasingly prohibit the use of well-established materials. The food industry has experienced bans on saccharine, cyclamates, food dyes, preservatives, and many other items. The detergent industry has been forced to reformulate its products several times because of ecological effects of phosphorous and enzyme additives. Government regulation of fuels to control auto emissions is another instance. Government regulations call for a small percent of all vehicles sold in 1998 to have zero emissions (as with electric cars). Thereafter, the percentage is expected to increase significantly. Such government restrictions can be expected to continue to grow in number and severity.

If proper value analysis of alternative materials have been done, then materials alternatives do not have to be hastily sought when either shortages occur or government edicts forbid the use of specific materials, and/or change the way that they can be used. The procedures for value analysis should be applied to established products during their mature life cycle stages, as well as to new products during their start-up and growth life cycle stages.

Successful value analysis requires a well-structured approach. This is reflected by the consistent application of a set of relevant questions. For example:

1. How do materials affect what this product or service is intended to do?
2. How do materials affect the cost of this product or service?
3. Have there been prior changes in the materials used?
4. What other materials could be used?
5. How do alternative materials affect Questions 1 and 2 above?

The value analysis approach has been designed to increase insights by providing a structural framework to encourage the development of alternative materials management strategies. The starting point is the methods analysis of existing outputs and processes.

Often, the most important materials in the product and process are labeled primary classes; others are called secondary classes. Major efforts are placed on the primary classes which are then related by analogy to other products, processes and materials that are thought to provide similar properties.

TECHNOLOGY

Inco Ltd.'s "Vendor Rationalization Process"

Inco Ltd. of Sudbury, Ontario, is undergoing what it calls a "Vendor Rationalization Process" to choose suppliers based on a list of formal criteria. Inco operates "the world's largest mining complex" in Sudbury, where 6,700 employees work in ten mines, two refineries, a smelter (which separates impurities from metals), and an office center. Inco's selection criteria for major suppliers now include commitment to quality, security of supply, commercial impact, technical support, and commitment to environmental concerns. The supplier's capability to receive purchase orders from Inco through electronic data interchange (EDI) also is taken into account under the technical support criterion.

About three years before implementing Vendor Rationalization, Inco began using EDI with some of its suppliers. The driving forces behind Inco's implementation of EDI were "compressing the cycle time, eliminating the paper, and getting the material to the customer more efficiently," said Wayne Smith, manager of Purchasing, Warehousing and Traffic. The company issues about 15,000 purchase orders a month; 84 percent of Inco's purchase orders are sent electronically.

Inco purchases a wide variety of items for use in the mines and refineries such as drills, trucks, explosives, construction timber, steel rods, cement, tools, clothing, and safety supplies. The company also purchases large quantities of business equipment and office supplies for its 2,000-member office staff. Before implementing the Vendor Rationalization Process, Inco dealt with nearly 3,000 suppliers. One of the by-products of the process will be a decrease in the number of suppliers the company will deal with.

"Multi-Functional Commodity Teams" within Inco evaluate suppliers based on the criteria. The teams include representatives from purchasing, warehousing, traffic, maintenance, engineering, accounting, and research as well as the end users of the products. For example, the team responsible for determining suppliers for blasting products (explosives), includes miners, the people who use the product in Inco's underground operation.

Some of Inco's Sudbury operations have earned ISO 9002 certification and other Inco locations are "well on their way" to achieving certification this year, according to Jerry Rogers, manager of Public Affairs for Inco. Rogers said the company encourages its suppliers to become involved with the ISO 9000 series of quality standards as well. Sudbury is a North American leader in ISO with more than 100 suppliers to Inco having been certified. Inco worldwide provides about 25 percent of the world's supply of nickel. In addition to mining nickel, the Sudbury plant produces nickel by-products such as nickel powder used in rechargeable batteries.

Sources: Fiona Christnsen, "Electronic Ordering Part of Inco's Push for Quality," *Northern Ontario Business*, July 1, 1994, and a conversation with Jerry Rogers, Inco's manager of Public Affairs.

Sometimes, by means of analogies, materials alternatives can be derived. Comparisons are developed between different joining methods—adhesion, cohesion, welding, brazing, and mechanical fastenings such as nuts and bolts, cotter pins, nails, etc. When things are to be held together it is wise to know all of the ways that fastening can be achieved.

Both methods analysis and value analysis are used to discover new tactical, efficient alternatives. But P/OM must be on guard against investments in efficiency studies before effectiveness issues have been thoroughly considered.

12.8
ABC CLASSIFICATION— THE SYSTEMS CONTEXT

Materials management—from the inception of the purchasing process all the way through production and the shipping of finished goods—can be improved by utilizing an important systems concept called the **ABC Classification**. The concept is that some materials are more important than others. Further, that it is prudent to do what is important before doing what is less important.

This concept can be applied in a number of different ways, two of which are discussed below.

ABC Classification
Concept that some materials are more important than others, and do what is important before doing what is less important.

Material Criticality

It is necessary for materials management to categorize the critical nature of parts, components and other materials. There are various definitions of critical that fit different situations. When a part failure causes product or process failure, that is a critical part. For an airline engine, most of the parts may be critical, in this sense.

The following type of scenario might be relevant. This airline considers most critical any part which has greater than a 10 percent probability of failing within one year. Membership in the top ranking group is called *A-critical*. Perhaps the top 25 percent of all critical parts can be considered to be A-critical.

A second set, called *B-critical*, is identified by the airline as any part which has greater than a 10 percent probability of failing within five years. A-critical items are excluded. Perhaps the next 25 percent of all critical parts can be considered to be B-critical.

The third set, called *C-critical*, might be associated with greater than 10 percent probabilities of failing after five years. Because 50 percent of the parts are in the A and B groups, the remaining 50 percent of the parts are in the C group.

As an alternative definition, part failure can have a probability (not a certainty) of stopping the process or product. Thus, a possible description of critical parts relates to the probability of process or product failure when the part fails.

The parts are rank ordered by the probability of causing process or product failure. Say that if the first part fails there is a 30 percent probability of product failure. The next part has a 20 percent probability. Perhaps the top 25 percent of critical parts has an 80 percent probability of causing (each independently) product failure. This kind of situation is pictured in Figure 12.2.

Some processes do not fail as a result of part failure but instead the production output is reduced by a significant amount. A curve similar to Figure 12.2 above could be created for this kind of situation.

Criticality

Crucial to performance or dangerous to use; reflects costs of failures.

Criticality is a coined term which can mean crucial to performance or dangerous to use. Thus, an alternative definition of criticality could apply to the danger involved in using materials. Flammability, explosiveness and toxicity of fumes could be crucial safety factors for materials management.

Whichever definition of criticality is used, the procedure is to list the most critical parts first. Then, systematically rank-order parts according to their criticality. Criticality should reflect the costs of failures including safety dangers, loss of life, and losses in production output.

FIGURE 12.2

ABC CURVE FOR THE CRITICALITY OF MATERIALS

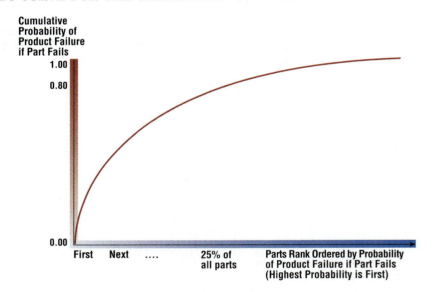

Spare parts and other backup materials should be provided to conform to remedial failure strategies. Also to be considered are replacement parts for preventive failure strategies. Further, spare parts must be inspected to insure their continued integrity over time.

A large South American refinery decided to double its rubber spare parts inventory. Within two years many of the parts had begun to deteriorate and could not be called upon to replace critical failed parts. The high penalty resulted in a return to the original policy. Organizations have found that poor spare parts policies can be the "Achilles' heel" for business.

Material Dollar Volume

The second set of ABC categories is based on sorting materials by their annual dollar volume. Dollar volume is the surrogate for potential savings that can be made by improving the inventory management of specific materials.

Accordingly, all parts, components and other materials used by a company should be listed and then rank ordered by their annual dollar volume. Thus:

$$DV_i \ (t = 1 \ yr.) = c_i \sum_{t=0}^{t=1} q_i \ (t)$$

where:

DV_i = the annual dollar volume of the ith item

c_i = the dollar cost per unit

$\sum_{t=0}^{t=1} q_i \ (t)$ = the number of units of the ith item ordered per year

Start with those items which have the highest levels of dollar volume DV_i and rank order them from the highest to the lowest levels. The top 25 percent of these materials will be called A-type items. The next 25 percent are called B-type items. The bottom 50 percent are called C-type items.

FIGURE 12.3

ABC CURVE FOR ANNUAL DOLLAR VOLUME OF ITEMS

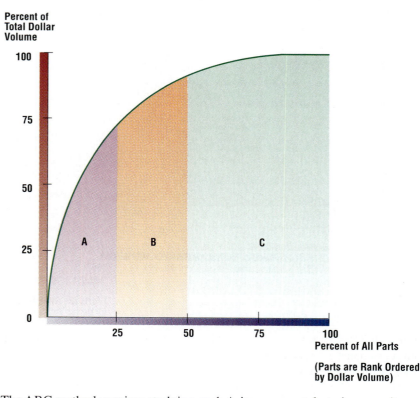

The ABC method requires studying each A-item separately to improve its performance. Details include: how much to order at one time which determines how often to order, who to order from, what quality standards to set, lead times for deliveries, the consistency of lead times, as well as all special agreements with suppliers. B-type items are studied in groups and with less attention to detail. Policies for C-type items are set to be as simple as possible to administer.

C-type items have low-dollar volume which means that they have low price per unit. They may have low volume too but most are driven by the low price. Small penalties are paid for overstocking such items so they do not have to be ordered as often. Still and all, it may turn out that a C-type item can lead to a critical situation if the CEO's bathroom runs out of toilet paper—which is a typical C-type item.

If MM does not use this ABC systems approach which sorts all materials into those that have more potential savings than others, then a major strategy for systematic and continuous improvement is being overlooked.

Companies differ with respect to what percent of all their items account for 75 percent of their total annual dollar volume. Generally, a small percentage of all items accounts for a large percentage of annual dollar volume as shown in Figure 12.3.

F I G U R E   1 2 . 4

STRAIGHTLINE ABC CURVE WITH EQUAL INCREMENTS

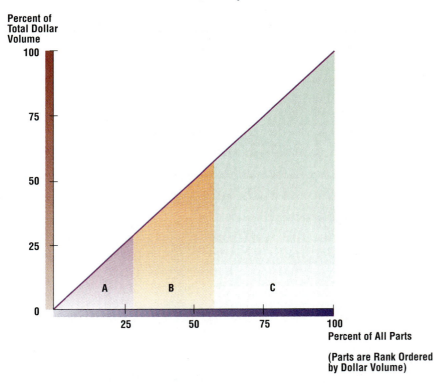

Figure 12.3 portrays a typical case where 20 to 30 percent of all items carried account for as much as 70 to 80 percent of the company's total dollar volume. Consider the hotel chain which stocks 1,000 different items for which it spends one million dollars annually. Twenty-five percent of these items, costing $750,000 constitutes the A-class of inventory items.

That is, 250 items control 75 percent of the annual dollar volume and conversely, 750 items control only 25 percent of the annual dollar volume. It is highly unusual for the ABC curve to look like the one drawn in Figure 12.4. There, the percent of items always equals the percent of dollar volume. Rank order means nothing because all items are the same.

Because the costs of inventory studies tend to be proportional to the number of items under consideration, MM always chooses to study and update A-class items. What is to be included in A-type will depend on β in the following equation:

$$y = x^{\beta}, (0 \leq x, \leq 1), (0 \leq \beta \leq 1)$$
$$y = percent\ of\ dollar\ demand$$
$$x = rank\text{-}ordered\ percent\ of\ all\ items$$

Beta is one in Figure 12.4. Beta is less than one half in Figure 12.3. The Problem Section at the end of this chapter suggests that a curve be drawn for $\beta = 0.5$.

There is no fixed convention that x-class breaks must occur at 25 and 50 percent. Some companies use only A and B classes. It is historically interesting that Del Harder of the Ford Motor Company is said to have developed the ABC concept in the 1940s and the curve which reflects the disproportionately large effect derived from a small number of items in the population.

12.9

CERTIFICATION OF SUPPLIERS

Certification of suppliers is a process for grading suppliers to ensure that suppliers' organizations conform to standards which are essential for meeting the buyer's needs. It is expensive and time consuming. The certification process should be reserved for suppliers of A-type items with respect to both criticality and dollar volume.

In addition to establishing minimum standards which every supplier must meet, certification is like a bidding process for long-term relationships. Companies use the certification process to choose the best in the class—just like they use various criteria to hire students based on grades, dean's list and personal evaluations. Most often, the requirements for suppliers are equivalent to the company's internal standards for itself with respect to excellence in quality and reliability.

The number of suppliers chosen can vary from one to several. Often, supplier organizations which do not make the grade are encouraged to improve. Many companies help potential suppliers upgrade those capabilities on which they are rated as deficient. Accepted suppliers are regularly reviewed to make certain that they maintain their "winning" status. Thus, while certification aims at long-term relationships, it is subject to reassessment. In numerous instances, the buyer raises the acceptance standards while assisting certified suppliers to meet the new and more stringent standards.

The rating procedures are formal evaluations of price, quality, delivery time, and the ability to improve all three. Suppliers' productivity improvement programs are expected to result in lower prices. Suppliers' total quality management (TQM) programs are monitored for expected improvements. ISO 9000 standards and the Baldrige Award criteria are used to provide the foundation. Lead time management programs track delivery time reduction. Time-based management concepts form the basis for evaluation.

The buyer's materials management information system (MMIS) has to be able to handle many suppliers and potential suppliers for hundreds and even thousands of A-type items. Who can do this successfully? The list is impressive. Chrysler, Ford Motor Company, General Motors, Hewlett-Packard, Honda Motor Company, IBM, Motorola, Texas Instruments, and Toshiba, are only a few of the companies that have made public their use of certification programs. Certification procedures demand the use of the systems approach.

Certification of suppliers

Process for grading suppliers to ensure that the supplier organizations conform to standards which are essential for meeting the buyer's needs; rated on quality, price, and delivery.

S U M M A R Y

The chapter starts with the query "What is materials management (MM)?" The answer is that MM is a set of connected systems functions which are parts of the organization's internal supply chain. These many interlinked components must be coordinated to be managed successfully. A taxonomy of materials is presented to cover the different kinds of materials to be managed. Then, the cost leverage of savings in materials and the penalties of errors in ordering materials are explored.

Materials management is unique for each of the various kinds of work configurations (flow shop, job shop, and project environment), as well as for flexible processing systems. All configurations require an intelligent information system which is based on designing systems parameters that provide timely decision making for the right variables—often in a global information systems context.

Purchasing is a crucial player on the MM team. It can take advantage of the global reach provided by telecommunications capabilities. The learning organization uses purchasing to gain twenty-first century opportunities for improving competitiveness. "Turnover" and "days of inventory" are important measures of competitiveness.

Also covered are the MM functions of receiving, inspection and storage. MM's use of bidding is discussed as well as the categorization of parts in terms of their criticality. A failure model for critical spare parts is presented. Then, it is shown how MM's capabilities are enhanced by the application of value analysis. To conclude, materials are categorized into ABC classes that rank order their criticality, and then, their importance in terms of the potential savings that can be made by improving the way they are managed.

▶ **Key Terms**

ABC Classification   *539*
criticality   *540*
expediting   *524*
logistics   *515*
materials management system (MMS)   *514*
purchasing agents (PAs)   *525*
turnover   *527*

bidding   *532*
cross-docking   *531*
intelligent information system   *523*
materials management (MM)   *514*
methods analysis   *536*
static inventory problem   *535*
value analysis (VA)   *536*

certification of suppliers   *543*
days of inventory (DOI)   *526*
internal MM system   *515*
materials management information
    system (MMIS)   *523*
trade-off models   *520*

▶ **Review Questions**

1. What is materials management (MM) and who does it?

2. Why is the systems approach important for MM?

3. Present the taxonomy of materials.

4. Why is it said that materials have the greatest leverage for making cost savings?

5. What kinds of errors can be made in ordering materials and what are the penalties?

6. Explain differences in MM according to work configurations.

7. Describe the materials management information system (MMIS) and explain why it plays such an important role for P/OM.

8. Explain the difference between centralized and decentralized materials management.

9. Discuss how the information system participates in determining the degree of centralization of MM.

10. From the MM point of view, what is a global information system? Is it centralized or decentralized?

11. Discuss the effect of the systems communication flows.

12. Describe the purchasing function and the role of purchasing agents (PAs).

13. Why are PAs said to be part of the twenty-first century learning organization?

14. Define turnover.

15. Define days of inventory (DOI).

16. What ethical problems must be resolved by purchasing?

17. Describe the international issues that may create ethical quandaries for purchasing agents.

18. Detail the functions for receiving, inspection, and storage. How are they a part of MM?

19. Explain why bids may be sought by P/OM.

20. Explain why materials costs often are a large percentage of both factory costs and operating costs of services.

21. Why does the low bid tend to decrease with more bidders?

22. Discuss the materials management of critical parts.

23. What is a static inventory problem? How often does it arise in the real world?

24. Explain the role played by value analysis for materials management. Why is value analysis part of the OM-team effort?

25. Explain the role played by methods analysis. Why is methods analysis part of the OM-team effort?

26. Describe ABC Classification as applied to material criticality.

27. Describe ABC Classification as applied to material dollar volume.

28. Explain why supplier certification requires the systems approach to include all aspects of materials management. (Note: do not overlook the use of bidding models, ABC concepts and benchmarking which permit comparison with the best in the class.)

▶ **Problem Section**

1. Assume that downsizing could save the Market Research Store $5,000 in annual labor costs. The downsizing study costs $1,000. If instead, the study focused on materials savings, what would be the net payoff for the first year—assuming average firm ratios for labor and materials costs?

2. If the ratio of labor costs to materials costs is 1/2 and labor costs are decreasing by 10 percent per year while material costs are increasing by 10 percent per year, what is the ratio at the end of the second year? (Note: this means that at the present time labor costs are 50 percent of materials costs.) Does the result seem reasonable?

3. If labor is 20 percent and materials 80 percent of COGS—say $20 and $80, then a 10 percent reduction in labor costs yields $18 added to $80 totals $98. This produces a 2 percent reduction in total costs, (i.e., $(100 - 98)/100 = 0.02$). Make the same comparison for a 10 percent reduction in materials costs.

4. Assume that the probability distribution has been changed for the decision matrix previously constructed for the complex machine spare parts failure problem. Thus, Table 12.2 above should be altered to conform to the following data:

For the complex machine, let $p_i \geq 0$ for $i = 0, 1, 2, 3$. Note that now $p_0 = 1/6$, and that there can be no more than 3 failures over the lifetime of the machine. The probability of failure is distributed as follows:

$$i = 0 \text{ failure; } p_0 = 1/6$$
$$i = 1 \text{ failure; } p_1 = 1/3$$
$$i = 2 \text{ failures; } p_2 = 1/3$$
$$i = 3 \text{ failures; } p_3 = 1/6$$

How many spare parts should be ordered initially?

5. Assume that the probability distribution has been changed again for the decision matrix previously constructed for the complex machine spare parts failure problem. Table 12.2 above should be altered to conform to the following data:

For the complex machine, let $p_i \geq 0$ for $i = 0, 1, 2$. Note that $p_0 = 3/4$, and that now there can be no more than 2 failures over the lifetime of the machine. The probability of failure is distributed as follows:

$$i = 0 \text{ failure; } p_0 = 3/4$$
$$i = 1 \text{ failure; } p_1 = 1/8$$
$$i = 2 \text{ failures; } p_2 = 1/8$$
$$i = 3 \text{ failures; } p_3 = 0.0$$

How many spare parts should be ordered initially?

6. Referring again to Table 12.2 above, what would the decision be if zero failures are expected with 0.97 probability and the other failure rates for $i = 1, 2$, and 3 are all 0.01?

7. Again turning to the problem shown in Table 12.2, if $c = 5$, $c' = 20$, and:

$$i = 0 \text{ failure; } p_0 = 0.0$$
$$i = 1 \text{ failure; } p_1 = 1/2$$
$$i = 2 \text{ failures; } p_2 = 1/3$$
$$i = 3 \text{ failures; } p_3 = 1/6$$

how many spares should be ordered? Note that the initial acquisition cost remains the same but is now only 1/4 the failure replacement cost.

8. It has been stated that 16 percent of the beer drinkers drink 40 percent of all the beer that is consumed. Is this reasonable if β is equal to 0.5? Draw the curve for the equation:

$$y = x^\beta, (0 \leq x, \leq 1), (0 \leq \beta \leq 1)$$

where:

$y =$ percent of dollar demand, $x =$ rank-ordered percent of all items, and $\beta = 0.5$

This is equivalent to

$$y = \sqrt{x}$$

9. For the β in Problem 8 above, determine the percent of total dollar demand that occurs when the A category is set at 25 percent of all items.

10. Calculate the inventory turnover rate when monthly net sales are $120 million and average inventory evaluated at the selling price is $240 million. What might this product be and comment on this level of turnover for such a product.

11. Continuing with Problem 10 above, calculate the days of inventory (DOI) if average inventory is $120 million (calculated in terms of costs) and the monthly cost of goods sold (COGS) is $60 million. Use 30 days per month. Once again, consider what product might be described by these data and comment on the level of days of inventory for such a product.

12. Compare the answers obtained in Problems 10 and 11 above. Explain how they relate to each other.

▶ **Readings and References**

D. N. Burt, *Proactive Purchasing*, Englewood Cliffs, NJ, Prentice-Hall, 1984.

D. W. Dobler, L. Lee, Jr., and D. N. Burt, *Purchasing and Materials Management*, NY, McGraw-Hill, 1984.

Lisa M. Ellram and Laura M. Birou, *Purchasing for Bottom Line Impact*, Vol. 4 of the National Association of Purchasing Management (NAPM), Homewood, IL, Irwin, 1995.

S. F. Heinritz, P. V., Farrell, L. Giunipero, and M. Kolchin, *Purchasing: Principles and Applications*, 8th Ed., Englewood Cliffs, NJ, Prentice-Hall, 1991.

Thomas K. Hickman and William M. Hickman, *Global Purchasing: How to Buy Goods and Services in Foreign Markets*, Homewood, IL, Business One Irwin/APICS Series in Production Management, 1992.

Kenneth H. Killen and John W. Kamauff, *Managing Purchasing*, Vol. 2 of the National Association of Purchasing Management (NAPM), Homewood, IL, Irwin, 1994.

Michiel R. Leenders, and H. Fearon, *Purchasing and Materials Management*, 10th Ed., Homewood, IL, Irwin, 1993.

Michiel R. Leenders and Anna E. Flynn, *Value-Driven Purchasing*, Vol. 1 of the National Association of Purchasing Management (NAPM), Homewood, IL, Irwin, 1994.

L. D. Miles, *Techniques of Value Analysis*, NY, McGraw-Hill, 1961.

Alan R. Raedels, *Value-Focused Supply Management*, Vol. 3 of the National Association of Purchasing Management (NAPM), Homewood, IL, Irwin, 1994.

R. Schonberger, and J. Gilbert, "Just-In-Time Purchasing: A Challenge for U.S. Industry," *California Management Review*, Vol. 26, No. 1, Fall 1983, pp. 54–68.

R. J. Tersine, *Materials Management and Inventory Systems*, 3rd Ed., NY, Elsevier North-Holland Publishing, 1987.

Notes...

1. An insight is available from Edward T. Hall, "The Silent Language in Overseas Business," *Harvard Business Review*, Vol. 38, No. 3, May–June, 1960, pp. 87–96.

Chapter 13

Aggregate Planning

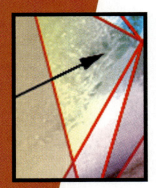

After reading this chapter you should be able to...

1. Explain the function of aggregate planning (AP).
2. Discuss the systems nature of AP in terms of classes of resources and product-mix families.
3. Explain standard units of work.
4. Relate the importance of forecasting to AP.
5. Compare constant (or level) production with a chasing policy—supply chases demand.
6. Detail the cost structure for aggregate planning.
7. Describe how to use linear programming for AP.
8. Describe how to use the transportation matrix for AP.
9. Explain why a nonlinear cost model can be used as a bench-mark for other AP models.

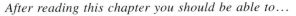

Chapter Outline

13.0 WHAT IS AGGREGATE PLANNING (AP)?
Aggregation of Units
Standard Units of Work
The Importance of Forecasts
Basic Aggregate Planning Model

13.1 THREE AGGREGATE PLANNING POLICIES
Pattern A—Supply Is Constant
Pattern B—Supply Chases Demand
Pattern C_1—Increase Workforce Size by $\pm$ One
The Cost Structure

13.2 LINEAR PROGRAMMING—(LP) FOR AGGREGATE PLANNING

13.3 TRANSPORTATION MODEL (TM) FOR AGGREGATE PLANNING
Network and LP Solutions

13.4 NONLINEAR COST MODEL FOR AGGREGATE PLANNING

Summary
Key Terms
Review Questions
Problem Section
Readings and References

The Systems Perspective

Aggregate planning matches classes of jobs with the supportive resources needed to produce each of them. The concepts of "what does it take to make this class of items" or "what skills and technologies are used to provide this class of services" are successful if based on broad systems thinking. Aggregate planning takes the expected demand for the job shop product-mix and assigns available resources in an optimal way (say minimum total costs). Also, from the systems point of view, it is sensible to extend the investigation to include alternatives to the product-mix and the available resources.

13.0

WHAT IS AGGREGATE PLANNING (AP)?

Aggregate planning (AP) is required to design a generalized production schedule. The generalized schedule is intended to meet demand which is expressed in standard units instead of specific units of various measures. The reason that the term aggregation is used is because the specific identities of all units are merged into a common pool of standard units or standard hours. Supply is stated in these aggregated units and demand also is estimated in the same aggregated units. Demand is generated by a variety of customers for the different kinds of products made by the job shop or the different kinds of services offered by job shops such as the Market Research Store.

It is important to note that aggregate planning is an internal production management function which leads to detailed production scheduling in later stages. The specific purpose of AP is to decide when to schedule work and under what conditions to schedule it. The conditions can vary such as second shift, overtime work, and subcontracting.

In Figure 13.1, aggregate planning is shown as following strategic business planning.

AP is the producer's general plan for getting work done. It is not detailed scheduling which follows aggregate planning.

F I G U R E 1 3 . 1

AGGREGATE PLANNING FOLLOWS STRATEGIC PLANNING

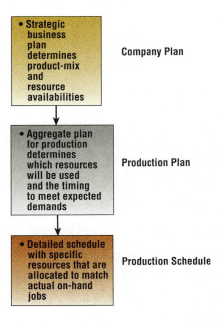

- Strategic business plan determines product-mix and resource availabilities — **Company Plan**

- Aggregate plan for production determines which resources will be used and the timing to meet expected demands — **Production Plan**

- Detailed schedule with specific resources that are allocated to match actual on-hand jobs — **Production Schedule**

Aggregate planning starts a chain reaction in the supply chain of suppliers-producer-customers. Once the internal assignments that relate to the factory and the office are scheduled, then the external flows begin to function. These are flows of materials from the factory to the warehouses by trucks and other transport systems which must be coordinated. Transport from the warehouses to the customers must be organized. External flows from suppliers to the production system must be activated. Equipment may have to be rented or bought and people hired and trained or workforce reductions may have to be initiated.

The driver of AP is *forecasted customer demand*. This can take the form of customer orders and/or inventory plans. The forecasts are translated into production plans. The goal is to be prepared to make the product and deliver it when the actual demands are on-hand. The same applies to delivering services to the customer at the optimum time.

Material flows inside the company can be scheduled in detailed specific items, or in categories of specific items that have been aggregated. Before trying to do detailed scheduling, aggregate planning (generalized) should be used to avoid costly mistakes that arise from not preparing enough resources at the right time.

Aggregation of Units

When the specific production items to be made are called apples, oranges and pears, there is much more detail required for planning n-periods ahead. There is too much detail to schedule the production of each of these items. It is much simpler to schedule the production of one aggregated item, called "fruit."

Standard units are used as the common denominator for aggregated units in the production scheduling problem. Forecasts of specific items such as sweaters of different colors, cotton, wool, and mixtures, are converted into the standard machine hours required to make each type of sweater. This is referred to as aggregation of the mixed-model product line of the job shop. It is the aggregation of finished goods used as the basis for planning what will be made in the period starting n-months ahead.

Standard units

The common denominator for aggregated units in the production scheduling problem.

Aggregate planning enables Bayway Refining Company to move 60,000 barrels of gasoline, diesel, heating oil, and liquified petroleum gas per day using this modern truck-loading rack.
Source: Tosco Corporation

If the planning period is for one quarter, then it starts at the beginning of t_1 and it finishes at the end of t_3. If that planning period starts one month ahead (lead time $n = 1$), and it is now January 1st, then the planning period starts February 1st and ends April 30th.

Aggregate planning is achieved by collecting and lumping all items to be produced together. The idea is to strip away the specifics while retaining the aggregate properties of the product. The purpose is to match aggregate capacity against aggregate demand. This results in generalized determination of workforce requirements including the possible use of multiple shifts, overtime and part-time work, to satisfy demand across a great variety of stockkeeping units.

Consider the paint manufacturer with a product line that includes water-based, oil-based and acrylic paints. They come in many colors and can sizes. For aggregate planning, the standard product unit might be gallons of paint; the time period could be monthly; the planning horizon could be one year ahead. Alternatively, it might have been decided to separate aggregate planning for each type of paint, water-based, oil-based and acrylic, on a week-by-week basis, for half a year.

The idea is to aggregate demands that use essentially the same resources. This allows effective utilization of resources across competing demands. By the same resources, it is meant that essentially the same equipment, materials, people, space, and experience are transferable between the different items to satisfy demand as capacity allows in an optimal manner.

Aggregate-planning methods can derive better solutions when P/OM can find:

1. ways to expand the number of products that can be made by a specific class of resources
2. ways to expand the number of resources that can make a class of products

Points 1 and 2 provide increased resources and flexibility to satisfy demand. Also, planning results can be improved by investing in more capacity of the right kind (scarce). Changing the product line and the order mix can result in other benefits. These are systems-oriented concepts for improvements using AP.

A paint manufacturer may use gallons of paint for the standard product unit for aggregate planning.
Source: The Sherwin-Williams Company

Also, to be effective, the forecast interval has to provide sufficient lead time to allow the resource mix to be changed in accordance with the plan. The paint company might decide to do aggregate planning using monthly time buckets with a planning horizon of six months and updating the conclusions once every three months. Figure 13.2 illustrates these time-planning concepts with the numbers just given.

FIGURE 13.2

AGGREGATE PLANNING WITH PLANNING HORIZON AND UPDATING INTERVAL

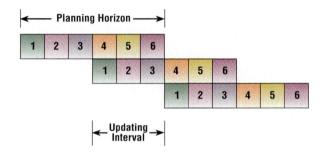

Standard Units of Work

It was said that "aggregate planning is achieved by collecting and lumping all items to be produced together." This should be amended to "… lumping all items together that share the use of common resources." This point, which was previously made, is emphasized by the process of creating standard units.

This process requires that different parts, activities, products, and services all be described and accounted for in terms of an arbitrarily chosen, but agreed upon standard unit of work. Thus, a pint of deluxe external white paint might require two standard units of work. These two standard units will be added to the five standard units required to make a gallon of special green latex paint. These seven standard units will be added to the standard units of work required for other products or services.

The aggregated total will be used for workforce planning. Seven standard units of workforce time is required to make the pint of white, and the gallon of green, paint. Workforce planning is of great importance to all service industries. It applies to crew scheduling for airlines, hospitals and banks. Because many job shops provide services and not goods, it pays to emphasize the fact that AP is useful to all kinds of job shops. It also can apply to intermittent flow shops which have lumpy demands over time.

The Blackfeet Indian Writing Company in Browning, Montana, is a small company that sells a variety of writing instruments such as markers and pencils. Demands are sporadic and aggregate planning is essential. Book publishers (like boyd & fraser) have to plan for erratic demand patterns which sometimes tax their production capacities. Theme parks, hotels and resorts have shifting demands as a function of seasons that lend themselves to AP. They cannot inventory low-season, unused capacities to accommodate peak demands. Nevertheless, they can use AP for workforce scheduling.

In addition to workforce planning, AP treats inventory and equipment availability. The AP model cannot always satisfy demand. Sometimes capacity is insufficient. At other times, some part of the systems resources will be idle because supply is greater than demand. As the AP models are developed, note in particular how certain kinds of supply (like overtime) provide protection against not being able to deliver the goods. However, P/OM may be happy to see that overtime supply is seldom utilized.

The direct approach for understanding standard hour computations is to work through an example of how planners aggregate demands for different products or services. This means changing actual hours of work into standard units of required production capacity. For the example that follows, the standard unit is selected as a standard machine hour (SMH) based on using a standard machine (SM) for the standard of comparison.

The total production capacity consists of four machines (people, departments, etc., also can be used when appropriate). The four machines called M1, M2, M3, and M4, work at different rates. It should be noted that the rate differentials apply across all of the jobs that these machines are assigned.

That is: M2 is fastest—for all of the jobs, and by the same amount in comparison with the other machines. For this reason it is convenient to choose M2 as the standard machine (SM). It will be assigned an SM index of 1.0. Then, all of the other machines will have fractional SM indexes. Thus:

Machine Number	SM Index	Description
1	0.5	Half as fast as M2
2	1.0	Standard Machine
3	0.8	80% as fast as M2
4	0.6	60% as fast as M2

Each machine, department, or person is ranked by an index number which when multiplied by the actual machine hours available per week, yields the standard machine hours (SMH) available per week. Thus, Table 13.1 provides the actual machine hours that are available per week converted by means of the SMH index into standard machine hours available per week.

TABLE 13.1

SUPPLY OF ACTUAL AND SMH HOURS AVAILABLE PER WEEK

Machine Number	SMH Index	SMH per Week	Actual Hours
1	0.5	18	36.0
2*	1.0	54	54.0
3	0.8	64	80.0
4	0.6	20	33.3
	Total SMH	156	

Note the asterisk (*). M2 was chosen as the standard machine (SM). If M1 had been chosen as SM, the indexes would have been (1.0, 2.0, 1.6, 1.2) which is the result of multiplying the SMH index above by 2. This kind of relationship is not unusual between machines. It often applies to skills. Some variation from perfect index relations can be tolerated as providing a reasonable approximation.

There is a total of 156 standard machine hours (SMH) available per week. How many standard hours are in demand? The fact that there are 203.3 actual machine hours available per week has no significance for the resolution of the problem. Actual hours must first be converted to standard hours for application to the jobs that need to be done. This will be understood by examining the job requirements in Table 13.2 which represents a forecast of demand for next week.

TABLE 13.2

A FORECAST OF DEMAND FOR NEXT WEEK

JOB	A	B	C	D	E
Units demanded	300	210	240	1800	400

Table 13.3 provides the production output rate (also called *productivity rate*) of the standard machine for each job.

TABLE 13.3

PRODUCTIVITY RATES OF STANDARD MACHINE

JOB	A	B	C	D	E
Production rate of the Standard Machine in units per SMH	6	7	6	30	25

The computations are guided by the following equation where dimensions are shown in *italics* and units cancel to yield standard machine hours (SMH).

$$D\,(\text{units}) \div PR \left(\frac{\text{units}}{\text{SMH}} \right) = D\,(\text{SMH})$$

where:

$D\,(\text{units})$ = demand in units for each kind of job

PR = the production rate in units per SMH

$D\,(\text{SMH})$ = demand in aggregated standard machine hours

For Jobs A through E, Table 13.4 converts actual units of demand into standard machine hours (SMH) of demand, as follows:

TABLE 13.4

DEMAND CONVERTED INTO STANDARD HOURS

SMH of Demand – by Jobs

A	300/6	=	50
B	210/7	=	30
C	240/6	=	40
D	1800/30	=	60
E	400/25	=	16

To satisfy demand . . . . . .196 standard machine hours are needed!

The transformation of the dimension *units* to *SMH* should be noted. It makes the generalized nature of standard machine hours understandable.

Both supply and demand have been converted to the common terms of SMH. There are only 156 standard machine hours available. Therefore, $196 - 156 = 40$ SMH of demand will not be met.

The same reasoning applies to standard workforce hours (people hours) and to combinations of people and machines working together. Various weighting schemes can be used to bring different kinds of estimates of supply and demand into a common framework.

The Importance of Forecasts

Figure 13.3 shows the connection between the three activity levels of production planning. It is to be noted that the first level—aggregate planning—is the only one that requires a forecast. This will be used in addition to orders that are on-hand. Activity level two is loading and activity level three is sequencing, all to be explained shortly.

FIGURE 13.3

**THREE LEVELS OF PRODUCTION PLANNING—
LEVEL 1 IS AGGREGATE PLANNING
WHICH REQUIRES A FORECAST**

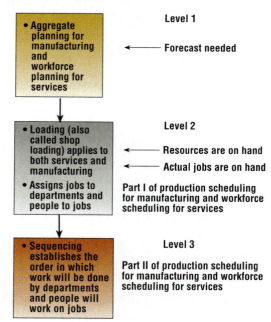

If all three levels of production planning are well managed, job shop profitability is significantly enhanced. Thus:

First Level: Set the **planning horizon** for forecasting, as required by AP, equal to or greater than the longest lead time. If it takes eight months to have an order filled or to hire five computer programmers, no plan can be carried out until five months have elapsed. The lead times include the interval required to train the appropriate workforce, obtain the needed facilities, and be able to do the jobs that are forecast.

Planning horizon

Used in the first level of production planning (which is aggregate planning); the forecast ≥ the longest lead time.

The planning horizon for the pickle producer must include lead time for the cucumber harvest and pickling process.
Source: Dean Foods Company

Second Level: Assign jobs to departments (called **loading**) which have been equipped and staffed according to the aggregate plan. The jobs are bound to deviate in number and kind from the AP forecast. The estimate of total standard hours required might be fairly good, but how they get allocated to specific jobs will have greater variability. As usual in these situations, the errors tend to cancel out. The second level of production planning permits a comparison to be made of forecast versus actual demand. Feedback concerning accuracy may be helpful in improving future forecasts. Further, at the second level, the real resources that were collected and prepared, based on the aggregate plan, must now be used to deal with the actual demand on-hand.

Third Level: **Sequence** the work according to a good shop floor control policy. This determines the order in which jobs will be done by each department. The interval between loading and sequencing is short (days or hours) whereas the interval between AP and loading is long (months). Sequencing decisions have an important effect on customer satisfaction. They are determinants of delivery schedules and affect costs as well. The quality of sequencing decisions is always a consequence of AP and loading decisions, and sequencing skills and knowledge.

Integration of the Three Levels: If the three levels of the job shop are not recognized, a number of problems and possible crises can result. Job shop "old timers" call it "organized chaos." Lack of careful attention to the three levels leads to random management of the system. Job-shop management involves so many details (different jobs, machines, workers) that strong information management capabilities are essential.

At the AP level, forecasting plays such a critical role that it is useful to review forecasting material. In particular, note exponential smoothing which is powerful and widely used.

If management is not in control of the situation, losses occur because of neglect at all three levels. The costs are cumulative. They impair an organization's competitiveness. Errors made at the first level create unnecessary costs at the second and third levels. Errors made at the second level create unnecessary costs at the third level. That is what is meant by cumulative costs.

Loading

Used in the second level of production planning assigns jobs to departments and people to jobs.

Sequence

Used in the third level of production scheduling; establishes the order in which work will be done by departments and people working on jobs.

Ken Bogard
· · ·
Registrar, Miami University

Ken Bogard, registrar of Miami University in Oxford, Ohio, deals with the aggregate plan that touches the lives of all college students—the master class schedule. Bogard's office uses a "demand analysis" approach for planning each year's master schedule. Miami University is a state-assisted, residential university enrolling approximately 16,000 students per year. Miami offers undergraduate degrees in nearly 100 academic areas, master's degrees in 64 areas, and the doctoral degree in 10 areas. In a recent semester Miami offered 2,122 different courses for a total of 3,937 sections.

With the demand analysis approach, a master plan for each academic year is set up in spring of the preceding year. Miami starts with an advance registration process conducted in April for two weeks. Students preregister for fall courses based on an academic planning guide published by the registrar's office. This guide contains information on two semesters of course offerings.

After the information is collected (including information from students who attend orientation during the summer), it is sent to the departments and academic divisions. The departments analyze the demand for their courses and add or shift faculty as necessary, e.g., from a low enrollment course such as Federal Income Tax Accounting to a higher enrollment course such as Principles of Accounting.

According to Bogard, the biggest challenges in achieving the master plan are "spreading the classes over the entire day so that you maximize the availability of courses for students," and "placing those courses into the classrooms." Planning for next year's classrooms begins with the plan for the current year as a base. Adjustments to the plan are based on variations in the number of sections offered.

Bogard says that demand for some popular "over enrolled" courses could never be satisfied. For example Anthropology 175, "Peoples of the World," typically is requested by 700 students per semester, but only 400 students can be accommodated. Bogard said a major consideration in determining the number of course alternatives to offer is to have enough "so that students can make normal progress toward their degree requirements."

Source: A conversation with Ken Bogard.

Information is essential for forecasting and planning at level one and for decision making at levels two and three where the aggregate units have become specific jobs. Most companies maintain the necessary information about processing in the form of blueprints, bills of materials, operations sheets, and routing sheets. Some of these are on paper and others are available as computerized databases.

Methods for ordering materials required for the job shop follow aggregate planning. These include developing master production schedules which are precise commitments of what is to be made in the shop or done by the service organization.

Aggregate planning gains utility from the advantage of forecasting for aggregate phenomena as compared to specific components of the aggregation. Aggregate forecasts are always better than the set of component forecasts because the standard deviation of the total of *n* components is equal to the square root of *n* multiplied by the standard deviation of the individual components. Thus:

$$\sigma_T^2 = \sigma_1^2 + \sigma_2^2 + \ldots + \sigma_n^2$$

and assuming that the variances for $i = 1,2,3,\ldots,n$ are all about equal:

$$\sigma_T^2 = n\sigma_i^2$$

and

$$\sigma_T = \sigma_i\sqrt{n}$$

The square root effect is a significant decrease in the otherwise linear result of summing variation to yield $n\sigma$.

Basic Aggregate Planning Model

Job shops require a strong methodological approach for planning ahead. There is so much detail and variety to capture. Aggregate planning provides just the approach that the job shop requires. Demand and production output are treated in aggregation across a variety of different work facilities and output jobs. The aggregate is treated as one job made by one facility operating under several different modes, e.g., regular and overtime production, with and without subcontractors.

The organization's facilities are used to satisfy varying demand levels over time *and* in whatever way promises to minimize total costs. These total costs vary according to the production schedule used. Demands for different outputs are aggregated by considering them all to be a unified demand for the output capacity of the facility.

Figure 13.4 shows the demand side of the system. Varying demands for different items (1, 2 and 3) have been aggregated into a single demand called S_t. The demand resulting from sales during a period T is called S_T.

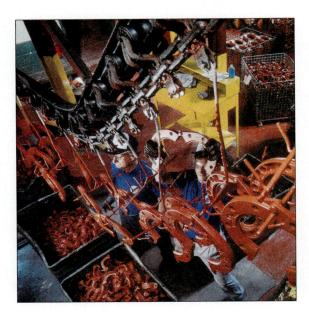

Job shops require a strong methodological approach for planning ahead.
Source: Tyco International Ltd., Exeter, NH

AGGREGATE DEMAND AS THE SUM OF THE DEMAND OF 3 ITEMS IN STANDARD UNITS OVER TIME t

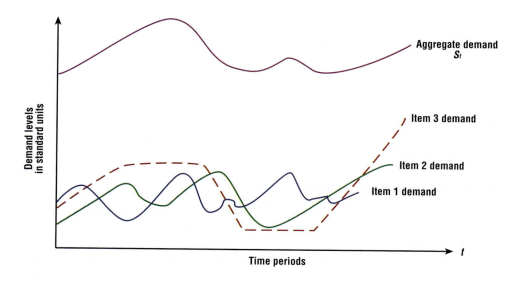

Figure 13.5 shows the production side of the system.

AP'S PRODUCTION SIDE

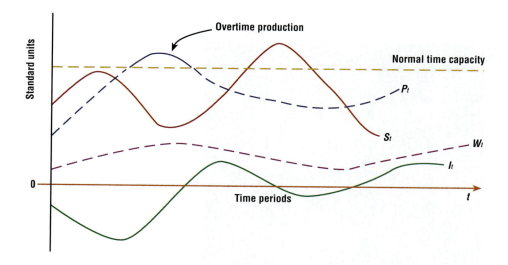

The factors of aggregate planning are demand (or Sales) = S_t, supply (or Production) = P_t (Note overtime use), Workforce level = W_t, and Inventory level = I_t (normal time capacity is given)—all in standard units.

The aggregate supply resulting from production during period t is P_t. In other words, the amount produced during time period t is called P_t. Both the facilities used and the organization's workforce are likely to vary over time. The workforce level during time t is called W_t. For the prior period, called $t - 1$, the workforce level is called W_{t-1}. The same kind of time considerations apply to supply, demand and inventory levels. An additional variable that the P/OM wants to control over time is the inventory level, I_t, I_{t-1}, I_{t-2}... which also is shown in Figure 13.5.

Using the forecast for aggregate demand S_t over time, the aggregate planning problem to solve is:

> *How should P_t, W_t, and I_t be varied so as to optimize the system's performance?*

More specifically, the aggregate planning problem requires period by period solutions that will optimize the total system's performance. There are interperiod dependencies which mean that one period might not be as good as it could be in order that the total result for the entire planning horizon is best. This is classic systems thinking in which component performance may have to be suboptimized to obtain the system's optimal. Note Tables 13.5, 13.6, and 13.7 below, as examples of aggregate systems planning.

13.1

THREE AGGREGATE PLANNING POLICIES

Three policies suggest themselves.

1. Do not vary the workforce. This means keeping P_t constant over time. By formula: $P_t = k$ for all times, t. Keeping a constant workforce is called a level policy. Figure 13.6 labels this Pattern A.

F I G U R E   1 3 . 6

AGGREGATE PLANNING—P_t IS A CONSTANT

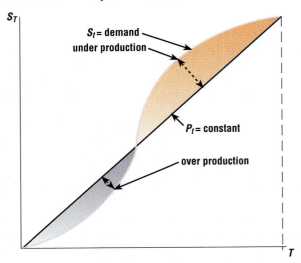

There is overproduction and underproduction with Pattern A.

Pattern A's policy provides a constant, fixed, and level volume of production. Routine production sequences can be followed every day which saves money because process alterations are avoided. However, there are likely to be costs associated with the fact that the "level policy" does not match demand.

2. Vary W_t so that P_t matches the demand S_t as closely as possible. In Figure 13.7 this is labeled Pattern B and called a chasing policy because P_t strives to match S_t at all times.

AGGREGATE PLANNING—CHASING POLICY

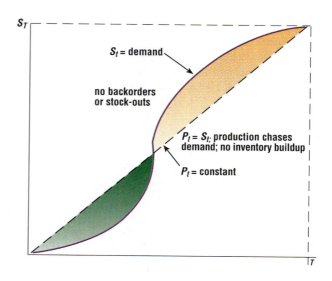

Pattern B is called a chasing policy because P_t matches S_t. The objective is to use up and not build up inventory.

Production Pattern B matches supply with demand—as exactly as possible—for every period in time, i.e., $P_t = S_t$. In this situation, workers are hired, fired or furloughed, in each time period, according to the expected demand of the particular time period. The workforce changes could be made daily, weekly, monthly, etc.

Fixed-volume, level production shops—as in Pattern A above—create a more stable job shop environment than the variable-volume, chasing production policy shops of Pattern B. Greater investments in training can be made. As a result the level shop has less start-up and learning waste. It is associated with higher quality product and service. A drawback is that it also is identified with more idle time.

Idle time is minimized by the variable-volume job shop. When there is a large amount of demand fluctuation, a variable workforce capability is appealing but there are offsetting costs which include information systems to keep track of workforce changes. That is the bookkeeping part of the problem. Being able to hire and layoff involves human resources management with a lot of detail. There are unpleasant aspects to layoffs. The use of temporary employees for peak loads, and subcontracting, can reduce the need for layoffs.

   The models for AP can provide guidance about minimum cost policies using tangible cost factors. The less tangible factors must be included qualitatively. This may involve spreading management too thin. The systems approach enables OM to do aggregate planning with costs that reflect human resources management's limitations. Each company can find a mixture of policies that comes closer to meeting its needs than the pure policies of Patterns A and B.

3. Find superior combinations of Policies A and B. Note that there is a boundless number of variable-volume lines that could connect zero and S_T in Figure 13.7. Each is a partial chasing policy. Examples of combination policy lines are shown in Figure 13.8.

FIGURE 1 3 . 8

AGGREGATE PLANNING—COMBINATION POLICIES

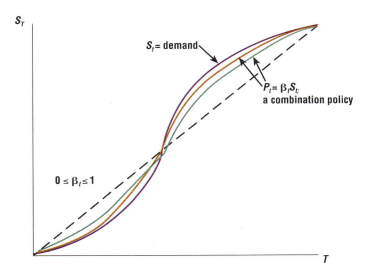

Combination (*C*) policies where production partially chases demand. Systems interrelatedness when P_t partially chases S_t.

Each line in Figure 13.8 has different costs. Certain costs disappear when Pattern *B* is followed instead of Pattern *A*. At the same time, other costs appear. There are mixtures of the costs for the lines in Figure 13.8. Each solution represents some combination of changing production rates, changing workforce size, varying degrees of overtime utilization, fluctuating inventory levels, overstocks and out-of-stocks.

   Note that a partial chasing policy represents a combination policy falling between the two pure policies. It has to satisfy OM, marketing, human resources managers, distribution and warehouse managers, and others. It must capture the system's interrelatedness.

Pattern A—Supply Is Constant

Consider the first case (Pattern A) where the workforce is maintained at a constant level ($W_t = 38$). Assume that each worker can produce $k = 10$ units per time period. Then, production will be constant at $P_t = 380$. This in turn means that inventory can and will fluctuate as demand exceeds and falls below the constant supply. Note that total supply equals total demand (2,280 units) for the six month planning horizon.

At the end of the first period, production has fallen short of demand by 40 units which is indicated by $I_t = -40$. This backordered work is shown as negative inventory. The cumulative inventory is $\sum I_t = -40$ units because it is assumed that prior to period 1 there was zero inventory. When the cumulative inventory changes signs from minus to plus, it means that backorders have been filled and accumulation of stock has occurred. The record is shown in Table 13.5.

T A B L E   1 3 . 5

PATTERN A

t	S_t	P_t	I_t	$\sum_{t=1}^{t} I_t$	W_t	$W_t - W_{t-1}$
1	420	380	−40	−40	38	0
2	360	380	+20	−20	38	0
3	390	380	−10	−30	38	0
4	350	380	+30	0	38	0
5	420	380	−40	−40	38	0
6	340	380	+40	0	38	0
Total	2280	2280				

It is not possible to tell which actual jobs will not be completed when backorders arise. This is because the inventory is stated in terms of aggregates. It is certain, however, that some jobs, or parts of some jobs, are going to be backordered. Cumulative inventory shows backorders occurring in four out of six periods.

Customers whose jobs are backordered are not going to be happy about that state of affairs. It is not unusual to find that some customers will cancel their orders. In certain product classes this is more likely than in others. This cancelling proposition is referred to as "fill or kill."

It is not the custom to cancel Boeing 777s—they are always backordered. On the other hand, it is customary to "fill or kill" orders for music tapes or CDs which are not available in the store. Customers are seldom willing to accept backorders for records, tapes and CDs. The same holds true for cosmetics. Lipsticks and nail polishes are fill or kill items.

Note that in the period 2, demand S_t falls 20 units behind production, P_t. This adds 20 units to inventory, I_t. These units will be shipped to reduce backorders from 40 to 20. The remaining 20 backordered units are shown in row 2 under the cumulative inventory column,

$$\sum_{t=1}^{t=2} I_t$$

as −20 units.

In summing up Pattern A, cumulative backorders equal 130 units and there are no overstocks to report with the constant workforce of 38.

Pattern B—Supply Chases Demand

Now consider the second case, Pattern B, where supply chases demand perfectly. What happens to costs when the workforce level is changed in order to allow production to match demand?

Since $(W_t - W_{t-1})$ will not always equal zero, as in Table 13.5, the production Pattern B $(P_t = S_t)$ can be followed by altering the workforce levels. Monthly variation of the workforce is used to match supply and demand. There will be no inventory accumulation if demand can be perfectly predicted so that the workforce size can be perfectly adjusted. If perfection is elusive, then production always will be chasing demand trying to reduce overstocks and decrease backorders. If perfection is obtainable, then backordering and overstocking costs will not occur.

In general, the forecast will not be perfect. Unexpected orders and cancellations are usual. Also, workforce adjustments cannot always be made. New workers cannot be found and trained immediately.

Running the numbers, again, assume that each worker produces 10 units per time period. Demand in the first period is 420 units and 42 workers have been employed. Perfect forecasts of demand can be made. The results are shown Table 13.6.

TABLE 13.6

PATTERN B

t	S_t	P_t	I_t	$\sum_{t=1}^{t} I_t$	W_t	$W_t - W_{t-1}$
1	420	420	0	0	42	0
2	360	360	0	0	36	-6
3	390	390	0	0	39	$+3$
4	350	350	0	0	35	-4
5	420	420	0	0	42	$+7$
6	340	340	0	0	34	-8
Total	2280	2280				

Production in the first period is 420 units, i.e., $P_1 = 420$. Production exactly matches demand, so there is no inventory, i.e., $I_1 = 0$. Cumulative inventory,

$$\sum_{t=1}^{t=1} I_t$$

also is 0.

No change in workforce size $(\Delta W_t = W_t - W_{t-1})$ is assumed for the first period. Had there been fewer than 42 workers, then backorders would have arisen and the workforce size in the next period would have been increased to catch-up with demand. Whenever a forecast error causes overstocking or backordering, the workforce will be adjusted to catch-up and produce zeros in the inventory columns.

In the second period, Because the demand is 360, the workforce contracts to 36, which is a reduction of 6. Thus, $\Delta W_t = -6$. Then, the workforce expands to 39 in the third period. The remainder of the record is read from the table in the same way. W_t goes from 35 to 42 to 34. $P_t = S_t$, so there are neither inventories nor stock-outs and, under usual circumstances, all orders are filled with minimum delay. The table portrays production perfectly chasing demand.

The supply chain is value adding continuously and in synchronization with demand. There are no unfilled orders and there is no inventory sitting around waiting for demand to increase. On the other hand, the costs for changing workforce size can be substantial.

In addition to costs for changing W_t (i.e., related to ΔW_t), there are other costs which include alterations of space requirements, and services (such as cafeteria and restrooms) for the workforce. More people need more room and more services. Further, $-\Delta W_t$ causes group morale problems and teamwork failures. These correspond to expenses that are difficult to estimate, but they are there nevertheless.

Contrasting the 2 cases of A and B, the constant workforce has produced 3 instances of supply being less than demand. The total shortfall is 90 units. However, back-orders never exceed 40 as seen from the cumulative inventory column,

$$\sum_{t=1}^{t=6} I_t$$

It should be remembered that cumulative inventory is the measure of actual stock on-hand which is net stock on-hand.

In the constant workforce case, there is no positive accumulation of inventory. That is by chance because there is nothing to have prevented it from occurring. If the first period had been deleted, there would have been positive inventory to report. Note that for the constant workforce of 38, total demand for 6 periods equals total supply which is 2280 units.

The same totals apply to the figures for the chasing policy. In trade for no inventory or backorders, the organization is required to hire 10 workers and lay off 18. The costs of workforce changes need to be contrasted with the savings that result from having no backorders and no inventory accumulation.

Which approach has the lowest expected cost, Patterns A, B, or C? Recall that C-type patterns allow many different combinations of A and B. Thus, it is important to consider configurations that are logical mixtures of Patterns A and B.

Pattern C_1—Increase Workforce Size by ± One

To begin with a simple change, test the workforce level, namely, using the stable workforce size of 38 as a basis:

1. increase the number of workers from 38 to 39 whenever demand is greater than 380
2. decrease the number of workers from 38 to 37 whenever demand is less than 380

It is an entirely sensible way to begin. The plan *slightly* chases demand when $S_t > P_t$, and thereby reduces backorders. The plan *slightly* reduces overstocks when $S_t < P_t$.

Workforce size is variable in a
business like gold exploration.
Source: FMC Corporation

TABLE 13.7

PATTERN C_1

t	S_t	P_t	I_t	$\sum\limits_{t=1}^{t}$	W_t	$W_t - W_{t-1}$
1	420	390	−30	−30	39	+1
2	360	370	+10	−20	37	−2
3	390	390	0	−20	39	+2
4	350	370	+20	0	37	−2
5	420	390	−30	−30	39	+2
6	340	370	+30	0	37	−2
Total	2280	2280				

The result of this *slightly* chasing policy (C_1) for six periods is:

$$\sum_{t=1}^{t=6} \sum_{t=1}^{t=6} I_t = -100 \text{ for total backorders and } +0 \text{ for total stock on-hand}$$

$$\sum_{t=1}^{t=6} \Delta W_t = +5 \text{ for hires and } -6 \text{ for workforce reductions}$$

Note: there is no additional labor charge because the average number of work-ers for the six periods in C_1 is 38.

The comparable figures for Pattern A—the *stable* workforce policy—of $W_t = 38$ are:

$$\sum_{t=1}^{t=6} \sum_{t=1}^{t=6} I_t = -130 \text{ for total backorders and } +0 \text{ for total stock on-hand}$$

$$\sum_{t=1}^{t=6} \Delta W_t = 0 \text{ for hires and for workforce reductions}$$

The comparable figures for Pattern *B*—the *chasing* policy where no inventory costs occur—are:

$$\sum_{t=1}^{t=6} \sum_{t=1}^{t=6} I_t = -0 \text{ for total backorders and } +0 \text{ for total stock on-hand}$$

$$\sum_{t=1}^{t=6} \Delta W_t = +10 \text{ for hires and } -18 \text{ for workforce reductions}$$

The Cost Structure

The costs that apply must be used to decide which is best. Comparing the stable policy *A* with the slightly chasing policy C_1, there is a reduction of backorders by 30, an increase of +5 for hires, and an increase of 6 for workforce reductions.

Say that backorders cost \$100 per occurrence, costs for carrying inventory are \$25 per unit, hires cost \$200 per instance, and reductions cost \$300 per event—all per unit time period. Then, for the C_1 policy, the total variable cost, TVC(C_1) is:

$$\text{TVC}(C_1) = (100 \times 100) + (0 \times 25) + (5 \times 200) + (6 \times 300)$$

$$= \$12,800$$

Compare this to the total variable costs of the stable workforce of 38 workers. For policy *A*, the total variable cost, TVC*(A)* is:

$$\text{TVC}(A) = (130 \times 100) + (0 \times 25) + (0 \times 200) + (0 \times 300)$$

$$= \$13,000$$

Preference is for C_1, the *slightly* chasing policy, which has reduced costly backorders at the expense of increased hires and layoffs. The difference is only \$200.

Because backorders are driving the preference, it is logical to examine option *B* which completely eliminated inventory costs by chasing demand. Total variable cost TVC*(B)* for the *B* policy is:

$$\text{TVC}(B) = (0 \times 100) + (0 \times 25) + (10 \times 200) + (18 \times 300)$$

$$= \$7,400$$

This is clearly the preferred way to go for the particular costs that have been assumed. Other costs could lead to entirely different conclusions.

Cost trade-off analysis

Comparisons using only the variable portions of total costs; variable costs are those costs which differ between alternative choices.

The comparisons are examples of **cost trade-off analysis** using only the variable portions of total cost. The variable portions are those costs which differ between alternative choices. Thus, the total variable costs do not include the costs of the workforce size, W_t, so long as there are no additional wages to be paid. If a policy required hiring a 39th worker, the cost of that individual's wages would have to be included as a variable component.

Another factor to consider is the difference in the costs of increasing the workforce size and of decreasing the workforce size. There is no reason for them to be equivalent. Hiring involves orientation and training. There are many forms of workforce reduction—some of which are temporary and others which are permanent.

Also, note that when supply (P_t) varies with demand (S_t), workforce adjustment costs could represent overtime costs with a constant size workforce—or a fluctuating workforce size, without the use of overtime. The cost structures must apply to the policies that are followed.

FIGURE   13.9

TRADE-OFF MODEL

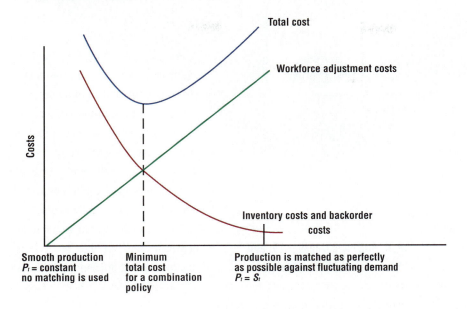

Trade-off model to find optimal balance between level production and perfect chasing policy as a function of workforce adjustment costs and inventory/backorder costs—minimum total cost point indicates the optimal degree of chasing.

There is a trade-off between workforce adjustment costs and inventory/backorder costs. The trade-off model is shown in Figure 13.9 where the x-axis runs from smooth or level production at the left to perfect chasing where production matches demand at the right.

Workforce adjustment costs and inventory/backorder costs are added together to create the total cost curve which has a minimum point. That minimum total cost is associated with the concept of a combination policy which is not specific. The idea of a fifty percent chasing policy is ambiguous.

The purpose of Figure 13.9 is to note that if workforce adjustment costs rise, the line pivots around the 0,0 point up and to the left. The effect is to move the optimal combination closer to level production. If the inventory and backorder costs rise, that curve moves upward and to the right. The effect is to move the optimal combination closer to perfect matching of production and demand.

13.2

LINEAR PROGRAMMING (LP) FOR AGGREGATE PLANNING

Figure 13.9 illustrated the way in which total cost reaches a minimum value for some particular production schedule. The key is how to find that **minimum cost assignment schedule**. This is equivalent to identifying the production policy for the minimum cost point on the x-axis of Figure 13.9.

Minimum cost assignment schedule

The equivalent to identifying the production policy for the minimum cost point.

**Planning Services in the
Health Care Industry**

Changing practices and breakthroughs in technology have dramatically affected the planning of operations in the health care industry. With the strong encouragement of insurance companies, health care providers have been employing new processes and technology in streamlining their services.

The goal of everyone involved in the health care process, including providers, insurance companies and consumers, is to have the patient receive the highest quality care in a safe and cost-effective manner, said Mary Jane Klarich, Jr., director of Sub-Acute Services at Parma Community General Hospital near Cleveland, Ohio. Klarich said that Parma Community is currently implementing a Health Care Information System to facilitate meeting these goals and to handle the planning and scheduling of services. This planning currently is done by individual departments at the hospital.

One way in which health care providers can keep costs down is to keep the length of stays in an "acute care setting" to a minimum. The length of stays for some patients can be minimized through the use of more precise surgical tools and improved anesthesia. For some treatments, such as tonsillectomies, patients are operated on and released the same day.

Another technique to minimize hospital stays is the "observation bed" status, which is used with patients whose conditions are too complicated for an outpatient visit but not serious enough for inpatient admission. Hospitals also are changing the process of administering tests to determine the exact status of medical conditions by performing more of the tests "bedside" rather than wheeling patients to many different locations for the tests. They also are cross-training technicians to perform multiple testing procedures.

Klarich points out that glitches in the system may occur in "hand-offs between levels of care." For example, if an elderly person is released from an acute care setting to a nursing home bed too early, the person may not receive the appropriate care at the nursing home. Another problem arises when there are many physicians involved on one case and each is looking at a different part of the patient's care. "If they're not talking to each other and coordinating their care well, then that affects the quality and efficiency of the care the person receives," Klarich said.

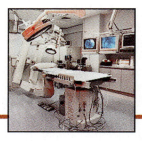

Sources: Charles Stein, "Same-day service to cut costs, insurers push hospitals to discharge patients faster. Is it too fast?" *Boston Globe*, April 24, 1994; conversation with Mary Jane Klarich, Jr., director of Sub-Acute Services at Parma Community General Hospital.

   A linear equation can be written for the minimum total cost objective. Also, a set of constraints will be needed to characterize patterns such as those previously described as A, B and C_1. The advantage of LP is that computer programs can quickly solve problems with a large number of variables and many constraints. The nature of the variables and the constraints now will be explained in terms of symbols, many of which have been introduced already in this chapter.

Data input that is given and not a variable:

$$S_t = \text{sales in period } t \text{ (sales demand is forecast or on-hand)}$$

$$t = \text{the period of time which is often a month or a quarter}$$

Variables that can be controlled:

1. $W_t =$ the number of workers at the beginning of period t
   $P_t = kW_t =$ normal production output in period t (k is the output per worker in period t). P_t is derived from W_t as long as k is a constant.
2. $A_t =$ additions to the workforce (number of people) at the beginning of period t.
3. $R_t =$ reductions to the workforce (number of people) at the beginning of period t.
4. $I_t =$ inventory (number of units in stock or out-of-stock) at the beginning of period t.
5. $O_t =$ overtime production (units of output) in period t.

Reasonable constraints for each period:

1. $W_t \leq W$ which places limits on the workforce because of space, payroll and other resource availabilities.
2. $A_t \leq A$ which places limits on additions to the workforce because of constraints imposed by training facilities and the ability to absorb additional workers.
3. $R_t \leq R$ which places limits on reductions to the workforce because of constraints on ability to process layoffs and company policy regarding the number of layoffs and furloughs that will be permitted in a time period.
4. $W_t = W_{t-1} + A_t - R_t$
   This equation describes the size of the workforce at the beginning of each time period t. It is subject to Constraints 1, 2, and 3 above.
5. $I_{\min} \leq I_t \leq I_{\max}$ places limits on how large or small the inventory can become. Sometimes a specific number is given for the inventory level at time t. Thus, $I_t = I$.
6. $O_t \leq O$ which places a limit on how much overtime is acceptable.
7. $I_t = I_{t-1} + P_t + O_t - S_t$
   $\quad = I_{t-1} + kW_t + O_t - S_t$
   This describes the amount of inventory (net stock or out-of-stock) at the beginning of each time period t imposed by Constraints 5 and 6 above.

There are five variables and seven constraints for each time period. Say that the time period is months. Then, with a planning horizon of six months there would be 30 variables and 42 constraints. In practice, some of the constraints that have been listed as possibilities would not be exercised. Say that half of them are real constraints. This would result in 30 variables and 21 constraints. As described below, only 21 variables will have values greater than zero. This means that $30 - 21$ variables will have zero values. They will not be active variables.

Fundamental theorem of linear programming

There cannot be more active variables than real constraints.

The **fundamental theorem of linear programming** states that there cannot be more active variables than real constraints. There are six variables for each of the following: W_t, A_t, R_t, I_t, and O_t. Because workforce will be positive, and inventory levels are likely to be positive, this means that at least nine uses of overtime, or additions and reductions to the workforce will be set at zero. In other words, they will not be used.

To obtain the LP solution, costs must be assigned as follows:

c_w = earnings of worker per time period t

c_A = cost of adding a worker

c_R = cost of reducing the workforce—removing a worker

c_C = cost of carrying a unit for the time period t

c_O = cost of production for an overtime unit

The total cost objective function to be minimized is:

$$TC = \sum_{t=1}^{t=T} (c_w W_t + c_A A_t + c_R R_t + c_c I_t + c_o O_t)$$

The most common objection to use of the LP model is that it has a linear structure for the costs of the objective function and for the constraints. Below, a pioneering aggregate planning model will be briefly explained (call HMMS) that allows nonlinear costs—which require much more work to use.

OM considers how output can be increased by introducing new technology like this machinery that continuously processes 400 apples per minute.
Source: FMC Corporation

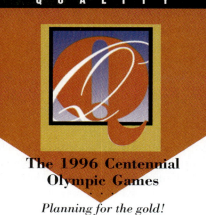

QUALITY

The 1996 Centennial Olympic Games
• • •
Planning for the gold!

The Olympic Games epitomize the quest for quality and excellence. When more than 10,000 Olympic athletes compete in 26 sports over a 16-day period in July and August, spectators give little thought to the planning effort that makes it all possible. This is the scenario for the 1996 Centennial Olympic Games held in Atlanta, Georgia. More than 2 million visitors will purchase approximately 11 million tickets to attend the 271 planned Olympic events.

With the support of the city of Atlanta, a small group of volunteers waged a three-year campaign to host the 1996 Olympic Games that began in 1987 and continued until the city's selection by the International Olympic Committee (IOC) in September 1990. The Atlanta Committee for the Olympic Games (ACOG), a private, not-for-profit corporation established in January 1991, is responsible for planning and staging the 1996 Olympic Games in coordination with the IOC, the United States Olympic Committee (USOG), the city of Atlanta, and the Metropolitan Atlanta Olympic Games Authority.

The ACOG consists of many departments including communications, construction, corporate services, finance and management services, games services, international relations, licensing, marketing, operations, sports, technology, and venue management. The venue management department is responsible for developing the operations plans and ultimately manages the operations of the games at all venues (or sites).

The site plan is the foundation for the development of the operations plan. In developing the plans, venue management considers the needs of anyone who will use, or is affected by, the facility. The venue management department works with all departments within ACOG and the facility's current management, then coordinates all plans into one integrated operations plan. This plan is guided by public safety issues and standard event management practices.

Olympic events are staged at 30 different venues. Eleven competition venues are located within the "Olympic Ring" which is an imaginary circle with a radius of 1.5 miles emanating from the Georgia World Congress Center in downtown Atlanta. In planning which venues are appropriate for a given event, location and seating capacity, among other factors, must be considered.

Sources: Atlanta Committee for the Olympic Games Press Guide, February 1995 with special assistance from Lyn May, communications director/ACOG Operations.

Another important factor is the production output represented by k. OM carefully considers how k can be increased by introducing new technology and by improving training methods. This is not part of the LP solution. Note, however, that by increasing k, different solutions will be obtained. Overtime costs can be scaled back, workforce additions can be decreased and workforce decreases may be possible. The minimum total cost can be reduced considerably by using improved process methods. From the systems perspective, coordination with human resources management for training improvements, and marketing for better product-mix alternatives, can significantly decrease costs.

The solution to the problem requires a great deal of calculation which only computer programs for LP can handle. It also should be noted that fractional values of some of these variables (i.e., a part of a person to be added to the workforce) may not be feasible if part-time work is not used. When a problem requires integer solutions for the variables, integer programming (IP) software is available for the computer. This kind of application makes sense for airline crew scheduling.

13.3

TRANSPORTATION MODEL (TM) FOR AGGREGATE PLANNING

The transportation matrix provides a convenient representation of the aggregate-planning problem. It can provide an optimal aggregate solution based on relatively simple assumptions. Computer software can solve these transportation problems using network algorithms[1] or linear programming (LP).

Two versions of transportation matrices are presented. The first one allows no backorders and is easy to solve by hand. The second one permits backorders and their costs. It is laborious to solve by hand so computer software is preferred.

Table 13.8 presents the first version of the aggregate planning transportation matrix which does not allow backorders.

TABLE 13.8

TRANSPORTATION MODEL FOR AGGREGATE PLANNING— NO BACKORDERS ALLOWED

	Sales Periods					
	1	2	3	Final Inv.	Slack	Supply
Initial Inv.	0	c	$2c$	$3c$	0	I_0
Regular 1	r	$r+c$	$r+2c$	$r+3c$	0	R_1
Overtime 1	v	$v+c$	$v+2c$	$v+3c$	0	O_1
Regular 2	X	r	$r+c$	$r+2c$	0	R_2
Overtime 2	0 X	v	$v+c$	$v+2c$	0	O_2
Regular 3	X	X	r	$r+c$	0	R_3
Overtime 3	0 X	c X	v	$v+c$	0	O_3
Demand	D_1	D_2	D_3	I_f	SL	Grand Total

where:

c = carrying cost per unit for the interval of time
r = regular production cost per unit
v = overtime production cost per unit
I_0 = initial inventory
I_f = final inventory

Demand is shown as the column totals. The first three column totals are demand levels in the first, second and third periods (D_1, D_2, and D_3). Final inventory (called I_f) is specified at the bottom of the fourth column. Slack (SL)—which balances supply and demand—is listed at the bottom of the fifth column.

Supply is shown as the row totals. The first row is initial inventory (called I_O). Thereafter, rows alternate in presenting production supply for regular time and for overtime, called R_1 and O_1, R_2 and O_2, R_3, and O_3.

The objective is to minimize total cost. Constraints reflect the availability of regular and overtime production capacity in the period. Alternatives can be examined where capacity is added or removed by additional hires and/or new technology.

Standard units are required to aggregate different kinds of product—if the problem is applied to situations where the same resources are used to make diverse models such as small and large cans (aggregated as "cans"), or white and green paint (aggregated as "paint"), or 40, 60, and 75 watt bulbs (aggregated as "bulbs").

The TM is flexible. It does not require a square matrix. Each matrix applies to a specific period of time. Table 13.8 has eight rows and six columns—one of which is a Slack column. In this case, a Slack column is added instead of a Slack row because total supplies are assumed greater than total demand. Thus, $S > D$, so $S - D > 0$, and $S - D = SL$. Then: $S = D + SL$.

TABLE 13.9

TRANSPORTATION MODEL FOR AGGREGATE PLANNING— BACKORDERS ALLOWED

	Sales Periods					
	1	2	3	Final Inv.	Slack	Supply
Initial Inv.	0	c	$2c$	$3c$	0	I_0
Regular 1	r	$r+c$	$r+2c$	$r+3c$	0	R_1
Overtime 1	v	$v+c$	$v+2c$	$v+3c$	0	O_1
Regular 2	$r+b$	r	$r+c$	$r+2c$	0	R_2
Overtime 2	$v+b$	v	$v+c$	$v+2c$	0	O_2
Regular 3	$r+2b$	$r+b$	r	$r+c$	0	R_3
Overtime 3	$v+2b$	$v+b$	v	$v+c$	0	O_3
Demand	D_1	D_2	D_3	I_f	SL	Grand Total

where:

v = overtime or second-shift production costs per unit made

c = inventory carrying costs, per dollar for a given time period

r = normal production labor costs (regular shift) per unit made

b = backorder costs per unit and per time period backordered

On the other hand, the model is linear. This means that unit costs or profits are not able to be changed as a function of volume. Profit saturation and nonlinear cost increases cannot be represented.

Normal transportation methods are used to obtain a solution. In this case, however, backorders are prohibited (by the *x*'s in the six cells). Each *x* represents a situation where production units made in the *nth* period are intended for delivery in the *n−1st* period. This is equivalent to making units in February that are intended to satisfy January demands that were backordered. For the computer, backorders are prohibited by putting very large costs per unit in the *x*-marked cells.

Table 13.9 presents the second version of the aggregate planning transportation matrix where backorders are allowed.

This second transportation matrix has new costs, *b,* which are backorder costs per unit backordered. The new cost is integrated without disruption and the model can be solved by conventional transportation techniques including computer software using network algorithms or linear programming. Several simple numerical examples are assigned in the Problem Section at the end of this chapter.

Calling the time period months, the transportation matrices in Tables 13.8 and 13.9 cover a planning horizon of one quarter. Note how the carrying costs are zero for the first month. For the second month *c* is charged per unit carried in that month. For the third month, $2c$ is charged because inventory has been carried two months. Backorder costs are like carrying costs, growing with each additional month that a unit is backordered. Slack costs are all zero.

Network and LP Solutions

The AP problem in Table 13.8 above is unique because no backorders are allowed. It can be solved by assigning as many units as possible to the lowest cost cells in each column. Start with column 1. When the supply in a row is used up, go to the next lowest cost in the column and assign as many units as possible in that row. The constraints that must be met are the row and column totals—more than either cannot be assigned to the lowest cost cells.

When column 1 allocations equal the column 1 total, move to column 2. Proceed in the same way making sure not to assign more units to lowest cost cells than the row and column totals can support. The lowest cost assignments in column 2 may be blocked by assignments already made in column 1. Blocked assignments cannot be violated. Continue until all columns have been assigned in order. The optimal solution is the result.

It is necessary to understand the equations for solving the transportation problem as an LP model in order to input data correctly for the software. The constraining equations or inequalities (also called inequations) are:

$$\sum_{j}^{M} x_{ij} \leq S_i$$

where i = rows (supply); i = 1, 2, ..., N

$$\sum_{i}^{N} x_{ij} \leq D_j$$

where j = columns (demand); j = 1, 2, ..., M

$$\sum_{i}^{N} \sum_{j}^{M} x_{ij} = \text{Grand Total } (\sum_{i}^{N} S_i \quad \text{or} \sum_{j}^{M} D_j; \text{ whichever is greater.})$$

The objective function to be minimized is

$$TC = \sum_i^N \sum_j^M c_{ij} x_{ij}$$

where all cells *ij* are included.

Data requirements include the row totals S_i, the column totals D_j, the number of rows (N) and columns (M), and the costs c_{ij} for all *i* and *j*. The software has built-in methods for finding the first feasible solution. Most software permits **sensitivity analysis** which allows investigation of the effect of altering supply capacities, demand levels and costs on a "what if" basis.

Sensitivity analysis

Allows investigation of the effect of altering supply capacities, demand levels and costs on a "what if" basis.

13.4
NONLINEAR COST MODEL FOR AGGREGATE PLANNING

The linear limitations of LP and the transportation model are so often cited that it is important to explain what the nonlinear structure is like. A *nonlinear* method, called **HMMS** after its four developers, *Holt, Modigliani, Muth,* and *Simon,* was developed at PPG's paint factory.[2] The HMMS model addresses four separate issues.

1. What are nonlinear costs?
2. How complex is HMMS?
3. How much of a difference will it make?
4. In practice, how much is it used?

HMMS

Model which addresses four issues:

• What are nonlinear costs?

• How complex is HMMS?

• How much of a difference will it make?

• How much is it used?

A linear cost equation is $y = cx$. A quadratic or second order cost equation is $y = cx^2$. HMMS used both linear and quadratic cost equations to approximate the actual cost system of the PPG job shop. Examples of quadratic costs are shown in Figure 13.10.

The solid lines are the shapes of the cost curves that are believed to define the actual cost structure. The dashed lines are the HMMS functions used to approximate the actual cost curves. Thus, payroll is linear and described by $C_1 W_t$. The cost of hiring and layoffs is made into one quadratic cost function $C_2(W_t - W_{t-1})^2$ instead of two lines with different slopes. Similar descriptions apply to the other costs. Note that the HMMS model includes costs for overtime payroll, setups, and back-ordering. Also, demand S_t was based on forecasts.

To evaluate the complexity of the HMMS model, a good starting point is the total cost equation.

$$\textit{Minimize } TC = \sum_{t=1}^{t=T} C_t$$

(*T* is the last month or end period of the planning horizon.)

$$C_t = C_1 W_t + C_2(\Delta W_t)^2 + C_3(P_t - C_4 W_t)^2 + C_5 P_t - C_6 W_t + C_7(\sum I_t - C_8 - C_9 S_t)^2$$

subject to the logical balancing of inventory:

$$\sum_{t=1}^{t=t^*} I_{t-1} + P_t - S_t = \sum_{t=1}^{t=t^*} I_t$$

where cumulative inventory through period *t* equals inventory at the end of the previous period plus production added minus shipments made, during the period. The equality applies for all values of *t*, called *t**.

FIGURE 13.10

COST FUNCTIONS–USED FOR THE HMMS AGGREGATE PLANNING MODEL

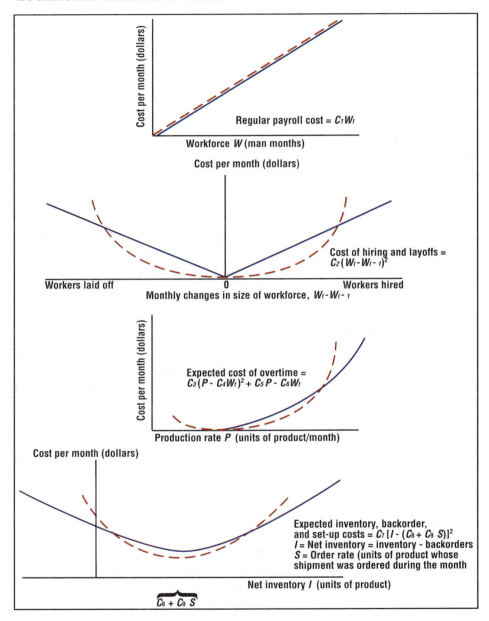

Approximating cost function – – – –

From Charles C. Holt, Franco Modigliani, and Herbert A. Simon, "A Linear Decision Rule for Production and Employment Scheduling," *Management Science*, Vol. 2, No. 1, October 1955.

The costs in the equation are identified as follows:

Costs

$$C_1 = \text{costs related to payroll (i.e., size of the workforce)}$$

$$C_2 = \text{hiring and layoff costs, in terms of workforce changes}$$

$$C_3, C_4, C_5, C_6 = \text{various forms of overtime costs}$$

$$C_7, C_8, C_9 \quad = \text{different kinds of inventory costs}$$

$$\textstyle\sum C_t \qquad\quad = \text{total costs over } t \text{ periods } (t = 1, 2, \dots T)$$

Given by forecast

$$S_t = \text{demand in units for period } t$$

Variables

$$\textstyle\sum I_t = \text{on-hand inventory minus backorders—end of period } t$$

$$P_t \quad = \text{aggregate unit production rate—period } t$$

$$W_t \quad = \text{workforce size—beginning of period } t$$

Methodology

The method for obtaining the optimal solution is fully detailed in the HMMS text (Note 2 above). It is a procedure of calculus using partial derivatives that surpasses the level of mathematics required by this text. Optimum values of W_t, P_t, and I_t must be obtained for each time period. The computational work is significantly greater than when this is done using LP.

HMMS is complex. It is costly to use. Solving the system of equations is mathematically arduous. Using the computer would help. However, finding the correct quadratic cost functions is a large and tedious project in itself. If the cost equations are only approximate, LP provides a better alternative.

How much difference will the use of HMMS make? There is no hard data to go on, but using LP properly allows good linear approximations of nonlinear functions. To illustrate, note how the LP model treats workforce additions (hiring) and reductions (layoffs) separately. The HMMS approximation, in this case, is less accurate. Also, there are additional techniques that can be used with LP to further the cause of approximating nonlinear cost functions.

Addressing the final question that was raised above, no organizations are known to be users of the HMMS model. This is probably because it is more work than it is worth—especially bearing in mind that LP and the transportation model can be used. However, the HMMS model provides important insight into the proper use of the alternative LP and TM models.

The HMMS model stresses the need to get correct cost functions. It provides a basis for discussion of costs. It offers a *benchmark* for comparison of models. On ease of use the TM gets first place, LP gets second place, and HMMS is a distant third. On getting the cost structure into perspective, HMMS wins first place, LP is second, and TM is third. In conclusion, it is useful for P/OM to suggest to aggregate planners that they examine all cost functions for linearity. When significant curvature characterizes a cost curve, the LP model should be designed—as much as possible—to take that into account.

S U M M A R Y

Aggregate planning (AP), an internal production management function, matches classes of jobs with the supportive resources needed to produce each of them. Its relationship to business planning and production scheduling is discussed. The methods for aggregation are explored. Then three aggregate planning policies are developed and compared. These are level production, chasing sales, and combinations. Then, linear programming is explained. The linear programming (LP) model is an effective means of achieving optimal aggregate plans but the linear cost structure may be a limitation. The transportation model (TM) is then introduced for AP. It has the same linear cost constraints. The transportation model for aggregate planning without allowing backorders can be solved by simple network rules. With backorders allowed, the TM can be solved as an LP model. The insightful nonlinear HMMS model is introduced and its second order equations to describe cost are explained. Then, the three AP models are benchmarked against each other.

▶ **Key Terms**

aggregate planning (AP) *550*
HMMS *577*
planning horizon *556*
standard units *551*

cost trade-off analysis *568*
loading *557*
sensitivity analysis *577*

fundamental theorem of linear programming *572*
minimum cost assignment schedule *569*
sequence *557*

▶ **Review Questions**

1. Why is aggregate planning used? What does it do?

2. What is meant by aggregation of units?

3. Explain standard units of work.

4. Explain the use of backordering for goods.

5. Give an example of how backordering can be used for services.

6. Explain the significance of "fill or kill" and give some examples.

7. Describe the system's nature of aggregate planning from the point of view of classes of resources and product-mix families.

8. Explain why aggregate planning follows strategic planning.

9. Explain the statement that aggregate planning starts a chain reaction in the supply chain of suppliers-producer-customers.

10. Discuss the importance of forecasting for AP.

11. Explain the planning horizon and the updating interval.

12. It has been stated that: "For the average job shop product, the best planning interval would range from 3 to 6 months." Why might this statement provide a reasonable rule of thumb?

13. Why do forecasts for aggregate planning have an advantage over forecasts for individual (disaggregated) jobs?

14. How can the effects of seasonal demands be taken into account for aggregate planning? Explain.

15. In some job shop industries, a smooth production rate is the preferred choice. Explain what this means.

16. In some job shop industries, the workforce size is altered to chase the expected demands. Explain what this means.

17. Compare smooth or level-aggregate-production policies with chasing policies. Explain when each are likely to be preferred.

18. At one time, the canning industry was totally dependent on harvest dates. As a result, major workforce alterations occurred sporadically. After careful study, steps were taken to smooth the demand patterns. What measures might have helped?

19. Explain the meaning of this question:

How should P_t, W_t, and I_t be varied so as to optimize the systems performance?

20. What is a combination aggregate policy and why is it called partially chasing?

21. Explain how the trade-off model for workforce adjustment costs and inventory costs yields an optimal aggregate planning policy.

22. What is meant by a *slightly* chasing policy?

23. Describe the use of linear programming for AP.

24. What does the fundamental theorem of linear programming state? How does this theorem provide important guidance?

▶ **Problem Section**

1. Seven jobs will be in the shop next week. The demand in units for each job (called d_j), and the production rates of the standard operator in pieces per standard operator hour for each type of job (called PR_j) are given as follows:

Job	A	B	C	D	E	F	G
d_j	600	1000	500	50	2000	20	800
PR_j	60	20	25	10	40	2	40

What production capacity in standard operator hours is required to complete all of these jobs?

2. Six service calls are on-hand for next week. The number of steps required for each has been determined and is listed below as d_j. The supervisor is listed as the standard operator and his output rates in steps per standard operator hour for each type of job are listed as PR_j below.

Job	A	B	C	D	E	F
d_j	500	400	200	150	2000	48
PR_j	100	20	25	75	400	12

What workforce capacity in standard operator hours is required to complete all the service calls?

3. It has been estimated that annual demand for five types of soup that are made by The Big Soup Company are as follows:

Soup	A	B	C	D	E
d_j	900	630	240	1800	1200

Also, there are three plants located in the U.S. The most productive plant has been chosen as the standard plant. Its output is listed below in standard plant output per day.

PR_j	6	7	2	10	25

The other two plants have indexes of 0.9 and 0.7. There are 250 working days in the year and all numbers are given in thousands of cases.

Is it likely that the three plants can handle the annual demand?

4. Using the information in Problem 3 above, assume that the two other plants have indexes of 0.8 and 0.7.

Is it likely that the three plants can handle the annual demand?

5. Complete Table 13.10 for the six month period shown:

TABLE 13.10

AN ALTERNATIVE WORKFORCE PATTERN

t	S_t	P_t	I_t	$\sum I_t$	W_t	$W_t - W_{t-1}$
1	340	380			38	0
2	420	380			38	0
3	350	380			38	0
4	390	380			38	0
5	360	380			38	0
6	420	380			38	0

What kind of aggregate planning policy is this?

6. Compare the results obtained from Table 13.10 in Problem 5 above with the results obtained from Table 13.5 in the text.

7. Using the information in Problem 5 above, what would be the effect of adding another person to the workforce? Specifically, increase the number of workers from 38 to 39. Is this sensible?

8. Using Table 13.5 in the text, what would be the effect of adding another person to the workforce? Specifically, increase the number of workers from 38 to 39. Is this sensible?

9. Using the information in Table 13.10 in Problem 5 above, calculate the total variable costs for the six month period. Backorders cost $100 and carrying inventory costs $25 per unit per time period. Hires cost $200 per person added and layoffs cost $300 per person. The cost of an additional person working is $2,000 per month. Fractional payroll amounts can be calculated and used as well.

10. Using the information in Problems 7 and 9 above, what is the total variable cost? Compare this total variable cost with that derived in Problem 9 above. Present your recommendations.

11. Using the information in Problems 8 and 9 above, what is the total variable cost? Compare this total variable cost with that derived in the text for Table 13.5. Present your recommendations.

12. Complete Table 13.11 below:

AN ALTERNATIVE WORKFORCE PATTERN

t	S_t	P_t	I_t	$\sum I_t$	W_t	$W_t - W_{t-1}$
1	340	380			38	
2	420	390			39	
3	350	360			36	
4	390	370			37	
5	360	360			36	
6	420	400			40	

What kind of aggregate planning policy is this?

13. Write the linear programming equations for aggregate planning using $k = 500$. What additional data is needed to solve the LP aggregate planning problem?

14. There is sufficient data in Table 13.12 to solve this aggregate planning transportation model. Assume that carrying costs are stated as dollars per month. *Note:* because no backorders are allowed, this particular TM can be solved directly.

Specify the solution and explain it.

TRANSPORTATION MODEL FOR AGGREGATE PLANNING— NO BACKORDERS ALLOWED

	Sales Periods 1	2	3	Final Inv.	Slack	Supply
Initial Inv.	0	c	$2c$	$3c$	0	I_0
Regular 1	r	$r + c$	$r + 2c$	$r + 3c$	0	$R_1 = 100$
Overtime 1	v	$v + c$	$v + 2c$	$v + 3c$	0	$O_1 = 200$
Regular 2	x	r	$r + c$	$r + 2c$	0	$R_2 = 300$
Overtime 2	x	v	$v + c$	$v + 2c$	0	$O_2 = 200$
Regular 3	x	x	r	$r + c$	0	$R_3 = 100$
Overtime 3	x	x	v	$v + c$	0	$O_3 = 50$
Demand	$D_1 = 200$	$D_2 = 300$	$D_3 = 200$	I_f	SL	Grand Total

where:

c = carrying cost per unit for the interval of time = \$1
r = regular production cost per unit = \$2
v = overtime production cost per unit = \$3
I_0 = initial inventory = 50
I_f = final inventory = 200

15. For the aggregate planning TM presented in Problem 14 above, assume that carrying costs per month rise from \$1.00 to \$1.50 and overtime costs decrease from \$3.00 to \$2.50 per unit. Specify the solution and comment on the following question: How sensitive is the solution to these changes in costs?

16. Table 13.13 is equivalent to Table 13.12, in Problem 14 above, and uses the same data. However, in this case, the aggregate planning transportation matrix *allows backorders*. Start with the result obtained for Problem 14 and see whether allowing backorders will change the result. What is the answer? (Note: b = \$1.)

TABLE 13.13

TRANSPORTATION MODEL FOR AGGREGATE PLANNING— BACKORDERS ALLOWED

	Sales Periods					
	1	2	3	Final Inv.	Slack	Supply
Initial Inv.	0	c	2c	3c	0	$I_0 = 50$
Regular 1	r	r + c	r + 2c	r + 3c	0	$R_1 = 100$
Overtime 1	v	v + c	v + 2c	v + 3c	0	$O_1 = 200$
Regular 2	r + b	r	r + c	r + 2c	0	$R_2 = 300$
Overtime 2	v + b	v	v + c	v + 2c	0	$O_2 = 200$
Regular 3	r + 2b	r + b	r	r + c	0	$R_3 = 100$
Overtime 3	v + 2b	v + b	v	v + c	0	$O_3 = 50$
Demand	$D_1 = 200$	$D_2 = 300$	$D_3 = 200$	$I_f = 200$	SL	Grand Total

17. Refer to Figure 13.10 (earlier in the text) which presents four kinds of nonlinear (second order) costs used by the HMMS AP model. Make a table or plot on graph paper the linear payroll cost for C_1 = \$100 per day and between 35 and 45 workers.

18. Refer to Figure 13.10 (earlier in the text) which presents four kinds of nonlinear (second order) costs used by the HMMS AP model. Make a table or plot on graph paper the costs of hiring and layoffs where C_2 = \$200 for between 35 and 45 workers.

19. Refer to Figure 13.10 (earlier in the text) which presents four kinds of nonlinear (second order) costs used by the HMMS AP model. Make a table or plot on graph paper the expected cost of overtime where C_3 and C_5 = \$20, C_4 and C_6 = \$100. The production rate varies between 350 units and 450 units. Remember the relationship between P_t and W_t.

20. Refer to Figure 13.10 (earlier in the text) which presents four kinds of nonlinear (second order) costs used by the HMMS AP model. Make a table or plot on graph paper the expected sum of inventory, backorder and set-up costs where $(C_8 + C_9 S)$ = 40 when $I = 50$. Treat $(C_8 + C_9 S)$ as a constant. Also, C_7 = \$30. Let I range from −50 to 150.

▶ Readings and References

(Aggregate planning models were fully developed by 1970. Therefore, a few citations are early, going back to the basics. What is new is the software that can be used to solve large AP problems. These are cited below. In addition, some more recent references with present-day perspectives about mathematical programming are given.)

M. Attaran, *MSIS: Management Science Information Systems*, NY, Wiley, 1992.

M. Attaran, *OMIS: Operations Management Information Systems*, NY, Wiley, 1992.

E. H. Bowman, "Production Scheduling by the Transportation Method of Linear Programming," *Operations Research*, Vol. 4, No. 1 February, 1956.

C. C. Holt, F. Modigliani, and H. A. Simon, "A Linear Decision Rule for Production and Employment Scheduling, *Management Science*, Vol. 2, No. 1, October, 1955.

C. C. Holt, F. Modigliani, J. F. Muth, and H. A. Simon, *Planning Production Inventories and Work Force*, Englewood Cliffs, NJ, Prentice-Hall, Inc., 1960.

Interfaces, "Special Issue: The Practice of Mathematical Programming," (15 articles covering various aspects of linear and mathematical programming), Vol. 20, No. 4, July–August, 1990.

L. Lasdon, and A. Waren, "Gino: General Interactive Optimizer" (software), and J. Liebman, L. Lasdon, A. Waren, L. Schrage, Modeling and Optimization with GINO, Danvers, MA, boyd & fraser publishing company, 1986.

G. L. Nemhauser and L. A. Wolsey, *Integer and Combinatorial Optimization*, Wiley, NY, 1988.

Donald R. Plane, *Management Science: A Spreadsheet Approach*, Danver, MA, The Scientific Press Series, boyd & fraser publishing company, 1994. (Solvers apply to LP Spreadsheets).

What's Best!, Version 2.0 for DOS, boyd & fraser publishing company, 1993.

Quattro Pro, Version 4 for DOS, 1994; Quattro Pro, Version 6 for Windows, 1995.

Lotus 1-2-3, Version 2.x for DOS, 1994.

Donald R. Plane, *Management Science: A Spreadsheet Approach for Windows*, Danver, MA, The Scientific Press Series, boyd & fraser publishing company, 1995. (Solvers apply to LP Spreadsheets).

What's Best!, The Spreadsheet Solver with Users' Guide, boyd & fraser publishing company, 1995. Quattro Pro, Version 6 for Windows, 1995.

Lotus 1-2-3, Version 4 for Windows, 1995.

Excel, Version 5 for Windows, Microsoft, 1995.

L. Schrage, Lindo: An Optimization Modeling System, 4th Ed., (text and software), Danvers, MA, boyd & fraser publishing company, 1991.

T. E. Vollmann, W. L. Berry, and D. Clay Whybark, *Manufacturing Planning and Control Systems*, 3rd Ed., Homewood, IL, Irwin, 1992.

Notes...

1. For archival purposes, one of the first citations was E. H. Bowman, "Production Scheduling by the Transportation Method of Linear Programming," *Operations Research*, Vol. 4, No. 1, February 1956, pp. 100–103.

2. C. C. Holt, F. Modigliani, J. F. Muth, and H. A. Simon, *Planning Production Inventories and Work Force*, Englewood Cliffs, NJ, Prentice-Hall, Inc., 1960. This is also cited for archival purposes.

Chapter 14

Inventory Management

After reading this chapter you should be able to...

1. Explain what inventory management entails.
2. Describe the difference between static and dynamic inventory models.
3. Discuss demand distribution effects on inventory situations.
4. Discuss lead time effects on inventory situations.
5. Describe all costs that are relevant to inventory models.
6. Differentiate inventory costs by process types.
7. Explain order point policies (OPP) and when they are used.
8. Discuss the use of economic order quantity (EOQ) models for determining the optimal order size for batch delivery.
9. Discuss the use of economic lot size (ELS) models for determining the optimal production run for continuous delivery.
10. Explain the operation of the perpetual inventory model and explain why it is the most widely-used inventory control system.
11. Explain the operation of the periodic inventory model and describe the special circumstances that make its use desirable.

Chapter Outline

14.0 TYPES OF INVENTORY SITUATIONS
    Static versus Dynamic Inventory Models
    Make or Buy Decisions—Outside or Self-Supplier
    Type of Demand Distribution—Certainty, Risk, and Uncertainty
    Stability of Demand Distribution—Fixed or Varying
    Demand Continuity—Smoothly Continuous or Lumpy
    Lead-Time Distributions—Fixed or Varying
    Independent or Dependent Demand

14.1 COSTS OF INVENTORY
Costs of Ordering
Costs of Set Ups and Changeovers
Costs of Carrying Inventory
Determination of Carrying Cost
Costs of Discounts
Out-of-Stock Costs
Costs of Running the Inventory System
Other Costs

14.2 DIFFERENTIATION OF INVENTORY COSTS BY PROCESS TYPE
Flow Shop
Job Shop
Project
Flexible Process Systems (FPS) including FMS

14.3 ORDER POINT POLICIES (OPP)

14.4 ECONOMIC ORDER QUANTITY (EOQ) MODELS—BATCH DELIVERY
Total Variable Cost
An Application of EOQ

14.5 ECONOMIC LOT SIZE (ELS) MODEL
The Intermittent Flow Shop (IFS) Model
An Application of ELS
From EOQ through ELS to Continuous Production
Lead-Time Determination
Expediting to Control Lead Time
Lead-Time Variability

14.6 PERPETUAL INVENTORY SYSTEMS
Reorder Point and Buffer Stock Calculations
Imputing Stock-Outage Costs
Operating the Perpetual Inventory System
Two-Bin Perpetual Inventory Control System

14.7 PERIODIC INVENTORY SYSTEMS

14.8 QUANTITY DISCOUNT MODEL

Summary
Key Terms
Review Questions
Problem Section
Readings and References

The Systems Viewpoint

Inventory is the storage of all materials used or made by anyone in the organization for the direct or indirect purposes of offering the finished products or end services to customers. This chapter focuses on part of the total inventory used by companies. That part relates to items that are used continuously and consistently over a long period of time. They can be ordered independently instead of in conjunction with other parts. There are many such items required by all companies to support the production of goods and the delivery of services. With respect to such items, a multinational company with a centralized inventory management system has to keep track globally of where everything is, where and when it will be needed, when to order it, and where to store it. It is an enormous system to manage. Without the systems perspective, everyone, everywhere will be ordering what they need as they need it. Lack of coordination diminishes buying power and loses the knowledge-based benefits of the centralized system. The systems perspective is required to optimize the production plans of suppliers and producers to best meet the needs of their customers throughout the supply chain. In that way the lowest costs and best deliveries can be achieved for mutual benefit.

14.0

TYPES OF INVENTORY SITUATIONS

Inventory

Those stocks or items used to support production, supporting activities, and customer service; storage of all materials used or made by anyone in the organization for the direct or indirect purposes of offering finished products or end services to customers.

Inventory is: those stocks or items used to support production (raw materials and work-in-process items), supporting activities (maintenance, repair, and operating supplies), and customer service (finished goods and spare parts).[1] (The above quote from the APICS DICTIONARY is just part of the definition that is written there. The rest is more technical and requires understanding issues that will be discussed in this chapter—which is devoted to in-depth explanation of inventory. The American Production and Inventory Control Society (APICS) is a professional society which has played an influential role in the inventory management area.)

Who manages inventory? APICS has established the fact that the management of inventory is a major P/OM responsibility. How to manage inventory is dependent on the type of inventory that is involved. Most types of inventory situations are best handled by well-designed computer systems that utilize as much centralization of record keeping and order placement as is feasible. The manifold advantages of the systems perspective with centralized buying includes the fact that larger quantities provide stronger supplier relationships, bigger discounts, a more informed choice of suppliers and less chance for mistakes.

Each of the seven classes of inventory situations described below requires its own type of management even though they can all be centralized or decentralized according to the dispersion of use for producers, suppliers and customers.

1. Order repetition—static versus dynamic situations.
2. Make or buy decisions—outside or self-supplier.
3. Demand distribution—certainty, risk, and uncertainty.
4. Stability of demand distribution—fixed or varying.
5. Demand continuity—smoothly continuous or sporadic and occurring as lumpy demand; independent.
6. Lead-time distributions—fixed or varying. (**Lead time (LT)** is the interval between order placement and receipt.)
7. Dependent or independent demand—when components are dependent on one or more end items, the information system must be able to calculate linked demands.

Lead time (LT)

Interval between order placement and receipt.

Static versus Dynamic Inventory Models

To explain order repetition, **static inventory models** have no repetition. They portray "one-shot" ordering situations, whereas **dynamic inventory models** place orders repetitively over long periods of time. A few examples underscore the practical nature of this distinction between *one order only*, or a *repetitive stream of orders* for the same item placed over time.

The pure static case also is called a one-period model even though under some circumstances a corrective "second shot" may be allowed. The "Christmas tree problem" is a good illustration of the static situation. The owner of a tree nursery that sells Christmas trees locally said that she placed her orders with a Canadian tree farm north of Montreal back in July. Reasoning that it would be a good year—because people were feeling more prosperous than the prior year—she had bought the maximum number of trees that her organization could truck and accommodate.

Unfortunately, the two weeks before Christmas were unusually rainy. This discouraged people from buying trees for the holiday. In the last week before Christmas, the owner had posted sale signs slashing prices. She felt that helped but nevertheless, 25 percent of the trees remained unsold on Christmas eve. Five percent of them were live trees which could be saved, but 20 percent would have to be scrapped.

The seller's problem is: how many trees to order in July for next December? Most of the sales take place a few days before Christmas. There is no time to take corrective action.What is to be done if too few trees were stocked? Driving up to Canada to replenish supplies is impractical. Also, there may not be any trees left there to sell. If sales are unexpectedly strong, buying locally will cost too much.

If too many trees were stocked, the best alternative is to advertise discounts hoping to get anyone who was going to buy a tree to buy it from the overstocked dealer. The dealer never really knows whether the order size was over or under, or just right, until Christmas day.

Other static examples include the storekeeper who has to decide how many *Wall Street Journal* newspapers to buy for each day. Only one decision can be made to buy *n* newspapers. In its purest form, there is no opportunity to correct that decision based on later information. Another example is the problem of the hot dog vendor at the ballpark.

Static inventory models

Within the order repetition class of inventory situations; these models have no repetition.

Dynamic inventory models

Within the order repetition class of inventory situations; these models place orders repetitively over long periods of time.

Consider the department store buyer who places an order in July for toys to be sold at Christmas time. If the toy is a dud, there is severe overstock. If the toy is hot, there will not be enough stock to meet demand. Both types of situations occur regularly. Another example is the spare-parts order for the complex machine. When placed with the original order, the parts are relatively inexpensive. When required later, because of unanticipated failure, the costs are exorbitant. This spare-parts model is a static decision problem.

In the case of overestimated demand, salvage value is sometimes available. For example, a department store that overbuys on toys, shipped from abroad in time for the holiday season, may be able to sell those toys at a discount after the selling season is finished.

Dynamic situations require different considerations because the demand for such items is constant. Orders are placed repetitively over time. The problem becomes one of adjusting inventory levels to balance the various costs so that total variable costs are minimized. Variable, in this case, means these costs change with order size.

Dynamic models apply to inventories that are used for flow shops, intermittent flow shops, batch work that occurs fairly regularly and supplies that are used for ongoing support systems—no matter what work configurations are typical of the process. The models developed in this chapter address dynamic situations. Service systems use many kinds of supplies with dynamic demand patterns (as in hospitals and hotels).

Make or Buy Decisions—Outside or Self-Supplier

Make or buy decisions

One of the seven classes of inventory situations; the decision of whether inventory should be purchased externally or made internally; based on cost and quality considerations.

Inventory is either purchased externally or made internally. The decision about which way to go is dictated by many factors. Cost and quality are among the most important considerations. This often is called the **make or buy decision**. Make is self-supply. One of the reasons for self-supply is that profit paid to the supplier can be saved by being your own supplier. This requires being as efficient as the best of the suppliers. If the volume that the self-supplier will produce is significantly lower than the volumes produced by the leading suppliers, it is doubtful that self-supply will be justified.

Weyerhauser, a wood and paper products manufacturer, is its own supplier of wood and fiber.
Source: Weyerhauser Company

Costs can be compared using breakeven analysis. Making the product has significant fixed costs—much larger than the buy option. The variable costs have to be low enough to offset the fixed costs of the investment. With a new process, it is reasonable to assume that variable costs will start high and gradually decrease. They may never decrease enough to warrant choosing the make option on a strictly cost basis.

A car company that has never made windshield glass faces an awesome task if it decides to learn how to do it. An enormous amount of experience is required. During the education period, which can last a long time, costs will be excessive and quality will be erratic. Vertical integration could be used which entails buying a windshield producer. Vertical integration means buying companies that provide components which are otherwise purchased from suppliers. The other option is to buy from an expert supplier.

Learning about new process technology—to be up to par or better—is another reason for reaching the decision to make the product rather than to buy it. Eventually, cost and quality advantages are expected to make the buy decision worthwhile. By learning, the company may be better able to deal with suppliers.

Type of Demand Distribution—Certainty, Risk, and Uncertainty

Decision problems that have certain outcomes do not have a probability distribution. Signing a contract to supply a given number of units converts demand from a risk to a certainty. Locating a supplier down the block reduces delivery time uncertainty. Generally, there is a cost for certainty. Because contracts reduce the producer's risk, the buyer expects the price will reflect this fact. The supplier that locates across the street from the buyer expects compensation for being at that location.

Sometimes certainty is a reasonable assumption. This only works when the degree of variability will not affect the solution. Then, certainty is assumed for convenience and it does not violate the spirit of the model. Linear programming methods and transportation models make this assumption. Lead time for delivery is often treated in this way. Say that two weeks is the average for a specific case. The variability around the average will determine if two weeks can be used as if it were certain. It is important that users of inventory models know when the certainty assumptions are allowable. When lead time variability could cause a stock-out, it cannot be treated as constant.

Buffer stock is stock carried to prevent outages when demand exceeds expectations. Basic methods of inventory dealing with order point policy (OPP) make assumptions about the demand distribution that get translated into buffer-stock levels.

Uncertainty means that there is no good forecast for the probabilities of demand and/or lead-time distributions. When uncertainty exists, the probabilities for various levels of demand occurring are speculative. However, when there is a known risk, some planning can be done. Delivery of critical materials from the port city of Kobe in Japan always was totally reliable until the January 18, 1995 earthquake caused serious delays. Some companies factor such possibilities into their planning and can react quickly.[2] There is a finite probability for various catastrophes so having a team ready to assess the situation and find solutions may make a difference.

While less of a disaster, the risk that an order in process will be cancelled has been experienced by almost every production department. Contingency planning for cancellation manages this type of situation as best as can be handled.

Buffer stock

Stock carried to prevent outages when demand exceeds expectations; also called reserve and/or safety stock.

When forecasting is difficult or impossible, systems are developed to search for advance warning. They are tuned in to whomever might have information about orders. Every effort is made to find the key players who originate the orders and to keep them in the communication loop at all times. Other factors that might trigger orders also are tracked. Efforts are made to gain some control over the uncertain occurrences.

Stability of Demand Distribution—Fixed or Varying

Uncertainty includes the possibility that the demand distribution is changing over time. If it is known how it is changing, then the risk levels may be higher but forecasts are possible. OPP methods are applicable to stable demand distributions. They can be modified if it is known how the distributions are changing. Otherwise, the system is searching for advanced warning. Other methods than OPP should be used to manage these situations.

Demand Continuity—Smoothly Continuous or Lumpy

Most of the remarks above apply to demand continuity. OPP needs demand continuity. It also requires stability of the demand pattern, or special knowledge about how it changes. Alternative inventory methodology can deal with the lack of smoothly continuous demand. It should be noted that assuming smooth and continuous demand is akin to converting a known risk situation into one of certainty. The assumption is usually valid and can be tested by simulating different patterns that are more or less smooth and continuous and measuring the extra costs incurred for assuming perfect smoothness. Such testing is most readily done with computer simulation programs. It is called **sensitivity analysis** whether it is done by hand or computer.

Sensitivity analysis
Testing the effect of varying conditions.

Lead-Time Distributions—Fixed or Varying

Lead-time (the interval between order placement and receipt) variability may be a factor in setting the size of the buffer stock. As lead times get longer, inventory systems become more sensitive to variations that increase replenishment time. There also is the matter of how critical the materials are for production. For materials that are not critical, the assumption of fixed lead times is sustainable. When materials are critical, planning had best proceed based on forecasts for the lead-time distribution. In that event, the lead-time distribution will add to the buffer stock which is held to provide safety against longer than average lead times. Simulation testing for sensitivity is advised for this situation.

Independent or Dependent Demand

This chapter focuses on independent-demand systems. This means that orders can be placed for these items without considering what end products they are used with. Alternatives to OPP should be used when dependent demand systems are involved. This is most applicable for components and subassemblies that are used as parts of one or more finished products. The dependency is greatest when the end-product demands are sporadic.

Q U A L I T Y

Auto Industry Outsourcing

Auto parts suppliers uphold strict quality standards in the design, manufacture, and delivery of car parts.

Auto makers worldwide rely heavily on outside suppliers to produce the thousands of parts required to build a car. And they expect suppliers to do more than just fill orders for parts. Suppliers are expected to participate in the design of cars and car parts, minimize defectives in their products, be online with manufacturers via electronic data interchange (EDI), and provide just-in-time (JIT) delivery of car parts.

Toyota, for example, sets standards for quality, cost and delivery that suppliers must meet and comes to the aid of suppliers who have problems meeting these standards. The company is involved in training its suppliers in new manufacturing techniques through the Toyota Suppliers Support Center in Lexington, Kentucky. Thomas Zawacki, assistant general manager of purchasing for Toyota, said the company considers suppliers to be "an extension of our own company."

One Canadian-based automotive supplier, the A. G. Simpson Co., recently expanded its metal stamping operations to a new Dickson, Tennessee, facility. One of the reasons the company located there was its proximity to the Nissan plant in Smyrna, Tennessee. Simpson, a just-in-time supplier of sheet metal stamped parts and welded assemblies for three Nissan models, ships parts to the Nissan plant daily. The company has a "10-minute window for delivering parts" to the Smyrna plant, said Steve Quarles, Simpson's quality and sales manager.

Overseas, the Czech auto maker Skoda has taken just-in-time a step further by giving auto part suppliers space for manufacturing the car parts and components in its own assembly plant. One of the suppliers located within the Skoda plant is Milwaukee-based Johnson Controls, which manufactures car seats. Skoda sends a computer message to the seat-making area 160 minutes in advance of when the seat installations will be needed on the assembly line. The message indicates to Johnson Controls' seat assemblers the model required and they in turn prepare the seats in the proper sequence. The seats are then shipped to the assembly line via a 25-minute monorail journey.

Sources: David Heath, "A good car is the sum of its parts," (Louisville, Kentucky) *Courier-Journal*, February 7, 1993; Cyrus Afzali, "Dickson plant allows firm to put stamp on new markets," *Nashville Business Journal*, November 14, 1994; and Jane Perlez, "Skoda Gives Its Suppliers a Place in the Auto Plant," *New York Times*, November 19, 1994.

14.1
COSTS OF INVENTORY

The core of inventory analysis lies in finding and measuring relevant costs. Six main kinds of costs are discussed and then in a seventh category, a few others are mentioned.

The six costs that will be discussed are:

1. costs of ordering
2. costs of set ups and changeovers
3. costs of carrying inventory

4. costs of discounts
5. out-of-stock costs
6. costs of running the inventory system

Costs of Ordering

The best way to determine the cost of an order is to do a systems study of the elements that go into making up and placing an order. The elements are then given costs related to materials and labor used, space required and equipment charges. It is easy to say how to do the cost analysis, but difficult to get the correct cost allocations. Assume that the purchasing department can process 100 orders per day but things are slow and so it processes only 50 orders per day. Does that mean that the cost of an order is twice as great when things are slow? It does mean that. If things stay slow, redesigning the purchasing department makes sense.

What is needed is the determination of the variable costs of ordering. How long does it take and what skill levels are involved? What should an order cost? The fixed costs of ordering have to be separated from the variable costs. Fixed costs are not changed by the order size or frequency of ordering, so they are not included in the inventory policy model. They are included in the cost of the purchasing department.

It is possible to evaluate the fixed costs to determine if purchasing is carrying too much capacity. Activity-based costing can help to ascertain an optimal size for the order department. If union regulations allow, purchasing people can do other things when they are not occupied with purchasing. This converts fixed costs to variable costs.

With variable costs defined, the systems study of the ordering process holds. It consists of writing up the purchase requisition form, phone calls made in connection with specifications and ordering, faxing or mailing purchase orders to the supplier. All costs that increase as a function of the number of purchase requisitions and the size of the orders qualify and should be included.

Finally, a specific example helps to illustrate how the ordering cost relates to undercapacity of service orders. Assume that the order department now processes 100 orders per week. A new inventory policy requires that 150 purchase requisitions be processed in a week. It is agreed that the ordering department must be enlarged. The increase in labor costs, equipment and overhead is considered to be an addition to variable costs. Thus, the ordering cost is determined on top of a base-ordering system.

Costs of Set Ups and Changeovers

When self-supply is being used, the ordering cost is replaced by the cost of setting up the equipment to do the run. This requires cleaning-up from the prior job which also is called taking down. The entire process is known as changeover, and the costs can be considerable.

A number of parts may have to be made before the set up is complete. The cost of learning is involved and defectives play a role. When acquiring equipment, changeover times and costs can be as important to consider as output rates. Flexible manufacturing systems (FMS) are notable for their ability to provide fast changeovers, but the expense of the equipment makes the costs of the changeovers high.

Inventory carrying costs include the expense of storage until delivery.
Source: Avery Dennison Corporation

Costs of Carrying Inventory

Good inventory policy maintains the *minimum* necessary stock on-hand. When demand slackens, companies cut back further on their inventory. They do this because of the belief that the minimum necessary level is going down.

Also, inventory is a form of investment. Capital is tied up in materials and goods. If the capital were free, alternative uses might be found for it. The alternatives include spending for R & D, new product and/or process development, advertising, promotion, and going global. Some firms even put the money into financial instruments, the stock market, or the savings bank.

Expanded capacity and diversification are typical opportunities which, when ignored, incur the cost of not doing that much better with the investment funds. By holding inventory the company foregoes investing its capital in these alternative ways. Such opportunity costs account for a large part of the costs of carrying inventory.

Inventory carrying costs are variable costs which depend upon the average number of units stocked ($Q/2$), the cost per unit (c), and the carrying cost interest rate that is applicable (C_c). These costs vary with Q, which is the number of units per order.

Inventory carrying costs also include the expense of storing inventory. As was the case for the ordering department, costs should be measured from a fixed base. The key is to find the variable cost component that is associated with storage.

These are the costs over which P/OM can exercise control using inventory policies. Thus, if a company has shelf space for 1000 units, but can get a discount if it purchases at least 2000 units, then to get this discount it must expand storage capacity. It can buy or rent additional space.

There are options for ordering large quantities to obtain discounts without requiring expanded storage capacity. One of these is *vendor releasing* whereby the supplier agrees to deliver small increments of the larger order over time. It will be mentioned again in connection with inventory deliveries for the flow shop.

Another method for reducing storage space requirements is the use of *cooperative storage*. Commonly used supplies are ordered at discount in large quantities, stored in cooperative warehouses, and dispensed to participating hospitals, in the same metropolitan area, on an as-needed basis. Airlines share the storage and investment carrying costs of commonly used parts such as jet engines. Cooperative sharing reduces storage costs and increases the availability of expensive components which require large dollar expenditures.

Items carried in stock are subject to the costs of pilferage, obsolescence, and deterioration. These costs represent real losses in the value of inventory. Pilferage, which is petty theft, is characteristic of small items such as tools. Department stores suffer from stolen merchandise. Hotels lose ashtrays and towels; pencils and stamps disappear in offices.

Obsolescence may be the most important component of carrying cost because it happens so often and so fast. Obsolescence occurs quite suddenly because a competitor introduces technological change. Also, it can be the kind of loss that is associated with style goods, toys, and Christmas trees. Out-of-season and out-of-style items can lose value and must be sold at a special reduced rate. The determination of how much inventory to carry will be affected by the nature of the inventories and the way in which units lose value over time.

Deterioration affects the carrying cost of a broad range of products. Industrial products that deteriorate include adhesives, chemicals, textiles and rubber. Weather deteriorates iron and wood. Rubber gaskets in pumps can fail.

Food and drugs are vulnerable to spoilage. Companies put spoilage retardants in foods, but increasingly customers avoid additives. Adding ingredients to prevent spoilage increases material and production costs while decreasing the carrying cost for product deterioration. The net market effect must be factored in as well. Spoiled milk and stale bread are quickly perceived but customers can avoid buying products before they spoil by noting freshness dating information as required for milk and bread. Whatever cannot be sold because of real or dated deterioration is added to the carrying cost.

Freshness dating increases
inventory and production costs.
Source: Pepsi-Cola, N.A.

Some products that are not required to use freshness dating are testing its effect on their market share. Freshness dating not only increases the carrying cost component, but it also increases demands on operations to make and deliver product as much before the freshness date as possible. Delays in getting dated products to market will decrease productivity and increase carrying costs. It remains to be seen whether freshness dating has enough consumer appeal to compensate for the added production and inventory costs.

An additional component of carrying cost includes both taxes and insurance. If insurance rates and taxes are determined on a per unit basis, then the amount of inventory that is stocked will determine directly the insurance and tax components of the carrying costs.

Determination of Carrying Cost

Table 14.1 is furnished as a guide. Numbers have been provided that are similar to the carrying cost computations of most companies in the U.S.A. Each situation is different, so the accountants and operating managers must assess those costs that apply to their particular situation.

TABLE 14.1

CARRYING COST (C_C) EXPRESSED AS A PERCENT OF THE PRODUCT'S COST PER YEAR

Category	Percents of Cc	
	(High)	(Low)
Loss due to inability to invest funds in profit-making ventures, including loss of interest	15.00	9.00
Obsolescence	3.00	0.00
Deterioration	3.00	1.00
Transportation, handling, and distribution	2.00	1.00
Taxes	0.25	0.10
Storage cost	0.25	0.10
Insurance	0.25	0.10
Miscellaneous	0.25	0.10
Theft	6.00	0.10
C_c Totals	30.00	11.50

Costs of Discounts

Accepting discounts by buying at least a certain amount of material involves extra costs which may make taking the discount unprofitable. An appropriate inventory cost analysis must be used (and is furnished below) to determine whether or not a discount that is offered should be taken. The extra costs for taking the discount are compared to the savings obtained from the discount. Extra costs include additional carrying costs, part of which are incurred for additional storage space. Table 14.1 above can be consulted for other variable cost components associated with holding extra inventory.

Out-of-Stock Costs

When the firm cannot fill an order, there is often some penalty to be paid. Perhaps the customer goes elsewhere, but will return for the next purchase. Then the penalty is only the value of the order that is lost. If the customer is irritated by the out-of-stock situation and finds a new supplier, the customer may be lost forever. The loss of goodwill must be translated into a cost which is equivalent to the termination of the lifetime value of that customer.

If the buyer is willing to wait to have the order filled, the company creates a backorder. This calls for filling the order as soon as capacity is available or materials arrive. Backorder costs include the penalties of alienated customers. To avoid this penalty, some mail-order companies prefer to fill customers' orders with a more expensive substitute instead of creating a backorder. The outage cost is related to the decrease in profit margin. However, the goodwill generated by the gesture is an intangible addition to long-term profit. Depending on the system that is used (i.e., backordering, substitutions, fill or kill, and so on), various costs of being out of stock will occur. The lost-goodwill cost is the most difficult to evaluate.

Organizations which are not close to the customer, and there are many, frequently ignore lost goodwill because it cannot be measured. Many bureaucracies are identified with this characteristic of neglecting customer satisfaction.

Costs of Running the Inventory System

Processing costs that are associated with running the inventory system are referred to here as systemic costs. This category of costs is usually a function of the size of the inventory that is carried and the importance of knowing the correct stock levels up-to-the-minute. Costs are related to the number of terminals accessing the computer and people that are operating the inventory system. Operating costs also include the amount of systems assistance and programming that is needed. There are training costs. The amount of time that the system operates (round-the-clock) and the number of locations that are networked into the centralized data system (round-the-world) will have to be factored into this cost.

This 512,000 square-foot distribution center has a computerized inventory-tracking system that insures orders are shipped within 24 hours.
Source: Cooper Industries, Inc.

Systems involving many stockkeeping units (SKUs) are dependent upon having an organized information system. There are different SKU part numbers for each model type, as well as for each size and color. Part numbers for SKUs often identify the specific suppliers and where the inventory is stored. When there are frequent online transactions with many SKUs, the amount of detail is great and the systems are expensive to operate.

It makes sense to focus on the SKUs for materials and items which are critical for production. Also, it makes sense to pay special attention to items which have high dollar volume (called A-type) because they can waste the most money if handled badly. If the A-category of the ABC inventory system is large and if there are many critical items to manage, then keeping all the information up-to-date is important and costly.

The status of work in process (WIP) can be monitored by means of bar codes and optical readers. Often, this technology is combined with the concept of a paperless factory. There are significant advantages but the systems problems are formidable. Design and equipment choices interact with decisions about what kind of inventory system to use.

Other Costs

The six costs discussed above are generally among the most relevant ones in determining inventory policy. However, other costs can play a part in specific cases. The costs of delay in processing orders can take on great significance when a heart transplant is involved. Not quite as dramatic, every manufacturer recognizes the costs of delay when lengthy set ups are required. Service organizations are sensitive to the costs of delay. Organizations set goals for the maximum time that can be allowed to elapse before answering a ringing phone.

The costs of production interruptions has previously been related to the critical nature of inventory items. Salvage costs can play an important role as can the costs required for expediting orders. The costs of spoilage for food might be better handled as a separate cost instead of as a part of the carrying cost related to obsolescence. The costs of central warehousing as compared to dispersed warehousing can be crucial. In various circumstances, one or more of these costs can dominate the inventory policy evaluation.

14.2

DIFFERENTIATION OF INVENTORY COSTS BY PROCESS TYPE

Significantly different inventory costs characterize each type of work configuration. These cost differences play an important part in OM planning and in the selection of process types.

Flow Shop

With serialized production, the cost of ordering is low. This is because orders for the same items are placed with regularity. The carrying cost of inventory reflects industry profitability. The reason for this relationship is that the carrying cost rate

C_c is larger where reinvestments in the process or in marketing and distribution can bring excellent returns on investment. Out-of-stock costs tend to be severe. The line may have to shutdown or run at some fraction of total capacity. For this reason, large buffer stocks are usually held in the inventory, in spite of the fact that carrying costs are high. Because of the large quantities of similar materials, purchases can qualify for discounts.

On the other hand, many companies—striving to be lean and mean—have made efforts to move toward just-in-time (JIT). Using trusted relations with fewer suppliers, the reliability of the supplier is viewed as if the supplier were part of the company being supplied. Serialized processes lend themselves to being well-balanced and having synchronized flows of materials.

There is a loophole for JIT enthusiasts who want discounts. A large amount of material is purchased to be delivered in small quantities—JIT. This agreement which is called **vendor releasing** is not available from all suppliers. It is up to the buyer to negotiate with the supplier to achieve it.

Vendor releasing

A large amount of material is purchased to be delivered in small quantities—JIT, or nearly so.

Job Shop

The cost of ordering is often high for the job shop. There are many jobs and a great variety of materials used in small quantities. The right vendors have to be located for new work. Quantities are too small to build trust relations with a few suppliers. Vendors may not value the business and take liberties including not delivering when bigger and better opportunities come their way.

It is difficult to get engineering and R & D's attention to develop new materials and components because the order sizes are too small. Carrying costs will tend to be lower than those of the flow shop to the degree that the job shop is less profitable. Overordering to provide protection against defectives leaves an increasing accumulation of small amounts of materials that are difficult to sell as salvage. This increases the costs of obsolescence.

Many companies have moved toward just-in-time suppliers in an effort to reduce the quantity and carrying cost of inventory.
Source: Emerson Electric Co.

Projects

Ordering costs for projects are likely to be high. Bidding, which is expensive, is often used. Project managers may have to buy from a variety of vendors with whom they have had little experience. Projects are unique and project purchasing is often not repeated.

For small items, that are essentially one-shot, minimum effort is made to bring efficiency and control to the purchasing function. Carrying costs are relatively unimportant, because the inventory is quickly used up. Time minimization drives project management.

The importance of being on time with the project has led to the development of critical path methods which facilitate project planning and reporting of project status. These reports to OM permit a great deal of control over out-of-stock items. Though these project-control methods are costly, they are warranted by the high out-of-stock costs. They help minimize (or entirely eliminate) the number of outages to which the high out-of-stock costs can be applied.

Flexible Process Systems (FPS) including FMS

Flexible manufacturing systems (FMS) have low ordering costs. This is because there is considerable homogeneity of materials used for products by processes. The variety of materials used more nearly matches the flow shop than the job shop. FPS systems have high carrying cost rates because of the large capital investment and the profitability of the technology when it is properly managed.

Out-of-stock costs are less significant than for the flow shop because FMS run small batches. Other products can be made while waiting for delivery of out-of-stock materials. FMS lot sizes are small, allowing air delivery. Obsolescence of small amounts of inventory are less likely than for the job shop because materials controls are monitored by computers that run the show.

The FPS situation is a cross between flow shop and job shop characteristics. However, these processes have additional unique advantages that are directly attributable to the flexibility of FPS technology. FPS permits better cost control than is available for more traditional work configurations.

14.3

ORDER POINT POLICIES (OPP)

Order point policies (OPP) define the stock level at which an order will be placed. In other words, as withdrawals decrease the number of units on-hand, a particular stock level is reached. That number of units, called the reorder point (RP), triggers an order for more stock to be placed with the appropriate supplier. These OPP models specify the number of units to order. The time between orders varies depending on the order quantity and the rate at which demand depletes the stock on-hand. So the order size is fixed and the interval between orders varies. This discussion is about one type of order point system called the **perpetual inventory system.** It continuously records inventory withdrawals. Most often it is used online and in real-time.

Next, another type of order point policy will be described. It is called a **periodic inventory system**. On consecutive specific dates separated by the same amount of time, the item's record is called up and an order is placed for a variable number of units. The order size is determined by the amount of stock on-hand when the

Periodic inventory system

One type of OPP where on consecutive specific dates separated by the same amount of time, the item record is called up and an order is placed for a variable number of units.

record is read. It is the date that triggers the review and the order being placed. Therefore, in this case, the interval between orders is fixed while the ordered amount varies. The order amount is a function of the rate at which demand occurred in the period between reviews of the item's records.

This type of OPP has become outmoded in many companies because powerful computer systems enable the perpetual inventory system to be used which has advantages that will be explained. Nevertheless, the periodic system will be briefly described for two reasons. First, it is used by companies that want to order certain items together from the same supplier. This may be done by small- and medium-sized organizations to obtain carload freight rates because they do not have sufficient volume to fill a freight car with one item. Second, some more advanced inventory models mix the use of the perpetual and the periodic systems with the result that costs are further reduced. Therefore, it is useful to understand the mechanism of the periodic model.

Economic order quantity (EOQ)

The fixed amount to order; a type of OPP; also called square root models.

Economic lot size (ELS)

The fixed amount to produce; a type of OPP

Models such as the two mentioned above are based on the concept and mathematical structure of an **economic order quantity (EOQ)**—the fixed amount to order, or an **economic lot size (ELS)**—the fixed amount to produce. These concepts were first developed in 1913 by an engineer named Harris. They have been expanded in many ways to give OM a powerful set of OPP inventory models, but the basic structure remains the same.

There is an alternative approach to OPP inventory models. It is called materials requirements planning (MRP) and is mentioned and contrasted with OPP below. At this moment, the only point to make is that MRP is used when the assumptions about order continuity and consistency of demand that are required by EOQ and ELS are not met. Table 14.2 indicates when OPP models apply.

TABLE 14.2

CONDITIONS FOR OPP

Continuity Requirement	Model Type
static order	one-shot, single period
dynamic orders	OPP
continuous smooth demand	OPP
sporadic demand in clusters	MRP

Demand Dependency Requirement	Model Type
independent	OPP
components dependent on end item	MRP

The most important of the requirements for EOQ and ELS (which are OPP models) are consistency of independent demand over time with relatively smooth and regular withdrawal patterns of units from stock. MRP works for sporadic dependent demand by using information systems control of messy demand patterns. MRP requires the information processing capacity of computers. It did not appear until the 1960s. When OPP is online, dealing with the perpetual inventory management of large numbers of SKUs, computers also are required.

Baird Information Systems

Managing inventory in the perishable food industry presents companies with the problem of an extremely short (average 72-hour) shelf life for products. Baird Information Systems (BIS), a wholly owned subsidiary of Mrs Baird's Bakeries, Inc. of Fort Worth, Texas, has developed a computer-based route information system that helps bakeries deal with the perishable nature of their products.

The system is based on a "stores on wheels" concept, with route salespeople using handheld computers to record point-of-sale transactions and to keep track of inventory. The goal is to have no inventory left on the truck at the end of the day. When the route salespeople complete their daily deliveries, the transaction information is imported into a database that has about three years of information on each stop. This information helps salespeople with forecasting on a daily basis how much inventory to carry on their route. Sales patterns

that emerge are "highly day-oriented," explains Jerry Baird, president of BIS. "It really matters what you do on Monday as opposed to what you do on Tuesday," he said.

Jerry Baird and his staff originally developed the software solely for Mrs Baird's Bakeries use. After gaining control of the deluge of information generated by the handheld computers for Mrs Baird's, he saw an opportunity in providing the same service to other bakeries that needed help in this area. Rather than buying the system from BIS, the bakeries pay one fixed monthly fee for the software, hardware, operating costs, training, and support. BIS supports

800 handheld devices in the field with twice-daily consultations and daily information processing.

Jerry Baird, who spent four years as manager of Mrs Baird's downtown Fort Worth plant, stresses that BIS is an operating business. "This is not programmers programming and walking out at the end of the day. People are depending on us for their business. We're on call and on the line all the time," he said. His goals for BIS are to continue to work to improve the information systems, to increase the number of outside routes, and to expand internationally.

Sources: Cathleen Cole, "No half-baked ideas for Jerry Baird's BIS," *Business Press* (Fort Worth, Texas), December 24, 1993, and a conversation with Jerry Baird.

14.4

ECONOMIC ORDER QUANTITY (EOQ) MODELS—BATCH DELIVERY

EOQ models, also called square root models, form the basis of most order point policies. They can be related to Japanese inventory methods which stress small lot production and just-in-time (JIT) deliveries. EOQ calls for JIT when there is:

1. extreme reduction of the cost of ordering
2. extreme reduction of the cost of setting-up jobs
3. very high costs of carrying inventory

The Japanese inventory methods also emphasize supplier relationships to accomplish goals which are not directly represented by the inventory models—namely, high product quality, communication of impending problems, short lead times and delivery reliability. These have all been converted into international methods for dealing with inventory.

How do the costs defined in Sections 14.1 and 14.2 above operate in an inventory system where deliveries of purchased supplies are made at regular intervals? The quantities will be delivered in batches of Q units (the order quantity). Also, it is assumed that the production system uses up the inventory at a *constant rate* and that there is *no variability* in this rate or in the delivery intervals. Because of these assumptions, all units that are ordered will be consumed in a fixed period. There will be no salvage costs for excess units, and no stock-outages. Later on, variability in the rate of inventory usage and in the lead-time replenishment interval will be included.

The proposed conditions apply to many P/OM inventory systems. They are applicable to FPS and the flow shop, which receive materials in batches that can be small and delivered just-in-time. They also are relevant for the job shop, or the project or the intermittent flow shop, where the question arises: should all the inventory that will be needed to meet the demand be purchased at one time, or in several lots, over time?

Although it is assumed that consumption occurs at a uniform, continuous rate, it is seldom a serious problem if the conditions are not exactly met. The objective is to balance the time stream of opposing costs so that an optimal inventory procedure can be designed.

Figure 14.1 shows the relationship of the order quantity, Q, with carrying costs having the linear form aQ which is called Line A. The figure also shows the order cost curve having the nonlinear form b/Q which is called Line B. Note that a and b are abstract constants which will be defined for the EOQ and ELS cases.

F I G U R E   1 4 . 1

VARIABLE CARRYING COST LINE (*A*) AND
VARIABLE ORDERING COST CURVE (*B*)

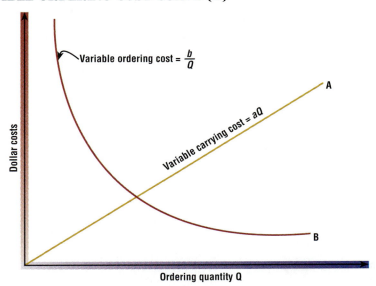

Variable ordering cost = $\frac{b}{Q}$

Variable carrying cost = aQ

Dollar costs

Ordering quantity Q

The lines vary as a function of Q.

As the number of units (Q) purchased per order increases, the carrying costs rise. This is Line A. As the number of units per order increases, the number of orders placed per year will decrease lowering the ordering costs as shown in Line B.

If the demand (D) for a particular item amounts to 1,000 units per year, all of the units could be ordered at one time. $Q = 1,000$ and only one order would be placed per year. The 1,000 units gradually decrease to zero from the beginning to the end of the year so that the average annual stock would be 500 units. Two orders per year cuts the average annual number of units in half to 250.

$$\overline{Q} = \text{Average inventory with smooth withdrawal} = \frac{Q}{2}$$

Figure 14.2 illustrates the continuous withdrawal pattern that OPP assumes.

FIGURE 1 4 . 2

CONSTANT WITHDRAWAL FOR Q = 1,000 & Q = 500 ORDER POLICIES

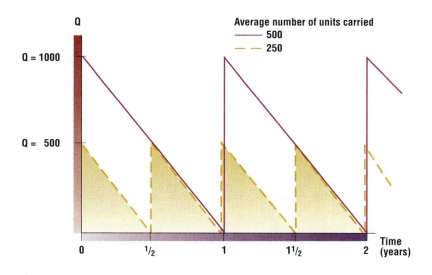

Demand *(D)* = 1000 units per year.

It is shown for two order size policies both of which satisfy demand of 1,000 units per year. The first policy places a single order of $Q = 1,000$ units once a year. $\overline{Q} = 500$ units. The second policy orders twice a year so $Q = 500$ and $\overline{Q} = 250$ units.

Note that demand per year *(D)* divided by the number of units per order *(Q)* equals the number of orders per year *(n)*.

$$\frac{D}{Q} = n \quad \text{or} \quad D = nQ$$

The progression is evident: as Q is halved, n is doubled. Figure 14.3 illustrates two additional continuous withdrawal patterns for OPP where the order frequency is increasing.

CONSTANT WITHDRAWAL FOR $Q = 250$ & $Q = 125$ ORDER POLICIES

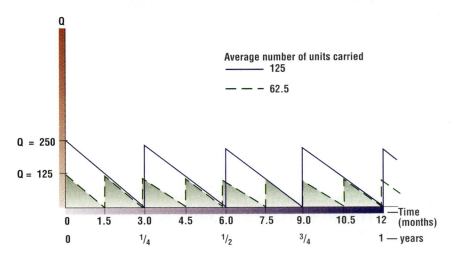

Demand $(D) = 1,000$ Units Per Year

For $Q = 250$, $\overline{Q} = 125$ and $n = 4$ orders per year. For $Q = 125$, $\overline{Q} = 62.5$ and $n = 8$ orders per year. Although the ordering cost is doubling, the carrying cost is being cut in half because the average inventory is cut in half.

Total Variable Cost

Total variable cost is the sum of the ordering cost and the carrying cost. Thus:

$$TVC = CC + OC$$

where:

TVC = total variable cost; CC = carrying cost; and OC = ordering cost

The goal is to write the total variable cost equation and then minimize it with respect to the order size Q. Figure 14.4 shows TVC as the sum of the two cost factors (CC and OC) previously shown in Figure 14.1 above. This is the sum of the carrying cost (Line A) and the ordering cost (Line B).

Minimum total variable cost (min TVC) is associated with an order quantity $Q = Q_O$, where Q_O is referred to as Q optimal. Some components of the equation are already familiar.

1. The average number of units carried in stock is $Q/2$, where Q is the number of units purchased per order.
2. The average dollar inventory carried is $cQ/2$, where c is the per unit cost of the item.
3. The total variable carrying cost per year is $(cQ/2)C_C$, where C_C is the carrying cost rate, previously defined. This is the first term of the total variable cost equation.

FIGURE 14.4

TOTAL VARIABLE COST:
TVC = $\sum$ (CARRYING COSTS + ORDERING COSTS) IN DOLLARS
PER YEAR AS A FUNCTION OF Q

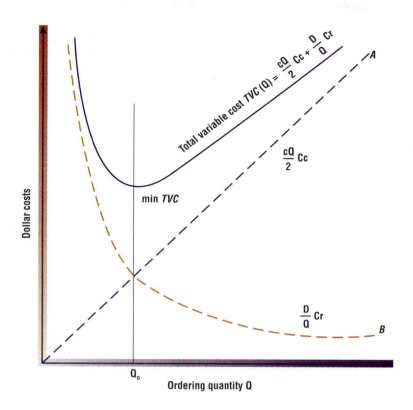

4. The number of orders placed per year is D/Q, where D is the total demand per year.

5. The total variable ordering cost per year is $(D/Q)C_r$, where C_r is the cost per order. This is the other term of the total variable cost equation TVC.

6. The time interval most often used is a year. This is consistent with the stability of the flow shop and the conceptualization of carrying cost rates and interest rates per year. Adjustments can always be made if needed. It is essential that all time periods in the equation be the same.

Using math symbols, the total variable cost TVC is described as follows:

$$TVC = (cQ/2)C_C + (D/Q)C_r$$

This equation appears in Figure 14.4 above where the costs are shown individually and summed.

The minimum value of TVC can be obtained by trial and error methods. Finding the minimum can be done visually using Figure 14.4. There is no need to use trial and error because the *minimum total variable cost always occurs at the crosspoint of Lines A and B*. This statement applies to all equations of the form $y = aQ + b/Q$ which is the structure of the TVC equation shown above and in Figure 14.4.

The minimum total variable cost also can be derived by taking the derivative of y, setting it equal to zero, and solving for the optimal value of $Q(Q_O)$ as follows:

$$\frac{dy}{dQ} = a - \frac{b}{Q^2} = 0 \quad \text{or} \quad Q_0^2 = \frac{b}{a} \quad \text{and} \quad Q_0 = \sqrt{\frac{b}{a}}$$

By showing it in this way with just a, b, y and q, there are far fewer letters and the structure can be readily seen. Then, it can be applied to the derivation of optimal order quantity, Q_O, substituting TVC for y.

$$\frac{d(TVC)}{dQ} = \left(\frac{c}{2}\right) C_c - \left(\frac{D}{Q^2}\right) C_r = 0$$

Solving:

$$Q_0^2 = 2DC_r/cC_c \quad \text{and} \quad Q_0 = \sqrt{\frac{2DC_r}{cC_c}}$$

The same result is obtained by setting the two terms of the equation equal to each other. They are equal at the cross-point. Thus: Set $(cQ/2)C_c = (D/Q)C_r$, and solve for Q to obtain the economic order quantity (Q_0). The result is:

$$Q_0 = \sqrt{\frac{2DC_r}{cC_c}}$$

This square root inventory model is basic. It is the foundation equation for all inventory modeling of the OPP type and applies to flow shops, job shops, flexible processing, and projects. Note that human intuition tends to extrapolate linearly and does not empathize with square root relationships.

If one of the variables under the square root is doubled, what happens to Q_O? As a specific illustration, if demand (D) in the numerator is increased to $2D$, it is not instinctive that Q_O should increase by 1.414.

When risk exists in the system because of demand or lead-time variability—a version of the EOQ model can be built in which buffer stock is added to Q_O. Other alterations of the basic model permit discount policies to be examined to ascertain whether the company should take or refuse a discount. When inventory is self-supplied instead of purchased from an external vendor, the model can be converted to indicate optimal run size. This variant is called the economic lot size (ELS) model. In all of these cases, which will be addressed shortly, the foundation is the square root model.

An Application of EOQ

Assume that the carrying cost rate is six percent per year. This is low which means that the interest rate on money is dominating the carrying cost. Table 14.1 above shows that even that is conservative.

C_C = 0.06 per year or 6 percent per year

c = $0.005 per unit (this is price paid for a unit of a bulk item—1,000 units cost $5.00—i.e., nails, napkins, and noodles)

D = 3,000 units per month = 36,000 units per year

C_r = \$2 cost per order

The total annual variable cost equation is:

$$TVC = (0.005)(Q/2)(0.06) + (36,000/2) 2 = 0.00015Q + 72,000/Q$$

$$Q_o = \sqrt{\frac{(2)(36,000)(2)}{(0.005)(0.06)}} = \sqrt{(480)(10)^6} = 21,909$$

This indicates that 21,909 units should be ordered at one time. This number could be rounded off to 22,000 without disturbing anything.

$$n_o = \frac{D}{Q_o} = 36,000/21,909 = 1.643 \text{ orders per year, or}$$

$$t_o = \frac{Q_o}{D} = 21,909/3,000 = 7.3 \text{ months}$$

$t = 1/n$ is introduced for the first time. It is the period of time that elapses between placing or receiving orders. t could be calculated in years but it is shown in months between orders. D per month has been used. Accordingly, n in months is calculated:

$$n_O \text{ (annual)}/12 = n_O \text{ (monthly) or } 1.643/12 = 0.13692 \text{ orders per month}$$

Assume that the present ordering policy is to order 3,000 units every month to just match demand and approximate a just-in-time ordering policy. Call this demand matching policy TVC_{3000}. How does the demand matching policy compare with the optimal policy, $TVC_{21,909}$?

$$
\begin{aligned}
TVC_{3000} &= 0.00015(3000) + 72,000/(3000) \\
&= \$0.450 + \$24.00 = \$24.45 \\
TVC_{21,909} &= 0.00015(21,909) + 72,000/(21,909) \\
&= \$3.286 + \$3.286 = \$3.286(2) = \$6.57
\end{aligned}
$$

The optimal policy always has equal carrying and ordering costs. In this case they are both \$3.286. The optimal TVC saves a considerable amount, namely, \$24.45 − 6.57 = \$17.88. If the same kind of savings can be made on many other purchased items, total savings can be substantial. Optimal TVC is 270 percent lower than the demand matching policy. This is because Q for demand matching is small which associates it with the steeply rising portion of the total variable cost curve in Figure 14.4 above.

However, it should be noted that the total cost curve is steep only where the optimal ordering quantity, Q_O, is small. The curve flattens out as Q_O gets larger. Also, there is less difference between the minimum cost at Q_O and the costs associated with small changes (ϵ) in Q_O such as $Q_O \pm \epsilon$ than large changes. This low sensitivity is illustrated in the Problem Section at the end of this chapter.

The application of the systems point of view leads to questions: where should the 7.3 months' supply of these bulk items be stored? Will these bulk items age and deteriorate, or become obsolescent? Why is it that matching demand does not provide the just-in-time cost benefits that are so popular?

Small sums are involved. It may well be that the use of EOQ is not worth the bother. For this particular item, the annual expense is 36,000(0.005) = $180.00 a year. This is a C-type item in the ABC categories. The savings for going optimal are tremendous on a per unit basis but trivial from the systems point of view.

Now, to answer the questions raised above. Where should the 7.3 months' supply of the bulk items be stored? The answer depends on the size, weight and volume of the items to be stored. Thousands of paper clips can be placed in a very small space. Are bulk items likely to age, deteriorate, or become obsolescent because of the way that they are stored? The answer depends on the kind of item. Some items are more vulnerable than others, and the carrying cost calculation should reflect this fact. Paper clips and pencils will be taken home. Cherries will deteriorate. Computer chips will become obsolete.

Why is it that matching demand with monthly orders does not provide the just-in-time benefits that are so popular? The answer is twofold. The carrying cost as reflected by cC_c is too low. The cost of half a cent per unit is so small that JIT is not needed. The combination of low per unit cost and a 0.06 rate for carrying stock make matching demand with supply costly. Even the low ordering cost operates against JIT. Large set-up costs and high carrying costs are foundations for JIT.

Instead of EOQ batch delivery at fixed intervals a serialized process which delivers product (or services) on a continuous basis now will be explored. It will be seen that batch production is one extreme of this more general model of production. At the other extreme is continuous production which keeps on running. In between are run lengths determined by the same factors that drive the EOQ model.

14.5

ECONOMIC LOT SIZE (ELS) MODELS

Developing the ELS model is best accomplished by comparing it with the EOQ model. ELS provides continuous delivery of product (say hourly or daily) instead of delivery in batches (say every three days or once a week). The two options are pictured in Figure 14.5 where a batch delivery of 18 units is made every third day (called A). Continuous delivery of two units every day (called B) is the other option.

Generally, batch delivery is the result of batch manufacture where work is done in stages within the job shop, as shown in Table 14.3 where six units are delivered every three days. All work is completed in stages. Not one unit can be delivered until the entire batch is completed on days 3, 6, and 9.

TABLE 14.3

BATCH PRODUCTION

Day	Number of Units	Stage of Completion
1	6 units started	six units are one-third finished
2		six units are two-thirds finished
3		six units are completed and delivered
4	6 units started	six units are one-third finished
5		six units are two-thirds finished
6		six units are completed and delivered
7	6 units started	six units are one-third finished
8		six units are two-thirds finished
9		six units are completed and delivered
Total	18 units	delivered in 9 days

FIGURE 14.5

CONTRASTING BATCH AND CONTINUOUS DELIVERY SYSTEMS

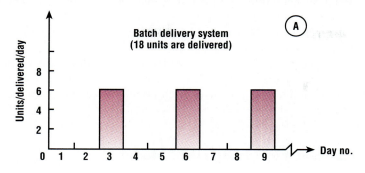

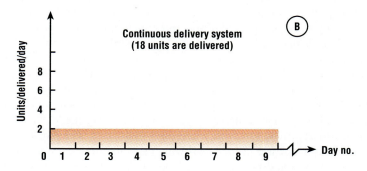

A portrays batch and *B* portrays continuous delivery.

The other possibility is that work is continuous using a flow-shop process and deliveries are made almost continuously as shown in Table 14.4.

TABLE 14.4

CONTINUOUS PRODUCTION

Day	Number of Units	Stage of Completion
1	2 units started	two units finished and delivered
2	2 units started	two units finished and delivered
3	2 units started	two units finished and delivered
4	2 units started	two units finished and delivered
5	2 units started	two units finished and delivered
6	2 units started	two units finished and delivered
7	2 units started	two units finished and delivered
8	2 units started	two units finished and delivered
9	2 units started	two units finished and delivered
Total	18 units	delivered over the 9-day period

The continuous output case (B) in Figure 14.5 and in Table 14.4 above could be delivered in batches as in A of Figure 14.5 and Table 14.3 above. Various reasons account for the producer storing continuous output for batch delivery. Among these are a JIT agreement between buyer and supplier, lack of storage space at the buyer's site, delivery economics where less than fully-loaded shipping containers, trucks or railroad cars are costly.

Some of these points hold true whether the units are delivered by an outside supplier or produced internally. The pull system of production does not allow deliveries until a work station calls for it and often permits a batch of no more than one unit to be delivered.

The Intermittent Flow Shop (IFS) Model

A flow shop which does not have sufficient demand to warrant running it continuously can be converted with new set ups and changeovers to run other products. It is an intermittent flow shop where run time is called online and nonrun time is called offline.

Figure 14.6 pictures the way that this IFS functions for one product. It shows the length of time of the production run (t_1), the time that the system is free to run other kinds of units (t_2), the average number of units that are carried in stock, and the size of the storage facility that is required to accommodate units at the peak. The peak storage quantity occurs just before the system goes offline.

Note that Figure 14.6 shows the production rate which has a greater slope than the stock on-hand line because daily withdrawals are being made. The ELS model portrays intermittent production with continuous deliveries. The stock level continuously increases until the apex is reached which sets the maximum storage requirement. Then, stock on-hand begins to decrease because production for the item is offline. Eventually, SOH reaches zero (or some protected level of stock if buffer units are held for service safety reasons).

FIGURE 14.6

THE OPERATION OF THE ECONOMIC LOT SIZE (ELS) MODEL FOR INTERMITTENT FLOW SHOP PLANNING

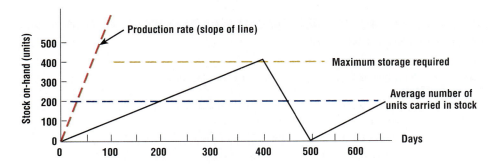

Next comes the formulation of the economic lot size (ELS) model. Because the production run quantity is called a lot, the model gets its name. It can be used to determine the optimal production lot size which translates into run time as a function of the production output rate. The model also can be used to study the effect of varying the production rate.

FIGURE 14.7

THE ELS TRIANGLE IS DIVIDED INTO TWO
RIGHT TRIANGLES CALLED *A* **AND** *B*

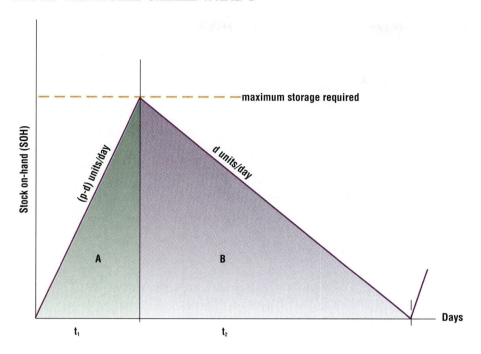

A shows stock accumulation the rate of $(p - d)$ and *B* reflects continuous deliveries at the rate of (d) units per time period.

Note that if the production output rate per unit time (p) is increased, then the slope of the line rising to the apex increases. In other words, the left side of the SOH triangle rises faster. Also, the production rate line (drawn in Figure 14.6) rises faster. This means that run time could be shortened leaving longer free times on the line for other products.

Furthermore, note that as the slope of the left side of the SOH triangle increases—rising faster—this triangle looks more and more like the right triangles that characterize EOQ batch models (see Figures 14.2 and 14.3 above).

Figure 14.7 shows the ELS triangle divided into two right triangles called *A* and *B*. The upward slope of the left side of the *A* triangle represents the net rate of accumulation of the stock on-hand. It is the production rate per day (p) minus the demand rate per day (d) which is $(p - d)$. The downward slope of the right side of the *B* triangle represents the demand per day (d). There is no production going on during that interval called t_2.

The ELS model's ability to determine optimal production run size seems of primary interest to suppliers and producers. The EOQ model's calculation of the optimal order quantity would seem to be of primary interest to buyers and customers. The systems perspective refutes this self-interest. The ability of suppliers and producers to obtain minimum cost solutions and adequate production runs operate to the advantage of buyers and customers. The benefits are lower costs and satisfactory delivery times. The buyer's EOQ solution and the supplier's ELS solution should be complementary and congruent.

The ELS model can provide continuous JIT-type delivery with intermittent production. The discontinuity in production is behind the scenes. This characterizes flexible production systems and intermittent flow shops which are designed to run one item and then change over to run another item using the same or similar equipment, the same work stations, and the same workers. The ELS solution is increasingly encountered with flexible NC and CNC equipment. Because of the ability to change over rapidly and at low cost, the economic lot size can be small.

The ELS model can run the gamut from the continuous production of the flow shop to short production runs. When the supplier delivers the economic order quantity to the buyer, the production of the required batch can be thought of as being instantaneous. The ELS math model developed below shows these properties.

The cost of preparing to produce product or provide a service (run the lot) includes the set-up cost, C_S. Set-up costs are producers' costs. They are usually significantly larger than order costs experienced when buying from suppliers. Because FPS are pre-engineered for rapid set ups, C_S is considered to be negligible.

Set-up costs are composed of two main parts:

1. the cost of labor that is required to prepare the facility for the new production run
2. the cost of lost production created by the facility's downtime while being prepared for the new job

Optimal run size is derived in a manner similar to the derivation of the economic order quantity. Many of the variables are the same. Two new variables include p and d, previously mentioned:
Thus:

$$p = \text{production rate in units per day}$$

$$d = \text{demand rate in units per day}$$

The total variable cost equation is:

$$TVC = \left(\frac{cQ}{2}\right)\left(\frac{p-d}{p}\right)C_c + \left(\frac{D}{Q}\right)C_s$$

where

$$\left(\frac{p-d}{p}\right)\left(\frac{Q}{2}\right) = \text{the average number of units carried in the inventory}$$

As with the EOQ model, the minimum total cost can be derived by taking the derivative of TVC for ELS, setting it equal to zero, and solving for the optimal value, Q_O as follows:

$$\frac{d(TVC)}{dQ} = \left(\frac{c}{2}\right)\left(\frac{p-d}{p}\right)C_c - \left(\frac{D}{Q^2}\right)C_S = 0$$

$$Q_O^2 = \left(\frac{2DC_S}{cC_c}\right)\left(\frac{p}{p-d}\right) \text{ and } Q_O = \sqrt{\left(\frac{2DC_S}{cC_c}\right)\frac{p}{p-d}}$$

Alternatively, setting the two terms of the ELS equation for *TVC* equal to each other produces the same result because minimum *TVC* occurs at the Q_O value where the two lines in Figure 14.4 above cross.

The production run size $Q_O = pt_1$ (where t_1 = the run time or online time). Use $Q_O = pt_1$ to solve for t_1 as follows:

$$t_1 = \frac{Q_O}{p}$$ **Equation 14.1**

Time t is required for each total cycle where $t = t_1 + t_2$ and t_2 is the offline time. Because $Q_O = dt$ (the run size must satisfy demand for the entire period t), it follows that:

$$t = \frac{Q_O}{d}$$

Figure 14.6 above will provide graphic meaning to the online and offline relationships. Equations 14.1 and 14.2 compute online and offline production times.

$$t_2 = t - t_1 = \frac{Q_O}{d} - \frac{Q_O}{p} = Q_O\left(\frac{p - d}{pd}\right)$$ **Equation 14.2**

An Application of ELS

Units of different colors are to be made on new equipment using an intermittent flow shop which runs 250 days per year. Different colors are run successively. Blue has the largest market share.

Data have been assembled for blue as follows:

Yearly demand $= D = 200{,}000$ per year $= 250d$

Daily demand $= d = 800$ per day $= D/250$

Daily production $= p = 1800$ per day

Setup cost $= C_S = \$20$ per setup

Carrying cost rate $= C_c = 0.15$ of the unit cost per year

Cost $= c = \$0.06$ per unit

Calculating the optimal lot size:

$$Q_O = \sqrt{\frac{(2)\,(200{,}000)\,(20)\,(1800)}{(0.06)\,(0.15)\,(1000)}} = \sqrt{\frac{(144)\,(10)^8}{(9)}} = \frac{(12)\,(10)^4}{3}$$

$$= 40{,}000 \text{ units}$$

Production run time for blue is:

$$t_1 = \frac{Q_O}{p} = \frac{40{,}000}{1{,}800} = 22.22 \text{ days}$$

Cycle period is determined:

$$t = t_1 + t_2 = \frac{Q_O}{d} = \frac{40{,}000}{800} = 50 \text{ days}$$

Offline period between runs is:

$$t_2 = Q_0\left(\frac{p-d}{pd}\right) = 40{,}000 \left(\frac{1000}{800 \times 1800}\right) = 27.78 \text{ days}$$

This means that 22 days (rounding) are required for the blue run and $50 - 22 = 28$ days (rounding) are available for other colors. The problem of fitting mixed-model systems on the same production line is seldom easy. An approach that is used is to start with the individual product optimal assignments. Then modify them with the "cut and try" method. More sophisticated mathematical models also can be used.

When can all of the items of a mixed-model set achieve optimal runs that minimize each of their *TVC*s? The answer is when there is enough extra capacity so that none of the items (say colors) are constrained. When the capacity is overtaxed by demand levels, then it is rare that each item of a mixed-model set can have an optimal run.

Imagine what would happen if in addition to blue, red and yellow had very similar characteristics. How could all three items be run on the same system? The only way that they can all be run is by cutting down on the run sizes. This means that each of the items will have *TVC*s that are suboptimal. Thus, overall costs for the system will be minimized but each item would have lower costs if there were no resource constraints. The best that can be done is to achieve the systems' optimal.

From EOQ through ELS to Continuous Production

The ELS equation for optimal run size:

$$Q_0 = \sqrt{\left(\frac{2DC_S}{cC_c}\right)\left(\frac{p}{p-d}\right)}$$

lends itself to analysis of extremes. First, if d is almost equal to p, then $(p - d)$ approaches zero. This means that Q_0 becomes very large.

$$Q_0 \to \infty \quad \text{as} \quad (p - d) \to 0$$

The conclusion is that if the demand rate is as great as the production rate, then run the process continuously. There is no inventory buildup. The slope of the line in Figure 14.8 is near zero. The stock on-hand runs almost flat against the time axis.

If p is much greater than d ($p >>> d$), then

$$Q_0 = \sqrt{\frac{2DC_s}{cC_c}}$$

which is the EOQ model with C_s instead of C_r.

Increasing stock on-hand rises perpendicularly, as in Figures 14.2 and 14.3 above, or almost so, as in Figure 14.7 above. The producer can supply the order quantity, Q_0, on demand. This is similar to ordering from the outside supplier.

Between the two extremes described above, many online and offline combinations of production levels are feasible.

FIGURE 14.8

THE ELS MODEL INDICATES ALMOST CONTINUOUS PRODUCTION SHOULD BE USED AS ($p - d$) APPROACHES ZERO

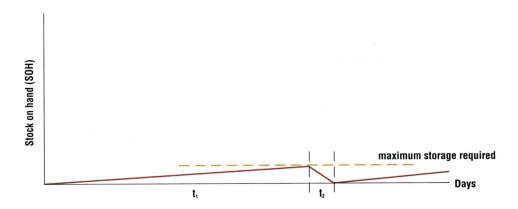

Lead-Time Determination

Lead time (LT) is the interval that elapses between the recognition that an order should be placed and the delivery of that order. For both EOQ and ELS models, the required lead time can be identified by studying their respective graphs, as shown in Figure 14.9.

First, note how the diminishing stock level reaches a threshold called Q_{RP} on the diagram. Q_{RP} stands for the stock level of the reorder point. That threshold triggers the order for replenishment. The level of stock that is at the *RP* is based on determining the lead time. Figure 14.9 below shows the graphical relationship between Q_{RP} and *LT*.

FIGURE 14.9

LEAD TIME SETS THE REORDER POINT FOR BOTH EOQ AND ELS MODELS.

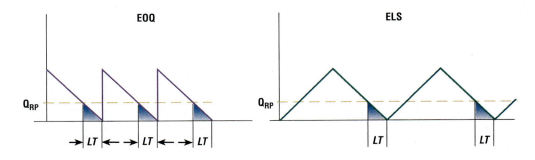

Legend:

Lead Time = *LT*

Reorder Point = Q_{RP}

Les Waltman
· · ·
GE Aircraft Engines

Les Waltman is a manager of inventory control at GE Aircraft Engines' Production and Procurement Division headquartered in Evendale, Ohio. With recent annual revenue of $5.7 billion, GE Aircraft Engines manufactures jet engines for military and commercial aircraft.

Waltman's responsibilities as inventory control manager include: (1) coordinating the budgeting and forecasting process for inventory management; (2) working with various client sites in developing "inventory reduction action items": (or initiatives taken to reach inventory reduction goals); (3) developing strategic plans to support top-level management inventory objectives; and (4) disposition of "excess inventory" which includes finished goods or raw materials that manufacturing no longer needs.

Over the last several years GE Aircraft Engines has "totally revamped" its thinking on how the manufacturing shops are laid out, Waltman said. Instead of shops being batch process shops or process

centers, "we've made them into manufacturing cells where each cell is somewhat self-contained for a given product or products," he said. This means that instead of a part coming into a centralized inventory area and then being sent to many different operations all over the plant or different plants, inventory areas are set up in a horseshoe "cell" arrangement for each product or group of similar products. "Instead of going 10,000 feet the part might go a couple hundred feet; instead of having a lot size of 50, you have a lot size of one; instead of having a cycle time of 15 weeks, we have a cycle time of 2 weeks," he said.

Waltman cites developing and executing "stretch targets" to show significant improvements over past years as the biggest challenge of his job—in the corporate environment where "quantum change" rather than minimal change is expected. The "inventory team" has helped the company achieve quantum changes in inventory level. "Over the last three years we've reduced our inventory about 50 percent," Waltman said, and this means about a billion dollars in reductions. "We've made changes over the last few years that people thought would take ten years," he said. To Waltman, who has been with GE Aircraft Engines for 30 years, the quantum changes occurring in manufacturing today "really make things exciting."

Source: A conversation with Les Waltman.

Consider eight components of lead time for the EOQ case.

1. The amount of time that is needed for recognition of the fact that it is time to reorder. If the reorder point is monitored continuously, then this period is as close to zero time as the system permits. If the stock level is read at intervals, then the average interval for noting that the reorder point has been reached is part of the LT interval.

2. The interval for doing whatever clerical work is needed to prepare the order. This includes determining how much to order and from whom to order. It might even include preparing for bidding. If multiple suppliers are to be used, determining how to divide the order.

The Campbell's Soup Company's Continuous Product Replenishment program replenishes the retailer's inventory at the same rate as the consumer takes it off the shelf by using an electronic ordering system.
Source: Campbell's Soup Company

3. Mail or telephone intervals for communicating with the supplier (or suppliers) and placing the order.
4. How long will it take for the supplier's organization to react to the placement of an order? This necessitates communication with everyone who participates in filling the order. How long does it take to find out if the requested items are in stock? If the items are not in stock, how long will it take for the supplier to set up the process to make them? How long will it take to produce, pack and ship?
5. Delivery time includes loading, transit, and unloading times. If transshipment is needed the period must be extended to take that into account. This period begins when the product leaves the supplier's control and ends when the buyer takes control of the items.
6. Processing of delivered items by the receiving department and determining storage locations.
7. Inspection to be sure that items match specifications is generally required. Checking quantity as well as quality is part of the inspection. Time may be required to deal with problems uncovered by the inspection.
8. Record-keeping computer time is needed to enter items in the warehouse stock records. It also is needed to transport the items to the warehouse and to store them properly.

All eight of these lead-time components add together to form the total lead time. Determination of lead time requires awareness of all relevant systems factors. Put another way, proper lead-time determination requires a systems study that traces out all contributing factors including their averages and variances. A similar set of steps applies to ELS for resuming production.

Expediting to Control Lead Time

Expediting is part of the OM function. It is used to control and improve lead times in the plant, in the office and with outside suppliers and shippers. Expediting is the process of keeping track of the state of an order. It includes reminding and following up with anyone who could be delaying order processing.

Delivery time includes loading,
transit, and unloading times.
Source: Campbell's Soup Company

Expediters are accountable for making certain that due dates are met. Often, there are many dates and places to be tracked. For example, CarTrack used by DuPont provides information about all railroad boxcars. This is the first step—knowing where things are geographically. Tracking orders in a supplier's factory permits knowledge about the degree of completion of the order.

Expediters also are trained to prevent and remedy problems. Delays are addressed as to causes and corrections. Using information systems to track and evaluate situations, the expediters know how to take action and when to use special capabilities (such as using FedEx) to speed things up.

There is specific contingency training required in the use of equipment and techniques to take effective action. Such expediting techniques are frequently used in project management where slippage along time lines is considered to carry very high monetary penalties.

Lead-Time Variability

Lead times are usually variable. There are delays of varying length in each of the eight lead-time components above. When variance is significant, estimates must be made of the degree of variability. Appropriate steps should be taken to include the effects of variability in the analyses. Expediting can be considered to decrease variability.

Expediters deal with the causes of variability and attempt to keep them under control so that due dates can be met. Expediters also keep management informed about slippage and due dates that will be overrun.

Although lead times are usually variable, they are often treated as if they were fixed, single values. There are two reasons for this:

1. the amount of variability is considered to be small and relatively negligible
2. it is complex to assume variable demand and variable LT simultaneously

Then, to take variable lead time into account in a quick and easy way, the idea of increasing protection against stock-outs is used. This is accomplished by assuming a larger (worse case) average than the actual average for the LT. To illustrate, if lead time is known to average 10 days, and it is also known to range ± two days almost all of the time, then an estimate of 12 days for LT might be tried out.

How this affects the stock levels will now be shown as the perpetual inventory system uses extra stock protection against surges in demand and delays in delivery.

14.6

PERPETUAL INVENTORY SYSTEMS

Perpetual inventory systems continuously record inventory received from suppliers and withdrawn by employees. Most perpetual inventory systems are used online, in real-time with some demand variability. The EOQ and ELS models discussed so far do not include demand variability. In many practical situations, this assumption is unrealistic. With modifications, the EOQ and ELS models can deal with demand variability.

Variability of demand arises from variations in order sizes. Variability is passed along the supply chain affecting everyone. It is caused by customers but there are other causes of variability as well. These include changes in the number of defectives produced by the process.

Consider what occurs when a machine setting changes creating a large number of rejects which cannot be reworked. Production must run larger numbers of items to compensate for the scrap. More materials must be withdrawn from inventory to be used by the process. The supplier has furnished the usual number of units per order which was based on the estimate of the expected demand (D). Because demand is larger than expected, the reorder point Q_{RP} will be reached sooner.

The perspective that will be used to explain how the EOQ model can be modified to protect the buyer will assume batch delivery of the quantity:

$$Q_O = \sqrt{\frac{2DC_r}{cC_c}}$$

for processes that are on-going so that demand is relatively constant over time. Such modeling does not apply to job shops with small orders because they lack continuity of demand.

The focus now is on items that are stocked regularly, where the demand can vary. The buyer orders the economic order quantity, but has on-hand some buffer stock (which is sometimes called reserve stock or safety stock). The buffer stock is extra stock so that when demand is heavier than expected, orders can still be filled.

An option is to design a perpetual inventory system. It is perpetual because it is an online system, tracking the stock on-hand (SOH) at each transaction of withdrawal or stock entry. To design the system it will be useful to answer the following questions:

1. **Q:** *How many buffer stock units should be carried?*
   A: The answer involves balancing the added carrying costs for extra stock against the costs of running out of stock. As one cost goes up, the other comes down.

2. **Q:** *When should an order be placed for the buffer stock?*
    A: The buffer stock should be created as soon as it is determined that it is beneficial to have it. It does not have to be replenished when it has been drawn on to avoid stock-outs because the use of buffer stock is expected to average out at zero.

3. **Q:** *How often should an order be placed for buffer stock?*
    A: Once, and then whenever the buffer-stock level needs to be updated.

4. **Q:** *How many units should be ordered to meet expected demand?*
    A: Use the same economic order quantity, Q_O.

5. **Q:** *How often and when should the orders be placed for units to meet the demand?*
    A: An order is placed whenever the reorder point, Q_{RP}, is reached.

6. **Q:** *How does this perpetual system work?*
    A: Withdrawal quantities are entered in the computer each time one or more units are taken out of stock. These quantities are subtracted from the previous stock-level balance to determine the new balance quantity of stock on-hand.

Reorder Point and Buffer Stock Calculations

The reorder point quantity is designated for each item and is entered in the computer program. When the reorder point has been reached, the program recognizes this fact and an order is placed for the economic order quantity Q_O.

The stock level of the reorder point (Q_{RP}) is equal to the expected demand in the lead-time period called $\overline{D}_{LT}$, plus the buffer stock *(BS)* quantity. Thus:

$$Q_{RP} = \overline{D}_{LT} + \text{BS} \qquad\qquad \textbf{Equation 14.3}$$

Figure 14.10 shows the distribution of demand in the lead-time period. It also provides a physical interpretation of the components of Equation 14.3.

F I G U R E   1 4 . 1 0

DETERMINATION OF THE REORDER POINT Q_{RP}.

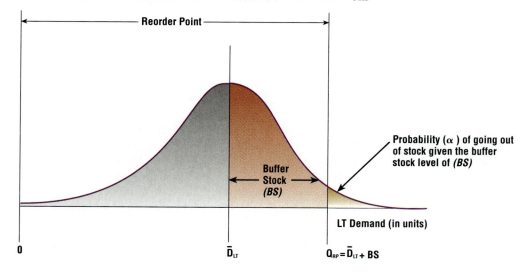

Q_{RP} is the Sum of Expected Demand in the Lead-Time Period and the Buffer Stock *(BS)*.

The first part of the reorder point goes from zero (at the left side of the diagram) to the mean of the distribution, which is $\overline{D}_{LT}$. The second part of the reorder point is the buffer stock. It spans a distance $+k_a\sigma_{LT}$, which is k standard deviations of the lead-time distribution. Thus, adding $k_a\sigma_{LT}$, which equals the buffer stock, to the mean of the distribution defines the reorder point Q_{RP}. The reorder point is fully specified by Equation 14.4 where the value of k_α in $+k_\alpha\sigma_{LT}$ represents a probability design decision regarding stock-out frequencies.

$$Q_{RP} = \overline{D}_{LT} + (\text{BS} = + k_a\sigma_{LT}) \qquad \textbf{Equation 14.4}$$

The tail of the distribution (in yellow) at the right-hand side of Figure 14.10 above represents the probability α of going out of stock. Outages happen whenever actual demand in the lead-time period exceeds $\overline{D}_{LT} + k_a\sigma_{LT}$. The probability α in the tail area determines how often that happens. That probability can be adjusted by increasing or decreasing the value of k_α.

When the tail area is small, then the buffer stock is large and the likelihood of going out of stock is small. The large buffer stock means that the *carrying cost of stock is high to make sure that the actual cost of stock-outages is small.* When the tail area is large, then the amount of buffer stock that is being carried is small and the carrying cost is small but the outage penalties are going to have to be paid more often.

The best way to determine the lead-time distribution is by observation. First, take as large a number of past lead times as the computer records make available and calculate the average lead time (say it is 10 days). Then, again by observation, record the demand over successive 10 day intervals. Plot the distribution and determine the standard deviation.

If the successive 10-day periods are independently distributed, as they usually are, the lead-time distribution obtained by observation will be a reasonable approximation. This assumes the same supplier and delivery system over the sample period. If a change has occurred, then the new lead-time parameters will have to be estimated some other way.

Enough intervals should be collected in the sample to provide a smooth nearly normal shape for the lead-time distribution when it is plotted. If the distribution appears bimodal or skewed, further statistical investigation is needed. It is best to state the average lead time in small periods of time, such as days, to expedite creating the lead-time distribution.

Imputing Stock-Outage Costs

Buffer stock is designed to prevent some percent of out-of-stock situations. The size of the alpha tail of the distribution in Figure 14.10 above indicates the level of protection being sought. The buffer-stock level that is chosen is based on balancing the costs of stock-outs with those of carrying costs. Thus:

$$\text{Carrying Cost } (BS) = cC_c(k_a\sigma_{LT}) \qquad \textbf{Equation 14.5}$$

where

c = cost in dollars per unit

C_C = the carrying cost rate as a percent per dollar per year

cC_C = carrying cost in dollars per unit per year

$k_\alpha \sigma_{LT}$ = number of units carried in the buffer stock (k_α has been chosen to limit the probability (α) that demand in the lead time will exceed $k_\alpha \sigma_{LT}$)

$cC_C(k_\alpha \sigma_{LT})$ = carrying cost in dollars per year for buffer stock that offers an α level of protection

The cost of stock-outages relates to their frequency. As k_α gets larger, the frequency of stock-outages decreases. Say that the tail of the distribution with $k_\alpha = 1.64$ provides protection against demand levels exceeding buffer stock 95 percent of the time. That means that five out of 100 lead-time intervals (or one out of 20 *LTs*) will experience a stock-outage so $\alpha = 0.05$.

The number of lead-time intervals is the same as the number of orders placed. Say that $n = 5$ orders are placed per year. Then, an outage is expected once every four years. Stated in an alternative way, it is expected that four years will elapse between stock-outs. Finally, assume that the carrying cost per year for buffer stock is $2,500 obtained by using Equation 14.5 above. In four years this sums to a value of $10,000 without discounting.

The value of k_α chosen has an *imputed stock-out cost* of $10,000. Imputed means attributed in the sense that an action attributes value. It is now up to management to decide whether that implied cost seems sensible and is acceptable. In many instances, it is very difficult to determine stock-out costs directly. The direct measure requires finding all of the consequences of the stock-out and then measuring the impact of those consequences. Therefore, the use of imputation of cost provides an appealing alternative. If the imputed cost is accepted, then:

$$\text{Outage Costs } (BS) = \$10,000$$

Operating the Perpetual Inventory System

The fastest route to understanding how the perpetual inventory system works is by studying the graphic representation of it. Figure 14.11 illustrates the two parts that have been put together. The EOQ model sits on top of the buffer-stock part.

Follow the stock on-hand line as it forms an irregular sawtooth pattern. When demand becomes greater, the line moves down faster and vice versa. In other words, when the slope of the line becomes steeper, the reorder point (Q_{RP}) is reached more quickly.

Next, note that the interval between orders is variable depending on whether the demand has been faster or slower than the average rate. When the demand is consistently at the average rate the spacing between orders will be constant and unchanging at:

$$t_0 = \frac{Q_0}{D}$$

Note in Figure 14.11 above, the first of the replenishment cycles stops right at the buffer-stock line. At that point, SOH is zero and no buffer stock is withdrawn. Demand dips into the buffer stock region *only* for the second replenishment cycle. In other words, buffer stock is called upon once in three replenishment cycles. That buffer stock is the first to be replaced when the next order is filled.

The third of the three replenishment cycles shows an order arriving when the stock level is above the buffer stock level. If the demand distribution is stable over time, then the buffer stock level will average out to $k_\alpha \sigma_{LT}$ units.

FIGURE 14.11

A PERPETUAL INVENTORY SYSTEM—SOH IS COMPUTED WITH EACH WITHDRAWAL

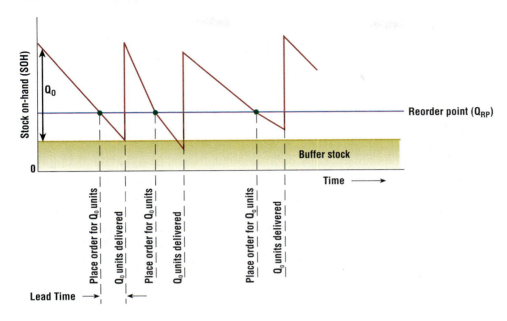

Two-Bin Perpetual Inventory Control System

The two-bin system is a smart way of continuously monitoring the order point. It is a simple self-operating perpetual inventory system. Figure 14.12 shows the two bins with familiar labels.

FIGURE 14.12

THE TWO-BIN SYSTEM IS A SELF-OPERATING PERPETUAL INVENTORY SYSTEM WHERE Q_{RP} = CONSTANT

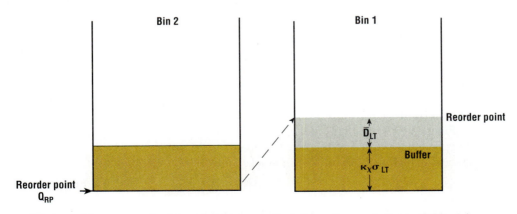

The two bins are marked 1 and 2. Assume that Q_O units are delivered. Bin 1 is filled to the reorder-point level. The remainder of the units are put into Bin 2. Figure 14.12 shows the construction of the reorder point in Bin 1.

Withdrawals are made from Bin 2 first. When Bin 2 is depleted, an order is placed for the EOQ, and further units are taken from Bin 1. Each time bin 2 is emptied, a new order is placed—it is equivalent to reaching the reorder point.

The two-bin system is not feasible for many kinds of items. When applicable, much clerical work is eliminated. This two-bin system is well-suited to small items like nuts, bolts, and fasteners. These are things that are too small and too numerous to make withdrawal entries for each transaction. The same reasoning applies to recording withdrawals of liquids.

14.7

PERIODIC INVENTORY SYSTEMS

Periodic inventory systems were more popular than perpetual inventory systems before inventory information was digitized and put online. Periodic inventory systems are based on regular fixed review periods which were ideally suited for manual entries. Periodic systems based on that convenience were outmoded by computers. Nevertheless, there are reasons why periodic systems operated by computers continue to be used. These include requirements of suppliers concerning when they will accept orders, requirements of shippers about when they will schedule deliveries, and the need to combine orders to obtain volumes sufficient for shipment discounts. Some organizations have central warehouses that will only accept orders from their regional distributors once a week on a specific day. Also, some industries prefer the regularity of the periodic method, which can be linked to changeover intervals for production processes.

Periodic inventory systems also should be understood because advanced inventory models (call Ss policies) combine the ordering rules of perpetual and periodic order systems to obtain lower total costs. These can be encountered in big inventory systems installations such as the Armed Forces use.

The optimal interval is based on the familiar square root relationship:

$$t_O = \frac{Q_O}{D} = D\sqrt{\frac{2DC_r}{cC_c}} = \sqrt{\frac{2C_r}{DcC_c}}$$

High-cost items and large carrying costs increase the desirability of short review periods. High-ordering costs and large set-up costs call for long review periods.

To use the periodic model, first fix the review interval. At each review, determine SOH by adding total receipts and subtracting total withdrawals. An order is placed for a variable quantity contrasting with the perpetual inventory model where it was a fixed quantity at variable intervals.

Figure 14.13 illustrates the way that the periodic order system functions. Note the upper line marked M. It is set by summing the order quantity Q and the stock on-hand, which includes the buffer stock. Thus:

$$M = Q + SOH \qquad\qquad \textbf{Equation 14.6}$$

As seen in Figure 14.13 above, the SOH includes the buffer stock and the expected demand in the lead-time period. This is:

$$SOH = \overline{D}_{LT} + BS \qquad\qquad \textbf{Equation 14.7}$$

THE PERIODIC INVENTORY SYSTEM WHERE SOH IS COMPUTED AT FIXED INTERVALS AND Q IS VARIABLE

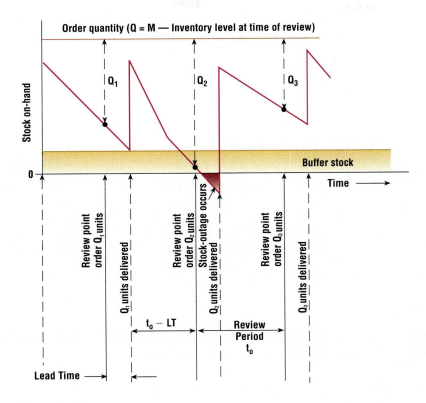

Observing that the order quantity Q is the expected demand in t_O (the interval between reviews), Equation 14.6 above is rewritten using Equation 14.7 above, as follows:

$$M = \overline{D}_{t_o} + \overline{D}_{LT} + BS = \overline{D}_{(LT + t_o)} + BS \qquad \textbf{Equation 14.8}$$

The size of the buffer stock for the periodic review system must include protection against excessive demand in a review period plus one lead-time interval. Thus:

$$BS = k_\alpha \sigma_{(LT + t_0)} \qquad \textbf{Equation 14.9}$$

M is calculated as shown in Equation 14.10.

$$M = \overline{D}_{t_0} + \overline{D}_{LT} + k_\alpha \sigma_{(LT + t_0)} = \overline{D}_{(LT + t_0)} + k_\alpha \sigma_{(LT + t_0)}$$

$$\textbf{Equation 14.10}$$

Periodic buffer stock is calculated using the distribution of $(LT + t_O)$ because after an order is placed, variations in demand can be experienced during the period t_O. Then a review occurs and an order is placed. Demand variations can continue during LT. Therefore, exposure to variable demand before a correction can be made is over a review period and a lead time. This requires extra stock for protection which is a major drawback of the periodic model.

The operation of the periodic model is to determine the review dates. Calculate *SOH* and order an amount $M - SOH = Q_O$. Thus:

$$M - (\overline{D}_{LT} + k_\alpha \sigma_{(LT + t_o)}) = Q_O \qquad \textbf{Equation 14.11}$$

It is worth noting that $(\overline{D}_{LT} + k_\alpha \sigma_{(LT + t_o)})$ in Equation 14.11 is the formula for the reorder point (Q_{RP}) of the perpetual inventory model. Therefore, $M - Q_O = Q_{RP}$. However, there is a notable difference in buffer stock, $k_\alpha \sigma_{(LT + t_o)} > k_\alpha \sigma_{LT}$.

14.8

QUANTITY DISCOUNT MODEL

Accept a *quantity discount* only when it lowers the total costs. Therefore, to determine when the discounted order quantity should replace the optimal undiscounted order quantity Q_O, it is necessary to derive the total cost curves. Note that it is necessary to compare total costs instead of total variable costs because for discount analysis the cost of goods purchased must play a significant role in the decision to accept or reject a discount. Lowering the price paid for goods is the purpose of taking the discount but the true criterion must be total costs.

The discounting situation is directly reflected by the following schedule:

T A B L E   1 4 . 5

DISCOUNT SCHEDULE WITH 3 PRICE BREAKS

Quantity (Q units)	Cost per unit	Total Cost
$Q = 0$ up to Q_1	c	$TC(c)$
$Q = Q_1$ up to Q_2	c'	$TC(c')$
$Q = Q_2$ and greater	c''	$TC(c'')$

(There can be any number of price breaks.)

Total cost is *TC* using *TVC*, as previously defined, plus dollar demand (*cD* and *c'D*) for undiscounted and then discounted costs. To simplify the discussion, only two price breaks will be used.

$$TC(c) \ = \ TVC(c) \ + cD = (cQ/2)C_c \ + (D/Q)C_r + cD \quad (0 \le Q < Q_1)$$
$$TC(c') \ = \ TVC(c') \ + c'D = (c'Q/2)C_c \ + (D/Q)C_r + c'D \quad (Q \ge Q_1)$$

These total cost curves are drawn in Figure 14.14.

FIGURE 14.14

QUANTITY DISCOUNT MODEL: TOTAL COST CURVES

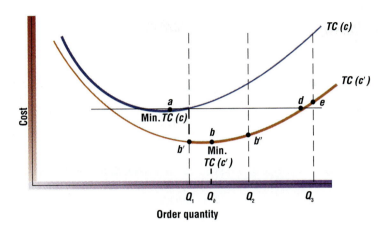

The quantity discount model involves more than one total cost equation but only one is applicable in each cost break's range c, c'.

Each curve includes its own respective dollar volume which is cD for the first curve and $c'D$ for the second curve. The top curve $TC(c)$ is based on an undiscounted cost c. The bottom curve $TC(c')$ is applicable when a discount is available, but it is only applicable at and above the quantity needed to obtain the discount.

Let $Q \geq Q_j$ be the specified quantity required to obtain the discount. Q_j is called the discount break point. If Q_j is Q_1 in Figure 14.14 above then the discount must be taken because the TC at b' is lower than the TC at a which is the minimum TC without the discount. In fact, the order quantity should be increased from Q_1 to Q_b, which is the order quantity at point b. Point b is the minimum total cost that can be obtained in the discount region. It is the lowest cost that can be acquired. Note that the top TC curve applies from $0 \leq Q \leq Q_1$; the range of the bottom curve is for $Q \geq Q_1$. Figure 14.15 magnifies this discontinuity of the two cost curves.

The top curve is valid left of the vertical line at Q_1 and the bottom curve is valid at and to the right of it. The cost of point b is lower than that of point b'.

In Figure 14.16 when Q_j is specified as being at Q_2, then point b'' provides a lower cost than point a. Therefore, take this discount at break point Q_2. Point a is the minimum total cost without the discount. Point b'' is the lowest total cost that is available in the discount region.

Now refer to Figure 14.14 (above) again. If the discount break point $Q_j = Q_3$, then the intersected cost point is e, which is a greater cost than a. Therefore, the order quantity that is related to point a should be used. However, if Q associated with cost point d had been the break point then there is a tie between the order amounts associated with cost points a and d.

FIGURE 14.15

DISCONTINUITY OF TWO COST CURVES

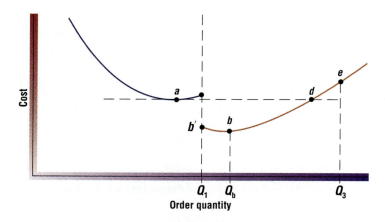

A specific discounting situation where Q_O is the minimum point of the discounted total cost curve. The discount should be taken with a larger order quantity b than the minimum amount b' needed to obtain the discount.

FIGURE 14.16

DISCONTINUITY OF TWO COST CURVES

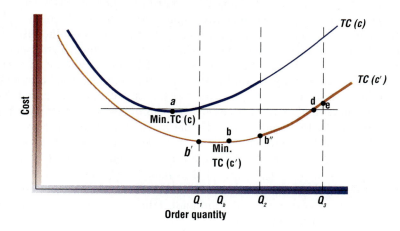

Another discounting situation where Q_2 is the break point quantity at the cost b'' which is the lowest cost that can be obtained so the discount should be taken at that quantity. Note how the two TC curves are each applicable in their own regions.

Theoretically, the inventory planning team would be indifferent about either buying the small amount at point a or buying the large amount at point d. In fact, they would consider a variety of additional factors such as: could there be a shortage of the material, is there a possibility of a strike, and can the purchase of a large quantity make it difficult for others to buy it? These notions might favor point *d*. On the other hand, does this item spoil easily, does it require a large amount of storage space, does it experience obsolescence? These notions are more likely to favor purchasing the smaller quantity associated with point *a*.

The reasoning process that has been applied, and the curves that have been drawn can be extended to any number of price breaks for quantity discounts. What is recommended is to draw the total cost curves, mark the price break ranges and examine where the minimum cost points fall. A purely mathematical approach can be programmed for rapid decision making when faced with many discount offerings.

S U M M A R Y

This chapter addresses "independent demand" inventory management. It explains order point policy (OPP) models in terms of demand continuity. The costs of inventory are treated. These include costs of: ordering, setups and changeovers, carrying inventory, discounts, stock-outages, and running the inventory system. Costs are differentiated by process types.

Then, economic order quantity models (EOQ) are developed for batch delivery. The total variable cost equations are set up and solved. Economic lot size models (ELS) for continuous delivery of product are contrasted to EOQ models. The ELS are applied to intermittent flow shops (IFS). There is discussion of lead-time variability.

Next, the perpetual inventory system with its reorder point and buffer-stock calculations is detailed. This includes operation of the two-bin perpetual inventory control system. The periodic inventory system is explained including why this system is being outmoded by computer capabilities—yet it still is needed in certain situations. The chapter ends with an explanation of how quantity discount models work.

▶ Key Terms

buffer stock *593*	dynamic inventory models *591*	economic lot sizes (ELS) *604*
economic order quantity (EOQ) *604*	inventory *590*	lead time (LT) *591*
make or buy decisions *592*	order point policies (OPP) *603*	periodic inventory system *603*
perpetual inventory system *603*	sensitivity analysis *594*	static inventory models *591*
vendor releasing *602*		

▶ Review Questions

1. It has been stated that as a rule of thumb, "the best inventory is no inventory." Discuss this heuristic.

2. For a toothpaste manufacturer, how is the decision made concerning how many caps should be ordered? Could it be a different number than the number of tubes that are ordered at one time?

3. How often should an order size be updated?

4. What method should be used to determine the order quantity for a raw material that is used continuously within a flow shop?

5. What method should be used to determine the order quantity for a raw material that is used continuously within an intermittent flow shop?

6. Who are the people that are responsible for placing orders?

7. Electronic data interchange (EDI) makes communication between connected supply chain participants immediate. How does the use of EDI reduce the order quantity?

8. Using the information in Problem 7 above, how does the use of EDI reduce the buffer stock?

9. A manufacturer of costume jewelry has a salesforce that uses handheld computers with modems. Salespeople take stock at each retail establishment and then call the factory with stock-level data. How does this procedure affect the manufacturing schedule of the company? Explain your answer in terms of ELS models.

10. Benetton is a well-known manufacturer and retailer of clothing all over the world. The Benetton factories are tied in with retailers so that demand information is relatively immediate and complete concerning what is selling and what is not. How does this information affect lot-size planning?

11. Salespeople use handheld telecommunications devices to communicate inventory status to the warehouse in a major toy company. Why is this system needed and what does it affect?

12. What is carrying cost composed of and what is the range of values that will be found for it?

13. What is the logic for a buyer accepting a discount?

14. What is the logic for a seller offering a discount?

15. Why is there an ordering cost and of what is it composed?

16. What is the difference between an ordering cost and a set-up cost?

17. When is it likely that an ordering cost will be larger than a set-up cost?

18. When is it likely that a set-up cost will be larger than an ordering cost?

19. Relate the number of orders placed and the order quantity.

20. Why are order point policies (OPP) called by that name?

21. Why are total variable cost equations written for OPP models instead of total cost equations?

22. When discounts are being considered, total cost equations must be used instead of total variable cost equations. Why is this so?

23. Differentiate between EOQ and ELS models.

24. When is lead-time variability a problem and what can be done about it?

25. How can lead-time variability be modeled?

26. What is a two-bin system? When is it applicable?

27. Describe a perpetual inventory system.

28. When is a perpetual inventory system preferred?

29. Describe a periodic inventory system.

30. When is a periodic inventory system preferred?

31. How can a quantity discount model be used by a buyer and supplier to negotiate a price break point schedule that benefits both of them?

32. Why is it that multiple price break points for many discount levels can be examined in the same fashion as one price break point for a single discount?

33. Is the two-bin inventory system perpetual or periodic?

34. Assume that a new IFS has been set up to make a baby stroller. The ELS analysis is completed and the run size and run time are determined. Discuss what to do with the time, t_2, when the process is not running the product.

35. Distinguish between inventory problems under certainty, risk and uncertainty.

▶ **Problem Section**

1. Water testing at the Croton reservoir requires a chemical reagent that costs $500 per gallon. Use is constant at 1/3 gallon per week. Carrying cost rate is considered to be 12 percent per year, and the cost of an order is $125.

   What is the optimal order quantity for the reagent?

2. Continuing with the information about the Croton reservoir given in Problem 1 above, the city could make this reagent at the rate of 1/8 gallon per day, at a cost of $300 per gallon. The set-up cost is $150. Use a seven-day week.

   Compare using the EOQ and the ELS systems. What course of action do you recommend?

3. Water testing at the Delaware reservoir requires a chemical reagent that costs $400 per gallon. Use is constant at 1/3 gallon per week. Carrying cost rate is considered to be 10 percent per year, and the cost of an order is $100.

   What is the optimal order quantity for the reagent?

4. Continuing with the information about the Delaware reservoir given in Problem 3 above, the town could make this reagent at the rate of 1/4 gallon per day, at a cost of $500 per gallon. The set-up cost is $125. Use a seven-day week.

   Compare using the EOQ and the ELS systems. What course of action do you recommend?

5. The Drug Store carries Deodorant R which has an expected demand of 15,000 jars per year (or 60 jars per day with 250 days per year). Lead time from the distributor is three days. It has been determined that demand in any three-day period exceeds 200 jars only once out of every 100 three-day periods. This outage level (of 1 in 100 LT periods) is considered acceptable by P/OM and their marketing colleagues. The economic order quantity has been derived as 2,820 jars. Set up the perpetual inventory system.

6. Use the information given in Problem 5 above, plus the fact that it has been determined that demand in any 50-day period exceeds 4,000 jars only once out of every 100 50-day periods. This outage level (of 1 in 100 LT periods) is considered acceptable by the P/OM and their marketing colleagues. Set up the periodic inventory system.

7. Compare the results derived in Problems 5 and 6 above. What do you recommend doing? Explain how you have taken into account all of the important differentiating characteristics of perpetual and periodic inventory systems.

8. Consider the recommendation made in Problem 5 above, taking into account the fact that The Drug Store must combine orders for Deodorant R with other items in order to have sufficient volume to qualify for the distributor's shipping without charge. With this constraint, what are your recommendations?

9. The information required to solve an EOQ inventory problem is as follows:

   c = $10 per unit, C_C = 16 percent per year, D = 5000 units per year, and C_r = $10 per order

   What is the optimal order quantity?

10. Using the information in Problem 9 above, instead of buying from a supplier the decision is to make the item in the company's factory. The new equipment is able to produce $p = 30$ units per day. Costs are changed to $c = \$6$ per unit and $C_S = \$150$ per setup.

    What is the optimal run size?

11. Using the information in Problems 9 and 10 above, which is better: make or buy?

12. Using the information in Problems 9, 10, and 11 above, what factors that are not in the equations might shift the decision?

13. The quantity discount schedule below has been offered for the situation described in Problem 9 above.

$$c = 10 \quad Q = \text{up to } 300$$
$$c' = 9 \quad Q = 300 \text{ to } 499$$
$$c'' = 8 \quad Q = 500 \text{ and up}$$

Should either of these discounts be accepted?

14. The quantity discount schedule offered in Problem 13 above prompted a competitor to offer the following discount schedule:

$$c = 10 \quad Q = \text{up to } 250$$
$$c' = 9 \quad Q = 251 \text{ to } 599$$
$$c'' = 7 \quad Q = 600 \text{ and up}$$

Should any of these discounts be accepted?

15. Compare the answers to Problems 13 and 14 above and discuss these results making appropriate recommendations.

16. Murphy's is famous for their coffee blend. The company buys and roasts the beans and then packs the coffee in foil bags. It buys the beans periodically in quantities of 120,000 pounds and assumes this to be the optimal order quantity. This year it has been paying $2.40 per pound on a fairly constant basis.

    The company ships 1,200,000 one-pound bags of its blended coffee per year to its distributors. This is equivalent to shipping 24,000 one-pound foil bags in each of 50 weeks of the year. This can be considered to be constant and continuous demand over time.

    What carrying cost in percent per year is implied (or imputed) by this policy if an order costs $100 on the average? Discuss the results. Use:

$$Q_O = \sqrt{\frac{2DC_r}{cC_c}}$$

17. Using the information in Problem 16 above, suggest a better ordering policy.

18. What production run lengths are individually optimal for each item in the table below?

i	D_i (per year)	C_{s_i}	c_i	p_i (per day)
1	200	100	6	10
2	400	50	10	12
3	600	20	15	14
4	800	80	9	20

where $C_c = 0.24$ per year and $D_i = 250d_i$ for changing daily demand into yearly demand.

19. Using the information in Problem 18 above, can all of these items be scheduled to run on the same equipment? Discuss.

20. Using the data in Problem 18, employ cut and try methods to determine a reasonable set of production run lengths so that all items can be scheduled to run consecutively on the same equipment. Compare these results with the individual optimal values obtained in Problem 18.

21. The manager of the greeting card production department has been buying two rolls of acetate at a time. They cost $200 each. Card production requires 10 rolls per year. Ordering cost is estimated to be $4 per order. What carrying cost rate is imputed? Is it reasonable?

22. Using the information in Problem 21 above, if the cost of rolls of acetate increases to $250 each, what happens to the imputed carrying cost rate? Is this reasonable?

23. Refer to Figure 14.1 in the text and the example upon which it was based. Now, to illustrate the flatness of the total cost curve when Q_O is large, determine the TVC for $Q = 20,000$. Recall that the optimal value presented in the text and related to Figure 14.1 was $6.57 associated with $Q_O = 21,909$. Calculated as follows:

$$TVC_{21,909} = 0.00015(21,909) + 72,000/(21,909)$$
$$= \$3.286 + \$3.286 = \$3.286(2) = \$6.57$$

Determine the percent increase in TVC and decrease in Q.
Discuss your findings.

24. Using the information in Problem 23 above, calculate TVC for $Q = 25,000$. What percent increase has occurred in TVC? Calculate the percent increase in Q from the optimal value of $Q_O = 21,909$. Does this back up the statement that: "The total variable cost curve flattens out as Q_O gets larger. There is less difference between the minimum cost at Q_O and the costs associated with small changes (ϵ) in Q_O such as $Q_O \pm \epsilon$."

25. Masters of Travel, Inc. (MTI) give customers who have purchased travel arrangements a soft canvas shoulder-strap bag. Bags Limited, the company that furnishes these bags to MTI has suggested that if the order quantity is doubled, it would discount the cost per bag by 20 percent. At present, MTI buys 600 bags imprinted with MTI at a time for $1.50 per bag. This represents a 2 month supply.

Should the discount be taken?

26. Using the information in Problem 25 above, a competitor to Bags Limited has made the following offer:

Number of Shoulder Bags	Price
600	at $1.45 each
800	at $1.35 each
1,000	at $1.25 each

What should MTI do?

▶ **Readings and References**

Lisa M. Ellram and Laura M. Birou, *Purchasing for Bottom Line Impact*, Vol. 4 of the National Association of Purchasing Management (NAPM), Irwin, Homewood, IL, 1995.

D. W. Fogarty, and T.R. Hoffmann, *Production and Inventory Management*, Cincinnati, OH, South-Western Publishing Co., 1983.

Thomas K. Hickman and William M. Hickman, *Global Purchasing: How to Buy Goods and Services in Foreign Markets*, Business One Irwin/APICS Series in Production Management, Homewood, IL, 1992.

Kenneth H. Killen and John W. Kamauff, *Managing Purchasing*, Vol. 2 of the National Association of Purchasing Management (NAPM), Irwin, Homewood, IL, 1994.

Michiel R. Leenders and Anna E. Flynn, *Value-Driven Purchasing*, Vol. 1 of the National Association of Purchasing Management (NAPM), Irwin, Homewood, IL, 1994. E. McTighe, Ed., Nikkon Kogyo Shimbun, Ltd., *Factory Management Notebook Series: Mixed Model Production*, Cambridge, MA, Productivity Press, 1991.

R. Peterson, and E.A. Silver, *Decision Systems for Inventory Management and Production Planning*, 2nd Ed., NY, Wiley, 1984.

Donald R. Plane, *Management Science: A Spreadsheet Approach*, Danvers, MA, boyd & fraser publishing company, 1994.

Alan R. Raedels, *Value-Focused Supply Management*, Vol. 3 of the National Association of Purchasing Management (NAPM), Irwin, Homewood, IL, 1994.

R. J. Tersine, *Principles of Inventory and Materials Management*, 3rd Ed., NY, Elsevier-North Holland Publishing, 1987.

T. E. Vollmann, W.L. Berry, and D. C. Whybark, *Manufacturing Planning and Control Systems*, 3rd Ed., Homewood, IL, Irwin, 1992.

Notes...

1. J. F. Cox, J. H. Blackstone, Jr., and M. S. Spencer, Eds., APICS DICTIONARY, 7th Ed., APICS Educational and Research Foundation Publishers, 1992.

2. "Sea-Land Beats the Odds in Kobe, Company is Ready to Resume Shipping," *New York Times*, February 4, 1995.

Chapter 15

Material Requirements Planning (MRP)

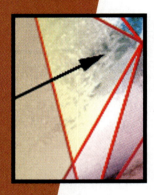

After reading this chapter you should be able to...

1. Explain the difference between dependent and independent demand systems.

2. Describe when MRP is preferred to the OPP approach for materials planning.

3. Explain the importance of the master production schedule (MPS).

4. Discuss the role of the bill of material and describe different forms that are used for the BOM.

5. Detail the data inputs and information outputs of MRP.

6. Talk about lot sizing methods used for ordering with MRP.

7. Describe capacity requirements planning (CRP).

8. Explain distribution requirements planning and relate it to replenishing inventory at the branch warehouses.

9. Explain distribution resource planning (DRP) and relate it to planning for the key resources of the supply chain.

10. Discuss how information systems using coding connect parts from more than one product into a cohesive planning whole.

11. Distinguish between closed-loop MRP and MRP II.

Chapter Outline

15.0   THE BACKGROUND FOR MRP

15.1   DEPENDENT DEMAND SYSTEMS CHARACTERIZE MRP
        Forecasting Considerations

15.2   THE MASTER PRODUCTION SCHEDULE (MPS)
        Total Manufacturing Planning and Control System (MPCS)
        MPS: Inputs and Outputs
        Changes in Order Promising

15.3   THE BILL OF MATERIAL (BOM)
        Explosion of Parts

15.4   OPERATION OF THE MRP
        Coding and Low-Level Coding

15.5   MRP BASIC CALCULATIONS AND CONCEPTS

15.6   MRP IN ACTION
        Scenario 1
        Scenario 2
        Scenario 3
        Scenario 4

15.7   LOT SIZING

15.8   UPDATING

15.9   CAPACITY REQUIREMENTS PLANNING (CRP)
        Rough-Cut Capacity Planning (RCCP)

15.10   DISTRIBUTION RESOURCE PLANNING (DRP)

15.11   WEAKNESSES OF MRP

15.12   CLOSED-LOOP MRP

15.13   STRENGTHS OF MRP II (MANUFACTURING RESOURCE PLANNING)

Summary
Key Terms
Review Questions
Problem Section
Readings and References
Supplement 15: Part-Period Lot-Sizing Policy

The Systems Viewpoint

Material requirements planning (MRP) is inherently a systems approach to the management of the entire supply chain. The method addresses a broad range of P/OM situations for inventory planning using information systems to connect parts from many products in an overall plan of action. MRP employs detail about customers, suppliers and producer's capabilities to provide guidance for actions of when to order, and how much to order, among other things. Time-dependent factors are an integral part of the analysis and solution development.

Systems that require MRP include interdependencies between final assembly products (parents) and parts (components which include subassemblies). In other words, the same part occurs in various products and so when planning production runs, over time the different demand patterns for the parts (or components) in each product must be linked to determine the overall demand for the part. Good systems analysis is essential for achieving the goals of having what is needed, when needed, and where needed.

The ultimate systems plan is manufacturing resource planning (MRP II) which links OM production planning with marketing goals, financial policies, accounting procedures, R & D plans, human resources management constraints, and so on. Put in an alternative context, MRP II is responsive to the strategic business plan of the company. Thus, new product introductions are most likely to be successful if the total organization is communicating and coordinated. The organization as a whole can mobilize to handle serious problems as they arise, such as large numbers of defectives and major equipment failures.

The reactions to predicaments can include adding additional physical capacity as well as coproducing which means subcontracting work to firms having the requisite capabilities. The production budget will have to be changed. From the systems point of view, MRP, in its various forms, helps to achieve varying degrees of the functionally-coordinated organization.

MRP II operating at the highest level of coordination is difficult to achieve. It is a goal worth striving for because it provides one of the most powerful platforms for finding best solutions to the dynamics of change.

15.0

THE BACKGROUND FOR MRP

There always has been a need for a method to deal with the great complexity of inventory control for job shops that have many parts that are interdependent in many different ways. The method that was needed had to be able to organize all of the data on stockkeeping units (SKUs). It had to keep track of stock levels and order timing, while coordinating process scheduling with delivery dates for customers. The challenge of organizing all of this information continued to grow with the increasing size of markets and production capabilities.

The first book on MRP by Joseph Orlicky was published in 1975. In the Preface to this book, Orlicky states that the development and installation of computer-based MRP systems began in 1960—over 36 years ago.[1] Also, he points out that, "the number of MRP systems used in American industry gradually grew to about 150 in 1971, when the growth curve began a steep rise as a result of the 'MRP Crusade,' a national program of publicity and education sponsored by the American Production and Inventory Control Society (APICS)." Today, there are tens of thousands of companies using various kinds of MRPs, and they are located all over the world.

MRP could not work in a practical way without the computer capabilities that were emerging by the 1960s. The need to be able to manage the massive amount of information required for ordering and scheduling in job shops grew. The ability to meet that need grew even faster. MRP has drawn broad national attention and support as computer power became increasingly available at significantly lower prices. It continues to be true that the pros and cons of specific MRP software must be discussed separately from the principles that drive the methodology.

15.1

DEPENDENT DEMAND SYSTEMS CHARACTERIZE MRP

MRP applies to a dissimilar class of problems than order point policy (OPP) models which are used under distinctly different circumstances. OPP models (which are represented by EOQ and ELS) are powerful as long as the assumptions for their applicability hold true. MRP is used when the OPP assumptions below do not hold.

1. Demand is *independent*. This is true of end products like cars and VCRs whose demand is driven by the marketplace. In contrast, the demand for the raw materials, subassemblies and components used in the automobiles and VCRs are *dependent* on the number and types of cars and VCRs that are made. **End products**, which also are called parent products, are considered to have independent demand systems.

2. Demand for the item has to be regular and relatively continuous over time for order point models and policies to be applicable. An end item can be ordered sporadically which makes it an MRP application.

End products

Also called parent products; considered to have independent demand systems where demand is driven by the marketplace.

MRP should be used for any items that do not satisfy both of the two OPP assumptions above. Alternatively, MRP is used when:

1. Demand is dependent because it will organize and control bringing together the kit of interdependent parts and subassemblies to make up the end product. The more complex the dependencies, the more critical is the use of MRP.
2. Demand for the item is sporadic and lumpy. This can occur for end products which have demand patterns that are not smooth and regular. However, quite often sporadic and lumpy conditions occur for parts which are dependent on end products which have regular and smooth demands. The various reasons why this is true include the fact that the part appears in several end products.

Sometimes the part is used by different end products that are produced simultaneously on parallel lines. In other circumstances, the common part is used in different end products that are scheduled to be made at widely separated times. Part demands can change according to which special options of the parent product are being produced.

To illustrate this last point, say that VCRs with different options (one model has two heads—H2, and the other model has four heads—H4) are run on the same line. The line switches from one model to the other on alternating weeks. The demand is fairly constant for VCRs so that 200 are made everyday. Note how the demand for the two-head components and the four-head components changes on a weekly basis. Every other week there is no need to have stock on-hand for both the H2 and the H4. Table 15.1 illustrates the sporadic character of the dependent demands for H2 and H4.

TABLE 15.1

DEMAND FOR END-PRODUCT VCRS IS SMOOTH WHILE DEMAND FOR COMPONENTS H2 AND H4 PULSES

Week	1					2					3→
Day	1	2	3	4	5	1	2	3	4	5	1
VCRs	200	200	200	200	200	200	200	200	200	200	200...
H2	200	200	200	200	200						200...
H4						200	200	200	200	200	
Week 1	Need 1000 H2s					Need 0 H2s					
Week 2	Need 0 H4s					Need 1000 H4s					
Week 3	Need 1000 H2s					Need 1000 H2s					
Week 4	(the pattern continues)					(the pattern continues)					

Although VCR production is smooth and continuous, the demands for components H2 and H4 are discontinuous—fluctuating between needing 1000 units one week and none the next week.

Dependencies are seldom as regular as in Table 15.1 because the components are used in different end products each of which has its own demand pattern. Most are only relatively continuous and not as stable as the VCR pattern shown here. The comparison with regard to stock on-hand for independent and dependent demand systems is shown in Figure 15.1.

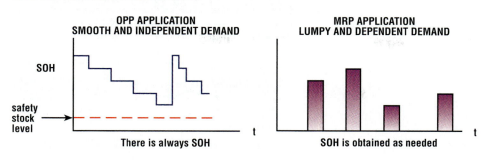

FIGURE 15.1

STOCK ON-HAND PATTERNS

When reorder point models are used to control stock on-hand for linked parts, parent demand systems often produce inventory swings that lead to overstocks and out-of-stocks. This occurs because the information about the products and the parts are not properly merged. The result is that the inventory levels are mismanaged. Lump withdrawals of parts to supply their parent products create demand surges which overpower the steady demand pattern required for reorder point models.

MRP reflects the systems viewpoint by connecting the parent products and the dependent parts together. This linkage is made by managing the combined information concerning dependent demands, SOH, deliveries and lead times.

Forecasting Considerations

Reorder point models require that demand distributions exist for their forecasting requirements. When such reliable, long-term forecasts are not feasible, MRP is indicated. This accounts for a broad range of P/OM situations that require MRP because they have irregular demands—and do not have stable distributions.

Even though MRP demands are sporadic and lumpy, they can be predicted as discrete order events when there is advance knowledge about probable order placements. Predictions about the demand for parts is often based on forecasts for the end product. This is the result of dependent demand. The distinction that is being made is between predictions that are generated by customer notification of intentions to buy (use MRP), as compared to forecasts based on demand distributions that exist in a steady state over time (use OPP).

Actual customer orders (CO) on-hand provide relatively solid information about future demands. However, cancellations of orders are not uncommon so risk factors still play a part in the determination of future events. Often, there are many customers with a variety of types of products (associated with the job shop). This creates numerous interdependencies and sources of change long after order entry for the customer's order. Change also can take place after orders are placed with suppliers for the parts needed to make the end products.

Ollie Wight wrote in 1970, "…the number of pages written on independent demand-type inventory systems outnumbers the pages written on material requirements planning by well over 100 to 1. The number of items in inventory that can best be controlled by material requirements planning outnumbers those that can be controlled effectively by order point in about the same ratio."[2] The ratio of papers written for OPP as compared to MRP is now closer to 50/50, but the number of items in inventory that fit MRP instead of OPP is not much changed.

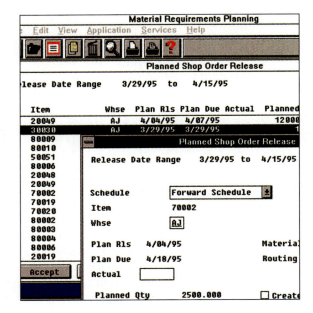

Predictions about the demand for parts is often based on forecasts for the end product.
Source: SSA (System Software Associates, Inc., Chicago, IL)

15.2
THE MASTER PRODUCTION SCHEDULE (MPS)

Master production schedule (MPS)

One of three inputs to the MRP information system— other two are bill of material (BOM) and inventory stock-level reports; time-phased plan which indicates product specifications and production quantity.

The **master production schedule (MPS)** is one of three inputs to the MRP information system. The other two are the bill of material (BOM) which describes what constitutes the end product and inventory stock-level reports.

Aggregate production planning determines the resource capability and capacity to produce generic product in standard hours or in some other generalized dimension. For example, the aggregate plan might specify the number of gallons of paint to be made in January. The master production schedule converts this number into a time-phased plan which indicates exactly when each type and color of paint and size of can should be made, and how much of it should be made. Table 15.2 shows the planning only for January. Similar plans will be made by the MPS for February, March and all succeeding months.

The basis for the assignments of MPS—quantity to be made of specific models at chosen times—are customer orders, and forecasts of orders to come that are considered highly probable. The MPS makes the assignments in response to sales department commitments which are called **order promising**.

Order promising

The process of making a delivery commitment; involves a check of uncommitted material and availability of capacity.

Order Promising is defined as: "The process of making a delivery commitment, i.e., answering the question, 'When can you ship?' For make-to-order products, this usually involves a check of uncommitted material and availability of capacity. Syn. customer order promising, order dating."[3]

Figure 15.2 relates order promising to the timeline that characterizes all production scheduling. It shows that the drivers (or inputs) of the MRP system are the customers' orders for parent products which have been scheduled by the master production schedule (MPS). These are drawn above the parent product process timeline.

Figure 15.2 also shows that the MRP system determines what parts are used by the parent product (*Parts 1* through *m*). It indicates when they are required by the process. It physically represents the lead times (LT) between placing orders and receiving them. MRP assignments are shown below the parent product process timeline.

TABLE 15.2

CONVERTING THE AGGREGATE PLANNING SCHEDULE INTO THE MASTER PRODUCTION SCHEDULE—JANUARY 130,000 GALLONS

	January Master Production Schedule Paint Code (Color, Size, Type) (in 000)							
Week	WGF	WHGF	WGO	RGGL	RHGF	WPGL	YGF	SUM
1	10	10	10					30
2	10		10	15	5			40
3			20			10		30
4			10			10	10	30
SUM	20	10	50	15	5	20	10	130

Legend for paint codes:

First letter is color: W = white, R = red, Y = yellow

Second letter(s) is (are) size: G = gallon, HG = half gallon, P = pint

Third letter(s) is (are) type: F = flat, O = outdoor, GL = glossy

FIGURE 15.2

THE TIME-PHASED MPS FUNCTION

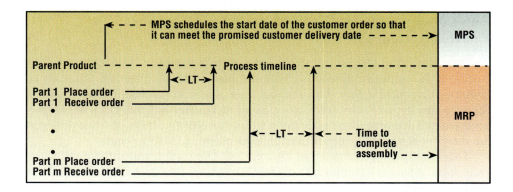

(Note: Place order   = planned-order release;

      Receive order = planned-order receipt)

Examples of parent products include VCRs, computers, paint, farm equipment and industrial machines.

Total or cumulative lead time (LT) is defined by APICS as follows: "In a logistics context, the time between recognition of the need for an order and the receipt of goods. Individual components of lead time can include order preparation time, queue time, move or transportation time, and receiving and inspection time. Syn. total lead time."[4]

Total or cumulative lead time (LT)

Time between recognition of the need for an order and the receipt of goods. Total time to do what needs to be done to get the order delivered.

The cumulative lead time for a parent product is found by looking at the longest lead-time path to obtain all of the parts that go into it as the end item. To assemble A in Figure 15.3 requires ordering B one week before the assembly operation is to begin.

CUMULATIVE LEAD-TIME TREE

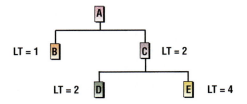

Part C must be ordered two weeks before assembly is to begin on part A. Parts D and E are necessary components for part C. Part D has $LT = 2$ and part E has $LT = 4$. The longest path results from $A + C + E = 6$, and it dominates cumulative lead time. However, if part E is a standard item which means it is stocked, then E should not be considered when calculating total lead time. If D is a special item which means there is no stock on-hand for D, then $A + C + D = 4$ dominates cumulative lead time. If both D and E are standard items, then $A + C = 2$ has a longer path than $A + B = 1$. If C also is a standard item, the longest lead-time path is $A + B = 1$.

Total Manufacturing Planning and Control System (MPCS)

The manufacturing planning procedure started with resource evaluation and forecasting at the aggregate scheduling stages of manufacturing planning. Based on an authorized production plan, it moved to master production scheduling (MPS) which reacts to specific customer order expectations and orders on hand. Once the MPS is accepted and authorized, the planning process proceeds to details of scheduling that are the province of the material requirements planning (MRP) system.

These steps are flowcharted in Figure 15.4. The system called the total manufacturing planning and control system (MPCS) captures what takes place in job shops all over the world.

The MPS stage needs to be documented. In Figure 15.5, an MPS chart is shown which is created for every parent product. It represents the middle step—MPS—in the Total MPCS shown in Figure 15.4.

This MPS chart goes nine weeks into the future with parent-product requirements scheduled for Weeks 2, 5, and 9. The MPS must project Gross Requirements far enough ahead to permit actions to be taken that are needed for the success of parent-product production.

FIGURE 15.4

THE TOTAL MANUFACTURING PLANNING
AND CONTROL SYSTEM (MPCS)

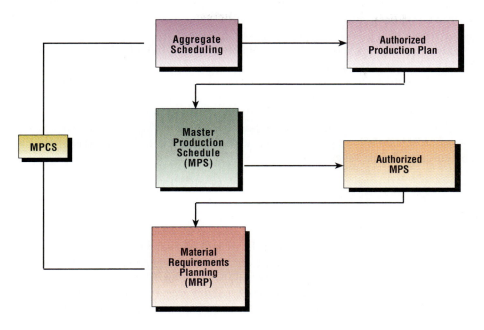

FIGURE 15.5

MPS CHART FOR SPECIFIC PARENT PRODUCT

Weeks Ahead	1	2	3	4	5	6	7	8	9
Gross (amount) Requirements		20			45				60

Weeks are the intervals, or time buckets, most often used. However, other periods can be required. Some organizations use short- and long-term master production schedules where the short-term schedules are in days or weeks and the long-term schedules are in months or quarters. The time buckets used reflect the rapidity of changes to schedule, lead time lengths, and how far ahead it is possible to forecast.

The MPS chart in Figure 15.5 has 20 units of parent product to be made in the second week ahead from *now*. If it takes 4 weeks (total lead time, $LT = 4$) to order parts that are needed for the parent product, then the order should have been placed already. Note that this chart is updated each week. The gross amount required moves from the *n*th week ahead to the *n* − 1 week ahead of *now*. Thus, the second week from now at one time appeared as Week 5. When it was in the Week 5 position, the parts were ordered during Week 1. Therefore, it is expected that the necessary order for parts already has been placed for the *present Week 2 Ahead* Gross Requirements of 20.

There are 45 units required for the *present Week 5 Ahead*. The planned order should be released now, in Week 1. There are 60 units required for the *present Week 9 Ahead*. Perhaps there should be a planned-order release in *Week 5 Ahead* for these 60 units. However, the ordering policy, which will be discussed shortly, may determine that the order for *Week 5 Ahead* (which is released in Week 1) should cover some (or all) of the *Week 9 Ahead* requirements. The point to note is that: The master production schedule is required to have a planning horizon as long as the longest cumulative lead-time function.

MPS: Inputs and Outputs

Master production schedulers set down the make-build schedule for specific parent products. Their scheduling function moves beyond the generic units of time and money that were the limitations of aggregate scheduling. They deal with real-time and specific SKUs. The master production schedules are partially driven by the sales forecast which is one of the usual inputs to the production plan. It also is driven by actual orders on-hand. It takes into account priorities for orders based on urgency and customer importance. Master production schedulers always seek to make the best use of actual capacity and resources. These considerations are brought together in Figure 15.6.

FIGURE 15.6

INPUTS AND OUTPUTS OF MPS

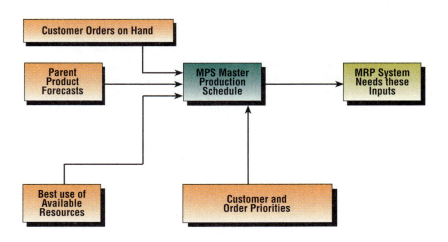

TECHNOLOGY

FastMAN Software Systems, Inc.

Managers of materials, production, and procurement use simulation and "what-if" tools in analyzing the impacts of potential opportunities or problems that may arise in their manufacturing environments. FastMAN Software Systems Inc. has developed a simulation and "what if" analysis software package that works in conjunction with a manufacturer's existing MRP system.

Most MRP systems in use today are mainframe-based technology, which often have the capacity to perform simulation and "what-if" analyses, but execution is often cumbersome and time-consuming. The PC-based FastMAN software allows managers to perform analyses "offline"—permitting experimentation without tying up a company's scarce computer resources. It reduces the time it takes for a typical MRP run from hours to less than five minutes, according to company President Martin Horne.

FastMAN software was developed to meet the needs of manufacturers who deal with a great many parts and

frequent changes in supply and/or demand. It can provide answers to critical business questions like: can we deliver a potential new order by the date the customer wants it? Can we get the materials? Do we have the production capacity? If not, when could we deliver? When is the best time to implement an engineering change to minimize excess or obsolete inventory? What customer orders will be impacted by a vendor delivery delay? What would be the overall impact on product cost if we changed vendors?

FastMAN Software Systems Inc. was founded in 1994 by Horne, who conceived the core of the software

while responsible for materials management and MIS at Digital Equipment Canada and DY-4 Systems. Several leading manufacturers including Hewlett-Packard, Rockwell International, and Measurex participated in its development over a three-year period. FastMAN software also is being used by Motorola, Panasonic, Lexmark, and Microsoft Corp. which used the software to consolidate materials worldwide.

Sources: FastMAN Software documents provided by Brett McAteer, director of marketing, FastMAN Software Systems Inc., Ottawa, Ontario, Canada, and "AGRA sets new standard in 'what-if?' MRP technology; engineering leader offers affordable, PC-based manufacturing planning software; announces multi-part agreement with Microsoft Corp.," Business Wire, November 16, 1993.

Changes in Order Promising

Order promising by sales and marketing and order fulfillment goals of production are matched and coordinated by the MPS. If marketing promises dates that production cannot deliver, changes are required either in the due dates promised to customers or in the production schedules for the plant.

Detail report by date provides report sharing requirements and scheduled receipts by date for each parent item. This shows all "in" and "out" activities.
Source: Macola Software (Marion, OH)

Make-to-stock (MTS) environment

Emphasis is on having sufficient stock on-hand to provide excellent service with minimum delivery times.

In a **make-to-stock (MTS) environment**, there is some latitude in what is scheduled. Also, there may be sufficient inventory in stock to help fulfill order promising. The emphasis is on having sufficient stock on-hand to provide excellent service with minimum delivery times. The MRP emphasis is on finished product which is usually smaller in number than the number of components which make up the finished product. This is typical of automobiles, home owners' tools and VCRs.

Make-to-order (MTO) inventory environment

No finished goods inventory to call upon; end items are produced to customers' orders.

In a **make-to-order (MTO) inventory environment**, there is no finished goods inventory to call upon. End items are produced to customers' orders. Consequently, greater dependency exists between sales which promises delivery and OM which has to produce what is to be delivered. Customers expect finished goods to be delivered on time. This characterizes small- and medium-sized job shops—often called machine shops—which work from blueprints and do not have product lines. Carpenters and plumbers live in an MTO environment. On a larger scale, auto replacement part manufacturers including windshield glass have a significant percent of their business as MTO.

Assemble-to-order (ATO) environment

End items are built from subassemblies made available to meet demand within promised due dates.

The **assemble-to-order (ATO) environment** is one where end items are built from subassemblies. Master production schedules ensure that these subassemblies are available to meet demand within promised due dates. Often the number of components is smaller than the number of finished products and master production schedules operate at the components or subassembly level. This means that parts and subassemblies are scheduled instead of parent products. In this case, the MPS is only secondarily concerned with parent end products and primarily with making the parts to complete the subassemblies. The master production schedule is focused on the subassemblies much like the modular BOM concentrates on options (see Section 15.3 below). The heavy equipment industry uses assemble-to-order.

15.3

THE BILL OF MATERIAL (BOM)

The bill of material (BOM) is required for the use of MRP.

The APICS definition is quite complete. A **bill of material (BOM)** is: "a listing of all of the subassemblies, intermediates, parts, and raw materials that go into a parent assembly showing the quantity of each required to make an assembly. It is used in conjunction with the master production schedule to determine the items for which purchase requisitions and production orders must be released. There is a variety of display formats for bills of materials, including the single-level bill of material, indented bill of material, modular (planning) bill of material, transient bill of material, matrix bill of material, and costed bill of material. It may also be called the "formula," "recipe," "ingredients list," in certain industries." [5]

BOMs include every material that goes into or onto a product. This includes screws, nails, rivets, glue, paint, and packaging. Figure 15.7 displays several of the commonly used BOMs. Each of the different kinds of bills of materials mentioned in the APICS definition captures some important aspects of the MRP process. BOMs are the starting point for actual use of MRP and the varieties are consequently instructive.

F I G U R E   1 5 . 7

SINGLE-LEVEL, MULTILEVEL, AND INDENTED BOMS

Single-Level BOM: Parent Product with Parts One Level Down. The number of parts needed for the level above are shown in parentheses.

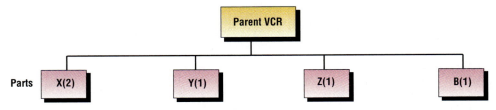

Multilevel BOM: Parent Product with Parts at All Levels Down. Often called a product-structure tree, where the number of parts needed for the level above are shown in parentheses.

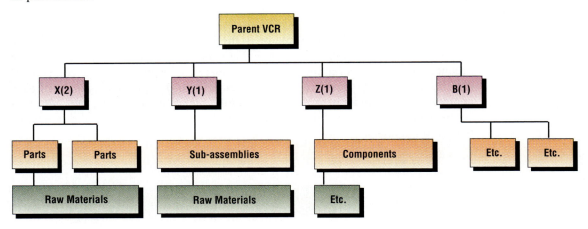

or, more specifically:

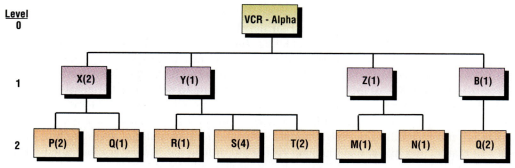

The Multilevel Indented BOM: Parent Product and Parts Are Carried at Levels (0, 1, 2, and 3) for Two VCR Models (the Alpha and the Beta) as shown below.

The number of parts needed for the level above are indicated in parentheses.

VCR Alpha		Level 0	VCR Beta		Level 0
X(2)		1	G(2)		1
	P(2)	2		P(2)	2
	Q(1)	2		B(1)	2
Y(1)		1	Y(1)		1
	R(1)	2		R(1)	2
	S(4)	2		S(4)	2
	T(2)	2		T(2)	2
Z(1)		1	V(1)		1
	M(1)	2		C(1)	2
	N(1)	2		D(1)	2
B(1)		1	B(1)		1 (Code 2)
	Q(2)	2		E(2)	2 (Code 3)

(See Figure 15.11 for the VCR-Beta product structure and note that because B(1) is low-level coded at level 2, E(2) is at the third level. Low-level coding will be explained shortly.)

The multilevel BOM (as a tree or in the indented form) is essential for MRP purposes because it captures everything that must be ordered to complete the parent product. The APICS definition is precise, namely: "a display of all the components directly or indirectly used in a parent, together with the quantity required of each component. If a component is a subassembly, blend, intermediate, etc., all of its components will also be exhibited and all of their components, down to purchased parts and materials."[6]

Modular BOMs are organized so that product options can serve in place of single-product parents. This BOM is associated with products that can be assembled in many different ways. Thus, assemble-to-order is common with computers that can be configured in many ways using different boards and drives. It also is characteristic of automobile assembly with different packages of features. The options are fewer than the end products which can combine options in many different ways. Thus, by using options, fewer bills of materials are required to control the ordering and inventories.

The matrix BOM is a special chart constructed by analyzing the BOMs of a family of similar products. It arranges parts in columns and parent products in rows. This is useful where the objective is to make the greatest number of products with the minimum number of parts. This matrix characterizes what is called the modular product design objective.

Cost BOMs show complete parent-part hierarchical product structures with the latest costs per part appended. The costs must be regularly updated to prevent incorrect and misleading information. The cost BOMs are useful for purchasing decisions when alternative suppliers are being considered. They also are an important means of determining the costs of parts including raw materials, subassemblies, and components.

Transient or phantom BOMs are used for subassemblies that are not kept in stock. It will be noted here that phantoms are assigned zero lead times and lot-for-lot order quantities (concepts that will be explained below). This permits the MRP system to bypass the subassemblies and go directly to their components.

Explosion of Parts

The multilevel product structure was called an explosion by Orlicky who wrote, "In executing the explosion, the task is to identify the components of a given parent item and to ascertain the location (address) of their inventory records in computer storage so that they may be retrieved and processed."[7]

15.4

OPERATION OF THE MRP

The **inventory record files** are the third major input to the MRP computer system. This is shown in Figure 15.8 along with the outputs from the MRP system that are about to be discussed.

The stock on-hand levels need to be up-to-date. So should all ordering information including the suppliers' lead times. The role of the MPS and the BOM already have been discussed. It is to be noted that the preparation of the MPS involves knowledge of the BOMs as well as the inventory and supplier information.

Inventory record files

Third major input to the MRP computer system; includes up-to-date stock on-hand levels and ordering information including suppliers' lead times.

FIGURE 15.8

THE INPUTS AND OUTPUTS OF THE MRP SYSTEM

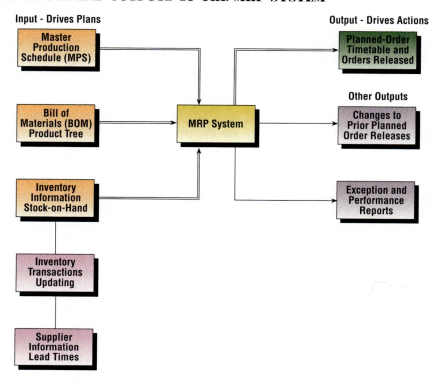

Legend:
Double lines indicate major inputs and outputs of the MRP system.

Coding and Low-Level Coding

When various parent items have had their parts explosion accomplished, coding provides a means of identifying the fact that some parts belong to more than one parent. In Figure 15.7 above, note that the indented BOMs of the Alpha and Beta VCR Models share *Y(1)* subassemblies at level one. All components of *Y(1)* are similarly shared at the next level down. These are *R(1)*, *S(4)*, and *T(2)*. Further, *P(2)* is used by both models at the second level.

When the master production schedules call for various amounts of Alpha and Beta VCRs, the MRP must combine the demands of commonly shared parts. For the specific example that is being discussed (Figure 15.7), the MRP program must be able to recognize that one Alpha VCR and one Beta VCR call for 2 *Y(1)s*, 2 *R(1)s*, 8 *S(4)s*, 4 *T(2)s*, and 4 *P(2)s*. The increased demand for parts shared in common will affect the order quantities in different ways, depending upon the method used to determine order lot sizes.

Coding techniques capture other relevant information about parts, such as the supplier(s), storage locations and substitutabilities. Coding of part names is the foundation for an effective information system to link the components and the parents together. It also is the best means of identifying families of parts which can be made with minimal set-up changes. Such part-family-oriented processes are the subject that is identified by the name "group technology."

Part numbers should not be chosen randomly. There is so much useful information that can be instantly gleaned from carefully coded part numbers. It is best to code part numbers so that they reflect the level of the part, and the product parents of the part. Additionally, as suggested above, codes show who are present suppliers of the part, who might be future suppliers of the part, alternative locations for storage, and the type of materials of which the part is made.

The method called **low-level coding** is commonly used to organize order quantity calculations for parts with combined demands. Top-level parents, such as the VCR models, are called the level zero. Parts and components that are one level down are coded level one. Those two levels down are coded level two, etc. This coding scheme has the ability to show how distant parts are (in their linkage) from their parent products.

"Once low-level codes are established, MRP record processing proceeds from one level code to the next, starting at level code 0. This ensures all gross requirements have been passed down to a part before its MRP record is processed. The result is planning of component parts coordinated with the needs of all higher-level part numbers. Within a level, the MRP record processing is typically done in part number sequence." [8]

Low-level coding uses the principle that the lowest position of a part should determine where that part is located for computer scanning. Part *B* of VCR Beta appears as *B(1)* at both the 1st and 2nd levels of the multilevel indented BOM in Figure 15.7. The MRP software counts down from the level zero to determine order quantities and when to release the orders for parts. Thus, for the Beta VCR, it will carry *B(1)* at level one down to level two where it will combine it with the level two *B(1)*. This is shown in Figure 15.11 below.

Figure 15.9 shows the product structure for a hall coat rack tree where a *G* part appears at both levels one and two. Figure 15.10 redraws to accommodate the low-level coding requirement. The graphics aside, low-level coding is simply an organizing rule for computing materials requirements.

Low-level coding

Organizing rule for computing material requirements; lowest part position determines where the part is located for computer scanning.

FIGURE 15.9

PRODUCT-STRUCTURE TREE FOR A HALL COAT RACK WITH UNFULFILLED LOW-LEVEL CODING REQUIREMENT

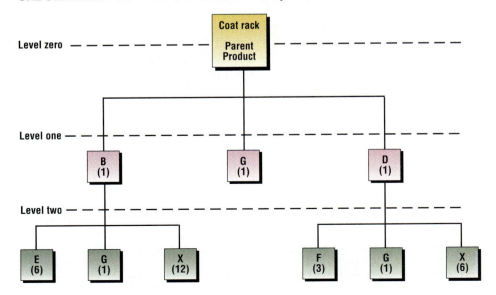

PRODUCT-STRUCTURE TREE FOR A HALL COAT RACK WITH LOW-LEVEL CODING REQUIREMENT FULFILLED

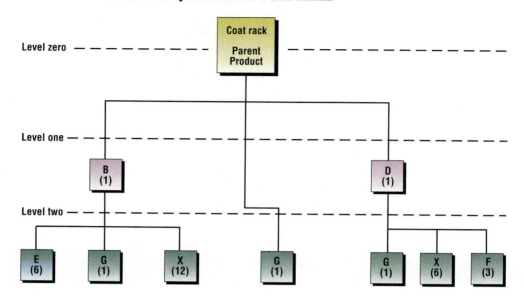

Exception codes are used to identify important items. This is equivalent to adding on to the part number an exception code indicator to reflect the fact that checks should be made for the accuracy of the data and the status of the part. Ten percent of all items might receive such special coding which depends on the importance of the customer, the difficulty of the job, or the fact that MRP cannot schedule completion of a level zero product because something has gone awry.

15.5

MRP BASIC CALCULATIONS AND CONCEPTS

Gross requirements (GR_t) for end items for week t are stated by the MPS. Inventory information is on file for the prior week's stock on-hand (SOH_{t-1}). There may be some open orders which will be received at the beginning of week t and these are called scheduled receipts, SR_t.

$$\text{Projected stock on-hand at } t = SOH_{t-1} + SR_t$$

The gross requirements less the projected stock on-hand at time t are called the net requirements, NR_t, thus:

$$NR_t = GR_t - (SOH_{t-1} + SR_t)$$

The process of determining net requirements is called netting. Based on net requirements and order policies, planned-order receipt times and planned-order release times can be determined, as explained below.

If projected stock on-hand is larger than the gross requirements:

Valendrawers, Inc.

Drawer maker ships all orders within 48 hours

Valendrawers Inc. of Lexington, North Carolina, manufactures make-to-order drawers for the furniture and kitchen cabinet industries. Valendrawers, which began U.S. operations in 1983, imports generic unfinished drawer pieces from its Bologna, Italy, factory. For each U.S. order, the Lexington plant cuts the drawer pieces to the appropriate length and custom mills the joinery pieces. Valendrawers's U.S. president, Donn K. Wilbur, said the system is unique because it mills the pins right onto the components in one step using high speed boring technology. This technology replaces the older process of boring a hole into the drawer piece, then inserting glue and a pin into the hole.

According to Wilbur, outsourcing drawer subassemblies makes sense for furniture manufacturers because making the drawers is labor intensive, thus not profitable, for them. The company's "Quick-Ship" program assures customers that any size order will be shipped within 48 hours of receipt. This has a positive affect on the MRP efforts of Valendrawers's customers. Wilbur credits this just-in-time service as the prime reason for the company's success.

Even though Valendrawers is located in North Carolina, a center for residential furniture manufacturing, most of the company's customers are office furniture manufacturers. Residential furniture manufacturers tend to produce their own drawers and other furniture components and have not been as receptive to specialty suppliers. Outsourcing is more common among European residential furniture manufacturers.

Valendrawers's Bologna operation began almost a half century ago. Together with the U.S. plant, the company has produced more than 250 million drawers. Wilbur, a former professional basketball player in the U.S. and Italy, said the company's initial successes in the U.S. were with "companies that understand the values of a vendor network and lean manufacturing techniques."

Source: Karl Kunkel, "The Italian Connection," (Greensboro, North Carolina) *Triad Business*, October 31, 1994.

$$(SOH_{t-1} + SR_t) > GR_t$$

then, there would be stock on-hand at the end of period t. Thus:

$$SOH_t = SOH_{t-1} + SR_t - GR_t > 0$$

Net requirements would be zero in the above equation because there was more stock on-hand than gross requirements. However, if:

$$GR_t > (SOH_{t-1} + SR_t)$$

then, $NR_t > 0$, and an order must be placed early enough so that with the lead time, delivery can be made at the beginning of period t. This order is called a planned-order receipt, PR_t. In the equation below, $PR_t = 0$ when projected stock on-hand is greater than gross requirements.

$$SOH_t = SOH_{t-1} + SR_t + PR_t - GR_t$$

Lot-for-lot ordering

Planned-order receipt is equal to the net requirements.

When there is a shortfall, the size of PR_t must be decided. When **lot-for-lot ordering** is used, this means that the planned-order receipt is equal to the net requirements.

$$PR_t = NR_t \text{ with lot-for-lot ordering}$$

If other methods are used for ordering, then the order will exceed net requirements, and the planned-order receipt, PR_t will be greater than the net requirements.

To keep the charts simple in the examples that are worked out below, scheduled receipts are set to zero. Therefore, projected stock on-hand is the same as SOH_{t-1}.

15.6

MRP IN ACTION

The easiest way to understand MRP is to simulate the action. Here are some scenarios to go through that provide the best explanation of the workings of MRP.

First, create the MPS for the Alpha and Beta Models. Use rough-cut (approximate) capacity planning (RCCP) to make certain there is enough capacity to produce Table 15.3 specifications.

TABLE   1 5 . 3

MASTER VCR SCHEDULE—ALPHA AND BETA MODELS

	Week									
Net Requirements	1	2	3	4	5	6	7	8	9	10
Alpha Model	—	60	—	—	—	80	—	—	40	—
Beta Model	—	—	50	—	—	20	—	—	—	60

Second, develop the product structure for both Alpha and Beta Models, as in Figure 15.11.

FIGURE   1 5 . 1 1

PRODUCT-STRUCTURE TREES FOR ALPHA AND BETA VCRS

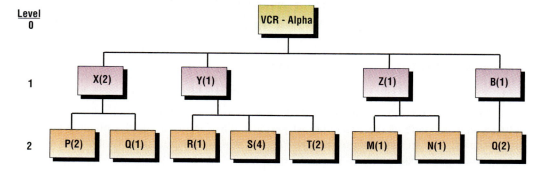

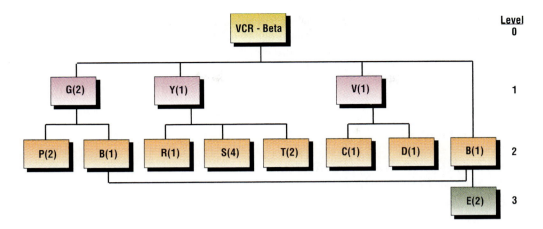

Note that because of low-level coding *B(1)* is at level 2 placing *E(2)* at level three. See also the Indented BOM in Figure 15.7.

Scenario 1

Lot-for-Lot Ordering with No Parts Sharing Parents—Level One

MRP methods will be used to determine order timing for part *Z* in the Alpha Model of the VCR. *Z(1)* occurs only in the Alpha Model, and at level one. There is one unit per VCR. Lead Time (LT) = 2 weeks which means that LTs for both *M(1)* and *N(1)* are less than two weeks. There must be time for the *Z* component to be assembled after receiving the parts *M* and *N*.

The initial *SOH* = 80. Lot sizing follows a lot-for-lot policy. This means that both the Planned-order receipt and the Planned-order release exactly match the Net requirements number of units.

TABLE 15.4

MRP PLAN FOR VCR ALPHA—Z(1)

	End Week									
	1	2	3	4	5	6	7	8	9	10
Gross requirements	—	60	—	—	—	80	—	—	40	—
Stock on-hand (SOH)	80	20	20	20	20	—	—	—	—	—
Net requirements	—	—	—	—	—	60	—	—	40	—
Planned-order receipt	—	—	—	—	—	60	—	—	40	—
Planned-order release	—	—	—	60	—	—	40	—	—	—

The conclusion is to have an order release (place the order) for 60 units of Z in Week 4 for receipt in Week 6. Also, order 40 units of Z in Week 7 for receipt in Week 9. There are two orders and once the initial stock on-hand is used up, there are no carrying costs. This is due to the fact that lot sizing is done on a lot-by-lot basis. The short lead time makes this feasible.

Scenario 2

Lot-for-Lot Ordering with No Parts Sharing Parents—Level Two

Part $N(1)$ is dependent on part $Z(1)$. One N and one M are needed to make one Z. MRP methods will be used to determine order timing for part N in level two of the Alpha Model of the VCR. Lead time is one week and SOH is 40 units.

TABLE 15.5

MRP PLAN FOR VCR ALPHA—N(1)

	End Week									
	1	2	3	4	5	6	7	8	9	10
Gross requirements	—	—	—	60	—	—	40	—	—	—
Stock on-hand (SOH)	40	40	40	—	—	—	—	—	—	—
Net requirements	—	—	—	20	—	—	40	—	—	—
Planned-order receipt	—	—	—	20	—	—	40	—	—	—
Planned-order release	—	—	20	—	—	40	—	—	—	—

Forty units of stock are carried for three weeks. Two orders are placed. One order is in Week 3 and the other order is in Week 6. Planned receipt is in Weeks 4 and 7.

Scenario 3

Lot-for-Lot Ordering with No Parts Sharing Parents—Level Two

Part $M(1)$ is dependent on part $Z(1)$. The difference between the scenarios for N (above) and M (now) is that part M has no SOH (SOH = 0 units). Again, MRP methods will be used to determine order timing for part M in level two of the Alpha Model of the VCR. LT is one week as it was for part N.

TABLE 15.6

MRP PLAN FOR VCR ALPHA—M(1)

	End Week									
	1	2	3	4	5	6	7	8	9	10
Gross requirements	—	—	—	60	—	—	40	—	—	—
Stock on-hand (SOH)	—	—	—	—	—	—	—	—	—	—
Net requirements	—	—	—	60	—	—	40	—	—	—
Planned-order receipt	—	—	—	60	—	—	40	—	—	—
Planned-order release	—	—	60	—	—	40	—	—	—	—

There is no stock carried. Two orders are placed in Weeks 3 and 6 with planned receipt in Weeks 4 and 7.

Table 15.7 summarizes the first three scenarios which are related in the VCR hierarchy.

TABLE 15.7

SUMMARY OF MRP ORDERS PLACED

SOH	Name	End Week									
		1	2	3	4	5	6	7	8	9	10
0	VCR	—	60	—	—	—	80	—	—	40	—
80	Z(1)	—	—	—	60	—	—	40	—	—	—
40	N(1)	—	—	20	—	—	40	—	—	—	—
0	M(1)	—	—	60	—	—	40	—	—	—	—

Having stock on-hand for Z(1) deferred the need for an order until Week 4. Thereafter, the lot-for-lot ordering policy took over. On the other hand, the fact that N(1) had some stock on-hand did not defer an order. It just made it a smaller one. The pattern of dependency is evident in that M and N need to be ordered in Weeks 3 and 6 to satisfy Z's need for stock in Weeks 4 and 7.

The next, and final, scenario assumes that an economic order quantity Q_O has been determined to set the order size. Note it is used because it minimizes the sum of ordering and carrying costs. Also, it is not the basis for a continuous (dynamic) ordering system with a reorder point.

Scenario 4

Economic Order Quantity for a Part that Shares Parents

$Y(1)$ subassemblies occur in both the Alpha and the Beta Model. Therefore, the demands in the MPS are combined. There is one unit used whichever model is made. $LT = 1$ week which means that $R(1)$, $S(4)$, and $T(2)$ all have lead times that are one week or less.

The initial $SOH = 80$. Lot sizing is done using the economic order quantity (EOQ) model. The optimal lot size has been determined to be $Q_O = 90$ units.

TABLE   1 5 . 8

MRP PLAN FOR VCR ALPHA AND BETA—Y(1)

	End Week									
	1	2	3	4	5	6	7	8	9	10
Gross requirements	—	60	50	—	—	100	—	—	40	60
Stock on-hand (SOH)	80	20	—	60	60	—	50	50	10	—
Net requirements	—	—	30	—	—	40	—	—	—	50
Planned-order receipt	—	—	90	—	—	90	—	—	—	90
Planned-order release	—	90	—	—	90	—	—	—	90	—

$Y(1)$ total demand (which combines demand for both the Alpha and the Beta Models) drives the planned-order releases and receipts. There are three orders and 230 units are carried over the 7 week period. This period, which stretches from end of Week 3 to end of Week 10, starts once the initial inventory of 80 units is used up. The average number of units that are carried $= 230/7 = 32.9$, or 33. For the ten-week period, it is the same ($330/10 = 33$). Also, there will be 40 units carried during Week 11.

Note: the chapter supplement compares the total costs of all scenarios. That comparison is necessary for OM to make a choice between alternatives.

15.7

LOT SIZING

Lot-for-lot sizing is among the simplest ordering methods but it places more orders than other methods which order less frequently in larger quantities. Lot-for-lot is as close an approximation to just-in-time as it makes sense to get with an MRP model. However, if there are set ups required then the costs associated with lot-for-lot are likely to be too high and alternative methods which order less frequently will prevail.

The cost of this policy can be obtained by multiplying the number of orders that are made using lot-for-lot by the cost of an order or set up. As revealed by the simulations (scenarios), with lot-for-lot sizing and no order cancellations, the amount of inventory carried would be zero. MRP systems are not immune to last minute changes which result in stranding inventory that was earmarked for make-to-order. When orders are cancelled, there is stock left, and this might have to be carried for quite a while.

Grouping periods require summing the demands to cover various fixed periods of time, the length of which is selected based on the concept of an optimal period t_O as shown below. The period also can reflect accounting conveniences and policies of the supplier concerning how often it is able to deliver.

The idea is to decrease the number of orders. If a four week fixed period is chosen, then there will be 13 orders placed per year. It is not difficult to keep track of average inventory carried and to put a cost on carrying stock. If this cost is close to the cost of an order or set up multiplied by 13, then, a good balance has been established. This approach, called the part-period model using an economic part-period criterion, is examined in the supplement to this chapter.

Economic order quantity is calculated for items using the traditional OPP method:

$$Q_O = \sqrt{\frac{2DC_r}{cC_c}}$$

The annual demand D, can be estimated. The cost per unit c, is known. The carrying cost C_c, and the ordering cost C_r, both must be estimated. The economic order quantity sets the order size. The equivalent to the reorder point is that positive net requirements signal the need to place an order. It is expected that the number of such orders placed will average out to be t_O, as calculated here:

$$t_O = \sqrt{\frac{2C_r}{cDC_c}}$$

The sensible approach for important items is to compare the costs obtained by using the various methods. The methods that have been discussed for economic part period balancing (ppb) are only able to approximate the optimal. It should be pointed out that over a long planning horizon the dynamic programming approach based on the Wagner-Whitin algorithm comes closest to minimizing the total costs which are the sum of the carrying and ordering costs. Computing time, which has been a stumbling block to implementation of this model for large inventory systems, has been gradually reduced to a practical level.[9] This is for two reasons: better computing algorithms and faster computers.

15.8

UPDATING

There are two approaches for updating the MRP files and records. The first of these is called *regeneration MRP* and the second is called *net change MRP*. APICS definitions should be used as a standard for the field. They are precise and concise.

Regenerating materials planning builds a work file from Customer Order Processing, Purchase Order, Bill of Material Processor, and Shop Floor Control modules, which will determine what requirements are needed for all outstanding areas.
Source: Macola Software (Marion, OH)

Regeneration MRP

MPS is totally reexploded down through all bills of material, to maintain valid priorities; new requirements and planned orders are then recalculated.

Net change MRP

MRP is continually retained in the computer; whenever a change is needed, a partial explosion and netting is made only for affected parts.

"**regeneration MRP**—An MRP processing approach where the master production schedule is totally reexploded down through all bills of material, to maintain valid priorities. New requirements and planned orders are completely recalculated or 'regenerated' at that time. Ant. net change MRP, requirements alteration."[10]

"**net change MRP**—An approach in which the material requirements plan is continually retained in the computer. Whenever a change is needed in requirements, open order inventory status, or bill of material, a partial explosion and netting is made for only those parts affected by the change."[11]

The MRP records are usually regenerated once a week. However, the variation that exists in practice can be traced to the degree to which change affects the system and how frequently it occurs. Because net change systems respond to transactions and only calculate the effects of that transaction, it is a simpler system requiring less time. Net change systems can be programmed to be activated by exception reports. This allows the system to react quickly to problems that arise.

It is not common practice to rely upon the net change system for all transactions over time. There are too many runs required for small and local effects. The many runs can lead to accumulation of errors instead of corrections. The systems for correction may not have received the same amount of thought and attention as the regeneration program. Consequently, it is often the case that net change is reserved for exceptions and regeneration is used on a regular basis.

15.9

CAPACITY REQUIREMENTS PLANNING (CRP)

Capacity requirements planning (CRP) is an extension of MRP. Sometimes it is necessary to readjust capacity when the MRP system delivers a plan that is not feasible because of capacity limitations.

Capacity requirements planning (CRP) is: "the function of establishing, measuring, and adjusting limits or levels of capacity…(it) is the process of determining how much labor and machine resources are required to accomplish the tasks of production. Open shop orders, and planned orders in the MRP system are input to CRP, which 'translates' these orders into hours of work by work center by time period. Syn: capacity planning."[12]

There is the short-range picture of shifting capacities and the long-range view of commitments to strategic alterations. CRP addresses the former which can be called "the more immediate problems." The latter, which is treated in Section 15.13 below, is called MRP II.

CRP is generally applied to short-term modifications for capacity problems that are encountered as a result of infeasible MRP plans. Specifically, capacity changes can be affected by using overtime, longer and extra shifts, renting more space, and hiring part-time workers. Subcontracting often is used when backlogs arise.

Bottlenecks can be removed by methods studies. Changes can be made in the way that purchasing functions. Priorities can be altered and sales can be asked to emphasize certain products and not take orders for others. For the most part, these are considered to be stop-gap measures which can be employed until the problems are solved in a more permanent fashion.

Figure 15.12 illustrates the way in which capacity requirements planning (CRP) is used to modify the MPS and the MRP when supply—Resource Availabilities (RA)—and demand—Material Requirements (MR)—do not match.

Capacity requirements planning (CRP)

Function of establishing, measuring, and adjusting limits or levels of capacity to determine how much labor and machine resources are required to accomplish production tasks.

FIGURE 15.12

THE CAPACITY REQUIREMENTS PLANNING SYSTEM

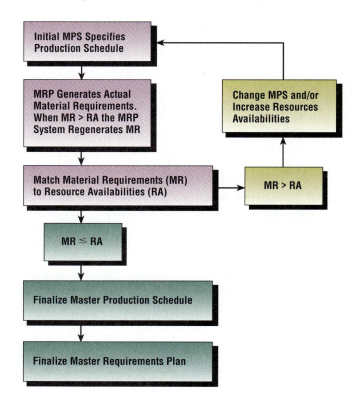

The fact that CRP operates at a higher level than MRP is an important first step in realizing potentials for the organization. Thus, capacity planning permits consideration of alternative sources of product. What is being made can be shifted to what will be bought, and vice versa. Nevertheless, as will be noted when MRP II is discussed, CRP looks outside the boundaries of MRP on a relatively short-term basis.

Rough-Cut Capacity Planning (RCCP)

Rough-cut capacity planning (RCCP) uses approximation to determine whether there is sufficient capacity to accomplish objectives. It serves as a precursor of the CRP process. RCCP occurs when the aggregate plan is being transformed into the master production schedule. In order to authorize the MPS (see Figure 15.3 above), it is essential to check that there is enough capacity to do the job that the MPS describes. RCCP is used to be certain that realistic objectives are being transmitted to the MRP process. If there is insufficient capacity to produce the MPS, then steps are taken to bring in additional key resources and/or to use subcontracting and coprocessing.

Bills of resources (BOR) are used to accomplish this step. The **bill of resources (BOR)** is a listing of the key resources needed to make one unit of the selected items (or family of items as in group technology). Resource requirements are identified with a lead time offset to show the impact of the requirements on resource availability with respect to the timeline. RCCP makes use of BOR to determine the expected capacity requirements of the MPS. The BOR also is called the product load profile or the bill of capacity.

Capacity requirements planning is post-MRP planning. However, capacity planning is going on before this stage in the form of rough-cut capacity planning. It is used to disaggregate the aggregate schedule in order to develop the master production schedule. It also should be noted that the MPS is the basis for the manufacturing budget. If there is not enough production capacity to fulfill orders promised, capacity has to be changed. This will alter the manufacturing budget.

15.10
DISTRIBUTION RESOURCE PLANNING (DRP)

When the acronym DRP is used, it refers to distribution resource planning which is akin to MRP II (described below). DRP is the essence of supply-chain management. A total systems point of view prevails. The basic distribution system is extended to consider not only how to keep the warehouse supplied but, also, how to configure the key resources of the distribution supply chain including the manufacturing supply.

Distribution resources can be modified once the underlying distribution requirements planning system (defined below) is charted and understood. Note that the distribution requirements planning system is akin to MRP, thus **distribution requirements planning** is defined as follows:

"1) The function of determining the needs to replenish inventory at branch warehouses. A time-phased order point approach is used where the planned orders at the branch warehouse level are "exploded" via MRP logic to become gross requirements on the supplying source. In the case of multilevel distribution networks, this

---

Rough-cut capacity planning (RCCP)

Uses approximation to determine whether there is sufficient capacity to accomplish objectives; precursor to CRP.

Bill of resources (BOR)

Listing of the key resources needed to make one unit of the selected items; also called product load profile or bill of capacity.

Distribution requirements planning

A time-phased order point approach for determining the needs to replenish inventory at branch warehouses.

Glenda Paquin

• • •

Mitel Corporation

Mitel Corporation is a Canadian-based international manufacturer and distributor of telecommunications systems. With annual sales of just over $500 million, Mitel also is developing products for the emerging computer telephony applications markets. "In operations, Mitel's number one challenge is to be as flexible as possible in reacting to changes in demand, while minimizing inventory investment and maintaining 100% delivery to promise dates," said Glenda Paquin, Mitel's director of Purchasing and Materials for North America.

Paquin's consolidated Purchasing and Materials function provides purchasing support services to Mitel's product-based business divisions. The group is responsible for all the purchasing functions that support manufacturing, as well as procurement of capital equipment and services. The main emphasis of Paquin's job is on strategic planning with the business divisions, but she is involved in operational details as well, because a large portion of dollars spent by the company is on materials purchases.

Mitel implemented its MRP system in the early 1980s. Then a typical MRP run took more than a

day and was performed over the weekend. When demand for Mitel's products began to increase, the company needed a tool that would allow it to deal on a daily basis with demand fluctuations. "Batching up daily demand changes for the weekly MRP run was not acceptable," said Paquin. "We don't make decisions weekly around here; we need to react quickly and intelligently to changes in market demand." She said Mitel needed the ability to identify changes and incorporate them into the schedule. They also needed to be able to isolate problem parts, simulate potential alternatives, and then select the alternative that minimized costs and optimized sales opportunities.

To solve these problems, Mitel considered several software products available on the market and also considered developing its own programs to integrate with the MRP

system. They chose software available on the market from FastMAN Software Systems Inc. of Ottawa, Ontario, Canada (see "Technology" box in this chapter). Paquin said they chose FastMAN in part because it was PC-based and object-oriented. Mitel uses this software as a simulation tool to evaluate demand increases and decreases, supply problems, and inventory levels.

Paquin said Mitel is continuously searching for tools and methodologies that improve responsiveness to customers. "We don't blame a missed revenue opportunity on bad forecasts," she said. "In operations we are focused on improving our delivery performance to all customer requests. All aspects of cycle time and response turnaround are taken very seriously."

Source: FastMAN Software Systems Inc. document provided by Brett McAteer and conversations with Glenda Paquin

explosion process can continue down through the various levels of regional ware-houses, master warehouse, factory warehouse, etc., and become input to the master production schedule. Demand on the supplying source(s) is recognized as dependent, and standard MRP logic applies. 2) More generally, replenishment inventory calculations may be based on other planning approaches such as "period order quantities" or "replace exactly what was used" rather than being limited to solely the time-phased order point approach."[13]

The analogy to MRP is evident. Material requirements connect the supply-chain participants. Lead-time dependencies exist as factories feed central warehouses and these, in turn, supply regional warehouses. It will be noted that shipping quantities are much like lot sizing. Lot-for-lot replacement is a potential strategy.

DRP is a higher order planning system than the distribution requirements planning system because it also plans for resource availability such as storage space in warehouses, trucking capacity, railroad boxcar availabilities, distribution workforce commitments and training, cash flow needed to keep the distribution system moving. Thus, DRP operates as an extension of distribution requirements planning by addressing what resources are needed in the distribution system.

15.11

WEAKNESSES OF MRP

MRP is vulnerable to variability in lead times. The parent product cannot be made if any one of the items on the lower levels is not delivered on time for the scheduled production runs. This raises the point that a sensible policy might require some safety stock.

MRP is supposed to do away with the need for safety stock. However, OM must evaluate the risk of delayed supplier deliveries and take adequate protective measures. This is done by increasing the lead time by an amount that should cover delivery contingencies. Such "safety time" translates into added stock being carried.

MRP Software, BPCS, helps identify critical needs for both production and purchasing activities in response to day-to-day events. With these tools, users keep the right items in stock at the right time to meet the flow of operations and reduce the financial burden of overstocking inventories.
Source: SSA (System Software Associates, Inc., Chicago, IL

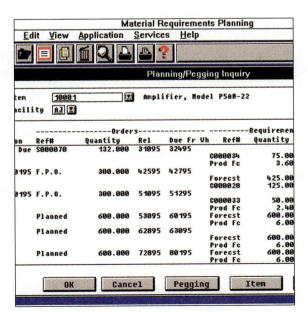

The fact that a production or delivery problem at any level in the hierarchy can shut down the system cannot be treated as an MRP weakness. Whether MRP is used or not, the lack of materials will result in stopping the process. MRP helps to keep problems from arising by documenting the time-phasing for all parts that are needed. Follow-up calls can be made to ascertain the status of orders. Other forms of expediting can be used when slippage seems to be occurring. The weakness of MRP, in this regard, is that it does not automatically check status of orders and expedite them. That remains an organizational matter related to how information is used. Expediting should be treated as an integral part of the MRP organization.

There are problems evaluating MRP software. Many kinds of MRP software are available off-the-shelf. Some companies buy these and then modify them to suit their own particular systems' needs. Large firms often build their own systems. Failures with MRP can often be traced to having chosen software that was not well-suited to the applications.

MRPs are information systems that are vulnerable to data errors. Data entry must be checked carefully for original records, and then for updates. Errors can creep into net-change systems which are difficult to detect. Many organizations have so many parts that it is difficult to be sure that all changes are recorded. Engineering design changes occur so frequently in some companies that it is hard to keep up with them. With MRP, the system for keeping up with changes had better be as good as the MRP system.

One of the worst problems that MRP faces is last minute changes in orders. It is inevitable that the size of orders will be altered, and that cancellations will occur. To illustrate, assume that the order for 90 *Y(1)* subassemblies has been placed in Week 2 for Scenario 4. A week later, sales cancels the Beta Model VCR run of 50 units. It is possible that no more Betas will be made. There is a frantic effort to cancel the 90 *Y(1)* subassemblies to no avail. Fortunately, the *Y(1)* subassemblies also are used in the Alpha Model. The dependent nature of shared parts adds to the complexity of resolving errors and to the severity of potential problems.

15.12

CLOSED-LOOP MRP

Closed-loop MRP is: "A system built around material requirements planning that includes the additional planning functions of sales and operations (production planning, master production scheduling, and capacity requirements planning). Once this planning phase is complete and the plans have been accepted as realistic and attainable, the execution functions come into play. These include the manufacturing control functions of input-output (capacity) measurement, detailed scheduling and dispatching, as well as anticipated delay reports from both the plant and suppliers, supplier scheduling, etc. The term **closed-loop MRP** implies that not only is each of these elements included in the overall system, but also that feedback is provided by the execution functions so that the planning can be kept valid at all times."[14]

The closed-loop MRP has material requirements planning as one of its steps. It is replete with feedback loops that permit planning and replanning. It extends the MRP concept to include capacity requirements planning, shop scheduling and supplier scheduling. The first step in the planning hierarchy is production planning which indicates that master production scheduling is based on more than open orders and sales forecasts.

Closed-loop MRP

System built around MRP which includes the additional planning, manufacturing control and execution functions as well as feedback thereby ensuring planning validity at all times.

The essence of the closed-loop MRP systems approach is represented by Figure 15.13. Some diagrams of closed-loop MRPs do not include distribution requirements planning, but good systems thinking endorses its inclusion. MRP becomes more powerful with the additions of closed-loop MRP.

FIGURE 15.13

CLOSED-LOOP MRP

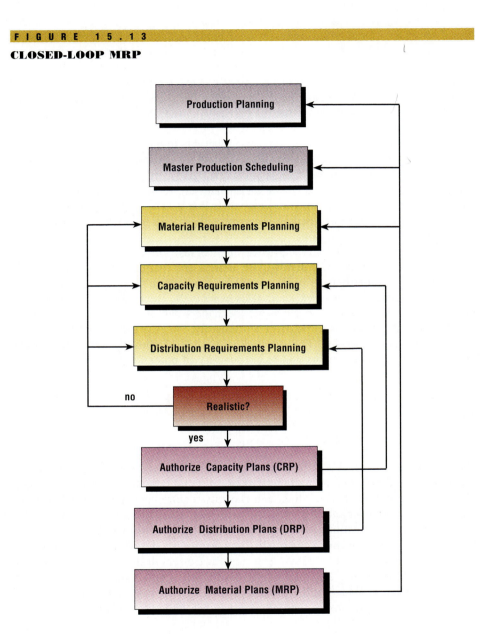

Closed-loop MRP becomes the more powerful MRP II with the addition of business planning as discussed below.

15.13

STRENGTHS OF MRP II (MANUFACTURING RESOURCE PLANNING)

MRP II entails total manufacturing resource planning as part of the strategic business plan. Thus, minor capacity adjustments can be made on a short-term basis (CRP), however, longer-term and broader systems perspectives should prevail. For example, marketing considerations raise the possibility of changing the master production schedule to allow for the introduction of new products and the alteration of the company's product mix.

Financial planning can access and control what is being done and determine what will be done in the future. Accounting is linked directly to the MRP II cost systems so that accounts payable, accounts receivable and cash flow are online with production. R & D's agenda is linked to long-term plans for the company and the architecture of its MRP II.

Technology is constantly changing. MRP II encourages and supports reevaluation of the way that work is being done. It permits assessment of the product mix and the input-output processes that are being used.

Competitive shifts in strategy produce forces to alter what is being done. This can be brought up for consideration by any of the systems players—finance, marketing, R & D, HRM, OM and purchasing. Concretely, MRP II puts business planning into a box on top of production planning. Compare Figures 15.13 and 15.14. The Business Planning box is in a feedback loop with the Production Planning box.

Because strategic business plans drive the MRP II system, this allows consideration of the effect of buying new machines, changing computer equipment, building new plants, altering the product mix, and modifying advertising and promotion campaigns, among others. Based on systems thinking, changes to any part of the planning process shown in Figure 15.13 should be in line with strategic business plans as in Figure 15.14. Making certain that the actions of these two figures are coordinated will assure achievement of the best plan for manufacturing resources and thereby, materials requirements. The long-range viewpoint in search of optimal configurations is the essence of MRP II.

Ollie Wight said, "...the theme of MRP II (is) managing *all* of the resources of a manufacturing company more productively." In his book[15] he then proceeds to develop eight chapters as follows:

Chapter 8—The CEO's Role in MRP II
Chapter 9—MRP II in Marketing
Chapter 10—MRP II in Manufacturing
Chapter 11—MRP II in Purchasing
Chapter 12—MRP II in Finance
Chapter 13—MRP II in Engineering
Chapter 14—DRP: Distribution Resource Planning
Chapter 15—MRP II in Data Processing Systems

The scope of MRP II is fully revealed by the intent of this founder of the MRP II system. Each of the eight chapters represents considerations that affect what will be made and what will be ordered. The titles reveal the broad systems scope and appeal of MRP II.

MRP II

Entails total manufacturing resource planning as part of the strategic business plan.

FIGURE   1 5 . 1 4
MRP II

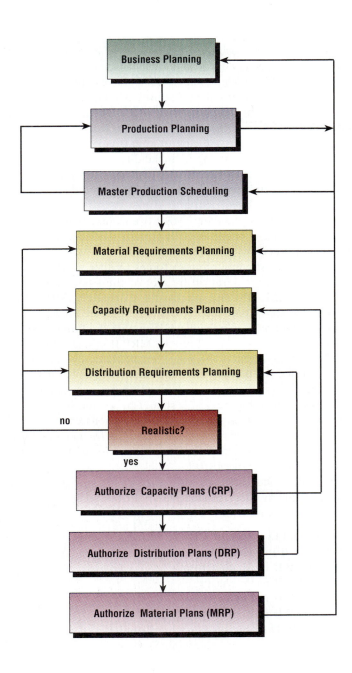

S U M M A R Y

This chapter presents the background for understanding and appreciating material requirements planning widely referred to as MRP. Demands for components and subassemblies that are dependent on the demands for the end items to which they belong characterize MRP. The first major tool of MRP is the master production schedule (MPS) which fits within the framework of the total manufacturing planning and control system (MPCS). The differences between make-to-stock (MTS), make-to-order (MTO), and assemble-to-order (ATO) are discussed. The means for converting gross requirements into net requirements is explained.

The next major component of the information system is the bill of material (BOM) which leads to a product-structure tree reflecting the explosion of parts. Coding and low-level coding are detailed for operating the MRP system. Lot-sizing scenarios are described. Updating the system is discussed. Then higher orders of MRP systems are presented. These are CRP (capacity requirements planning), DRP (distribution resource planning), closed-loop MRP, and finally, MRP II which is manufacturing resource planning.

▶ **Key Terms**

assemble-to-order (ATO) environment *652*	bill of material (BOM) *653*	bill of resources (BOR) *668*
capacity requirements planning (CRP) *667*	closed-loop MRP *671*	distribution requirements planning *668*
distribution requirements planning *668*	end products *643*	inventory record files *655*
lot-for-lot ordering *660*	low-level coding *657*	make-to-order (MTO) inventory environment *652*
make-to-stock (MTS) environment *652*	master production schedule (MPS) *646*	net change MRP *666*
MRP II *673*	regeneration MRP *666*	rough-cut capacity planning (RCCP) *668*
order promising *646*		
total or cumulative lead time (LT) *647*		

▶ **Review Questions**

1. How does inventory planning differ for the flow shop and the job shop?

2. Explain how MRP relates to the job shop and the flow shop.

3. What is meant by time-phased requirements planning for MRP?

4. What is the difference between updating by "regeneration" and updating by the "net change" method for MRP? Discuss the pros and cons of each method.

5. Explain net requirement, planned-order receipt and planned-order release. When are they equal and when are they not equal?

6. What is CRP?

7. What is DRP? How does it relate it to distribution requirements planning?

8. Distinguish between MRP, closed-loop MRP, and MRP II.

9. What characterizes dependent demand systems?

10. How do forecasting considerations apply to MRP?

11. Explain the master production schedule (MPS).

12. What is meant by the total manufacturing planning and control system (MPCS)?

13. Explain the bill of material (BOM). How does it relate to the product-structure tree?

14. What purpose is served by the explosion of parts?

15. How does coding and low-level coding impact order policies?

16. What is lot sizing?

17. What are some of the different methods that are used for lot sizing? Describe those that you know.

18. What is the difference between gross and net requirements?

19. Describe the make-to-stock (MTS) environment.

20. Explain the make-to-order (MTO) inventory environment.

21. Explain the assemble-to-order (ATO) inventory environment.

22. How is rough-cut capacity planning (RCCP) used?

23. Explain bills of resources (BOR).

▶ **Problem Section**

1. The master schedule for end product P is as follows:

Week	Gross Requirements
1	400
2	200
3	200
4	300
5	500
6	100
7	600
8	400

Stock on-hand for p_1, a component of P, is 700. Develop the planned-order release schedule for p_1, assuming a lot-for-lot sizing policy and lead time, $LT =$ two weeks.

2. Use the information in Problem 1 above to address the following issue: the SOH of 700 units is maintained because there is unreliability of delivery. How does the lot-for-lot sizing policy deal with this issue?

3. With further reference to Problem 1 above, the cost of a unit of p_1 is $12. The carrying cost rate is 0.25 percent per week (0.0025). The set-up cost to produce p_1 is $48. Demand per year can be estimated from the table in Problem 1. Use the economic order quantity to set the order size whenever an order is triggered by positive net requirements. Describe the performance of this system.

4. Using the information in Problem 3 above, what is the expected interval between orders?

5. The master schedule for parent product *M* is as follows:

Week	Gross Requirements
1	3400
2	4200
3	5200
4	6300
5	7500
6	3100
7	4600
8	5400
9	6600
10	7800

The policy is to carry no stock on-hand for m_1, which is a subassembly of *M*.

Develop the planned-order release schedule for m_1, assuming a lot-for-lot sizing policy and lead time, $LT =$ one week.

6. Use the information in Problem 5 above to address the following issue: the MRP manager wants to try two-period ordering. She requests that you compare your lot-for-lot plan with the two-period plan. Also, she asks, "Would it help to have some SOH?" Make the comparison and answer the question.

7. With further reference to Problem 5 above, the cost of a unit of m_1 is $1. The carrying cost rate is 24 percent per year. The ordering cost for m_1 is $25. Demand per year can be estimated from the table in Problem 5. Use the economic order quantity to set the order size whenever an order is triggered by positive net requirements. Describe the performance of this system.

8. Using the information in Problem 7 above, what is the expected interval between orders?

9. Develop the lot-for-lot ordering scenario for VCR-Alpha Part Q, using Table 15.3 in the text and the product-structure tree given by Figure 15.11 above. Note that *Q* appears more than once in the product-structure tree. Also, observe the number of units of *Q* that are required for each unit of *X* and for each unit of *B*. The lead times for both *X* and *B* are zero because they can be assembled immediately. The lead time for *Q* is one week. Part Q has no other parents or end items than VCR-Alpha.

In addition to providing the quantitative ordering instructions, evaluate the performance of the ordering system. *Hint:* include in the planned-order release chart levels 0, 1, and 2.

10. Develop the lot-for-lot ordering scenario for VCR-Alpha Part Q, using Table 15.3 in the text and the product-structure tree given by Figure 15.11 above. As in Problem 9, note that *Q* appears more than once in the product-structure tree. Also, observe the number of units of *Q* that are required for each unit of *X* and for each unit of *B*. The lead times for both *X* and *B* are one week to assemble. The lead time for *Q* is one week. Part Q has no other parents or end items than VCR-Alpha.

In addition to providing the quantitative ordering instructions evaluate the performance of the ordering system. *Hint:* include in the chart for planned-order release, levels 0, 1, and 2.

11. Develop the lot-for-lot ordering scenario for VCR-Beta Part B using Table 15.3 in the text and the product-structure tree given in Figure 15.11 above. Note that *B* appears more than once and on different levels of the product-structure tree. Also, observe that only one unit of *B* is required in both cases. The lead time for *G* is zero and the lead time for *B* is two weeks. Assume that Part B has no other parents or end items.

    Provide quantitative ordering instructions and an evaluation of the behavior of the ordering system. *Hint:* include in the chart for planned-order release levels 0, 1, and 2.

12. Develop the lot-for-lot ordering scenario for VCR Beta Part B using Table 15.3 in the text and the product-structure tree given in Figure 15.11 above. Note that *B* appears more than once and on different levels of the product-structure tree. Also, observe that only one unit of *B* is required in both cases. The lead time for both *G* and *B* is one week. Assume that Part B has no other parents or end items.

    Provide quantitative ordering instructions and an evaluation of the behavior of the ordering system. *Hint:* include in the chart for planned-order release levels 0, 1, and 2.

13. Determine what happens to *Y(1)* subassemblies if 50 VCR-Beta units ordered for Week 3 are cancelled. Use Tables 15.3 and 15.8 above. Note that *Y(1)* subassemblies are used in both Alpha and Beta end items.

14. What is the maximum cumulative lead time for parent assembly *A* where subassemblies *B, C, D,* and *E* are all special items with lead times as marked? The usage quantities shown in parentheses are all one-for-one meaning one *B* and one *C* subassembly are required for *A*, and one *D* and one *E* are required for *C*.

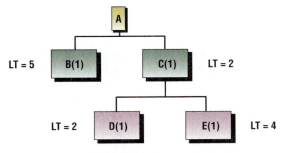

15. Using the graphics and information in Problem 14 above, what is the maximum cumulative lead time for parent assembly *A* if subassembly *E* is a standard item and there is SOH? *Note:* subassemblies *B, C,* and *D* are all special items.

16. Using the graphics and information in Problem 14 above, what is the maximum cumulative lead time for parent assembly *A* where subassembly *B* is a standard item with SOH? All the other components are special.

17. The product structure BOM drawn below does not account for low-level coding. Redraw this graphic so that it is consistent with low-level coding.

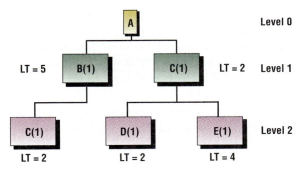

18. Using the graphic in Problem 17 above, what is the maximum cumulative lead time for parent assembly *A* where subassemblies *B, C, D,* and *E* are all special items with lead times as marked? The usage quantities are shown in parentheses.

19. Using the graphic in Problem 17 above, what is the maximum cumulative lead time for parent assembly *A* where subassembly *C* is a standard item for which there is ample SOH. Subassemblies *B, D,* and *E* are all special items. The usage quantities are shown in parentheses.

20. To which subassembly in the BOM below does the planned-order release chart apply?

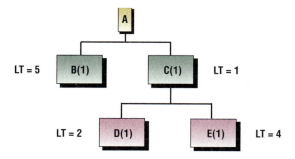

Weeks Ahead	1	2	3	4	5	6	7	8	9
Gross (Amount) Requirements: A		20			45				60
Planned-Order Release Date	20			45				60	

21. Complete the planned-order release chart below so that it applies to subassembly *C* which is two-for-one as shown in the BOM below.

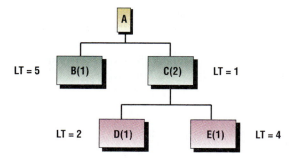

Weeks Ahead	1	2	3	4	5	6	7	8	9
Gross (amount) Requirements: A		20			45				60
Planned-Order Release: C(2)									

22. Using the graphic in Problem 21 above, complete the planned-order release chart below so that it applies to component *D*.

Weeks Ahead	1	2	3	4	5	6	7	8	9
Gross (amount) Requirements: A			25	35	40		20		60
Planned Order Release: C(2)									
Planned Order Release: D(1)									

▶ Readings and References

AGRA Software, FastMan (MRP software), (613-596-3344), Ottawa, Ontario, Canada, 1993.

K. R. Baker, "Requirements Planning," in S. C. Graves, A. H. G. Rinnooy Khan, and P. Zipkin, eds., *Logistics of Production and Inventory*, Amsterdam, North-Holland Publishing Company, 1993, pp. 571–628.

William Berry, Thomas E. Vollman, and D. Clay Whybark, *Master Production Scheduling: Principles and Practice*, Washington, DC, American Production and Inventory Control Society, 1979.

J. A. Buzacott and J. G. Shanthikumar, "Safety Stock versus Safety Time in MRP Controlled Production Systems, *Management Science*, Vol. 40, No. 12, December, 1994, pp. 1678–1689.

James F. Cox, John H. Blackstone, Michael S. Spencer, *APICS Dictionary*, 7th Ed., APICS Educational and Research Foundation, 1992.

R. B. Heady, and Z. Zhu, "An Improved Implementation of the Wagner-Whitin Algorithm," *Production and Operations Management Journal*, Vol. 3, No. 1, Winter, 1994.

Andre Martin, *Distribution Resource Planning*, Williston, VT, Owl Publications, Inc., 1982.

Joseph Orlicky, *Material Requirements Planning*, NY, McGraw-Hill, 1975.

Thomas E. Vollman, William L. Berry, and D. Clay Whybark, *Manufacturing Planning and Control Systems*, 3rd Ed., Homewood, IL, Irwin, 1992.

Oliver W. Wight, *The Executives Guide to Successful MRP II*, Williston, VT, Owl Publications, Inc., 1981.

Oliver W. Wight, *MRP II, Unlocking America's Productivity Potential*, Boston, MA, CBI Publishing Co., 1982.

S U P P L E M E N T   1 5

Part-Period Lot-Sizing Policy

The lot-for-lot policy—without errors, cancellations or changes in order size—exactly matches orders to requirements. Therefore, there would be no carrying cost and many orders. If placing orders is expensive, because a new set up is required in each case, then there would be a preference for fewer orders and some carrying costs. Consequently, a lot-sizing policy that is worthy of consideration is one which balances set-up costs with carrying costs.

Ordering more than lot-for-lot results in parts being carried over periods of time. This statement is the basis for the part-period lot-sizing policy. It is an attempt to achieve the minimum total cost policy which occurs when set up, or ordering, costs are equal to carrying costs. An estimate can be made of the economic part-period, thus:

$$\text{Economic Part-Period} = \frac{\text{Total Set-Up Costs}}{\text{Unit Carrying Cost per Period}}$$

The dimensions of this measure are unit months, or whatever time period the carrying costs embody.

Assume that the following costs have been estimated: the unit cost of the part is $10, the carrying cost rate is 1 percent per month, and the unit carrying cost for one month is then $10(0.01) = 0.1. The set-up cost to produce this component is $15.

$$\text{Economic Part-Period} = \frac{\$15}{0.1} = 150 \text{ Unit Months}$$

Now calculate the number of unit months (part-periods) of each possible ordering policy.

Table 15.9 lists monthly net requirements. These are summed in the column titled "Cumulative Lot Size." Thus, if the order is to produce enough for the first month only, then the lot size will be 20 units. If the order is to produce enough for the first 4 months, then the lot size will be 150 units, etc. The carrying cost per unit calculations are given below the table.

TABLE 15.9

DATA REQUIRED FOR PART-PERIOD LOT SIZING

Month	Net Requirements	Cumulative Lot Size	Months Carried	Carrying Cost/Unit	Policy Is: Order for Month
1	20	20	0	$0	1
2	40	60	1	$4	1+2
3	60	120	2	$16	1+2+3
4	30	150	3	$25	1+2+3+4

The carrying cost is determined for each policy. If the policy is to produce or order enough for only the first month, then there are no carrying costs:

20 units carried for 0 months = 0 part-periods or unit months

Producing or ordering for 2 months entails no carrying charges for the first 20 units. They will be used for production in Month 1. There are carrying costs of $4 per unit for the second month. Thus:

20 units $\times$ \$10/unit $\times (0.01) \times 0$ months carried $= \$0$

40 units $\times$ \$10/unit $\times (0.01) \times 1$ month carried $= \$4$

Total units ordered for 2 months = 60 units

Part-periods $= (40 \times 1) = 40$ unit months

Total cost = $4

Producing or ordering 120 units for three months requires:

20 units $\times$ \$10/unit $\times (0.01) \times 0$ months carried $= \$0$

40 units $\times$ \$10/unit $\times (0.01) \times 1$ month carried $= \$4$

60 units $\times$ \$10/unit $\times (0.01) \times 2$ months carried $= \$12$

Total units ordered for 3 months = 120 units

Part-periods $= (40 \times 1) + (60 \times 2) = 160$ unit months

Total cost = $16

This ordering policy (3 months) comes close to matching the prior determination of the economic part-period which was 150. It is necessary to see whether the next ordering policy (4 months) can come as close.

To determine the carrying cost of producing 150 units for 4 months' requirements, simply add the 3-month total to the carrying cost of the fourth month's net requirements.

$$30 \text{ units} \times \$10/\text{unit} \times (0.01) \times 3 \text{ months carried} = \$9$$

$$\text{Carried forward from 3-month lot-size policy} = \$16$$

$$\text{Total units ordered for 4 months} = 150 \text{ units}$$

$$\text{Part-periods} = (40 \times 1) + (60 \times 2) + (30 \times 3) = 250 \text{ part-periods}$$

$$\text{Total cost} = \$25$$

The 3-month ordering policy is indicated in two related ways. First, its economic part-period policy is closest (150 versus 160). Second, the set-up cost of $15 is closest to the carrying-cost charges of $16 for a lot size of 3 months' requirements. That meets the criterion of balancing these two opposing costs. Consequently, produce or order 120 units.

The deviation between the economic part-period measure and the chosen order size will vary as a function of the actual Net Requirements, as shown in Table 15.9 above.

Problems for Supplement 15

1. Determine the production or order quantity for Table 15.9 above when the set-up, or ordering, cost is $24.

2. Assume that the following costs have been estimated: the per unit cost of the part is $50, the carrying-cost rate is 2 percent per month, and the set-up cost to produce this part is $75. What is the economic part-period?

3. Using the information in Problem 2 above, calculate the number of unit months (part-periods) of each possible ordering policy for the table below.

Month	Net Requirements	Cumulative Lot Size	Months Carried	Carrying Cost/Unit	Policy Is Order for Month
1	30		0		1
2	50		1		1+2
3	20		2		1+2+3
4	40		3		1+2+3+4
5	30		4		1+2+3+4+5
6	25		5		1+2+3+4+5+6

Then, determine the production or order quantity to be used.

4. When the order is placed in Problem 3 above, start the calculations again from that point. Derive the next order quantity and continue in this way until the data are used up.

5. Use the part-period lot-sizing method for the situation described in Problem 1 of the general Problem Section above.

Notes...

1. Joseph Orlicky, Material Requirements Planning, NY, McGraw-Hill, 1975, p. ix.

2. Oliver W. Wight, "Designing and Implementing a Material Requirements Planning System," *Proceedings of the 13th International Conference of APICS*, 1970. The quote appears on page 1 of *Material Requirements Planning* by Joseph Orlicky, NY, McGraw-Hill, 1975.

3. Jon F. Cox, John H. Blackstone, and Michael S. Spencer, *APICS Dictionary*, 7th Ed., APICS Educational and Research Foundation, 1992.

4. *APICS Dictionary*, 7th Ed., 1992.

5. *APICS Dictionary*, 7th Ed., 1992.

6. *APICS Dictionary*, 7th Ed., 1992.

7. Ibid, Orlicky, p. 56.

8. Thomas E. Vollman, William L. Berry and D. Clay Whybark, *Manufacturing Planning and Control Systems*, 3rd Ed., Irwin, Homewood, IL, 1992, p. 34.

9. R. B. Heady, and Z. Zhu, "An Improved Implementation of the Wagner-Whitin Algorithm," *Production and Operations Management Journal*, Vol. 3, No. 1, Winter, 1994, pp. 55–63.

10. *APICS Dictionary*, 7th Ed., 1992.

11. Ibid, *APICS Dictionary*.

12. *APICS Dictionary*, 7th Ed., 1992.

13. *APICS Dictionary*, 7th Ed., 1992.

14. *APICS Dictionary*, 7th Ed., 1992.

15. Wight, Oliver W., MRP II: *Unlocking America's Productivity Potential*, CBI Publishing Co., Inc., Boston, MA, 1982.

Chapter 16

Production Scheduling

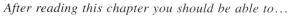

After reading this chapter you should be able to…

1. Explain why production scheduling must be done by every organization whether it manufactures or provides services.
2. Discuss the application of loading to the job shop (often called job-shop loading or shop loading).
3. Draw a Gantt load chart and explain its information display.
4. Describe how to resolve the loading problem where teams compete for permanent assignment to the best facilities.
5. Explain why "best assignments" require the systems approach.
6. Relate best-loading assignments to opportunity costs in terms of lowest cost, greatest profit or fastest completion time.
7. Detail the strengths and weaknesses of the assignment model.
8. Describe how the transportation model resolves the "no splitting" problem of the assignment model.
9. Detail strengths and weaknesses of the transportation model.
10. Describe the role of sequencing and how to apply sequencing rules for one facility and for more than one facility.
11. Explain the purpose of priority sequencing rules and the means for using them.
12. Describe the critical ratio and other priority rules for sequencing.
13. Explain how finite scheduling and the OPT concept take bottlenecks into account.
14. Explain queue control and synchronized manufacturing.
15. Discuss changing the capacity of bottlenecks.

Chapter Outline

16.0   PRODUCTION SCHEDULING: OVERVIEW

16.1   LOADING
Plan to Load Real Jobs—Not Forecasts

16.2   GANTT LOAD CHARTS

16.3   THE ASSIGNMENT METHOD FOR LOADING
Column Subtraction—Best Job at a Specific Facility
Row Subtraction—Best Facility for a Specific Job

16.4   THE TRANSPORTATION METHOD FOR SHOP LOADING

16.5   SEQUENCING OPERATIONS
First-In, First-Out (FIFO) Sequence Rule

16.6   GANTT LAYOUT CHARTS
Evaluatory Criteria

16.7   THE SPT RULE FOR n JOBS AND $m = 1$ FACILITIES

16.8   PRIORITY-MODIFIED SPT RULES: JOB IMPORTANCE—DUE DATE—
LATENESS—OTHER PRIORITIES
Critical Ratio for Due Dates

16.9   SEQUENCING WITH MORE THAN ONE FACILITY ($m > 1$)

16.10 FINITE SCHEDULING OF BOTTLENECKS—OPT CONCEPT
Flow Shops and Continuous Processes Do Not Have Bottlenecks
Queue Control
Synchronized Manufacturing
Changing the Capacity of the Bottleneck

Summary
Key Terms
Review Questions
Problem Section
Readings and References

The Systems Viewpoint

Department scheduling decisions rarely achieve optimal assignments for every job at every facility. Best assignments can seldom be made on a one-by-one basis. Instead, the problem must be looked at as a whole. As an analogy, NASA found that if every component of a space vehicle is optimized with respect to its function and only its function, the "bird will not fly." Instead, the vehicle must be designed together as a coordinated system of components. The same applies to scheduling for the job shop. The assignment of work and its timing need to be orchestrated. Bottlenecks need to be taken into account and the goal of synchronized manufacturing requires coordination of the entire job shop with respect to the mix of jobs that is being orchestrated.

The systems approach is called for so that the total set of assignments is optimized. The system's total costs are minimized or the total profits are maximized. Total quality and/or total productivity are maximized. The nature of the goal is to select assignments which—although less than best individually—result in the *overall* system's best.

Production scheduling is always a system's problem because jobs, people and teams compete with each other for best facilities. Jobs (as surrogates for customers) also compete with each other concerning which gets done first. Facilities contend with each other for jobs. Departments that do similar work compete with each other for preferred work. Orders placed with suppliers are tied to priorities with respect to jobs and customers—so suppliers' orders also rival each other in terms of importance and how they are treated.

The company must achieve systems optimization to rationalize the various preferences in a way that is not self-defeating. When the systems point of view prevails, the solutions are company-wide optimizations even though people and facilities experience suboptimal assignments. Over time, the facilities need to be altered in such a way as to minimize the degree of suboptimization. That too is a systems principle.

16.0
PRODUCTION SCHEDULING: OVERVIEW

Production scheduling

Assigns actual jobs to designated facilities with unambiguous stipulations for completion at specific times.

The final step of **production scheduling** assigns actual jobs to designated facilities with unambiguous stipulations that they be completed at specific times. The steps in scheduling will be reviewed below moving from generic resource planning to actual assignments at workstations.

1. *Aggregate scheduling* developed resource plans based on forecasts of orders in generic units such as standard hours.
2. Later, with actual orders on-hand, or with reasonable predictions about orders—the *master production schedule (MPS)* assigned jobs to time slots—to permit orders to be placed for required materials using *material requirements planning (MRP)*. These time period assignments are so well defined that they are referred to as *time buckets*.

3. The next step in production scheduling is to *load facilities* which means taking the actual orders and assigning them to designated facilities. **Loading** answers the question: which department is going to do what work? **Sequencing** answers the question: what is the order in which the work will be done?

To summarize, the third step in production scheduling does both loading and sequencing. Loading assumes that the material requirement analysis has been done and that orders have been properly placed for required materials and for needed parts and subassemblies. Further, that the parts will be on-hand, as planned. A problem in supplier shipments will delay scheduling production of the affected order.

Specifically, loading assigns the work to divisions, departments, work centers, load centers, stations, and people. Whatever names are used in a given organization, orders are assigned to those who will be responsible for performing the work. Loading releases jobs to facilities.

At this point, sequencing will be explained so that the entire process of production scheduling will be clear to the reader. Sequencing models and methods will follow the discussion of loading models and methods. Sequencing involves **shop floor control** which consists of communicating the status of orders and the productivity of workstations. It assigns priorities that determine the order with which work will be done. Say that Jobs x, y, and z have been assigned to workstation 1 (loading). Jobs x, y, and z are in a queue (waiting line). Sequencing determines which job should be first in line, which second, etc.

Loading

Takes actual orders and assigns them to designated facilities; decides which department is going to do what work.

Sequencing

Decides the order in which the work will be done; specifies precise timing of assignments thereby allowing customers' notification of delivery dates.

Shop floor control

Part of sequencing; communicates the status of orders and the productivity of workstations; assigns priorities that determine order with which work will be done.

16.1

LOADING

Loading, also called shop loading, is required to assign specific jobs or teams to specific facilities. Loading is needed for machine shops, hospitals and offices. While loading assigns work to facilities, it does not specify the order in which jobs should be done at the facility. Sequencing methods (described below) determine the order of work at the facility.

Plan to Load Real Jobs—Not Forecasts

Aggregate scheduling used standard hours based on forecasts to determine what resources should be assembled over the planning horizon. Loading takes place in the job shop when the real orders are on-hand. If the aggregate scheduling job was done well, then the appropriate kinds and amounts of resources are available for loading. The master production schedule also made resource assignments which could be modified if capacity was not adequate. Planning actual shop assignments is a regular, repetitive managerial responsibility. Releasing the jobs, as per assignment, is another.

Each facility carries a backlog of work which is its "load"—hardly a case of perfect just-in-time in which no waiting occurs. The backlog is generally much larger than the work in process which can be seen on the shop floor. This is because work not assigned yet is waiting but not visible. One major objective of loading is to spread the load so that waiting is minimized, flow is smooth and rapid, and congestion is avoided.

These objectives are constrained by the fact that not all workstations can do all kinds of jobs. Even though the job shop uses general purpose equipment, some workstations and people are better suited for specific jobs than are others. Some stations cannot do jobs that others can do. Some are faster than others and tend to be overloaded. The scheduling objective is to smooth the load with balanced work assignments at stations.

16.2

GANTT LOAD CHARTS

Gantt charts

Provide a graphic system that is easy to follow by visualizing the progress of jobs and the load on departments; method for trying to optimize assignments.

The use of Gantt charts for loading dates to the early 1900s when Henry L. Gantt (1861–1919) developed them as a formal means for assigning jobs to facilities and charting job progress. **Gantt charts** provide a graphic system that is easy to follow by visualizing the progress of jobs and the load on departments. Often these two purposes are combined on one chart. Too much job waiting (overloading), facilities not working at full utilization (underloading) or load imbalances between stations (some busy, others idle) could be seen immediately by someone familiar with these charts.

Gantt charts deal with the systems realities of loading for production scheduling. These are the issues like: which jobs have not started that should have started, how much progress toward completion has been made, which departments are overloaded, and which ones are underloaded?

The ideal situation is creative customization (of load scheduling charts) that meets the needs of the process managers and the sales department. Proper charts reflect the interests of customers and the impact of suppliers. Figure 16.1 illustrates the Gantt load chart and the discussion that follows explains it.

FIGURE 16.1

GANTT LOAD CHART

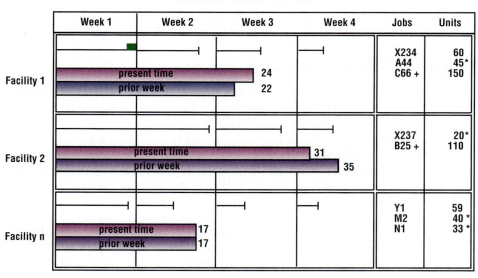

This is a traditional Gantt load chart. A legend is used to explain lines, bars and other chart marks.

The Gantt load chart in Figure 16.1 has a time scale running along the top. The rows represent resources: people, machines, stations, departments, or whatever facilities will be required to do the job. The time scale is labeled in either dated calendar time or with general intervals ahead, such as weeks.

The left-hand side of the chart is today. Depending on the time scale, a specific date can be used, or time 0. Bars and lines, running from left to right, convey various kinds of information in order to fit a department's particular needs.

Gantt Chart Legend

- For each facility, and week, there is a box which contains a horizontal line and two bars under that line. The top bar is purple. The lower bar is dark purple. The lines and bars for each week are divided into ten units of time. For all four weeks there are 40 units of time.
- The single horizontal line is the workload percent by the week. All lines start at the left. Moving to the right represents higher percentages of utilization. If the line runs from one side of the box to the other side, then that facility is 100% loaded for the week.
- Sequencing decisions have been reached with respect to weekly assignments as per the master production schedule. Orders have been placed with suppliers. Lead times have been calculated so that materials will be received in time to permit production to begin the job in that week. The order in which jobs will be done within each week has not yet been determined. *Note:* if the time buckets are monthly, then all times are monthly to be consistent.
- The Jobs column describes all of the jobs that are waiting at that facility. To the right of the job name is the number of units to be made for that order. The load consists of on-time jobs and backlogged jobs. Work is on time when it is scheduled for production so that delivery can be made on or before the promised due date. Work is backlogged when it is going to be late. This is shown by an asterisk in the Units column. An additional column can be added to the Gantt Load Chart to explain the causes and extent of delays.
- The top bar is the cumulative workload at the present time.
- The bottom bar (crosshatched) is the cumulative workload at the beginning of the prior week.

The cumulative load at each facility is presented in two ways. First, there is the load that exists at the beginning of the present week. Second, there is the load that existed at the beginning of the prior week. Comparing these can be useful to determine where and when the load is increasing, decreasing or staying the same. This helps the scheduler to assign load where it can be best processed.

Which departments are underutilized can be seen at a glance. That is one of the primary purposes of the Gantt chart which makes no pretense at being a method for optimizing assignments. In management science terms, *Gantt graphics facilitate heuristic methods*—which rely on "rules of thumb" and sound generalizations developed from experience. OM decisions must often be made with fast data scans ("eyeballing") that rely on the assistance of charts such as the Gantt load chart.

The charts can deliver additional information as well. They can show that certain departments have reserved time for preventive maintenance. This would be indicated by blocking off time along the horizontal line. Facility 1 has blocked 10% of its time in Week 1 with the symbol ■ for preventive maintenance.

Creating a chart can help visualize
the progress of jobs and the loads
on departments.
Source: Caterpillar, Inc.

The charts can show that particular jobs cannot be split up. This is shown by a + sign in the Jobs column. Running one job in two departments would mean two setups which can take too much time, cost a lot, and adversely impact quality control. Two different setups cannot have the same statistical fingerprints. For some jobs this lack of homogeneity can be a serious problem. Good Gantt charting prevents split assignments that impair productivity and quality.

When jobs cannot be split, for the economic and quality reasons mentioned above, there are some percentages of time available in departments which cannot be assigned work. Note that Facility 2 has 10% idle time in the second week. If it does not require scheduled maintenance, this time could be used for process improvement, training and cross-training exercises.

When meeting due dates is a problem, it is useful to add cumulative backlog of jobs to the Gantt chart. As with total load, the cumulative backlog data should be shown for the present week and the prior week. This information will indicate whether the backlog is increasing, decreasing, or staying about the same. *Note:* at Facility 1 load has increased from 22/40 to 24/40. At Facility 2 load has decreased from 35/40 to 31/40. However, it could be that backlog has decreased at the first facility and increased at the second one. Thus, when due date problems are a concern, the Gantt chart should reflect that fact.

As an illustration, the load scenario for Facility 1 with due date considerations might be as follows:

Week Number	Load Level Percent of Capacity	Cumulative Load		Cumulative Backlog	
		Present	Prior	Present	Prior
1	90	9/10	10/10	3/10	2/10
2	70	16/20	15/20	5/20	4/20
3	50	21/30	19/30	6/30	6/30
4	30	24/40	22/40	6/40	7/40

The load at Facility 1 by Week 4 has increased from 22/40 to 24/40. Due date problems (backlog) have diminished going from 7/40 to 6/40. Accompanying notes can indicate which backlog situations are considered to be the most serious.

Studying load charts can reveal whether there is a long-term capacity problem. It highlights bottlenecks and problem areas. When work that is scheduled is not completed on time, either the scheduler underestimated the time required for completion, or there is a process problem to be determined.

Constant matching of estimated and actual times improves the database for estimation, monitors process productivity, and reflects on the knowledge and skill of the schedulers. It is important to strive for the load being equally distributed among facilities. Overloaded departments suffer morale problems, especially when they see other departments that have idle capacity. Long-term solutions may be called for to provide capacity balance.

This kind of information must be fed back to the MRP departments and the master production schedulers. Often, the original scheduling concepts were applicable but the character of the jobs changed. Shifting job requirements can lead to an imbalance in load among departments. The changes can favor different materials, machines and skills. The variability of the order mix is a factor in planning capacity and taking corrective actions during aggregate scheduling. Variability in the loading mix is a difficult problem because the time for resource planning is long past.

Without knowing what the jobs are and the capabilities of the departments, it is not possible to evaluate how good a scheduling job has been done—both in the present week and in the weeks ahead. In a company where management knows its processes and its people, everyone in the organization should be aware of the schedule. Teamwork is the key to successfully adjusting to variability.

Employing the systems perspective, anyone who has some special knowledge that affects quality and/or productivity should feel empowered to communicate with those doing the scheduling. Inputs from suppliers regarding schedules can be critical and customer considerations may require special communication.

16.3

THE ASSIGNMENT METHOD FOR LOADING

The **assignment method** provides production scheduling with a decision model for matching jobs with facilities in an optimal way. Such loading of facilities determined by this model are subject to a variety of conditions described below.

The first scheduling question is always which facility, person, or department is best-suited to do the job. That raises another question: if the best-suited facility is occupied, or for any other reason unavailable, what facility is next best?

Using systems thinking, it becomes clear that these questions beg the issue, which is: what set of assignments will produce the overall best schedule? Best can be defined in many ways including highest output quality, greatest productivity, lowest cost, fastest completion or greatest profit. Best assignments (on a pair-by-pair basis) are unlikely to add up to optimal systems-wide solutions. The assignment problem must be visualized as a collection of paired assignments that interact as a whole.

The assignment model provides solutions when jobs compete for facilities where it is uneconomic to split the job. No splitting means that whichever facility, person or team gets the job does the whole job. Usually this is because split job set-up costs are prohibitive and/or quality control at two facilities is not manageable.

Assignment method

Provides production scheduling with a decision model for matching jobs with facilities in an optimal way.

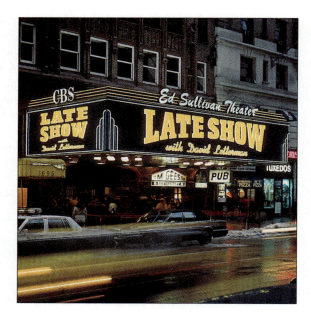

No splitting means that whichever facility gets the job does the whole job, and no other job.
The Ed Sullivan Theater in New York City is exclusively dedicated to, and the only facility where, The Late Show with David Letterman is taped.
Source: James Morse/CBS

There may be tooling or equipment for only one location. Many factory jobs cannot be worked on at two locations simultaneously. Project assignments are generally of this singular-assignment type.

It is uncommon for department-loading decisions to achieve optimal assignments for every job at every facility. Therefore, it is the set of assignments that is to be optimized. The concessions—which make some assignments that are less than the best—should result in an overall systems best.

Opportunity costs

Costs of not making the best possible assignments of jobs to facilities and vice versa.

The assignment model is based on **opportunity costs** which are the costs of not making the best possible assignments of jobs to facilities and vice versa. The reason that the best possible assignments cannot be made for every job is that jobs compete with each other for available time on those facilities for which they are best-suited. Thus, consider Table 16.1 which shows the cost per part for Jobs 1, 2, and 3, at Facilities A, B, and C.

T A B L E   1 6 . 1

COST PER PART FOR JOBS AT FACILITIES

		Facilities		
		A	B	C
Jobs	1	0.10	0.09	0.12
	2	0.08	0.07	0.09
	3	0.15	0.18	0.20

Jobs 1 and 2 have their lowest cost per part when made at Facility B. They are in direct competition with each other for the use of this facility. Job 3 has the lowest cost per part when assigned to Facility A. It is not competing with any other job for that facility. However, this assignment cannot be made until the conflict between Jobs 1 and 2 has been resolved. The reason is that whichever job is not assigned to Facility B might be better assigned to Facility A than Job 3.

All the numbers in the matrix are interrelated. This is a systems problem. Assignments should not be made on an individual basis. The assignment model develops comparisons of penalties (opportunity costs) for making an assignment other than the best possible one. To make the necessary comparisons, subtract the smallest number in each row from all other numbers in that row. This yields Table 16.2 which is a matrix of the opportunity costs for not assigning each job to the best possible facility for that job. There is a zero in each job row.

TABLE 16.2

JOB OPPORTUNITY COSTS

		Facilities		
		A	B	C
	1	0.01	0	0.03
Jobs	2	0.01	0	0.02
	3	0	0.03	0.05

All of the zeros in Table 16.2 are best possible job assignments. Because of the conflict between Jobs 1 and 2 at Facility B, the solution is stymied. Also, there is an assumption that the orders are of the same size. Otherwise, the matrix should represent total costs, not costs per unit as in Table 16.1 above. This means that the assignment model would begin with a matrix of total costs.

Column Subtraction—Best Job at a Specific Facility

Column subtraction yields the opportunity costs at a given facility for assigning jobs that are not the best jobs. These are facility opportunity costs. Using Table 16.2 above, subtract the smallest number in each column from all other numbers in that column. The opportunity costs of each facility with respect to jobs are shown in Table 16.3. There is a zero in each column.

TABLE 16.3

FACILITY OPPORTUNITY COSTS

		Facilities		
		A	B	C
	1	0.02	0.02	0.03
Jobs	2	0	0	0
	3	0.07	0.11	0.11

Job 2 has all zero opportunity costs in its row. It is the preferred job at each facility. The facilities are in competition with each other for Job 2. Therefore, thus far, there is no feasible assignment.

Row Subtraction—Best Facility for a Specific Job

The next step is to subtract the smallest number in each row of Table 16.3 above from all other numbers in that row. The instruction is applicable only for rows 1 and 3 because row 2 has zeros in it. This provides the combined opportunity cost matrix. It shows the penalty for each job being assigned to less than the best facility—and for each facility being assigned to less than the best job. Table 16.4 presents the total opportunity cost matrix.

TABLE 16.4

ROW AND COLUMN—COMBINED TOTAL OPPORTUNITY COST MATRIX STARTING WITH ROW SUBTRACTION

| | | Facilities | | |
		A	B	C
	1	0	•	0.01
Jobs	2	0	0	•
	3	•	0.04	0.04

(• = assigned zeros)

All jobs can be assigned to Facility A with zero opportunity cost. If this were acceptable (assuming that there was other work to be done at Facilities B and C), then a sequencing problem exists at Facility A. That is, which job goes first, which goes second, and which goes third?

Given that a job must be assigned to each facility without splitting, there is a feasible set of 0 opportunity cost assignments. They are shown in Table 16.4 as filled-in zeros, thus, • = assigned zeros.

The logic of this assignment starts with the fact that there is only one 0 in column C. Job 2 must be assigned there. There is only one 0 in row 3. Job 3 must be assigned to Facility A. That forces the assignment of Job 1 to Facility B.

Note that Table 16.2 above was derived by using row subtraction first. This yielded the opportunity costs of each job being assigned to less than the best facility. Instead start with column subtraction first and derive Table 16.5.

TABLE 16.5

ROW AND COLUMN—COMBINED TOTAL OPPORTUNITY COST MATRIX STARTING WITH COLUMN SUBTRACTION

| | | Facilities | | |
		A	B	C
	1	0.01	•	0.01
Jobs	2	0.01	0	•
	3	•	0.03	0.03

(• = assigned zeros)

Digital Equipment Corporation developed software it calls the "Factory Scheduler" for use in manufacturing printed circuit board assemblies. The software improves productivity by organizing circuit boards into compatible groups, thus reducing the time spent on machine set up. Digital manufacturing engineers have seen a significant increase in quality due to reduced feeder (or line transport) changeovers. Allan Fraser, a software engineer with Digital's Solutions Infrastructure Engineering Group, said that implementation of the new product groupings has reduced set-up time by 60 percent.

The Factory Scheduler software was developed by Digital's Modules Assembly Automated Production Planning (MAAPP) team. Input to the software consists of MRP data such as parts availability and product build forecasts; product description data such as part numbers, quantities and

Digital's "Factory Scheduler"

Increases quality with reduced changeovers

package types; and factory descriptions such as existing set ups, machine configurations and number of production lines. Factory Scheduler output groups products and assigns them to production lines. In doing so, the system assures that all material, feeder, and time constraints are met and the lines are loaded evenly. The software output also provides information on the utilization of machine and feeder slots and component usage by

product, machine, product group, line, and factory.

Digital has entered into an agreement with Mitron Corp. of Beaverton, Oregon, authorizing Mitron to sell Factory Scheduler to other manufacturers. Mitron has integrated Factory Scheduler into its own line of CIMBridge Framework productivity and quality management software. Doug McDougall, president and CEO of Mitron, said that "Without a tool like the Factory Scheduler, manufacturing organizations do not have the time to do effective operational planning with so many variables and such dynamic conditions. The Factory Scheduler handles the complexity and provides a means to optimize the manufacturing process and maximize return on factory assets."

Source: "Mitron to Sell DEC Factory Optimization Software for Circuit Board Manufacturing," Business Wire, March 7, 1994.

. . . .

The same set of assignments as in Table 16.4 above results. Although the final matrices are not identical, it is always equally correct to use either row or column subtraction first. However, sometimes it turns out to be easier to use one first instead of the other to reduce the matrix to an assignable set of zeroes. There is no way of prejudging which will be easier beforehand. Also, there are other complications that can be encountered. To deal with them requires studying the assignment method or the use of appropriate "assignment method" software.

It should be noted that if profit maximization is the goal, then all profit numbers should be subtracted from a number larger than any one of them. To illustrate, if the profit per part, given by the matrix in Table 16.6 is subtracted from $1.00, the result is Table 16.1 above. Thus, the problem solved in Tables 16.4 and 16.5 (above) maximizes profit.

TABLE 16.6

PROFIT PER PART

		Facilities		
		A	**B**	**C**
	1	0.90	0.91	0.88
Jobs	2	0.92	0.93	0.91
	3	0.85	0.82	0.80

Using the assignment method demonstrates the importance of opportunity costs for loading. The weakness (or strength when it applies) of the assignment model is that only one job or team can be assigned to each department, facility or machine, at a time. This condition accurately describes the permanent post of managers to stores. It is acceptable when every job takes the same amount of time. However, when jobs take different amounts of time to complete, depending on the pairing, the assignment method is not appropriate.

16.4

THE TRANSPORTATION METHOD FOR SHOP LOADING

When it is alright to split assignments, the transportation model can be used. The transportation model permits assigning more than one job to a machine or team. In Table 16.7, Machine 1 is assigned three jobs called A1, B1, and C1. Loading involves matching the supply of available machine time, $S(i)$ with the demand of jobs for machine time $D(j)$.

The second weakness of the assignment method is that no job can be divided. In Table 16.7, Job C has been split into three parts: C1, C2, and C3.

While the transportation method overcomes assignment method difficulties, it has certain restrictive assumptions of its own. Split assignments are easy to make, but when demand, $D(j)$ is loaded against supply $S(i)$ the amounts of supply and demand must be expressed in terms of standard hours.

TABLE 16.7

SPLIT ASSIGNMENTS OF TWO KINDS WITH THE TRANSPORTATION MODEL USED FOR JOB-SHOP LOADING

		JOBS (j)			
		Job A	**Job B**	**Job C**	**Supply $S(i)$**
	1	A1	B1	C1	
Machines	2			C2	
(i)	3			C3	
Demand D(j)					

To create standard hours it is necessary to assume that strict proportionality exists between the productivity rates of the machines. To review this concept, developed when explaining aggregate scheduling, note that in Table 16.8, Machine 1 is half as fast as Machine 2—which has arbitrarily been chosen to be the Standard Machine (SM). Similarly, M3 is 0.8 as fast as the SM.

TABLE 16.8

RELATIVE PRODUCTIVITY AND THE EQUIVALENT
SMH INDEX (FOR ALL *I*)

Machine (i)	SMH Index	Relative Productivity
M1	0.5	Half as fast as SM
M2	1.0	Standard Machine (SM)
M3	0.8	80% as fast as SM

The restrictive assumption is that the SMH Index for each machine applies to all jobs. It is called the presumption of productive proportionality. Thus, *M2* is fastest—for all of the jobs—by the same amount, in comparison with the other machines. The index is derived in more rigorous terms than before, as follows:

$$\text{SMH Index of } M\,(i, j) = \frac{\text{Production Output of } M(i,j)}{\text{Production Output of } M(SM,j)} \quad \text{(for all } j\text{)}$$

For each machine, when the SMH Index is multiplied by the actual machine hours available per week, the standard machine hours (SMH) available per week are obtained. Thus, Table 16.9 provides the actual machine hours that are available per week

A quick changeover packaging machine enables Merck & Company to assign more than one job to a machine.
Source: Merck & Co., Inc.

converted by means of the SMH Index into standard machine hours available per week. Table 16.9 concludes that Total SMH = 136 per week. Note that a week was chosen as a convenient period which could as readily have been a day or a month.

DEPARTMENT TIME CONVERTED INTO *SMH*

Machine	SMH Index	Actual Hours/Week	SMH per Week
1	0.5	36	18
2	1.0	54	54 (SM)
3	0.8	80	64
Total:	Actual Hours/Week =	170	SMH = 136 Hours/Week

The fact that there are 170 actual machine hours available per week has significance after the resolution of the problem. The transportation model is solved using the 136 standard hours that are available per week.

Table 16.10 provides production output rates (in pieces per hour) for all the machines for each job. The entries in the matrix are called $PR(i,j)$ where i = machine and j = job, which is consistent with the notation throughout this section. The production rate of the standard machine is denoted by $PR(SM,j)$.

PRODUCTION RATES *PR(i, j)* IN PIECES PER HOUR FOR MACHINES *I* AND JOBS *J*

	Jobs →	A	B	C	SM Index
	M1	3	3.5	15	0.5
Machines	M2	6	7	30	1.0 (SM)
	M3	4.8	5.6	24	0.8

Demands are initially stated as the number of units per order and can vary greatly in type for each order. The orders can put loads on the facilities for various lengths of time. An order for 300 *Cs* would take 10 hours on the standard machine and 20 hours on M1.

The demands for goods or services, called $D(j)$, also must be turned into standard hours. To accomplish the transformation divide the demands by the production rates of the standard machine. This yields demand in standard hours $D'(j)$ instead of demand in units $D(j)$. The equations are shown on the left side of Table 16.11. On the right side of Table 16.11 is unit demand divided by the production rates of the SM for each job.

TABLE 16.11

CONVERT JOB DEMANDS FROM UNITS TO STANDARD HOURS

$$D(j) \div PR(SM, j) = D'(j)$$

$D(A) \div PR(SM,A) = D'(A)$	300 A units $\div$ 6 =	50
$D(B) \div PR(SM,B) = D'(B)$	210 B units $\div$ 7 =	30
$D(C) \div PR(SM,C) = D'(C)$	1800 C units $\div$ 30 =	60
	Total standard hours of demand	= 140

Now, both supply and demand have been converted to the common terms of SMH. There are only 136 standard machine hours available per week and 140 hours demanded to complete the orders. Therefore, $140 - 136 = 4$ SMH of demand will not be met this week.

A dummy machine must be created to pick up the slack. It is called MD and has zero output for all jobs. Some part of whichever job is assigned to the dummy will not be done. The balanced supply and demand system with all numbers in standard hours including the output rates per standard hour in the upper right-hand corners of each assignment cell is given in Table 16.12.

TABLE 16.12

TRANSPORTATION MODEL: ALL NUMBERS IN STANDARD HOURS
INNER BOXED VALUES ARE PRODUCTIVE OUTPUT PER HOUR

	Job A	Job B	Job C	Supply S(i)
M1	3	3.5	15	18
M2	6	7	30	54
M3	4.8	5.6	24	64
MD	0	0	0	4
Demand D(j)	50	30	60	140

The profit per piece π_j is calculated for each Job j. When it is multiplied by the product output rates $PR(i,j)$ the result is $\Pi_{ij} = \pi_j PR(i,j)$ which is profit per standard hour. This formulation assumes that the costs per piece are the same on all machines.

When that assumption is incorrect, the revenue r_{ij} less the cost c_{ij} is multiplied by the product output rates $PR(i,j)$ for each cell in the matrix. Thus, where π_{ij} = profit per piece for each cell i,j, and Π_{ij} = profit per standard hour for each cell i,j:

$$\Pi_{ij} = \pi_{ij} PR(i,j) = (r_{ij} - c_{ij}) PR(i,j) \qquad \textbf{Equation 16.1.}$$

Often, the revenue is simply r_j because it is a function of the item and not the machine. However, if quality differences occur as a result of the machine used which affect price charged, then Equation 16.1 should be used.

When the profit per piece is as shown in Table 16.13, then the optimal load assignments are as indicated in that table. Note that the product output rates per hour have been adjusted to reflect the profit per standard hour in the boxes in each cell.

TABLE   16.13

TRANSPORTATION MODEL: MATRIX ADJUSTED FOR PROFITS, Π_{ij}

Profit per Piece	$2.00	$1.00	$0.50	
	Job A	Job B	Job C	Supply S(i)
M1	18 [6]	[3.5]	[7.5]	18
M2	[12]	[7]	54 [15]	54
M3	32 [9.6]	26 [5.6]	6 [12]	64
MD	[0]	4 [0]	[0]	4
Demand D(j)	50	30	60	140

The goal is to find the loading arrangement that uses the resources available to maximize total profit. The optimal loading schedule is shown in Table 16.13 in standard machine hours with the:

Maximum total profit $(\sum_i \sum_j \Pi_{ij})$ = $(18 \times 6) + (54 \times 15) + (32 \times 9.6) +$
$(26 \times 5.6) + (6 \times 12) + (4 \times 0) = \$1,442.80$

There are four rows and three columns which means that $M + N - 1 = 6$ assignments should be made for the transportation model. The requisite six are in place. The optimal has been found for this example using the standard techniques employed to solve the transportation model. Software is readily available to solve transportation problems involving many variables. It also can be solved as a linear programming problem.

The solution in terms of standard hours must now be converted back into actual machine hours and actual job units. Divide the standard hour assignments by the SMH Index for each machine. This is shown in Table 16.14.

TRANSPORTATION MODEL: ACTUAL ASSIGNMENTS

		Job A	Job B	Job C	Supply S(i)
Machines	M1	36 3	3.5	15	36
	M2	6	7	54 30	54
	M3	40 4.8	32.5 5.6	7.5 24	80
	MD	0	4 0	0	4
Demand Filled		300	182	1800	140
Shortages		0	28	0	
Demand D(j)		300	210	1800	

- For Job A: The load is established:

  18 SMH ÷ SMH Index of 1/2 equals 36 actual hours assigned on Machine 1 Job A which yields 36 × 3 = 108 units of A

  32 SMH ÷ SMH Index of 0.8 equals 40 actual hours assigned on Machine 3 for Job A which yields 40 × 4.8 = 192 units of A

  Total 108 + 192 = 300 units as required for Job A

- For Job B:

  26 SMH ÷ SMH Index of 0.8 equals 32.5 actual hours assigned on Machine 3 for Job B which yields 32.5 × 5.6 = 182 units of B

  4 SMH are assigned to the dummy machine MD so Job B will not be completed

  There is a shortage of (210 − 182) = 28 units for Job B

- For Job C:

  54 SMH ÷ SMH Index of 1.0 equals 54 actual hours assigned on Machine 2 for Job C which yields 54 × 30 = 1620 units of C

  6 SMH ÷ SMH Index of 0.8 equals 7.5 actual hours assigned on Machine 3 yield 7.5 × 24 = 180 units of C

  Total = 1620 + 180 = 1800 units as required for Job C

This transportation model has been solved for total profit maximization. It is as simple to use costs and solve for total cost minimization. Total time minimization, productivity maximization and other goals can be sought as well.

The criterion that must be satisfied for using this approach to shop loading is that reasonable proportionality exists between machine output rates. When the SMH Index does not apply, then a heuristic modification of the transportation method should be used. If one machine is especially efficient for Job A and another machine is best for Job B, the heuristic could make those assignments more profitable or less costly. Also, the loading problem can be treated by linear programming without the transportation model constraints.

16.5

SEQUENCING OPERATIONS

Loading assigns work to facilities without regard to the order in which the jobs will be done. Sequencing establishes the order for doing the jobs at each facility. Sequencing reflects job priorities according to the way that jobs are arranged in the queues. There are different costs associated with the various orderings of jobs.

Loading and sequencing in a typical job shop are done over and over again. Each time, doing it the right way provides a small savings. This has been demonstrated for loading where doing it the right way can provide better profit or lower costs.

Good sequencing provides less waiting time, decreased delivery delays and better due date performance. There are costs associated with waiting and delays. Total savings from regularly doing it right can accumulate to substantial sums. When there are many jobs and facilities, sequencing rules have considerable economic importance.

First-In, First-Out (FIFO) Sequence Rule

FIFO
"First-in, first-out" sequencing rule; the first jobs into the shop get worked on first; sometimes referred to as "first-come, first-served" (FCFS).

LIFO
"Last-in, first-out" rule. Usually true on an elevator.

The most natural ordering for doing work is in the order that the jobs are received. That means that the first jobs into the shop get worked on first. This is called **FIFO** for "first-in, first-out." Supermarkets like to use FIFO for their late-dated products (do not use after 6/6/98). There is a cost advantage in getting older products to be purchased first.

LIFO which is "last-in, first-out" can cause spoiled milk problems. On the other hand, LIFO can save warehouse handling costs where the product date does not matter; to move things around to get the first one in to be the first one out can cost a great deal.

FIFO is an appealing sequencing policy because it seems to be the fairest rule to follow. Sometimes—to emphasize the fair treatment sense—it is called "first-come, first-serve" (FCFS). Customers can get angry when someone seems to jump to the head of the line. The cost of angry customers is not to be trivialized. However, by at least one measure, FIFO is unfair because it penalizes the average customer. The penalty is extra waiting time for the set-up and processing time of the average order. This means that all regular customers will wait longer even though FIFO may get specific orders finished sooner.

Customers who regularly submit orders with short set-up and processing times will benefit if the job shop does not employ the FIFO rule, but uses instead a shortest processing time (SPT) priority rule. Customers who regularly submit long orders may be discriminated against by the shortest processing rule. If so, compensatory steps can be taken at the discretion of those doing the sequencing. Note Section 16.8 below on modified SPT rules.

16.6

GANTT LAYOUT CHARTS

It is useful to visualize the conversion of loading information to sequencing decisions. Each department has a list of assigned jobs. The Gantt layout chart arranges the list in order of processing.

Gantt, the master chart maker, developed this chart to show sequence assignments. Called a **layout or sequencing chart**, it reserves specific times on the various facilities for the actual jobs that have been assigned. It appears in a number of different forms depending upon the sequencing applications.

Figure 16.2 shows how the Gantt layout chart assigns the specific jobs to the particular facilities, over some given period of time.

Layout or sequencing chart

Reserves specific times on the various facilities for the actual assigned jobs; Gantt-type chart.

FIGURE 16.2

THE GANTT LAYOUT CHART (A RESERVED TIME-PLANNING SYSTEM)

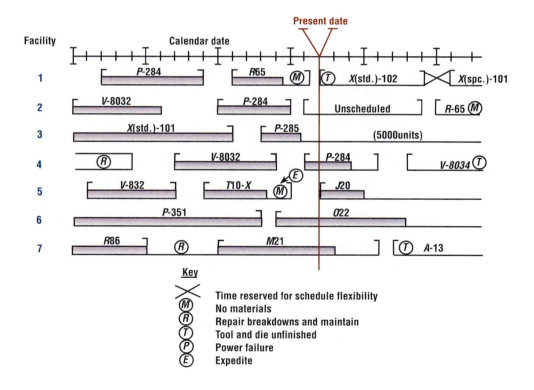

Key

✕ Time reserved for schedule flexibility
Ⓜ No materials
Ⓡ Repair breakdowns and maintain
Ⓣ Tool and die unfinished
Ⓟ Power failure
Ⓔ Expedite

Status: (1) job P-284 is ahead 2 days; (2) job J20 is ahead 3.0 days; (3) job O22 is ahead 6.0 days; (4) job M21 is ahead 1.1 day; (5) job R-65 is 2 days short of completion and 2.5 days late waiting for materials (M); (6) job P-285 is 1.5 days late with 5000 units; (7) job T10-X is 1.5 days short and about 3.5 days late (M) and (E); (8) Facility 2 is unscheduled for 7 more days.

The FIFO (First-In, First-Out) Sequence Rule or first-come, first-served seems to be the fairest rule to follow in the service sector. Source: Supercuts and Gensler & Associates

Concurrently, past sequences of work can be monitored to discover the state of completion of those jobs that were scheduled to be run in prior time periods. Thus, while Gantt layout charts provide work-schedule control they offer little help in determining the best-work sequences.

The chart shows the job schedule at each facility and the state of completion of all jobs. The present date is shown by the arrow and its associated vertical line on the chart. Thereby, the chart is divided into time past, present, and future. It is easy to see which jobs have been finished and which should have been finished but are not. It is simple to look ahead to observe which jobs are coming up and in what order. Thus, the difference between loading and sequencing is that sequencing specifies the precise timing of assignments. Thereby, it allows customers to be notified of delivery dates.

Once an assignment is made it blocks other assignments from being made. Because sequencing charts are revised regularly, assigned time can be unblocked if it appears to permit a better schedule. A correct amount of time is allowed between jobs, to account for machine maintenance, to absorb divergences from estimates and to allow for set ups and takedowns.

Frequently, additional symbols are attached to Gantt charts to indicate why a job has not been completed and which jobs are being expedited. Some of these symbols are shown on Figure 16.2. A lot of useful information can be conveyed by means of succinct shorthand notations. Each organization develops its own conventions. Consultants learn a great deal about a company by studying its Gantt charts.

The Gantt layout chart is continually updated. Jobs must be rescheduled. A good chart must be easy to alter for contingencies such as machine breakdowns, materials that do not arrive as scheduled, and customers suddenly demanding delivery because of special circumstances. Sequencing flexibility is encouraged when it is easy to redraw the charts. There are many commercial products that permit on-the-wall Gantt charting.

Jobs are indicated by i and the symbol n is used to refer to the number of jobs that are waiting to be sequenced through the facility ($i = 1, 2, \ldots, n$). In the typical job shop, n varies a great deal. Also, it is necessary to specify the number of facilities, m, through which the jobs must pass. The character of the sequencing problem is often specified as the n by m problem and represented by $n \times m$.

It is necessary to know how much time each job must spend at each facility. For a given facility, this is called t_i, for the ith job. It is usual for t_i to include set-up time and processing time. (From here on, the combination of set-up and operations times will be referred to as processing time.) It does not include waiting time W_i.

The potential for varying flow patterns in the job shop must be understood. When a set of jobs follows a fixed ordering, the conditions exist for an intermittent flow shop. Knowledge of technological orderings is required for job-shop sequencing. The sizes of orders, the number of alternative facilities for doing each job, the length of production runs, the set-up costs and times, among other things, are the determinants of what type of shop exists.

Evaluatory Criteria

Many criteria exist for evaluating production schedule sequencing. To determine how good a sequence is, evaluation measures called total flow time and mean flow time are used. These are defined in the following way.

Total flow time is the cumulative time required to complete a group of jobs. It is composed of the sum of the complete-to-ship times (including waiting time) for each job in the group. **Mean flow time** is the average amount of time required to complete each job in the group. It is the average of the wait-to-start and processing times for every job in the group.

$$\sum_{i=1}^{i=n} C_i$$

is the equation for total flow time and

$$\sum_{i=1}^{i=n} C_i / n$$

is the equation for mean flow time.

To sequence a group of jobs at a specific facility, define for each Job i a waiting time W_i, and a processing time, t_i. If Job i begins at time zero, then its completion time is:

$$C_i = W_i + t_i = 0 + t_i = t_i$$

Beginning at time zero is the same as saying that the release time, called r_i, equals zero. This might be construed to be the start of a day or a week, etc. Table 16.15 is constructed for three Jobs A, B, and C. They are processed in that order with $r_A = 0$.

Total flow time

Cumulative time required to complete a group of jobs; composed of the sum of the complete-to-ship times for each job in the group.

Mean flow time

Average amount of time required to complete each job in the group; average wait-to-start and processing times for every job in the group.

TABLE 16.15

THREE JOBS TO BE SEQUENCED

Order	Job i	W_i	t_i	C_i
1	A	0	5	5
2	B	5	3	8
3	C	8	4	12
	Sums	13 $+$	12 $=$	25

$\sum_i C_i$, the total flow time is, in this case, 25. Mean flow time with $n = 3$ is

$$\sum_i C_i / n = 25/3 = 8\,\tfrac{1}{3}$$

Part of flow time and mean flow time is job waiting time W_i. The target is to make the total waiting time as small as possible on the assumption that all customers want to have their orders filled as quickly as possible. Therefore, if mean flow time is minimized, this is equivalent to minimizing the mean waiting time because the processing times t_i are fixed.

If the average completion time $\sum_i C_i / n$ is as small as possible, the average customer's order is delayed the minimum necessary time. This is a primary objective in determining rational sequence priorities.

Figure 16.3 shows the three Jobs (A, B, and C) which have been sequenced in the order that they arrived (FIFO).

FIGURE 16.3

THREE JOBS SEQUENCED IN THEIR ORDER OF ARRIVAL (FIFO)

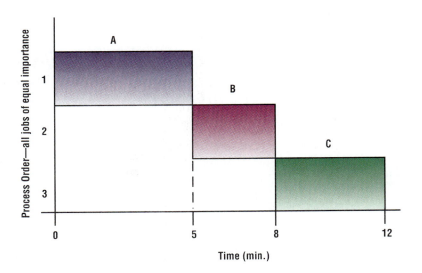

The x-axis is elapsed time. The y-axis shows the sequence used. It is the same information as in Table 16.15 above. A goes first and starts at the top. B is the next step down. The steps are all of unit value because all jobs are considered to be of equal importance. In the section on priority-modified jobs, the step size will be greater for important jobs. Job A is completed after 5 minutes. Jobs B and C have waited 5 minutes. Then Job B is done in three minutes while Job C waits. Next Job C is completed at 12 minutes. The completion times of 5, 8, and 12 are added together to obtain total flow time for the group of three jobs.

Other measures are used in evaluating the performance of differing sequences. There are promised delivery dates. Define d_i as the due date of a job. Then, $L_i = C_i - d_i$ is a measure of the lateness of Job i if $d_i < C_i$. To minimize average lateness, minimize mean flow time. It is assumed that promised dates are reasonably determined.

The degree of facility idleness also may play a critical role in sequence determination. The fact is, however, that sequencing models become increasingly difficult to handle with optimizing models, as the number of facilities through which the jobs must pass becomes large. Because the jobs take different times at each facility, what is the best sequence at one location may not be best at another.

The degree of data complexity is large. Consequently, heuristics have been developed to cope with the level of detail involved in making highly repetitive sequencing decisions required by many job shops.

16.7

THE SPT RULE FOR *n* JOBS AND *m* = 1 FACILITIES

The common objective of facility sequencing—to minimize mean flow time—is usually equivalent to minimizing average job waiting times, as follows:

$$\min \frac{\sum_i^n C_i}{n} = \min \frac{\sum_i^n (W_i + t_i)}{n} = \min \frac{\sum_i^n W_i}{n}$$

A simple rule achieves this objective. Namely, if a set of n jobs in one facility's queue ($m = 1$) are arranged so that the operations having the shortest processing times (SPT) are done first, then mean flow time, and mean waiting time will both be minimized. Sequencing jobs according to their shortest processing times is called the **SPT rule**.

Using SPT to sequence Table 16.15 for Jobs A, B, and C results in the following decrease in total flow time.

SPT rule

Sequencing jobs according to their shortest processing times (SPT); helps to minimize average delivery lateness and maximize fulfillment of delivery promises.

Order	Job i	W_i	t_i	C_i
1	B	0	3	3
2	C	3	4	7
3	A	7	5	12
	Sums	10 +	12 =	22

There is a 12 percent reduction in total flow time. For further contrast, note the effect of sequencing by longest processing time, called LPT.

Red Pepper's "ResponseAgents"

Red Pepper Software was founded in 1993 to provide manufacturers with a new generation of advanced planning and scheduling software. The company's products evolved from the Ground Processing Scheduling System developed for NASA to schedule after-flight refurbishment of the space shuttle. Red Pepper's founder, Monte Zweben, was Deputy Director of the Artificial Intelligence (AI) Division at NASA Ames Research Center. Red Pepper developed an AI-based, object-oriented system it calls the "ResponseAgent" that functions as a manufacturing manager's "intelligent assistant" in establishing plans, monitoring critical production variables, flagging problems as they arise, and optimizing and recommending solutions in real-time.

ResponseAgents are designed to integrate with a manufacturer's existing information systems like MRPII to capture basic data. The ResponseAgents system includes a "Databridge" which is a data exchange vehicle that integrates ResponseAgents with transactional systems and transfers information to and from those systems. Once the links are established, ResponseAgents independently collect transactional data and communicate changes in plans and schedules.

ResponseAgents consider both materials constraints (as with MRP systems) and physical constraints (as with finite scheduling techniques) and integrate these with customer demand to generate optimized plans that minimize changeovers, balance lines, and efficiently allocate scarce inventories. The system employs "exception-based optimization" to accommodate schedule changes. The model is built on the basis of constraints, exceptions (or violations), and repairs (to the exceptions). The system employs a scoring model that seeks to minimize the negative scores associated with the exceptions in order to achieve the best schedule. It goes through an iterative process and scores the schedule in terms of the number of constraints that are violated, always striving for the minimum score. The system must check "repairs" to make sure existing constraints are not violated.

ResponseAgents can be customized for an individual plant or a worldwide enterprise. A virtual model of the factory is constructed, through which data is processed during the examination of constraints. Because of the product's object-orientation, the software takes only one to three months to customize. ResponseAgents were beta-tested by Cisco Systems, a manufacturer of network routers. In addition to Cisco Systems, ResponseAgents are being used at 3Com Corporation, Sun Microsystems Inc., and other leading manufacturers.

Source: "Red Pepper Software founded to provide new class of manufacturing planning and scheduling systems compatible with existing MRP, shop floor control systems," Business Wire, November 4, 1994; and special assistance from David Obershaw, consultant to Red Pepper Software.

Order	Job *i*	W_i		t_i		C_i
1	A	0		5		5
2	C	5		4		9
3	B	9		3		12
	Sums	14	+	12	=	26

With respect to the SPT result, there is an increase of 18 percent in the total flow time.

Figure 16.4 provides a larger example of SPT, where:

$$t_4 < t_5 < t_2 < t_6 < t_3 < t_1$$

JOBS ORDERED BY SHORTEST PROCESSING TIMES

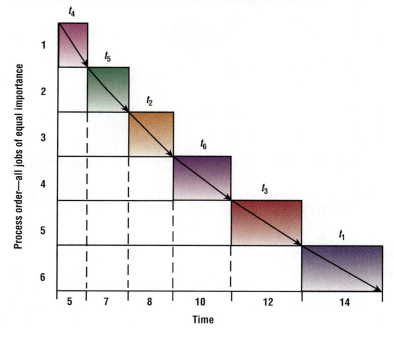

There are *n* = 6 jobs and *m* = 1 facility as shown in Table 16.16.

TOTAL WAITING TIME AND FLOW TIME WITH SPT

Order	Job *i*	W_i		t_i		C_i
1	4	0		5		5
2	5	5		7		12
3	2	12		8		20
4	6	20		10		30
5	3	30		12		42
6	1	42		14		56
		109	+	56	=	165

Note that the curve in Figure 16.4 has its convex (curving outward) surface down which minimizes the area of waiting time under the curve.

In Figure 16.5 the same jobs shown in Figure 16.4 and Table 16.16 (above) are processed according to FIFO.

FIGURE   16.5

JOBS ARE PROCESSED FIFO

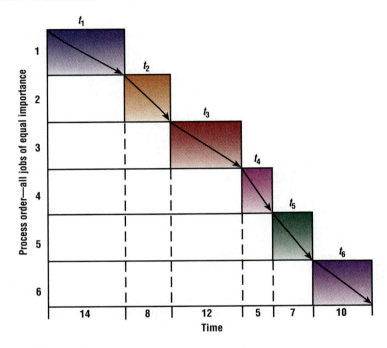

In terms of processing times this is a random sequence without regard to SPT. Any length job is as likely to enter the shop first as any other. The orders are processed in FIFO ordering. Neither the area under the curve nor the mean flow time is minimized. Table 16.17 presents waiting time, processing time and flow time calculations.

TABLE   16.17

FIFO PROCESSING ORDER

Job i	W_i	t_i	C_i
1	0	14	14
2	14	8	22
3	22	12	34
4	34	5	39
5	39	7	46
6	46	10	56
	155 +	56 =	211

The waiting time areas under the curves in Figures 16.4 and 16.5 above can be determined by multiplying each job's processing time t_i by the number of jobs waiting for it to finish, and summing them all.

In each figure, there are columns of blocks which represent the number of jobs waiting plus being processed. Each job carries its stack of waiting jobs, except the last one. When the stack size is multiplied by the processing times, the result is total job time spent in the system. It is equal to flow time. Table 16.18 shows the computations for Figure 16.4 above based on SPT.

TABLE 16.18

TOTAL JOB TIME SPENT IN THE SYSTEM WITH SPT

i	t_i	×	**Number of Jobs Waiting and Working**	=	**Job Time Spent in the System**
4	5	×	6	=	30
5	7	×	5	=	35
2	8	×	4	=	32
6	10	×	3	=	30
3	12	×	2	=	24
1	14	×	1	=	14
Total waiting time in the system				=	165

This is the identical value obtained for flow time in Table 16.16 above. Note how the larger values of t_i are multiplied by the smaller values of the number of jobs waiting and working. That is the SPT effect.

Compare the results of this stack analysis with that applicable to Figure 16.5 above where FIFO sequencing was used.

TABLE 16.19

TOTAL JOB TIME SPENT IN THE SYSTEM WITH FIFO

i	t_i	×	**Number of Jobs Waiting and Working**	=	**Job Time Spent in the System**
1	14	×	6	=	84
2	8	×	5	=	40
3	12	×	4	=	48
4	5	×	3	=	15
5	7	×	2	=	14
6	10	×	1	=	10
Total waiting time in the system				=	211

This is the same value that was obtained earlier in Table 16.17 for flow time. In this case, the largest value of t_i happens to be multiplied by the largest value of the number of jobs waiting and working. FIFO is not a logical sequence if it is used to minimize waiting time or total time spent in the system.

Sequencing jobs by the SPT rule can help to minimize average delivery lateness L_i, and maximize fulfillment of delivery promises. It cannot overcome delivery promising that is physically unrealizable. The beneficial effects follow from the fact that wasted time W_i is minimized under SPT. Total processing time,

$\sum_i^n t_i$, is fixed. When flow time, $\sum_i^n C_i$, is minimized so is average lateness, L_i.

16.8

PRIORITY-MODIFIED SPT RULES: JOB IMPORTANCE—DUE DATE— LATENESS—OTHER PRIORITIES

The SPT rule may need to be modified to take into account the fact that some jobs are more important than others. This is done by dividing each job's time t_i by the relative importance of that job w_i.

Thus, t_i/w_i are rank ordered. The job having the smallest value of t_i/w_i is placed first. The job having the next smallest value of t_i/w_i is placed second, and so on.

An important job will have a large value of w_i resulting in a small number t_i/w_i. This will place the job earlier in the modified SPT sequence than would otherwise have been the case. Importance may be assigned a number from 1 to 10 depending on the potential lifetime value of the customer, the profitability or margin contribution of each job, the probability of making an occasional customer into a long-term loyal one, or who yells the loudest.

Alternative interpretations of w_i include lateness such that t_i/L_i is the ratio used for SPT. A job which is not late carries an $L_i = 1$. Another system is to let w_i equal the number of days that the job has been waiting to be processed.

Critical Ratio for Due Dates

Due date

Number of days remaining before promised delivery.

As another variant, **due date** d_i is the number of days remaining before promised delivery. A sequencing rule—called critical ratio—rank orders the ratios d_i/t_i. Jobs are sequenced according to the SPT rule applied to these ratios. The smallest ratio is first in line, etc. The first in line will have some combination of least time available until due date divided by longest processing times which obstruct on-time delivery.

Figure 16.6 illustrates the modified SPT rule for priority ratios of t_i/w_i but it applies as well to d_i/t_i.

FIGURE 16.6

JOBS SEQUENCED IN THE ORDER OF THE SMALLEST RATIOS OF WEIGHTED PROCESSING TIME

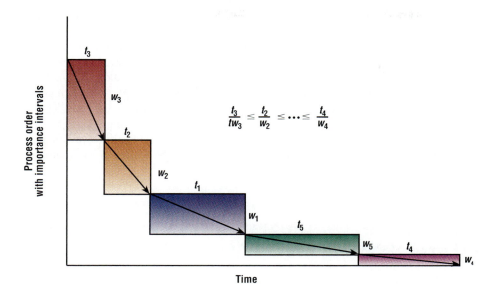

$$\frac{t_3}{tw_3} \leq \frac{t_2}{w_2} \leq \ldots \leq \frac{t_4}{w_4}$$

The importance weight, w_i alters the height of the blocks which is the step size, and thereby the slope of the arrows. Previously, in Figures 16.4 and 16.5, the w_i's were all equal to one. Now, the most vertical arrow (which has the largest slope) will belong to the job with the smallest ratio t_i/w_i. The same reasoning applies to the critical ratios.

As shown in Figure 16.6, the ratio of t_3/w_3 is smallest. Therefore, Job 3 with processing time t_3 is placed first in the sequence. This ratio rule assures a smooth (convex) function that will minimize the area under the curve.

By using a ratio, the production scheduler doing the sequencing gains greater control over the factors. Processing time is an important sequencing factor, but the other factors drive priorities as well.

16.9

SEQUENCING WITH MORE THAN ONE FACILITY ($m > 1$)

The SPT character of the sequencing problem is easy to understand with n jobs and 1 facility. The optimal sequence through the single facility can be calculated. There are $n! = (n)\,(n - 1)\,(n - 2)\ldots(1)$ ways to sequence n jobs through one facility.

The complexity of finding the minimum flow-time sequence becomes increasingly difficult as larger numbers of facilities are considered. When there are n jobs with $m = 2$ facilities, the method for finding the minimum waiting-time sequence is fast and concise.

With the $n \times 3$ problem, the optimal sequence cannot always be derived. There are certain conditions that must be met. When m is four or greater, heuristic methods must be used to find a satisfactory sequence. Optimization methods cannot be used to minimize flow time. The number of sequencing combinations $(n!)^m$ grows large quickly. Thus, when $n = 5$ and $m = 4$, $(n!)^m$ is larger that 200 million.

The SPT concept still applies when jobs must pass through 2 facilities in a given technological ordering. For example, Facility 1 (F1) drills the part. This is followed by Facility 2 (F2) which deburrs the part after the drilling.

Instead of using another manufacturing example, shift to a health care situation. The HMO clinic has five people requiring diagnosis to be followed by treatment. Table 16.20 presents the data (in minutes).

TABLE 16.20

HMO SEQUENCING PROBLEM—TIME GIVEN IN MINUTES

Job (Person)	F1 (Diagnosis)	F2 (Treatment)
a	6	3
b	8	2
c	7	5
d	3	9
e	5	4

S. M. Johnson's algorithm derives the minimum completion times for all "no passing" cases.[1] "No passing" means that the order of processing jobs through the first facility must be preserved for all subsequent facilities.

To use the modified SPT sequencing rule for Table 16.20 where F1 must be done first, select the job (or person) with the shortest processing time in either the F1 or the F2 columns. If this minimum value is in the F2 column, place that person last in sequence (here in fifth place). If it is in the F1 column, put that person in first place. For this example, person b with 2 in the F2 column will be treated last. Remove person b from further consideration, then continue in the same way.

Select the smallest number remaining in the matrix. If it is in the F2 column, assign that person to the last place—if it is available, or the next to last place if not. If the smallest number is in the first column, that person is given first place—or next to first place.

Resolve ties by randomly selecting either position for assignment. For the example above: person b goes last, person d goes first, person a goes next to last, person e is treated next to next to last, and person c fills the remaining slot. The minimum total completion time is 31, as shown by the Gantt chart in Figure 16.7. There is idle time at F2 of five minutes.

FIGURE 16.7

GANTT CHART FOR SEQUENCING *n* × 2 PROBLEM

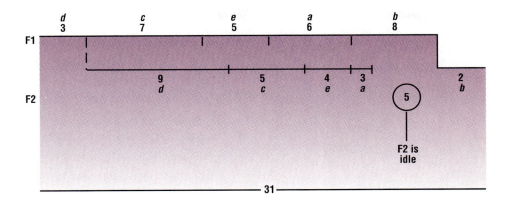

16.10

FINITE SCHEDULING OF BOTTLENECKS— OPT CONCEPT

Bottlenecks (BN) are well described by their name. The narrowing of the bottle's neck constricts the flow of output. In terms that are more germane to production, the slowest machine sets the pace of the process flow. When the pace of a person or machine is exceeded by the demand for his, her or its time, waiting lines begin to grow.

Everyone has experienced the feeling of keeping people waiting because other assignments must be completed first. People who are bottlenecks in their organizations (often because of the design of the system and through no fault of their own) are placed under stress. Workstations that are bottlenecks create tension. Job shops always have one or more bottlenecks. These change, according to the product mix in the shop.

Bottlenecks (BN)
The slowest machine sets the pace of the process flow; when the pace is exceeded by the demand, waiting lines grow.

Flow Shops and Continuous Processes Do Not Have Bottlenecks

Properly designed flow shops and continuous processes, such as a refinery, have no bottlenecks. They are designed to eliminate all bottlenecks by pre-engineering. Their process flows are fully balanced. In the flow shop machines are timed to supply each other at just the right rate. When people and machines interact at workstations, a paced conveyor belt carries the work between stations. The timing of the conveyor (how fast it moves between stations and how long it stays at the station) is the drum beat of the processing system.

Dr. Eliyahu M. Goldratt
. . .

Dr. Eliyahu M. Goldratt, renowned worldwide for developing new production control and management philosophies and systems, challenges companies to "break outside of the box" of conventional practices in their ongoing efforts to achieve business goals. In his early work, Goldratt developed the concept and technology for bottleneck-based finite scheduling. From this he concluded that optimizing production scheduling in isolation from other business variables usually does not lead to higher productivity or achievement of company goals. As Goldratt's work evolved, he began to call his new, more encompassing ideas and concepts the Theory of Constraints (TOC).

Based on systematic thinking processes, TOC studies the cause and effect relationships between dependent aspects of a given problem or project. It provides an overall framework for determining (1) *what to change* (not everything is broken); (2) *to* what to change (what are the simple, practical solutions); and (3) *how* to implement change (overcoming the inherent resistance to change). TOC enables the determination of core problems, construction of detailed solutions, and devising of implementation plans. TOC has major applications in production, project management, distribution, finance, marketing and sales, and management strategy but

can be used as well to solve any problems that arise in life.

Dr. Goldratt brings his finite scheduling theory and TOC framework to life in his books, *The Goal: Excellence in Manufacturing* and the sequel, *It's Not Luck*. He wrote these two business textbooks as novels, "to give the content in a way people can absorb it," he said. The books give students and practitioners of business a "real-life" context for applying Goldratt's ideas for achieving business excellence. (He does not accept credit for inventing this creative genre, however, attributing inspiration to novelist/philosopher Ayn Rand.) Goldratt has written several traditional texts as well, including *The Haystack Syndrome, The Race*, and the *Theory of Constraints*. In *The Haystack Syndrome*, Goldratt presents evidence that "cost accounting is public enemy number one of productivity." In *It's Not Luck*, he asserts that balance sheets are "nothing more than the liquidation value of the company ... and should

not be used to judge an ongoing operation."

Application of Goldratt's theories have proven "that the core problem right now in most of the industry is a lack of a systems approach," he said. This core problem is referred to throughout his work as "striving to run the whole system based on local optimums." Goldratt teaches that one area of a business may need to sacrifice its own local optimums for the greater good of the entire system. Goldratt sees an eagerness among the business community to embrace systems thinking, but finds major resistance to this approach in academia. He believes that academics who do not present a systems perspective are "sending graduates into the market totally unprepared." Goldratt's theories have met with widespread acceptance, as evidenced through sales of over two million copies of *The Goal*—with no advertising expense incurred! When he's not consulting with managers of the world's largest corporations, Goldratt lives in Israel with his wife and two youngest children.

Source: A conversation with Dr. Eliyahu M. Goldratt

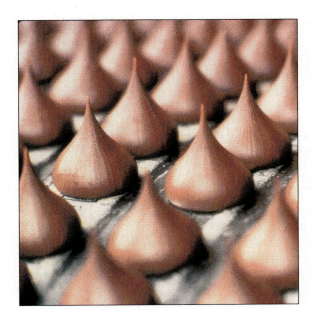

Properly designed flow shops and continuous processes have no bottlenecks.
Source: Hershey Foods Corporation

Bottlenecks are caused by the mismatching of process rates which the flow shop purposefully, and often at great expense, eliminates. The mismatch is not the result of bad management. It occurs because of the mixture of orders that are received and in process for a particular shop configuration. If the job mixture changes, the shop configuration can be altered but it takes time.

However, because bottlenecks are a fact of life for the job shop, it is worth learning to control where and when they occur. This makes bottleneck control an expected and routine part of job-shop management. It turns attention to the causes of scheduling problems instead of dealing with the symptoms which are inability to fill orders on time.

Queue Control

Multiple bottlenecks usually exist within any process. When processes are being altered from hour to hour, as is the case with batch work, the potential for clogging the pipeline is large. The damaging effects of queues multiply as the number and length of the waiting lines grow. Orders get misplaced and materials are lost. The chaos level crescendos. The need for queue control is increasingly apparent. Studying the queuing situation at the bottlenecks results in improved production scheduling.

To reduce the wastefulness of queues, the OPT[2] (Optimum Production Technique, also known as Optimized Production Technology) system recommends feeding the bottleneck. This is equivalent to saying that all prior operations should be able to supply the bottleneck with what it needs, when it needs it. Having identified the BNs removes part of the problem. Seeing that they are kept working, removes another part of the problem. Finally, when the bottleneck-capacity constraints are unacceptable, they can usually be removed. Then, other bottlenecks emerge.

The bottleneck's processing rate is called (in OPT terms) the drum because it sets the production rate for all of the upstream and downstream activities. For synthetic processes (which assemble items), work begins upstream at the tributaries and flows downstream to the trunk encountering various bottlenecks along the way. This is called the OPT V process because it looks like a V moving from upstream to downstream.

Upstream activities (which occur prior to the bottleneck) should be slowed down or stopped to synchronize with the drum. In the stream analogy, if dams cannot slow the flow, then flooding occurs. Levees must be built to contain the backed up waters. Similarly, work in process backs up and needs a place to stay. Containment of excess work to which value is not being added is simply less desirable than slowing the flow to the bottleneck. Flow control can result in smaller lot sizes being made more often.

This is called finite scheduling and it follows two important rules of OPT. These are that the transfer batch may be smaller than the process batch which provides greater flexibility and less cost. Also, the process batch should be variable and not fixed. Relatively small transfer batches can be used effectively to provide inventory buffers for the bottleneck (drum). The span of protection (called the time buffer) is varied to protect the plant against disruptions. For example, the time buffer can be increased from four to five days when there is greater risk of not being able to supply the bottleneck.

Downstream activities cannot expect to receive more work than the fully utilized bottleneck can deliver. The bottleneck sets the course for the rest of the process—that is until the next bottleneck is encountered—as the work flow moves downstream. In process terms, *excess capacity* located downstream from the bottleneck may be *wasted investment* in capacity. If the BN rate is increased, it may not be wasted investment.

Synchronized Manufacturing

Synchronization

Requires control of the timing of flows; includes how much is made at one time and transferred to the next station.

Synchronization requires control of the timing of flows. This includes how much is made at one time and transferred to the next station. Transfer amounts and storage amounts are determined by control mechanisms which can be viewed as valves or gates that can be opened and closed by pulling on ropes to control the flow of materials to the bottlenecks and to the operations that follow the bottleneck. The ropes are like communication links that synchronize the flow rates throughout the system. The controlled flow rates establish what are called the time buffers.

Figure 16.8 illustrates this drum-buffer-rope concept which is the queue-control mechanism designed by OPT. Queue control over time and between stations is why OPT is called synchronized production-scheduling methodology. It applies to **finite scheduling** which means that production is scheduled in batches of finite size for a variety of jobs in the shop—which might coincide with lot-for-lot scheduling.

Finite scheduling

Production is scheduled in batches of finite size for a variety of jobs in the shop; feed the bottleneck and protect it with days of buffer stock.

Variability can disrupt the smooth flow of product. Consequently, it is necessary to provide bottlenecks with assurance that their suppliers will increase production when there are signs that one or more BNs could be idled. When any bottleneck is idled, the existing system is not producing to its maximum.

OPT uses time buffers. Time buffers are different from stock buffers. The latter specify the number of units of stock that should be kept in reserve whereas the former specify the amount of time that the BN can keep working if any of its suppliers are shut off. Time reserves are maintained by means of the communication links that OPT calls ropes (see Figure 16.8).

FIGURE 16.8

OPT SYSTEM'S DRUM-BUFFER-ROPE MODEL

Legend

▭ - Raw Materials

● - Operations - Order-Mix I

● - Operations - Order-Mix II

◉ - Bottleneck Constraint - Order-Mix I   Drums for I

◉ - Bottleneck Constraint - Order-Mix II   Drums for II

— - Timing Ropes Operate for Appropriate Bottlenecks

▭ - Time Buffers Operate for Appropriate Bottlenecks

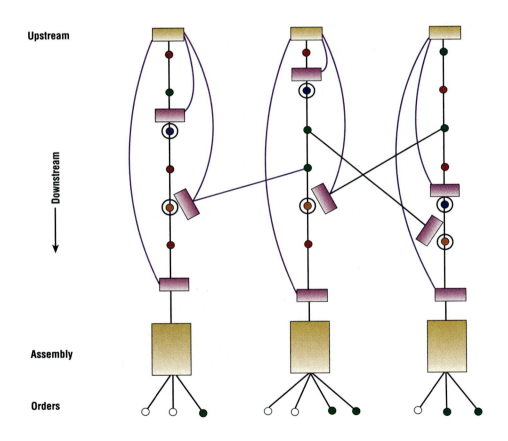

The bottleneck-feeder stations are aware of the need to monitor and allow for statistical variation in the process. The OPT time buffers are maintained by regulating the outputs of all feeder stations in line with the status of each time buffer. Being constantly in communication with each other assures that the bottleneck and its feeder systems are integrated. The rope—similar to a ship's telegraph which signals the desired speeds of the engine—signals the production output rate required to prevent idling the bottleneck facilities.

This is symbolically represented in Figure 16.8, above, which shows shifting bottlenecks for two different order mixes. Order Mix I consists of five jobs and Order Mix II consists of five different jobs. There are time buffers before three bottleneck operations in each case. Also, there are time buffers before the three batch assembly operations. Notice how this job shop is in communication with its bottleneck-constraint system. As the order mix changes, it shifts BNs (new drums), calculates new time buffers and communication controls upstream production (new ropes).

Changing the Capacity of the Bottleneck

It is sensible to consider the effect of increasing the capacity of every bottleneck facility. Sometimes by using new technology, or more people, a greater volume of work can be done. This may result in shifting the bottleneck to another part of the process. Flow rates can be studied, as in Figure 16.9, to permit bottleneck analysis of specific situations.

FIGURE 16.9

THE BOTTLENECK IS OPERATION B

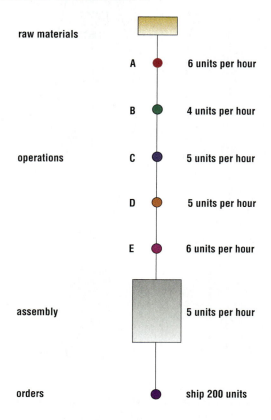

Bottleneck *B* (the drum) has a processing or throughput rate of four units per hour. It constrains the system's output rate to four units per hour. The order is for 200 units so if there are 40 work hours per week, then 160 units will be made in the week. That leaves 40 more units to be made which requires 10 more hours.

Operation *A* is instructed to feed the bottleneck and the time buffer is set to provide an hour of protection in the event that Operation *A* breaks down. *B* needs four units every hour which means that between *A* and *B* there should be a reserve of four units. This is equivalent to *A*'s working ⅔ of an hour. The first two hours *A* will work at full speed and create the necessary time buffer. Thereafter, if *B* works faster and draws down on the hour of buffer stock, *A* will work longer than 40 minutes to make up the difference. It will be noted that *C*, *D*, and *"Assembly"* all can work at five units per hour. *E* can produce six units per hour. None of them will be able to work faster than *B* as the bottleneck *B* sets the pace.

Complex flow patterns are more difficult to analyze than the simple example above, but the basic idea is the same.

1. Locate the bottleneck. This is called prospecting for bottlenecks. It involves developing detailed flow-process charts for all important orders in the shop and measuring or estimating the processing rates. Every operation must be included and associated with its processing rate.
2. Determine appropriate time buffers for the bottlenecks. Then control the transfer lot sizes and the frequency of delivery from all upstream operations (in this case only Operation *A* needed to be controlled).
3. Decide on the cost and desirability of increasing the throughput rate of the bottleneck.

If the use of new technology, costing $100,000, could increase Operation *B*'s output to five units per hour, *B*, *C*, *D*, and *"Assembly"* all become bottlenecks. A time buffer must now be built up between the many bottlenecks so that *B* can feed *C*, *C* can feed *D*, and so on. It is really easier to have only one bottleneck to tend.

To evaluate the change in bottleneck configuration, the system can now produce 200 units per week. The additional 40 units times 50 weeks per year yields 2000 extra units per year. Assume that they can be sold without incurring carrying costs. Then if each unit brings a $50 profit, the investment in bottleneck reconfiguration will pay for itself in one year. Thus:

$$V = \text{investment to increase production by X}$$

$$X = \text{extra production per year}$$

$$\pi = \text{profit per unit}$$

Then, alter the BN when

$$\frac{V}{\pi X} \leq n$$

where *n* is the chosen payback period.

$$\text{For the numbers above, } \frac{\$100,000}{\$50(2000)} = 1$$

S U M M A R Y

Production scheduling is the culminating series of steps that determines *when* orders are to be worked on and *by whom*. The function goes back to the earliest days of systematic production and assignment of service jobs to work crews. The Gantt load chart was developed at that time, and it is still used. Loading decisions concern which jobs are to be assigned to which teams or facilities. Jobs on-hand are scheduled; jobs that are forecasted are not scheduled.

It is explained that the assignment method for loading is powerful but limited because job splitting is not permitted. The model can be used to maximize profits or minimize costs. Then, the transportation method is developed for loading. It permits job splitting and multiple job assignments at one facility. The drawback of the transportation model is the need to use standard units and the presumption of productive proportionality.

Loading is always followed by sequencing which is the production-scheduling step that determines the order for job processing. Which job goes first, second, and so on? Servicing people often requires first-come, first-serve so that they do not get upset. With inventory that is called first-in, first-out (FIFO). However, processing orders first with shortest processing times (SPT) provides benefits to everyone except those who regularly have orders that take a long time to process.

Gantt layout charts are used to organize sequencing assignments. The sequencing situation depends on the number of jobs *(n)* and the number of machines *(m)* that are available to work on the jobs. Solution methods differ according to the numbers n and m for the $n \times m$ problem. SPT rules are developed for $n \times 1$ and $n \times 2$. The SPT rule can be modified to reflect various forms of priorities including: job and customer importance, due date, and lateness.

The chapter concludes with a discussion about identifying bottlenecks and using the OPT approach of drum-buffer-rope for synchronized manufacturing. Bottlenecks set the pace of production. They are protected from being idled by statistical variability using time-based stock buffers. The rest of the system is synchronized to match the input-output performance of the bottleneck.

▶ Key Terms

assignment method *691*	bottlenecks (BN) *715*	due date *712*
FIFO *702*	finite scheduling *718*	Gantt charts *688*
layout or sequencing chart *703*	LIFO *702*	loading *687*
mean flow time *705*	opportunity costs *692*	production scheduling *686*
sequencing *687*	shop floor control *687*	SPT rule *707*
synchronization *718*	total flow time *705*	

▶ Review Questions

1. The Gantt load chart can be used to determine whether the "load" is equally distributed among facilities. To whom is balanced loading important?

2. Explain the statement: "Load real jobs, not forecasts."

3. Opportunity costs are instrumental for loading models and decisions. Are these real costs? Explain.

4. What is the assignment method of loading? What special conditions apply?

5. Distinguish between the maximization model for the assignment problem and the minimization model. Explain the differences.

6. What is the transportation method of loading? What special conditions apply?

7. Distinguish between the conditions necessary for using the assignment model and the transportation model.

8. How do Gantt load charts differ from Gantt layout charts?

9. When does FIFO make sense as a sequencing rule?

10. When does LIFO make sense as a sequencing rule?

11. When should SPT be used?

12. What are priority-modified SPT rules? Give some examples and explain when they should be used.

13. Explain the use of critical ratios for due dates.

14. What is the significance that sequencing charts are convex (curving downward) when the sequencing rule is SPT?

15. What is the significance that sequencing charts are not generally convex (curving downward) when the sequencing rule is FIFO?

16. What is the $n \times 1$ sequencing problem?

17. What is the $n \times m$ sequencing problem?

18. What is the $n \times 2$ sequencing problem? How can it be resolved?

19. What are time buffers?

20. Explain the meaning of drum-buffer-rope for scheduling.

21. What is meant by synchronized manufacturing?

22. Why is OPT called the scheduling theory of constraints?

23. Why is finite scheduling concerned with bottlenecks?

24. Discuss the reasons that job shops and not flow shops are associated with the problem of bottlenecks.

25. When does it pay to remove a bottleneck? What is entailed?

26. Explain why the removal of a bottleneck creates a new one.

27. Describe what is meant by "feed the bottleneck."

▶ **Problem Section**

1. The matrix of total costs per day for Jobs *1, 2*, and *3* if assigned at Facilities *A, B*, and *C* of the Rivet and Nail Factory is:

| | | **Facilities** | | |
		A	**B**	**C**
	1	$1000	$ 900	$1200
Jobs	2	800	700	900
	3	1500	1800	2000

What relatively permanent assignments will minimize total costs per day?

2. The matrix of costs per part for *Jobs 4, 5,* and *6* if assigned at Facilities *A, B,* and *C* of the Rivet and Nail Factory is:

		Facilities		
		A	**B**	**C**
	4	$0.10	$0.19	$0.12
Jobs	5	.16	.14	.18
	6	.30	.36	.40

The jobs are about the same size and duration. What relatively permanent assignments will minimize total costs?

3. The matrix of costs per part for *Jobs 7, 8,* and *9* if assigned at Facilities *A, B,* and *C* of the Rivet and Nail Factory is:

		Facilities		
		A	**B**	**C**
	7	$0.10	$0.09	$0.12
Jobs	8	.08	.07	.09
	9	.30	.36	.40

The jobs are about the same size and duration. What relatively permanent assignments will minimize total costs?

4. The executive offices for five vice presidents of an airline are on the tenth floor of the airline's building. The decision is made to assign the offices in such a way as to maximize total satisfaction. Each VP is asked to rank his or her preferences for the available offices. One is the most preferred location and five is the least preferred. The data are as follows:

Office	**VP1**	**VP2**	**VP3**	**VP4**	**VP5**
O1	1	1	2	3	2
O2	5	3	1	2	3
O3	4	2	5	1	1
O4	3	5	4	3	4
O5	2	4	3	5	5

a. What is the best assignment plan from the company's viewpoint?

b. How does this problem relate to production schedules?

c. How does this problem relate to service scheduling?

5. Use the assignment model to achieve an optimal shop-loading arrangement for an intermittent flow shop. The per unit profits are given in the matrix below. Relatively continuous production can be expected for each assignment over the next quarter.

Machines

Jobs	A	B	C	D	E
1	19	17	15	15	13
2	12	30	18	18	15
3	13	21	29	19	21
4	49	56	53	55	43
5	33	41	39	39	40

6. For the data in Problem 5, a new machine F has become available that can only work on *Jobs 1, 2,* or *3*, with unit profits of 14, 11, and 12, respectively.
   a. Should one of the present machines be replaced with F?
   b. What information is lacking that could help make the recommendations more definitive?

7. The matrix below has productivity per standard machine hour $[PR(i,j)]$ filled into each box within a machine-job cell. The revenue per piece is given by column and the cost per piece is given by row.

Revenue per Piece		$2.00	$1.00	$0.50	
Cost per Piece		Job A	Job B	Job C	Supply S(i)
$0.20	M1	3	3.5	15	
$0.10	M2	6	7	30	
$0.30	M3	4.8	5.6	24	
$0.00	MD	0	0	0	
Demand D(j)					

Determine the profit per standard hour entries that should be used. Draw up some blank matrices so that the results can be entered, and used for later calculations. It also is possible to make entries directly onto the text pages. Thus, these numbers can be filled into the empty boxes within each machine-job cell in the matrix in Problem 8 below.

8. The matrix below has the supply of available standard machine hours in the right-hand column. It has demand in standard hours in the bottom row. Fill in the Northwest Corner (NWC) assignments which means start at the upper left-hand box and enter as many units as the row or column sums allow. Then move to the east or to the south depending upon whether it is the row or column sum that still needs to be satisfied. Continue in this way.

Revenue per Piece		$2.00	$1.00	$0.50	
Cost per Piece		Job A	Job B	Job C	Supply S(i)
$0.20	M1				18
$0.10	M2				54
$0.30	M3				64
$0.00	MD				4
Demand D(j)		50	30	60	140

What is the total profit associated with the NWC assignment?

9. The matrix below has been derived after a number of shifts from the NWC assignment aimed at improving profit.

Revenue per Piece		$2.00	$1.00	$0.50	
Cost per Piece		Job A	Job B	Job C	Supply S(i)
$0.20	M1		12	6	18
$0.10	M2			54	54
$0.30	M3	50	14		64
$0.00	MD		4		4
Demand D(j)		50	30	60	140

What is the total profit associated with the "improved" assignment? By how much has it improved the NWC solution?

10. Convert the solution obtained for Problem 8—the NWC solution—into actual assignment hours and units.

11. Convert the solution obtained for Problem 9—the optimal solution (?)—into actual assignment hours and units.

12. The Door Knob Company has four orders on-hand, and each must be processed in the sequential order:

Department A—press shop

Department B—plating and finishing

The table below lists the number of days required for each job in each department. For example, Job IV requires one day in the press shop and one day in the finishing department.

	Job I	Job II	Job III	Job IV
Department A	8	6	5	1
Department B	8	3	4	1

Assume that no other work is being done by the departments.

Use a Gantt sequencing chart (see Figure 16.7 above) to show the best-work schedule. (Best-work schedule means minimum time to finish all four jobs.)

13. Using the information in Problem 12 above, find the best sequence for each department separately.

14. How do the solutions obtained for Problems 12 and 13 above, compare?

15. The Market Research Store has four orders on-hand, and each must be processed in the sequential order:

   Department A—computer analysis

   Department B—report writing and printing

   The table below lists the number of days required by each job in each department. For example, Job IV requires two days during computer analysis and one day in report writing and printing.

	Job I	Job II	Job III	Job IV
Department A	3	6	5	2
Department B	8	3	4	1

Assume that no other work is being done by the departments.

Use a Gantt sequencing chart to show the best-work schedule. (Best-work schedule means minimum time to finish all four jobs.)

16. Using the information in problem 15 above, find the best sequence for each department separately.

17. How do the solutions obtained for problems 15 and 16, above, compare?

18. Examine the sequences for the following $n \times 1$ system. There are six jobs with processing times t_i. These jobs arrived in the alphabetical order ($a, b, c,...$).

i	t_i	i	t_i
a	5	d	9
b	4	e	12
c	6	f	8

Derive the SPT solution. What is the SPT sequence? What are the total and mean flow times for that sequence?

19. Using the information in Problem 18 above, derive the LPT (longest processing time) solution. What is the LPT sequence? What are the total and mean flow times for that sequence?

20. Using the information in Problem 18 above, derive the first-come, first-serve (or FIFO) effect. What is the FIFO sequence? What are the total and mean flow times for that sequence?

21. For Problem 18 above, what effect does the information that Jobs *a, b,* and *c* are half as important as Jobs *d, e,* and *f,* have on the sequence solution?

22. For the following $n \times 1$ problem, use a priority-modified SPT rule where the priority is based upon reducing lateness L_i. Work times t_i and lateness are given in the table below.

i	t_i	L_i	i	t_i	L_i
a	5	1	d	9	3
b	4	2	e	12	1
c	6	1	f	8	1

What is the optimal sequence and what are its total and mean flow times? Compare the answer with the one obtained for Problem 18.

23. For the following $n \times 1$ problem, use a priority-modified SPT rule where the priority is based upon the critical ratio using due dates d_i and work times t_i as given in the table below.

i	t_i	d_i	i	t_i	d_i
a	5	10	d	9	18
b	4	16	e	12	36
c	6	18	f	8	25

What is the optimal sequence and what are its total and mean flow times? Compare the answer with the one obtained for Problem 18.

24. For the following $n \times 1$ problem, use a priority-modified SPT rule, where the priority is based on the additional information about customer importance called w_i (given in the table below).

i	t_i	w_i	i	t_i	w_i
a	5	6	d	9	3
b	4	5	e	12	2
c	6	4	f	8	1

Important customers get larger w_i numbers.

What is the optimal sequence and what are its total and mean flow times?

25. The problem faced by Information Search, Inc. (ISI) has the processing times shown below:

Jobs (*j*)

	a	b	c	d	e
Look Up Reference	4	8	9	4	6
Write Report	3	2	5	7	4

What is the optimal sequence and what are its total and mean flow times?

26. Workstation *B* is the bottleneck (see Figure 16.9 above). *B* has a throughput rate of four units per hour which constrains the system's output rate to four units per hour. An order has been received for 240 units which led to the decision to buy an automatic feeder machine. This machine can raise *B*'s output rate to six units per hour. There is a 40-hour work week. What output rate will be achieved? How long will it take to fill the order?

27. Using the information in Problem 26 above, what time buffers should be set for workstation *A*? Are there any other time buffers to be set?

28. Combine the information in Problem 26 above with the fact that the feeder machine will cost $25,000. The profit per unit for extra units will be $2.00. Do you agree with the decision to buy the new machine? Explain your response.

29. Combine the information in Problem 26 above, with the fact that for $50,000 all of the units can be connected with a paced-conveyor belt and run at six units per hour. Will this investment pay off? Discuss the problem of feeding multiple bottlenecks.

▶ **Readings and References**

R. W. Conway, "Priority Dispatching and Job Lateness in a Job Shop," Journal of Industrial Engineering, Vol. 16, No. 4, July, 1965.

Digital Equipment Corporation, Factory Scheduler, licensed to the Mitron Corp. of Beaverton, Oregon, (503-690-8350), 1994.

Eliyahu M. Goldratt and Jeff Cox, The Goal: A Process of Ongoing Improvement, 2nd Revised Edition, North River Press, 1992.

Eliyahu M. Goldratt and Robert E. Fox, The Race, North River Press, 1986.

Eliyahu M. Goldratt, The Theory of Constraints Journals, Volumes 1–6, Great Barrington, MA, 800-486-2665, North River Press, Inc., 1990.

Eliyahu M. Goldratt, The Haystack Syndrome, Great Barrington, MA, 800-486-2665, North River Press, Inc., 1990.

Eliyahu M. Goldratt, It's Not Luck, Great Barrington, MA, 800-486-2665, North River Press, Inc., 1994.

Red Pepper Software, ResponseAgents (Optimizing Production Scheduler), San Mateo, CA, (415-960-4095), 1994.

M. Michael Umble and M. L. Srikanth, Synchronous Manufacturing: Principles for World Class Excellence, Cincinnati, OH, South-Western Publishing Co., 1990.

1. S. M. Johnson, "Optimal Two- and Three-Stage Production Schedules with Set-up Times

Notes...

1. S. M. Johnson, "Optimal Two- and Three-Stage Production Schedules with Set-up Times Included," *Naval Research Logistics Quarterly*, Vol. 1, No. 1, March 1954, pp. 61–68. This article laid the foundation for later work in sequencing theory. It is historically of interest.

2. Eliyahu Goldratt, and R. Fox, The Race, North River Press, 1986. The OPT system was developed by Eliyahu Goldratt as a commercial product for best scheduling practice.

Chapter 17

Cycle-Time Management

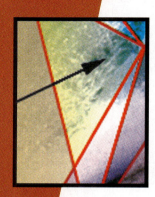

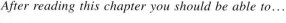

After reading this chapter you should be able to...

1. Explain the nature of cycle-time management.
2. Illustrate the relationship between the line balancing of serialized production systems and cycle-time management.
3. Explain why line balancing is important.
4. Detail the different kinds of line-balancing situations that exist with respect to labor-intensive and technology-oriented workstations.
5. Describe the difference between deterministic, stochastic and heuristic line-balancing situations.
6. Explain what needs to be done to balance a production line.
7. Relate cycle-time management to time-based management (TBM) and distinguish between tactical and strategic TBM.
8. Draw a precedence diagram and calculate output rates for different numbers of stations.
9. Describe perfect balance, measure balance delay and line efficiency, and explain why they are useful measures.
10. Demonstrate a heuristic line-balancing algorithm and explain why the use of computer software is desirable.
11. Relate stochastic line balancing and queuing models.
12. Explain why all service systems can be viewed as queuing systems which must be balanced to satisfy customer expectations.
13. Demonstrate how single-channel queuing models can be used to estimate average length of waiting lines and waiting times.
14. Describe the role of suppliers in cycle time management.
15. Discuss the advantages and conversion of push-production systems to pull-production systems.

Chapter Outline

17.0   CYCLE-TIME MANAGEMENT
       Time-Based Management and Synchronization of Flows

17.1   LINE-BALANCING METHODS
       Basic Concepts for Line Balancing
       The Precedence Diagram of the Serialized Process
       Cycle Time and the Number of Stations
       Constraints on Cycle Time
       Developing Station Layouts
       Perfect Balance with Zero Balance Delay
       Rest and Delay

17.2   HEURISTIC LINE BALANCING
       Kilbridge and Wester's (K & W) Heuristic
       Other Heuristics

17.3   STOCHASTIC LINE BALANCING
       Queuing Aspects of Line Balancing
       Single-Channel Equations
       Multiple-Channel Equations

17.4   JUST-IN-TIME DELIVERY

17.5   PUSH- VERSUS PULL-PROCESS DISCIPLINE

Summary
Key Terms
Review Questions
Problem Section
Readings and References
Supplement 17: Simulation of Queuing Models

The Systems Viewpoint

Cycle time provides a measure of the output rates of different parts of the system. For the system as a whole, it is a common denominator that interrelates the timing of activities and flows of materials. Timing is as important to the performance of repetitive production systems as it is to the performance of a symphony orchestra. Cycle time is not only the interval that work stays at the workstations or the rate at which items are finished and leave the production line. The bigger systems perspective connects customers' demand rates and suppliers' delivery rates with management's control over the production cycle time.

Cycle-time management can be applied to many different work configurations including that of relatively permanent flow shops or the shorter durations of intermittent flow shops (IFS) which produce large batches of work. Cycle-time management will reflect decisions about how much to make to ship and how much to make to stock under normal and peak demand conditions. The systems view requires inclusion of the cycle times of internal functions that support the production process, such as production scheduling, materials inspection and maintenance. It requires inclusion of the external cycles related to delivery of materials and distribution along the supply chain. The systems viewpoint relates time-based management, cycle-time management and just-in-time management for interdependent functions. Synchronization and orchestration are systems-wide issues of cycle-time management.

17.0

CYCLE-TIME MANAGEMENT

Cycle time

Interval that elapses between successive units of finished goods coming off the production line; also length of time a unit spends at each workstation.

Cycle time is defined as the interval that elapses between successive units of finished goods coming off the production line. If a completed Subaru comes off the production line every 2.5 minutes, then 2.5 minutes would be the cycle time for that assembly line.

Because of the above definition, cycle time also must be the length of time that a unit spends at each workstation. This means that if the process can be altered so that work spends less time at a station, then the interval between successive units of finished goods can be shortened and the production rate increased.

If the production rate is altered up or down, this will change the cycle times for many activities that support the production system. More or less materials may be needed to feed the process. Fewer or more workers doing their own inspection may be needed to deal with quality and quantity checks. More or fewer trucks may be wanted to carry the finished goods to distributors' warehouses. Cycle time of the process is interconnected with many other activities. For a smooth working production system, the timing of many cycles must be understood and controlled.

In recent years, cycle time has become one of the OM topics that has received major attention from the organization's top executives. This is because cycle-time improvements can provide great financial leverage. There can be big improvements in output and revenue—gained at low cost. Thus, decreasing cycle time by working smarter and not harder has the effect of increasing productivity, decreasing the breakeven point, and increasing the system's effective capacity.

Decreasing cycle time by improving the process design raises output. The extra input costs are minor in comparison to the output gains. Consider the example of Rolls Royce which cut the cycle time for making a car from 50 to 30 days. This reduced the breakeven point from 2,600 cars in 1992 to 1,300 cars in 1994. The result was profitable operations in place of serious losses from operations. In contrast, many auto-assembly plants in the U.S., Europe and Asia finish a car a minute.

To apply cycle time to airline flights, treat each time the same plane takes off as the next finished unit coming off the production line. The interval between takeoffs is then the cycle time. U.S. Air cut the ground time between flights from 45 minutes to 20 minutes. This was done by reducing the number of crew changes, speeding up the check-in procedures, and loading planes with enough food and drinks for several trips. The resulting productivity improvement increased the available seating capacity for the fleet of planes by 3,500 seats per day.

This airline example is particularly useful because it changes perspective. It broadens the applicability of the concept of cycles from the traditional and entirely correct notion that cycle time applies to repetitive operations such as an assembly line. Thus, cycle time also is appropriate for any set of operations that are linked together as a chain of repetitive activities such as supporting airplane takeoffs.

In setting time standards, there are short and long cycles that comprise the job. Having to go to the storeroom to get materials, such as rolls of tape to seal boxes, may have to be done once every 1,000 boxes sealed, but it is a legitimate cycle that deserves to be studied. Perhaps it could be done less frequently which might decrease the cost of the job and increase the productivity of the worker.

Cycles exist in the repetitive intervals between placing an order and receiving the supplies. Inspection cycles exist between receiving and accepting materials for use on the production line. Just-in-time (JIT) deliveries require shorter cycle times for less materials. This affects the frequency with which truckers make deliveries (shippers' cycles). Just-in-time can alter the production cycles of suppliers as well. JIT deliveries of sealing tape would free the workers at the stations from interrupting their normal procedures which could increase the productivity of the process.

Time-Based Management and Synchronization of Flows

Cycle-time management is one facet of **time-based management (TBM)**. It is the aspect of TBM that applies to running a better production process. This is generally regarded to be tactical in scope. The reason that it is tactical is that it entails management of the existing production process. With TBM, units are produced in an improved way by a faster process.

The concept of continuous process improvement, instead of reengineering (or radical change) often, but not always, applies to cycle-time reduction. For instance, Rolls Royce's success in cutting cycle time was based on the "greenfield" concept which means literally building on a vacant lot—or starting from scratch when associated with reengineering.

Time-based management (TBM)

Cycle-time management and production scheduling management are facets of TBM; units are produced in an improved way by a faster process.

The other facet of TBM which is always dealt with by OM is project management. It has to do with developing a new product or process as quickly as possible. It is considered strategic because it represents a change in use of resources, and often of company policy. It would apply to the way in which Rolls Royce handled the project of building, installing and starting-up its new production process.

Tactical timing goals must synchronize cycles of all participants. The flows of materials used on the line must get to where they are needed on time. Shipping schedules of finished goods must be synchronized so that inventory pileups do not occur on the shipping docks. When the cycle times of parts of processes are reduced in one place, other places in the process must be adjusted so that balance is maintained and the flows are synchronized. This aspect of cycle-time management is inherent in proper control of the supply chain.

The management of the broad range of activities that can be defined as cycles puts the spotlight on the timing. Substantial reduction of intervals between successive outputs from the production line can usually be obtained by rethinking the way that things are done. Cycle-time management and quality management have a lot of similarities. The reduction of cycle time and the reduction of defectives can result from being clever about process design. Conversely, poor processes result from a lack of teamwork and a shortage of team members' suggestions. Thus, as it is said that some part of quality improvement is free, some part of cycle-time improvement also is free.

In building confidence that one can improve cycle times, the place to start is with line-balancing serialized work configurations. This includes intermittent flow shops. There is often reluctance to spend as freely on line-balancing temporary flow shops as compared to permanent ones. Line balancing is the main cycle-management technique in both cases—whether the system of supporting machines, conveyors and training are for dedicated or temporary production purposes.

17.1

LINE-BALANCING METHODS

Line balancing (LB)

Effort to nearly equalize the output rate capabilities of successive workstations on the production line; also effort to reduce and match excess idle time at the workstations; three LB classes include: deterministic, heuristic, and stochastic.

Line balancing (LB) is the effort to nearly equalize the output rate capabilities of successive workstations on the production line. LB also is the effort to reduce and match excess idle time at the workstations. Line balancing has to take into consideration constraints which include the order in which work can be done and the impossibility of splitting some operations between different stations. The overriding objective of LB is to achieve a balance of production resources which can deliver the right amount of production to satisfy demand requirements with minimum waste of resources.

This goal can be directly translated into supply chain terms. LB synchronizes material flows by properly assigning required resources. These include people, production equipment, storage space and transportation facilities. When production equipment is part of a dedicated flow shop (as in automobile assembly), engineers provide perfect line balance for such mechanized manufacturing systems.

Unlike the engineering of dedicated flow shops, the use of LB for intermittent flow shops (IFS) is common when batch sizes are large enough to warrant a temporary serial form of production. A manufacturer of baby strollers runs different models

The use of line balancing is common for intermittent flow shops.
Source: Union Camp Corporation/Elizabeth Arden

on the same production line for varying times according to demand. Each model has unique characteristics that change the line-balance requirements. A manufacturer of electric switch boxes makes most models on a job shop basis in batches. When an order is received for batch sizes that tie-up machines for more than two days, an intermittent flow shop set up is used. The Market Research Store uses IFS processing only when a report has more than five regions and/or three products being compared. The intermittent flow shop is the process of choice for growth firms and product situations where short life cycles characterize the market. They are so prevalent that line balancing for cycle management is a necessity for competitive capability.

With IFS, people often are an important part of the production system. The participation of people at the workstations in manufacturing, service systems, and along the supply chain, increases the variability of output rates. The effect of worker variability is particularly important for cycle management using line balancing of service systems. Variability of service systems is noticeable in both the arrival rate of people and work for servicing, and in the rates at which services are administered. People and work wait in line which is inefficient. Then, at other times, the service system is idle which is inefficient. This aspect of line balancing will be covered in the section on stochastic line balancing which is another way of saying line-balancing systems that have random arrivals and/or service times.

Serialized processes are comprised of operations that are repetitive and sequential. They are linked together—usually by a conveyor that moves the pieces and then stops—so that work can be done at each station. Line balancing requires assigning operations $(i = 1, 2, \ldots, k)$ to workstations $(j = 1, 2, \ldots, n)$ so that all of the workstations are about equally busy. This terminology can be applied to supply chain and service systems.

Some workstation activities are labor intensive. Others are mechanized. Glasstech's Automotive In-Line Windshield Forming and Annealing System is entirely mechanized. Most windshield manufacturing is a combination of both human labor and robot equipment. While the trend is toward automation in volume manufacturing systems it remains labor intensive in services.

Volume production of semiconductors reflects a unique combination of labor and technology in manufacturing. The production process takes place in a clean room. Workers in full containment garb are part of the process. The slightest bit of dirt on computer chips ruins circuitry. Clean-room technology constantly filters away the smallest particles. Workers interact with machines at workstations within this special environment.

Automotive-assembly line balancing reflects a broad range of technology and workers. Installing window glass and upholstery is relatively labor intensive whereas, painting and welding are primarily computer and machine driven. Timing is set for the assembly-line conveyors that travel at rates carefully set by engineering and management design. When there are unions, they participate in agreeing to the rate of the paced-conveyor system.

The history of line balancing deserves mention. In the 1940s at the Louisville, Kentucky plant of General Electric which manufactures appliances, a group called Operations Research and Synthesis (OR & S) decided to investigate the line-balancing problem. Their studies began a stream of activities that led many researchers to contribute their results. By the 1960s a great body of work existed in all three classes of line balancing which are deterministic, heuristic and stochastic line balancing.

The three classes are described below.

Deterministic line balancing is based on the assumption that each total process takes an exact amount of time which is known and not variable. The total process is made up of operations. The operations' times when added together constitute the total process or total job time. The operations' times are all fixed. If they were not, then the total process time would not be deterministic.

Heuristic line-balancing methods employ "rules of thumb," common sense, logical thinking, and experience to find near-optimal solutions for large problems. These are problems that involve too many operations to allow enumerating and testing all of the possibilities. The number of possible configurations in line balancing grows rapidly with operations and stations. Only heuristic methods are economically feasible for many real problems.

Heuristics are used when complexity of size is an issue and the flow shop is temporary. When the flow shop is relatively permanent, as in computer-chip manufacturing and automobile assembly, engineering efforts can be justified to design perfectly balanced flow shops. That level of expense is not often acceptable for intermittent flow shops which have sufficient complexity to require heuristic methods and sufficient volume to pay for the application of the heuristic. Under these circumstances, the line-balancing problem offers one of the best examples of a real need for heuristic methods.

Stochastic line balancing also must be addressed. It exists when the assumption of fixed and deterministic times for operations and total process time is not correct. As previously noted, this might be especially relevant for workstations where

Deterministic line balancing

Each total process takes an exact amount of time which is known and not variable.

Heuristic line-balancing methods

Employ "rules of thumb," common sense, logical thinking, and experience to find near-optimal solutions for large problems.

Stochastic line balancing

Exists when the assumption of fixed and deterministic times for operations and total process time is not correct.

labor-intensive operations dominate. It also can be critically important where the variability makes it impossible to complete the work at a station. When the conveyor carries defective work to the next station, it often shuts off production for the rest of the downstream line. Thus, bottlenecks emerge which limit the benefits to be gained from serialized flow-shop configurations.

Some assembly lines permit workers to stop the conveyors from moving work further when a serious problem arises at their station. This will be discussed later. Also, it is worth noting that the stochastic line-balancing problem is related to queuing models. These models describe the length of queues and the average waiting times that can occur when lines are not balanced. In effect, the demand for service becomes greater than the supply. Queue length is greater when requests for service occur randomly than when requests are evenly spaced.

If demand volume justifies it, the high volume, properly balanced flow shop, is the best work configuration to obtain lowest costs and finest quality. The smooth running, conveyor-driven flow shop, with its well-timed stream of product, has low WIP (work-in-process) levels and enjoys economies of scale. It is a production system which moves just the right amount of material between stations.

Serialized production systems, without conveyors to set the pace, can push uncontrolled amounts of materials from one station to another. Line imbalances get translated into uneven workloads. Stations are not designed to handle inventory overloads. The result of worker overloads is the production of poor quality work as the operators speed up to compensate for the extra materials they must process. Push-type production systems create bottlenecks which cause disruptive backups of costly WIP to form. When stations push materials, there is no consistency in what is being delivered for processing by adjacent stations. On the other hand, pulling product along the line with a goal of being just-in-time is part of proper line balancing.

Basic Concepts for Line Balancing

Step one is to detail the process to be balanced. This means listing all of the operations that need to be done to make the product or deliver the service. Detailing requires specifying the order in which all operations are to be done. This includes stating which operations must be done before others can be started. A precedence diagram, used to chart the sequence, will be described shortly. Knowledge of the process technology is essential to successfully carry out step one.

It is useful to have a number of people review the list of operations and their sequence. Even great chefs have different recipes for baking the same kind of cake. One recipe might be better than another for different volumes of production. Team responsibility for accepting the list of operations and the order for doing the operations is desirable. That team may be called upon to revise the set of operations and the precedence diagram in order to better balance the line.

This first step is often taken for granted. How the job will be accomplished must be precisely stated. The "devil" is in the details. For making toothpaste, it is not enough to state "mix the slurry for 20 minutes." The details include the mixing rate, the quantity mixed, the way the materials to be mixed are loaded and unloaded, and how and when other materials are added.

Other examples of deciding in detail how each of the operations is to be done include: which adhesive will be used and how much at what temperature? How will the adhesive be applied? Or what software will be used and which version on which computer? Such a set of decisions constitutes the first feasible enumeration of the process. Accepting exactly how the job is to be done is the inception of intelligent line balancing. There is ample proof that it pays to question every facet of the stated procedure before beginning the next step.

Step two requires estimating how long the operations will take. Getting these operation times right is important and will be discussed shortly.

Step three follows acceptance of the process, and estimation of the operation times. *Step three is to assign the stated operations to workstations.* As a criterion, perfect balance—which can seldom be achieved—is when every one of the stations has the same amount of work to do and there is no idle time at any station. Another way of saying this is that the sum of the operation times for each station would be the same, and all available time is completely utilized at each station. Note that workstation assignments cannot violate precedence conditions.

Having equal sums of operation times at each station is not easily achieved because:

1. Precedence constraints limit flexibility. Some operations must be done in specific sequences. Others must be done together at the same station (see the section on zoning constraints below). The exact arrangements are detailed using precedence diagrams as shown in Figure 17.2 below and discussed in the next section.
2. Operation times can be increased and decreased in various ways. The degree of flexibility is limited by the physical process, management policies and practices, technology and training. When some operations are speeded up, there is a loss in quality. Other operations cannot be done faster unless an entirely new technology is available to be used.
3. Human performance is variable. Individuals and teams shift rates over time. These problems, which are pronounced for the intermittent flow shop, can often be circumvented by the use of technology and proper design of the dedicated flow shop.

The step three goal is to assign operations so that the sums of operation times at all stations are about equal. Thus, starting-up the process: raw materials enter station 1 at 9:00 A.M. Stations 2, 3, and 4 are idle. Between 9:00 and 9:03 A.M., the raw materials are worked on at station 1. Precise operational details must be specified.

The materials (now work in process) move to station 2 at 9:03 A.M. Station 2 adds value to the work in process it has received from station 1 by using its set of operations. Precise operational details must be specified. At the same time, new raw materials enter station 1 which repeats the same operations every three minutes.

At 9:06 A.M., station 3 begins to operate on WIP passed from station 2. Continue in this way until all stations are actively engaged. Every 3 minutes raw materials enter at station 1 and finished goods leave the final station. Cycle time is $C = 3$. Table 17.1 provides the start-up log and Figure 17.1 pictures the situation.

LOG OF FOUR STATION START-UP

9:00 A.M.   Station 1 begins working.

Station 1 is scheduled to take 2.9 minutes to complete its operations. Idle time at station 1 is 0.1 minutes.

9:03 A.M.   Station 1 repeats its operations.

Station 2 begins working. Station 2 is scheduled to take 2.7 minutes to complete its operations. Idle time at station 1 is 0.1 minutes and at station 2 is 0.3 minutes.

9:06 A.M.   Stations 1 and 2 repeat their operations.

Station 3 begins working. Station 3 is scheduled to take 2.8 minutes to complete its operations. Idle time at station 1 is 0.1 minutes, at station 2 it is 0.3 minutes, at station 3 it is 0.2 minutes.

9:09 A.M.   Stations 1, 2, and 3 repeat their operations.

Station 4 begins working. Station 4 is scheduled to take 3.0 minutes to complete its operations. Idle time at station 1 is 0.1 minutes, at station 2 it is 0.3 minutes, at station 3 it is 0.2 minutes, and it is 0.0 minutes at station 4.

9:12 A.M.   All Stations repeat their operations until shutdown.

After start-up has been accomplished, each of the $n = 4$ stations have operations that take approximately three minutes to perform. This is shown in Figure 17.1. However, this system had to go through start-up where first only station 1 is busy; then stations 1 and 2, and so on. Finally all stations are operating as shown in Figure 17.1. Because deterministic operation times are assumed, the operation times at each station are constant as are the idle times at each station.

A GRAPHIC INTERPRETATION OF WORK AT FOUR STATIONS WHERE CYCLE TIME C = 3 MINUTES

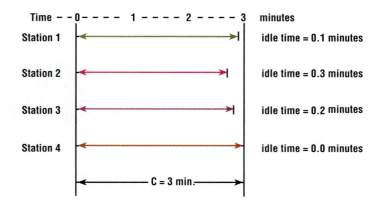

P E O P L E

Lucinda Heekin
· · ·
The Last Best Place

Dreams of the American west inspired Lucinda Heekin's mail-order catalog, The *Last Best Place*. Heekin describes her business, which began operations in August 1993, as a 100 percent mail-order driven store "by and about the American West … hallmarking the American manufacturer and artist." She acknowledges the many cycles that must be managed in the mail-order business: "I feel like I'm always juggling the Olympic rings," she said. Heekin sees the catalog business as governed by three basic cycles. These are the production cycle for the "store" or catalog, the purchasing cycle, and the cycle involving distribution of goods to the customer. Satellite cycles involve managing mailing lists, shipments of catalogs to customers, and sales data analysis.

The production cycle for the catalog begins anew with each printing. In its first two years of operations, *The Last Best Place* was printed three times each year. (Beginning in 1996 it will be printed four times a year.) Production for each printing generally takes six to eight weeks, with the majority of work done in the month prior to printing. Most of the production activities for the spring catalog, for example, take place in January and the first two weeks of February. Major activities to be managed entail finalizing items to include in the catalog, photographing the items, copy writing, page layout,

digital color separation, and final proofing. "We're always in production," Heekin said, "because some part of this always needs to be produced better, smarter, and faster… and that comes with doing the compilation and analysis of the product sales numbers." Each product is analyzed in terms of square-inch, category, dollar performance, and vendor cooperation.

As a result of the analysis, products with proven sales patterns for the season are retained. This allows Heekin and her merchandising staff to begin each purchasing cycle (attending trade shows and arts and crafts fairs) knowing exactly what product categories to look for and how many spaces to allocate to each. *The Last Best Place* contains 12 product categories with subtypes under each. Once goods are selected, they flow from the manufacturer to the fulfillment house to the customer. The company initiated a TQM program in January 1995 and is vigilant about having vendors buy in

to the importance of quality. "I've got to get them to see that Ms. Smith is their customer and she will come back looking for their products through us only if it's a high-quality product." Heekin also stresses the importance of vendors meeting the "start ship, end ship, and cancel" dates. "The key is to get the inventory in just within days of the drop (or catalog shipment) hitting the first wave of homes across the U.S.," she explains.

Heekin refers to the third basic cycle—involving the distribution of goods to customers, total customer service, and returns—as the "back end" of the business. The company manages this phase through its fulfillment house, which receives and inspects the merchandise, receives and fills customer orders, then provides after-sale service, if necessary. Heekin said the quality control staff at the fulfillment house is zealous about quality and will not inventory any product that doesn't pass inspection without a written okay from her company.

Source: A conversation with Lucinda Heekin.

Note that no allowance has been made for rest and delay. There is little slack, in general, and none at station 4. If there are people working at the stations, some person must be available to relieve employees who want to use the restrooms.

Having developed a way to visualize the operation times at stations and their comparative idle time, the best way to understand the concept of line balancing is to study a small problem. It is useful to bear in mind that real line-balancing problems are seldom so small.

The Photo Lab is a mail-order film processor. It has developed a computer-controlled color film process that promises to improve the quality of work while speeding up film developing and printing time. The process designers have created a flow shop that starts with mail receipt of film and money. Each cycle ends with an envelope of photo prints and negatives ready for mailing and control of payments for deposit to The Photo Lab's account.

The Precedence Diagram of the Serialized Process

The first version submitted by the process designers to the operations team was changed after detailed discussion of contingencies that might arise. The next version was rejected by the plant manager as being difficult to service. The final process version is shown by the precedence diagram in Figure 17.2 which shows the order of operations. **Precedence diagrams** are critically important graphic displays which show operations that must be done before others.

Precedence diagrams
Graphic displays which show operations that must be done before others.

F I G U R E 1 7 . 2

PRECEDENCE DIAGRAM FOR THE PHOTO LAB

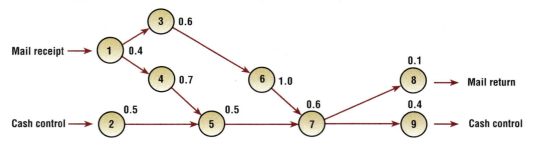

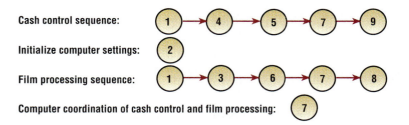

This sequence specifies operations that are antecedents and successors. The precedence diagram specifies the progression such that A follows or precedes B. However, the numbers on the nodes do not specify the order in which work must be done. Thus, (1) must precede (3) and (4), but (4) might be done before (3). Similarly, (2) could be done before (1) or vice versa.

Precedence diagrams are helpful in sequence specific operations like this operation where the gloss of a wood coating is measured after it has been applied.
Source: Air Products and Chemicals, Inc.

The operations that constitute the process are labeled i where $i = 1, 2, \ldots, k$. For each operation, an operation time t_i has been estimated. These times are shown in the precedence diagram of Figure 17.2 and in Table 17.2.

OPERATION TIMES FOR THE PHOTO LAB

Operation (i)	Operation Time (t_i), Minutes
1	0.4
2	0.5
3	0.6
4	0.7
5	0.5
6	1.0 t_{max}
7	0.6
8	0.1
9	0.4

The longest operation time is $t_{max} = 1.0$ minute.

Operation times are estimates because the line has not been set up yet. The actual times cannot be measured. The estimates can be based on simulations that come close to emulating the situation that will hold after the line is running. Having correct estimates is crucial for line balancing.

The design of the process, and thereby of the precedence diagram, must be right. Then, the estimates of the operation times must be good ones.

Next, the operations team must estimate how many orders will have to be processed per hour. Say that The Photo Lab presently processes 40 orders per hour which is 320 orders per eight-hour day or 1600 orders per week. If more orders than that are received, there will be a waiting line, which will begin to disappear when less than 320 orders are received per day.

With a processing capacity of 320 orders per day, if 400 orders are received on a particular day, a waiting line of 80 orders will result. If on the following day, 260 orders are received, the backlog of 80 will be reduced to 20 orders waiting.

Also, as an alternative to a waiting line, the line can be run for more than eight hours per day.

Mail-order film customers do not like to wait too long. Working overtime is costly. Consequently, if the probability that more than 320 orders will be received per day is significant, then the process should be reexamined and adjusted. If 320 orders per day is reasonable, then it is sensible to proceed with line balancing to find out if the processing rate of 40 orders per hour is feasible under line-balanced conditions.

Cycle Time and The Number of Stations

Product output rate O is equal to the time period T divided by the cycle time C. Cycle time has been defined before as the work time at each station. It will be defined by an equation shortly. First, note that T is the time period during which the order will be filled whereas t_i is the time required to do the ith operation at the station to which it is assigned. The output to cycle time relationship is:

$$O = T/C \quad \text{and} \quad C = T/O \qquad \qquad \textbf{Equation 17.1}$$

With $O = 40$ orders per hour, $T = 60$ minutes per hour, the indicated cycle time is:

$$C = 60/40 = 1.5 \text{ minutes per order}$$

This is the desired cycle time if the demand rate of 40 orders processed per hour is to be achieved. The process will be designed to realize this output time as a lower bound. The actual cycle time chosen will be shorter to allow for some idle time and to provide workers with an allowance for rest time.

Cycle time has been calculated here which will satisfy the demand, assuming perfect balance. Next, the number of stations is chosen to deliver the required cycle time. Then, the technology of the process will determine how feasible it is to divide operations evenly among the workstations.

Reference to Figure 17.2 and Table 17.2 (above) indicates that the selected output rate of 40 orders per hour may be possible. Every operation time is shorter than the cycle time of 1.5 minutes per order. This includes $t_{max} = 1$ as shown in Table 17.2. Therefore, one or more operations can be completed at a station.

Physically, the number of stations n must always be an integer. For each integer number of stations, there is an associated cycle time. The calculation is dependent upon total work content which is defined in Table 17.3. There the operation times have been added and found to sum to 4.8 minutes. This sum of operation times $\sum_i t_i$ is called total work content.

TABLE 17.3

THE PHOTO LAB'S PROCESSING TIMES

Operation (i)	Operation Time (t_i), Minutes
1	0.4
2	0.5
3	0.6
4	0.7
5	0.5
6	1.0 t_{max}
7	0.6
8	0.1
9	0.4
	$4.8 = \sum_i t_i$

Total Work Content = 4.8 minutes

When total work content is divided by the desired cycle time, it is possible to derive a noninteger value for n (called n^*). This case is illustrated by using the data in Table 17.3.

$$n^* = \frac{\sum_i t_i}{C} = \frac{4.8}{1.5} = 3.2$$

This result of 3.2 stations can be viewed as a minimum (ideal) number of workstations. The actual number of stations n will have to be larger to accommodate all operations in a physically feasible system.

When total work content is divided by the larger integer number of stations the result is a feasible cycle time, as shown in Equation 17.2.

$$nC = \sum_i t_i \quad \text{and} \quad C = \frac{\sum_i t_i}{n} = \frac{4.8}{4} = 1.2 \qquad \textbf{Equation 17.2}$$

Dividing the total work content time (in minutes) by the integer number of stations determines the number of minutes spent by an order at each station which is C. The cycle time $C = 1.2$ in Equation 17.2 also is the number of minutes elapsing between completion of orders.

Line balancing divides up the total work content making assignments of various operations i to each workstation. This has the effect of assigning differing amounts of t_i to the workstations.

Summing up the effects of Equations 17.1 and 17.2 above, the desired output rate O is used to determine C. Cycle time C in conjunction with $\sum_i t_i$ is used to determine the number of stations n. The number of stations working on different operations to fill the orders is n. Each order is at different stages of work in process and completion. The greater the number of stations, the shorter the cycle time and the larger the output rate.

Constraints on Cycle Time

Limits on cycle time also are established. Cycle time can be no less than the longest operation time; no more than total work content time. Thus, as shown in Equation 17.3:

$$t_{max} \le C \le \sum_i t_i \qquad \text{Equation 17.3}$$

In Tables 17.2 and 17.3, the sixth operation has been marked t_{max} because it is the longest operation time. Cycle time must be equal to or longer than t_{max}. If it were shorter, then the t_{max} operation must be completed at more than one station. In this case, $t_6 = 1.0$ would have to be broken into two segments.

It is reasonable to ask if any t_i can be further divided, but especially the long ones. If t_{max} can be divided now, it should have been done when the process was under review. It always is beneficial to have shorter operations which can be assigned to workstations more readily than longer ones. Large t_i are difficult to assign and they reduce line-balancing flexibility.

Cycle time must be equal to or less than total work content. This follows from Equation 17.2 above ($C = 4.8/n$) using the smallest integer value of $n = 1$. The Volvo method of production in Sweden assembles each car at one station. The cycle time that Volvo uses to assemble the car is that of the total work content. By using parallel "lines" more than one car can be assembled at the same time.

Table 17.4 presents The Photo Lab's chart of all possible cycle times C and hourly output rates O for all possible integer number of stations n. *Warning:* this is based on perfect balance (PB) which means no idle time. Figure 17.1 does not have PB and PB may not be able to be achieved!

T A B L E   1 7 . 4

THE PHOTO LAB'S CHART OF POSSIBLE CYCLE TIMES, C, HOURLY OUTPUT RATES, O = T/C, FOR INTEGER NUMBER OF STATIONS, n

n	$C = \sum_i t_i/n$	$O = T/C = 60/C$	Total Idle Time
1	4.8 minutes	12.5 orders/hour	0
2	2.4	25.0	0
3	1.6	37.5	0
4	1.2	50.0	0
*5	0.96	62.5	0
*6	0.80	75.0	0
*7	0.69	87.5	0
*8	0.60	100.0	0
*9	0.53	112.5	0

Several points need to be discussed related to Table 17.4.

1. When $n \ge 5$, $C < 1.0$ which means that it is not a feasible arrangement. (See asterisks (*) in Table 17.4.) As stated by Equation 17.3, $C \ge (t_{max} = 1.0)$.
2. The desired output $O = 40$ cannot be achieved directly with less than four stations. $O = 37.5$ with $n = 3$ whereas, $O = 50$ when $n = 4$. If three stations are used, some overtime would be required, but not much and it might provide a suitable solution.
3. From Table 17.4, the cycle time is 1.2 minutes when there are four stations and perfect balance. The line is poorly balanced in Figure 17.3 with $n = 4$ and $C = 1.5$.

FIRST EFFORT AT LINE BALANCING WITH FOUR STATIONS FOR THE PHOTO LAB, C = 1.5 MINUTES, O = 40 ORDERS PER HOUR

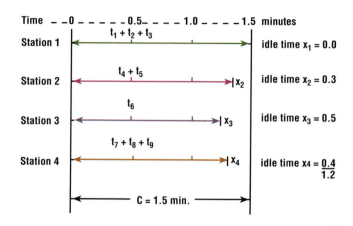

4. Let the idle time at each station be denoted by x_j. Total idle time in Figure 17.3 is

$$\sum_{j=1}^{i=n} x_j = 1.2$$

   This line cannot be balanced with $C = 1.2$ because total idle time is 1.2 minutes.

5. This arrangement with a cycle time of $C = 1.5$ produces an output of $O = 40$ orders per hour which satisfies demand.

6. The question arises: is there a better arrangement of the operations at the stations that conforms to the precedence diagram in Figure 17.2 above?

7. Note that the arrangement shown in Figure 17.4 is feasible with respect to the precedence diagram. Cycle time has been increased to 1.6 minutes and there are now $n = 3$ stations instead of $n = 4$. Output is decreased to 37.5 orders per hour which falls short of the required 40 but there is no idle time. Some overtime is indicated.

8. Another possibility is to set up parallel production lines. For example, with perfect balance, two parallel production lines with two stations each would be able to process 50 orders per hour. This is shown in Figure 17.5.

9. A systems interaction to consider! If it turns out that two parallel lines—each with two stations—lowers production costs while raising the capacity of the process, then OM and marketing could set a reduced price strategy for film processing which might increase the volume of business and total profit.

10. The variable costs of production include: the number of stations being managed $c_1 n$, the use of t amount of overtime $c_2 t$, and the cost of total idle time

$$c_3 \sum_{j=1}^{i=n} x_j$$

FIGURE 17.4

SECOND EFFORT AT LINE BALANCING WITH THREE STATIONS FOR THE PHOTO LAB, C = 1.6 MINUTES, O = 37.5 ORDERS PER HOUR

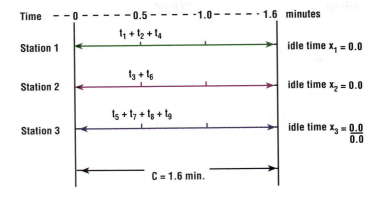

FIGURE 17.5

PERFECT BALANCE WITH 2 PARALLEL PRODUCTION LINES

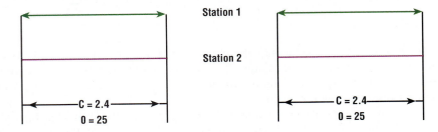

Cycle Time is 2.4 minutes and output of each line is 25 units per hour.

11. It is usually assumed that operation times cannot be shortened, having been chosen as the smallest reasonable components into which the total process or job can be subdivided. This assumption can be relaxed when technological factors permit alternative arrangements.

Developing Station Layouts

The procedure for developing station layouts is based on the use of precedence diagrams. Place in station I (the first column) all operations that need not follow others. Reference to Figure 17.2 above shows that operations (1) and (2) can be placed in station I.

Then, place in station II operations that must follow those in station I. These are operations (3) and (4). Note that (5) must follow (4) and, therefore, cannot be in the second station. Continue to the other columns in the same way. Thus, in Figure 17.6 The Photo Lab's operations have been placed into the maximum number of stations required by operational sequence, not by operation times.

FIGURE   17.6

THE PHOTO LAB'S PRECEDENCE DIAGRAM WITH FIVE STATIONS

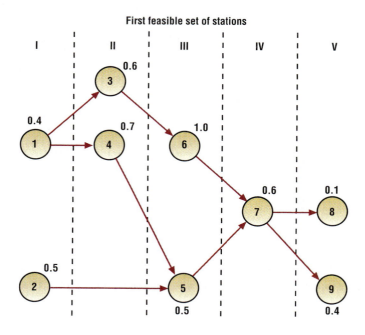

Sequence is fully specified but column position is not fixed. For example, operation (2) could be done in either station I or II. This first feasible set of five stations is determined by the longest chain of sequenced operations, not in time, but by the number of operations in the chain.

Various orderings exist that can satisfy the precedence requirements. Intracolumn movement is totally free between operations that are mutually independent (not connected by arrows). Thus, at station II, operation (3) could be done before operation (4) or vice versa. This kind of flexibility with respect to the order with which operations are completed at a station is called **permutability of columns**.

Also, operations can be moved sideways from their columns to positions to their right without disturbing the precedence restrictions. Operations (3) and (6) could be done at the third workstation, but (3) must always precede (6). Flexibility to move sideways is called **lateral transferability**.

With this five station layout, the cycle time would be set at 1.5 minutes (or more) by the assigned operation times at station III. The total station operation times and idle times are shown in Table 17.5.

Permutability of columns

Flexibility with respect to the order with which operations are completed at a station (with respect to precedence diagrams).

Lateral transferability

Flexibility to move operations sideways and into different stations (with respect to precedence diagrams).

TABLE   17.5

THE PHOTO LAB'S FIVE STATION LINE-BALANCED LAYOUT WITH C = 1.5 AND TOTAL IDLE TIME = 2.7

Station	I	II	III	IV	V
Operations	(1), (2)	(3), (4)	(5), (6)	(7)	(8), (9)
Total operation times	0.9	1.3	1.5	0.6	0.5
Idle times	0.6	0.2	0.0	0.9	1.0

There is a great deal of idle time at stations I, IV, and V. Another way of representing this assignment pattern is illustrated in Table 17.6.

TABLE 17.6

THE PHOTO LAB'S FIVE-STATION LINE-BALANCED LAYOUT

Station	i	t_i	Station Sum	Cumulative Sum	Idle Time
I	1	0.4			
	2	0.5	0.9	0.9	0.6
II	3	0.6			
	4	0.7	1.3	2.2	0.2
III	5	0.5			
	6	1.0	1.5	3.7	0.0
IV	7	0.6	0.6	4.3	0.9
V	8	0.1			
	9	0.4	0.5	4.8	1.0

Using trial and error methods, efforts can be made to improve upon this line-balancing arrangement. The most promising approach is to turn to heuristic methods for line balancing which will be explained in the next section. First, however, this section dealing with concepts of line balancing is concluded with consideration of perfect and imperfect balance, and a new term which is balance delay.

Perfect Balance with Zero Balance Delay

With cycle time C fixed by design (1.5 minutes for The Photo Lab), the number of stations under perfect balance would be $\sum_i t_i/C = n$, or 4.8/1.5 = 3.2 stations which is not possible. The integer number of stations, $n = 3$ or $n = 4$ might qualify. Three stations yield a lower production rate. Four stations provide a challenge to use more capacity.

Perfect balance is not a realistic objective for a number of reasons. First, it is difficult to group operations into stations such that the total operation times at all stations are equal. Technological factors exist that do not permit operations to be split. Such physical restrictions are referred to as zoning constraints.

The system's inefficiency (or imperfect balance) is measured by a quantity d, called the **balance delay**. This term refers to the percent of unproductive time at the stations.

Balance delay
Percent of unproductive time at the stations.

$$d = 100 \ (nC - \sum_i t_i)/nC \qquad \text{Equation 17.4}$$

Balance delay d equals zero for all cycle times listed in Table 17.4 above. That is because:

$$nC = \sum_i t_i \ \text{for all values of } n \qquad \text{Equation 17.5}$$

For The Photo Lab's cycle time of 1.5 minutes, $n = 3.2$ was derived for perfect balance. Even though $n = 3.2$ is not physically feasible, the calculations demonstrate that $d = 0$.

$$d = 100[(3.2)(1.5) - (4.8)]/(3.2)(1.5) = 0 \text{ percent}$$

If the integer value of $n = 4$ is used with the cycle time of 1.5 minutes, balance delay would be:

$$d = 100[(4)(1.5) - (4.8)]/(4)(1.5) = 20 \text{ percent}$$

The illustration in Figure 17.7 aims to show that balance delay is the total idle time of all stations as a percentage of total available working time of all stations. Thus:

$$\sum_{j=1}^{i=n} x_j/nC = \text{(Total Idle Time)/(Total Available Time)}$$

F I G U R E   1 7 . 7

THE PHOTO LAB'S FLOW SHOP WITH BALANCE DELAY OF 20 PERCENT

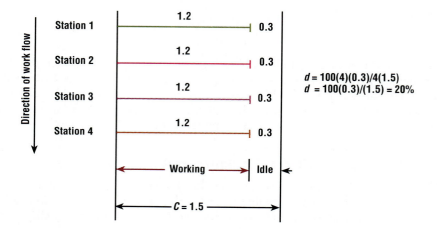

Tables 17.5 and 17.6 (above) show the five station line-balanced layout for The Photo Lab. Total idle time is 2.7 minutes out of a total work time of $5 \times 1.5 = 7.5$ minutes which is balance delay, $d = 36$ percent.

Balance delay is characterized as a systems measure of idle time because it relates total idle time to the total work content of the entire process. For cycle-time management, the cost of balance delay $c_4 d$ may be a better measure than the cost of total idle time

$$\sum_{j=k}^{i=n} x_j$$

because it describes the percent of processing time that is idle.

Line efficiency (Λ) is a related measure used to assess how well the line is balanced. It is:

$$\Lambda = \frac{\sum_i t_i}{nC} \text{ where } 0 \le \Lambda \le 1$$

Balance delay and line efficiency have a linear relationship with a negative slope as follows: $d = 100(1 - \Lambda)$. When line efficiency is one, balance delay is zero and both describe the condition of perfect balance.

As an aspect of the tradeoff between technology and people, human inspection of these transmitter units produced on an advanced robotics line is an important aspect of quality control at TRW.
Source: TRW, Inc.

Rest and Delay

Perfect balance without idle time for rest is expected for a serialized system of interconnected machines, but people require time for rest and delay. Between 10 and 20 percent of station work time normally is added to a perfectly balanced line. Thus, from a systems perspective, cycle-time management has trade-offs to consider between longer cycle times using people and shorter cycle times using technology. Also, with technology it is easier to achieve perfect balance and to maintain greater control over variability.

17.2
HEURISTIC LINE BALANCING

The term heuristic has been used by Simon and Newell[1] to describe a common sense approach to problem solving and decision making. Heuristic models utilize logic derived by observation and introspection. Heuristic approaches replace mathematical optimization efforts when optimization is not sensible or feasible. Heuristic line balancing uses rules of thumb to set near-equal workloads at successive stations.

Complex formulations of linear programming have been devised to obtain solutions to the line-balancing problem, but in practice they are not acceptable. Heuristic approaches are faster and offer more opportunities for testing alternative solutions. For practicality, they require computer software.

The essence of the heuristic approach is in the application of selective routines that reduce the complexity of a problem. Thus, the assembly line-balancing problem can be treated by simulating the decision-making pattern of managers when faced with this problem. The advantages of this approach are consistency of the decision rules, speed of application, and by using the computer, the ability to cope with large systems.

Boeing Co.
· · ·
*Boeing cuts cycle time
and costs—to increase
customer satisfaction*

Boeing Co., the jet aircraft manufacturer headquartered in Seattle, Washington, has set concurrent goals of cutting cycle time by 50 percent, cutting production cost by 25 percent, and increasing customer satisfaction. These goals were set by the company as a response to a global downturn in the market for aircraft and customer needs for faster delivery of aircraft. Boeing's customers include passenger and cargo airlines, the military, and NASA.

To cut cycle time, Boeing started with identifying periods in the production schedule when work-in-process aircraft sections sit idle. Periods of idle time were trimmed in part through the use of just-in-time techniques for ordering materials used in the manufacture of the aircraft. Boeing teams also achieved time savings through examination of steps in the aircraft assembly process that cause delays. One such delay was identified as the

installation of galleys (or kitchen areas) of passenger planes. Boeing workers who installed galleys pointed out that different airlines required different kinds of galley areas. The problem was solved by introducing a modular flooring design, so different modules of flooring could be installed as needed.

Boeing plans to shorten production cycle time for some of its jet aircraft to as few as six months. This doesn't mean that airlines will place an order and six months later, receive an

airplane. It does mean that once an airline places an order, and obtains a spot in Boeing's production line schedule, they will receive a plane in a much shorter period of time.

Boeing's new production facility in Frederickson, Washington, utilizes the streamlined production processes. The facility's state-of-the-art production line manufactures new composite materials which are used in place of aluminum for the skin of the new 777 jetliner. The composites are made up of carbon fiber tapes and resins which are combined and baked under pressure. Boeing worked with customers and suppliers in developing both the "paperless" design and manufacturing process for the 777.

Sources: John Davies, "Boeing Slashes Output Cycle; Tougher Cuts Lie Ahead," *AirCommerce,* March 28, 1994; "Boeing equipment is installed," *Pierce County Business Examiner,* July 12, 1993; Polly Lane, "Condit: Boeing needs to take some risks president cautions, 'You can't stay static,'" *Seattle Times,* May 3, 1994.

The key is to trace out and then embody the thinking process that an intelligent decision maker uses to resolve the line-balancing problem. A few popular heuristics will be described. References will be given to other heuristics with brief descriptions of what they do.

If the product or service output of the flow shop is expected to have a long and stable life cycle, then engineering design will be the preferred method for line balancing. On the other hand, when the output life cycle is not stable over a long time, shorter-term flow shops (including intermittent flow shops) are needed. These are ideally suited for heuristic resolution.

Kilbridge and Wester's (K & W) Heuristic

Kilbridge and Wester[2] proposed a heuristic procedure that assigns a number to each operation describing how many predecessors it has. This is done using the appropriate precedence diagram. The operations are rank ordered according to the number of predecessors each has. Zero is ranked first in line followed by 1, then 2, etc. The first operations assigned to stations are those with the lowest predecessor numbers. Zero is first, then 1, then 2, etc.

The procedure is illustrated in Tables 17.7 and 17.8 where The Photo Lab is experimenting with the K & W heuristic for a three station design with cycle time $C = 4.8/3 = 1.6$ minutes. The number of predecessors for each operation is obtained using Figure 17.2 above and listed in Table 17.7.

TABLE 17.7

OPERATIONS RANKED BY THE NUMBER OF PREDECESSORS

Operation	Number of Predecessors	t_i
1	0	0.4
2	0	0.5
3	1	0.6
4	1	0.7
6	2	1.0
5	3	0.5
7	6	0.6
8	7	0.1
9	7	0.4

Operations are assigned to stations in the order of the least number of predecessors. For station I, select operations (2) and (1) in that order because of the rule for ties below. They have a total operation time of 0.9 which leaves 0.7 free time in station I. This follows from C = 1.6. Either introduce operation (3) with $t_i = 0.6$ or operation (4) with $t_i = 0.7$ into station I. They are tied with respect to number of predecessors.

When ties exist, another rule applies. Choose first the longest operation times that can be used. Short operations are saved for ease of fitting them later. In this way, earliest stations are given the least idle time possible. As a result of this rule, in Table 17.8 operation (2) is assigned before (1).

Operation (4)'s time of 0.7 is longer than operation (3)'s time of 0.6. Therefore, operation (4) is assigned to station I. This results in total time of 1.6 for station I. It is now fully assigned and has no idle time.

Attending to station II, of the nonassigned operations, (3) has the least number of predecessors. Next comes operation (6) with two predecessors. Station II with (3) and (6) has a total operation time of 1.6. Therefore, station II also is fully assigned with no idle time.

If operation (6) required more time than was available at station II, then operation (5) would have been chosen. Operations (7), (8), and (9) could not be used. They must be saved for a later station assignment because of precedent constraints.

When an operation with the smallest number of predecessors has too long an operation time to be included in the station, the heuristic is programmed to look for the operation with the next smallest number of predecessors which fits the available time at the station and precedent constraints. Thus, operations (5), (7), (9), and (8) are assigned at station III, which is fully assigned and has no idle time. Note that operation (9) is assigned before (8) because of the rule concerning ties.

Perfect balance has been achieved using the K & W heuristic. This is shown in Table 17.8.

TABLE 17.8

K & W HEURISTIC WITH CYCLE TIME = 1.6 MINUTES

Station	i	t_i	Station Sum	Cumulative Sum	Idle Time
I	2	0.5			
	1	0.4			
	4	0.7	1.6	1.6	0
II	3	0.6			
	6	1.0	1.6	3.2	0
III	5	0.5			
	7	0.6			
	9	0.4			
	8	0.1	1.6	4.8	0

The precedence diagram in Figure 17.8 is subdivided in accordance with Table 17.8 assignments.

FIGURE 17.8

THE PHOTO LAB'S THREE-STATION ASSIGNMENT USING KILBRIDGE AND WESTER'S HEURISTIC WITH C = 1.6 MINUTES

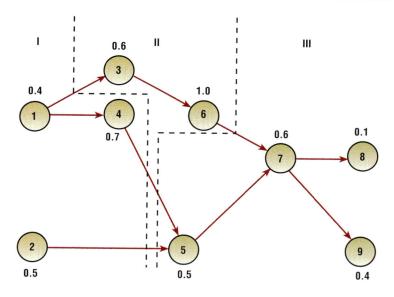

Three perfectly balanced stations can be used if the technology permits. With people working, add (say) 0.2 for rest and delay. This adds 12.5 percent to the cycle time. Perfect balance output would be $60/1.6 = 37.5$ orders per hour (see Table 17.4 above). With rest and delay, the output would be further reduced to $60/1.8 = 33\ 1/3$ orders per hour.

At the end of the day, instead of 320 orders being processed, there would be only 266 2/3 orders completed. That is 53 1/3 orders shy of expected demand. Dividing 53 1/3 by 33 1/3 yields 1.6 hours of overtime to correct the situation. There are lots of other alternatives including use of the $n = 4$ station line-balance solution.

Other Heuristics

Helgeson and Birnie's (H & B) heuristic suggested first assigning to stations those operations whose followers have the largest total time. This is the sum of all successors' operations times.[3]

$$\sum_{i = \text{all followers}} (t_i)$$

H & B called their approach the "ranked positional weight" method.

In Table 17.9 each operation of The Photo Lab's example is associated with a weight equal to the sum of all operation times that follow it. The operations are then ranked by descending order of the weight measure.

Check Figure 17.2 above to develop the set of weights that appear in Table 17.9. As one example, operations (3), (4), (5), (6), (7), (8) and (9) (*not* 2) follow operation (1). The sum of these operations' times is 3.9. This is the weight given to operation (1) in Table 17.9.

Similarly, operations (8) and (9) follow operation (7). The sum of their operations' times is 0.5, which is the weight for operation (7) in Table 17.9.

The operations have been assigned to the stations in accordance with the weights. The largest weight is 3.9 for operation (1) and it goes first. Next is operation (3) with a weight of 2.1, and so on. Note that the cycle time of 1.6 minutes was selected to allow comparison with the results of the Kilbridge and Wester heuristic. Four stations are required to use $C = 1.6$ minutes.

TABLE 17.9

H & B HEURISTIC WITH CYCLE TIME = 1.6

Station	Operation in Ranked Order	Weight	Station t_i	Sum	Idle Time
I	1	3.9	0.4		
	3	2.1	0.6		
	2	1.6	0.5	1.5	0.1
II	4	1.6	0.7		
	5	1.1	0.5	1.2	0.4
III	6	1.1	1.0		
	7	0.5	0.6	1.6	0.0
IV	8	0	0.1		
	9	0	0.4	0.5	1.1
				4.8	1.6

Figure 17.9 shows the precedence diagram divided into four station sectors.

FIGURE 17.9

THE PHOTO LAB'S FOUR-STATION ASSIGNMENT
USING HELGESON AND BIRNIE'S HEURISTIC

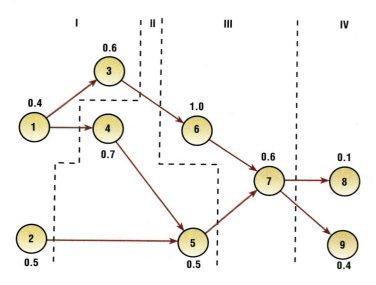

Cycle time = 1.6 minutes.

With $n = 3$, the cycle time is 1.6 minutes or greater. Operations are assigned without violating precedence or zoning constraints. When the total time for a station is exceeded, the attempt is made to find a feasible operation further down the list that can be included. This was not possible with station I.

The heuristic requires that operations (1) and (3) be included as part of station I. Operations (2), (4), and (6) are the only ones that might then be assigned without violating precedence. The total time for operations (1), (3), and (4) is 1.7 minutes. For (1), (3), and (6) it is two minutes. For (1), (3), and (2) it is 1.5 minutes, and so that was used.

The H & B approach does not yield three stations but results in four stations operating under the cycle time of 1.6. The fourth station's unused capacity is particularly poor. Balance delay is high:

$$d = 100(4 \times 1.6 - 4.8)/(4 \times 1.6) = 25 \text{ percent}$$

This same result is obtained by adding up the total idle time in Table 17.9, dividing it by total station time, and making it a percent, i.e., $100[1.6/(4 \times 1.6)]$.

Tonge's heuristic develops a learning procedure[4] that rewards success and penalizes failure by increasing and decreasing the probability of selecting a heuristic randomly from a catalog of heuristics. Each heuristic assigns operations to stations on its own basis. Those heuristics which require the least number of stations are rewarded. Those which require a greater number of stations are penalized.

Among the set of heuristics employed to choose the next operation are:

A. longest operation time, t_i

B. largest number of immediate followers

C. operation i chosen at random

All assignments are made subject to nonviolations of precedence and zoning constraints.

The choice of heuristic is determined by $p(A)$, $p(B)$, and $p(C)$. These probabilities sum to one. Initially they are all equal, i.e., $\frac{1}{3}$. Then, according to the success of the heuristic chosen, the associated probabilities are increased or decreased.

Tonge concludes that this probabilistic approach results in fewer workstations than any one heuristic or by random choice of operations.

Arcus's heuristic[5] sets the pattern for random generation of feasible sequences. All assignments are made subject to not violating any constraints based on precedence, zoning, and feasibility relations. The computer program generates 1000 line-balance arrangements. The line-balance arrangement chosen is that one out of a 1000 which requires the minimum number of stations.

On computers in the 1990s—rapidly approaching 2000—large problems take a few minutes to generate thousands of sequences. It is historically interesting to note Arcus's experience in 1966 with a relatively small problem—namely, he reported that the generation of 1000 sequences for a system of 70 operations on an IBM 7090 required 30 minutes.

It should be noted that for very large problems, the objective of minimizing the number of workstations—with cycle time given—makes good sense. Also, in spite of its antiquity, E. J. Ignall's article[6] is one of the best summaries of line-balancing approaches.

17.3

STOCHASTIC LINE BALANCING

Another facet of the line-balancing problem occurs when operation times at the workstations are variable. Most machines have relatively constant operation times and workers do not. Without planning, bottlenecks can arise. Then a WIP buildup occurs. If a paced conveyor is being used, it may have to be stopped or other steps must be taken to keep the line moving.

When operation times are randomly distributed over time, they are said to be stochastic. To illustrate, The Photo Lab's operation (6) might vary 95 percent of the time between 0.8 minutes and 1.2 minutes with the average time being 1.0 minute.

When work at a station can exceed the allotted time, a plan is needed. The alternatives are:

1. Let the unfinished or defective item move on to the next station. Depending on the nature of the problem, successive stations may or may not continue to work on the item.
2. Develop a group of special workers who can shift from station to station to provide extra help as needed.
3. Label this item so it can be removed at the end of the line and reworked then. Stations will not continue to work on items so labeled.

4. Stop the line and correct the problem. This applies not only to the product but also to the part of the process that created the problem.

5. Remove the defective item as soon as it is discovered and let the line move on without it. Downstream stations will have to wait until good items start to reach them again. Take the item to a rework area.

Each plan has a cost which can be determined with analysis. Other plans can be developed which might be more appropriate.

As far as line balancing is concerned, operation times should be increased to reflect the longest time that might be experienced with each operation. The number of standard deviations used will depend on the severity of the penalty that is paid for having exceeded the cycle time of the paced conveyor. Very severe penalties might require three or more sigmas.

The longer operation times used for planning will lead to more stations being required in order to deliver the specified output. The cost for protection against extra long operation times is more stations to manage and increased idle time at those stations. The trade-off analysis must be done with reasonable estimates of the costs to determine what policies to follow.

Generally, methods of statistical process control can be used effectively to control, and sometimes reduce, the variability of the operation times. Programs of Total Quality Management can be started with emphasis on seeking suggestions for improvement from those working on the process.

Queuing Aspects of Line Balancing

Queues can form when stochastic behavior characterizes any t_i. The word is commonly used to describe a waiting line for service. Say that t_2 and t_4 in Figure 17.6 above have variable operation times. This means that operation (5) in station III is sometimes idle while at other times there is a queue made up of different numbers of completed (2)s and (4)s. In the same way, assume that t_5 of operation (5) is variable. This affects station IV with fluctuations of arrivals leading to queues sometimes and idleness at other times. Queuing theory has the ability to describe what happens in such systems. It begins by characterizing arrival rates as random (Poisson distributed) and service intervals as random (exponentially distributed). Cycle-time management is required when variability characterizes operation times.

The paced conveyor is not practical when variability exists for either arrival times or service times. Then, line balancing is achieved by putting the right amount of service power in place to handle the arrivals. Such balance changes over time as demand cycles from peak loads to normal and below normal loads.

The behavior of a single workstation can be described as an input-output system with variable arrivals and variable service times. It is illustrated in Figure 17.10 with an arrival rate λ and a service rate μ. A queue of four has formed in front of the station.

A SINGLE-CHANNEL WORKSTATION

Assume that the workstation has an average output capability of four units per hour ($\mu = 4.0$) and μ is the average of a Poisson distribution. The arrival rate of orders averages three per hour ($\lambda = 3.0$). λ is the average of a Poisson distribution.

The use of this queuing model makes certain specific assumptions. First, the source of arrivals is large enough to be considered infinite. Second, there are unlimited accommodations for a waiting line. Third, service is granted on a first-come, first-served or FIFO basis. Fourth, it already has been stated that this analysis and the equations that follow apply to single-service channels. Fifth, the arrival and service distributions are Poisson. (If intervals between arrivals and between services are used, instead of numbers of orders, then both distributions are exponential.)

Single-Channel Equations [7]

$$\rho = \frac{\lambda}{\mu} = \text{the systems utilization factor}$$

The **systems utilization factor**, for the single channel, is the average number of orders being serviced over time. With a **single-channel system** *rho* is the average percent of time that the station is providing service. *Rho* must be less than one because if it is one or greater, then the system always is occupied and the waiting line will grow indefinitely. Thus: $\rho < 1.0$ is required.

$L = L_q + \rho = $ the average number of orders in the system which comprises the average number of orders waiting in line (L_q) and the average number of orders being served.

$$W_q = \frac{L_q}{\lambda} = \text{the average time that an order spends on line}$$

L_q was defined above as the average number of orders waiting in line. An equation for the L_q of a single-channel station will be given shortly. $W = W_q + 1/\mu =$ the average time that an order spends in the system and consists of the average time spent waiting plus the average service time, $1/\mu$.

Figure 17.10 above shows the single-channel situation where $\lambda = 3$ and $\mu = 4$. Although the station's output capacity is greater than the demand rate, a waiting line can be expected to develop from time to time. This occurs when either demand becomes greater than the average lambda (λ) and/or the service rate falls below the average mu (μ). If both occur simultaneously, then the resultant queue can be formidable.

Equation 17.6 describes the average number of units in the queue where:

$$L_q = \frac{\rho^2}{(1 - \rho)} \qquad\qquad \textbf{Equation 17.6}$$

where:

$$\rho = \frac{\lambda}{\mu} = \frac{3}{4} = 0.75 \text{ system's utilization factor}$$

Systems utilization factor, ρ (rho)

Average number of orders being serviced over time for single-channel systems.

Single-channel system

Rho is the average percent of time that the station is providing service.

(the average number of orders being serviced)

whence:

$$L_q = \frac{9}{4} = 2.25 \text{ orders}$$

(average number of orders waiting for service)

and:

$$L = L_q + \rho = 2.25 + 0.75 = 3.00 \text{ orders}$$

(average number of orders in the system)

Note also:

$$L = L_q + \rho = \frac{(\rho)}{(1 - \rho)} \qquad \textbf{(Using Equation 17.6)}$$

Further:

(the average time that an order spends in line)

$$W_q = \frac{L_q}{\lambda} = 2.25/3 = 0.75 \text{ hour}$$

Also:

(the average time that an order spends in the system)

$$W = W_q + \frac{1}{\mu} = 0.75 + 0.25 = 1.00 \text{ hour}$$

Note that as long as there is a waiting line, the average cycle time between successive completed orders is 0.25 hour or 15 minutes. This means that when there is a waiting line, the station's output rate is four orders per hour. However, output drops to zero when there are no orders in the system. To find out what percent of the time the station is not working, it is useful to determine the probability distribution of system occupancy. Using that distribution some additionally interesting insights can be gained.

The probability distribution is denoted by (P_n), where (P_n) equals the probability that the total number of orders in the system is (n). When $n = 0$ there are no orders and the station is idle. When $n = 1$ the station is working but no orders are waiting. When $n = 2$ the station is working (set-up times included) and one order is waiting. The pattern continues for $n > 2$.

The values of P_n are derived from Equation 17.7.

$$P_n = (1 - \rho)\rho^n \qquad \textbf{Equation 17.7}$$

Table 17.10 shows the calculations for $\rho = 0.75$.

TABLE 17.10

DISTRIBUTIONS FOR SINGLE-CHANNEL QUEUING MODEL WITH ASSUMPTIONS LISTED ABOVE AND $\rho = 0.75$

n	P_n		Probability Density	Cumulative Probability
0	$P_0 = (1 - \rho)$	=	0.2500	0.2500
1	$P_1 = (1 - \rho)\rho$	=	0.1875	0.4375
2	$P_2 = (1 - \rho)\rho^2$	=	0.1406	0.5781
3	$P_3 = (1 - \rho)\rho^3$	=	0.1055	0.6836
4	•	=	0.0791	0.7627
5	•	=	0.0593	0.8220
6	•	=	0.0445	0.8665
7	•	=	0.0334	0.8999
8	•	=	0.0250	0.9249
9	•	=	0.0188	0.9437
10	•	=	0.0141	0.9578
•	•	•	•	•
•	•	•	•	•
•	•	•	•	•
n	$P_n = (1 - \rho)\rho^n$	etc.	etc.	etc.

FIGURE 17.11

PROBABILITY DENSITY DISTRIBUTION FOR $\rho = 0.75$

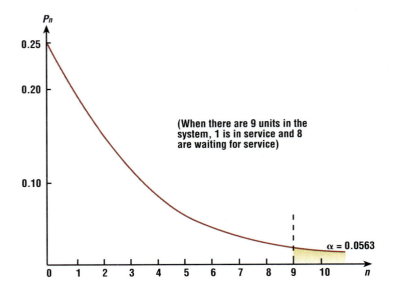

(When there are 9 units in the system, 1 is in service and 8 are waiting for service)

$\alpha = 0.0563$

Figure 17.11 shows the probability distribution for n orders being in the system. The area of the tail of this distribution (yellow) is the probability (α) that the queue length will exceed ($n - 1$). The graph states the probability that the queue will be longer than eight is 0.0563, which is ($1 - P_9$).

Thus, the work waiting can be expected to be greater than eight more than five percent of the time. If only eight units can be stored in front of the station, then at least five percent of the time the system will turn away arriving orders and/or shut down. Because this is clearly undesirable, steps will be taken to decrease the probability of exceeding eight units in queue or of providing more storage space. These problems typify cycle-time management for stochastic systems.

Other performance characteristics of this system are apparent from Table 17.10 above. The station will be idle 25 percent of the time because $P_0 = 0.25$. The workstation's output is given by Equation 17.8.

$$\text{Expected output} = (1 - P_0)\mu = (0.75)(4) = 3 \text{ units per hour} \qquad \textbf{Equation 17.8}$$

The expected output is $3(8) = 24$ units per day. Cycle-time management can take the form of increasing μ. This might be done by adding an additional station.

Multiple-Channel Equations[8]

Multiple-channel stations

Total systems utilization must be less than one; the single-line forms in front of M stations waiting for a free station to provide service; more complex than a single-channel system.

Figure 17.12 shows a queue of four with arrival rate λ waiting for service by M **multiple-channel stations** which have service rates of $\mu_1, \mu_2, ..., \mu_M$. This kind of waiting system is used by many banks, airline ticket windows, and the post office.

MULTIPLE CHANNEL WORKSTATIONS

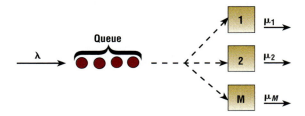

The system utilization factor when there are M servers is

$$\rho = \frac{\lambda}{M\mu}$$

To avoid infinite waiting lines $\lambda < M\mu$, which means that rho must be less than one. If the decision is made to use two parallel stations, then $M = 2$ and

$$\rho = \frac{\lambda}{2\mu} = \frac{3}{8}$$

when $\lambda = 3$ and $\mu = 4$, as before.

The multiple-channel equations for L, W, etc. are much more complex than for single-channel systems, so only the result is now shown of adding another station. Table 17.11 compares the total number of orders (L) that are in the system—waiting and being serviced. It also compares P_0 which for $M = 1$ means one station is idle and for $M = 2$ means both stations are idle.

TABLE 17.11

COMPARISON OF ONE AND TWO STATIONS WITH $\lambda = 3$ AND $\mu = 4$

	$M = 1$	$M = 2$	Percent Change
L	3.00 orders	0.85 orders	-72
P_0	0.25	0.39	$+56$

An additional strategy is available for cycle-time management. If the single-channel workstation can be mechanized so that the service intervals are constant, then L will be reduced by about 35 to 45 percent. Any reduction in the variability of service time produces significant reductions in waiting lines and in cycle time.

Finally, the kind of system that is being managed can vary greatly. The service units can be part of a job shop. Each station stands alone as an individual, stochastic input-output system providing batch service. When the flow shop configuration is used, then the various service units are interrelated, and the outputs of one system become the inputs to another. This requires careful coordination and matching of the stochastic behaviors of systems that are depending upon each other.

Queues can be comprised of customers' orders waiting for service, WIP waiting for the next machine, airplanes waiting to be cleared to land, or cars waiting at the toll booth. The waiting line is invisible but present when a busy signal is encountered on the telephone line. Express checkout lanes in supermarkets alter the FIFO priority and would require other models. Parking garages put up the sign "Sorry Full" which turns away arrivals.

Queuing models do not provide optimum solutions for output rates and cycle times. They provide descriptive measures. OM evaluates various designs using these descriptive measurements of queuing criteria. Design changes are made based on test results, and the system is retested. This goes on until a satisfactory production line is obtained.

Many different models exist and most of them have been fully explored in the literature. When the situation does not lend itself to mathematical analysis, then simulation can be used to determine the behavior of the service system (see the supplement to this chapter).

17.4

JUST-IN-TIME DELIVERY

In line-balance terminology, each station is a customer for the "supplier" station that precedes it. JIT is a cycle-time management principle which states that delivery to the next station of a specified amount of material is to be made when and only when that next station is ready to receive that delivery. Just-in-time (JIT) delivery requires agreement between suppliers and buyers of what, when, where and how much. It should be noted that JIT puts production at risk in the event that something happens to the supplier or to the supplier's shipper. Thus, risk reduction is achieved by carrying extra stock. This policy is referred to as just-in-case.

A **kanban** is any method that calls attention to the fact that the next station wishes to receive a JIT delivery. Kanban is a card used in Japanese manufacturing plants to signal that a delivery will be needed. The word has been adopted all over the world to describe an array of signaling devices which include colorful squares which signal when uncovered, Ping-Pong balls dropped through tubes, and lights flashing.

Kanban

Any method that calls attention to the fact that the station wishes to receive a JIT delivery.

T E C H N O L O G Y

Quality Control Color

Keeping abreast of technology, Quality Control Color (QCC) has evolved from a color separation trade shop to providing one-stop graphics communications services for customers. QCC began operations in 1987 as a supplier to the printing and advertising industries. The company's specialty was color separation, the process used for preparing a color photograph or transparency for reproduction on a printing press.

Widespread use of desktop publishing technology by advertisers and printers, however, has all but eliminated the need for physical separation of colors by specialists. The function is now a part of software programs that perform all of the "prepress" activities including typesetting, page makeup, and color separation. These activities are now done on a personal computer whose output is read by the printing press.

Desktop publishing is used in the production of all types of printed materials, from brochures and business cards to textbooks.

QCC now specializes in producing color brochures and catalogs. QCC's technology includes Indigo E-Print 1000 Digital Offset Color printing presses which are used to take projects from concept to printed page. Finished products are achieved in hours rather than days. One of the strengths of QCC's system is its capability to store pictures digitally. This is important for customers such as Thomasville Furniture, which has QCC store thousands of photographs of its products digitally for rapid retrieval. With QCC's system any one photograph is available at the push of a button. QCC owner Jim Sprinkles looks toward the future of his industry. "The new technology is going to open up some avenues we never dreamed of," he said.

Source: Jack Scism, "Quality Control Color thrives with changes," *Greensboro News & Record,* November 7, 1994.

Pull-production systems

Production systems that signal for delivery from upstream stations (including outside suppliers); materials are pulled as needed.

Push-production systems

Earlier era production systems which push materials on the downstream stations; when not carefully controlled, they can impede productivity

The same concept applies to outside suppliers making deliveries as required of what is needed by the producer, when, where and how much. Various plants and distributors' warehouses are built with walls that open anywhere along the production line. In this way, supplies can be delivered where they are needed on the production line. Automobile-assembly plants reduce JIT delivery risks by having suppliers located on adjacent land right outside the plant.

Production systems that signal for delivery from upstream stations (including from outside suppliers) are known as **pull-production systems**. They pull materials as they need them to work on. In contrast, production systems which push materials on the downstream stations are regarded as impeding productivity.

The smooth flows and balanced cycle times of pull-production systems are gradually replacing the **push-production systems** of an earlier era.

What accounts for workstations pushing as much WIP as possible onto the next workstation? The motivating factor for such performance is often the accounting system which records the utilization of workers and equipment, and rewards people for using equipment and producing product. The focus is on keeping busy. The result is process bottlenecks, and storage space filled with unfinished product which has carrying costs. Smooth running JIT-type pull systems have multiplicative advantages over the management accounting model which encourages full utilization of all resources and facilities.

Serialized production systems which are not connected by paced-conveyor belts must often compensate for variable arrival and service rates. These systems are allowed to work free with kanban serving as the communication link to coordinate the interdependent activities. Such systems are line balanced so that there is the same average amount of work at each station. They also are balanced as much as possible with respect to the amount of variability that characterizes each station.

The above description also applies to balancing the flows in the supply chain. When strong relationships have been developed with suppliers, the effect of the externality of the supplier to the producer is minimized. There is communication, trust, and a sense of mutual responsibility. Also, there is less difference between adjacent stations that are entirely within the plant, and adjacent stations that are in two different plants.

There is shared confidence that the quality control system of the supplier will ensure that standards are met or bettered. Deliveries are regular and reliable. Problems at either end are immediately shared. Improvements are suggested by both parties for the other's plants. Suppliers operating in this environment are regarded as part of the buyer's team. The supplier's cycle-time management is the buyer's concern and vice versa.

Another form of cycle-time management that connects suppliers and buyers is vendor releasing. This is an agreement that the buyer makes to purchase a quantity that entitles the buyer to a discount. However, the supplier is in a position to make

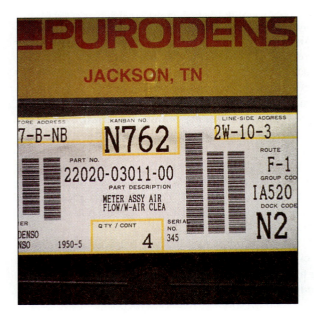

Kanban cards are used as part of
the manufacturing process
at Toyota.
Source: Toyota Motor
Manufacturing, U.S.A., Inc.

regular deliveries of parts of the total quantity purchased. To illustrate, six months of supply can be purchased at one time with weekly deliveries of specific amounts which can be changed as per agreement.

Suppliers that offer vendor releasing often can do so because they have other customers which allow them to produce in larger volumes. By producing either continuously, or in substantial economic lot quantities, shipments are permitted directly from the line and only occasionally from stock. Steel and aluminum companies use this delivery cycle-control technique to sell big quantities of product to customers who could not otherwise buy large amounts to earn discounts because they lack storage space.

17.5

PUSH- VERSUS PULL-PROCESS DISCIPLINE

As an example of the conversion from a push-production system to a pull-production system consider the following situation. Say that Joe toasts hamburger buns and Josephine puts the burger between the buns and adds tomatoes, pickles, relish and sauce. Further, assume that Joe's toaster produces ten finished crowns and heels per minute while Josephine's production is six burgers per minute.

If both work at their top capacities, Joe is pushing four extra buns per minute on Josephine and her queue will grow to 240 buns in an hour and to almost 2000 by the end of the eight-hour shift. Push production processes are relentless in their overfeeding of the bottleneck process, which in this case is Josephine's operation. To prevent absurd queues from growing, Joe shifts to the french fries department whenever there is no more space left to push the finished buns onto.

Josephine is a student of OM. She has learned about a manufacturing process that uses kanban to signal the need for supplies. From Josephine's point of view, the simplest kanban would be a yellow square on which Joe placed toasted buns whenever the square had no buns on it. The ideal quantity to be called for would be one bun at a time.

To do this, Joe's conventional toaster which delivers all finished goods in a batch should be replaced with a circular toaster, as shown in Figure 17.13.

Like a ferris wheel which stops every so often to let passengers depart and new ones to embark, the circular toaster deposits a finished bun (crown and heel) and is reloaded to start toasting every 10 seconds. This permits Joe to feed buns to Josephine as they are needed. When the demand drops off, because the peak period is past, Josephine slows down and makes fewer burgers per minute. Joe must then adjust the bun loading pattern to feed the bottleneck activity at the rate that it needs to be fed. Josephine has succeeded in pulling buns as needed.

If demand exceeds the capacity of Josephine to make burgers, then a swing person (such as the manager) starts to work alongside of Josephine. The setting is turned up on the toaster and its speed of rotation is increased. This effectively increases its throughput rate without jeopardizing quality. The flexibility of the pull system is dependent upon the speed control of the feeding systems.

The pop-up toaster that was formerly used produced a batch of ten finished buns every minute and these were delivered en masse to Josephine. Joe thought that he was doing a great job when he really loaded up Josephine's workstation because that is the criterion of high productivity when using a push system. Customers are

FIGURE 17.13

SERIALIZED TOASTING PERMITS PULL PROCESS

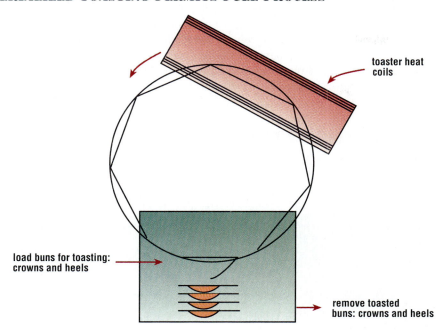

toaster heat
coils

load buns for toasting:
crowns and heels

remove toasted
buns: crowns and heels

much happier with the pull one-at-a-time system initiated by Josephine. The buns are fresher. Josephine is happier too because she sets the pace and is not always working to reduce the bun inventory that swamps her workplace.

There is more than ample evidence that pushing output on the next workstation is counterproductive. The pull number does not have to be one. Josephine might have preferred to have two buns delivered to her every time her kanban square was empty.

Pushing production is based on two concepts:

1. the independence of adjacent workstations in the process
2. the objective of having everyone, and every facility, work at maximum capacity

Both points represent nonsystems thinking. Adjacent workstations in any process should be viewed as interdependent. They are connected by material transfer systems and successive stages of work entail product design dependencies. Workstations always are connected by common systems objectives which include waste reduction of all kinds including space, WIP and time.

Hewlett-Packard has made an exceptional videotape (entitled "Stockless Production") at their Greeley Division in Colorado. It has been used internally to demonstrate to their employees the effectiveness of moving from push- to pull-type production systems.

An intermittent flow shop was set up to make a mock product. The production line consisted of five stations each with one person doing simple operations to a styrofoam insert that was then put into a cardboard box. The line did not appear to be balanced but each of the five stations had about the same amount of work to do.

Three different production arrangements were tested. The first pushed six completed units from each station to the next one. The second pulled three units and the third pulled one unit.

The five evaluation measures used were space required, work in process, cycle time, rework quantity and problem visibility. HP measured cycle time as the interval between start work on a part and its completion. It is shown to be correlated with both work in process and the rework quantity. To determine rework quantity, defective material inserted into the line was detected at the fifth and final station. All work in process queued up between stations was counted to determine how much rework would be required once the problem was spotted.

Table 17.12 shows the results of these push versus pull comparisons.

TABLE 17.12

RESULTS OF THE PUSH VERSUS PULL COMPARISON

	Push 6	Pull 3	Pull 1
Space Used	2 tables	2 tables	1 table
Work in Process	30 units	12 units	4 units
Cycle Times	3'17"	1'40"	0'19"
Rework Quantity	26 units	10 units	3 units
Problem Visibility	hidden	hidden	visible

There is a decrease of 90 percent in cycle time between Push 6 and Pull 1. The decrease in WIP and rework quantity also is about 90 percent. Space needs are cut in half because WIP is down to four units. Problems can be spotted immediately because the WIP clutter is removed. Everyone who views this tape agrees, the push- versus pull-pilot experiment is impressive. Process visibility had improved so much that the production manager commented, it might be possible to consider automating the line.

JIT processing was a by-product of the switch from push to pull-one production. Cycle-time management is enhanced by this transformation of the process. Line balancing is a natural outgrowth of the change. No investment was required other than the will to change. There were no costs and the savings are evident and significant. The applicability of this method to real production systems and supply-chain functions is immense.

SUMMARY

Cycle-time management is the management of the interval between finished goods coming off the production line or the interval between completion of servicing operations. Cycle-time management also is the management of how long work remains at workstations and how much idle time there is at the stations. Line balancing is used to provide near-equal assignments to all workstations. This prevents some stations from being overloaded while others are idle. Cycle-time management is a form of time-based management related to synchronization of material flows including those from suppliers.

The different approaches to line balancing consist of the deterministic, heuristic and stochastic models. All require designing the best possible processes using precedence diagrams of serialized flow systems. The general objective is to obtain the desired production output with the minimum number of stations and the least amount of idle time.

Perfect balance, which has zero balance delay, still needs rest time added if people are working on the line. Heuristic models include Kilbridge and Wester's (K & W) line-balancing techniques. The heuristic approaches are needed for complex intermittent flow-shop systems because the problems are too large for optimization models and they are too important to ignore the benefits of line balancing.

Stochastic line balancing deals with the affects of variability in cycle management. This includes consideration of a variety of queuing models. It should be pointed out that a primary function of queuing models is to facilitate the design of balanced-service systems (requests for service and servicing capabilities) throughout the supply chain. There is discussion of suppliers' participation in effective cycle-time management.

The chapter concludes with two examples of the conversion of push-production systems to pull-one-unit production systems. The advantages of pull-one unit are detailed. A supplement deals with simulation of stochastic line-balanced queuing models.

▶ Key Terms

balance delay   *751*

heuristic line-balancing methods   *738*

line balancing (LB)   *736*

precedence diagram   *743*

single-channel system   *761*

time-based management (TBM)   *735*

cycle time   *734*

kanban   *765*

multiple-channel stations   *764*

pull-production systems   *766*

stochastic line balancing   *738*

deterministic line balancing   *738*

lateral transferability   *750*

permutability of columns   *750*

push-production systems   *766*

systems utilization factor   *761*

▶ Review Questions

1. What is cycle-time management?

2. What is line balancing?

3. How does line balancing relate to cycle-time management?

4. Do line balancing and cycle-time management apply to services as well as material products?

5. What is the difference between deterministic and stochastic line balancing? Can a paced conveyor be used for either one? Explain your answer.

6. Why is it said that heuristic line balancing is a practical approach to achieving a reasonably good intermittent flow shop? Can a paced conveyor be used for the resulting production line? Explain your answer.

7. Explain why heuristic line balancing is associated with computer software to run the algorithms.

8. Why is perfect balance often regarded as an unrealistic goal? When is that not the case?

9. Explain permutability of columns.

10. What does lateral transferability mean for line balancing?

11. What are zoning constraints for line balancing ?

12. Why is it essential to design the best possible process before starting line balancing?

13. Why is it important to get good estimates for operation times before starting line balancing?

14. An automatic teller machine (ATM) corresponds to a stochastic input-output flow shop system. As such, queuing theory can be used to study the behavior of the system and to provide cycle-time management. Explain.

15. The flow shop is designed to remove stochastic behaviors. Explain this statement and comment on how this objective can be accomplished. When is this objective unrealistic?

16. How do queuing models relate to cycle-time management?

17. For single-channel systems, why is $\lambda > \mu$ a violation?

18. For single-channel systems, why is it necessary to assume that the source population is very large?

19. For single-channel systems, why is it necessary to assume that the storage space can accommodate any number of units?

20. What difference characterizes single-channel configurations from multiple-channel configurations?

21. Would the FIFO order of entry make sense for an automatic teller machine?

22. For multiple-channel systems, why is $\lambda > M\mu$ a violation?

23. What are Poisson arrivals?

24. What is meant by exponential service times? How can this be stated in terms of the Poisson distribution?

25. Does FIFO hold for supermarket checkout counters?

26. What kind of stochastic situation occurs in a supermarket with n different checkout lanes?

27. What constitutes cycle time in the supermarket checkout situation?

28. What is the supplier's role in cycle-time management?

29. Explain vendor releasing from the point of view of both the buyer and the supplier.

30. When is a push-production system desirable?

31. Explain how to use a kanban to convert a push-production system to a pull-one-unit production system.

▶ Problem Section

1. Assume that you are going into business selling replacement tires. Draw a precedence diagram for changing tires. Discuss the way in which this job could be done with a flow-shop configuration. Why would a flow shop be preferred? Suggest a possible division of labor that could produce a reasonable line balance.

2. Draw a precedence diagram for the way in which tellers handle withdrawals at the bank. Can this job be accomplished by tellers in flow-shop configurations?

3. Line efficiency (Λ) is a measure used to assess line balance. It is:

$$\Lambda = \frac{\sum_i t_i}{nC}$$

Determine line efficiency for The Photo Lab where $n = 4$, $C = 1.5$ minutes, and total work content is 4.8.

4. Determine line efficiency Λ (as described in Problem 3 above) and relate it to the balance delay measure:

$$d = 100(nC - \textstyle\sum_i t_i)/nC$$

when total work content is three hours and there are 180 stations operating with a cycle time of one minute? Compare the results and meanings of d and Λ.

5. For The Photo Lab what happens if t_{max} can be reduced from 1.0 to 0.7 by improving the technology of the film development step? Refer to Figure 17.2 and Tables 17.2 and 17.3 in the text.

6. Using the information in Problem 5 above, what happens if t_{max} is reduced from 1.0 to 0.6 by installing two photo developing units in parallel for operation (6)? Develop a table for this answer similar to Table 17.4 above.

7. Use The Photo Lab's precedence diagram in Figure 17.2, but change the operation times in Table 17.2 as follows:

Operation (i)	Operation Time (t_i), Minutes
1	1.4
2	1.5
3	1.6
4	1.7
5	1.5
6	2.0
7	1.6
8	1.1
9	1.4

What productivity rates are possible with perfect balance?

8. Using the information in Problem 7 above, develop a feasible station layout for $n = 5$. What are the characteristics of this arrangement, i.e., cycle time, productivity and balance delay?

9. Using the information in Problem 7 above, use Kilbridge and Wester's heuristic line-balancing algorithm with $n = 4$. What are the characteristics of this arrangement, i.e., cycle time, productivity and balance delay?

10. Drop operation (9) from The Photo Lab's precedence diagram in Figure 17.2 above, and alter Table 17.2 as follows:

Operation (i)	Operation Time (t_i), Minutes
1	0.4
2	0.5
3	0.6
4	0.7
5	0.5
6	1.0
7	0.6
8	0.1

What productivity rates are possible with perfect balance?

11. With the information in Problem 10 above, develop a feasible station layout for $n = 3$. What are the characteristics of this arrangement, i.e., cycle time, productivity and balance delay?

12. For the information in Problem 10 above, use Kilbridge and Wester's heuristic line-balancing algorithm with $n = 4$. What are the characteristics of this arrangement, i.e., cycle time, productivity and balance delay?

13. Solve The Photo Lab's problem, but multiply all operation times in Table 17.2 above by 10. Use the Kilbridge and Wester's heuristic to achieve a productivity level of 20 finished units per eight-hour day.

14. Find the solution for Problem 13 discussed above using the Kilbridge and Wester's heuristic with $n = 4$.
    a. What is the cycle time?
    b. What is the productivity rate?
    c. How much idle time results?

15. Using the information from Problems 13 and 14 above, what do you recommend, three or four stations?

16. Using the information from Problems 13, 14, and 15 above, would you consider working nine hours per day with $n = 3$?

17. Using the information from Problems 13 and 14 above, plus the fact—just known—that the system is totally mechanized, would it be worthwhile to try to speed up certain operations so that an output rate of 40 units per hour is achieved?

18. Using the information from Problem 17 above, is it possible to design a balanced flow shop that can deliver hourly output rates of 37.5, 40.0, and 42.5 at the flick of a switch?

19. Figure 17.14 depicts the present line balance used by the Baby's Stroller Company for an intermittent flow shop that makes convertible strollers. These strollers, which can be made into baby car seats, have become increasingly popular. The IFS is now run twice a month for four to five days.

F I G U R E   1 7 . 1 4

BABY'S STROLLER LINE BALANCE

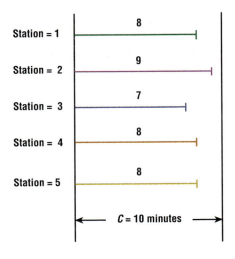

All the work is done with five stations.
    a. What is the productive output per hour?
    b. What is the balance delay?

20. Viewing the automatic teller machine (ATM) as a stochastic, single-channel, input-output system with Poisson arrivals and exponential service times, determine the values of W_q and L_q for $\lambda = 8$ and $\mu = 10$. *Note:* FIFO holds as well as the other assumptions required for the single-channel queuing model.

21. Viewing a vending machine as a stochastic, single-channel, input-output system with Poisson arrivals and exponential service times, determine the values of W and L for $\lambda = 8$ and $\mu = 10$. *Note:* FIFO holds as well as the other assumptions required for the single-channel queuing model.

22. Viewing the painting booth in the automobile-assembly plant as a stochastic, single-channel, input-output system with Poisson arrivals and exponential service times, determine the values of P_0 and P_1 and $P_{n > 1}$ for $\lambda = 8$ and $\mu = 10$. *Note:* FIFO holds as well as the other assumptions required for the single-channel queuing model.

23. For the painting booth described in Problem 22 above, what kind of in-process (WIP) storage will be needed? Where should it be located?

24. If an assembly line is mechanized with two workstations such that the first station's output rate is three units per minute and the second station's output rate is four units per minute, what is the output rate of the system—assuming deterministic performance?

25. At The Burger Place, Jane's toaster produces 20 finished crowns and heels per minute while Jim dresses 10 burgers per minute. They have been trained to use push-production methods. When Jane sees that Jim has no more space she helps out at the drive-by window. What is the waiting line if they both work at rated speed for 15 minutes? How big a storage space would Jim need? Is quality affected?

26. Using the information in Problem 25 above, convert Jane's operation to a pull-five system. What delivery rate will apply?

27. Using the information in Problems 25 and 26 above, what kind of a kanban do you recommend for times when demand for burgers drops off to less than 10 per minute? Would it make sense for both Jane and Jim to combine the toast and dress operations and work in parallel? What maximum demand rate could they satisfy?

▶ **Readings and References**

Reza H. Ahmadi and Hernan Wurgaft, "Design for Synchronized Flow Manufacturing," Management Science, Vol. 40, No. 11, Nov. 1994, p. 1469–1483.

W. M. Chow, Assembly Line Design, Amsterdam, Marcel Dekker, 1990.

R. B. Cooper, Introduction to Queueing Theory, 2nd Ed., NY, Elsevier North-Holland Publishing, 1981.

S. Ghosh, and R. Gagnon, "A Comprehensive Literature Review and Analysis of the Design, Balancing and Scheduling of Assembly Systems," International Journal of Production Research, Vol. 27, No. 4, 1989.

S. C. Graves, "A Review of Production Scheduling," Operations Research, Vol. 29, No. 4, July-August, 1981.

D. Gross, and C. H. Harris, Fundamentals of Queuing Theory, 2nd Ed., NY, Wiley, 1985.

Hiroyuki Hirano, JIT Implementation Manual: The Complete Guide to Just-In-Time Manufacturing, Cambridge, MA, Productivity Press, 1990.

Edward J. Ignall, "A Review of Assembly Line Balancing," Journal of Industrial Engineering, Vol. 16, No. 4, July-August, 1965.

Japan Management Association (Ed.). Kanban and Just-in-Time at Toyota: Management Begins at the Workplace, Trans. by D. J. Lu, Cambridge MA, Productivity Press, 1989.

A. M. Law, and W. D. Kelton, Simulation Modeling and Analysis, NY, McGraw-Hill, 1982.

D. R. Plane, Management Science, A Spreadsheet Approach, Danvers, MA, boyd & fraser publishing company, 1994.

A. A. B. Pritsker, C. E. Sigal, and R. D. J. Hammesfahr, SLAM II, Network Models for Decision Support, The Scientific Press, 1994.

T. E. Vollmann, W. L. Berry, and D. C. Whybark, Manufacturing Planning and Control Systems, 3rd Ed., Homewood, IL, Irwin, 1992.

S U P P L E M E N T   1 7

Simulation of Queuing Models

It has been proposed that a two-station system such as the one shown in Figure 17.15 be used to sort bulk mail.

FIGURE   1 7 . 1 5

MAIL SORTING WITH STORAGE OF SIZE N

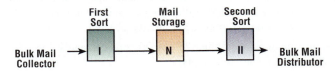

The output rates of the first and second stations are expected to vary as shown in Table 17.13.

T A B L E   1 7 . 1 3

STATION II—ARRIVAL AND SERVICE RATE DISTRIBUTIONS

	Output Rate (units/min)	Interval between Inputs to Station II (minutes)	Probability Distribution Density	Cumulative
	2	0.500	0.10	0.10
Station	3	0.333	0.50	0.60
I	4	0.250	0.30	0.90
	5	0.200	0.10	1.00
		Service Duration at Station II (minutes)		
	2	0.500	0.10	0.10
Station	3	0.333	0.30	0.40
II	4	0.250	0.50	0.90
	5	0.200	0.10	1.00

The average output rate for station I is 3.4 units per minute and for station II it is 3.6 units per minute. Although II processes at a faster rate than I, a queue will develop and the storage facility between I and II is needed. Station I must stop working when the storage queue gets larger than N.

Output rates are converted to intervals between inputs from station I and to service durations at Station II by taking the reciprocals. This is column 2 of Table 17.13. Column 3 gives the probability density distribution values. These are converted into cumulative probability distribution values in column 4.

Monte Carlo Numbers (MCNs) are specified in Table 17.14. They translate the cumulative probability distribution estimates of column 4 in Table 17.13 into number clusters that are like labels for the different output rates and their associated intervals. MCNs can be used to generate random sequences of arrivals with different input intervals from Station I and service intervals at Station II. These generated patterns conform to the respective probability distributions of outputs and inputs at the stations.

TABLE 17.14

MONTE CARLO NUMBER (MCN) ASSIGNMENTS

	Station I			Station II	
Input Interval (min.)	Monte Carlo Numbers	Quantity	Service Interval (min.)	Monte Carlo Numbers	Quantity
0.500	00-09	10	0.500	00-09	10
0.333	10-59	50	0.333	10-39	30
0.250	60-89	30	0.250	40-89	50
0.200	90-99	10	0.200	90-99	10

Using a random process for selecting numbers, there are 10 chances to pick a Monte Carlo Number 00-09 which is associated with an input interval from Station I of 0.500 minutes. Similarly, there are 50 chances of picking a MCN 10-59. This is the right proportion of Station I input intervals of 0.333 minutes.

An abbreviated table of random numbers is used to generate a series of random events. Table 17.15 shows a small set of random numbers produced by the computer. The character of the numbers in that table is such that any digit 0 through 9 is as likely to appear as any other. Because the two-digit MCNs are associated with particular events, the occurrence of the events is similarly random. Random number tables are useful for instructive purposes but computer programs are available which generate the numbers as needed for the various algorithms that use them.

A systematic pattern of reading the numbers in the table (i.e., vertically, diagonally, horizontally) must be used so that no bias is introduced in the way that the numbers are selected. Reading successive pairs of digits from Table 17.15, the first pair is equivalent to an input interval from Station I, and the second pair is equivalent to a service interval at Station II. (*Note:* Successive triplets of digits 000-999 would be used if the probabilities are stated in three places, etc.)

AN ABBREVIATED TABLE OF RANDOM NUMBERS

05621	64483	38549	62908	71579	19203	83546	05917	51905
10052	03550	59144	59468	37984	77892	89766	86489	46619
50263	91130	22188	81205	99699	84260	19693	36701	43233
62719	53117	71153	63759	61429	14043	49095	84746	22018
19014	76781	61086	90210	55006	17765	15013	77707	54317

In reality, the simulation of the two-station flow shop for sorting bulk mail would be done by computer. The details are given here as though the simulation was done with pencil and paper solely to provide understanding of the method.

The simulation consists of choosing successive pairs of random numbers and matching these against the input interval and the service interval Monte Carlo Number assignments specified in Table 17.14.

The first pair of random numbers indicates the interval until the first arrival at Station II. The second pair of random numbers specifies how long servicing will take. Four more random numbers are drawn. The first pair of these random numbers is used to specify the interval until the next arrival at Station II. The second pair of these random numbers details service time.

When a sufficient sample is drawn, the behavior of this system can be evaluated. The maximum and average length of the waiting line can be estimated. The bottleneck effect of limited storage between stations can be assessed. The idle time of Station II can be estimated, and the overall quality of this stochastic flow shop configuration can be observed to determine whether an acceptable line balance has been achieved.

Cycle-time management will depend on what strategies are available to alter λ and μ. Can Station II be speeded up? Can another Station II sorter be added at peak load times? What new technology might be brought to bear? Perhaps some of the load from Station I can be directed to another sorter location.

Using a left to right scan of the random number table, only the top line is shown here.

05621	64483	38549	62908	71579	19203	83546	05917	51905

The first two pairs of digits are 05 and 62, representing an input interval of 0.500 and a service duration of 0.250.

The simulation continues in Table 17.16. Especially note Arrival Time, Completion Time, Idle Time, and Queue Size.

All of the different performance characteristics of the system can be simulated. Complex assumptions can be made that defy mathematical analyses. To illustrate, it has been assumed that the input and service intervals are independently distributed. If this is not so, and they are conditional upon each other or on prior states, then these conditional dependencies can be modeled in a simulation. Math analysis of complex dependencies is very difficult. With a computer simulation program, is not difficult to simulate stochastic line-balancing problems involving hundreds of activities.

TABLE 17.16

EIGHT STEPS OF THE SIMULATION

Sample Number	Random Numbers	Input Interval	At Station II			Idle Time	Queue Size
			Arrival Time	Service Duration	Completion Time		
1	05, 62	0.500	0.500	0.250	0.750[a]	0	0
2	16, 44	0.333	0.833[a]	0.250	1.083	0.083[a]	0
3	83, 38	0.250	1.083	0.333	1.416	0	0
4	54, 96	0.333	1.416	0.200	1.616[b]	0	0
5	29, 08	0.333	1.749[b]	0.500	2.249[c,d]	0.133[b]	0
6	71, 57[e]	0.250	1.999[c,d]	—	—	0	1[c]
7	91, 92[f]	0.200	2.199[d,f]	—	—	0	2[d]
6			2.249[d,e]	0.250[e]	2.499[d]	0	1[e]
7			2.499[f]	0.200[f]	2.699	0	0[g]
8	03, 83	0.500	2.699	0.250	2.949	0	0

.

.

etc.

a—Station II is idle from 0.750 until 0.833 (0.083 minute).

b—Station II is idle from 1.616 until 1.749 (0.133 minute).

c—Station II is busy so arrival number six must wait from 1.999 until 2.249 (0.250 minute).

d—A second unit, arrival number seven, joins the line at 2.199. Assume that there is storage space for it. Otherwise, Station I would have to stop at 2.199 and wait until 2.499 before it could begin again. One unit waits from 1.999 until 2.249 (0.250 minute). The other unit waits from 2.199 until 2.249 (0.050 minute).

e—The random number 57 in the sixth sample indicates a service duration of 0.250. So, the sixth arrival begins to be serviced at 2.249.

f—The seventh arrival waits from 2.199 until 2.499 to be serviced. The random number 92 indicates a service duration of 0.200.

g—The line is cleared of waiting units. To this time, Station II has been idle a total of 0.216 minute. The waiting line has been occupied by a single unit for 0.200 minute, by two units for 0.050 minute (together 0.100), then by a single unit for 0.250 minute, for a total waiting time of 0.550. There has been no shutdown of Station I.

Problems for Supplement 17

1. Use the following random numbers to simulate the behavior of the two-station mail-sorting system.

   10052   03550   59144   59468   37984   77892   89766   86489   46619

2. Use the following random numbers to simulate the behavior of the two-station mail-sorting system.

   50263   91130   22188   81205   99699   84260   19693   36701   43233

Notes...

1. H. A. Simon, and A. Newell, "Heuristic Problem Solving: The Next Advance in Operations Research," *Operations Research*, Vol. 6, No. 1, January–February, 1958, pp. 1–10.

2. M. D. Kilbridge, and L. Wester, "A Heuristic Model of Assembly Line Balancing," *Journal of Industrial Engineering*, Vol. 12, No. 4, July-August, 1961, pp. 292–99.

3. W. B. Helgeson, and D. P. Birnie, "Assembly Line Balancing Using the Ranked Positional Weight Technique," *Journal of Industrial Engineering*, Vol. XII, No. 6, November-December, 1961, pp. 394–398.

4. Fred M. Tonge, "Assembly Line Balancing Using Probabilistic Combinations of Heuristics," *Management Science*, Vol. 11, No. 7, May 1965, pp. 727–735.

5. A. L. Arcus, "Comsoal: A Computer Method of Sequencing Operations for Assembly Lines," see Elwood S. Buffa, Ed., *Readings in Production and Operations Management*, NY, Wiley, 1966, pp. 336–360.

6. E. J. Ignall, "A Review of Assembly Line Balancing," *Journal of Industrial Engineering*, Vol. 16, No. 4, July-August 1965, pp. 244–254.

7. For complete coverage of queuing equations see: D. Gross and C. H. Harris, *Fundamentals of Queuing Theory*, 2nd Ed., NY, Wiley, 1985.

8. David G. Dannenbring and Martin K. Starr, *Management Science: An Introduction*, NY, McGraw-Hill, 1981, pp. 603–607. See also, D. Gross and C. H. Harris, note 7 above.

Industry Perspective on Smart Technology

AT&T's LITTLE ROCK REPAIR AND DISTRIBUTION CENTER

The AT&T Repair and Distribution Center in Little Rock, Arkansas, supports the AT&T Multimedia Group's Global Business Communications Systems (GBCS) business unit. Headquartered in Basking Ridge, New Jersey, GBCS specializes in the design, manufacture, sales, and repair of business telephone systems and other communications equipment. GBCS has annual revenue in excess of $4 billion. Manufacturing for GBCS is done at facilities in Denver and Shreveport and several international locations. The Little Rock Repair and Distribution Center is the primary repair facility for GBCS. Repair operations involve refurbishing and repairing equipment for AT&T business telephone systems under warranty or maintenance contract or for remarketing. Equipment to be repaired includes telephone sets, controllers for small telephone systems, and miscellaneous equipment such as fax machines. Exhibit 1, "Pareto Analysis of Little Rock Repair Unit Volume," shows the breakdown of products that typically flow through the facility for repair.

The Little Rock facility, which had been a manufacturing and distribution center for AT&T computers, inherited equipment repair operations from a large West Chicago facility in summer 1994. (A smaller repair and distribution operation in Denver also was

EXHIBIT 1

PARETO ANALYSIS OF LITTLE ROCK REPAIR UNIT VOLUME

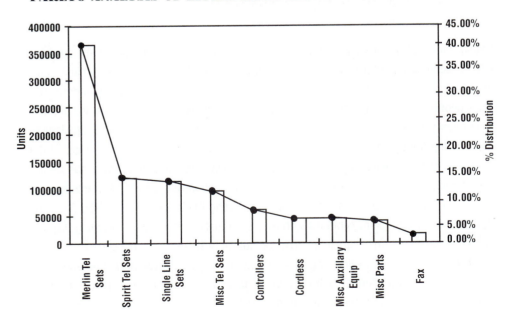

Moving GBCS repair operations to Little Rock afforded the availability of a trained work force and the opportunity to structure a new relationship with union labor.
Source: AT&T

consolidated into Little Rock.) AT&T's computer manufacturing facility and warehouse in Little Rock were no longer needed as a result of AT&T's merger with NCR in 1992. The merger created the AT&T Global Information Solutions (GIS) business unit, and all computer operations were moved to existing NCR facilities. The company decided to move GBCS repair to the unutilized space in Little Rock. This would afford advantages such as an available, trained workforce; the lower southern wage rates (as compared to those in West Chicago); and the opportunity to develop a new relationship with the union, which involved establishing the groundwork for a self-directed work force with worker pay based on performance rewards.

Process Reengineering

Rather than duplicate the West Chicago operation in Little Rock, GBCS seized the opportunity to reengineer the business processes with the goal of achieving "best-in-class" performance. This allowed GBCS to introduce state-of-the-art materials tracking and personnel scheduling systems to enhance the handling and logistics processes. The use of these Smart systems allowed efficiency gains to better utilize people, space, and capital investment within the operation. In addition to revamping processes, a complete redo of the computer platform (including hardware, software, and systems service) also was necessary. A major bottleneck that existed at the West Chicago facility (and to be overcome with the reengineered processes in Little Rock) was sorting the incoming materials. The Little Rock center began repair operations with over 8,000 pallets of materials awaiting disposition—that is, the returned equipment that had not yet been identified and sorted.

A major challenge in relocating repair to Little Rock was keeping the operation running during the transition. The transition management effort began with rapid movement of existing repair process support systems and resources from West Chicago into

Little Rock. During this initial stage, the repair processes and systems support were fragmented with no integrating strategy in place. To address this situation, a partnership between GBCS Global Information Technology Solutions (GITS) and GBCS Materials Logistics (MLS) was established to integrate process improvements and system support. GITS is offering information solutions to its internal customer, GBCS.

The original three-person project team was assembled in September 1994. By May 1995 the team consisted of about 19 members from within MLS and GITS, internal consultants from other AT&T business units (i.e., Bell Labs, Network Systems and GIS), and consultants from academia. (About half of the team members were academics or consultants hired on a temporary basis for their particular expertise.) Team members represented the following areas of specialization: Material Requirements Planning (MRP), Material Logistics, Repair Operations, Process Reengineering, Network Systems Development and Integration (hardware and software), and Project Management. The team's fundamental approach was to focus on the business processes and the associated key measures (or benchmarks) of success. Computer systems were then designed to support the reengineered process. The GBCS reengineering project for repair is a "laboratory setting" for GITS. Successful process solutions will be rolled out to other GBCS and AT&T locations.

The team began the planning phase of the reengineering project in September 1994 by defining the master process in AT&T's standard Process Quality Management & Improvement (PQMI) block diagram format. The master process (as shown in Exhibit 2, "Little Rock Repair Planning and Execution Process Current/Future") provides a framework in defining both the "current" process (as it existed at West Chicago and initially transferred to Little Rock) and the "future" (or reengineered) process. The planning phase was complete by October 1994. Implementation of the new processes began in December (with the "production pilot" of the reengineered repair process for Merlin telephone sets), with completion projected by early summer of 1995. As of May 1995 the reengineering process was 70 to 80 percent complete. According to Marc Barnett, Little Rock reengineering project manager, "We've gotten to the point where receiving materials and supplying the lines are in pretty good shape—now we're looking at optimizing the processes."

Smart Systems

The computer system used in the West Chicago facility was a mainframe-based information system which relied heavily on manual data input. Reengineered systems are run on a UNIX System V operating system. This multitasking hardware is used as a server and to control software applications. The new strategy was to use as much "off-the-shelf" software as possible to take advantage of upgrades in these packages. Software used in the system includes off-the-shelf MFG-PRO for MRP; off-the-shelf MLMS and AT&T-developed RMS for inventory; and AT&T-developed ATIS (Assembly Test Information System) for quality control. An AT&T-developed "Interface Control

EXHIBIT 2

LITTLE ROCK REPAIR PLANNING AND EXECUTION
PROCESS—CURRENT/FUTURE

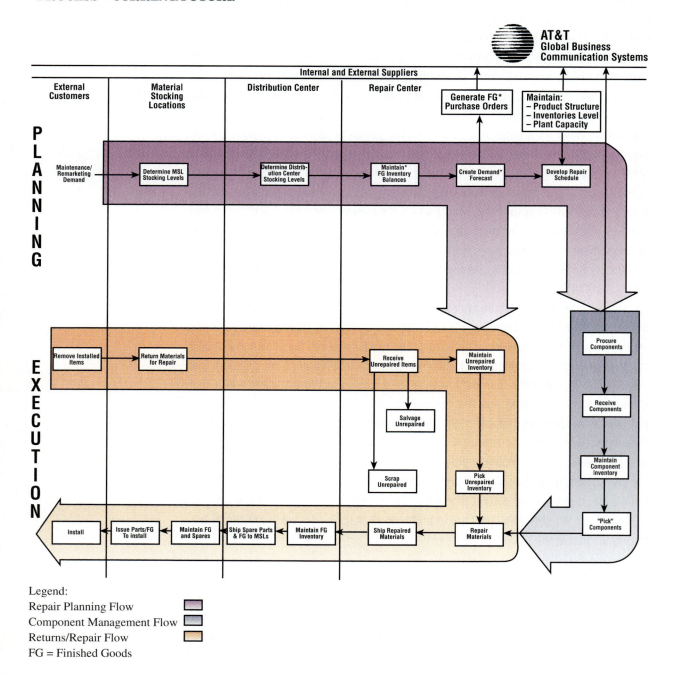

Legend:
Repair Planning Flow
Component Management Flow
Returns/Repair Flow
FG = Finished Goods

* Currently performed in Denver.

Environment" (ICE) mediates the conversation between multiple software packages for demand, inventory, MRP, financial, and support applications. A series of about 500 terminals is connected via fiber optic network to the system, including bar coding terminals, computer terminals, and radio frequency (RF) terminals. RF technology is used to communicate work directions to all mobile material handlers (vehicles are equipped with computer and RF terminals). As many as 200 users interface with the computer system at any given time. At the heart of the reengineered computer system is the bar code-driven Material Logistics Management System (MLMS) software developed by Scanning Technologies, Inc. Use of bar codes in the process to track materials replaces the need for keyboarding information into the mainframe computer, which enhances accuracy. (In keyboarding, the error rate is 1 in 300 characters keyed as compared to the bar code scanning error rate of 1 in 2.7 million scans.) The MLMS is made up of three modules: (1) forward picking; (2) material tracking; and (3) personnel scheduling.

Smart systems are evident in reengineered processes in the interactions between people and technology. Materials handlers, for example, may receive up to ten job assignments at a time through the MLMS from a terminal onto which they log at the beginning of their work day. The computer gives job assignments starting with high priority activities. During periods of slack time the system directs the materials handlers to perform lower priority activities, keeping them busy all day. The MLMS system has 14 levels of priorities and can move the people to complete highest priority jobs first. These smart systems result in a largely self-directed workforce.

Smart systems for the reengineered processes are based on benchmarks established through research by The Logistics Institute (TLI), a consortium of universities and leading corporations. GBCS worked closely with TLI in deriving the algorithms on which the smart systems are based. Key to the system are the algorithms for physical volumetrics, which inform the materials handlers of capacities of materials handling devices and material storage devices for given quantities of materials. One new algorithm used by the Material Logistics Management System (MLMS) is the Dynamic Stock Location Assignment Algorithm (DSLAA) which outperforms by about 30 percent previously used algorithms (such as the Global Assignment and the CPOI or Cube Per Order Index). The basic difference between these is that the older algorithms look at historical data, while the DSLAA bases forecasts and bills of material on anticipated demand and volumetrics.

Don Stuart, a GBCS/GITS manager and one of the three original reengineering team members, emphasizes that the systems at the facility are "real-time systems that enable us to have very precise control of our materials movement, personnel scheduling, and inventory control." The MLMS controls all materials movement, personnel scheduling, and inventory tracking to optimize processes.

The Repair Process

The process begins with customers calling the Technical Support Center to report problems with their AT&T business telephone equipment. A customer service associate takes the "trouble report" and either directs it to the National Parts Sales Center (NPSC) or a local service technician. The NPSC air ships the customer the needed part or a refurbished phone overnight, and either instructs the customer to scrap the broken phone or component or send it back to the NPSC. (The NPSC consumes or redirects 55% of the phones refurbished at the Little Rock facility.)

On a typical service call, the technician replaces a malfunctioning unit with a refurbished set stored on his or her truck. The service technician returns the broken set to a "Material Stocking Location" (MSL). There are 38 MSLs geographically spread throughout the U.S. which send materials to the Little Rock repair facility. The NPSC returns materials to Little Rock as well.

Replenishment of stock at the MSLs and NPSC is done in a traditional MRP "push" environment now. For example, with Merlin phone sets, service technicians' trucks might carry 7 telephone sets, with 2 pallets kept at the MSL, and 36 pallets at the distribution center in Little Rock. When the levels of phone sets fall below these standards, they are automatically restocked.

Returned equipment of all sorts (the database of potential products to be returned includes over ten thousand items) is bundled up at the MSL in a 4′ by 4′ box called a "gaylord" or shrink wrapped together onto a wooden pallet. The material is shipped to Little Rock and delivered to an "awaiting disposition" receiving dock. The facility receives between 100 and 150 pallets of returned materials a day. About 40 percent of the incoming material is scrapped. At present all material is sent to Little Rock for control purposes: GBCS doesn't want AT&T telephone equipment being sold by unauthorized dealers or showing up at flea markets. It does want Little Rock to be a clearinghouse for recapturing the equipment for recycling purposes. The facility processes (refurbishes and repairs) approximately 80,000 units per month.

When a carton is full, an operator scans the bar code on the flow rack lane for the product. At this point the computer system directs disposition of the cartoned units to one of five paths.
Source: AT&T.

As a result of reengineered processes, throughput for a telephone set (from receipt at the dock to final refurbishment and repair) is less than a day. This is four times faster than the throughput at the West Chicago facility, achieved with just 70% of the work force,

The Sorting Process

The reengineered process for sorting the equipment differs significantly from the previous method whereby units were sorted manually and identification numbers keyboarded into "unrepaired inventory" on the mainframe computer. In the new process, unrepaired incoming units from gaylords or shrink-wrapped pallets are put on conveyors. Operators identify the units (heuristically or by standard AT&T identifiers called "comcodes"), pick them off, and put like units into cartons. The sorting task is particularly difficult because the units do not always have identification numbers on them and different sets may look identical on the outside. There are 12 different paths the items are sorted into, including four paths for telephones.

The cartons hold a fixed number of like units, such as six Merlin phone sets. When a carton is full, the operator scans the bar code on a lane in a dedicated "flow rack" for that product. Fixed location bar codes for each lane in the flow rack identify the type of product, quantity, dollar value, and location. This data is stored in the Material Logistics Management System (MLMS). At the point of this initial scan, the MLMS interfaces with the Returns Management System (RMS) and directs disposition of the cartoned units to one of five paths. The RMS directs the operator to (1) "Keep this" and the carton is put into unrepaired stock; (2) "Move this to Shreveport or Denver or to some other location" and the carton will be put back on a conveyor and loaded on a truck. (3) The RMS might indicate that the sets contain parts no longer made, so they are put into a location for disassembly to salvage parts for use in refurbishing other units. (4) The RMS could direct the units to the "remarketing" unit that sells refurbished equipment. (5) Or, the RMS could direct the material to be scrapped. (All scrapped materials, including the circuit boards, precious metals, and plastic shells are recycled.)

If the operator is directed to "keep" the material, he or she scans the bar code location again and pushes the carton down into the flow rack lane. The MLMS knows when a flow rack lane is filled (through calculation of volumetric data) and transmits a radio frequency transaction to a forklift truck operator. For example, it will send a transaction that tells the forklift operator to go to location XYZ and palletize all the material in that lane, and then take it to the point of use (such as at the repair line) or overflow in the warehouse. To minimize material movement, the units are delivered to point-of-use whenever possible. Locations to which the unrepaired material is taken are bar coded as well (such as in the warehouse and the kanban bin flow racks at the repair lines).

Another receiving operation handles components used in repairing the units. As the component material flows in from the dock, it is delivered to a kanban location directly on the line or to a reserve location in the warehouse. The two receiving flows (units to be repaired and components used in repairing them) are merged at the repair line.

Shop Floor Logistics

Shop floor logistics is in place for components and equipment used in the repair process. Materials are categorized into A, B, or C categories. C items are low-cost, non-critical components (such as nuts, bolts, clips, and screws) with short lead times. These items are managed outside MRP and bypass the component storeroom; they are delivered directly from the dock to the kanban bins at the repair lines. B items are non-critical items that require longer lead times than C components; B items respond well to traditional MRP methods; they go from dock to component storeroom to kanban bin. A items are the most critical to the process (an example of an A item is an unrepaired telephone set); they are more complex, higher cost items that require long lead times and extra effort in handling; A items also go from dock to component storeroom to kanban bin. Because material managers don't have to deal with C items, they have more time to handle A and B items.

The Repair Line

Steps in the repair process are (1) refurbishing the units—replacing the plastic shell, cord and handset, and putting a serial number bar code label on the unit (the bar codes are produced using Zebra Technology printers located at the end of every two or three repair lines); (2) testing the units; (3) repairing the units—replacing components such as circuit cards as necessary (about 15% of the units require repair at this point); (4) retesting the units to make sure the problem was corrected; and (5) random quality checking the units. In the first pass quality check, the target defect rate is .01 (the benchmark in the telecommunications industry). This quality check runs the unit through the same visual and technological tests as those in steps (2) and (4) above.

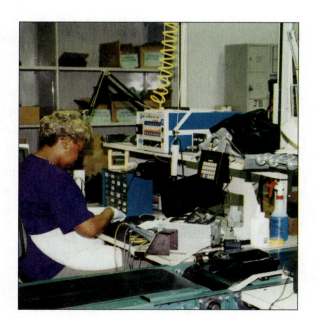

Each telephone set is tested twice during the repair process, then quality checks are performed at random on repaired sets, with a target defect rate of .01.
Source: AT&T

With the bar coding system, each unit is identified with the year and week of its visit to the Little Rock repair facility and its serial number. This will allow tracking of each unit's repair history. Units repeatedly sent to Little Rock are identified as "loopers." Al Taylor, Quality Engineer at the repair facility said that much of the quality documentation in place in West Chicago was lost during the move to Little Rock. New documentation is being patterned around ISO9000 guidelines.

The team is in the process of collecting frequency data for components usage during the repair process. (For example, 20% of all phones require new dials.) Frequencies are being established for all components, including circuit cards, dials, keypads, keypad membranes, etc. Each operator repairs about 65 phone sets per day.

During the repair process, bar codes for all components not replaced 100% are scanned into the system. The data is used to support the MRP and materials purchasing process. In an effort to keep inventory levels down, this information is used to refine the bill of material, that interfaces with the kanban system for component ordering purposes.

A dual-bin kanban system—with a work-in-process bin physically backed up by an identical reserve bin on a flowrack lane—is used to ensure continuous restocking of repair materials. When one bin is empty, the bar code is scanned. This is the kanban or signal to the MLMS to tell the next available fork-lift operator to replenish it. (Bins are stocked with either new or recycled components.) For work centers dedicated to one high-volume product (i.e., Merlin telephone sets), the kanban-bin flow racks are permanently located at the end of the repair line. Work centers that work on multiple products make use of mobile or portable kanban flow racks. These flow racks are filled at the receiving dock, then trailered to the work center when needed. When the line is changed to another product, the flow racks are taken back to the receiving area, filled with components appropriate for the next product, then returned to the location. Movement of the portable kanban flow racks is triggered via RF smart assignments.

Bar coded, serialized repaired units are placed into individual boxes. Appropriate documentation, depending on the end-use for the item, is placed into the box as well. Each box is then bar coded (with the "child" bar code). Four boxes are placed in a larger carton, which also receives a bar code (the "parent" bar code). These cartons are palletized, receive a "grandparent" bar code, then shipped to the distribution centers.

GBCS works closely with Bell Laboratories, an AT&T division providing engineering support for the repair operation, in designing the products for disassembly and reuse. Don Stuart said that the company is in the "process of closing the loop" between the repair facility and Bell Labs in terms of feedback of information collected at Little Rock about why certain products fail and how long they are in service before they fail, etc.

The Next Level

According to Marc Barnett, the next level of reengineering at the facility will be to optimize the processes by exploring what causes some lines to be more productive than others. They also plan to reduce warehouse inventory by another 65%. (The space necessary in the warehouse for storage already has been reduced to 1/6 of what was utilized when materials were initially shipped from West Chicago and Denver, due to use of volumetrics and inventory reductions.) When successful processes have been exported to GBCS manufacturing facilities in Shreveport and Denver, Stuart anticipates that all locations, including the MSLs, will be networked together and run on real-time information systems. Just as on the shop floor, equipment going to the MSLs will be restocked in a "pull" rather than a "push" environment. Don Stuart looks toward a "virtual warehouse" in the future of GBCS's repair and distribution facilities utilizing a combination of radio frequency and satellite technology. This will create "one warehouse management system for the world," he said, "where a service technician's truck is just another warehouse location."

Case Questions

1. How did GBCS keep the repair operation running during the transition from West Chicago to Little Rock? Suggest alternative ways the transition could have been accomplished.

2. Approximately what percentage of materials coming in to the Little Rock plant are processed (if 80,000 units are processed per month and 40 percent of all incoming units are scrapped)?

3. Why is managing materials on a repair line more difficult than managing materials on a manufacturing production line?

4. How will replenishment of stock to the MSLs differ in a "pull" as opposed to a "push" environment?

5. Suggest and discuss ways in which processes in the repair lines can be optimized.

6. How could GBCS be even more efficient in its processes? What elements of the process can be further improved or changed?

7. What happens to the incoming unrepaired Merlin telephone sets? Detail their path step-by-step in the reengineered process from the time they are received at Little Rock until they are cartoned and shipped to the distribution centers.

8. Discuss the role of technology in the processes. What other technologies can be utilized to further improve the system?

References

Little Rock Repair and Distribution Process/Systems Investment Analysis, January 19, 1995, AT&T Global Business Communication Systems.

Deb Navas, "AT&T Reengineers Business Processes," *IDSystems*, May 1995.

"Affordable Innovation" product literature on Bar Code Systems by Scanning Technologies Incorporated, 2315 Durwood Road, Little Rock, Arkansas 72207.

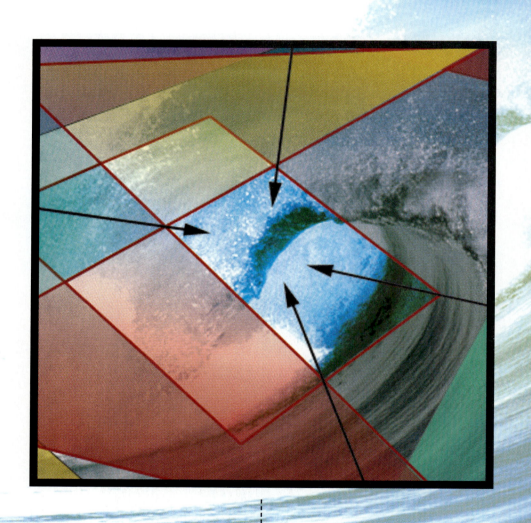

TRANSITION MANAGEMENT

Preparing for the Future

Two chapters, 18 and 19, comprise PART V which deals with transition management. Seven aspects of transition management are highlighted. Three of these are in Chapter 18 and four are in Chapter 19.

Chapter 18—Rapid-Response Project Management

1. Change may be forced from the outside by competitive actions or by customer dissatisfaction. It may originate from the inside where R & D, creative product development and innovative process management suggest new ideas that will provide competitive advantage. In either case, change should be planned and managed as a project. Without project planning, major transitions will be less successful.

2. The time scale for planning and completing projects has been altered dramatically by telecommunications technology, computers and television. Transmission of ideas and information leading to successive stages of approval which used to take months and years now takes days and weeks. Every part of the project process has been speeded-up. As a result, rapid-response project management is a competitive necessity. Among the important methods for rapid response in project

management is concurrent engineering—also called concurrent planning.

3. The goal of rapid-response using well-structured projects is realizable through teamwork which embodies every function's point of view, goals and objectives. The essence of concurrent planning and engineering is the systems approach. It is being used by an increasing number of organizations that take pride in their excellence.

Chapter 19—Change Management

1. P/OM has a major impact on the environment. By judicious choice of what it does P/OM can help restore unpolluted ground, rivers and air. It can produce products that can be disassembled so that many parts can be reused. It can choose materials which do not contribute to toxic waste.

2. Ethical practice is important for P/OM. It includes honesty about materials and the qualities of products and projects. Ethics is a systems factor which links P/OM to the customers, the employees and the suppliers. Government regulation of food products has been increasing but quality of output is P/OM's responsibility. Safety of the workplace is not legislated into place. It is designed and managed. Ethical benchmarks are changing with worldwide convergence on what is deemed to be acceptable ethical practice.

3. Going global requires teamwork. P/OM is on that team. Suppliers are as likely to be users of one currency as another. Customers are as likely to use one current as another. The system is a global one. Time-based management and the broader sense of capabilities management are discussed.

4. Re-engineering is the way to manage transitions from a regional company to a transnational company; from a company that follows local ethical practice to a leader that raises ethical standards; from a company that follows the letter of the law with respect to the environment to one that protects the environment.

Change management requires flexibility and leadership. Both re-engineering and continuous improvement are used to achieve successful change. Technology is driving companies to change. Technological developments such as waterjetting, mobile service robots and the integrated information systems are altering the way that everyone is accustomed to doing work. As Fortune Magazine said on the cover of its October 1994 issue, "The Job is Dead." The job description is whoever is free will do whatever needs to be done.

Chapter 18

Rapid-Response Project Management

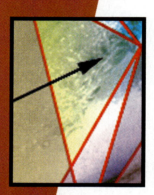

After reading this chapter you should be able to…

1. Explain the unique work configurations known as projects.
2. Classify projects by their various types.
3. Describe the life cycle stages of projects.
4. Explain how project managers differ from process managers.
5. Discuss project-management leadership and teamwork.
6. Explain the basic rules for project management.
7. Describe pros and cons of parallel-path project management.
8. Describe Gantt Project-Planning Charts.
9. Explain the advantages and disadvantages of Gantt Project-Planning Charts.
10. Describe how critical-path methods differ from Gantt chart procedures.
11. Explain how to use the forward-pass calculation to determine the shortest feasible time to complete the project.
12. Explain how to use the backward-pass calculation to determine which activities are on the critical path.
13. Describe what slack means and explain how to derive it.
14. Explain when deterministic and probabilistic estimates for activity times apply.
15. Show how to use optimistic and pessimistic activity time estimates to obtain a variance measure for activity times.
16. Set up a COST/TIME trade-off analysis to determine when to "crash" an activity as compared to doing it in normal time.
17. Describe the penalty-reward system that is used for motivating on-time (or better) project completion.
18. Discuss the analysis of cost/time trade-offs for crashing projects and explain the relationship of crashing and quality.
19. Determine when to use resource leveling to reduce project-completion time.
20. Explain why the computer is an essential facility for adequate project management.
21. Explain the concurrent-engineering approach for speeding up project completion.
22. Discuss time-based management (TBM) of project-cycle time and the ideas of continuity, concurrency and synchronicity.

Chapter Outline
18.0 DEFINING PROJECTS
18.1 MANAGING PROJECTS
　　Project Managers as Leaders
18.2 TYPES OF PROJECTS
18.3 BASIC RULES FOR MANAGING PROJECTS
18.4 GANTT PROJECT-PLANNING CHARTS
18.5 CRITICAL-PATH METHODS
　　Constructing PERT Networks
　　Deterministic Estimates
　　Activities Labeled on Nodes (AON)
　　Activities Labeled on Arrows (AOA)
　　Activity Cycles
18.6 CRITICAL-PATH COMPUTATIONS
　　Forward Pass
　　Backward Pass
　　Slack Is Allowable Slippage
18.7 ESTIMATES OF TIME AND COST
18.8 DISTRIBUTION OF PROJECT-COMPLETION TIMES
　　The Effect of Estimated Variance
　　Why Is Completion Time So Important?
　　Progress Reports and Schedule Control
18.9 DESIGNING PROJECTS
　　Resource Leveling
18.10 COST AND TIME TRADE-OFFS
　　PERT/COST/TIME/(and QUALITY)
18.11 RAPID-RESPONSE TIME-BASED MANAGEMENT (TBM) OF PROJECTS
　　Concurrent Engineering (CE)
Summary
Key Terms
Review Questions
Problem Section
Readings and References

The Systems Viewpoint

Projects have life cycle stages which mean that the project environment is dynamic and changing. It is like a race through an obstacle course where there is planning to "get ready," steps to be taken to "get set," and at the command "go," the race begins. It may take months or years to complete the project, but time to completion is a crucial factor. The project to bring out a new product is being raced against the competition. As competitors became faster, the time factor became more than an economic issue of how long will it be until return on investment begins. It became an issue of whoever is first into the marketplace with a new quality product gains significant competitive advantage. It was discovered that systems-wide cooperation speeded-up the project while improving the quality of the new product. Companies which do not have the systems viewpoint lack the ability to achieve the level of cooperation necessary to win the race.

Projects require cycle-time management which involves simultaneous attention to strategic and tactical issues. Not being fast enough involves risks of losing large sums of money which provide ample incentives for learning how to use the systems approach. The project system is so big that it includes information from every part of the organization as well as many suppliers. It is not possible to keep track of it all or to use project management without a computer tied in to all parts of the system. Suppliers that are not part of the team can cause the project system to fail.

18.0

DEFINING PROJECTS

Projects

Special work configurations designed to accomplish singular or nearly singular goals

Projects are special work configurations designed to accomplish singular or nearly singular goals such as putting on one play, writing new software, creating a mail-order catalog, and constructing a building. Bringing out a new product, building a factory and developing a new service belong to the same category of unique activities and qualify as projects. Each of these projects consists of a set of goal-oriented activities which end when the goal is achieved. Such undertakings have a finite planning horizon. This is in contrast to the character of batch and flow shop production. Projects have many attributes that are similar to custom work. However, the scale of projects is much greater involving many participants and resources.

Projects can include some repetitive activities. Building several houses on one land subdivision is a project. Software programming is a project even though use is made of modular components (object-oriented programming). Projects may entail some batch work and even some intermittent flow shop work. However, the project itself integrates activities as it moves towards completion much as each additional chapter is written for a book or floors are added to buildings.

Projects can be classified as relatively simple. Many engineering design changes which result in engineering change orders (ECOs) appear to be minor alterations in the product design. However, even simple changes require alterations of the process that can lead to systems complexities. A small design change can destroy the ability of fixtures to hold the parts for all downstream activities. Also ECOs can multiply in number and lead to severe quality problems. These problems are especially noticeable if there is insufficient time to test the interactions of the proposed changes. Having too many ECOs can disrupt the normal business of an organization. In a well-known business case, a large computer manufacturer fell behind on promised deliveries as a result of trying to correct both errors and steps taken to correct errors. The problem arose in the first place because of the desire to deliver products before they were thoroughly tested. Then, conflicts began to appear when too many ECOs were attempted at the same time.

Projects can be classified as frequently done. NASA has launched many shuttle flights. Housing developments consist of the same house design being built many times. There are benefits from having repetitive activities within a project. Parts can be purchased with quantity discounts. Training for repetitive activities is justified. The same activity plan (charts) can be used. As the project frequency increases, the project mind-set must remain in place.

That mind-set is goal oriented with completion planned for a specific time. If the activities begin to be treated as a repetitive system, then the project orientation has been replaced by one of repetitive scheduling as used by job shops and intermittent flow shops. Note that even though many houses of the same design are being built, there are unique site considerations which must be taken into account. Nevertheless, some builders have produced houses in volume to reduce the costly project factors and replace them with lower manufacturing costs.

Projects can be classified as complex meaning that the number of issues to consider is very large. Building a new factory is very complex. It requires doing a great variety of things that have not been done before. The same applies to bringing a new product to the marketplace.

Projects can be classified by the fact that they have never been done before. Examples of such projects might include: NASA building a space station, construction of a monorail train using new technology, and construction of Eurotunnel (called the Chunnel) under the English Channel connecting England to France and the European continent.

Bridge design and construction is a complex project.
Source: HNTB Corporation

18.1

MANAGING PROJECTS

Good project-management methods keep track of what has been done and what still needs to be done. Also, they point to activities that are critical for completion and expedite those activities that seem to be slipping. These points are part of the four project life cycle stages:

1. *Describing the goals* requires developing and specifying the desired project outcomes. (The architect lays out the plans for the building and thereby sets the goals of the project.)
2. *Planning the project* requires specifying the activities that are required to accomplish the goals. It involves planning the management of the project including the timing of the activities. (The project manager lays out the charts of sequenced activities and estimates how long it will take to do them. The time frame sets in motion the execution of the plan. The builder is usually the project planner.)
3. *Carrying out the project* requires doing the activities as scheduled. (Getting the building permit, ordering the materials, assembling the different kind of work crews required at the right times, and constructing the building. The builder is usually the project manager.)
4. *Completing the project* can mean disbanding the work groups and closing down the project-management team. However, firms that are in the business of project management, such as companies that build refineries, move their crews from project to project. Each project is goal specific and finite. That is the mission of project-management companies as compared to organizations which need to use project management from time to time. The latter should not avoid the fact that an ECO is a project and needs to be managed as such. Companies that are not in the project business might bring out a new product and then disband the project teams when the job is done. As will be discussed later, increasingly organizations opt to maintain continuous project capability.

Project Managers as Leaders

Organizations encounter the need for project management whenever they consider introducing a new product or service. Often, they turn to their process managers and appoint them to deal with the project over its lifetime. The kinds of problems encountered in projects are different from those encountered in the job shop and flow shop. Time is money in several ways.

First, until the project is completed there is seldom any return on investment (ROI). Second, when projects are new products, the first into the marketplace with a quality product gets a substantial market advantage. In the same way, when the project is a major process improvement, there may be a cost or quality advantage which also translates into a market differential.

The project manager is constantly trying to reduce the cycle time from inception to completion of the project. This is quite different from the job shop manager who is trying to reduce the cycle time of batches of work waiting to be delivered to customers. It also is different from the intermittent flow shop manager who sets the process-cycle time to deliver the required output to satisfy demand. While one person can be good at both, a different hat should be worn for each mode.

The project manager is guided by strategic planning which is tuned to windows of opportunity in the marketplace. This often means putting more resources to work to speed-up project completion. Problems arise which slow the project. The costs of such delays can be in the many millions of dollars whereas the batch shop manager can accept delays which cost much less and are correctable the next time around. Usually, there is no next time around for the project manager.

The ability to manage under pressure and crises is a leadership issue that should be recognized when selecting project managers. Tracy Kidder's description of the enormous stress that was experienced by Tom West's project group when they developed a new computer for Data General[1] makes memorable reading.

Project managers are accustomed to living with great risk and the threat of large penalties. Their goals are strategic and usually vitally important to top management and the success of the company. Often, their goals are the change-management plans for the company. Thus, the profile of a successful project manager is different from that of job shop and flow-shop process managers. Further, project managers often require rapid systems-wide cooperation to resolve their problems quickly. This is a different kind of leadership than that required by process managers who are control oriented.

18.2

CONTRASTING PROJECTS AND PROCESSES

Two main types of P/OM activities are:

(α) **change management projects** which require planning, designing, redesigning, and implementing the system

(β) **control management processes** which entail running the system and shielding it from external disturbances

α-type activities are the domain of project managers. β-type activities are the domain of process managers. Some organizations, such as Bechtel, Computer Science Corporation, EG&G, BE&K, and E.D.S., devote a lot of their resources to managing projects. They develop managers with expertise in α-type activities. These organizations and their project management knowledge are often hired by β-type companies that require projects to remain competitive.

Change management projects—called α

Require planning, designing, redesigning, and implementing the system; domain of project managers.

Control management processes—called β

Entail running the system and shielding it from external disturbances; domain of process managers.

18.3

BASIC RULES FOR MANAGING PROJECTS

The following basic rules apply to project management:

1. State project objectives clearly. They should be reduced to the simplest terms and communicated to all team members. There often are many participants in a project, and knowledge about objectives should be shared.

2. Expertise is required to outline the activities of the project and sequence them correctly. These activities are what must be done to achieve the goals. As a simple example of what happens if the right steps and sequences are not known, when the walls are plastered and painted before the electrical wiring and plumbing are done, the house will have to be unbuilt (going backwards) and then rebuilt to achieve goal completion.

3. Accurate time and cost estimates for all project activities are essential. Slippage from schedule often means real trouble, whereas at other times it can be tolerated because there is sufficient slack. Slack is a time buffer and will be precisely defined in a later section. Project management requires knowing which activities should be monitored and expedited.

4. Duplication of activities, in general, should be eliminated. Under some circumstances, however, parallel-path project activities are warranted. Namely:

   a. If a major conflict of ideas exists and there is urgency to achieve the objectives, then it is sometimes reasonable to allow two or more groups to work independently on the different approaches. Preplanned evaluation procedures should exist so that as soon as it is possible the program can be trimmed back to a single path.

   b. At the inception of a program (during what might be called the exploratory stage) parallel-path research is frequently warranted and can be encouraged. All possible approaches should be considered and evaluated before large commitments of funds have been made.

   c. When the risk of failure is high, for example, survival is at stake. When the payoff incentive is sufficiently great with respect to the costs of achieving it, then parallel-path activities can be justified for as long a period of time as is deemed necessary to achieve the objectives.

   d. If Rule 4 a, b, or c are not applicable, the management of parallel path projects is costly and unnecessary.

5. One systems-oriented person should be responsible for all major decisions. The project manager must be able to lead a team that understands technological, marketing, and production constraints. Multiple project leaders are not feasible.

6. Project management methods are based on information systems which utilize databases that are updated on a regular basis.

   a. Project methods categorize and summarize a body of information that relates to precedence of activities, and their time and cost.

   b. Project methods can assess the effects of possible errors in estimates.

18.4

GANTT PROJECT-PLANNING CHARTS

The great chartist, Henry L. Gantt, developed a planning graphic in the early 1900s that is still used in much the same form. Gantt's chart permits a basic project plan to be set down. Then, using trial and error, improvements in project scheduling can be attempted. The Gantt chart does not computerize complex project situations nor does it permit the mathematical analysis of project paths. PERT and CPM methods which have these capabilities will be described after Gantt charts are covered.

A Gantt chart is good for small projects and for visualizing the basic structure of any project. Large projects require other techniques where the power of computer driven project-planning programs can be used. Many computerized, project-planning models are in use.

Figure 18.1 shows the structure of a Gantt project-planning chart. Figure 18.2 shows a chart which describes in general terms the activities required to launch an entirely new automobile. Figure 18.1 has detail lacking in Figure 18.2.

FIGURE 18.1

GANTT PROJECT-PLANNING CHART WITH LEGEND

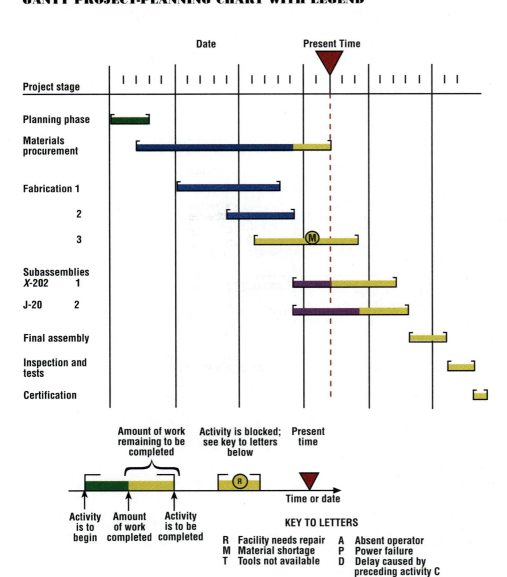

GANTT PROJECT-PLANNING CHART FOR TIME REQUIRED
TO BRING A NEW CAR TO MARKET

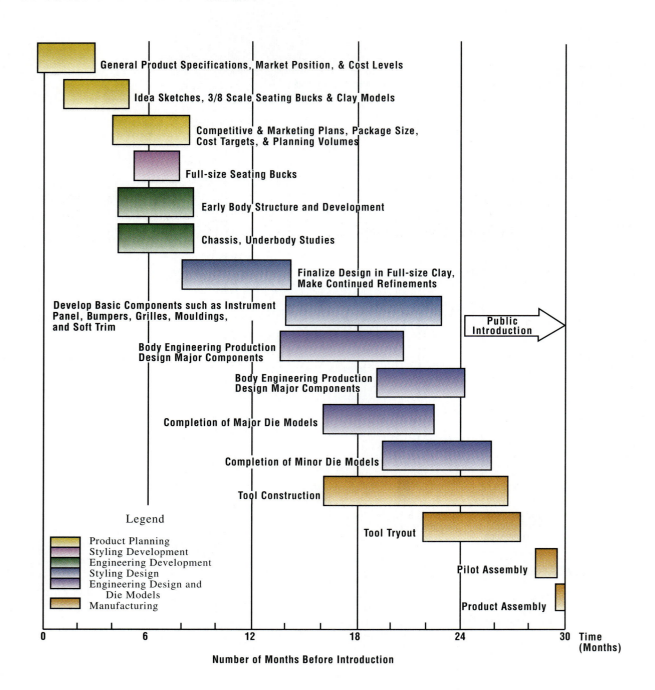

The project planner/manager uses the Gantt chart in two ways: first, to set down the necessary activities; second, to track the status of the project's activities over time. Referring to Figure 18.1, the conventional symbols are bars that show planned start and finish times. The bars are all gold initially. As work proceeds, the colored bar moves over the gold one to show the activity's percent of completion. Note the red vertical line that shows present time. Each day it moves to the right.

Project stages are listed along the left side of the chart. The planning phase has been completed as scheduled. Materials procurement is behind time. It was supposed to have been completed today. Presumably, the project manager knows what is missing, when it is expected, and what is being affected. Apparently, the materials for Fabrication activity 1 arrived because it has been completed on time. The same is true of Fabrication activity 2.

Fabrication activity 3 has not been started and the letter *M* indicates the problem is material shortage. The delay started six days ago and might seriously compromise the completion time of this project. However, the two subassemblies do not seem to be affected. One is on time and the other is ahead of schedule. Final assembly is scheduled in six days ahead of now. Fabrication activity 3 requires eight days. Unless some way is found to speed it up, the project will be delayed two days. Inspection and certification will be delayed.

The Gantt chart in Figure 18.1 has been helpful in tracking a project's progress. The Gantt chart in Figure 18.2 is useful in a different way. It shows that the car requires 30 months from general product specifications and market positioning to public introduction. This is a major reduction in the time required from start to finish. The cycle took four to five years just a short time ago. Detailed attention will be paid to cycle-time reduction for project development in the concluding section of this chapter.

Figure 18.2 shows the broad categories of activities that need to be completed to achieve product assembly and launch. Each bar on the Figure 18.2 chart would be detailed with many specific charts of the kind shown in Figure 18.1.

Technology accounts for the activity structure of the project, i.e., the activities and their sequences. The variables that account for bar length are related to the amount of resources used. Certain activities can be accomplished faster or slower, depending upon the number of people employed, the kinds of facilities that are utilized, and so on. Resource allocations determine time and cost. There are various options. Therefore, for any project there is at least one best sequence to be followed and one best use of resources with respect to cost and time objectives.

Gantt project-planning methods cannot lead to optimal resource utilization or provide sufficient project-tracking capability for the complex projects that now are undertaken by large organizations. Because critical-path methods (CPM) can, there is good reason to study these methods.

18.5

CRITICAL-PATH METHODS

Starting about 1957, two similar approaches to large-scale project planning and tracking were begun at separate locations and for different reasons. These were:

PERT — Program Evaluation Review Technique

CPM — Critical-Path Method

PERT was developed by the U.S. Navy Special Projects Office in conjunction with Booz, Allen and Hamilton for the Polaris submarine launched missile project. This cold war project was considered urgent by the government and time was a critical variable. There were about a hundred thousand activities divided amongst thousands of suppliers. PERT set up activity networks, ideal for large projects, which could be systematically analyzed by computers.

CPM was a similar method developed by DuPont and Remington Rand which later became Unisys. It was used to design and coordinate chemical plant operations. Even at the time of development, computers were essential.

Both applications were very successful in reducing project time. Before network methods existed, project slippage was a fact of life. Projects often took 20 percent more time than expected and cost 20 percent more than budget estimates. With PERT and CPM, 20 percent reductions in expected values were experienced. The adoption of the new project methods was immediate within the U.S. Many different kinds of software were developed which could be used for very large projects including year-end budget preparation. (See the Readings and References section at the end of this chapter.)

PERT and CPM differed only in details. Because both methods share the notion of a **critical path**, this section is called critical path methods for the property of the network and not the name of the program. Further, PERT is the most familiar method to project managers. It was adapted by NASA as NASA-PERT and is required by government agencies for participation in U.S. government projects. Therefore, the discussion refers to the PERT variant of critical-path methods.

Constructing PERT Networks

Three steps are required to utilize network models.

1. Detail all of the activities that are required to complete the project.
2. Draw a precedence diagram for the precise sequencing to be used based on technological feasibility, administrative capabilities, equipment and workforce constraints, and managerial objectives. The rationale for sequential ordering should be documented so that all teammates can share it and the historical record is permanent and explicit.
3. The time to perform each task or activity must be estimated. The method of estimation for time must be detailed and related to project quality. More time is needed to use double-error checking in order to be sure that no project defects occur. Two options will be considered with respect to Step 3.

First, *deterministic estimates* for activity times will be used.
Second, *probabilistic estimates* for activity times will be used.

The critical path is defined in terms of time. PERT is a time-based method. Time and cost estimates will be related at a later point. For now, time is the crucial parameter. The goal is reduction and then control of project-cycle time.

Deterministic Estimates

After the project has been planned, the activities and their sequence is known so the project-precedence diagram can be drawn. Figure 18.3 shows a construction project-precedence diagram used for successive floors of a multistory building.

FIGURE 18.3

A PRECEDENCE DIAGRAM FOR CONSTRUCTION
OF EACH FLOOR OF AN OFFICE BUILDING

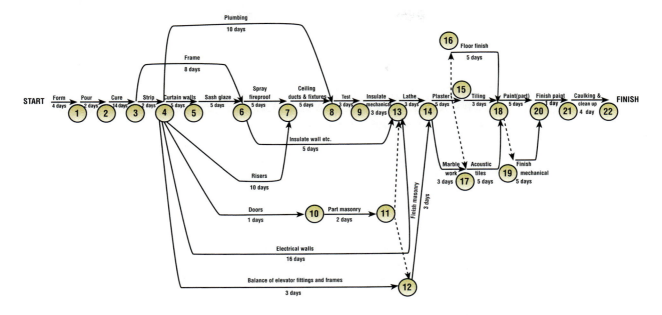

Each activity has a time estimate written on its arrow. That number could have been derived by requesting an engineering estimate. Sometimes it is an average based on historical records of the time required for that activity. Because it is one number, there is no way of inferring variability. Later, two parameters (the average and the variance) for probabilistic estimates will be discussed.

PERT charts are introduced with deterministic estimates to keep the discussion focused on their construction and on the determination of the critical path. Two kinds of charts can be drawn. The first is where activities are represented by circles called nodes.

FIGURE 18.4

ACTIVITIES ARE DRAWN AS NODES, CALLED
ACTIVITIES-ON-NODES (AON)

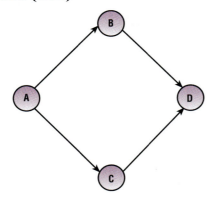

Activities Labeled on Nodes (AON)

All networks consist of arrows and nodes. One form of network chart makes each activity a node as in Figure 18.4. Then, the arrows describe the order or precedence of the activities. This is called an activity-on-nodes (AON) network.

The AON chart in Figure 18.4 shows that activity A precedes activities B and C. Activity D cannot be done until both B and C are completed. These relationships are summarized in Table 18.1.

TABLE 18.1

AON FOR FIGURE 18.4

Activity	Activities Which Must Immediately Precede
A	none
B	A
C	A
D	both B and C

There is no ambiguity about these instructions. However, when the activities are drawn on the arrows, there can be confusion about the precedence instructions.

Activities Labeled on Arrows (AOA)

When the activities are represented as arrows, then the node circles are called *events*. Each node marks either the starting event of one or more activities, or the finishing event of one or more activities, or both. This is shown in the event-oriented, activities-on-arrows (AOA) network of Figure 18.5.

FIGURE 18.5

ACTIVITIES ARE DRAWN AS ARROWS CALLED ACTIVITIES-ON-ARROWS (AOA) CREATING AMBIGUITY ABOUT ACTIVITY D

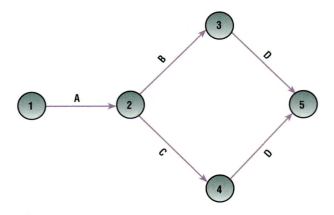

This chart is called ambiguous because there are two arrows labeled D. One arrow for activity D has to follow the arrow of activity B. The second D arrow follows the arrow of activity C.

This situation cannot be permitted for computational reasons. The computer (using the PERT-network algorithm) cannot distinguish between these two D arrows. The ambiguity is removed by redrawing this network.

FIGURE 18.6

THE AMBIGUITY OF THE AOA-PROJECT NETWORK IN FIGURE 18.5 CAN BE REMOVED BY USING A DUMMY ACTIVITY

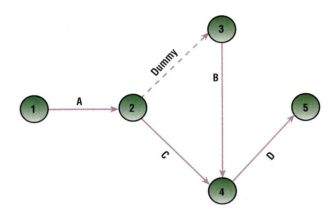

A dummy activity, represented by the dashed line in Figure 18.6, is inserted. The dummy activity takes zero time to do. It allows B and C to join at a single node which precedes activity D. It will not affect time computations along the network paths. However, it allows the computer to keep track of the nodes as shown in Figure 18.6 and Table 18.2.

TABLE 18.2

COMPUTER RECORD OF PROJECT DATA FOR START AND FINISH TIMES

Node	Event That Is Tracked	Time	Activity Time
1	Start of Activity A	t_1	
2	Finish Activity A	t_2	$t_A = t_2 - t_1$
	Start of Activity C	t_2	
	Start of Dummy activity	t_2	
3	Finish Dummy activity	t_3	$t_{dummy} = 0$
	Start of Activity B	t_3	
4	Finish Activity B	t_4	$t_B = t_4 - t_3$
	Finish Activity C	t_4	$t_C = t_4 - t_2$
	Start of Activity D	t_4	
5	Finish Activity D	t_5	$t_D = t_5 - t_4$

Look at Figure 18.5 and note that nodes 3 and 4 are both starting nodes for activity D. Thus, without the dummy, computations would stop when this ambiguity was encountered. Also, observe that $t_3 = t_2$.

Note that in Figure 18.6 the dummy activity could as well have preceded activity C instead of activity B. (It is requested that this alternative be drawn in the Problem Section at the end of this chapter.) Also, there is no need for dummy activities when AON is used. However, the computer cannot use the information in the AON diagram because it lacks the start and finish times. AOA supplies the start and finish times and is therefore required for the PERT method. The need to create dummy activities arises only when there is ambiguity about start or finish nodes for an activity—as in Figure 18.5. Dummy activities can be developed by trial and error.

The PERT algorithm (rules for calculation) will use the nodes and the activity times to compute four different kinds of time. Thus, the data in Table 18.2 is the starting point for further computations. This will be explained right after some other network conditions are stated.

Activity Cycles

In planning a project, some activities go through a cycle of steps. For example, a test on a new product design can result in pass or fail. If the product fails the test, then rework is required, and a retest is scheduled. Looping back and forth between test, rework and retest, as shown in Figure 18.7, is not permitted in PERT networks.

F I G U R E   1 8 . 7

PERT NETWORKS DO NOT PERMIT LOOPING BACKWARDS

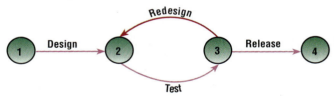

The test/redesign cycle shown in Figure 18.7 could occur more than once before the project is completed. Project planners often cannot prespecify the amount of rework and redesign that will be needed. Therefore, an estimate is made of likely rework requirements and the PERT network is drawn in extensive form as shown in Figure 18.8. Extensive means always moving forward with no arrows allowed to loop backwards. Variability can be expected.

F I G U R E   1 8 . 8

PERT NETWORKS MUST BE DRAWN IN EXTENSIVE FORM

If redesign occurs twice, <u>this must be treated</u> as follows:

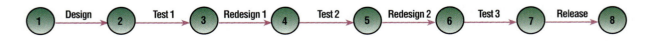

Figure 18.8 presents the project network in extensive form with two test/redesign occurrences followed by product acceptance and release after a third test. This is the project planner's best guess of what will be required. When more is known about the activities, the network will be redrawn and the PERT computations will be recalculated.

Projects of realistic size have thousands of different activities. Whenever an activity starts or finishes, an event node must be created. The project designer may use few or many activities to represent the same project elements. For effective project design, more activities are better than few because that allows greater activity control. However, activities should not be divided which are inherently interdependent and indivisible.

18.6

CRITICAL-PATH COMPUTATIONS

Figure 18.9 depicts a project network with three main branches called A, B, and C.

FIGURE 18.9

BASIC PERT CHART WITHIN FEASIBLE NETWORK BECAUSE A DUMMY ACTIVITY CONSTRAINT IS USED

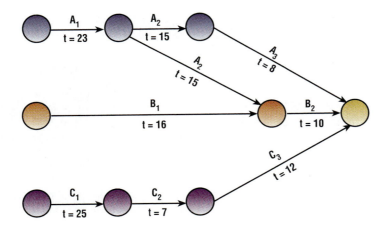

The project network cannot be used as drawn here. There are two arrows labeled A_2. They were both drawn to show that neither A_3 nor B_2 could begin until A_2 was finished. Figure 18.10 corrects the situation by constructing a dummy activity.

TECHNOLOGY

Rapid-Response Machine Tool Manufacture

For projects involving the introduction of new automobiles, rapid-response management is required not only of the automobile manufacturers but also the producers of the tools and dies that make the autos. "Automobiles must go from concept to delivery in three years, when it used to take five years," said Jack Boudrie, president of the Grand Rapids chapter of the National Tooling and Machining Association (NTMA). "That puts a lot of pressure on tool makers to use innovative ways in making tools. A typical tool that used to take 20 weeks to build is now being done in 14 weeks, or 12." He said auto makers would like this time to be reduced even further.

Auto manufacturers work closely with machine tool producers at the front end of projects in designing the tools and dies used to produce a new automobile model. "They recognize that the person who is making the tool and the person who is going to manufacture the part both understand how it is going to be used, so they have input in how that part is being built. It's about doing it on the front end instead of

straightening it out on the factory floor," said Don Klein, a professor at Grand Valley State University and researcher in the field of continuous improvement and world-class manufacturing. Some of the tool and die manufacturers outsource the design and development function to companies that specialize in these areas.

The machine tool industry has developed innovative ways to deal with the extensive use of JIT delivery in the auto industry. Machine tool companies experiencing overloads may outsource work to companies that specialize in providing these services. NTMA-member machine tool manufacturers also have established innovative cooperative

efforts such as making equipment not currently in use available to other member companies. Boudrie said a major challenge is to keep the expensive equipment busy all the time to help pay for the investment. "The expense of machines [is] very, very high, and they depreciate in value much more quickly now because of the rapid technology change... that makes it almost impossible to keep up," Boudrie said. "When we get a particular job and it needs work on a particular machine, [we] can get it done at another shop."

Source: Don VanderVeen, "Area Design Work, Outsourcing Grows," (Michigan) *Grand Rapids Business Journal,* August 15, 1994.

FIGURE 18.10

BASIC PERT CHART WITH ACTIVITIES NAMED ON ARROWS AND TIME ESTIMATES GIVEN

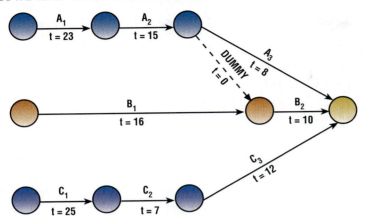

In a time-based project system, the activities A_1, A_2, and A_3 might be marketing activities needed to launch a new product. Branch B might be production process activities B_2 and B_2. Branch C might be activities C_1, C_2, and C_3 used to develop distribution channels. The branches are moving ahead together which will decrease time to market. The PERT network has no way of showing that when used every team member is in communication with every other team member.

All activities are labeled and each has an activity time t (in days). In addition to the dummy constraint which insures that A_2 will be completed before either A_3 or B_2 can start, it should be noted that A_3, B_2, and C_3 must be completed before product launch (at the last node) can occur. Otherwise, the usual precedence constraints apply.

There are four unique paths leading to the launch node.

(Marketing)	Top path: A_1, A_2, A_3
(Marketing and Production)	Mixed path: A_1, A_2, DUMMY, B_2
(Production)	Middle path: B_1, B_2
(Distribution)	Bottom path: C_1, C_2, C_3

Forward Pass

Every event node in the PERT network will be identified with the following four-part scorecard of earliest and latest starts and finishes. Activity time t accompanies the scorecard.

t = activity time

ES	EF
LS	LF

$EF = ES + t$

$LS = LF - t$

ES = earliest start, dependent on preceding activities
EF = earliest finish
LS = latest start, without delaying project
LF = latest finish, without delaying project

The node scorecards are now explained in terms of Figure 18.11 which starts scoring ES, the earliest start, and EF, the earliest finish for the first nodes along each path. For every arrow the left-hand node is its start time and the right hand node is its finish time.

F I G U R E   1 8 . 1 1

**PERT CHART WITH INITIAL CALCULATIONS
FOR EARLIEST FINISH (EF)**

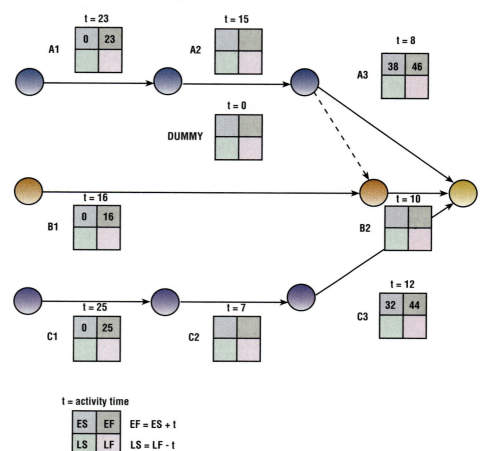

The earliest start for each beginning node is zero which is therefore entered in the upper left-hand box of the scorecard. Next, note that the earliest finish EF is equal to ES plus the activity time along the arrow that connects two nodes. Adding $t = 23$, the activity time for A1, to zero yields 23 which is entered into the upper right-hand box of the scorecard. Doing the same for branches B and C produces pairs of numbers (ES, EF) of (0, 16) and (0, 25) respectively.

Figure 18.12 completes the calculations for the top two boxes (ES, EF) of every node.

FIGURE 18.12

PERT CHART WITH ALL CALCULATIONS COMPLETED FOR EARLIEST FINISH (EF)

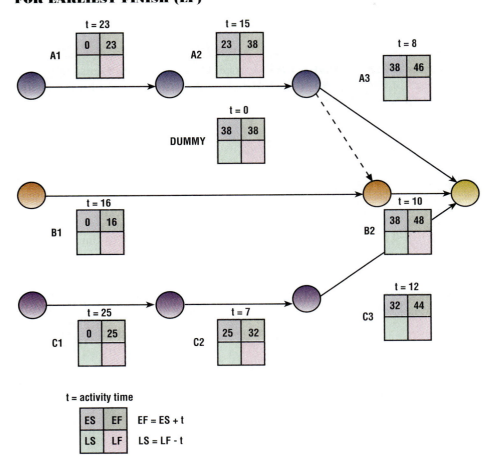

The value of EF for the preceding node becomes the value of *ES* in the next node. For activity A2, ES is 23, $t = 15$, and so EF is 38 which is entered in the upper right-hand scorecard for A2. The same operation is performed for A3. Carry the previous EF to the new ES. Add the activity time and the entry pair is (38, 46). This is consistent for the dummy activity which carries forward the same EF of 38 for earliest finish time of A2 and puts that number into ES for the dummy. Then, 38 plus the dummy's activity time of zero makes EF for the dummy equal to 38 (unchanged).

Skip the B-path for a moment because a second scoring rule applies there. Instead, calculate the values for the C-path. For C2, 32 = 25 + 7. For C3, 44 = 32 + 12. The top two boxes of all scorecards have been completed except for the B-path.

When two arrows converge at a single node, which occurs at the start node for B2 (which is also the finish node for the dummy and for the B1 arrow), always carry forward to the next ES the largest prior value of EF. Figure 18.13 focuses on just this one part of the network.

FIGURE   18.13

WHEN MULTIPLE ARROWS CONVERGE AT A NODE AS THEY DO HERE (THE DUMMY AND B1), THEN THE LARGEST VALUE FOR EARLIEST FINISH (EF) IS CARRIED FORWARD

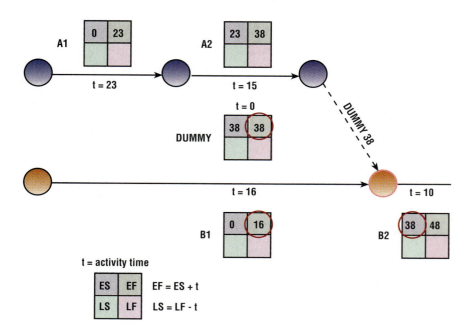

Figure 18.13 shows EF circled for both the dummy and B1. EF is 38 for the dummy and 16 for B1. The largest value is 38 and so that is entered as ES for B2. *The forward pass rule for arrows converging at a node is: when two or more arrows converge at a single node, choose the largest value of EF among the converging arrows for the ES value of the next arrow.* All further accumulation proceeds with this larger number, which is the earliest possible starting time for the next event in the network, and successive events thereafter. That is why the definition for ES is: ES = *earliest start, dependent upon preceding activities.*

Forward pass rule

When two or more arrows converge at a single node, choose the largest value of EF among the converging arrows for the ES value of the next arrow.

Note that in Figure 18.14 three arrows converge on the last node of the network.

FIGURE 18.14

PERT CHART WITH COMPLETED FORWARD PASS; THE EARLIEST FINISH (EF) POSSIBLE IS THE LARGEST NUMBER AT THE LAST NODE

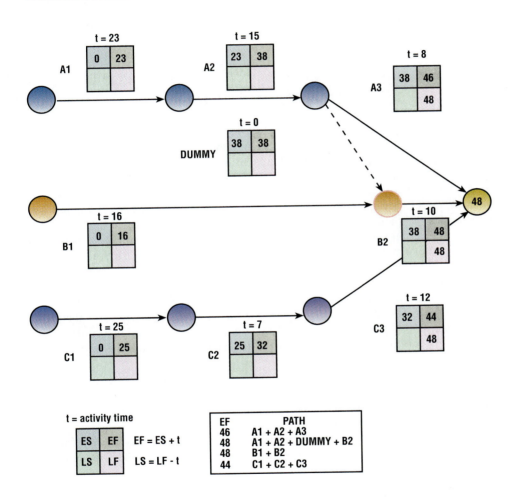

These are A3, B2, and C3. Their EF values are 46, 48, and 44 respectively. There is no next node to carry them to and therefore the largest one specifies the earliest finish time for the project. That number is 48 for the example in Figure 18.14. This project requires 48 days for completion.

The procedure just followed is called the *forward pass* through the project network to determine the earliest start and earliest finish times at all nodes. The largest value of the earliest finish (EF) at the last node also is the latest finish time possible for the last node. Thus, EF = 48 = LF and 48 is entered for LF in the bottom right-hand box of the scorecard *for all three of the arrows* converging at the last node in Figure 18.14.

This largest value of EF and LF at the last node in the network represents project-completion time assuming that nothing unexpected occurs. It also is a measure of the maximum cumulative time of any path *n* through the network, i.e., the longest time sequence of consecutive linked activities through the network. In the example, it represents the earliest possible time that the product launch can begin, which is 48 working days.

Backward Pass

Now to determine the activities that constitute the critical path, start with the final node of the last activity and move backwards. For example, which of the four unique paths that were previously listed accounts for the longest time sequence of 48 days? That path is called the critical path. There is at least one path that constitutes the critical path. For a small project network, it is easy to find out which path (or paths) of activities is responsible for the final EF and LF. With real project complexity, the PERT computations (forward and then backward) cannot be done without a computer. The backward pass part of the algorithm is as follows.

Start with the largest value of latest finish. In the example, this is LF = 48. Move backwards through the network subtracting activity times in accordance with LS = LF − *t*. The latest start time (LS) of the (*n* + 1) node becomes the latest finish time (LF) of the prior *n*th node. LF is the bottom right-hand box of the scorecard. LS is the bottom left-hand box of the scorecard. Thus, for activity B2, 38 = 48 − 10.

Note the B2 scorecard in Figure 18.15. It shows that LF was entered as 48 and *t* = 10 was subtracted yielding LS = 38. This becomes LF for the DUMMY. In Figure 18.15 two arrows emanate from the finish node of activity A2. This means that in going backwards, these two arrows converge on that node.

There is a backwards convergence rule that is similar and opposite to the forward convergence rule. It is opposite because it carries back the smallest value of LS. Thus: *the **backward pass rule** for arrows converging at a node is: when two or more arrows converge at a single node, choose the smallest value of LS among the converging arrows for the LF value of the preceding arrow.* All further accumulation proceeds with this smaller number, which is the latest possible finishing time for the preceding activity without delaying the project.

Backward pass rule

When two or more arrows converge at a single node, choose the smallest value of LS among the converging arrows for the LF value of the preceding arrow.

In Figure 18.15, the dummy calculation for LS is 38 minus zero equals 38. LS for A3 is 48 minus eight equals 40. Because 38 is smaller than 40, 38 becomes LF for A2. Figure 18.16 shows all node scorecards filled in with latest starts. The backward pass has been completed.

FIGURE 18.15

WHEN MULTIPLE ARROWS EMANATE FROM A NODE AS THEY DO HERE (THE DUMMY AND A3) THEN ON THE BACKWARD PASS THE SMALLEST VALUE FOR THE LATEST START (LS) IS CARRIED BACK. THIS IS THE VALUE OF 38 INSTEAD OF 40

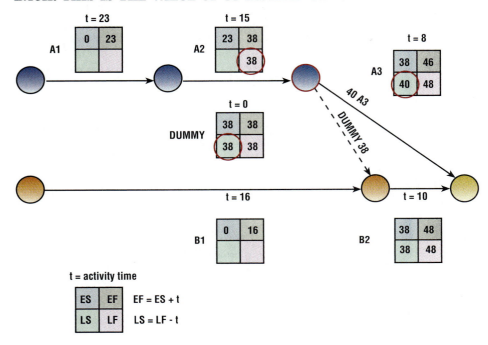

Table 18.3 tabulates the remaining results for all three branches as well as the others previously calculated on the backward pass.

TABLE 18.3

LATEST START FROM LATEST FINISH FOR ALL ACTIVITIES

Activity	LS = LF − t
A3	40 = 48 − 8
A2	23 = 38 − 15
A1	0 = 23 − 23
B2	38 = 48 − 10
B1	22 = 38 − 16
C3	36 = 48 − 12
C2	29 = 36 − 7
C1	4 = 29 − 25
DUMMY	38 = 38 − 0

F I G U R E   1 8 . 1 6

PERT CHART WITH ALL LATEST STARTS COMPLETED

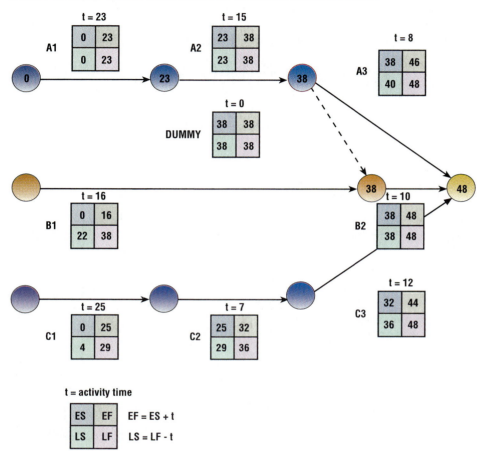

The results in Table 18.3 are used in two ways. First, to calculate slack which is slippage that can occur without changing project-completion time. Second, to identify the critical path which is the path of zero slack.

Slack Is Allowable Slippage

Table 18.4 calculates slack in two ways. Slack = LS − ES = LF − EF. Note that both subtractions produce the same result for each activity.

TABLE 18.4

CALCULATING SLACK IN DAYS

Activity	LS	ES	Slack	LF	EF
A3	0	0	0	23	23
A2	23	23	0	38	38
A3	40	38	2	48	46
B1	22	0	22	38	16
B2	38	38	0	48	48
C1	4	0	4	29	25
C2	29	25	4	36	32
C3	36	32	4	48	44
DUMMY	38	38	0	38	38

The slack of an activity describes the amount of time (here in days) that can be viewed as a safety buffer for slippage. The project manager has to view small amounts of slack and zero slack as the activities that demand constant attention because slippage there results in project delay. Shortly, it will be shown that reallocation of resources can affect slack which puts discretionary powers for what is critical, and what is not, in the hands of project planners.

Note the activities with zero slack follow the path A1 + A2 + DUMMY + B2. This is the critical path. It is the set of activities that define the shortest time in which the project can be completed without slippage. Figure 18.17 shows the critical path in red.

For all other activities, the amount of slack is indicated. Note that the C-path has Slack = 4 for the three C activities. If C1 starts four days late, all the slack is used up for C2 and C3. Contiguous slack along a path is shared slack. If activity A3 is delayed two days it becomes part of the critical path. Activity B1 has 22 days of slack. It can start late or take longer to be done. The project manager might think about transferring some people from B1 to one of the critical activities to relieve the stress.

Using the example of marketing, production and distribution brings up the question of best practice. Because production has 22 days of slack it might seem acceptable for OM not to get involved immediately; but it is not acceptable practice. Instead, production should get involved with marketing issues to help expedite work along the critical path and to share insights and systems perspectives. From a systems point of view note how the dummy activity shows an important systems linkage between A2 and B2. Production has the opportunity to work on marketing issues and then marketing has a little leeway to work on production issues.

FIGURE 18.17

CRITICAL PATH AND SLACK IS SHOWN

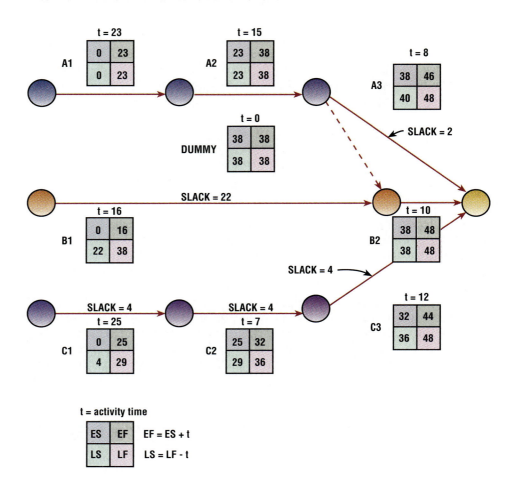

The slack situation is represented by Figure 18.18 in a concise way where both B and C teams can work with the A team at the start of the project. However, if B and C's slack is allowed to be used up, then all three paths of the project become critical. Only activity A3 can be allowed to slip without increasing project-completion time. Thus, more than one project path can be critical. Management is hard pressed to handle projects that have a large percent of critical activities.

Knowing which activities have slack is important. It is probably wasteful to expedite activities A3, B1, C1, C2, and C3. It is likely to be useful to expedite along the critical path. Note that such information about slack and the critical path are not available using the Gantt chart.

FIGURE 18.18

IF *B* AND *C* DELAY STARTING AS SHOWN, THEN THREE PATHS BECOME CRITICAL

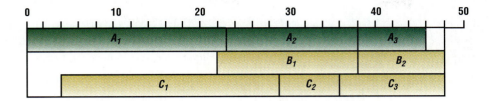

18.7

ESTIMATES OF TIME AND COST

Project management requires good estimates of time to manage the critical path and use slack resources in a proper fashion. Delivery of the finished project must be on time. As will be explained shortly, there can be penalties for being late.

As for cost estimates, when project bidding is used, underestimating the cost of a job results in lower bids with higher probabilities of the bid being accepted. The actual higher cost of the job reduces profit. A good estimator increases the organization's revenues by not overestimating costs, which produces bids that are too high, resulting in lower probabilities of winning the bid.

Management relies upon its ability to estimate before hard data are available. Organizations that deal in projects track the estimating capabilities of their employees by maintaining historical records of their estimation errors.

$$\text{Estimate} - \text{Actual Result} = \text{Error}$$

It is possible to evaluate estimators in terms of:

1. how close their average error is to zero
2. how dispersed are their errors
3. do their errors tend to be overestimates or underestimates

There are basically three different methods for obtaining estimates.

First is a person's opinion derived from experience.

Second is the pooling of several individuals' opinions derived from experience. Many different pooling techniques are available, the most obvious being to use the average value.

Third, several parameters of estimation are requested from an individual, and these are combined by a computing formula. Here too, the pooling of several persons' opinions can be achieved.

P E O P L E

Tom Walker

· · ·

*boyd & fraser
publishing company*

Publishing textbook projects in fast-cycle time means the difference between success and failure for Tom Walker and boyd & fraser publishing company. "If we can't fast cycle our products beyond what is generally accepted as the industry standard, we'll be out of business," declared Walker, the young, extemporaneous president and CEO of boyd & fraser.

boyd & fraser publishes educational materials in a number of formats (such as print, diskette, and CD-ROM) for disciplines including information systems and computer applications, quantitative methods and operations management, and others. Walker said that publishing in these rapidly evolving, technology-driven markets involves responding to "externalities" or forces beyond the publisher's control. He explains one such force that regularly affects boyd & fraser's line of computer applications products: "If Microsoft releases a new version of software, we know that it is only going to be available for a short, finite period of time. If we're going to publish learning materials on how to use that software, we'd better come out with them immediately after Microsoft releases the product. Being first to market in our business is often the determining factor between who wins and who's left sitting on unsaleable inventory at the end of the day."

Cycle time in the textbook publishing industry is measured for each project from the time the author signs a publishing agreement to write, to the time the product is delivered and available for purchase. The traditional textbook publishing process is described by Walker as being "archaic and batch-oriented" involving many hand-offs from author to editor, editor to pre-press production, production to manufacturing. Because of boyd & fraser's team approach to the publishing process, "what we don't have are these batch hand-offs," Walker said.

In the traditional publishing model, production on a textbook does not begin until the entire manuscript is deemed complete and final. With boyd & fraser's team-oriented fast-cycle time, the various activities are performed concurrently. This means that as the author completes the manuscript, earlier sections of the book are being typeset, digitized for film preparation, and may even be printed. Walker said the industry standard cycle time (from signing to product delivery) for traditional college textbook projects is between two to three years, while boyd & fraser's average cycle time is 15 months.

Walker acknowledges that boyd & fraser assumes substantial financial risks with fast-cycle time, because by compressing the cycle time, the company simultaneously accelerates cash outlay or investment in the project. If Microsoft announces that in four months it will release an upgrade for word processing software from 6.0 to 7.0, and a book on 6.0 is midway through the cycle, a decision on how to proceed with the project must be made. A major consideration is how much money has been invested already. For a fast-cycle project in its seventh month, that investment will have been much greater than for a traditionally cycled project in its seventh month. boyd & fraser must decide whether to scrap the project (treating the investment already made as a sunk cost), to recapture some of the investment by recycling as much as possible of the work in process for a book on the software upgrade, or to go ahead with the project as is.

Source: A conversation with Tom Walker.

For PERT, the prevalent estimation procedure is either the first or the third method. The U.S. Navy helped to develop multiparameter estimation which uses the Beta (three parameter) distribution. The PERT system requires three different estimates of time for each activity. These are:

1. an optimistic estimate, called a
2. a pessimistic estimate, called b
3. a most likely estimate, called m

The three estimates are combined to give an expected (mean) elapsed time, called t_e.

$$t_e = \frac{1}{6}(a + b) + \frac{2}{3}(m)$$

If a single number is given as an estimate it is the mode value, m. The computing formula adds variance information by providing a range between shortest likely and longest likely times. A possible distribution for these three elapsed time estimates is shown in Figure 18.19.

F I G U R E   1 8 . 1 9

A POSSIBLE DISTRIBUTION FOR ACTIVITY TIME ESTIMATES USING THE BETA DISTRIBUTION

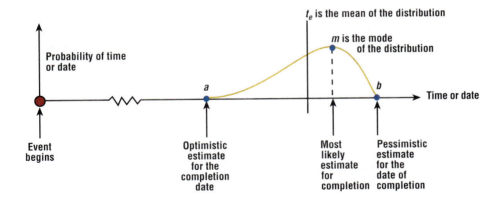

The three values a, b, and m are used to estimate the mean of a unimodal Beta distribution. For the unimodal Beta distribution, if a and b are equally spaced above and below m, then $m = t_e$. That is:

$$t_e = \frac{1}{6}[(a = m - x) + (b = m + x)] + \frac{4}{6}(m) = m$$

When a and b are not symmetric around m, the mean, t_e is moved in the direction of the greatest interval.

A Beta estimate of the variance (σ^2) associated with the mean value t_e is given by:

$$\sigma^2 = [1/6 (b - a)]^2$$

This method of estimating the variance is appealing because a single estimate of variance without historical data is difficult to make and hard to believe. Beta provides a logical basis for estimating the range.

The sum of the means of successive activities is the mean of the path. Thus, for the top (marketing) path:

$$t_{(A1 + A2 + A3)} = t_{A1} + t_{A2} + t_{A3}$$

The same kind of equation can be written for all the other paths.

The sum of the variances of successive activities is the variance of the path. Figure 18.20 illustrates the combining of mean and variance.

FIGURE   18.20

THE VARIANCE OF THE PATH IS EQUAL TO THE SUM OF THE VARIANCES OF THE INDIVIDUAL ACTIVITIES

The rule concerning the sum of the variances requires independent activities. This means that if one activity is late it will not affect any other activity's time. Then the following equation describes the variance of the top (marketing) path:

$$\sigma^2_{(A1 + A2 + A3)} = \sigma^2_{A1} + \sigma^2_{A2} + \sigma^2_{A3}$$

The variance of the mixed-marketing and production path is:

$$\sigma^2_{(A1 + A2 + DUMMY + B2)} = \sigma^2_{A1} + \sigma^2_{A2} + \sigma^2_{DUMMY} + \sigma^2_{B2}$$

The variance of the production path is:

$$\sigma^2_{(B1 + B2)} = \sigma^2_{B1} + \sigma^2_{B2}$$

The variance of the distribution path is:

$$\sigma^2_{(C1 + C2 + C3)} = \sigma^2_{C1} + \sigma^2_{C2} + \sigma^2_{C3}$$

Note that the paths share activities, so that the variances of the paths are not independent.

When the variance of a noncritical path (based on t_e) is larger than the variance along the critical path, it is possible that the critical path will switch between paths. Simulations can be run using the mean and variance information to determine the percent of time that each path becomes critical. Top attention can be paid to those paths which most often become critical.

18.8

DISTRIBUTION OF PROJECT COMPLETION TIMES

If activity times vary, then project-completion time varies. Assume that off the critical path there is enough slack to make variances along the critical path the main point of interest. An analysis of the variation of critical-path time provides risk estimates of major interest to project managers who promise to deliver the finished project at a specific time.

Figure 18.21 shows the mean and the variance (in parentheses) for each activity.

FIGURE 18.21

THE DISTRIBUTION OF PROJECT-COMPLETION TIMES AROUND THE EXPECTED TIME OF 48 DAYS IS SHOWN. THERE IS 0.05 PROBABILITY THAT THE ACTUAL COMPLETION TIME WILL NOT FALL WITHIN THE RANGE OF 40 TO 56 DAYS.

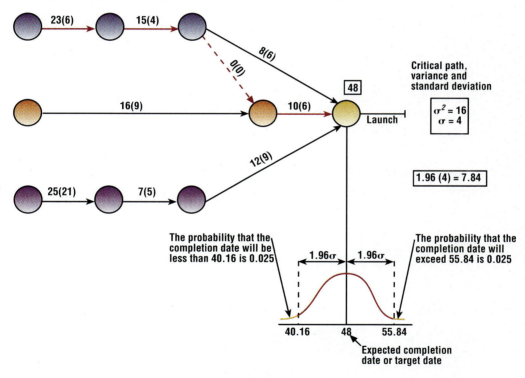

The variances are summed proceeding along the critical path. At the final node that signals project completion, the mean of the critical path (48) is shown with the estimated variance (16).

This is a normal distribution with both tails cut off at the limit of 1.96 standard deviations (plus and minus from the mean value). Each tail contains the probabilities of an event occurring approximately 25 out of 1000 times. The right-hand tail contains long completion dates. The left-hand tail contains short completion dates. There is a probability of 50 percent that the actual project-completion time will exceed 48 days.

There is a 95 percent probability that the actual completion date will fall within the range of $48 \pm 1.96\sigma$. This range also can be used to specify an earliest and latest project-completion date. For convenience, Note 2, at the end of this chapter, shows some other ranges for $k\sigma$.

The Effect of Estimated Variance

The analysis can be extended to reflect the effect of variance on all of the paths in any project network.

Path		Expected Completion Time	Total Variance	95% Range (1.96σ)
1	A_1, A_2, A_3	46 days	16	38.16–53.84
2	$A_1, A_2,$ DUMMY, B_2	48 days	16	40.16–55.84*
3	B_1, B_2	26 days	15	18.40–33.59
4	C_1, C_2, C_3	44 days	35	32.40–55.60

* Path 2 is the critical path.

The variance of Path 4 is so large that the upper limit of 55.60 is almost the same as the upper limit of the critical path (55.84). If the activities along the critical path remain as expected, but the Path 4 activities require excessive times, then completion will be dominated by a new critical path, C_1, C_2, and C_3, which could take as much as 55.6 days. This can occur within the 95 percent probability range. Path 4 is nearcritical with 44 days (versus 48 days) and its variance is twice that of Path 2.

This means that careful managerial attention should be paid to the distribution activities of Path 4. These activities can dominate the situation and cause the targeted completion date of the project to be missed.

The Woodstock '94 music festival was a large project that had to be completed by a pre-determined date.
Source: Amee Peterson

Why Is Completion Time So Important?

There are three dimensions for assessing projects. These are completion time, total cost, and quality of results. All three are important but it is often the case that the targeted date for finishing the project is the first factor considered.

This is not surprising. Certain products have marketing "windows of opportunity." Textbooks must be available in time for adoption, toys must be ready for Christmas, and roses need to bloom for Valentine's Day. Projects to launch planetary probes have specific "windows" defined by such factors as closeness to Earth, position of the Sun, and so forth. Seasonal factors produce a "window" for the Alaska pipeline and for new TV shows.

Even when the "window" concept does not apply, "as soon as possible" frequently does. Projects usually involve large investments that do not start paying off until the project is completed. When viewing project costs against when return on investment (ROI) begins, managers prefer to spend more now to decrease completion time so that a positive cash flow (or other benefits, i.e., hospital beds) can be obtained "as soon as possible."

In bidding situations, project-completion time is a major determinant for being awarded the job. Frequently, contract terms include stiff penalty clauses which are incurred if the organization fails to bring the project in on time.

Progress Reports and Schedule Control

At all stages of the project, the manager will ask for progress reports. Each activity, especially those with suppliers, should report status on a regular, predetermined basis. Usually, daily computer printouts indicate activities that are ahead, those that are behind, and by how much. Shifts in the critical path are reported. New slack values appear and are constantly monitored. Continually updated, the network reflects project progress and problems.

Team work requires that progress reports be shared by all. When problems arise, there is mutual concern. Shared efforts can be made to move resources from one part of the network to another to keep on schedule. The topic of allocating resources is relevant at the beginning design stage of the project. It also applies continuously during the project's lifetime when redesign is referred to as reallocation of resources.

One aspect of reallocation is resource leveling. Before resource leveling could be addressed, it was necessary that the PERT method be thoroughly understood. It also is now possible to address the trade-offs between time and cost which will be done after resource leveling is discussed.

18.9

DESIGNING PROJECTS

There are methods for designing projects so that resources can be used in the best way. Redesign can result in improving the performance of projects such as shortening the critical path, getting closer to target-completion time and meeting the budget. Stringent project quality goals have to be set, monitored and met. Steps to improve project-completion dates should not be permitted to damage quality and/or quality control.

Resource Leveling

Resource leveling is an important concept for project design. The fundamental idea of resource leveling is to balance resource commitments among activities over time. It also provides some control for time-based management (TBM) of project-life cycles and facilitates shortening the critical path. It is a method for pursuing the goal of rapid-response project management.

The design of many projects is such that they start slowly while many ideas are being considered. There is much slack in activities that are off the critical path as reports are written and approvals are sought. The critical path is itself stretched out by bureaucratic organization of the project. At some point, it is decided that time must be made up, and there is a surge in spending in order to speed-up the project. It is well-known that this pattern is not effective. Present-day project management avoids this damaging scenario.

In search of shorter project cycles, the newer approach to projects uses a multi-functional team with rapid communication to secure approvals. The cash-flow pattern is far more balanced from start to finish. Resources are assigned along all paths so that the critical path can be shortened and slack imbalances can be corrected.

Resource leveling seeks to move people from overstaffed activities to those which are understaffed. It attempts to reallocate money from where there is overspending to where there is underspending. These efforts at leveling must make sense in technological and process terms. Similarly, the project manager would prefer smooth demand for cash instead of sporadic cash outflows. If among a set of simultaneous activities, a few are receiving the greatest percentage of project expenditures, it often is desirable to level these allocations. Once the decision is made to use resource leveling, trade-off techniques can be designed to address the specific objectives.

A project that experiences fluctuating calls on its resources is hard to manage. If a large workforce is required for some project intervals and not for others, then an effort should be made to redesign activities so as to level the workforce requirements over time. It is costly and disruptive to make large changes in the project group size. Facilities are built or rented for specific occupancy levels. Overall, the task of managing a project is impeded by uneven resource-demand patterns.

The existence of slack is an important basis for resource leveling. To illustrate, three people have been shifted from activity B1 which has 22 days of slack to activity A1 on the critical path of Figure 18.17. The assumption is made that each added person decreases activity time t by one day. Similarly, it is assumed that each removed person increases activity time t by one day. The assumptions apply to the two activities, A1 and B1. The result of this resource leveling is shown in Figure 18.22.

Several effects are to be noted. The critical path remains unchanged with respect to nodes and activities. However, the length of the critical path has decreased from 48 to 45. Both project manager and clients will be delighted. Slack for B1 has decreased from 22 to 16. Because 22 seemed excessive, no one will object. Shared slack along the C-path has been reduced from 4 to 1. This may lead to some objections because the project manager now has more activities that have zero or near-zero slack.

If management is not up to handling the situation, a smaller resource shift could be tried (as is suggested in the Problem Section at the end of this chapter). Other options could be tried in addition to the change above, or in place of it. Resource leveling can apply to resources other than the workforce, including materials, equipment, suppliers, managerial and administrative time, and cash to pay both direct and indirect costs. It is worth pointing out that simple assumptions about the effects of trading-off resources (such as one-for-one) will have to be replaced with knowledgeable estimates of the effects of resource changes. This is accomplished in the next section on COST/TIME trade-offs.

RESOURCE LEVELING (COMPARE WITH FIGURE 18.17 ABOVE); A SHIFT OF THREE PEOPLE WAS MADE FROM B1 TO A1

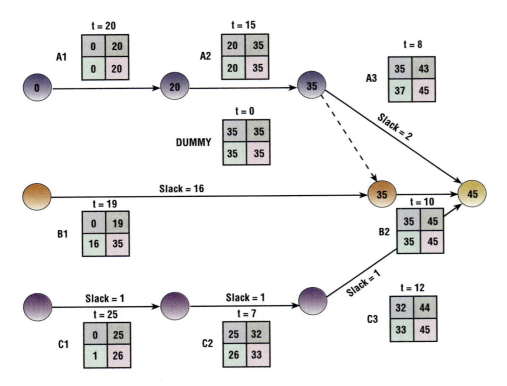

Trading-off resources between activities does not have to result in a shorter critical path. Instead, better project quality might sometimes be achieved by moving resources from the critical path to slack paths which would increase the length of time required to complete the project. Another possible advantage is that there would be fewer critical activities to manage. Sometimes it is possible to utilize resources for expediting and control to improve the variance along the critical path. Doing this does not change the target date, but reduces the risk of deviating from the target date and stabilizes the project.

The above discussion of resource leveling assumes that the amount of resources is held constant. It is only the allocation of resources which is altered. Another option is to withdraw resources or to add new resources. This approach changes budgetary constraints. That issue is addressed in the next section. The important point to make before finishing this discussion is that the resource-leveling activity is an important P/OM responsibility which affects the entire system without changing the budget. All team members are affected and should be kept informed and involved in decisionmaking.

Q U A L I T Y

Price Waterhouse

*Accounting firm reengineers
for quality!*

In struggling with how to apply the principles of TQM to its accounting and consulting business, Price Waterhouse decided to reengineer operations. The Pittsburgh and Los Angeles offices of Price Waterhouse are laboratories for reengineering the company's worldwide operations, according to Richard Lettieri, the company's director of total quality.

Lettieri said that reengineering a successful company to adopt the principles of TQM is like "trying to change the four tires of a [car] going down the road at 60 miles per hour." But the need for change is there. Price Waterhouse Managing Partner James Stalder points out that "public accounting is steeped in tradition," and the firm has "a way of falling into the habit of doing things the way [it has] always done them." Lettieri said the reputation Price Waterhouse has established over the last 100 years is perishable if not reevaluated.

Reengineering a service-based CPA firm is different from reengineering a manufacturing operation because efficiencies are harder to measure. "The biggest difference is a service firm's emphasis on client satisfaction," Lettieri said. "You can develop a better, more efficient process and still not deliver better value for the client." One operation Price Waterhouse has reengineered is the evaluation process accountants and consultants complete at the end of engagements. One problem identified was that the evaluations were not being completed in a timely manner. "With TQM in place, our percentage of completing them has dramatically improved," said Stalder. "TQM means a prioritization of things like this."

Communication and employee retention also should improve with the use of TQM principles. Senior managers are encouraged to take more seriously the suggestions of younger staff members. This also should enhance worker satisfaction and result in efficiency gains from lower turnover and training.

Source: Thomas Olson, "Price Waterhouse local office serves as laboratory for quality approach," (Pennsylvania) *Pittsburgh Business Times and Journal,* January 3, 1994.

18.10

COST AND TIME TRADE-OFFS

The approach thus far has dealt only with time and is called PERT/TIME. Another way to consider changing critical paths and slack time is by means of trade-offs of cost and time. It is often possible to reduce the critical-path duration by spending more money. Resource leveling shifts costs among activities but does not add new funds or reduce existing funds. The project manager transfers funds from activities with slack to activities on the critical path.

What must be known is how adding or decreasing the costs of resources affects the times required to complete activities. The comparison will be made between **"crash"** programs with peak resource requirements and "normal" programs with minimum resource needs.

Crash

Each activity will be done in the least possible or minimum time; entails additions to the project budget.

Assume that the project plan is to run all activities on a crash basis. This means that each activity will be done in the least possible or minimum time. Crash entails additions to the project budget. If the alternative objective to minimize cost is used, this will stretch out project time and reduce spending. The relationship of cost and time has received considerable investigation. The objectives of minimum cost and minimum time are negatively correlated. This means that *crashing for minimum time has maximum costs*. Minimum cost activities have maximum times associated with them.

Maximum quality (qualities) attainment is a third objective that project managers take very seriously. As the Challenger space shuttle tragedy showed, project-quality goals do not take care of themselves. Project quality is an OM responsibility. Flawed technology, faulty design standards, poor building techniques and improper materials can slip past as a result of disorganized project crashing.

It is necessary that crash-time spending should be sufficient to avoid cutting corners. Cost cutting with normal-time project procedures can inflict heavy penalties as well. Failures in bringing a new product or service to the marketplace often can be traced to the project instead of the product. This is a systems problem requiring extensive coordination between marketing, distribution and OM. The project team is accountable.

PERT/COST/TIME/(and QUALITY)

PERT/COST/TIME begins with the representative network of activities. However, in this case, two different estimates for each activity are derived. These are:

crash time: a minimum time estimate and its cost

normal time: a minimum cost estimate and its time

Figure 18.23 shows the relationships between these estimates.

The data for these COST/TIME relationships are presented in Table 18.5 for clarity.

TABLE 18.5

NORMAL AND CRASH TIMES t AND COSTS c

	Normal		Crash	
Activity	max t	min c	min t	max c
A	9	10	6	20
B	8	2	3	5
C	12	11	10	15
D	10	3	7	5

The best way to understand how to use these numbers is to apply them. This is done in Figure 18.24 where activities *A, B, C,* and *D* are illustrated in normal-time and crash-time networks.

The network on the left is based on the minimum cost, normal-time estimates for each activity. It is so simple a network that, for ease of explanation, the critical path is determined using only activity times. Earliest finish (EF) is shown in the two-box scorecard. EF for the upper branch of the network is 17. It is 22 for the bottom branch which is the critical path. Twenty-two is the shortest completion time for the network.

SOME REPRESENTATIVE COST/TIME RELATIONSHIPS WHERE THE ASSUMPTION IS MADE THAT LINEARITY EXISTS OVER THE SPECIFIED RANGES OF COSTS AND TIMES. THE END POINTS OF EACH LINE ARE ALSO ASSUMED TO BE LIMITS

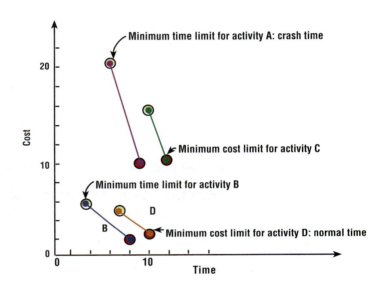

COMPARING NORMAL-TIME AND CRASH-TIME NETWORKS

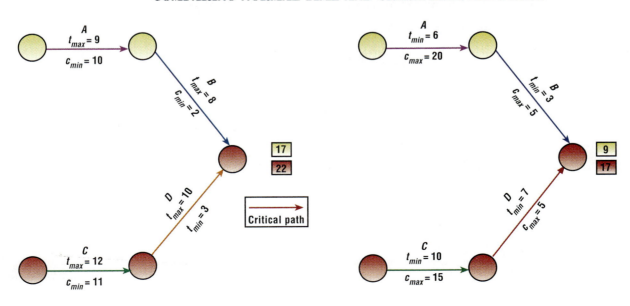

The network on the right is based on the maximum cost, crash-time estimates for each activity. Again, the critical path is determined using only activity times. Earliest finish in the two-box scorecards is nine for the upper branch and 17 for the lower branch which remains the critical path under crash-time conditions.

Table 18.6 sums the costs and times for normal and crash conditions.

TABLE 18.6

NORMAL- AND CRASH-TOTAL TIMES AND TOTAL COSTS

Activity	Normal		Crash	
	max t	min c	min t	max c
A	9	10	6	20
B	8	2	3	5
Upper-branch sums	17	12	9	25
C	12	11	10	15
D	10	3	7	5
Lower-branch sums	22	14	17	20

If 22 is unacceptable for project completion, then crashing C and D will increase costs from 14 to 20 which is a 43 percent increase. The decrease in time from 22 to 17 is a 23 percent decrease. It should be noted that there may be no advantage in crashing A and B which in normal time is 17. However, the entire network becomes the critical path if A and B remain at normal time while C and D are crashed. The slack of five shared by A and B in the all-normal configuration is removed when only C and D are crashed.

Assume the team goal is for project-completion time to be as close to 20 days as possible. Normal-time project planning misses the mark by +2 days, and the crash program overshoots the objective by −3 days. Accordingly, partial reductions can be made in activity times, requiring partial cost increases. Thus, activity C can be reduced from $t = 12$ to $t = 11$ which will mean that $c = 11$ will increase to $c = 13$. These numbers are obtained by interpolating the figures in Table 18.6 or by using the graphs in Figure 18.23 above.

Another alternative is to crash specific activities along the normal-time critical path but not all activities on that path. In this case, crash either C or D. In this way, the critical path can be shortened selectively until such time that:

1. another path becomes critical, and then it is selectively reduced, or
2. the modified critical path is acceptable

As a rule of thumb, use is made of those activities along the critical path where the ratio of cost increases to time reductions is smallest. Thus, select those activities on the critical path where

$$|\Delta \, Cost| \div |\Delta \, Time|$$

is smallest. Then the next-to-smallest ratio is used, and so on, until a satisfactory compromise between time and cost is achieved. If the critical path switches, the next alterations are made along the new path.

Using the data in Table 18.6, the smallest measure of the ratios (of the absolute values)

$$|\Delta\ Cost| \div |\Delta\ Time|$$

applied only to activities along the critical path is associated with activity D with a ratio equal to $\frac{2}{3}$ while C has a ratio of 2. Thus, as shown in Table 18.7:

TABLE 18.7

SMALLEST $|\Delta\ COST| \div |\Delta\ TIME|$ RATIO

Activity	Increase (Δ Cost)	Decrease (Δ Time)	$|\Delta\ Cost| \div |\Delta\ Time|$
C	+4	−2	2
D	+2	−3	$\frac{2}{3}$

For activity D, $t_{max} = 10$, $t_{min} = 7$. Using the normal-time critical path of 22, set $t = 8$ instead of 10. This produces the requested critical path of 20. Making that partial change for activity D, when $t = 8$, $c = 4\ 1/3$. This is derived by using interpolation of Table 18.6 or from the graph of Figure 18.23. The interpolation is done as follows:

$$\frac{c_{max} - c_{min}}{t_{max} - t_{min}} = \frac{2}{3} = \frac{c_D - c_{min}}{t_{max} - t_D} = \frac{c_D - 3}{10 - 8} = \frac{c_D - 3}{2} = \frac{2}{3}$$

whence:

$$c_D = \frac{4}{3} + 3 = 4\frac{1}{3}$$

The critical path remains C and D with total project-completion time of 20 and total cost of $15\frac{1}{3}$. The slack for the upper path of A and B is $20 - 17 = 3$. Figure 18.25 shows this trade-off solution with a two-box scorecard.

FIGURE 18.25

TRADE-OFF SOLUTION WITH CRITICAL PATH ADJUSTED TO BE 20

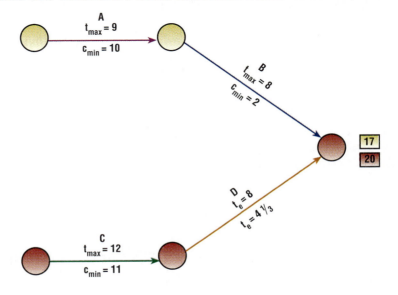

Trade-off models in project planning represent one of the most advanced aspects of P/OM capabilities. Projects are the means by which change is brought about and progress is achieved. Good project management abilities enable management to change the product line, organization, environment and itself.

18.11

RAPID-RESPONSE TIME-BASED MANAGEMENT (TBM) OF PROJECTS

An area of capability called time-based management (TBM) has captured attention because of the success that companies have had using it. TBM elevates time to a position that is equal in importance to cost, quality and productivity. Time is considered a surrogate for profitability, market share, satisfied customers, and so on. Being able to deliver product to customers quickly is one form of TBM. Cutting project duration by a significant amount is another valued form of TBM. Thus, whoever is able to be first into the marketplace with a quality product gains a significant advantage.

The need for speed in project completion has increased as new competitors emerge globally. New companies possess the advantage of not being encumbered with old plants and old ideas. The time between technological feasibility and new products that embody the new technology has been cut between a third and a half by practitioners of TBM. Technological diffusion concerning the state-of-the-art and what is feasible is nearly instantaneous among high-technology players many of whom are in Asia, Europe and North America.

The methods for shortening project life-cycle time include PERT analysis with particular emphasis on cutting the critical-path time without jeopardizing quality. Resource management includes resource leveling and COST/TIME trade-offs which previously have been discussed. Utilization of project funds is directed toward the goal of minimizing project duration.

The time between technological feasibility and new products that embody the new technology has been cut between one-third and one-half by practitioners of time-based management.
Source: Sony Electronics, Inc.

The way that the resources are managed has changed considerably as rapid-response project management has gained adherents because of its clear-cut competitive advantages. The use of more resources at the beginning of the project is encouraged. Project managers used to defer resource utilization until quite late in the project's life cycle.

A main reason for this delay in the early allocation of resources was the compartmentalization of authority which meant that each group would be left alone to conclude its assignment. Not until approval had been granted for that part of the project would the next stage be authorized to proceed.

Figure 18.26 shows the difference in timing of resource utilization under the time-based project management scenario.

Design for manufacture (DFM) and design for assembly (DFA) are illustrative of this early resource-allocation orientation of present-day project management. It is directed toward preventing the need for later changes by thinking everything through as early as possible. The concept requires many points of view being used in the initial stages of the project. The idea is to dream up contingencies leading to engineering design changes which could arise and delay the project. The management of resources to achieve these goals is called concurrent engineering.

FIGURE 18.26

MORE RESOURCES ARE ALLOCATED TO THE PROJECT INITIALLY TO PREVENT LATER CHANGES FROM BEING NEEDED

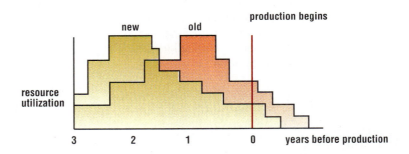

Concurrent Engineering (CE)

Concurrent engineering (CE)

Multifunctional parallel-path project management.

Concurrent engineering (CE) is defined as multifunctional, parallel-path project management. It is compared to a football team or a rugby team working together to carry the ball down the field. Using similar analogy, conventional project management is like a relay race where compartmentalized project groups complete their activities in isolation before handing the baton over to the next group. Everyone runs their own race, obtains separate approval, which allows the next one in line to run. Figure 18.27 aims to portray the managing team concept for concurrent engineering used to develop a new production system.

FIGURE 18.27

THE CONTINUOUS, MULTIFUNCTIONAL CONCURRENT-ENGINEERING TEAM USED TO DEVELOP A NEW PRODUCTION SYSTEM

Production Planning System	Marketing and Distribution	Input/Output Production Transformation System
	Accounting and Financials	
	Human Resource Management	
	Technology and Engineering	
	Operations Management	

The early use of resources is essential for the team effort that is an inherent part of the new project methods called concurrent engineering (CE) by many high technology and aerospace companies. Nonmanufacturing organizations which utilize the methods for new product development and service applications call the project system "concurrent planning (CP)."

Automotive and heavy industry companies call the approach "simultaneous engineering." The idea of technical teams working on different parts of the technology at the same time has been around for a long time. The focus was always on joint problem solving, not so much for speed as for technical exchange. Technological dependencies could be illuminated by simultaneous engineering which emphasizes another aspect of the need for the systems approach.

The terms "parallel planning" and "parallel engineering" are used because they highlight the fact that a multifunctional team is a permanent part of the project-resource plan. This team (or task force) is in many instances multidisciplinary when various branches of science and engineering are called upon to play a part. The team is expected to define the project goals in great detail and depth. The team is responsible for uncovering potential problems and for taking whatever steps are necessary to avoid them. Quality issues are therefore at the top of the task force's agenda. "CE requires a culture in which everyone is responsible for quality."[3]

Global competitors have used CE to develop continuous project-management systems. **Continuity** refers to project teams that complete one project and then move to another. In this way project skills are not dissipated when a project is completed. Because product and process improvement is an ongoing effort, the task force often supports the use of continuous improvement.

Continuity in project management also refers to the use of project teams in different parts of the world so that the work can progress as it follows the sun. Teams using distance learning technologies, such as interactive video, can effectively transfer information about the status of projects. Data is transferred from east to west as team specialists communicate what has been learned and what needs to be done.

Engineers at General Electric-Fanuc (a U.S.-Japan joint venture) design devices used to control factory operations that function around the clock, using a telelink with Japan.[4] Each day's accomplishments are downloaded from GE in the U.S. to Fanuc in Japan in the afternoon, and uploaded from Japan the next morning. Global companies are using round-the-clock shifts with CAD/CAM to design and manufacture textiles. Even two-shift project management is a major step forward in cutting project time.

Continuity

Complete one project and then move to another; use of global project teams enabling work to progress following the sun.

Concurrent Engineering is often used by high technology and aerospace companies in the development of a new product or system.
Source: Rockwell International Corporation/Tony Romero

CE is a way of assuring project quality. It strives to achieve team-based, seamless transfers of information. Successful teams have strong leaders who are experimental innovators. Teams often include customers' and suppliers' organizations. Principles of disassembly are used to accelerate learning. Benchmarking is a common practice to provide best-practice standards.

CE methods are known to be used by companies such as AT & T, Boeing, Chrysler, Digital Equipment Corporation, Eastman Kodak, Ford, GE, GM, Hewlett-Packard, Honda of America, IBM, Ingersoll-Rand, Komatsu, Nissan, Northrop, Siemens, Toyota, and the U.S. Air Force. This list of 18 organizations represents far less than one percent of the companies that are dedicated to using CE for project management.

S U M M A R Y

The chapter begins by explaining that projects are sets of activities designed to accomplish goals which, when achieved, terminate the project. Projects constitute a unique OM work configuration. Gantt project-planning charts are introduced and then critical-path methods and PERT are examined. Networks of activities are constructed leading to the determination of critical paths which are the set of activities that determine project duration being the longest time path from the beginning to the end of the project. Slack is calculated and used for resource leveling aimed at reducing project time.

Rapid-response project management also includes COST/TIME trade-off analyses for critical-path time reduction. Concurrent engineering is the time-based management approach which employs more resources early-on in a team-oriented framework to reduce project time and improve project quality. Project continuity also is described where competitive leverage occurs with significant decreases in project-completion times. Some of the details of project management that are covered include the difference between deterministic and probabilistic time estimates, different network-construction methods (activities labeled on nodes (AON) and activities labeled on arrows (AOA)), forward and backward passes for network calculations, the distribution of project-completion times, and the rugby versus relay race model of project management.

▶ **Key Terms**

backward pass rule *818*

continuity *839*

crash *832*

PERT *806*

change management projects – called α *801*

control management projects — called β *801*

critical path *806*

projects *798*

concurrent engineering (CE) *838*

CPM *806*

forward pass rule *816*

resource leveling *830*

▶ **Review Questions**

1. What is unique about project management? Frame your answer in terms of other OM work configurations.

2. Explain the construction and use of Gantt project-planning charts.

3. What are the strengths and weaknesses of Gantt project-planning charts?

4. Classify projects by type and describe project life cycles.

5. Why is the project manager considered a leader?

6. Why is teamwork repeatedly mentioned when discussing good project management?

7. The present air-traffic control system is one of two systems that were funded using the concept of parallel-path project development. Can this dual expenditure be justified?

8. What is PERT's relationship to critical-path methods?

9. What does the forward pass accomplish in critical-path methods?

10. What does the backward pass accomplish in critical-path methods?

11. What is the critical path and how does knowing it help project managers?

12. What is slack and how does knowing it help project managers?

13. What is the difference between deterministic and probabilistic estimates of activity time?

14. Evaluate the strengths and weaknesses of the Beta method for obtaining time estimates.

15. What are the strengths and weaknesses of PERT?

16. Explain how resource leveling is done and for what purposes.

17. Describe how COST/TIME trade-off methods can be used given the decision to spend an additional 10 percent on the project.

18. Describe project crashing and contrast it to normal time.

19. What advantages can be gained by, and what are the dangers of, using crashing?

20. What is the purpose of time-based management (TBM)?

21. How does concurrent engineering relate to project management?

22 Explain the resource-management principle that CE uses.

23. What is the special role of teamwork for CE?

24. Why is it said that computers are essential for project management?

25. What are the usual goals of rapid-response project management for reducing project time?

▶ Problem Section

1. Get a group to work together on the construction of the Gantt chart requested below. First, however, try to sketch out an approach. Then, note how working in a team improves the creativity of solutions and the quality of problem solving.

   Draw a Gantt project chart to implement the engineering change order: replace faulty motherboard in 500 computers. The steps, set-up times (SUT), and times per unit (t) are given below in minutes. One person is doing the work in batches of 500, i.e., Step a is completed on 500 computers before Step b is begun.
   a. Remove 4 screws and lift off case, SUT = 60, $t = 2$.
   b. Detach cables, SUT = 70, $t = 1$.
   c. Remove faulty motherboard, SUT = 15, $t = 0.5$.
   d. Replace with new motherboard, SUT = 15, $t = 0.5$.
   e. Attach cables, SUT = 45, $t = 3$.
   f. Put cover on with four screws, SUT = 60, $t = 4$.
   g. Inspection check: plug-in and turn on, SUT = 20, $t = 3$.
   h. Mark computers that do not pass test, SUT = 30, $t = 1$.

2. Use the team approach again on the more difficult two-person situation described below.

   Employ the information in Problem 1 to draw a Gantt project chart where two people are working at adjacent stations and the second person starts to work on Step b as soon as the first person completes Step a on 10 computers. Thereafter let the two people operate in a logical way. Note that the second person has 80 minutes to set up for Step b. Set-up time for Step a is 60 minutes and 10 units take two minutes each for an additional 20 minutes.

3 Convert the Gantt project-planning chart of steps required to bring a new car to market (Figure 18.2 in the text) into an appropriate critical-path diagram. Make whatever assumptions you require. Discuss the advantages and disadvantages of each form of representation.

4. Convert Figure 18.28 where activities are labeled on nodes, into an equivalent figure where activities are labeled on arrows.

F I G U R E   1 8 . 2 8

ACTIVITIES-ON-NODES (AON)

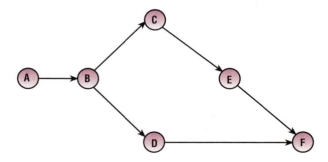

5. The Gantt project chart in Figure 18.29 below, has vertical lines drawn upward from the upper left-hand corner of each activity bar. Where these lines intersect the prior (upper) activity is the "milestone" which when reached allows that next activity to begin. To illustrate, the activity "order materials" is 13/20 finished when the activity "receive

materials" becomes active presumably because materials ordered earlier begin to be received. Note that the position of "Build A" is triggered by a milestone in the "Design" activity. There is no overlap with "Receive Materials." How can this be explained?

FIGURE 18.29

GANTT PROJECT CHART

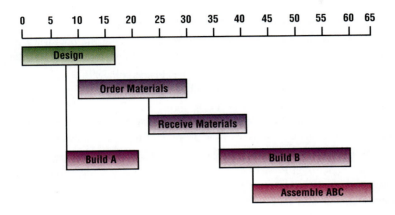

6. Using the information in Problem 5 above, draw a PERT diagram with activities-on-nodes (AON).

7. Using the information in Problems 5 and 6 above, draw a PERT diagram with activities-on-arrows (AOA).

8. The Delta Company manufactures a full line of cosmetics. A competitor recently developed a new form of hair spray that appears to be successful and potentially damaging to Delta's position in the marketplace. The sales manager has asked the OM what the shortest possible time would be for Delta to reach the marketplace with a new product packed in a redesigned container. The OM has drawn up the following table:

Activity	Initial Event	Terminal Event	Duration (Days)
Design product	1	2	30
Design package	1	3	15
Test market package	3	5	20
Distribute to dealers	5	6	20
Order package materials	3	4	15
Fabricate package	4	5	30
Order materials for product	2	4	3
Test-market product	2	7	25
Fabricate product	4	7	20
Package product	7	5	4

a. Construct the appropriate PERT diagram.

b. Apply the estimates of activity duration to the arrows.

9. Using the information in Problem 8 above, make the forward pass to fill out the ES and EF part of the four-box scorecard. As a reminder: ES = earliest start, EF = earliest finish.

t = activity time

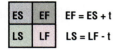

$EF = ES + t$

$LS = LF - t$

What is the shortest feasible completion time for this project?

10. Using the information in Problems 8 and 9 above, make the backward pass to fill out the LS and LF part of the four-box scorecard. As a reminder: LS = latest start, LF = latest finish. Identify the activities on the critical path.

11. Use the information in Problems 8, 9, and 10 above.
    a. Determine the slack for each activity.
    b. Neither the sales manager nor the OM is satisfied with the way the project is designed, but the OM insists that because of the pressure of time, the company will be forced to follow this plan. In what ways does this plan violate good practice?

12. Use the information in Problems 8, 9, 10, and 11 above. The decision is made to form a concurrent-engineering task force with people from sales, production, distribution, engineering, and R & D. The first recommendation of the CE team is to use resource leveling in the initial design-product and design-package phases. It is noted that one-for-one applies which means that if a person is moved from package to product design, the package-design activity time will increase by one day and the product-design activity time will decrease by one day. Up to 10 people can be shifted in this way. Do you recommend doing so?

13. The CE task force has redesigned the project for the Delta Company as follows:

Activity	Initial Event	Terminal Event	Duration (Days)
Design product	1	2	30
Design package	1	3	15
Test-market package	3	2	0
Distribute to dealers	5	6	20
Order package materials	3	4	15
Fabricate package	4	5	30
Order materials for product	2	4	3
Test-market product and package	2	7	25
Fabricate product	4	7	20
Package product	7	5	4

The product and package will be test marketed together which has necessitated adding a dummy activity for test-market package.

a. Construct the appropriate PERT diagram.
b. Apply the estimates of activity duration to the arrows.

14. Using the information in Problem 13 above, make the forward pass to fill out the ES and EF part of the four-box scorecard. What is the shortest feasible completion time for this project?

15. Using the information in Problem 13 and 14 above, make the backward pass to fill out the LS and LF part of the four-box scorecard. Identify the activities on the critical path.

16. Use the information in Problems 13, 14, and 15 above.
    a. Determine the slack for each activity.
    b. Evaluate the project design.
    c. If you have done Problems 8 through 12 compare those results with the answers for Problems 13 through 16.

17. Entertainment is a service area that has demonstrated growing awareness of the importance of OM for improving productivity and quality. Filmmaking uses OM for various scheduling activities as well as layout on location. Movies and films are costly and complex projects which benefit from the application of critical-path methods.

    The PERT network in Fig. 18.30 has been developed by the film's director.

F I G U R E   1 8 . 3 0

PERT CHART FOR FILM MAKING

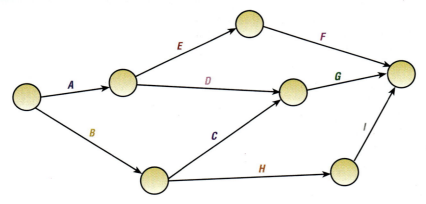

The table below provides crash and normal time, and cost information for the activities required to make the next major box-office winner.

Activity	t_{min}	t_{max}	c_{min}	c_{max}
A	5	9	5	12
B	10	12	4	10
C	3	8	9	15
D	6	7	3	8
E	4	14	12	20
F	5	7	9	12
G	2	3	6	11
H	7	10	2	9
I	3	5	7	9

Find the critical path if all activities follow normal time.

18. Using the information in Problem 17 above, find the critical path if all activities are crashed.

19. Use the information in Problems 17 and 18 above.

    a.   What slack exists in each case (normal vs. crash)?
    b.   What are the total costs of normal vs. crash?

20. Use the information in Problem 17. Assume that the director wants a target date halfway between the required total project times of crash and normal planning.

    a. Use resource leveling with cost and time trade-offs to achieve the objective.
    b. How can resource leveling over time and between activities be applied to the project of filming a movie? Treat both the over-time and between activities aspects of shifting resources. Note the degrees of freedom that exist for the time order of scenes.

21. While dummy arrows are a finer point of PERT construction, they play an important role when needed as in Problem 13 above where the CE task force recommended that test marketing the product and the package could be done together. When dummies were discussed in Figure 18.6 (in the text), it was stated that the dummy activity could as well have preceded activity C instead of activity B. Redraw Figure 18.6 to create this alternative.

22. Three people were shifted from activity B1 which has 22 days of slack to activity A1 on the critical path of Figure 18.17 above. The assumption was made that each added person decreases activity time t by one day. Similarly, it was assumed that each removed person increases activity time t by one day. The assumptions still apply to the two activities, A1 and B1. The results of this resource leveling are shown in Figure 18.22 above. Shift four people (instead of three) from B1 to A1. Discuss the results.

23. Using the information in Problem 22 above, shift two people from B1 to A1.

24. Set up the best possible schedule for project continuity using concurrent-engineering methods. The company and its suppliers have research offices in New York City (0), San Francisco (-3), Tokyo ($+14$), and London ($+5$). The number of hours that separate each city are within parentheses. Is there any way of maintaining perfect project continuity?

25. This problem requests that you make a model of the PERT network out of string. The string model of a project network requires that the length of the pieces of string be cut to scale. The length of string would be proportional to the time of the activity. The pieces are tied together in conformance with the precedence diagram. When the strings are pulled taut between the starting node and the final node, the critical path is revealedas the taut parts of the string model. The other strings show their slack by drooping.

▶ Readings and References

R. Burke, Project Management: Planning and Control, 2nd Ed., Wiley, NY, 1994.

J. R. Hartley, Concurrent Engineering: Shortening Lead Times, Raising Quality, and Lowering Costs, Productivity Press, Cambridge, MA, 1992.

H. Kerzner, Project Management: A Systems Approach to Planning, Scheduling and Control, 3rd Ed., Van Nostrand Reinhold, NY, 1989.

J. R. Meridith, and S. J. Mantel, Project Management: A Managerial Approach, Wiley, NY, 1995.

P. R. Scholtes, et al., The Team Handbook, Joiner Associates, (608-238-8134), Madison, WI, 1989.

H. Shaughnessy, *Collaboration Management: New Project and Partnering Techniques,* Wiley, NY, 1994.

Project Software:

AGS first case, 1995.

Harvard Total Project Manager, 1995.

Microsoft Project for Windows, 1995.

FlowMark by IBM, Workgroup Management Product, 1994.

allCLEAR III for Windows, Flowcharting Projects, 1994.

Notes...

1. T. Kidder, *The Soul of a New Machine*, Little, Brown & Co., Boston, MA, 1981.

2. This assumes a normal distribution, which would result if a number of Beta distributions were added together to form a single distribution for the project as a whole.

Number of Standard Deviations ($k\sigma$) $\pm$ from the Mean	Probability that the Actual Time Falls within the Specified Range
1.00 σ	0.680
1.64 σ	0.900
1.96 σ	0.950
3.00 σ	0.997

3. J. R. Hartley, *Concurrent Engineering: Shortening Lead Times, Raising Quality, and Lowering Costs*, Productivity Press, Cambridge, MA, 1992, p. 19.

4. C. R. Morris, "The Coming Global Boom," *The Atlantic*, October, 1989, p. 56.

Chapter 19

Change Management

After reading this chapter you should be able to...

1. Explain why organizations must be able to adapt to change
2. Describe what special contributions OM can make to change management for technological decisions and the timing of change.
3. Describe what special service OM brings to change management to assist in greening the environment.
4. Explain the special contributions that OM can make to maintain and raise ethical standards.
5. Describe the three factors that must exist for change to take place and explain how they combine into a powerful force for change.
6. Describe the operations manager as an important change agent and at the same time—a team player.
7. Explain why a "vision" of what the future can become must be understood by the team and supported by top management.
8. Describe various forms of benchmarking including by-process type and self-benchmarking.
9. Provide a list of "smart" benchmarks that capture essential OM characteristics of processes.
10. Relate benchmarking to both continuous improvement (CI) and reengineering (REE) aspects of change management.
11. Describe continuous improvement methods.
12. Explain the reengineering and business process redesign.
13. Discuss the principles of reengineering.
14. Give some examples of companies that live by the principles of continuous reengineering.
15. Explain the application of design for disassembly (DFD) and its role in environmental protection.
16. Explain the benefit of waterjetting for the ecology.
17. Explain the importance of recycling concrete in place.
18. Describe OM's interaction with ethics.
19. Relate quality integrity and the hazard analysis critical control point system (HACCP).
20. Provide examples of readying the organization for future technology.
21. Discuss P/OM's responsibility for going global successfully.
22. Explain ISO 14000/14001.

Chapter Outline

19.0 ADAPTATION TO EXTERNAL CHANGES

19.1 OM'S PART IN CHANGE MANAGEMENT

19.2 FACTORS THAT ALLOW ORGANIZATIONAL CHANGE

19.3 OM'S BENCHMARKING ROLE IN CHANGE MANAGEMENT

19.4 CONTINUOUS IMPROVEMENT (CI) FOR CHANGE MANAGEMENT
Planning Transitions

19.5 RE-ENGINEERING (REE) FOR CHANGE MANAGEMENT
Oticon Holding S/A
W. L. Gore & Associates, Inc.
Continuous Project Development Model

19.6 GREENING THE ENVIRONMENT
Waterjetting
Design for Disassembly (DFD)
Recycling Concrete in Place

19.7 SUPPORTING ETHICS
Quality Integrity
Hazard Analysis Critical Control Point (HACCP) System
Food and Drug Testing

19.8 READYING THE ORGANIZATION FOR FUTURE TECHNOLOGY
Mobile Robots
Service Robots
The Virtual Office

19.9 P/OM'S RESPONSIBILITY FOR GOING GLOBAL SUCCESSFULLY
Agribusiness–Consulting–Financial Services–Machinery
ISO 14000/14001
The Olympics as a Systems Symbol

Summary
Key Terms
Review Questions
Problem Section
Readings and References
Supplement 19: NPV for Net Present Value Model Applied to the Make or Buy Problem

The Systems Viewpoint

To be successful, change managers must adopt the broadest systems perspective. New forces arise continuously, and in response to them, strategic adaptation must occur. External change must be met by internal adaptation. Change is met by counterchange, and stimulus by response. Teamwork is essential because even small changes ripple through the system creating new conditions that have to be managed.

The change management team communicates with employees at every level from top to bottom attempting to determine how a change strategy might affect every activity of the organization. Tracing cause and effect is a systems task particularly well-suited to OM talents and abilities. Permission must be granted explicitly and implicitly that allows cutting across all boundaries.

Four of the most important factors that force change in this area are environmental, ethical, technological and global. The proper response of the organization to challenges and opportunities in each of these areas is the focus of change management. Change management uses both continuous improvement (CI) and reengineering (REE) which starts from scratch to bring about change. Benchmarking is the trigger for dissatisfaction when the existing system falls far short of best-practice accomplishments. Such dissatisfaction leads to both CI and REE. Managing change requires implementing modifications in a way that will have the most desirable cause and effect chains.

Bureaucracy is dedicated to protecting the status quo. When dissatisfaction with the status quo begins to grow because of social disapproval (environmental and ethical factors), and because of competitive disabilities (technological and global factors), OM (as part of the change management team) can provide a "vision" of what the future could be like. The "vision" is partly derived from benchmarking other companies that are used as models of excellence. Managers always are looking for companies to emulate. *Business Week, Fortune Magazine,* and other business publications regularly produce issues devoted to describing the "best managed" companies, and "The 100 Best Companies to Work for in America" (cited in the text) is a nationwide bestseller.

Benchmarks have to be germane (on target). OM, reflecting the systems viewpoint, is an essential member of the benchmarking team which decides what to measure and who to benchmark.

19.0

ADAPTATION TO EXTERNAL CHANGES

When **change management** is discussed, what is meant is that managers will reach strategic decisions leading to actions that will help their organizations adapt to significant changes in the competitive realm. Success demands systems thinking and cooperation. Figure 19.1 illustrates company adaptation to several types of external changes.

Change management

Managers reach strategic decisions which help organizations adapt to significant changes in the competitive realm.

F I G U R E   1 9 . 1

EXTERNAL CHANGES FORCE MANAGEMENT TO ADAPT BY ADOPTING STRATEGIES TO MANAGE TRANSITION (CHANGE MANAGEMENT)

External Changes — — — — ➔	Company Adaptation
New Technology Is Available	Company Strives to Adopt the New Technology Rapidly
New Environmental Regulations Are Created	Company Meets or Beats the New Regulations
Ethical Problems Receive Public Attention	Company Changes Procedures to Fulfill or Raise New Ethical Expectations
Company Adopts Higher Ethical and/or Environmental Standards	Other Companies Change Procedures to Meet or Raise Ethical and/or Environmental Standards
There Are New Global Competitors and New Competitive Methods	Company Changes to Become More Effective as a Global Competitor
Company Invents New Global Initiatives	Other Companies Respond with New Counter-Initiatives which Increase the Intensity of Global Competitiveness

New competitors appear with no old investment burdens. They start without the old designs and obsolete equipment their competitors have. Newcomers can put all of their investments into new technology deriving improved products from better processes. OM is able to evaluate the competitive advantage.

The old barriers to entry which kept competitors out of a market included:

1. Lack of experience with the technology and a learning curve to become as good as the existing suppliers.
2. Difficulty in taking customers away from existing suppliers.
3. Large investments to become as good as existing suppliers.

These three principles no longer apply. Technology changes continuously and rapidly. It transfers immediately. The owner of the latest technology has cost and quality advantages which the old supplier's customers find irresistible. There are global sources of capital that are constantly searching for best return on investment. Start-up companies have distinct advantages over established companies because experience with the old technology drags down innovative use of the new technology. OM is fully apprised of the technological advantages of start-ups and the importance of timing when up-grading old technology.

New technology forces established companies to find ways to adapt. The transition from old to new technology must be managed. Timing is crucial. Old facilities are modified or closed down. On a global scale, many of the old auto-assembly plants have been closed by Chrysler, Ford, and GM to keep pace with what new plants can achieve in quality and cost. OM's participation is critical on the technology management team.

New rules and regulations regarding environmental impacts of both processes (smokestack emissions) and products (auto-engine emissions) force change and adaptation. OM is the change manager in charge of meeting pollution controls and recycling requirements.

Customers change in what they expect. Health and safety considerations have forced entire industries (such as food and automobiles) to adapt to new customer levels of needs and wants. Customer-driven changes have occurred in what is considered to be acceptable quality. Raising standards of quality is part of a continuing dynamic which new competitors turn to their advantage. If OM cannot help the company adapt quickly enough, the market falters and the financial system withdraws its support.

Employees change in what they expect. They want a greater participatory role. They have greater contributions to make. OM is consistently weighing the trade-offs between using people and technology. This trade-off is constantly changing because both technology and employee expectations are dynamically evolving.

Suppliers change what they make, how they manufacture, what services they offer, their quality standards and what they charge. The buyer now goes to the global market. New organizations appear continuously. Start-ups have the new technology advantage which was discussed above in terms of "cost and quality advantages which … customers find irresistible." The advantages of dealing with established suppliers certified by OM for their present capabilities and likely future improvements must be weighed against the strategy of switching suppliers. OM is needed on that strategy planning team.

Community expectations change which the company calls home to its plants, warehouses and offices. Environmental expectations are real. The ethical conduct of the company is important. In the book, "The 100 Best Companies to Work for in America," one of the six rating characteristics is pride in work and company.[1] There can be no pride in working for a company considered lacking in ethics by employees.

Global activity is changing. It is increasing for every one of the preceding factors. Companies are becoming global in activities and mind-set. This forces continuous and substantial change. The communities mentioned above can be anywhere in the world. Customers, employees and suppliers speak many different languages and use a variety of currencies. Rules and regulations change by crossing borders. While teamwork is essential to manage such change, OM has a major role to play on that team.

19.1

OM'S PART IN CHANGE MANAGEMENT

OM brings special knowledge to the change management team which consists of three parts.

1. *Familiarity with technology* and the timing of technological change which includes technological forecasting and knowing when it is best to adopt a new technology and leave behind an old one. If a new start-up company has a technological advantage, there may be a benefit to not imitating that technology. Instead, by waiting for further technological development the company can "leapfrog the competition." To retain customers during the transition may require price cuts and full disclosure of the intended strategy. There are other scenarios which OM is uniquely suited to suggest and evaluate. Knowledge of timing is complex and critical. The OM advantage stems from experience with the technology of processes.

2. *Understanding the interaction between technology and environmental protection* takes the form of pollution controls for land, air, and water. It applies to processes and products. P/OM contains the body of knowledge that is essential for greening the globe. Whether the reference is to replenishment in forestry, preventing oil spills, or nuclear contamination, the issues arise from operations and technology which are both in the comfort zone of P/OM.

3. *Understanding the responsibility for ethical production practices*: A major area of ethics is related to the honest effort to deliver a safe and healthy product to customers. The quality of products is the responsibility of OM. Shipments of cans that might explode or foods that might be tainted often can only be stopped by operations managers. Other managers do not know the specifics of each day's production output.

If adequate standards are not maintained during production, and these problems are ignored, it is likely that deliveries will be made of unsafe food, toys that are dangerous, and cars that should not be driven. The design of the product also plays a part. Design is acknowledged to be a responsibility shared by OM and others. Design for manufacturability (DFM) and design for assembly (DFA) are subject to the conditions of design for safe use by customers and design for nonspoilage during manufacture.

The product must be safe to make. Ethics of the workplace reflect integrity to protect employees from noxious fumes, toxic substances, unsafe machines, nonsanitary facilities, jumbled storage and combustible materials. The factories and offices can be nice places to work. Care of the environment is free but takes caring from OM.

OM also encounters ethical issues when dealing with suppliers. In many places in the world, incentives to purchase from a particular supplier are not called bribes. There are ample examples of companies using inferior materials to save money. OM can oversee and enforce honest dealings with suppliers, maintain the integrity of government contracts and end the need for "whistle blowers." It may be impossible to put an exact cost on the damage done to workforce morale when management practice is unethical, but it is too high a cost to pay at any time.

19.2

FACTORS THAT ALLOW ORGANIZATIONAL CHANGE

Having established some of the reasons that OM has a special role in dealing with transitions and change management, it is important to point out that changing strategies for products and processes often require organizational changes. These organizational changes are not readily achieved by a bureaucracy which champions the status quo. There are specific conditions which must be fulfilled for change to occur. Various models describing the conditions for change have been proposed. The Gleicher model described below is an accepted change model.

Dr. David B. Gleicher's model was used to structure a meeting of the American Assembly aimed at change management. The three-day conference was convened to discuss reversing the declining competitiveness of the U.S.A.[2] The meeting addressed the issue of increasing the values of each of the three factors that promote change.

The model proposes that the product of three factors (*D, V, P*) must be greater than *C* to allow organizational change to take place. *C* is the cost of changing. Thus:

$$D \times V \times P > C$$

The factors are:

1. *dissatisfaction* with the present situation (*D*)
2. *vision* of what the future can become (*V*)
3. *practical first steps* to achieve the vision (*P*)

Note that because the three factors are multiplicative, if one of them is zero, then there can be no change. Further, when all three factors are greater than zero, their product must be greater than the cost (*C*) of changing the situation.

Increasing dissatisfaction with the present situation is often the result of comparing a company with its competitors. If it is learned from the comparison that "they" can build a car in 20 hours while it takes "us" 40 hours, the seeds for dissatisfaction (*D*) have been sown. Methodical comparisons called benchmarking are an OM function to be discussed shortly.

In the case above, the cause of dissatisfaction with the present situation also provides a "vision" (here, a target) of what the future can become. Benchmarking usually has the ability to indicate causes of dissatisfaction and targets to shoot for.

Practical first steps include: design studies aimed at using fewer parts and allowing quicker assembly, process improvement studies to replace workers with faster machines for tedious and repetitive work, and work simplification studies to redesign work stations so that teamwork can make the job more productive and less arduous. OM can be expected to be involved in this change management situation from beginning to end.

The American auto industry has changed a great deal. It benchmarked itself against the Japanese auto industry resulting in high *D* and clear "vision" as to where it must go. Various efforts were made to find practical first steps and over time, good answers were found. Problems still remain. More change management is needed. Industries that have used downsizing extensively have noted that it has the effect of eliminating dissatisfaction. Those that remain are satisfied to still be employed. Downsizing also obscures "vision" unless it is clearly identified as a means to the "vision" and not an end in itself.

The kind of "vision" that is obscured is the need for new product innovations and new state of the art plants. It also is significant that both unions and management in the U.S. auto industry have had different "visions," which in systems terms is equivalent to turning competition inward instead of keeping it external. The first practical steps would be to get everyone playing on the same team.

The components of the change model can be identified with OM activities as in Table 19.1.

TABLE 19.1

OM ACTIVITIES FOR CHANGE-MANAGEMENT FACTORS

Factor	OM Activity
Dissatisfaction (D)	Benchmarking
Vision (V)	Best-practice estimates derived from benchmarking or by proposing product and process innovations
Practical First Steps (P)	Project management often implementing new technology

There are other routes to be followed but the activities in Table 19.1 are basic OM contributions to change management. For example, "vision" is not useful unless it has real support from above.

OM can serve best as a change agent when it is part of a team. Many companies stress teamwork for change management. Concurrent engineering which is widely used for project management requires the team-oriented approach. Project management allows OM to provide a practical means for moving toward the realization of a goal.

19.3

OM'S BENCHMARKING ROLE IN CHANGE MANAGEMENT

Systematic comparative measurements made between similar processes (functions, departments, products, services, and so on) are benchmarks. The procedure for making comparative measurements is called benchmarking. Comparisons can be made between different industries. IBM, when developing mail and telephone-order services, benchmarked that process with Lands' End which is known to be a leader in mail-order service. Disney is frequently benchmarked for comparisons in customer relations.

A Xerox executive sent a group to Japan to determine quality goals. He stated, "This competitive benchmarking resulted in specific performance targets rather than someone's guess or intuitive feel of what needs to be done—which is the real power of the process."[3] The benchmarks were effective. Defects were decreased by 84 percent using the new targets. It is noteworthy how benchmarks permit global comparisons. Benchmarking is most useful when there is a clear correspondence of processes and their purposes and goals. The accounts payable function of Mazda and Ford led to improvements for the latter. Other processes that could benefit include the productivity of specific production processes. Anheuser-Busch regularly benchmarks all of its plants. Credit-approval, repair and maintenance, accounts receivable, performance appraisal, salary reviews, and so on, are other examples of processes that lend themselves to benchmarking.

Comparisons require at least two parties. Access must be available for the comparison to be made. Companies that would like to benchmark themselves against the competition can only use measures that are available through information in the public domain such as annual reports, newspaper and magazine articles. Data from newspapers and magazines must be used with care. Sometimes information is obtained from suppliers that are shared with competitors. There is an ethical question that arises. If competitors share information, there are legal issues about collusion. Benchmarks are sometimes available when there are joint ventures as was the case with Mazda and Ford.

Benchmarking similar departments in the same organization is a teamwork concept as long as it is used to encourage cooperative improvement and not as a basis for penalizing the lower-grade departments. In the Anheuser-Busch case, it is expected that the newer plants will have higher benchmark grades.

Self-benchmarking over time provides useful insights concerning self-improvement. Individual athletes in every kind of sport track themselves. Coaches use detailed team statistics to guide them in their efforts to design winning teams. Government departments track trade, productivity, unemployment benefits, the velocity of money, and so on.

Benchmarking can only be as good as the relevance of the measures that are used. There are three determinants. First, the measure must be appropriate. Second, measurement must be done properly. Third, the entities being compared must be comparable. Even if all three conditions are met, some benchmarks provide more insight than others. If they capture the competitive factor they are "smart."

Smart benchmarks
Capture significant competitive factors.

A list of **smart benchmarks** might include:

1. Cycle time; the inverse is output rate so short cycle times are chosen to increase output rates.
2. Net margins (after-tax income as a percent of revenue). Detroit's new-found obsession with net margins reflects its quest to improve the way it designs and builds cars and trucks. "It's a good management tool to keep everyone focused on where we need to go," according to Jack Smith, General Motors CEO.[4]
3. Return on equity; increasingly popular among chief financial officers (CFOs).
4. Inventory turns; per year; measured by net annual sales divided by average inventory level. It is increasingly used in preference to direct measures of profit.
5. Days of inventory; 365 divided by inventory turns.
6. How fast deliveries are made (fast-track deliveries are typically 3 days instead of a month).
7. Project life cycle intervals; rapid-response is often half the customary time.
8. Breakeven volume; comparison with industry norms and with the organization's BEP over time are both beneficial.
9. Breakeven time; Hewlett-Packard developed this measure to indicate how long a period is required for a new product to reach breakeven volume.
10. Other insight-building benchmarks include multidimensional quality mapping which shows the position of competing products on more than one dimension, team accomplishments, and levels of core competencies.

Some best-practice measures should reflect long-term survival of the enterprise. These measures can relate to environmental concerns, ethics, mastery of global business practices, and grasp of future technological developments.

Benchmarks have been criticized for various reasons. Government measures (such as unemployment, productivity, and the consumer price index) have been changed from time-to-time which makes them difficult to use. Company measures of resource utilization (machine and people) can be counterproductive. Measures of overhead can be misleading. Quarterly financial report statements are cited as a cause of short-term decisions. End-of-month shipments are used to meet quotas distorting production schedules and adversely impacting quality.

The systems complexity of benchmarking is exemplified by the U.S. government's monthly publication of flights that are delayed more than 15 minutes. The airlines that come out on top use this data in their marketing campaigns. Delays due to mechanical problems originally were excluded but later were made part of the on-time measure of performance. Various airlines have complained that this causes pressure to defer repairs to keep good grades.

An airline analyst says that "a better measure of reliability is the percentage of flight miles canceled, because a canceled trip irks travelers far more than minor delays."[5] Cancellations are measured by the percentage of scheduled miles that are actually flown. Criticism of a cancellation benchmark could be made if airlines failed to cancel flights (because of weather and mechanical problems) in order to maintain good grades. What starts out as a good idea to benchmark airline performance becomes complicated by issues of how these benchmarks are used and possibly abused. The power of benchmarking to influence behavior in unexpected ways must be taken into account.

19.4

CONTINUOUS IMPROVEMENT (CI) FOR CHANGE MANAGEMENT

Two approaches are used for change. OM is involved with both kinds. The first is **continuous improvement (CI)** which is a constant effort to find better ways to produce the products and deliver the services.[6] This is often local with conversions being done to the existing process. Because it is incremental, gradual and tactical, OM plays a large role in originating, implementing and evaluating the results.

Continuous improvement (CI)
Constant effort to find better ways.

Nevertheless, a CI program should be reviewed by the change-management team. Decisions to make simple changes can obscure opportunities for fundamental and sweeping changes associated with reengineering (REE). Adherents to the CI track have made commitments which place them in opposition to REE. One of the first decisions that change management faces is which track to take—the gradual or the extensive road to change.

The administration of specific improvements can be as simple as a new method for deciding when machine adjustments are to be made. It should not be as complex as the implementation of a new technology. A change of real magnitude deserves a total systems study which is characteristic of reengineering (examined in the next section).

Because CI changes are improvements to existing systems, a systems study to question goals and purposes is not undertaken. Benchmarking plays the guiding role in CI activities. It provides the basis for making course corrections on the journey to excellence. Work simplification, inventory reductions, material replacements and information system changes are typical of CI programs.

P E O P L E

Kevin Howard
• • •
Hewlett-Packard Company

Hewlett-Packard Company (HP) manufactures ink-jet printers at factories in the U.S. and Singapore. The printers are then shipped to five distribution centers (DCs) around the world. HP produced two models of printers which needed to be "localized" for their final destinations with a country-specific power module and language-specific software and manual. In Europe alone, there were 23 different versions of each printer. To better respond to fluctuating demand for specific localized combinations without increasing inventory levels, Kevin Howard, an HP packaging logistics engineer, was called upon to design a system to postpone the localization of printers until later in the distribution process. The problem involved how to move the printers from the factories to the DCs and when and where to localize the printers for their final destinations.

Howard's first approach to solving the problem was to ship the printers from the manufacturing plant to the DC in newly designed boxes. These boxes had a flap that could be opened and the localizing materials inserted at the DC without removing the boxes from the pallets. This solved the problem of localizing the units later in the distribution process. But this solution was not as economically sound as it might have been. "It was clear to me we were paying a lot for shipping this empty space in the boxes … and were spending almost $400,000 a year on repackaging boxes that had been damaged during distribution overseas," Howard said.

The next solution was to postpone both the packaging and the differentiation of the product until the DCs received it. Instead of palletizing half-empty printer boxes, printers were packed in bulk on cavitated foam trays with the same footprint size as a pallet. Elimination of the box, cushions, and empty space for the accessories allowed pallet density to increase from 32 units per pallet to 60 units per pallet.

Howard's final target in this reengineering effort was to replace the heavy, wood pallets used in moving the printers with plastic slip sheets. Elimination of the wood pallets allowed HP to get 25% more printers in the stack and save 25% of transportation, storage, and materials handling costs as a result. Howard believes the use of wood pallets to be "a tragic misuse of our natural resources." He said 42% of all hardwood cut down in the U.S. goes into the manufacture of 550 million wood pallets per year—360 million of which are thrown away after one use. Use of the plastic slip sheets is a much more environmentally sensitive way to ship printers. The high-density polyethylene (HDPE) sheets are made of 100% recycled milk jugs and are the thickness of a cardboard cereal box. HP returns used slip sheets to the manufacturer, and they are recycled into new slip sheets.

HP produces several million ink jet printers annually. The longest leg of distribution for most of these printers is between the manufacturing site and the DCs. Densifying the loads from 32 boxed printers on a pallet to 75 unboxed printers on a slip sheet saved tens of millions of dollars in logistics costs. Adding language-specific accessories at the DC rather than at the factory reduced inventory carrying costs about 20%, saving several million dollars annually.

Sources: Kevin Howard, "Postponement of packaging and product differentiation for lower logistics costs," invited paper presented to the Council of Logistics Management Conference, Michigan State University, May 1991, and conversations with Kevin Howard.

Planning Transitions

The prescription for successful CI changes includes total management involvement for implementation. When Chembank introduced new technology for its stock transfer system, senior management sat down with members of the workforce to provide hands-on training in the new procedures.

A lot of careful thought went into planning the transition. There was recognition of the fact that it is normal to resist change. Therefore, senior managers listened to the employees problems related to changing. Adjustments could be made when warranted because those that had planned the changeover were on the production line with those that do the work. That also signalled the importance of the changes being made.

When the (change) managers of the Royal Bank of Canada developed an expert system for monitoring risk with sophisticated technology, they stressed the importance of dialogue with members of the workforce. One of their purposes was to prevent employees from feeling threatened by the new technology. They did this by emphasizing employee ownership of the new technology and methods. They accented the need for teamwork to successfully facilitate making the changeover. The implementation team was aware of the need to listen, hear and make adjustments during the transition period.

19.5

REENGINEERING (REE) FOR CHANGE MANAGEMENT

Reengineering (REE)—also called business process redesign—seeks radical improvements over a short period of time. REE is at the opposite pole from CI. REE requires total replacement of existing systems with new systems that have been designed from scratch.

Reengineering has disturbed even its adherents because of extreme claims. Hammer and Champy wrote the original book in 1993 which started an REE crusade.[7] In 1995, Champy writes, "on the whole … reengineering payoffs appear to have fallen well short of their potential. *Reengineering the Corporation* set big goals: 70 percent decreases in cycle time and 40 percent decreases in costs; 40 percent increases in customer satisfaction, quality and revenue; and 25 percent growth in market share."[8] Studies done by Champy's company, CSC Index showed that "participants failed to attain these benchmarks by as much as 30 percent."

Such failures would be successes for many companies. In addition, the causes of failures can be traced to doing REE incorrectly. Champy's book strives to right that wrong by elaborating on the requirements for success. These have to do with identifying the core processes of the business (such as new product development for the pharmaceutical business and customer service for a bank). "Identify those key operational processes, reassemble the work that goes into them in line with the core mission of the business."[9] This is operations-management territory and OM knowledge is a crucial ingredient for success.

One important variant of REE highlights a different purpose using the same methodology. **Breakpoint business process redesign** "focuses on creating strategic-level competitive advantages through breakthroughs in the *core business processes that most affect customers and shareholders*."[10] The idea is to select functions to

Reengineering (REE)

Also called business process redesign; seeks radical improvements over a short period of time; opposite of continuous improvement.

Breakpoint business process redesign

Focuses on creating strategic-level competitive advantages.

reengineer which can create market response that is much greater than the resources required to achieve it. Emphasis is placed on competency-based competition which almost always involves OM and logistics-type capabilities.

Time-based management (including rapid-response project life-cycle management and fast-delivery cycle times) is one form of special capabilities. Four principles of capabilities-based management are stated by George Stalk, Philip Evans, and Lawrence E. Shulman in their seminal article as follows:

1. The building blocks of corporate strategy are not products and markets but business processes.
2. Competitive success depends on transforming a company's key processes into strategic capabilities that consistently provide superior value to the customer.
3. Companies create these capabilities by making strategic investments in a support infrastructure that links together and transcends traditional SBUs and functions. (SBUs are strategic business units.)
4. Because capabilities necessarily cross functions, the champion of a capabilities-based strategy is the CEO.[11]

OM is central to a capabilities-based strategy. What is happening is that fundamental axioms of business, which have been followed with great success for a hundred years, are no longer valid. Those axioms can be replaced by reengineering business processes. This change-management methodology entails great responsibility for OM. As Booz Allen & Hamilton state, "Where once the challenge was to find ways to be more competitive with the products or services you produced, it now lies in figuring out how to produce better products or services with which to compete."[12] The shift is from marketing and sales reliance to the systems approach with project teams composed of OM, marketing and sales, R & D, engineering, and everyone else involved.

General principles of reengineering and redesigning business processes are listed as a sequential set of steps:

1. *Identify special core qualities, competencies and capabilities* of the company. Some companies have developed unique ability to turn over inventory; others have developed a manufacturing skill in working with special materials. Rare process skills are required for making fine chocolate and the same applies to automotive windshield glass. A company that will be mentioned shortly has people with special knowledge and a mind-set about hearing-aid technology.
2. *Identify core processes* that deliver the core capabilities described in Step 1. Core processes are broad-based. For example, merchandising methods and pricing might be credited with the achievement of fast-inventory turnover. Alternatively, electronic data interchange (EDI) linking retail stores, distribution, and delivery systems might be the core process responsible for delivering fast turnovers.
3. *Trace cross-functional transactions* (between OM, marketing, distribution, suppliers, and so on) in detailed systems terms—for the core processes. To illustrate, assume that Fidelity Investments wants to examine how their computer systems are used by customers to make unassisted, discounted stock trades by telephone. OM's backroom operations handle the calls which would then be related to cash management, marketing activities, accounts receivable, accounts payable, customer services, and so on.
4. *Develop detailed process analyses* including the activities that are part of cross-functional transactions. The process components are made up of people, machines and workstations.

5. *Benchmark* cost, time, and quality for each process component. Include such measures as customer satisfaction, cycle times, productivity, complaints and errors. Some benchmarks will be broad-based and others detailed. Fidelity Investments will probably want to benchmark against Charles Schwab & Co., and vice versa. They might be able to do this on broad measures, such as customer satisfaction derived from market research. Data for operations are not likely to be shared by these competitors. Each can propose target-improvement levels for process components. This is done in the next two steps.

6. *Select processes to be redesigned.* Figure 19.2 (adapted from a Booz Allen & Hamilton figure[13]) provides a sensible way to prioritize processes with respect to their impact on the desired capabilities and the degree to which each process can be improved.

FIGURE 19.2

PRIORITIES USING EFFICIENCY FRONTIERS FOR REDESIGNING DIFFERENT PROJECTS (BASED ON *PROCESS ROLE IN DELIVERING CAPABILITIES* AND *PROCESS IMPROVEMENT ANTICIPATED*)

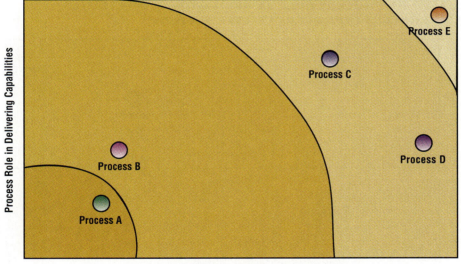

Process Improvement Anticipated

Various processes have been listed in Step 2 which can deliver the core capabilities; however, they do not deliver all equally. Process A has the least ability to deliver the core capabilities chosen in Step 1 whereas Process E has the best ability. Process C is next best and then Processes B and D are tied for the third best position. It is important to bear in mind that all five processes have been chosen because they deliver the core capabilities in varying degrees. In other words, the origin of the y-axis is not zero.

The x-axis reflects the degree to which processes can be improved. This evaluation is based on the information that was developed using Steps 4 and 5. Process E promises the greatest improvement, followed by Processes D, C, B and A, in that order.

The arcs shown in Figure 19.2 are called "**efficiency frontiers**" because they separate the regions with respect to their combined power to deliver improved capabilities. In effect, these frontiers represent combinations of delivery capabilities (*DC*) and improvements anticipated (*IA*) that are considered to be equally desirable. In equational form, they might be approximated by arcs of circles:

$$(DC)^2 + (IA)^2 = R^2 \qquad (R \text{ is the circle's radius})$$

Financial analysis uses risk-return efficiency frontiers in much the same way as REE uses DC-IA frontiers. Four regions have been demarcated. Process A is in the worst region. Process E is in the best region and should be chosen for reengineering. Note that efficiency frontiers drawn through Processes C and D do not provide much differentiation. If it is decided to reengineer an additional process, the choice between Processes C and D is not clear.

Figure 19.3 provides another way of prioritizing the projects which clears up the ambiguity for choosing Processes C or D.

FIGURE 19.3

PRIORITIES USING QUADRANT DIFFERENTIATION FOR REDESIGNING DIFFERENT PROJECTS (BASED ON *PROCESS ROLE IN DELIVERING CAPABILITIES* AND *PROCESS IMPROVEMENT ANTICIPATED*)

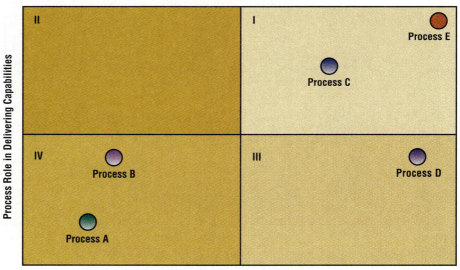

Using quadrant differentiation, processes in quadrant I are the most important ones for delivering the capability and have the greatest potential for improvement. That makes them ideal subjects for REE. Quadrant II is high on capabilities but low on improvement potential. As noted below, this may be because these processes have been already reengineered. Quadrant III is low in delivering the selected capabilities but high in improvement potential. Processes A and B in Quadrant IV are not attractive candidates for REE.

Figure 19.4 shows what happens to a process after it has been reengineered successfully.

FIGURE 19.4

PROCESS E HAS BEEN REENGINEERED

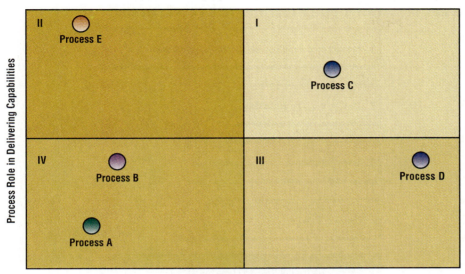

After REE, Process E will move from its position in Quadrant I of Figure 19.3 to Quadrant II as shown in Figure 19.4. It should be located high up in the left-hand corner indicating that little opportunity remains for this process to be further improved. Process E continues to play a major role in delivering core capabilities. Successful reengineering will result in a cluster of processes in the upper left-hand corner of Figure 19.4. The potential for process improvement will have been used up and the company's core capability is as good as it is going to get under the present set of systems conditions (being used to the best of the company's ability).

7. *Set target levels* for improved performance, such as reduce customer waiting time in half, reduce cycle time by 70 percent, decrease perceived difficulty for customer to input stock symbol by 30 percent, increase inventory turnovers from 6 to 10, set the maximum number of times that the phone should ring before being answered, and so on.

8. *Reengineer the process* (discussed below).

9. *Test, evaluate and replan* (stop when satisfied).

Figure 19.5 is a concise rendering of the REE process with the addition of feedback loops which are drawn from Step 9 to Steps 6, 7 and 8.

It is necessary to be able to reconsider what processes to redesign (6). It is usual practice to reset the target levels (7), and to continue reengineering the process originally selected (8). To accomplish Step 8 (reengineer the process) some additional heuristics (rules of thumb—from A to Z) should be considered. These are not intended to be applied sequentially and a number of the heuristics overlap.

A. Question the present task structure. For each task ask: why is this done and why is this done this way?

B. Before deciding to use reengineering check the benefits of using continuous-improvement methods instead. CI and REE should not be used simultaneously.

FIGURE 19.5

SEVEN STEPS LEADING TO REENGINEERING FOLLOWED BY TESTING AND REPLANNING WITH FEEDBACK TO STEPS 6, 7, AND 8 FOR FURTHER REENGINEERING

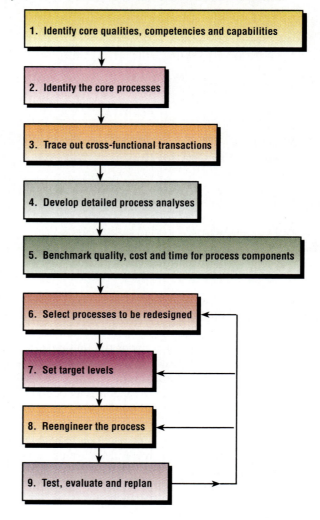

C. When the core processes, goals and subgoals are understood, (Steps 1 through 6) and the targets are set (Step 7), REE starts with a clean sheet of paper in an effort to achieve the targets. Because benchmarking is an ongoing activity, the processes selected and/or the targets may change.

D. For REE, focus on task design, not on task execution.

E. Search for radically different approaches.

F. Empower people to consider leading creative projects. *(Note:* the upcoming discussions of Oticon Holding S/A, Oticon, Inc., and W. L. Gore & Associates, Inc. will deal with this issue of creative leadership.)

G. Organize processes around outcomes not tasks. (If the process to be reengineered is "grant credit," then organize to grant credit and not to do a sequence of tasks such as check on credit history, request credit references, check on credit references, analyze references, analyze income, and so on.)

H. Using team collaboration, simultaneously work on as many activities as possible. Strive to avoid sequential activities where one must be finished before another can begin. Experience with concurrent engineering has shown how this approach speeds-up project completion. When team members work in parallel, individuals can coordinate tasks that are various parts of the job. Another way of stating this is: delinearize the process for rapid completion with the highest project quality.

I. When possible, make those who use the output of a process part of the group creating that output.

J. Combine information processing with output producing work. In other words, quantity and quality data should be recorded and analyzed by the same people that produce the product or service. Thus, someone who does credit accounting should be part of the process for setting credit ratings and granting credit.

K. Employ geographically dispersed resources as though they were centralized (using computer technology).

L. Make decisions where the work is done.

M. Place quality controls with the people who do the work.

N. Capture information only once and as close to its source as possible.

O. Develop detailed process flowcharts for the existing processes and for the proposed new processes. "An indispensable tool in Business Process Reengineering is process mapping. Competitive realignment … requires an extensive understanding of the activities that constitute core business processes and the processes that support them, in terms of their purpose, trigger points, inputs and outputs and constraining influences."[14]

P. Successful reengineering emphasizes balancing process flows—not maximizing production flow rates. Increasing output rates can be accomplished in many ways—after the balanced flow has been achieved.

Q. Stress every employee's right to choose to do what needs to be done—instead of fulfilling a job description.

R. Experiment with alternatives before finalizing new process configurations to avoid suboptimization.

S. Use **pilot studies** of proposed changes. These can be run in parallel to existing processes. Pilot plants are scaled-down models of the factory or service system.

T. Implement REE changes for only part of the system before expanding the application to the entire system. This requires: choosing a sample set of items to manufacture in the new way, a sample of accounts payable to be handled in the reengineered fashion, and a sample set of credit applications to be approved or rejected by the new methods. Then, if the experiment is successful, extend implementation to a larger subset, and eventually to the entire system.

U. Have patience! Efforts that do not work out can be learned from and should encourage further creative experimentation.

V. Institute a continuous reengineering program. This means that Steps 6 through 9 in Figure 19.5 above should be ongoing.

Pilot studies
Scaled-down models of new methods run in parallel to existing processes.

W. Continuous benchmarking is a learning process (Step 5 and heuristic rule C). Search for new benchmarks and continue to measure existing benchmarks over time. Translate benchmarks into targets.

X. Select new processes to be redesigned (Step 6). Successes and failures are indicators for new REE opportunities.

Y. Constantly reset the target levels (Step 7).

Z. Successful reengineering often requires reorganization.

Oticon Holding S/A

Lars Kolind, president of Oticon Holding S/A,[15] in 1991 reorganized this 89-year old Danish hearing aid company so that it fulfills the 26 heuristics and more detailed steps of reengineering. Oticon Holding S/A is home-based in Hellerup, Denmark. The company is organized to reengineer its business processes continuously. It embraces radical redesign using continuous project management without labeling it REE.

To compete with Siemens, Philips and Sony, Kolind (now CEO) freed everyone in the organization to generate and develop new ideas. He called his plan the "disorganization" of Oticon with the goal of achieving the "ultimate flexible organization." Departments and titles disappeared. All activities became projects initiated and pursued by informal groups of motivated individuals. Jobs were reconfigured to be fluid, matching individual abilities and company needs. Project leaders were empowered to find and hire the right people.

All offices were eradicated and replaced with open spaces filled with uniform workstations each consisting of a drawerless desk and computer. (In the section on technology, the growing use of the "hotel office" concept is described along with that of the "virtual office.") Everyone has access to all information with few exceptions. Incoming mail is optically scanned and then shredded. As a constant reminder of the no paper policy, the remains of the shredded paper plunge into a glass cylinder which is a centerpiece of the company's cafeteria.

Kolind also revolutionized how people communicate in the organization. Informal dialogue is the accepted mode of communication replacing all memos and other formal documentation. This includes electronic memo writing.

Employees think twice before sending e-mail to Kolind, lest he deem them "superfluous" by electronic return. Kolind believes that oral communication is "ten times more powerful, more creative, quicker, and nicer" than memo writing.

Because it has almost entirely done away with paper, Oticon requires employee computer fluency. A computer identical to those in Oticon workstations was offered to every employee for home use in exchange for a commitment to learn to use it.

The staff leaped at the opportunity, forming a "PC Club" to train each other outside of working hours. With computers installed at home, the concept of working hours disappeared. Mr. Kolind observed "that everyone works more and they're much more flexible."

Two aspects of this company's continuous reengineering process tell the tale. First, traditionally, hearing-aid companies focus on the technology required to make small hearing aids smaller. Oticon generated the vision to be the best in any hearing aid user-satisfaction survey in the world. One successful Oticon design was a bigger unit for hearing better. Aimed at looking like a modern communications system, it was not pink, but silver-gray like a mobile phone.

Second, Oticon is now the third largest hearing-aid manufacturer in the world. It grew 23 percent in a declining market and increased its gross profits by 25 percent. Price per unit was reduced 20 percent in two years. Startling new designs account for the excellent earnings growth. There are about 100 ongoing projects of various magnitude which have fine prospects.

This innovative company, which has earned ISO 9001 certification, epitomizes change management. *Industry Week* reports that Oticon is "now observed as a prototype for the company of the future," providing "vital strategies for survival" in a "knowledge-based era."

W. L. Gore & Associates, Inc.

W. L. Gore & Associates[16] has been practicing continuous reengineering long before that term was invented. The company makes Gore-Tex® brand fabric, a synthetic material which is popular for outdoor (and camping) use because of its waterproof and breathable qualities. It is made of an expanded form of polytetrafluoroethylene (PTFE) or Teflon. This company, founded in 1958 by Bill Gore, a former research chemist with DuPont in Newark, Delaware, makes many other electronic, industrial, and medical products with PTFE and Gore-Tex expanded PTFE.

Products have been developed by the 5,600 associates who work for the company. There are no employees—everyone is an associate. Gore uses a "lattice organization" which is intended to connote freedom to move in any direction; to lead, to grow, or to commit to new projects, for example. Many of the associates have developed ideas which have garnered supporters. Leaders arise when their ideas inspire others to follow them. Everyone is encouraged to develop projects leading to new products.

This continuous project-development method has achieved great results. The company has experienced steady growth with 1994 sales reaching $950 million. The company has generated 350 U.S. patents and over 1200 patents worldwide in its 37 years of operation which is more than 32 patents per year. It has hundreds of different successful products based on its core competencies with PTFE and Gore-Tex expanded PTFE.

There are 46 plants worldwide and, on average, 122 associates per location. This indicates another lattice organization principle which is to stay under 200 people per plant. Small plants with no authoritarian hierarchy encourage people to communicate with each other in creative ways. Because there are no bosses, sponsors act as advocates for associates in compensation matters. Advice flows freely in all directions because there are no titles and there is no formal organization. The basic principles of this global company with manufacturing sites located throughout the world can be summed up as fairness, freedom to grow and innovate, self- commitment, and communication.

Continuous Project-Development Model

The two companies provide useful models of ceaseless reengineering driven by means of continuous project development. Both companies have an internal focus that is insistent on pleasing customers. Both avoid hierarchies and grant employees freedom to be creative and to search for better ways to satisfy customers. Both firms are continuously reinventing themselves without needing slogans or names like REE. Figure 19.6 is an OM interpretation of the continuous project-management orientation.

FIGURE 19.6

CONTINUOUS PROJECT DEVELOPMENT

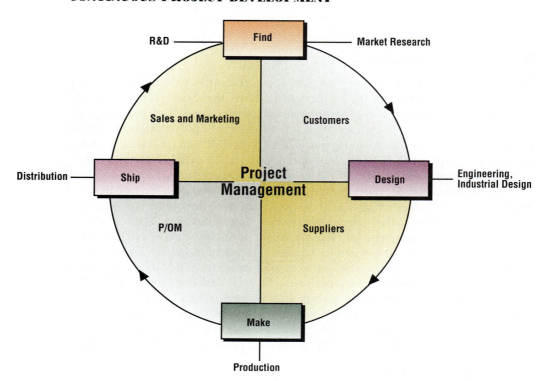

The project management team is multifunctional which facilitates the coordination of finding, designing, making, and shipping the product. The Find-Design-Make-Ship cycle must be supported by appropriate functions. All functions participate in each stage, but some of the main effects are as follows: Find and Design are best done while everyone "listens to the voice of the customer." Design and Make should be coordinated with suppliers. P/OM connects Make and Ship. Sales and Marketing drives Shipping. Market Research and R & D back up the creative impulses of the project-oriented management. Only a systems approach can relate all of these cross-functional considerations.

When one project leader finishes the cycle, another can take over the released resources. Alternatively, the same project leader can go through another product-development cycle with the purpose of making new contributions or further improvements. The same model applies to the dynamics of projects starting and finishing. A well-managed project wheel is constantly turning. The number of turns of this wheel measures the revenue generating energy of the system. Still some failures are bound to occur. When properly viewed, these provide learning experiences which improve the probabilities of future successes.

Project completions are modeled as a Ferris wheel in Figure 19.7.

FIGURE   19.7

THE FERRIS WHEEL MODEL OF CONTINUOUS PROJECT DEVELOPMENT FOR CHANGE MANAGEMENT

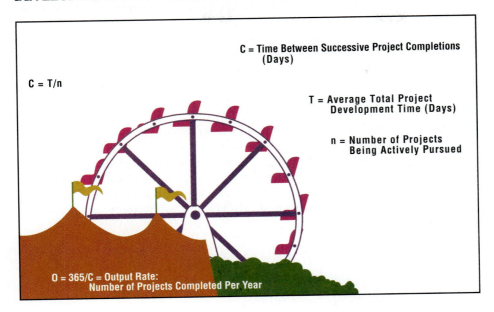

C = Time Between Successive Project Completions (Days)

C = T/n

T = Average Total Project Development Time (Days)

n = Number of Projects Being Actively Pursued

O = 365/C = Output Rate: Number of Projects Completed Per Year

The Ferris wheel, an amusement park ride, has n seating compartments. It makes a complete rotation in T minutes and stops every C minutes to let the riders get off and others get on.

To apply the model to projects, let T equal an average project's total development time (in days) and let n equal the number of projects that are actively being pursued. Then, the average interval between project completions (in days) C is given by:

$$C = T/n$$

Say that $T = 730$ days (which is two years) and $n = 100$ projects, then, on average, a project is completed every 7.3 days.

$$C = 730/100 = 7.3 \text{ days}$$

Also, O, the output per year (the number of projects completed per year) is given by:

$$O = 365/C = 365/7.3 = 50 \text{ projects completed per year}$$

When the numerator (365 days per year) is divided by C the result is projects completed per year. Say that 70 percent of the projects fail. Then, the probability of success:

$$P_{success} = 0.30$$

and the number of project successes would be:

$$nP_{success} = 50(0.30) = 15$$

Thus, there would be 15 project successes per year.

If each project has an average cost of $100,000 per year, then the annual load of 100 projects has a total cost of $10,000,000 per year.

Assume that the average gross margin for each successful project is $1,000,000 per year. (Gross margin is sales revenue less all manufacturing costs so it overlooks selling, administrative and development costs.)

Then, 15 successful projects will generate total gross margin of $15,000,000. Subtracting from this amount the total development costs of $10,000,000 yields a profit of $5,000,000. Even after administrative and sales costs are subtracted there is substantial profit from engaging in continuous project development which more than compensates for the cost of failed projects. Further, the development costs for successful products rapidly diminish, while the revenue stream grows. Also, the pool of successful products increases so that total gross margin continues to climb in successful companies.

Small increases in $P_{success}$ will improve the net profit figures substantially. The organizational models provided by Oticon and W. L. Gore & Associates are geared to have the kind of high values of $P_{success}$ that occur when people are motivated to use teamwork to continuously create new revenue-generating ideas, a pattern that is successful only in a nonauthoritarian change-management setting. (It is suggested in the Problem Section at the end of the chapter, that a sensitivity analysis be run by varying values of $P_{success}$.)

19.6

GREENING THE ENVIRONMENT

P/OM has capabilities for protecting the environment. Without P/OM's awareness and technical knowledge, the environment can be degraded irreversibly. With increasing ecological damage, proactive steps to repair the situation fall on OM shoulders. Using technology and good management, and driven by a sense of what is ethically correct, OM can help limit, if not prevent, harm.

Old polyurethane foam from automotive seating can be ground, reprocessed, and then molded into new seat cushions.
Source: Bayer Corporation

Waterjetting

Waterjetting is an example of a technology that has developed to a point where it can replace toxic methods that are used to remove paints and other coatings. Water-jet Systems, Inc. in Huntsville, Alabama, designs and markets ultra-high pressure (20,000–55,000 psi) waterjet systems with precision robotic controls for depainting and decoating large military vehicles such as aircraft and ships. Their mainstay product in widespread commercial use, however, removes corrosion and tenacious thermal barrier coatings such as plasma-sprayed ceramics and felt metals from jet engines and other industrial equipment components.

Waterjetting is a cost-effective and environmentally sound alternative approach to many conventional coating removal methods such as chemical stripping, incineration, machining, and abrasive blasting. The conventional technique for removing paint from aircraft skin is by applying methylene chloride, a toxic solvent which requires that workers use protective latex clothing and special breathing apparatus. Aircraft engine parts are dipped in toxic chemical baths whereas high pressure water-jet technology uses only water to achieve effective cleaning and decoating.

Special computer-designed waterjet end-effectors and nozzles coupled to a 6-axis manipulator mounted on a 21-foot arm, attached to a 2-axis, 30-ft. rotating vertical column and transported on a wire-guided AGV (automated guided vehicle) comprise the world's largest mobile robot. The depth of the decoating can be regulated precisely, allowing paint to be stripped one layer at a time. Particles are captured when the water is filtered during recycling by a closed-loop water reclamation system.

Thus, an OM decision regarding a function that applies to maintenance for many production systems can make a difference. A hazardous method that is widely used can be replaced by a harmless method.

The waterjet system has other applications as well. It is being used for cleaning airport runways, machine cutting tools, coal mining, cutting concrete, shoe leather and carpets, and for radioactive waste removal. An experimental application is surgical removal of basal cell carcinomas. Many new industries are emerging based on waterjet technology; Waterjet Systems, Inc., for example, is a NASA technology spinoff company.

Design for Disassembly (DFD)

The Vehicle Recycling Development Center is a joint effort of GM, Ford, and Chrysler. The purpose of the center is to learn how to design:

- cars that can be dismantled quickly and easily
- components that can be refurbished and reused
- parts that can be disposed of safely
- materials that can be recycled

Fortune Magazine states, "The men and women at the new center are riding the hottest new production trend in the world: design for disassembly (DFD)."[17] One of the factors that accounts for the growing importance of DFD is the increasing cost of the disposal of waste. In Europe **take-back laws** compel manufacturers to take back used product. As a direct result, manufacturers are forced to design cars, telephones, copy machines and refrigerators that are quick to disassemble and, as much as possible, reusable. BMW has set up experimental plants for disassembly and can recycle 80 percent (by weight) of some of its cars with a goal of 95 percent.

Take-back laws

Laws which compel manufacturers to take back used product; results in products designed for quick disassembly and reuse.

An equally important factor is the cost savings that can be realized. Economic and ecological benefits are so well correlated that in the U.S., without take-back laws, Detroit's car makers work with auto-part recyclers to reuse 75 percent of the weight of American cars. The effort to increase this percent will result in continuous innovations using new materials and principles of assembly-disassembly.

Such innovations are exemplified by the Saturn Corporation, a wholly-owned subsidiary of GM. A new disassembly line will be used by Saturn which is more energy efficient than the current method of shredding cars with monster machines. The new method of recycling has the capability to recycle 83 percent (by weight) of Saturn cars with a goal of 95 percent within five years.[18] The design of the car promotes recycling. Thermoplastic side panels for Saturn doors is an industry first. The panels do not dent which customers like. The plastic can be recycled which pleases environmentalists. Saturn is currently returning damaged accident parts from retailers nationwide using the roundtrip of its repair-parts trucks to Spring Hill, Tennessee, for recycling.

The methods and tools of life-cycle management are well thought out at Saturn. The company is a model of corporate commitment to environmental protection. The Saturn initiative spans design and construction of the manufacturing/assembly facility through the product life cycle which the Saturn Corporation describes as:

$$\text{Design} \rightarrow \text{Manufacture} \rightarrow \text{Consumer Use} \rightarrow \text{Post-Consumer Disposal}$$

Saturn has a permanent manager of Environmental Affairs who coordinates an initiative called "Designs for the Environment."

OM coordination of design for manufacturing, assembly and disassembly is paramount in many industries including Xerox (copy machines), Kodak (cameras), Caterpillar (tractors), and IBM (computers). IBM has take back and disassembly at its Engineering Center for Environmentally Conscious Products, and Digital Equipment Corporation has a Resource Recovery Center. Northern Telecom of Canada repackages phone components in new housings. The German engine manufacturer Deutz rebuilds thousands of engines for tractors and locomotives each year.

Hewlett-Packard (HP) has been a leader in the DFD area. It rebuilds and recycles every workstation that is returned to the company. HP's manager of product stewardship stated: "In the hierarchy of the three R's of design for the environment, the first two—reduce [the number of product parts] and reuse [the parts]—rank above recycling."[19] Table 19.2 expands these three R categories of design for disassembly into four objectives for OM.

T A B L E 1 9 . 2

THE FOUR PARTS OF THE DFD INITIATIVE ARE OBJECTIVES FOR OM

1. Rapid disassembly.
2. Reducing the number of parts needed.
3. Reusing, reworking and remanufacturing parts. If Step 3 can only be done n times, then
4. Recycle the material or dispose of it safely.

Eastman Kodak converted its throw-away camera concept to recyclable cameras because environmentalists were disturbed and planning to take action. Internal parts are now reused up to ten times and 87 percent of the camera (by weight) is reused or recycled. These products are Kodak's most profitable products.

TECHNOLOGY

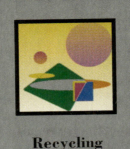

Recycling Technology

ndustry has developed various new environmentally sensitive technologies and processes for the manufacture of products. Two Oregon companies, for example, have received federal grants from the U.S. Department of Energy and the Environmental Protection Agency to develop new technology to reduce industrial waste and boost energy efficiency. Beta Control of Beaverton, Oregon, received $97,000 to complete development of a system for recycling hydrochloric acid used in metal finishing processes. The recycling system separates contaminates from the hydrochloric acid used by metal finishers to de-rust metal before galvanizing, plating, or other processing. The hydrochloric acid can then be reused, and ferrous chloride, the waste by-product, can be sold in a concentrated form for use in fertilizer, waste water treatment, and magnetic tape.

Alpine Technologies of Eugene, Oregon, received $348,000 from the federal government for the development of an optical scanning system to separate wastes from recycled glass. The optical sorting system scans raw glass for opaque materials such as metal, ceramics and cork. Contaminates are separated from the glass with high pressure blasts of air. The system can produce about 30 tons of contaminate-free crushed glass (called glass cullet) in about an hour. The cullet is sold to glass manufacturers who use it to produce glass products.

Another company involved with glass recycling is Vitreous Environmental Group Inc., of Vancouver, Canada. Vitreous makes a product from waste glass called GlasSand™. GlasSand is a sand-like substance that replaces silica sand and other materials in producing fibreglass, glass containers, glass beads and road paint. Other uses include end-use products such as abrasives, golf course construction and maintenance, and water filtration. In 1992 the company originally set out to sell or license the technology which produces GlasSand. But as Patrick Cashion, president of Vitreous Environmental explains, "We found there was a better return to the shareholders with the sale of the product, so there was a shift in focus over the first year." Cashion is optimistic about GlasSand's prospects. "Our problem isn't going to be finding markets, it will be keeping up with supplying those markets." The company has an ongoing need to find new sources of waste glass to recycle.

Sources: Robert Goldfield, "Grant helps waste reduction firms clean up," *Daily Journal of Commerce* (Portland, Oregon), August 11, 1993, and "Vitreous Environmental a massive appetite for waste," *Calgary Commerce* (Alberta, Canada), September 1, 1994.

For Xerox remanufacturing and recycling of parts is saving approximately $500 million per year. Xerox has a team called, Asset Recycle Management Organization, which works exclusively on DFD.

Design for assembly (DFA) has lowered production costs and is related to DFD. That which is easier to put together also is likely to be easier to take apart. With clever design DFA and DFD become symmetrical functions.

Recycling Concrete in Place

Environmentalists have adopted a cradle-to-grave approach in their evaluation of products and services. There are five phases to such an evaluation:

- raw materials extraction
- manufacture and processing
- distribution and placement
- use and maintenance
- disposal

Concrete, a mixture of cement with sand, rock, water, and different add mixtures (add mixtures include super plasticizers, and retarders, among other things) is affected by all five levels. Sources of aggregate from which cement is made are dwindling and thereby becoming more costly. The paving process is expensive and creates traffic disturbances that cost untold amounts including the costs of pollution resulting from cars caught in construction traffic jams.

The service life of roads can be extended in congested areas using new forms of concrete called HPC (high performance concrete). HPC provides environmental benefits and cost savings. There are six projects underway, financed by the Federal Highways to test the performance of HPC.

Another source of multiple benefits comes from the development of fast-track concrete paving machines. They have cut the process of resurfacing roads from two weeks to less than 36 hours for 22-foot wide stretches of road of up to one mile. The pollution relief and cost savings are significant motivators for additional road-building projects as described below.

Recycling concrete in service can provide an endless supply of raw materials for future use as aggregate. The problem is pressing because of the nation's deteriorating highway system. There is great incentive to recycle in place. Thus, instead of taking the pavement to the crusher involving costly transport in dump trucks, a train of crushers moves along the pavement site crushing the original pavement and laying it back down as subbase.[20]

This technology is still under development. With it, machines will lift concrete slabs off roadways, crush them, pull out steel reinforcing rods, process old concrete into a new mix and lay down fresh roadbeds. About 50 such machines could rebuild most of the nation's interstate concrete highways, at a rate of one mile a day, saving an estimated $850 million a year over the present method of ripping up roads with mechanized jackhammers.[21]

Road building is an OM project activity. The speed with which projects are completed has environmental impact that should be stated, either in general, or for each project. This would lead to increased investments in the rapid development of new road-building technology. It also would make companies that construct roads and bridges aware of the need to hire trained operations managers who know how to use rapid-response project management to reduce pollution and wasteful traffic jams.

19.7

SUPPORTING ETHICS

Greening the environment comes naturally to OM because health and safety start with workers on the line. That is why the Big 3 auto companies in Detroit are working together to reduce pollution caused by painting autos. Hazard reduction for workers, customers and communities is a matter of ethics—a set of acceptable social standards. These standards have been moving toward higher ideals in a global economy.

Ethics is not a separate part of business. When these precepts are treated as separate, it is not possible to build an ethical infrastructure. Ethics requires the systems point of view because the standards must be known and accepted by everyone who works for the company. It is not possible to conform to an ethical standard if some of the employees think it is alright to cut corners on quality, ship inferior product, overcharge some customers, solicit kickbacks, take bribes, and so on.

Quality Integrity

For OM, one of the most important ethical issues is to deliver only safe and healthy products and services. The standard is called **quality integrity** which means producing the product according to standards and not accepting or shipping anything to customers that deviates from those standards. One of the most sensitive areas to monitor for ethical adherence to standards is in foods and drugs.

Food and Drug Administration (FDA) activities are supported by information from the Centers for Disease Control and Prevention (CDCP) of the U.S. Public Health Service (PHS). CDCP data are used to identify trends and new concerns about diseases (such as an increase in botulism). In the CDCP system, seafood accounted for 4.8 percent of reported cases of foodborne illness from 1987 to 1994.

The FDA has taken numerous steps to provide consumers with protection against tainted foods and food poisoning. One of the new initiatives is the proposal for the application of the Hazard Analysis Critical Control Point (HACCP) principles to establish procedures for safe processing and importing of fish and fishery products.

Hazard Analysis Critical Control Point (HACCP) System

It is not necessary to restrict the discussion to the seafood industry which is comprised of organizations engaged in fishing, aquaculture, fish farming, fish and shellfish processing, and packing. It applies to all foods and drugs and can be extended to hazard analysis for all products and services. How much effort should go into the control of hazards is an ethical issue. The government sets standards that companies can elect to surpass.

Hazard Analysis Critical Control Point's application as a *preventive system for hazard control* was pioneered by the Pillsbury Company in the 1960s to create safe food for the space program. The program was applied successfully and since that time has been further developed.

It is an excellent example of total quality management where inspection and testing are not sufficient by themselves to achieve satisfactory quality.[22] Interactive quality monitoring is needed to assure that food poisoning does not occur at all. In the document cited in the footnote, the FDA proposes to make HACCP mandatory for the seafood industry. The seven steps for using HACCP's principles are:

1. *Identify the hazards.* These are the causes of food being unsafe for consumption. The principal seafood-related hazards that cause the foodborne illnesses are: bacteria, viruses, and natural toxins.
2. *Identify critical control points* in the process. These could be cooking, chilling, sanitation procedures, employee and plant hygiene. These are critical control points with respect to the hazards identified in Principle 1.

Quality integrity

An ethical standard to deliver only safe and healthy products or services.

Hazard Analysis Critical Control Point

Emphasizes documentation and updates changes in conditions.

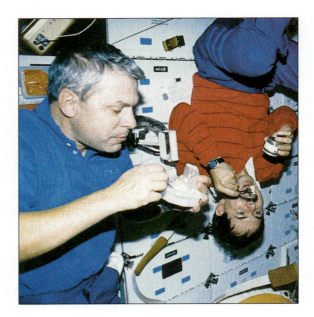

HACCP's application as a preventive system for hazard control was pioneered in the 1960s to create safe food for the space program.
Source: NASA

3. *Establish critical limits for preventive measures* that are associated with each identified critical control point. These critical limits are boundaries of safety for each critical control point. They specify thresholds for preventive measures to be taken for temperatures, times, moisture levels, available chlorine, pH, etc. The critical limits can be derived from sources such as regulatory standards and guidelines, experimental studies, literature surveys, and experts.

4. *Establish procedures to monitor critical control points.* This is a planned sequence of observations or measurements to assess whether a critical control point is under control and to produce an accurate record for future use in verification. Just like with quality control charts, this system monitors the process at critical prevention points to spot trends indicative of potential trouble or of real trouble. It is expected that corrective action will be taken to bring the process back into control before a real problem deviation occurs or to correct a real problem. Also important is the written documentation for use in verification of the HACCP plan.

5. *Establish the corrective actions* to be taken when monitoring shows that a critical limit has been exceeded. Three things need to be done:

   a. determine the disposition of food produced during a deviation
   b. correct the cause of noncompliance to ensure that the critical control point is under control
   c. maintain records of corrective actions

   Because food poisoning is an unacceptable quality deviation, the disposition of spoiled food becomes an obvious subject to be determined and documented. This is a good lesson for industry to copy. What happens to defective product should be decided as part of every quality-control system.

6. *Establish effective recordkeeping systems* that document the HACCP system. This principle requires documentation of hazards, critical control points, critical limits, and the maintenance of all records generated by the process during its operation.

7. *Establish procedures to verify that the HACCP system is functioning* as it is intended. This step includes verification that the critical limits are appropriate and working. It also is used to make sure that all aspects of HACCP are being followed. Finally, it ensures periodic revalidation of the plan to make certain that it is still relevant to raw materials, technology, and all other process factors and conditions.[23]

The implications of the HACCP system for processing technologies are significant. The design of process technology should lend itself to minimum costs for achieving the HACCP objectives. The attention to documentation and updating for changes in conditions makes HACCP a superior operational system.

All organizations entering the seafood industry and related businesses (as the use of HACCP grows for foods in general) will have to learn the HACCP regulations in order to be in compliance. This adds another dimension when planning and choosing process technologies. It is *design for compliance.* The rules must be understandable and compliance must be possible using reasonable technology and training.

Appropriate information technology will make the recordkeeping system a boon and not a burden. HACCP is a pragmatic system that has the potential to cure some basic problems. The shellfish industry is being destroyed by the fear of toxic poisoning. People who love to eat raw oysters and clams are giving them up because of misgivings. If HACCP can save the shellfish business it would mean jobs saved and consumers' health protected. The rewards are large and the costs are small.

This is an excellent example of the dynamics of ethics. Improvements in ethics are aided and abetted by governmental regulation. Unfortunately, some improvements are inhibited by government bureaucracy.

The HACCP model reflects good management of ethics. It might well be adopted by those who manage new food processes that use genetic materials and biotechnology. Hormones fed to cows to increase milk supplies and irradiation of food products to retard spoilage are two areas of concern. Once they are definitely established to be safe, there is distress about the control of process dosages. For environmental processes, it may be useful to consider the application of the HACCP principles and steps to pollution control and hazardous waste disposal management.

For process technologies in general, HACCP has relevance as an application of TQM under governmental regulation. Quality, a primary OM responsibility, always has been regulated, to some degree, by the federal government. Examples are weight, purity, and so on. New rules make change management essential.

Quality has been under constant discussion in the courts (product liability cases, etc). The difference is that HACCP regulates QC methodology, whereas previously, quality was only regulated as an end result. HACCP regulates quality at the process level.

The Occupational Safety and Health Act of 1970 (OSHA) provides precedence for governmental regulation of systems and processes. Inspectors are used to check workplace conformance to safety rules. There have been many complaints about bureaucracy and strenuous efforts have been made to have OSHA function as a positive complement and not antagonistically with business. It is hoped that HACCP regulators can learn from the OSHA experiences.

While OSHA is concerned with the safety of the process, HACCP deals with the quality of outputs from the process. Related to this subject is product liability which will be increasingly associated with HACCP-type rules. Even at the present time, the courts require documentation of efforts to "design for safety" and the use of QC methodology to assure safety.

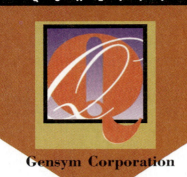

Gensym Corporation

*Gensym applies expert systems
to environmental quality
management.*

Headquartered in Cambridge, Massachusetts, Gensym Corporation supplies expert system software products worldwide for industrial, scientific, commercial and governmental applications. Gensym's family of intelligent software is built on its flagship product called G2®, which is a graphic, object-oriented environment that forms the foundation for building intelligent real-time systems. G2 can be used to develop expert systems applications for environmental quality management. The systems are used to monitor site conditions, identify and diagnose causes of unexpected process changes, and advise plant operators on how to minimize environmental contamination. The expert systems serve also to help minimize leaks, emissions, and waste and improve product quality, yield, and energy consumption.

Among the environmental applications of expert systems is minimization of air and water pollution. The South Coast Air Quality Management District (SCAQMD)—the air pollution control agency for the four counties in metropolitan Los Angeles—uses G2 to monitor power plant emissions and plants' compliance with complex air pollution rules. SCAQMD, which contends with the worst air pollution in the U.S., has implemented the most stringent pollution control requirements in the world. The agency's goal was to determine rule compliance on a daily basis. To adhere to emissions rules, utilities must monitor and measure boiler emissions throughout the day and at set intervals transmit the data to SCAQMD. The G2 system allows the agency to create rules representing the legislation's intent in order to model the data against regulations, and thus create compliance objectives, relationships, and attributes.

Yorkshire Water in Yorkshire, England, uses a G2 expert system in a continuing effort to improve the quality of its water supply. Yorkshire Water uses the G2 expert system to predict, and thus prevent, failures. One of the biggest problems is the unreliability of the sensors used to identify the quality of the water. The system detects which sensors are accurate and which are not. By knowing the water quality as it passes through the works, the process can be optimized and the quality of the water leaving the works can be improved.

Gensym Corporation, founded in 1986, serves more than 30 industrial sectors and has sold more than 3,000 licenses for its software products worldwide. Gensym has 14 U.S. offices and 11 offices overseas. The company offers a wide range of services to help customers design, develop, deploy and maintain their systems.

Sources: Alice H. Greene, "Environmental Protection Using Real-Time Expert Systems," *Quality,* January, 1994, and Gensym Corporation documents, with special assistance from Betsy Gartner, public relations manager for Gensym Corporation.

From these trends, it is evident that the systems perspective is necessary. It is required to deal with changes that are occurring in the regulatory and legal systems that promote ethical business in the U.S. and globally. OM has the lead role with HACCP. It must coordinate activities with R & D, process engineering, the legal departments and the government regulators. The customer must not be lost in a regulatory jungle.

Food and Drug Testing

Food and Drug Administration regulations for testing new drugs are extensive. Running counter to the rapid-response project management theme, pharmaceutical product cycles take many years. During that time, OM manufactures test materials with R & D and prepares suitable facilities and quality control methods for each stage.

New drugs go through years of laboratory studies followed by three phases of clinical testing with people. It takes the FDA 12 years (on average) to go from the initial application to final approval.[24] Further, it is not unusual for the FDA to refuse certification.

Some consider the FDA to have overly conservative testing standards and requirements. The main pros and cons associated with stringent test specifications, requirements and standards are captured by the following two points:

- Undertesting includes inadequate testing methods and deficient testing technology. Both can lead to disasters.
- Overtesting delays the acceptance of effective drugs. As a result, people die unnecessarily.

Two examples are presented for the first point. First, a vaccine was rushed into production to counter a predicted swine flu epidemic. One of the vaccine manufacturers, operating under crash project conditions, made process-technology errors which were not picked up in testing because that would have delayed the vaccination program. The result was that a far larger number of people died from the vaccine than from the swine flu which never became an epidemic. The coincidence of failures in this case can all be attributed to the axiom that "haste makes waste."

Second, the story about the drug thalidomide reflects the complicated trade-off between the risks and benefits of testing delays. European governments have organizations equivalent to the FDA in the U.S.A. Some of them had approved the use of thalidomide as a tranquilizer after a shorter period of testing with a smaller sample than was permitted in the U.S. The drug was never legally accepted by the FDA for the U.S.A.

It turned out that this drug adversely affected the fetus if taken by women in the early months of their pregnancy. European testing had overlooked the early months even though use of tranquilizers was common during that period. Thalidomide children were born with serious deformities which reinforced the conservative nature of the FDA in the U.S.A.

There are two kinds of potential errors that pose ethical quandries for OM and the organization. Should undertesting errors be preferred to overtesting ones? Some pharmaceutical and biotechnology firms would like FDA standards to be less stringent. They feel testing costs too much and takes too long. They argue that more funds would be invested in R & D for new drugs if testing standards were relaxed. Opponents assert that people are suffering and dying unnecessarily while the testing is proceeding. They propose delivering test products to those suffering serious illnesses.

The FDA, in recognition of this ethical issue, has speeded up the testing of AIDS drugs and cancer treatments. It has permitted their use for seriously ill patients before approval is granted. Aside from that, many reject the claim that the high cost of testing discourages the development of new drug technologies. They cite the fact that the pharmaceutical industry in the U.S.A. has done an excellent job of creating new drugs while making a great deal of money.

OM in the pharmaceutical industry emphasizes quality control as much as, or more than, any other industry. The cost of quality in pharmaceutical production is very high and there is pressure to find ways to cut these costs while increasing quality standards. The subject is an ideal candidate for reengineering. Because the FDA is known to accept change slowly and with great care, the topic of change management is totally relevant.

19.8
READYING THE ORGANIZATION FOR FUTURE TECHNOLOGY

Factories and offices of the future will not look like their present-day counterparts. OM will be practiced in different circumstances and, therefore, in different ways. The key is to be prepared for change. Predictions of what changes to expect are useful but hardly likely to hit the bull's-eye. The only way to benchmark these expectations is by noting present developments and then extrapolating. Some directions for change seem to be evident. Benchmarking against creative applications produces the following insights.

The September 19, 1994, cover of *Fortune Magazine* proclaimed, "The End of the Job: No longer the best way to organize work, the traditional job is becoming a social artifact. Its decline creates unfamiliar risks—and rich opportunities."[25] The traditional job is disappearing because of new technology, integrated information systems, a teamwork mind-set, and the global business olympiad where being better and better forces continuous change. This trend is evident on a global scale in best-practice companies.

Mobile Robots

Many repetitive, unpleasant and dangerous jobs can now be assigned to machines. Fixed machines can do some jobs, but they are limited by the lack of mobility. Robots[26] moving along in-floor guidewires are less constrained but still restricted compared to people. With better sight, robots play a greater role in factory and office designs. People have been freed-up for more creative working opportunities. Job descriptions are purposely vague to encourage "doing what needs to be done."

Past predictions about robots have fallen so far short of reality that it would be easy to overlook the true state of affairs which is that steady advances are being made in improving machine vision. Robots will be able to perceive objects in color and in three dimensions with speed and accuracy unmatched by present video systems. With high performance vision, robots will become mobile, thus transforming manufacturing, logistics, mining, space prospecting, and service operations.

"The combination of moving cameras and high-speed processing will give robots the ability to see objects around them in 3-D, and react quickly to avoid obstacles, find targets, recognize patterns, and provide intelligent human services for the elderly and handicapped. Additionally, they can provide remote "virtual reality" displays of actual environments which are too dangerous for human presence, for example in volcanic craters, nuclear reactors, the depths of the ocean, and space."[27]

HIGH PERFORMANCE ROBOT "EYE" SYSTEM

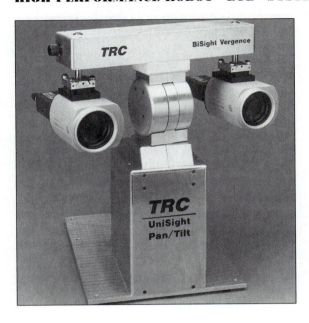

High performance robot "eye" system mimics the motions of human eyes and the head. The "head" rotates up to 1000 degrees per second to direct the view anywhere in the surrounding 360 degrees and can be zoomed and focused under computer control for close or distant viewing.
Source: Transitions Research Corporation

Figure 19.8 illustrates a high performance robot "eye"; Figure 19.9 shows how images are dissected into pixels. Figure 19.10 shows how a frog is seen by the system.

IMAGES ARE DISSECTED INTO PIXELS

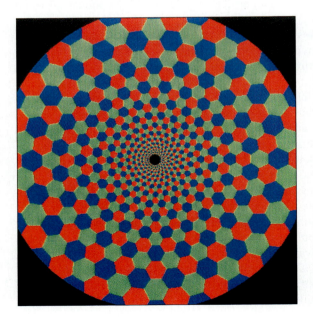

These hexagonal patches are necessary to process images fast enough to keep up with rapid camera motion. They are small in the center and large in the periphery—as with the human retina.
Source: Transitions Research Corporation

FIGURE 19.10

A FROG AS SEEN BY THE VISION SYSTEM

This image of a frog requires only about 5000 patches whereas video requires over 100,000 patches per image. Thus speed can be increased by 20 to 1 compared to video-image processing for robot vision systems.
Source: Transitions Research Corporation

Service Robots

Industrial robots are a multibillion dollar industry whereas the service robot industry is in its infancy. Few service robots are commercially available at present. "That will change significantly," says Joe Engelberger, chairman of the board at TRC Automation Technology, Danbury, Connecticut, and the pioneer of both industrial and service robotics. He believes that service robotics will pass industrial robot volume by 2005.

TRC presently manufactures various kinds of service robots including Sweepmate—a fully autonomous commercial floor sweeper, Autoscript II—linked to external computers to fill prescriptions in high-volume pharmacy operations, and HelpMate®—a hospital orderly.

Figure 19.11 shows two HelpMate robots being loaded with food trays which will be carried to designated delivery locations anywhere in the hospital.

HelpMate has laser vision and ultrasonic proximity sensors to avoid obstacles. These robots navigate from point to point using a map of the building to plan the best route and sensory feedback to follow that route. Elevators are modified to permit direct control by HelpMate as it totes x-rays, lab reports, food trays, and other items needed by doctors, nurses, and patients. Their verbal abilities include sixteen messages which are appropriate to specific situations.

Industrial robots replace humans working in fixed positions on the assembly line whereas service robots replace human couriers. HelpMates require extensive navigational abilities. Using walls and ceilings as guides, they access their internal maps. They are trackless as compared to industrial AGVS (automated guided vehicle system) which are confined to scheduled deliveries over preprogrammed routes. AGVS installations require extensive plant modifications. HelpMates have no such limitations.

FIGURE 19.11

HELPMATE SERVICE ROBOTS AT WORK IN IN A HOSPITAL

These two HelpMate robots are being loading with food trays. At the delivery point their trays will be unloaded.
Source: TRC

Glimpsing the future, production people will be accustomed to working with mobile robots that can navigate corridors and carry materials between stations. Mobile service robots will alter service industries providing rapid setups, maintenance, deliveries, repair, and cleaning services. They are particularly well-suited when second and third shift staffing is a problem. They have the potential to change the cost structure and the quality delivered by service industries.

Combine the transactional flexibility of ATMs (automatic teller machines) with the "potential creativity" of artificial intelligence (AI),[28] and the mobility of robots: the result is a plausible revolution for the service industries.

Another major step forward is underway. People need no longer be exposed to hazardous working conditions. Such problems are being analyzed and organized by the Association for Robotics in Hazardous Environments (RHE) which was created in 1994 by the Robotic Industries Association (RIA). RIA was founded in 1974 by the Society for Manufacturing Engineers located in Ann Arbor, Michigan. To be an effective substitute for people operating in hazardous environments, robots require excellent vision and mobility.

The Virtual Office

Offices become virtual when no space is dedicated to specific individuals. It has been called the nonterritorial office. "Sales representatives at American Telephone & Telegraph Co.'s Sacramento, California, office lost their desks. They were given laptop computers, cellular telephones and portable printers and told to create "virtual offices" at home or at their customers' offices."[29] The varieties of virtual offices include:

1. *Working at home.* This transfers the cost of space from the company to the worker. It is being used to decrease commuting by car. AT&T, IBM, Digital Equipment Company and other organizations are experimenting with this plan in California where increasingly tough regulations are expected to be passed requiring companies to limit the number of employees permitted to travel to a particular work site (for reasons of environmental pollution control).

2. *Working from cars, hotels and customers' offices.* These mobile offices are staffed by mobile workers. The move to mobile offices is associated with functions that require extensive travel to customers' sites. Thus, salespeople, accountants, auditors, consultants, and service team members are logical candidates for mobile-office assignments. The effect is to increase the amount of time that these kind of professionals spend with their customers which both parties find rewarding.

3. *Working in nonassigned spaces at the company site.* This is called hoteling, hotel offices and the check-in-and-out office. There also is doubling up (or better) the number of people assigned to one office. Ernst & Young increased the employee-to-desk ratio of two-to-one to five-to-one. In Chicago, the company consolidated their operations in the Sears' Tower and instituted "hoteling."

   "None of the 500 accounting and management consultants below the senior-manager level were given desks…. Now, when they want to spend more than a half-day at the home office … they must phone the firm's "concierge," who will reserve one of the 125 well-equipped offices, carry in the appropriate files, program their phone to ring through and hang their nameplate on the door."[30]

Many companies including Arthur Andersen & Co., Chiat/Day Inc., Dun & Bradstreet Corp., Ernst & Young, IBM, Price Waterhouse, and Travelers are reported to have reengineered their offices using combinations of the three points above. Link Resources in New York predicts there will be 25 million telecommuters by the year 2000.

All of IBM's seven regional U.S. markets are converting to virtual-office configurations. IBM's corporate headquarters is standardizing the process. IBM midwest has sent 700 executives into the field. Eventually all 2,000 employees at 30 midwest locations will be virtualized. Private offices are being replaced by "small team rooms" for use by the mobile executives…. Inside each room is a round conference table, plugs for local area network (LAN) communications, and, of course, telephones."[31] Eight executives at a time rotate in the use of these "work pods" used to solve problems. IBM has duly noted that change-management methods on a systems level are essential to make this major shift in working modes effective and productive.

Some question whether the virtual-office concept is a fad. Oticon uses nonassigned workstations. Price Waterhouse has started hoteling in Boston, Dallas, New York City, and Washington. The IBM and Ernst & Young stories are told above. It is reported that "the amount of office space needed for certain types of employees—those who travel a good deal—has declined to 48 square feet a person from 200 square feet."[32]

Technology makes the virtual office possible. Change management is needed to get the right timing for adopting the new technology and for getting people to accept and use the new systems effectively. The challenge to OM is far broader than the subjects that could only be covered briefly in this section.

19.9

P/OM'S RESPONSIBILITY FOR GOING GLOBAL SUCCESSFULLY

Supply chains are now global distribution systems which are as reliable and rapid as local systems. Telecommunications systems allow global decisions to be made as fast as local ones. The volume of products that are made in one country and then sold and shipped to another country are far greater than they ever have been and the trend is increasing.

Agribusiness–Consulting–Financial Services–Machinery

Agricultural business provides an excellent example of operations-intensive products which are being traded on a global scale. In addition to using a lot of labor, agricultural products require a great deal of energy. There are many companies in the supply chain. OM's role is critical at all stages in the process of growing, collecting, processing and distributing food. OM's contribution is high productivity, low cost and world-class quality.

"The war over the globalization of agricultural trade is escalating. With the stakes now exceeding US $300 billion annually, globalists are assaulting the defenders of agricultural self-sufficiency in their backyards by investing directly in local processing firms. They are also doing their best to shred the web of tariffs and subsidies that keep nearly 90 percent of the world's agricultural output from crossing a border."[33]

Countries are changing what they grow. Traditional farm products, which are staple items in the home market, are being replaced by crops that can be traded on international markets. Developing countries are trading great quantities of commodity products such as coffee, tea and cocoa which has lowered the prices to growers. The pressure on growers leads them to develop joint ventures with food processors.

"The fastest growing sector of the world's trade in agricultural products is the processed food industry. While the traders of traditional agricultural products struggle against the dogma of self-sufficiency and its accompanying barriers, processed foods share of the international market for agricultural products has increased from 58 percent in 1972 to 64 percent in 1990."[34]

The consulting business, always international, is now involved in every aspect of global competition. A Booz Allen & Hamilton, Inc. report states, "Global competition is increasing the number of competitors and minimizing geographic advantages. While companies around the world focus on new competitive imperatives, many have yet to achieve success."[35] The list of what must be done is an OM menu for change management.

Financial trading is an international business. Depression era banking laws which separate the businesses of commercial banks from brokerage houses and investment banks are being repealed. "As these walls crumble and the financial services industry becomes increasingly global, large foreign banks such as Union Bank of Switzerland … are putting down roots (in Wall Street)."[36] Operations makes internationalization of financial services possible.

"The intense drive to improve margins is leading companies globally to seek new efficiencies through restructuring. That process began in earnest in the U.S. in the late 1980s, and companies in all industrial economies have since followed the U.S.

lead by seeking to cut costs and reengineer operations."[37] Merrill Lynch goes on to state that the capital goods industry, which provides new production machines, will grow rapidly on a global scale.

ISO 14000/14001

Global environmental (and ethical) benchmarks will change the way that business is done worldwide. "Ready for ISO 14000? Just when you thought ISO 9000, the certification that shows manufacturers are observing international quality standards, was in place, the International Organization for Standardization in Geneva now aims for environmental management guidelines by 1996."[38]

"Many organizations have undertaken environmental reviews or audits to assess their environmental performance. On their own, however, these reviews and audits will seldom be sufficient to provide an organization with the assurance that its performance not only meets, but will continue to meet, its policy requirements. To be effective, they need to be conducted within a structured management system, integrated with overall management activity and addressing significant environmental impacts."[39]

ISO 14001

Standards for conformity assessment.

The ISO 14000 series includes **ISO 14001** which is being finalized to set standards for conformity assessment leading to certification. It is intended to provide the organizational and management structure and discipline for achieving excellence in environmental performance. ISO 9000 does not deal with specific qualities. The 14000 series does not deal with specific environmental qualities, such as emission levels. Like ISO 9000 standards, the 14000 standards will be coordinated by the American National Standards Institute (ANSI) in cooperation with the American Society for Quality Control (ASQC).

ISO 14001 is likely to start with an articulated environmental policy followed by management systems to plan the environmental program, provide for operations and implementation (including training, methods for communication, documentation and emergency preparedness). Also, there will be a provision for checking and taking corrective action. Monitoring and measuring will be detailed so that nonconformance can be prevented, but when it occurs, corrected. The methods will allow a certified environmental systems audit and review.

Reengineering and continuous improvement can have substantial environmental impacts. Every TQM and JIT program can effect the environment. Companies with ISO 14001 certification will not be forced to select which country's standards it will follow. Often this approach leads to conflicts of standards between countries. ISO will have provided unified global standards and certification procedures likely to be adopted by all countries—as has the ISO 9000 series.

The Olympics as a Systems Symbol

Olympics

Represent a system's symbol for "going global."

There is a lot to be learned about global business from the **Olympics**. The Olympic creed is to do the best that can be done. *Best* has to be defined with a global perspective because the best of the best come from all over the world to compete.

Choosing the right system to benchmark is vitally important.

The five Olympic rings linked together represent the interactions and connections of people from all over the globe: Africa, America, Asia, Australia, and Europe. The symbolism of these rings is global interdependency. Something that happens in one place can affect every other place. The rings provide an analogy for the global supply chain which depends upon use of the systems approach to achieve coordination required for best performance.

"After years of false starts, Washington is finally taking a solid crack at a sport long familiar to the Japanese and the Europeans: Commercial diplomacy."[40]

Best practice on a global scale for companies from any nation now requires governmental information and coordination of both public and private systems elements. The Export-Import Bank of the United States working together with the Overseas Private Investment Corporation (OPIC) helps to raise millions of dollars to finance various deals. In addition, OPIC, a financially self-sustaining U.S. government agency, provides political risk insurance.

OPIC is required by statute to conduct environmental assessments of every project and to decline to participate where there would be unreasonable adverse impact on the host country's environment. Similar considerations apply to worker rights and trade-related performance requirements.[41]

OPIC programs operate in 141 countries and areas. The corporation's net income exceeded $167 million in 1994, a four percent increase over 1993. Finance commitments quadrupled from $415 million to $1.7 billion. "Demand from American companies for OPIC support and services is growing at an unprecedented rate. OPIC welcomes project proposals from U.S. businesses of any size and from most industries and sectors."[42] Projects are supported in agribusiness, manufacturing, tourism, and so on.

OPIC's 1994 Annual Report goes on to say, "In many countries, the face of infrastructure development is changing ... government-owned facilities and natural resources are being privatized, opening the doors...." The investment areas covered include power, telecommunications, transportation and distribution, and natural resources. OPIC conducts educational outreach programs to inform small- and medium-sized businesses of global opportunities.

At the Commerce Department's "Advocacy Center," teams of trade and technology specialists work with corporations to develop strategies for successful trade missions. They are assisted with risk assessment and competitive analysis by the Central Intelligence Agency (CIA). The change from prior government practice is outstanding. The support policies of other countries provided useful (informal) benchmarks. Without labeling the process, the U.S. Commerce Department reengineered itself.

Peter Drucker observed, "Continuous improvement and benchmarking are largely unknown in civilian agencies of the U.S. government. They would require radical changes in policies and practices which the bureaucracy ... would ... fiercely resist."[43] This statement—which remains correct—in conjunction with the government's change of mind-set, could help reinvent government.

Operations managers play a major role in assessing global systems and in converting suitable opportunities into successful projects. Transition and change management abilities of OM are essential to success. Globalization has created careers for OM that are dramatically exciting. The teams that the U.S. now sends to the business Olympics must be world-class just like the sports teams that compete in the Olympics every four years.

Understanding and implementing world-class performance requires the systems point of view. It must be supported by a totally interconnected information system. The infrastructure is shown in Figure 19.12 where it is called the SME Manufacturing Enterprise Wheel. For service industries, a similar wheel of six concentric circles can be drawn, substituting customer services for manufacturing.

Start with the customer (1) who could be located everywhere in the world. Move to people organized for teamwork (2). This organization is designed to share knowledge by using the systems approach (3). The next concentric circle employs the systems approach to relate customer support, product/process factors and manufacturing (or customer services) (4). The next level (5) connects resources and responsibilities to monitor and guide all decisions that are made at the fourth level. The outer part of the wheel (6) encloses, connects, links, and shares everything in the system. Six is named the manufacturing (or service) infrastructure.

In the summer of 1993, the Computer and Automated Systems Association of the Society of Manufacturing Engineers (CASA/SME) announced this new version of what had previously been called the CIM Enterprise Wheel. CIM stood for computer integrated manufacturing. The center of that wheel focused on information systems architecture and databases.

FIGURE 19.12

SME MANUFACTURING ENTERPRISE WHEEL

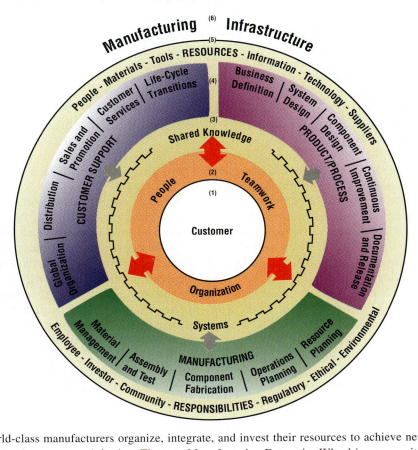

World-class manufacturers organize, integrate, and invest their resources to achieve new levels of customer satisfaction. The new Manufacturing Enterprise Wheel is a recognition of the significant progress to date, and the progress yet to be made, in manufacturing. It serves as an overview of today's best practice in total enterprise integration.

This new wheel centers on the customer. It considers "the virtual enterprise" which is the inclusive system of all of the people who are teamed together to achieve company objectives. The new wheel stresses: "The need to integrate an understanding of the external environment, including customers, competitors, suppliers, and the global manufacturing infrastructure."[44] The **SME wheel** provides a "vision" of a fully-integrated global company system.

SMG Wheel
World class performance must be supported by a totally interconnected information system.

S U M M A R Y

This chapter highlights the need for change management during periods of transition. Technology is advancing at an incredible rate creating new global competitors and opportunities for everyone to be one. Environmental and ethical factors are also changing the situation on a global scale.

The need to change is often apparent but the organization is not able to do so. Bureaucracy deflects change. The factors that permit organizational change are presented by a change-management model. It is then possible for P/OM and other functions to develop the factors that encourage change.

P/OM has mastery over continuous-improvement methods for gradual change. It also can provide leadership using reengineering (REE) methods—which is starting with a clean slate to redesign the system. The principles and heuristics for REE are presented along with a description of two companies that are modeled on the principles of reengineering. Both companies used the principles before REE was coined as a phrase to describe such activities. A company does not need to have an REE program to practice it.

Some important examples of OM processes that are environmentally sound are presented. With care and attention to ecological impacts, companies can save money and the environment at the same time. The same kind of results apply to ethical consideration for customers' welfare and employees' health and safety. In this regard, quality integrity is developed as an ethical issue. The HACCP program for seafood quality is a model of FDA regulation of total quality. Food and drug testing also is discussed.

OM is on the team which can prepare the organization for future technology. Several trends are examined including robots that can be mobile because of vision improvements. They are changing services and manufacturing. The virtual office is having a major impact on services. Dealing with such technology changes is a P/OM responsibility.

Agribusiness is going global, as are consulting, financial services, machinery and machine tools. P/OM has an important responsibility for preparing their organizations to compete globally. ISO 14000 and 14001 are explained as global certification procedures to indicate appropriate management structure for environmental concerns. At present, many different national standards exist.

The chapter concludes with the concept of the competitive "business olympics" modeled on the sports Olympics. Government resources are being used to help companies go global. Governments are in competition with each other in providing such supports. Companies that try to go it alone are not playing on a level playing field. The U.S. team consists of the Department of Commerce working with the Overseas Private Investment Corporation and other agencies; it has become as good as the best.

▶ **Key Terms**

breakpoint business process redesign   *859*

efficiency frontiers   *862*

olympics   *886*

reengineering (REE)   *859*

take-back laws   *871*

change management   *851*

hazard analysis and control (HACCP)   *875*

pilot studies   *865*

smart benchmarks   *856*

continuous improvement (CI)   *857*

ISO 14001   *886*

quality integrity   *875*

SME wheel   *888*

▶ **Review Questions**

1. What is OM's role in change management?

2. What eight factors are forcing the organization to change?

3. How do the four elements of the Gleicher change model work?

4. Use the Gleicher change model to explain why a new CEO might say that another "vision" statement is not needed.

5. What is the effect of downsizing on change management?

6. Why is it said that downsizing is not reengineering?

7. The elevators are kept locked at Oticon in Denmark to encourage meetings on the stairwells. What purposes are served?

8. What are the various kinds of benchmarking that are used?

9. What is continuous benchmarking?

10. How do smart benchmarks differ from other ones? Exemplify.

11. Why should continuous-improvement (CI) programs be regularly reviewed by a change management team?

12. It has been said that transition planning and training should include senior management. Explain this and give examples.

13. Provide nine reengineering principles and explain them.

14. What purpose is served by reengineering heuristics, A through Z. Pick the half you like the most, and explain why?

15. Why is it pointed out that Oticon Holding S/A and W. L. Gore & Associates, Inc. do not use the term reengineering even though both companies provide excellent examples of continuous reengineering?

16. What is the benefit of continuous project development?

17. What are the four aspects of DFD? Explain the role of DFD.

18. How does waterjetting qualify as an OM greening technique?

19. Why is recycling concrete in place used as an example of OM greening the environment?

20. What five phases are used for environmental evaluations?

21. Explain quality integrity as an ethical issue.

22. Describe the hazard analysis critical control point system in terms of the HACCP objectives and the seven recommended steps.

23. How does HACCP relate to TQC and TQM?

24. Why is HACCP called a custodian of social ethics?

25. Explain the problems associated with over- and undertesting the safety and effectiveness of food and drug products.

26. What is the impact of vision improvements on robots?

27. What are the uses of mobile robots with fine vision?

28. Explain the impact of the virtual office on operations.

29. Chiat/Day, an advertising agency, uses hoteling. Why is this office arrangement likely to be very sensible?

30. Relate the virtual office and the "end of the job" concept.

31. What are the responsibilities of P/OM with regard to helping the company successfully globalize its operations?

32. Why is agribusiness globalizing rapidly? Name other industry sectors that are likely to be doing the same.

33. Explain why ISO 14001 is important for OM to help the organization succeed in global opportunities.

34. 7-Eleven food stores in Japan have used JIT deliveries to such an extent that the company has been accused of creating road pollution. How would application of ISO 14001 help to sort this problem out?

35. Explain the SME wheel including each concentric ring.

36. Describe the competitive role of government in obtaining global business. What U.S. agencies participate?

37. How are the Olympic games symbolic of OM's efforts to apply the systems approach to global business success?

▶ Problem Section

1. Using the Gleicher model, assume that: $D = 9$, $V = 7$, $P = 1$ and $C = 40$. Will the organization be likely to accept change?

2. In Problem 1, there is a lot of dissatisfaction and a good vision of what the future could be like. There is much resistance to change. C is more than four times larger than D. Practical first steps are unknown ($P = 1$). Then, a consulting firm is hired which recommends a 15 percent workforce reduction. What changes might be expected in the original set of numbers?

3. The magazine, *Business Week*, benchmarks business schools every two years. There has been criticism of the benchmarks used. Develop a full set of criteria that you believe would permit an accurate comparison of business schools.

4. Make a chart for the benchmarks developed in Problem 3. Then rate your own school on each of the criteria from 1 to 7 where 7 is best. If you have friends attending other business schools, ask them to rate their schools using your benchmark criteria. Compare the ratings and discuss what they mean.

5. Referring to Problem 4, if you want to combine the numbers in some way to get a single number (as *Business Week* does), a scoring model will have to be developed. One possibility is to multiply the ratings as below. If there are n ratings for each school, then the value is $= R_1 \times R_2 \times R_3 \ldots \times R_n$. More complex scoring systems weight the ratings before multiplying them.

6. To improve the school, which factors should be changed first? (Use the information developed for Problems 3 and 4.)

7. For the economic analysis of continuous project development $P_{success} = 0.30$; $T = 730$ days (average total project development time in days); $N = 100$ projects (number of active projects). The average cost of projects is $100,000 per year. The average gross margins of successful projects is $800,000 per year.

   a. What is the estimated total gross margin?

   b. What percent increase is required for $P_{success}$ to increase the estimated total gross margin by 50 percent?

   c. What percent increase is required for $P_{success}$ to increase the estimated total gross margin by 100 percent?

8. For the economic analysis of continuous project development, $P_{success} = 0.50$. Also, $T = 500$ days (average total project development time in days), $N = 100$ projects (number of projects actively being pursued), projects have an average cost of $50,000 per year. Successful projects have average gross margins of $100,000 per year.
   a. How many successful projects will occur per year?
   b. Do you recommend this arrangement?

9. W. L. Gore & Associates, Inc., has generated over 1200 patents worldwide in its 37 years of operation—better than 32 per year.
   a. What is the average interval in days between patents granted?
   b. What kind of system could generate this number of projects for products that are unique enough to be awarded patents?

10. Various methods are used to determine the best candidates for process reengineering. "Efficiency frontiers" (Figure 19.2 in the text) are approximated by the equation $(DC)^2 + (IA)^2 = R^2$ where DC = Delivery capabilities and IA = improvements anticipated, and the radius, R, reflects the combined power of the efficiency frontier.

   Calculate the efficiency frontier values for the table below:

Project	A	B	C	D	E
R^2	4	4	9	9	16
DC	1.414		2.122	3	3
IA		2			

   Rank order your preference for these projects and explain.

11. Draw the chart indicating the appropriate efficiency frontiers and the points for each project as obtained in the table for Problem 10 above. Rank order your preference for projects.

12. Construct a quadrant differentiation chart using the data in Problem 10 above, and the quadrant center point of 2.2. Compare the results with those from the chart in Problem 11.

13. When office space is reengineered using "hoteling," the square feet needed per employee may decrease from 200 to 48.
   a. What is the percent decline in space required?
   b. At $12 per square foot per month what is saved per employee?
   c. If there are 1000 employees what is the total monthly saving?
   d. What business functions are best suited to hoteling?

14. Knowing that you have completed an OM course, the principal of your high school invites you to give a presentation about global careers in OM. Prepare an outline for your presentation.

Readings and References

D. K. Carr, K. S. Dougherty, H. J. Johansson, R. A. King, and D. E. Moran, *Breakpoint: Business Process Redesign*, Coopers & Lybrand Series, Arlington, VA, 1992.

James Champy, *Reengineering Management: The Mandate for New Leadership*, NY, Harper Business, 1995.

Food and Drug Administration (FDA), Proposal to Establish Procedures for the Safe Processing and Importing of Fish and Fishery Products, Federal Register, Vol. 59, No. 19, January 28, 1994. Companion Document: Fish and Fishery Products Hazards and Controls

Guide, FDA Draft Document, PB94-140-985.

Richard Tabor Greene, *Global Quality: A Synthesis of the World's Best Management Methods*, Burr Ridge, IL, Irwin, 1993.

Michael Hammer, and James Champy, *Reengineering the Corporation*, NY, Harper Business, 1993.

Michael Hammer, *Beyond Reengineering*, NY, Harper Collins, 1995.

Masaaki Imai, *Kaizen: The Key to Japan's Competitive Success*, NY, Random House Business Division, 1986.

Henry J. Johansson, Patrick McHugh, A. John Pendlebury, William A. Wheeler III, *Business Process Reengineering: Breakpoint Strategies for Market Dominance*, NY, Wiley, 1993.

Michael E. McGrath, *Product Strategy for High-Technology Companies: How to Achieve Growth, Competitive Advantage, and Increased Profits*, Burr Ridge, IL, Irwin Professional Publishing, 1995.

Jeffrey G. Miller, Arnoud De Meyer, and Jinichiro Nakane, *Benchmarking Global Manufacturing: Understanding International Suppliers, Customers, and Competitors*, The Irwin/APICS Series in Production Management, Burr Ridge, IL, Irwin, 1992.

Vladimir Puchek, Noel Tichy, and Carole K. Barnett, Eds. *Globalizing Management: Creating and Leading the Competitive Organization*, NY, Wiley, 1993.

G. Stalk, P. Evans, and L. E. Shulman, "Competing on Capabilities: The New Rules of Corporate Strategy," *Harvard Business Review*, March-April, 1992.

David A. Summers, *Waterjetting Technology*, E. F. Spon, Ltd., Division of Chapman & Hall, London, England, 1995.

S U P P L E M E N T   1 9

NPV for Net Present Value Model Applied to the Make or Buy Problem

The Make or Buy Model

Assume that the investment to make the product is $1,000,000 and the operating costs for making the product are $500,000 per year. The alternative is to buy the product from a supplier for $750,000 per year. The same volume is assumed in each case.

How do these plans compare? The simplest analysis is based on the fact that the *make* plan saves $250,000 per year. In four years, the savings will pay for the $1,000,000 investment. Is that soon enough?

The answer will depend on the product life cycle and the expected life of the investment. If it is more than the breakeven time of four years, then making the product can have cost savings as well as process-learning advantages. There also may be a quality benefit because the self-supplier is closer to the customer than the external supplier.

Disadvantages include complexity. There is only so much management time available. Risk always is a factor. In spite of the prediction that the mature phase of the life cycle extends beyond four years, there always is room for error and a potential loss of some part of a one million dollar investment.

There are those who insist that the make or buy model should discount the cash flows. To do so, an estimate for the interest value of money is required. For example, six percent per year might be the amount that could be obtained by investing a given sum of money in bank certificates. The estimate of this interest rate will vary depending upon the size of the company, its growth potential, capital requirements and going interest rates. Successful organizations can earn higher returns, simply by plowing back all available cash into their production output.

The six percent rate will be used for the illustration, recognizing that it is not easy, but necessary, to determine the appropriate rate of interest. What is the net present value of $750,000 per year for the cost payment stream as a function of the time period employed? The same question applies to $500,000. A time period of five years will be used.

The formula for discounting cash flow *(DCF)* to obtain net present value *(NPV)* is:

$$\text{NPV} = \sum_{n=1}^{n=n^*} CF_j \left(\frac{1}{1+i}\right)^n$$

where:

NPV = net present value

Option $j = 1$: $CF_1 = \$750,000$ is the annual cash flow for purchasing the product, or

Option $j = 2$: $CF_2 = \$500,000$ is the annual cash flow for operating costs to make the product.

$CF_j(n)$ = when cash flow differs for each period (n)

i = interest rate, six percent per year

n = the year number ($n = 0, 1, 2,..n^*$)

n^* = the planning horizon, five years in this case

$$\sum_{n=1}^{n=n^*} \left(\frac{1}{1+i}\right)^n = \left(\frac{1}{1+0.06}\right)^1 + \left(\frac{1}{1+0.06}\right)^2 + \ldots + \left(\frac{1}{1+0.06}\right)^{n^*}$$

This equation for NPV assumes that the costs of making and purchasing are delayed until the end of the period.[45] Thus, each dollar paid out at the end of the first year would have earned six percent interest and consequently costs less than a full dollar. At year end, the account would show a balance of $1.06. The bill that has to be paid is $1.00. The actual cost would be $1.00 − 0.06 = $0.94. That is why deferring payment to year end decreases the cost.

The calculation for the end of the first year is:

$$\left(\frac{1}{1+0.06}\right)^1 = 0.943$$

Table 19.3 shows how present value changes as a function of time periods.[46] Tabled values, available for different interest rates and time periods, make present value analysis easy. Also, calculators and computers are programmed to provide these figures.

TABLE   19.3

PRESENT VALUES FOR MAKE OR BUY EXAMPLE
(FOR i = 0.06; CF₁ = 750,000 AND CF₂ = 500,000)

n	*x*	*y*	Buying 750,000	750,000	Making 500,000	500,000
1	0.943	0.943	707,250	707,250	471,500	471,500
2	0.890	1.833	667,500	1,374,750	445,000	916,500
3	0.840	2.673	630,000	2,004,750	420,000	1,336,500
4	0.792	3.465	594,000	2,598,750	396,000	1,732,500
5	0.747	4.212	560,250	3,159,000	373,500	2,106,000

where:

$$x = \left(\frac{1}{1+i} \right)^n, \text{ and } y = \sum_n \left(\frac{1}{1+i} \right)^n, \text{ etc.}$$

The four year savings derived from making over buying—with discounting—is (2,598,750 − 1,732,500 = 866,250) and the five year savings is (3,159,000 − 2,106,000 = 1,053,000). Thus, with discounting almost five years of savings are required to pay for the investment in the equipment to make the product. Because of the risky nature of the decision, the discounting procedure adds a more conservative note than the nondiscounting method, but it still is fine-tuning an estimate that is rough in any case.

Given the decision to make the product, the paramount issue for P/OM is how to configure the system and how to run it. The approach to use is how to do the job better than the suppliers would have done it. If possible, gain an advantage from being a start-up with respect to newer technology and materials. An experimental and creative point of view is encouraged instead of the more typical "this is how it has always been done" attitude of specialists. Ultimately, the design must be determined and the conservative point of view that emphasizes reliability takes over.

Notes...

1. R. Levering and M. Moskowitz, *The 100 Best Companies to Work for in America*, Rev. Ed., A Plume Book, Penguin Group, NY, 1994.

2. "Running Out Of Time: Reversing America's Declining Competitiveness," a conference held at Arden House, Harriman, NY, November 19–22, 1987. For details, see M. K. Starr, ed., *Global Competitiveness: Getting the U.S. Back on Track*, 74th American Assembly, Norton, NY, 1988.

3. C. J. McNair, and K. H. J. Leibfried, *Benchmarking: A Tool for Continuous Improvement*, Harper Business, 1992, p. 20.

4. *USA Today*, March 1, 1995, p. 4B

5. The quote from Julius Maldutis at Salomon Brothers is reported in the *New York Times*, March 7, 1995, in an article by Adam Bryant, p. C5.

6. CI also is called Kaizen. See Masaaki Imai, *Kaizen: The Key to Japan's Competitive Success*, NY, Random House, 1986.

7. Michael Hammer, and James Champy, *Reengineering the Corporation*, NY, Harper Business, 1993.

8. James Champy, *Reengineering Management: The Mandate for New Leadership*, NY, Harper Business, 1995, p. 3.

9. Ibid, p. 112.

10. D. K. Carr, K. S. Dougherty, H. J. Johansson, R. A. King, and D. E. Moran, *Breakpoint: Business Process Redesign*, Coopers & Lybrand Series, Arlington, VA, 1992, p. v.

11. G. Stalk, P. Evans, and L. E. Shulman, "Competing on Capabilities: The New Rules of Corporate Strategy," *Harvard Business Review*, March-April, 1992, pp. 57–69.

12. Booz Allen & Hamilton, "Understanding the new environment," *Business Process Redesign: An Owner's Guide*, OPER 401, 13M, 1993, p. 4.

13. Ibid, p. 14.

14. Henry J. Johansson, Patrick McHugh, A. John Pendlebury, William A. Wheeler III, *Business Process Reengineering: Breakpoint Strategies for Market Dominance*, NY, Wiley, 1993, p. 209.

15. Information about Oticon has been derived directly from conversations with employees of the firm, and an article by Polly LaBarre, "The Dis-Organization of Oticon," *Industry Week*, July 18, 1994, pp. 23–28. Oticon, Inc., the U.S. daughter company of the Danish parent, manufactures and repairs in Somerset, New Jersey. It is not paperless but shares many of the innovations developed in Denmark. Among its creative products and programs are its services for children which include the Oticon 4 Kids Club and brightly colored hearing instrument decorations.

16. The information on W. L. Gore & Associates was derived by speaking directly with individuals who work for the firm and from the write-up of this company appearing in *The 100 Best Companies to Work for in America* by Robert Levering and Milton Moskowitz, Rev. Ed., A. Plume Book, Penguin Group, NY, 1994.

17. Gene Bylinsky, "Manufacturing for Reuse," *Fortune Magazine*, February 6, 1995, pp. 102–112.

18. This information was obtained in a conversation with John J. Resslar, Champion, Designs for the Environment, Saturn Corporation.

19. Bylinsky, "Manufacturing for Reuse," p. 110.

20. "Recycling concrete pavements on the run," *Concrete Pavement Progress*, Vol. 34, No. 1, 1994, p. 4.

21. Jon Anderson, "Concrete's Road to Success," *Chicago Tribune*, October 16, 1991.

22. Department of Health and Human Services, Food and Drug Administration, *Proposal to Establish Procedures for the Safe Processing and Importing of Fish and Fishery Products*, 21 CFR Parts 123 and 1240 [Docket Nos. 90N-0199 and 93N-0195].

23. For Steps 1–7, many of the phrases are taken directly from the government document cited in Footnote 22. At the same time, liberal paraphrasing has been used.

24. J. A. DiMasi, "Cost of Innovation in the Pharmaceutical Industry," *Journal of Health Economics*, Amsterdam, Elsevier Publishing Co., Vol. 10, No. 2, July, 1991, pp. 107–142.

Laboratory studies:	3.5 years
1st Phase —Safety	1.0 year
2nd Phase—Testing Effectiveness	2.0 years
3rd Phase —Extensive Clinical Testing	3.0 years
FDA Review	2.5 years
Total Time	12.0 years

25. William Bridges, "The End of the Job," *Fortune Magazine*, September 19, 1994, pp. 62–74.

26. The word *robot* was derived from the Czech words robota, meaning work, and robotnick meaning worker, by Karel Capek for his play, RUR (Rossum's Universal Robots) written in 1920.

27. Quote is from Carl F. R. Weiman, Ph.D., director of Vision Systems Research, TRC (Transition Research Corporation) in Danbury, CT.

28. Described in *Computer Models of the Fundamental Mechanisms of Thought*, by Douglas Hofstadter and the Fluid Analogies Research Group, NY, Basic Books, 1995.

29. Mitchell Pacelle, "Vanishing Offices: To Trim Their Costs, Some Companies Cut Space for Employees," *Wall Street Journal*, June 4, 1993, p. A1.

30. Ibid, June 4, 1993, pp. A1, A6.

31. Montieth M. Illingworth, "Virtual Managers," *Information Week*, June 13, 1994, pp. 44.

32. Susan Diesenhouse, "Price Waterhouse Tries Check-In-and-Out Office," About Real Estate Column, *New York Times*, Dec. 7, 1994.

33. John Madeley, "The Tide Turns for Global Agribusiness," *The World Paper*, December, 1993, p. 10.

34. Brandt Cameron and Yan Wenbin, "Out of the Fields, Into the Can: Global Food Processors on the March," *The World Paper*, December 1993, p. 11.

35. Booz Allen & Hamilton, Inc., *Business Process Redesign*, April, 1994, pp. 1-A-1, 1-A-2.

36. Laurence Zuckerman, "Look Where Wall Street Stars are Heading Now; As Fancy Bonuses Dry Up, Foreign and Regional Banks Make Hard-to-Resist Offers," Business Day, *New York Times*, March 17, 1995, p. C1.

37. Merrill Lynch, Pierce, Fenner & Smith, Inc., *Global Investing: Assessing the Investment Climate, Global Diversification—The Global Machinery Theme*, January, 1995.

38. Business Bulletin, *Wall Street Journal*, March 9, 1995, p. 1.

39. Quote from International Standards Organization (ISO) working paper ISO/TC 207/SC 1/N 47 for ISO/CD 14001 Environmental Management Systems—Specification with Guidance for Use, Geneva, Switzerland.

40. David E. Sanger, "How Washington Inc. Makes a Sale: Commerce Department's 'Economic War Room' Mixes Public and Private to Win Foreign Business," Business Section, *New York Times*, February 19, 1995, p. 1.

41. Overseas Private Investment Corporation, *Program Handbook*, April, 1994.

42. An Innovative Approach to a Global Market, Overseas Private Investment Corporation Annual Report, Washington, D.C., 1994.

43. Peter F. Drucker, "Really Reinventing Government," *The Atlantic Monthly*, February, 1995, pp. 49–54.

44. Computer and Automated Systems Association of the Society of Manufacturing Engineers (CASA/SME), *The New Manufacturing Enterprise Wheel*, 3rd Ed., publication of CASA/SME, Dearborn, MI, 1993.

45. If payments start at the beginning of the period, the first payment would be one dollar. In other words, the first period payment will be equal to 1.00 in Table 19.3. All other payment amounts will be dropped to the $(n + 1)$ row.

46. For example, Table of Present Values of an Annuity, R. S. Burington, *Handbook of Mathematical Tables and Formulas*, 5th Ed. McGraw-Hill Company, NY, 1973. (Current reprint available.)

Saturn Corporation

THE SATURN PROJECT

In June 1982 General Motors (GM) assigned its Advanced Product and Design Team to answer the question, "Can GM build a world-class-quality small car in the U.S. that can compete successfully with the imports?" This effort grew into what GM called the Saturn Project. The team used a "clean-sheet" approach—unbound by traditional thinking and industry practices—to set up the small car project. This approach afforded the opportunity to design the manufacturing process in parallel with the product and to adopt lean production philosophies and techniques. The Saturn Project's challenge in designing the process, product, and manufacturing facility was to eliminate the $2,000 per unit cost advantage that Japanese auto makers held over U.S. manufacturers.

With the Saturn Project underway, another GM group gathered to study ways to improve management-labor relations. This "Group of 99," composed of GM managers and staff personnel from 17 GM divisions and United Auto Workers (UAW) members from 140 regions, revealed the willingness of these two factions to work together as partners to accomplish business objectives. In 1985 the Saturn Project merged with the Group of 99 to become the Saturn Corporation. With a $5 billion commitment, GM established Saturn as a wholly owned subsidiary. A "Gang of Six" was appointed, which included a president and vice presidents for planning, engineering, manufacturing, sales, and finance.

The newly assembled group immediately separated into research teams to study intricate aspects of what Saturn could be. During a two-month period, members of the new team visited 49 GM plants and 60 benchmark companies worldwide to gather ideas in preparation for setting up the new system. According to Jay Wetzel, vice president for engineering, "We had the benefit of being able to start everything new, where other (GM) platforms had to work off their heritage and their existing facilities. We had the opportunity to take some risks in high technology that other units of General Motors may not have pursued."[1] GM planned to export successful processes and operations developed at Saturn to improve efficiency and competitiveness at all its manufacturing facilities.

In July 1985 the selection of Spring Hill, Tennessee, as the site for the new manufacturing facility was finalized. Five years later the first Saturn rolled off the assembly line. Saturn manufactured its 1 millionth car in early summer 1995. The company sold 267,450 units of its 1994 models, up from 228,833 of the 1993 models. Saturn's share of the total U.S. new-car market is more than 3%. In defining Saturn's mission and philosophy, team members established Saturn's commitment to protect and preserve the environment (see Exhibits 1 & 2). Saturn designed and built the mile-long, 4.5-million-square-foot manufacturing facility in Spring Hill around this commitment to protect the environment and to maintain the aesthetic value of the land.

EXHIBIT 1

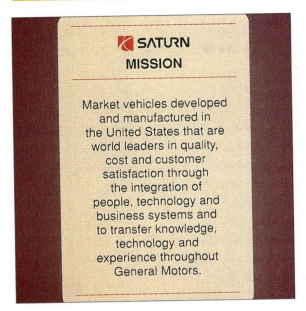

> # SATURN
> ## MISSION
>
> Market vehicles developed
> and manufactured in
> the United States that are
> world leaders in quality,
> cost and customer
> satisfaction through
> the integration of
> people, technology and
> business systems and
> to transfer knowledge,
> technology and
> experience throughout
> General Motors.

Saturn's Mission includes marketing quality vehicles and transferring technical knowledge gained throughout General Motors.
Source: Saturn Corporation

EXHIBIT 2

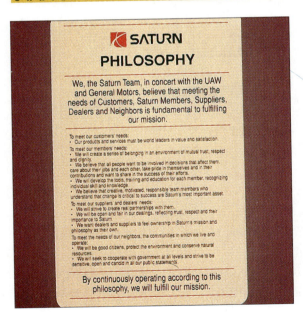

> # SATURN
> ## PHILOSOPHY
>
> We, the Saturn Team, in concert with the UAW and General Motors, believe that meeting the needs of Customers, Saturn Members, Suppliers, Dealers and Neighbors is fundamental to fulfilling our mission.
>
> To meet our customers' needs:
> • Our products and services must be world leaders in value and satisfaction.
>
> To meet our members' needs:
> • We will create a sense of belonging in an environment of mutual trust, respect and dignity.
> • We believe that all people want to be involved in decisions that affect them, care about their jobs and each other, take pride in themselves and in their contributions and want to share in the success of their efforts.
> • We will develop the tools, training and education for each member, recognizing individual skill and knowledge.
> • We believe that creative, motivated, responsible team members who understand that change is critical to success are Saturn's most important asset.
>
> To meet our suppliers' and dealers' needs:
> • We will strive to create real partnerships with them.
> • We will be open and fair in our dealings, reflecting trust, respect and their importance to Saturn.
> • We want dealers and suppliers to feel ownership in Saturn's mission and philosophy as their own.
>
> To meet the needs of our neighbors, the communities in which we live and operate:
> • We will be good citizens, protect the environment and conserve natural resources.
> • We will seek to cooperate with government at all levels and strive to be sensitive, open and candid in all our public statements.
>
> By continuously operating according to this philosophy, we will fulfill our mission.

The Saturn Philosophy is divided into three sections: meeting consumers' needs, meeting suppliers' and dealers' needs, and meeting the needs of the surrounding community.
Source: Saturn Corporation

The Saturn System

The Spring Hill plant is a "fully integrated" auto factory. This means all major parts for Saturn's coupes, sedans, and wagons are manufactured at one facility. The integrated nature of the plant facilitates interaction between business units and perpetuates quality-mindedness and information exchange throughout the system.

Saturn manufactures approximately 1150 cars per day. In a 20-hour day of 2, 10-hour shifts, that's roughly one car per minute. In 1995, the average price of a Saturn was $14,950; Saturn team members know that the company loses about $14,950 every minute the production line is down. As of spring 1995, approximately 8,000 team members work at the Spring Hill facility.

The three major "business units" at Saturn are Powertrain (engine and transmission), Body Systems, and Vehicle Systems (general assembly and interior systems). Components manufactured in Powertrain, Body Systems and Interior Systems are conveyed overhead into the Vehicle Systems assembly area. Exhibit 3 illustrates the basic layout of the Saturn manufacturing and assembly complex. Following is a brief summary of the production activities within the business units:

E X H I B I T   3

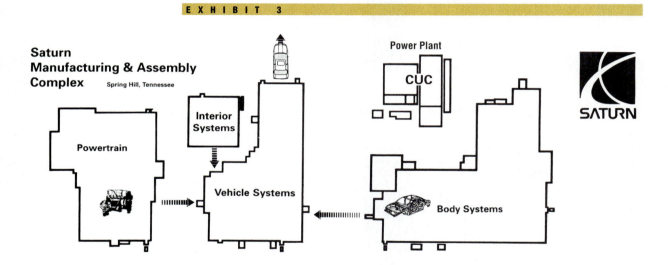

Powertrain

Die cast operations: Molten metal casting of parts including transmission cases, transmission parts, and engine covers.

Lost foam casting: Casting technology using polystyrene beads as patterns for molten metal casting. Parts produced here include engine blocks, engine heads, crankshafts, and differential housing.

Transmission assembly: Manual and automatic transmissions built on the same line. Flexibility to build up to 75% of either transmission to meet market demand.

Engine machining line: Flexible machining allows for one machining line instead of traditional two.

Final engine assembly: Assembles both high-performance and standard engines on the same line.

Body Systems

Steel body (or spaceframe) components: Rolled steel is cut to size on large transfer presses. These presses allow for a finished part to be manufactured using one press instead of the traditional five or six. Quick die changes are an integral part of Saturn's stamping plans. Saturn team members have achieved die changes of less than 10 minutes.

Body panels: Utilizing three different types of polymer pellets, composite panels are prepared in a 24 hour-a-day injection molding operation.

Body fabrication: Welding of the various components of the spaceframe including the body side, motor compartment, rear compartment and floor pan. The central point in body fabrication is the body framing and respot welding line.

Paint: Various components for Saturn vehicles are run through a phosphate and ELPO system which applies the rust-resistant undercoating, a robotic and manual sealer process, and a base coat/clear coat paint system. A completely automated panel storage and retrieval system has the capacity to store up to 900 complete sets of exterior panels.

Vehicle Systems

Skillet application: A unique ergonomic means of moving spaceframes and associates along the assembly line that emulates a stationary work station. Associates step on the moving wooden skillet platform, move with the car to perform work, then step off when they're finished.

Cockpit operation: All components are installed and tested for electrical functionality prior to installation in vehicle. Cockpit installation is traditionally a cumbersome procedure involving technicians getting down on their backs to do the necessary wiring on the instrument panel. At Saturn the cockpit is built outside the car so people don't have to assume the awkward positions

Trim installation: Employing the skillet system, all interior and exterior trim is installed.

Towveyor operation: Conveyor that tows the completed Powertrain, then installs it in spaceframe.

Exterior (body) panels scissor lift: Scissor lifts used to elevate car for ease of panel installation by team members.

Interior Systems: Injection molding and trim fabrication for interior.

Quality & Teamwork

Quality is paramount to success in the small car market. Meeting and exceeding customer requirements and expectations on a consistent basis is one of Saturn's key strategies for success. Quality at Saturn starts at the shop-floor level with the teams (Saturn refers to teams on the shop floor as "work units"). Teams are responsible for day-to-day car building and product-quality monitoring.

All Saturn employees are called "team members." Everyone at Saturn is associated with a team. All shop-floor UAW members are on a team. Each business unit (Power-train, Vehicle Systems, and Body Systems) has about 217 work units (or teams) for a total of 651. Business units are divided into modules which are subdivided into teams (or work units). (As an example, the "Door Finesse" team is in the "Front-end" module of the Vehicle Systems business unit.) Each team has from 8 to 15 members.

Within each team one person is responsible for gathering and analyzing information on quality issues. This person shares all quality information with team members. Module quality advisors look at outgoing (or between team) quality. The advisors pro-vide a communication bridge between the teams so each team knows what its internal customers upstream and downstream in the process require to make a quality, world-class vehicle. A series of Quality Councils, composed of team members and manage-ment meet on a periodic basis to set quality goals and provide general direction in terms of quality elements.

When problems arise, the team identifies the cause. The team may solicit help from engineering and operations (whose work areas also are located in the factory) in decid-ing what actions to take to solve the problem. A Total Quality Action Team (TQAT) may be formed to work on a specific problem. Saturn team members are not told by man-agement how to perform work or solve problems. They are made aware of what needs to be done, then the teams decide how to proceed. This includes: who performs the jobs; and whether a job is best done by a robot or a person. All decisions are made by consensus with the goal of promoting quality.

Quality is considered from the very beginning of a project in ascertaining what Saturn's target customers are looking for in a vehicle. Other quality inputs are obtained through warranty information and outside surveys done by organizations such as J. D. Power and Associates, which generates customer satisfaction comparisons. (In 1995, Saturn ranked #1 in J. D. Power's Customer Satisfaction Survey. In 1994, it was the only domestic nameplate within the top five.) All customer-related quality information gathered is analyzed and formatted to be understood by the teams. This helps team members understand what piece of the information they can affect to improve the prod-uct from the customer's perspective. In other words, customer feedback in this format is the means through which product quality is improved at Saturn.

Saturn's Environmental Initiatives

Saturn upheld its commitment to protect and preserve the environment in designing and constructing the Spring Hill production and assembly facility. The commitment continues in an effort Saturn calls "Designs for the Environment" that involves manag-ing environmental issues throughout the product life cycle, from design through post-consumer disposal of Saturn automobiles and parts. Bill Miller, Manager, Environmental Affairs, Manufacturing/Central Services said it has been Saturn's experience that "with very few exceptions, things that are environmentally correct are also things that make sense from an economic standpoint. ...If you think about it, pollution is nothing more than inefficiency in a process. The more efficient you make the process, the more environmentally correct it's going to be, and the more money you are going to make."

Impact of Facility on Physical Environment and Community

Saturn's approach in building in Spring Hill focused on how the facility would impact the residents' quality of life. Spring Hill residents were concerned that the plant would change the rural nature of the landscape. Saturn succeeded in maintaining the rural countryside. About half of the 2450 acres it owns are devoted to a working farm. To further preserve the landscape, Saturn used the same fencing materials as local farmers and painted the factory colors that blend in with the sky.

The company moved 6 million tons of dirt and rocks to build the factories. Rock from an on-site quarry was used to make the aggregate for the concrete used in construction. This eliminated having to truck hundreds of thousands of loads of materials to the site. The excavation site for the plant was 20 feet below ground level, and the dirt removed was used to sculpt hills to hide the plant from the road. The Saturn plant is barely detectable from the Saturn Parkway which links the facility to I-65 in Tennessee.

Finally, Saturn moved the trees from the excavation site rather than using the usual method of bulldozing and burning them. The company determined that it was not only more environmentally sensitive to save the trees, but more cost-effective as well. The uprooted trees were put in a nursery until they were used in landscaping around the office complex a few years later. On-site ponds were constructed to control storm water, provide water for fire protection, and provide areas for wildlife.

On-site ponds were constructed to control storm water, provide water for fire protection, and provide areas for wildlife.
Source: Saturn Corporation.

Impact of Manufacturing Process on the Environment

Saturn took various initiatives to protect the environment from pollution caused by the automobile manufacturing process. Saturn uses environmentally sound processes in painting and engine casting. Further environmental initiatives include designating loading docks for hazardous wastes, avoidance of underground tanks, and conducting stream surveys to measure water quality. Saturn reprocessed or recycled more than 40,000 tons of waste in 1994.

The painting process: One of the major environmental hazards associated with auto manufacture is painting the cars. Saturn was the first car manufacturer to use water-based rather than solvent-based paints. Water-based paints cause only $1/3$ to $1/5$ of the emissions of solvent-based paints during the painting process. Painting is done in a clean-room environment in paint spray booths where robots spray paint the car. Paint that misses the car flows into a water river under the paint spray booth. The water is taken into a large central room where the paint is separated from the water. (This is a challenge, because water-based paint is water soluble.) Chemicals that electrically charge the paint differently from the water are added to the mix so the paint will float. The paint is then skimmed off and run through a vacuum filter to get the remaining water out. The resulting non-toxic sludge is shipped to a landfill.

Airborne paint contaminates are captured in a carbon adsorption system. This system was developed to comply with the Clean Air Act of 1990 and to keep Saturn in compliance with future air standards (such as ISO 14000) which Miller predicts will become increasingly stringent. The $21 million carbon adsorption system is a series of stainless steel ducts and carbon beds. The system was installed by helicopter on the paint shop roof. The carbon adsorption system works like a carbon filter in an aquarium system by trapping pollution in the carbon. Then pollutants (volatile organic compounds in this case) are removed from five 85-foot long carbon beds with hot air. The air stream is directed through a thermal oxidizer (or incinerator) and then burned. This process destroys more than 99% of the remaining pollutants.

The lost foam process: Saturn uses a "lost foam process" for casting five large metal components including the engine block, cylinder heads, crankshaft, and differential transmission cases. The lost foam process is similar to the "lost wax" process used in ancient Egypt to make jewelry. With the lost-foam process, first a polystyrene mold is made of the part. This mold is placed into a container, then covered with sand. Molten aluminum (at 1400 degrees F) is poured into the sand. This melts the polystyrene which then dissipates into the sand. This process results in a significant reduction in the amount of contaminated waste sand generated than the traditional die-cast process. Waste sand from the lost foam process can be recycled up to 20 times. The process also requires less drilling than the traditional process; thus it saves approximately $15 million in machining tools. In addition, the process affords castings of greater precision and more flexibility than the traditional process.

Saturn uses a "lost foam process" for casting five large metal components including the engine block, cylinder heads, crankshaft, and differential transmission cases. Source: Saturn Corporation

Hazardous waste loading docks: In studying other GM manufacturing plants, Saturn found that loading docks are one of the major sites of chemical spills. Loading docks slope down toward the base of the building and typically, at the lowest point, have drains that go to the storm water system. Because of Saturn's extensive use of just-in-time delivery, the possibilities for spillage increased. To prevent contamination in the event of a spill, each business unit building at Saturn has a loading dock designated for hazardous materials. The drains at the low points of these docks lead to blind sumps, so if spills occur, the materials are contained.

Above-ground tanks: Tanks containing fluids used at an assembly plant, such as gasoline and windshield washer fluid, typically are contained in underground tanks. Saturn chose to locate all of its tanks above-ground. The tanks are surrounded by containment materials to prevent leakage.

Stream surveys: For assurance that the water supply is not being contaminated, Saturn tests the water in local streams about every other year. The company also participates regularly with the U.S. Geological Survey of water conditions. According to Miller, the water quality of the streams is better now than before the Saturn plant began operations.

Recycling of scrap material: All scrap materials resulting from the manufacturing process are recycled, including metal, plastic, casting sand, wood, corrugated paper, and office paper.

Impact of Product on the Environment

Recyclability of car parts: Recyclability of the plastic parts in a Saturn was considered in the design stage of the product life cycle—in anticipation of the post-consumer disposal stage. The exterior of a Saturn contains three different types of polymers, each of which is recyclable. Fascias (or bumpers) are made of a dent-resistant polymer that won't mark up in low-impact collisions. Fender and quarter panels are made of a nylon alloy which provides good bump resistance. The doors are made of a polycarbonate that can bounce back upon impact.

Saturn has instituted a recycling program wherein retailers return damaged plastic parts to Spring Hill. Retailers load damaged plastic car parts on Saturn trucks that have just delivered service parts, for the return trip to the factory. In a cross-division cooperation with General Motors Research and Packard Electric Division, Saturn has developed systems for reprocessing all painted thermoplastic body panel and fascia materials on the vehicle. The doors, rocker covers and decklids are reprocessed into Saturn rocker supports. Fascias are reprocessed into Saturn wheel liners. After being reprocessed into plastic pellets, fenders and quarter panels are used in bearing retainers at GM Saginaw Steering Division. (Saturn is working on a process to strip paint from parts using CO_2 pellets. This will allow the company to recycle plastic parts back to their original use.)

Sustainable Development

Saturn has demonstrated that achievement of a quality product and a commitment to protecting and preserving the environment are not mutually exclusive. The company chose to be proactive from an environmental standpoint with regard to customers, the legislative arena, and in order to be competitive globally. Saturn's environmental commitment can be summed up by the term *sustainable development,* which is popular in environmental discussions today. Bill Miller says that although the term is still developing, it essentially means "doing things right today to make sure that future generations have it at least as good as it is right now. A lot of what we've done, and are doing, here at Saturn fits right into that. ..."[2]

Sustainable development is doing things right today so that the world our children will inherit is at least as good as it is right now.
Source: Roy F. Weston, Inc.

Footnotes

1. Charles J. Murray, "Engineer on a Mission," *Design News,* Engineering Quality Award issue.

2. Gary S. Vasilash, "The Environment at Saturn," *Production,* April, 1994.

Case Questions

1. Discuss the "systems thinking" that went in to planning and building the Saturn facility and products.

2. Do you think simultaneous engineering was employed in designing the plant and products? Why or why not?

3. What was the importance and symbolism of the Group of 99 and Gang of 6 to the "clean sheet approach?" Discuss.

4. Explain how the Saturn Project in effect "reengineered" the automobile manufacture process?

5. If each business unit has 217 work units, and each work unit has 8 to 15 members, how many Saturn employees are "associated with" teams rather than "members" of teams? What kinds of functions do you think these Saturn employees perform?

6. How is quality ensured at Saturn? Discuss the role of Quality Councils and the TQAT.

7. What are the advantages of using the "lost foam" process for casting the large metal components over traditional casting operations?

8. Why does it make more sense economically for manufacturers to use processes that minimize pollution?

9. How does Saturn fit in the community? Discuss the company's concern for the environment.

10. Discuss the ways in which Saturn has demonstrated that achieving a quality product and environmental sensitivity are not mutually exclusive.

References

Gary S. Vasilash, "The Environment at Saturn," *Production,* April, 1994.

Raymond Sarafin, "The Saturn Story, *Advertising Age,* November 16, 1992.

Richard G. LeFauve and Arnoldo C. Hax, "Managerial and technological innovations at Saturn Corporation," *MIT Management,* Spring, 1992.

James P. Womack, "The Lean Difference: An International Productivity Comparison and the Implications for U.S. Industry," *Looking Ahead* (published by The National Planning Association, Washington, D.C.), April 1992, Vol. XIII, No. 4.

Michael Bennett, "The Saturn Corporation: New Management-Union Partnership at the Factory of the Future," *Looking Ahead* (published by The National Planning Association, Washington, D.C.), April 1992, Vol. XIII, No. 4.

Jack O'Toole and Jim Lewandowski, "Forming the Future: The Marriage of People and Technology at Saturn." Presentation at Stanford University Industrial Engineering and Engineering Management, March 29, 1990.

Numerous Saturn Corporation documents including "Important Dates in Saturn History;" "Saturn Fact Sheet—Sales and Marketing Highlights"; "Quality Information for Students;" and News Release on "Saturn's Environmental Initiatives," June 24, 1994.

James P. Womack, Daniel T. Jones and Daniel Roos, "The Machine That Changed the World: The Story of Lean Production," Harper Perennial, NY, 1991.

APPENDIX A

The Simplex Algorithm for Solving Linear Programming Problems

At one time it was considered essential to understand how to use the simplex algorithm to solve linear programming (LP) problems. Then, powerful software, such as LINDO[1] and Lingo were written. The software is versatile, fast and easy to use. Nevertheless, when time permits, it may be useful for P/OM students to know how the simplex method operates. It is the basis upon which computer programs have been written to solve linear programming problems.

First, an example is given of a problem that lends itself to an LP solution and a figure is presented to explain the geometry.

Second, the simplex method is explained as a set of logical operations in Steps I through V.

Third, the sequence of the steps is shown in a figure.

The Example: Greeting card designers have come up with two alternative card designs which use a special material. The production manager wishes to evaluate making one or both of these designs called A and B. The table below presents relevant data.

Number of cards to make	x_A	x_B
Time to print one card	2.4 min.	2.4 min.
Time to cut and fold	4.8 min.	1.6 min.
Material required	80 sq. in.	240 sq. in.
Estimated profit per card	$0.70	$0.80

Production works a 40-hour week which is 2400 minutes. There are 833 square feet of the material on-hand and no further orders for it are anticipated. There is no cutting waste and the job must be completed within one week. LP can obtain the optimal product mix.

There are three constraints as follows:

1. $(2.4)x_A + (2.4)x_B \leq (2400)$ print constraint
2. $(4.8)x_A + (1.6)x_B \leq (2400)$ cut and fold constraint
3. $(80)x_A + (240)x_B \leq (120,000)$ material constraints where $833 \times 144 = 120,000$ sq. in. of material.

The three lines are plotted in the figure below. Only the cross-hatched area and the lines that form it can contain the optimal solution. All else violates the constraints.

APPENDIX A FIGURE 1

THE OPTIMAL SOLUTION MUST OCCUR AT ONE OF THE VERTICES OF THE LINES THAT ENCLOSE THE FEASIBLE SOLUTIONS AREA

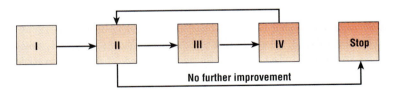

The profit line: $(0.70)x_A + (0.80)x_B = K$ also is drawn in the figure. The simplex method of LP locates the line with the largest value of K which has at least one point in common with the surrounding boundary of the feasible solution space. This must occur either at a vertex, or along the line that connects two vertices. In the example, this maximum point occurs at the intersection of the cut and fold constraint with the material constraint. The coordinates of this point are $x_A = x_B = 375$ cards. The maximum profit with this product mix is $K = \$562.50$. The line in the figure is $K = 600$ which does not quite touch the feasible space boundary.

The logic of the simplex method follows:

I.   Begin by selecting any vertex of the area or volume that is formed when two or more axes are involved. (The Cartesian coordinates of the two-dimensional geometric approach must be extended when more than two types of products or ingredients are to be mixed.)

II. Test to determine whether an improvement in profit can be made by moving to another vertex.

III. If improvement is possible, attempt to find the best possible change to make.

IV. Make the indicated change, i.e., choose the new, improved vertex solution. Operations II through IV are repeated until condition V occurs.

V.   Stop, because the test (in II) reveals that no further improvement is possible.

The sequence of steps is shown in the figure below.

A P P E N D I X   A   F I G U R E   2

THE OPTIMAL SOLUTION MUST OCCUR AT ONE OF THE VERTICES OF THE LINES THAT ENCLOSE THE FEASIBLE SOLUTIONS AREA

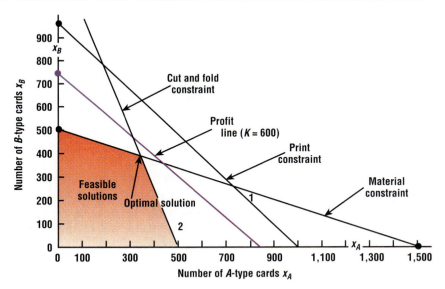

Notes...

1. Linus Schrage has created LINDO, Super LINDO, Hyper LINDO, Extended LINDO and Industrial LINDO, all published by The Scientific Press Series, boyd & fraser publishing company. LINDO Student's Edition, Release 5.3 handles up to 100 constraints and 200 variables, 1993.

Table of the Normal Distribution
AREAS UNDER THE NORMAL CURVE FROM K_α TO ∞

$$\alpha = \int_{K_\alpha}^{\infty} \frac{1}{\sqrt{2\pi}} e^{-x^2/2}dx$$

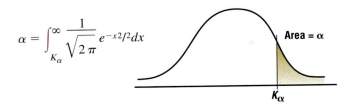

Area = α

K_α

K_α	0.00	0.01	0.02	0.03	0.04	0.05	0.06	0.07	0.08	0.09
0.0	0.5000	0.4960	0.4920	0.4880	0.4840	0.4801	0.4761	0.4721	0.4681	0.4641
0.1	0.4602	0.4562	0.4522	0.4483	0.4443	0.4404	0.4364	0.4325	0.4286	0.4247
0.2	0.4207	0.4168	0.4129	0.4090	0.4052	0.4013	0.3974	0.3936	0.3897	0.3859
0.3	0.3821	0.3783	0.3745	0.3707	0.3669	0.3632	0.3594	0.3557	0.3520	0.3483
0.4	0.3446	0.3409	0.3372	0.3336	0.3300	0.3264	0.3228	0.3192	0.3156	0.3121
0.5	0.3085	0.3050	0.3015	0.2981	0.2946	0.2912	0.2877	0.2843	0.2810	0.2776
0.6	0.2743	0.2709	0.2676	0.2643	0.2611	0.2578	0.2546	0.2514	0.2483	0.2451
0.7	0.2420	0.2389	0.2358	0.2327	0.2296	0.2266	0.2236	0.2206	0.2177	0.2148
0.8	0.2119	0.2090	0.2061	0.2033	0.2005	0.1977	0.1949	0.1922	0.1894	0.1867
0.9	0.1841	0.1814	0.1788	0.1762	0.1736	0.1711	0.1685	0.1660	0.1635	0.1611
1.0	0.1587	0.1562	0.1539	0.1515	0.1492	0.1469	0.1146	0.1423	0.1401	0.1379
1.1	0.1357	0.1335	0.1314	0.1292	0.1271	0.1251	0.1230	0.1210	0.1190	0.1170
1.2	0.1151	0.1131	0.1112	0.1093	0.1075	0.1056	0.1038	0.1020	0.1003	0.0985
1.3	0.0968	0.0951	0.0934	0.0918	0.0901	0.0885	0.0869	0.0853	0.0838	0.0823
1.4	0.0808	0.0793	0.0778	0.0764	0.0749	0.0735	0.0721	0.0708	0.0694	0.0681
1.5	0.0668	0.0655	0.0643	0.0630	0.0618	0.0606	0.0594	0.0582	0.0571	0.0559
1.6	0.0548	0.0537	0.0526	0.0516	0.0505	0.0495	0.0485	0.0475	0.0465	0.0455
1.7	0.0446	0.0436	0.0427	0.0418	0.0409	0.0401	0.0392	0.0384	0.0375	0.0367
1.8	0.0359	0.0351	0.0344	0.0336	0.0329	0.0322	0.0314	0.0307	0.0301	0.0294
1.9	0.0287	0.0281	0.0274	0.0268	0.0262	0.0256	0.0250	0.0244	0.0239	0.0233
2.0	0.0228	0.0222	0.0217	0.0212	0.0207	0.0202	0.0197	0.0192	0.0188	0.0183
2.1	0.0179	0.0174	0.0170	0.0166	0.0162	0.0158	0.0154	0.0150	0.0146	0.0143
2.2	0.0139	0.0136	0.0132	0.0129	0.0125	0.0122	0.0119	0.0116	0.0113	0.0110
2.3	0.0107	0.0104	0.0102	0.00990	0.00964	0.00939	0.00914	0.00889	0.00866	0.00842
2.4	0.00820	0.00798	0.00776	0.00755	0.00734	0.00714	0.00695	0.00676	0.00657	0.00639
2.5	0.00621	0.00604	0.00587	0.00570	0.00554	0.00539	0.00523	0.00508	0.00494	0.00480
2.6	0.00466	0.00453	0.00440	0.00427	0.00415	0.00402	0.00391	0.00379	0.00368	0.00357
2.7	0.00347	0.00336	0.00326	0.00317	0.00307	0.00298	0.00289	0.00280	0.00272	0.00264
2.8	0.00256	0.00248	0.00240	0.00233	0.00226	0.00219	0.00212	0.00205	0.00199	0.00193
2.9	0.00187	0.00181	0.00175	0.00169	0.00164	0.00159	0.00154	0.00149	0.00144	0.00139

K_α	0.0	0.1	0.2	0.3	0.4	0.5	0.6	0.7	0.8	0.9
3	0.00135	0.0^3968	0.0^3687	0.0^3483	0.0^3337	0.0^3233	0.0^3159	0.0^3108	0.0^4723	0.0^4481
4	0.0^4317	0.0^4207	0.0^4133	0.0^5854	0.0^5541	0.0^5340	0.0^5211	0.0^5130	0.0^6793	0.0^6479
5	0.0^6287	0.0^6170	0.0^7996	0.0^7579	0.0^7333	0.0^7190	0.0^7107	0.0^8599	0.0^8332	0.0^8182
6	0.0^9987	0.0^9530	0.0^9282	0.0_9149	$0.0^{10}777$	$0.0^{10}402$	$0.0^{10}206$	$0.0^{10}104$	$0.0^{11}523$	$0.0^{11}260$

Poisson Probability Distribution

$$P(r \mid \lambda) = \frac{\lambda^r}{r!} e^{-\lambda}$$

$$P(r = 1 \mid \lambda = 0.7) = 0.3476$$

r	0.10	0.20	0.30	0.40	λ 0.50	0.60	0.70	0.80	0.90	1.00
0	.9048	.8187	.7408	.6703	.6065	.5488	.4966	.4493	.4066	.3679
1	.0905	.1637	.2222	.2681	.3033	.3293	.3476	.3595	.3659	.3679
2	.0045	.0164	.0333	.0536	.0758	.0988	.1217	.1438	.1647	.1839
3	.0002	.0011	.0033	.0072	.0126	.0198	.0284	.0383	.0494	.0613
4	.0000	.0001	.0003	.0007	.0016	.0030	.0050	.0077	.0111	.0153
5	.0000	.0000	.0000	.0001	.0002	.0004	.0007	.0012	.0020	.0031
6	.0000	.0000	.0000	.0000	.0000	.0000	.0001	.0002	.0003	.0005
7	.0000	.0000	.0000	.0000	.0000	.0000	.0000	.0000	.0000	.0001

r	1.10	1.20	1.30	1.40	λ 1.50	1.60	1.70	1.80	1.90	2.00
0	.3329	.3012	.2725	.2466	.2231	.2019	.1827	.1653	.1496	.1353
1	.3662	.3614	.3543	.3452	.3347	.3230	.3106	.2975	.2842	.2707
2	.2014	.2169	.2303	.2417	.2510	.2584	.2640	.2678	.2700	.2707
3	.0738	.0867	.0998	.1128	.1255	.1378	.1496	.1607	.1710	.1804
4	.0203	.0260	.0324	.0395	.0471	.0551	.0636	.0723	.0812	.0902
5	.0045	.0062	.0084	.0111	.0141	.0176	.0216	.0260	.0309	.0361
6	.0008	.0012	.0018	.0026	.0035	.0047	.0061	.0078	.0098	.0120
7	.0001	.0002	.0003	.0005	.0008	.0011	.0015	.0020	.0027	.0034
8	.0000	.0000	.0001	.0001	.0001	.0002	.0003	.0005	.0006	.0009
9	.0000	.0000	.0000	.0000	.0000	.0000	.0001	.0001	.0001	.0002

r	2.10	2.20	2.30	2.40	λ 2.50	2.60	2.70	2.80	2.90	3.00
0	.1225	.1108	.1003	.0907	.0821	.0743	.0672	.0608	.0550	.0498
1	.2572	.2438	.2306	.2177	.2052	.1931	.1815	.1703	.1596	.1494
2	.2700	.2681	.2652	.2613	.2565	.2510	.2450	.2384	.2314	.2240
3	.1890	.1966	.2033	.2090	.2138	.2176	.2205	.2225	.2237	.2240
4	.0992	.1082	.1169	.1254	.1336	.1414	.1488	.1557	.1622	.1680
5	.0417	.0476	.0538	.0602	.0668	.0735	.0804	.0872	.0940	.1008
6	.0146	.0174	.0206	.0241	.0278	.0319	.0362	.0407	.0455	.0504
7	.0044	.0055	.0068	.0083	.0099	.0118	.0139	.0163	.0188	.0216
8	.0011	.0015	.0019	.0025	.0031	.0038	.0047	.0057	.0068	.0081
9	.0003	.0004	.0005	.0007	.0009	.0011	.0014	.0018	.0022	.0027
10	.0001	.0001	.0001	.0002	.0002	.0003	.0004	.0005	.0006	.0008
11	.0000	.0000	.0000	.0000	.0000	.0001	.0001	.0001	.0002	.0002
12	.0000	.0000	.0000	.0000	.0000	.0000	.0000	.0000	.0000	.0001

r	3.10	3.20	3.30	3.40	λ 3.50	3.60	3.70	3.80	3.90	4.00
0	.0450	.0408	.0369	.0334	.0302	.0273	.0247	.0224	.0202	.0183
1	.1397	.1304	.1217	.1135	.1057	.0984	.0915	.0850	.0789	.0733
2	.2165	.2087	.2008	.1929	.1850	.1771	.1692	.1615	.1539	.1465
3	.2237	.2226	.2209	.2186	.2158	.2125	.2087	.2046	.2001	.1954
4	.1733	.1781	.1823	.1858	.1888	.1912	.1931	.1944	.1951	.1954

POISSON PROBABILITY DISTRIBUTION

r	3.10	3.20	3.30	3.40	λ 3.50	3.60	3.70	3.80	3.90	4.00
5	.1075	.1140	.1203	.1264	.1322	.1377	.1429	.1477	.1522	.1563
6	.0555	.0608	.0662	.0716	.0771	.0826	.0881	.0936	.0989	.1042
7	.0246	.0278	.0312	.0348	.0385	.0425	.0466	.0508	.0551	.0595
8	.0095	.0111	.0129	.0148	.0169	.0191	.0215	.0241	.0269	.0298
9	.0033	.0040	.0047	.0056	.0066	.0076	.0089	.0102	.0116	.0132
10	.0010	.0013	.0016	.0019	.0023	.0028	.0033	.0039	.0045	.0053
11	.0003	.0004	.0005	.0006	.0007	.0009	.0011	.0013	.0016	.0019
12	.0001	.0001	.0001	.0002	.0002	.0003	.0003	.0004	.0005	.0006
13	.0000	.0000	.0000	.0000	.0001	.0001	.0001	.0001	.0002	.0002
14	.0000	.0000	.0000	.0000	.0000	.0000	.0000	.0000	.0000	.0001

r	4.10	4.20	4.30	4.40	λ 4.50	4.60	4.70	4.80	4.90	5.00
0	.0166	.0150	.0136	.0123	.0111	.0101	.0091	.0082	.0074	.0067
1	.0679	.0630	.0583	.0540	.0500	.0462	.0427	.0395	.0365	.0337
2	.1393	.1323	.1254	.1188	.1125	.1063	.1005	.0948	.0894	.0842
3	.1904	.1852	.1798	.1743	.1687	.1631	.1574	.1517	.1460	.1404
4	.1951	.1944	.1933	.1917	.1898	.1875	.1849	.1820	.1789	.1755
5	.1600	.1633	.1662	.1687	.1708	.1725	.1738	.1747	.1753	.1755
6	.1093	.1143	.1191	.1237	.1281	.1323	.1362	.1398	.1432	.1462
7	.0640	.0686	.0732	.0778	.0824	.0869	.0914	.0959	.1002	.1044
8	.0328	.0360	.0393	.0428	.0463	.0500	.0537	.0575	.0614	.0653
9	.0150	.0168	.0188	.0209	.0232	.0255	.0281	.0307	.0334	.0363
10	.0061	.0071	.0081	.0092	.0104	.0118	.0132	.0147	.0164	.0181
11	.0023	.0027	.0032	.0037	.0043	.0049	.0056	.0064	.0073	.0082
12	.0008	.0009	.0011	.0013	.0016	.0019	.0022	.0026	.0030	.0034
13	.0002	.0003	.0004	.0005	.0006	.0007	.0008	.0009	.0011	.0013
14	.0001	.0001	.0001	.0001	.0002	.0002	.0003	.0003	.0004	.0005
15	.0000	.0000	.0000	.0000	.0001	.0001	.0001	.0001	.0001	.0002

r	5.10	5.20	5.30	5.40	λ 5.50	5.60	5.70	5.80	5.90	6.00
0	.0061	.0055	.0050	.0045	.0041	.0037	.0033	.0030	.0027	.0025
1	.0311	.0287	.0265	.0244	.0225	.0207	.0191	.0176	.0162	.0149
2	.0793	.0746	.0701	.0659	.0618	.0580	.0544	.0509	.0477	.0446
3	.1348	.1293	.1239	.1185	.1133	.1082	.1033	.0985	.0938	.0892
4	.1719	.1681	.1641	.1600	.1558	.1515	.1472	.1428	.1383	.1339
5	.1753	.1748	.1740	.1728	.1714	.1697	.1678	.1656	.1632	.1606
6	.1490	.1515	.1537	.1555	.1571	.1584	.1594	.1601	.1605	.1606
7	.1086	.1125	.1163	.1200	.1234	.1267	.1298	.1326	.1353	.1377
8	.0692	.0731	.0771	.0810	.0849	.0887	.0925	.0962	.0998	.1033
9	.0392	.0423	.0454	.0486	.0519	.0552	.0586	.0620	.0654	.0688
10	.0200	.0220	.0241	.0262	.0285	.0309	.0334	.0359	.0386	.0413
11	.0093	.0104	.0116	.0129	.0143	.0157	.0173	.0190	.0207	.0225
12	.0039	.0045	.0051	.0058	.0065	.0073	.0082	.0092	.0102	.0113
13	.0015	.0018	.0021	.0024	.0028	.0032	.0036	.0041	.0046	.0052
14	.0006	.0007	.0008	.0009	.0011	.0013	.0015	.0017	.0019	.0022
15	.0002	.0002	.0003	.0003	.0004	.0005	.0006	.0007	.0008	.0009
16	.0001	.0001	.0001	.0001	.0001	.0002	.0002	.0002	.0003	.0003
17	.0000	.0000	.0000	.0000	.0000	.0001	.0001	.0001	.0001	.0001

r	6.10	6.20	6.30	6.40	λ 6.50	6.60	6.70	6.80	6.90	7.00
0	.0022	.0020	.0018	.0017	.0015	.0014	.0012	.0011	.0010	.0009
1	.0137	.0126	.0116	.0106	.0098	.0090	.0082	.0076	.0070	.0223
2	.0417	.0390	.0364	.0340	.0318	.0296	.0276	.0258	.0240	.0223
3	.0848	.0806	.0765	.0726	.0688	.0652	.0617	.0584	.0552	.0521
4	.1294	.1249	.1205	.1161	.1118	.1076	.1034	.0992	.0952	.0912
5	.1579	.1549	.1519	.1487	.1454	.1420	.1385	.1349	.1314	.1277
6	.1605	.1601	.1595	.1586	.1575	.1562	.1546	.1529	.1511	.1490
7	.1399	.1418	.1435	.1450	.1462	.1472	.1480	.1486	.1489	.1490
8	.1066	.1099	.1130	.1160	.1188	.1215	.1240	.1263	.1284	.1304
9	.0723	.0757	.0791	.0825	.0858	.0891	.0923	.0954	.0985	.1014
10	.0441	.0469	.0498	.0528	.0558	.0588	.0618	.0649	.0679	.0710
11	.0244	.0265	.0285	.0307	.0330	.0353	.0377	.0401	.0426	.0452
12	.0124	.0137	.0150	.0164	.0179	.0194	.0210	.0227	.0245	.0263
13	.0058	.0065	.0073	.0081	.0089	.0099	.0108	.0119	.0130	.0142
14	.0025	.0029	.0033	.0037	.0041	.0046	.0052	.0058	.0064	.0071

POISSON PROBABILITY DISTRIBUTION

r	9.10	9.20	9.30	9.40	λ 9.50	9.60	9.70	9.80	9.90	10.00
10	.1198	.1210	.1219	.1228	.1235	.1241	.1245	.1249	.1250	.1251
11	.0991	.1012	.1031	.1049	.1067	.1083	.1098	.1112	.1125	.1137
12	.0752	.0776	.0799	.0822	.0844	.0866	.0888	.0908	.0928	.0948
13	.0526	.0549	.0572	.0594	.0617	.0640	.0662	.0685	.0707	.0729
14	.0342	.0361	.0380	.0399	.0419	.0439	.0459	.0479	.0500	.0521
15	.0208	.0221	.0235	.0250	.0265	.0281	.0297	.0313	.0330	.0347
16	.0118	.0127	.0137	.0147	.0157	.0168	.0180	.0192	.0204	.0217
17	.0063	.0069	.0075	.0081	.0088	.0095	.0103	.0111	.0119	.0128
18	.0032	.0035	.0039	.0042	.0046	.0051	.0055	.0060	.0065	.0071
19	.0015	.0017	.0019	.0021	.0023	.0026	.0028	.0031	.0034	.0037
20	.0007	.0008	.0009	.0010	.0011	.0012	.0014	.0015	.0017	.0019
21	.0003	.0003	.0004	.0004	.0005	.0006	.0006	.0007	.0008	.0009
22	.0001	.0001	.0002	.0002	.0002	.0002	.0003	.0003	.0004	.0004
23	.0000	.0001	.0001	.0001	.0001	.0001	.0001	.0001	.0002	.0002
24	.0000	.0000	.0000	.0000	.0000	.0000	.0000	.0001	.0001	.0001

r	11.	12.	13.	14.	λ 15.	16.	17.	18.	19.	20.
0	.0000	.0000	.0000	.0000	.0000	.0000	.0000	.0000	.0000	.0000
1	.0002	.0001	.0000	.0000	.0000	.0000	.0000	.0000	.0000	.0000
2	.0010	.0004	.0002	.0001	.0000	.0000	.0000	.0000	.0000	.0000
3	.0037	.0018	.0008	.0004	.0002	.0001	.0000	.0000	.0000	.0000
4	.0102	.0053	.0027	.0013	.0006	.0003	.0001	.0001	.0000	.0000
5	.0224	.0127	.0070	.0037	.0019	.0010	.0005	.0002	.0001	.0001
6	.0411	.0255	.0152	.0087	.0048	.0026	.0014	.0007	.0004	.0002
7	.0646	.0437	.0281	.0174	.0104	.0060	.0034	.0019	.0010	.0005
8	.0888	.0655	.0457	.0304	.0194	.0120	.0072	.0042	.0024	.0013
9	.1085	.0874	.0661	.0473	.0324	.0213	.0135	.0083	.0050	.0029
10	.1194	.1048	.0859	.0663	.0486	.0341	.0230	.0150	.0095	.0058
11	.1194	.1144	.1015	.0844	.0663	.0496	.0355	.0245	.0164	.0106
12	.1094	.1144	.1099	.0984	.0829	.0661	.0504	.0368	.0259	.0176
13	.0926	.1056	.1099	.1060	.0956	.0814	.0658	.0509	.0378	.0271
14	.0728	.0905	.1021	.1060	.1024	.0930	.0800	.0655	.0514	.0387
15	.0534	.0724	.0885	.0989	.1024	.0992	.0906	.0786	.0650	.0516
16	.0367	.0543	.0719	.0866	.0960	.0992	.0963	.0884	.0772	.0646
17	.0237	.0383	.0550	.0713	.0847	.0934	.0963	.0936	.0863	.0760
18	.0145	.0256	.0397	.0554	.0706	.0830	.0909	.0936	.0911	.0844
19	.0084	.0161	.0272	.0409	.0557	.0699	.0814	.0887	.0911	.0888
20	.0046	.0097	.0177	.0286	.0418	.0559	.0692	.0798	.0866	.0888
21	.0024	.0055	.0109	.0191	.0299	.0426	.0560	.0684	.0783	.0846
22	.0012	.0030	.0065	.0121	.0204	.0310	.0433	.0560	.0676	.0769
23	.0006	.0016	.0037	.0074	.0133	.0216	.0320	.0438	.0559	.0669
24	.0003	.0008	.0020	.0043	.0083	.0144	.0226	.0329	.0442	.0557
25	.0001	.0004	.0010	.0024	.0050	.0092	.0154	.0237	.0336	.0446
26	.0000	.0002	.0005	.0013	.0029	.0057	.0101	.0164	.0246	.0343
27	.0000	.0001	.0002	.0007	.0016	.0034	.0063	.0109	.0173	.0254
28	.0000	.0000	.0001	.0003	.0009	.0019	.0038	.0070	.0117	.0181
29	.0000	.0000	.0001	.0002	.0004	.0011	.0023	.0044	.0077	.0125
30	.0000	.0000	.0000	.0001	.0002	.0006	.0013	.0026	.0049	.0083
31	.0000	.0000	.0000	.0000	.0001	.0003	.0007	.0015	.0030	.0054
32	.0000	.0000	.0000	.0000	.0001	.0001	.0004	.0009	.0018	.0034
33	.0000	.0000	.0000	.0000	.0000	.0001	.0002	.0005	.0010	.0020
34	.0000	.0000	.0000	.0000	.0000	.0000	.0001	.0002	.0006	.0012
35	.0000	.0000	.0000	.0000	.0000	.0000	.0000	.0001	.0003	.0007
36	.0000	.0000	.0000	.0000	.0000	.0000	.0000	.0001	.0002	.0004
37	.0000	.0000	.0000	.0000	.0000	.0000	.0000	.0000	.0001	.0002
38	.0000	.0000	.0000	.0000	.0000	.0000	.0000	.0000	.0000	.0001
39	.0000	.0000	.0000	.0000	.0000	.0000	.0000	.0000	.0000	.0001

POISSON PROBABILITY DISTRIBUTION

r	6.10	6.20	6.30	6.40	λ 6.50	6.60	6.70	6.80	6.90	7.00
15	.0010	.0012	.0014	.0016	.0018	.0020	.0023	.0026	.0029	.0033
16	.0004	.0005	.0005	.0006	.0007	.0008	.0010	.0011	.0013	.0014
17	.0001	.0002	.0002	.0002	.0003	.0003	.0004	.0004	.0005	.0006
18	.0000	.0001	.0001	.0001	.0001	.0001	.0001	.0002	.0002	.0002
19	.0000	.0000	.0000	.0000	.0000	.0000	.0001	.0001	.0001	.0001

r	7.10	7.20	7.30	7.40	λ 7.50	7.60	7.70	7.80	7.90	8.00
0	.0008	.0007	.0007	.0006	.0006	.0005	.0005	.0004	.0004	.0003
1	.0059	.0054	.0049	.0045	.0041	.0038	.0035	.0032	.0029	.0027
2	.0208	.0194	.0180	.0167	.0156	.0145	.0134	.0125	.0116	.0107
3	.0492	.0464	.0438	.0413	.0389	.0366	.0345	.0324	.0305	.0286
4	.0874	.0836	.0799	.0764	.0729	.0696	.0663	.0632	.0602	.0573
5	.1241	.1204	.1167	.1130	.1094	.1057	.1021	.0986	.0951	.0916
6	.1468	.1445	.1420	.1394	.1367	.1339	.1311	.1282	.1252	.1221
7	.1489	.1486	.1481	.1474	.1465	.1454	.1442	.1428	.1413	.1396
8	.1321	.1337	.1351	.1363	.1373	.1381	.1388	.1392	.1395	.1396
9	.1042	.1070	.1096	.1121	.1144	.1167	.1187	.1207	.1224	.1241
10	.0740	.0770	.0800	.0829	.0858	.0887	.0914	.0941	.0967	.0993
11	.0478	.0504	.0531	.0558	.0585	.0613	.0640	.0667	.0695	.0722
12	.0283	.0303	.0323	.0344	.0366	.0388	.0411	.0434	.0457	.0481
13	.0154	.0168	.0181	.0196	.0211	.0227	.0243	.0260	.0278	.0296
14	.0078	.0086	.0095	.0104	.0113	.0123	.0134	.0145	.0157	.0169
15	.0037	.0041	.0046	.0051	.0057	.0062	.0069	.0075	.0083	.0090
16	.0016	.0019	.0021	.0024	.0026	.0030	.0033	.0037	.0041	.0045
17	.0007	.0008	.0009	.0010	.0012	.0013	.0015	.0017	.0019	.0021
18	.0003	.0003	.0004	.0004	.0005	.0006	.0006	.0007	.0008	.0009
19	.0001	.0001	.0001	.0002	.0002	.0002	.0003	.0003	.0003	.0004
20	.0000	.0000	.0001	.0001	.0001	.0001	.0001	.0001	.0001	.0002
21	.0000	.0000	.0000	.0000	.0000	.0000	.0000	.0000	.0001	.0001

r	8.10	8.20	8.30	8.40	λ 8.50	8.60	8.70	8.80	8.90	9.00
0	.0003	.0003	.0002	.0002	.0002	.0002	.0002	.0002	.0001	.0001
1	.0025	.0023	.0021	.0019	.0017	.0016	.0014	.0013	.0012	.0011
2	.0100	.0092	.0086	.0079	.0074	.0068	.0063	.0058	.0054	.0050
3	.0269	.0252	.0237	.0222	.0208	.0195	.0183	.0171	.0160	.0150
4	.0544	.0517	.0491	.0466	.0443	.0420	.0398	.0377	.0357	.0337
5	.0882	.0849	.0816	.0784	.0752	.0722	.0692	.0663	.0635	.0607
6	.1191	.1160	.1128	.1097	.1066	.1034	.1003	.0972	.0941	.0911
7	.1378	.1358	.1338	.1317	.1294	.1271	.1247	.1222	.1197	.1171
8	.1395	.1392	.1388	.1382	.1375	.1366	.1356	.1344	.1332	.1318
9	.1256	.1269	.1280	.1290	.1299	.1306	.1311	.1315	.1317	.1318
10	.1017	.1040	.1063	.1084	.1104	.1123	.1140	.1157	.1172	.1186
11	.0749	.0776	.0802	.0828	.0853	.0878	.0902	.0925	.0948	.0970
12	.0505	.0530	.0555	.0579	.0604	.0629	.0654	.0679	.0703	.0728
13	.0315	.0334	.0354	.0374	.0395	.0416	.0438	.0459	.0481	.0504
14	.0182	.0196	.0210	.0225	.0240	.0256	.0272	.0289	.0306	.0324
15	.0098	.0107	.0116	.0126	.0136	.0147	.0158	.0169	.0182	.0194
16	.0050	.0055	.0060	.0066	.0072	.0079	.0086	.0093	.0101	.0109
17	.0024	.0026	.0029	.0033	.0036	.0040	.0044	.0048	.0053	.0058
18	.0011	.0012	.0014	.0015	.0017	.0019	.0021	.0024	.0026	.0029
19	.0005	.0005	.0006	.0007	.0008	.0009	.0010	.0011	.0012	.0014
20	.0002	.0002	.0002	.0003	.0003	.0004	.0004	.0005	.0005	.0006
21	.0001	.0001	.0001	.0001	.0001	.0002	.0002	.0002	.0002	.0003
22	.0000	.0000	.0000	.0000	.0001	.0001	.0001	.0001	.0001	.0001

r	9.10	9.20	9.30	9.40	λ 9.50	9.60	9.70	9.80	9.90	10.00
0	.0001	.0001	.0001	.0001	.0001	.0001	.0001	.0001	.0001	.0000
1	.0010	.0009	.0009	.0008	.0007	.0007	.0006	.0005	.0005	.0005
2	.0046	.0043	.0040	.0037	.0034	.0031	.0029	.0027	.0025	.0023
3	.0140	.0131	.0123	.0115	.0107	.0100	.0093	.0087	.0081	.0076
4	.0319	.0302	.0285	.0269	.0254	.0240	.0226	.0213	.0201	.0189
5	.0581	.0555	.0530	.0506	.0483	.0460	.0439	.0418	.0398	.0378
6	.0881	.0851	.0822	.0793	.0764	.0736	.0709	.0682	.0656	.0631
7	.1145	.1118	.1091	.1064	.1037	.1010	.0982	.0955	.0928	.0901
8	.1302	.1286	.1269	.1251	.1232	.1212	.1191	.1170	.1148	.1126
9	.1317	.1315	.1311	.1306	.1300	.1293	.1284	.1274	.1263	.1251

Figure and Table Credits

CHAPTER 1

Figure 1.2. Taxonomy of the Systems Approach: Reprinted with permission of the Institute for Operations Research and the Management Sciences from *Interfaces*, Vol. 24, No. 4, July–August, 1994, pp. 16–25, by Heiner Müller Merbach.

CHAPTER 4

Figure 4.2. Quality Assurance Certificate: Courtesy of Siemens

Figure 4.8. House of Quality: Reprinted by permission of the *Harvard Business Review*. An exhibit from "The House of Quality" by John R. Hauser and Don Clausing, May-June 1988, p. 72. Copyright © 1988 by the President and Fellows of Harvard College; all rights reserved.

Figure 4.9. ITI's House of Quality: Used with permission of International TechneGroup Incorporated, 5303 DuPont Circle, Milford, OH 45150.

Figure 4.10. Linked Houses Convey the Customer's Voice Through to Manufacturing: Reprinted by permission of the *Harvard Business Review*. An exhibit from "The House of Quality" by John R. Hauser and Don Clausing, May-June 1988, p. 78. Copyright © 1988 by the President and Fellows of Harvard College; all rights reserved.

CHAPTER 5

Figure 5.8. Matrix of Product Life Cycle Stages versus Process Life Cycle Stages: Reprinted by permission of the *Harvard Business Review*. An exhibit from "Link Manufacturing Process and Product Life Cycles" by Robert H. Hayes and Steven C. Wheelwright, Jan.–Feb. 1979. Copyright © 1978 by the President and Fellows of Harvard College; all rights reserved.

Figure 5.10. Matrix of Product Life Cycle Stages versus Process Life Cycle Stages (Adapted for Services): Reprinted by permission of the *Harvard Business Review*. An exhibit from "Link Manufacturing Process and Product Life Cycles" by Robert H. Hayes and Steven C. Wheelwright, Jan.–Feb. 1979. Copyright © 1978 by the President and Fellows of Harvard College; all rights reserved.

CHAPTER 8

Figure 8.7. Poka-Yoke Device: *From Zero Quality Control: Source Inspection and the Poka-yoke System* by Shigeo Shingo. English translation copyright © 1986 by Productivity Press, Inc., PO Box 13390, Portland, OR 97213-03289 (800) 394-6868. Reprinted by permission.

CHAPTER 9

Table 9.3. Time Value for Various Classifications of Motions: Reprinted with permission of the MTM Association for Standards and Research. © 1994 MTM Association, 1441 Peterson Avenue, Park Ridge, IL 60068.

Table 9.4. Computer Printout for 2M Analysis: Reprinted with permission of the MTM Association for Standards and Research. © 1994 MTM Association, 1441 Peterson Avenue, Park Ridge, IL 60068.

Figure 9.4. Constant Factors in Vehicle Seating: Diagram by Henry Dreyfuss Associates, "The Measure of Man and Woman," 1993. Used with permission.

CHAPTER 13

Figure 13.10. Cost Function—Used for the HMMS Aggregate Planning Model: From Charles C. Holt, Franco Modigliani, and Herbert Simon, "A Linear Decision Rule for Production and Employment Scheduling," *Management Science*, Vol. 2, No. 1, October 1955. Reprinted with permission of the Institute for Operations Research and the Management Sciences.

CHAPTER 19

Figure 19.12. CASA/SME Manufacturing Enterprise Wheel: Reprinted from the CASA/SME New Manufacturing Enterprise Wheel with permission from the Society of Manufacturing Engineers, Dearborn, Michigan, Copyright 1993, Third Edition.

Photo Credits

CHAPTER 1
John S. Reed
Courtesy of Citibank, N.A.

CHAPTER 2
Al Scott
Courtesy of Wilson Sporting Goods

CHAPTER 3
Doug Smith
Courtesy of Computervision Corporation

CHAPTER 4
Curt Reimann
Courtesy of National Institute of Standards & Technology

CHAPTER 5
Jeffrey Bleustein
Courtesy of Harley-Davidson, Inc.

CHAPTER 6
Ed Cornell
Courtesy of Motorola, Inc.

CHAPTER 7
W. Edwards Deming
Courtesy of MIT Center for Advanced Engineering Study

CHAPTER 8
Vickie Eckhert
Courtesy of Megatest Corporation

CHAPTER 9
Nancie O'Neill (Westin Hotels & Resorts)
Courtesy of Nancie O'Neill

CHAPTER 10
Scott Haag (Cinergy)
Courtesy of Scott Haag

CHAPTER 11
Jane Olson (FACS)
Courtesy of Jane Olson

CHAPTER 12
Wendell M. Kelly & Dennis J. Garrett
Courtesy of QualitiCare Medical Services, Inc.

CHAPTER 13
Ken Bogard
Courtesy of Miami University

CHAPTER 14
Les Waltman (G.E. Aircraft Engines)   Photo: Jeanne Busemeyer

CHAPTER 15
Glenda Paquin (Mitel Corporation)   Photo by Jeffrey deVries

CHAPTER 16
Eliyahu Goldratt
Courtesy of Avraham Y. Goldratt Institute

CHAPTER 17
Lucinda Heekin
Courtesy of The Last Best Place/Photo: Rod Walker, Boulder, CO

CHAPTER 18
Tom Walker
Courtesy of boyd & fraser publishing company

CHAPTER 19
Kevin Howard
Courtesy of Hewlett-Packard Company

Solutions

Chapter 1

1. a. Straight-grained slats are grooved a half lead deep. Graphite, clay, gums, and water are blended for #2 lead and laid into the grooves. Top slat fitted, glued and pressed. Cutters separate pencils which are then shaped, marked and crowned with an eraser.
   b. and c. -everyone is an expert.
3. See Figure 1.8.
5. Type of cooking, toasting, dressing and packaging links meat patties to burgers delivered to customers. Same approach for other items.
7. Profit = $4 million.
9. a. $0, b. loss of $250,000, c. profit of $250,000.

Chapter 2

1. Labor productivity (units shipped) = 8.57.
3. Total productivity = 1.07.
5. Cost of the quality problem is $350,000. There is a reduction in total output and total productivity.
7. Only part of the meaning is captured. Fixed costs are not reflected but the effects of investment in technology affect the variable costs and the prices that can be charged.
9. a. It is a possible description.
   b. $V_p = k$, so demand volume is constant (k) at any price (within a reasonable range where the relation holds).
   c. If k is large enough, b-type situations attract competitors which will create elasticity. Quality can be used to differentiate competitive products.
   d. For the p = 1.50 row:

β=0	β=0.25	β=0.50	β=1
5,000	4,518	4,082	3,333

11. a. When p = $1.00, vc_v = $0.51, profit = $39,000.
    When p = $2.00, vc_v = $0.72, profit = $54,000.
    b. Applicable for problem 10 but not for problem 11.

Chapter 3

1. Many possibilities exist such as the average of the four numbers which is near 25. Alternatively, the average change interval is [(9 + 5 + 6)/3] near 7 and 7 + 34 = 41. These are also stops on the New York subway. The next one is 42nd Street.
3. Multiply last year's actual sales by 1.5. For January the value would be 1500 × 1.5 = 2250, etc.
5. The 6-month moving average prediction for July is 166.7. The error is 200−166.7 = + 33.3. The absolute percent error = 16.7%. Then, MAD = 78.9 and CFE = -166.8.
7. The 6-month weighted moving average prediction for July is 183. The error is 200−183 = +17. The absolute percent error = 8.5%. Then, MAD = 70.7 and CFE = −184.
9. The solution is a = 11.68 and b = −0.455. The regression line is y_{t+4} = 11.68−0.455 x_t; r = −0.455 indicating poor fit.
11.

Day	1	2	3	4	5	6	7	SUM
Error ϵ_t	+1	-1	0	-1	-2	+2	+1	CFE=0
ϵ_t^2	1	1	0	1	4	4	1	12
$(\lvert\epsilon_t\rvert/x_t)100$	25	20	0	33.3	200	66.7	25	370

MAPE=52.86 AD=1.14 SE=1.71 σ=1.31

13. Magnitude of error in this time series appears to be decreasing which (if true) would violate the homogeneity assumption of linear regression. Advisable to get more data and look for the possibility of trends and/or cycles.

Chapter 4

1. D = 0 means perfect quality and total cost = 7,000.
3. Total cost = 10,480. The third point for plotting is 15,200.
5. Minimum total cost occurs at D = 0 (no defectives).
7. Warranty model for c. should include such considerations as: is it reliable, recyclable, caused by carelessness or product susceptibility. What are competitive warranties, cost of customer alienation? Similar reasoning can be used for a., b., and d.
9. Consider, for example, how weight and size interact with battery life, size of keyboard, size of screen and drives.
11. q gradually returns to s: q = 2, 3, 2.5, 2.25, 2.125, etc.

Chapter 5

1. a. Yes, it is the classic synthetic process. b. 1.92 min.
3. 1.92 versus 1.65 which is a reduction of 0.27 minutes.
5. New Jersey: Investment Cost per week = $1,923,070
   Labor Cost per week = 5,760,000
   Total Cost per week = 7,683,070
   Kentucky: Investment Cost per week = $2,307,696

Labor Cost per week   = 3,072,000
Total Cost per week   = 5,379,696

7. IFS Annual Cost: $367,800;
   Job Shop Annual Cost: $468,000;   Ratio = 0.79.

9. Set up a six by seven matrix and make notes in appropriate cells such as: complex products require more technology and the organization of the flow shop helps; highly varied productlines require job shops; small firms may not find high technology affordable; flow shops require volume; FMS are well-suited to families of products made in small lots. Profitable ventures allow experimentation, even jumping ahead of the competition — say from batch to flow shop.

11. Cycle time = 6 minutes.

Chapter 6

1. The idea is to draw a process flow chart for making burgers at McDonalds. To do that it helps to visit one at off-peak hours and talk to the manager about this assignment. Ask the manager if you can watch or have it explained.

3. The comment in problem 1 applies to process flow charting the way that chili is made and served at Wendy's.

5. Visit a travel agency or read about one. Role play what must take place with friends. Draw the process flow chart.

7. Really bake a cake, then draw a process chart for doing it.

9. Read the instructions for making concrete on a cement bag that you buy in the hardware store. Then create a process chart for producing concrete in volume — as a business venture. Try to set up an intermittent flow shop, say for paving roads.

11. The flow chart for apple cider vinegar should show: ferment apple juice, filter, add acid-making bacteria (called "mother") to convert alcohol in "hard" cider into vinegar. Then age 90 days in wooden tanks, reduce with water to 5% acidity, filter, pasteurize, fill bottles (16 and 32 oz. sizes), cap, label, ship to supermarkets. Use a flow shop configuration.

13. Create a process flow chart for routinizing preparation of research reports. Use a flow shop orientation.

15. a. The appropriate machine-operator chart shows that two operators are working at the same time (productively) but the system is not efficient because both make ready and put away have to be done as internal activities with the machine turned off. Has the SMED approach been tried?

Time	Operator	Machine	Operator 2
0	Make ready 1st unit	Idle	Idle
2			
4	Idle	Working 1st unit	
6			
	Make ready 2nd unit	Idle	Put away 1st unit
8			
10	Idle	Working 2nd units	Idle
12			
	Make ready 3rd unit	Idle	Put away 2nd unit
14			

The pattern from 8 to 14 continues until completion. The machine is 2/3 utilized. Both operators are 1/3 utilized. Compare this to the situation in Figure 6.8 (original process) where both the machine and operator were 50% utilized.

17. a. Light hours per replacement period are $8{,}760/4 = 2{,}190$ hours compared to expected lifetime of 3,000 hours presumed to be from a bell-shaped distribution. A reasonable standard deviation is 400 hours so light level will be excellent.
    b. $3,800,000.

19. The racing car is serviced while running. SMED ideas abound.

21. CST = $7,000 and USCT = $1,090 so uncontrolled storage wins.

Chapter 7

1. The scatter diagram seems to indicate a strong relationship.

3. Make 237 units.

5. For $\bar{x}$-chart: grand mean = 29.97, UCL = 30.51, LCL = 29.43; all points within limits. For the R-chart: $\bar{R} = 0.74$, UCL = 1.69, LCL = 0; all points within limits.

7. Reference Table 7.4, $\bar{R} = 0.67$, UCL = 1.53, LCL = 0; all points within limits. New 3pm data yield a better R-chart

9. $\bar{p} = 0.15$, for n = 9: UCL = 0.39 and LCL = 0.00, for n = 16: UCL = 0.33 and LCL = 0.00.

11. Use the data check sheet to track the Dow Jones Index and some other time series that might be related, such as the price of gold, other country stock indexes the day before, an index to reflect optimism-pessimism, etc., based on an Ishikawa analysis. Then, draw scatter diagrams to see if the causal factors might be related to the Dow.

Chapter 8

1. Use Box B.
5. 14.3 months. Payback period and cost recovery are synonymous.
7. Multiply possibilities.
9. 419,904.
11. 240/hour for cycle time of 15 seconds.

Chapter 9

1. $n! = 80$.
3. Yes, buy the simulator.

5.

7. Category   No. of Observations   Fraction of Total (p_i)
   Programming
   Data Input
   Analysis
   Report Writing
   Copying
   Printing
   Collating
   Binding
   Total*
   * You can also add in categories from Table 9.2 in the text.
   Observations should be made frequently at first, say once every hour in both morning and afternoons, every working day, but not at set times. If the proportions do not begin to stabilize, then conjecture on separate scenarios such as start-ups for new jobs, end-game for old reports, and look at data for each type of situation.
9. 49.6 seconds.
11.

Activity	No. of Obs.	Fraction (p_i)	Table 9.2 Obs.	(p_i)
1	212	0.625	359	0.620
2	90	0.265	154	0.266
A	10	0.030	19	0.033
B	8	0.024	14	0.024
C	7	0.021	14	0.024
D	12	0.035	19	0.033
Total	339	1.000	579	1.000

13. $7.
15. $2.

Chapter 10

1. The demand is sufficient to absorb the additional suits. Since this would result in an additional profit of $1800/week at a cost of $1200, you should do it.
3. Crank up U, the utilization like running above rated capacity.

5. Since the first-year demand exceeds BEP and the sales for the next four years are higher than capacity, then do it. Perhaps it might pay to over-produce in the first year in anticipation of the high demands.
7. The additional $5 million investment drops the BEP from 27,000 to 20,769 (see graph below) and reduces variable costs by $6 million, so you should introduce the automation.

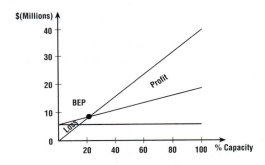

9. The forklift yearly cost is $19,000 while the conveyor yearly cost would be $20,000. Do not install the conveyor belt.
11.

n	$t(n) = 3\sqrt{n}$
1	3
2	4.2
3	5.2
5	6.7
10	9.5
15	11.6
20	13.4
25	15

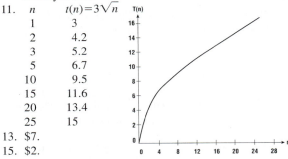

13. $7.
15. $2.

Chapter 11

1. A2, B1, C3.
3. a. The allocation is highly degenerate. There are only 5 positive flows here and 7 are needed for non-degeneracy.
   b. To create a non-degenerate allocation, perturb the numbers slightly, rerun the NWC algorithm and change the perturbations to zero allocations. For example, the first supply from 20 to 20.1 and the last demand from 50 to 50.1. After rerunning, change the 0.1 allocations to zero, but keep the cells active.
7. $2,060.
9. Ship 20 units to $P_1 M_D$. The revised TM is

	M_A	M_B	M_D
P_1	40		20
P_2		40	20
P_3			60

The profit is $2,120, a little more than for the NWC allocation.

11. Ship 20 units to P_3M_A. The revised TM is

	M_A	M_B	M_D
P_1	20	40	
P_2			60
P_3	20		40

The profit is $2,340, better than both of the previous allocations.

13. Since 4,000 total miles per year are eliminated, the cost is $27,000/4,000 or $6.50 per mile, which is a good investment.

Chapter 12

1. $15,000.
3. 0.08.
5. Buy 2 spares for an expected cost of $10.
7. Buy 3 spares for an expected cost of $15.
9. 50%.
11. 60 days. With automobiles, this is a reasonable turnover.

Chapter 13

1. 165 standard hours.
3. Since 588 standard days are needed to meet demand and a total of 650 standard hours are available at the 3 plants for the year, the demand can be met.
5.

t	S_t	P_t	I_t	ΣI_t	W_t	$W_t - W_{t-1}$
1	420	380	-40	-40	38	0
2	390	380	-10	-50	38	0
3	340	380	$+40$	-10	38	0
4	370	380	$+10$	0	38	0
5	360	380	$+20$	$+20$	38	0
6	400	380	-20	0	38	0
Total	2280	2280				

This is a fixed workforce plan.

7. The fixed production would go up by 10 units, resulting in an ending inventory of 60 but reduce backorders as shown below. The tradeoff between backorders and inventory plus additional workers is probably not sensible.

t	S_t	P_t	I_t	ΣI_t	W_t	$W_t - W_{t-1}$
1	420	390	-30	-30	39	0
2	390	390	0	-30	39	0
3	340	390	$+30$	$+20$	39	0
4	370	390	$+20$	$+40$	39	0
5	360	390	$+30$	$+70$	39	0
6	400	390	-10	$+60$	39	0
Total	2280	2340				

9. $10,500.

11. $17,750 for increased workforce scenario versus $13,000 for original scenario. Stick with the original scenario.

13. Min $\Sigma (c_w W_t + c_A A_t + c_R R_t + c_c I_t + c_o O_t)$ (costs)
ST $P_t = 500 (W_t + A_t - R_t) + O_t$ (production on regular time with k = 500 and overtime)
$S_t + I_t = P_t$ (produced units are sold or inventoried) with appropriate upper and lower bound on all the variables and the cost parameters designated.

15. The minimum cost solution of $2625 is

	1	2	3	Final Inv.	Slack	Supply
Initial Inventory	50					50
Regular 1	100					100
Overhead 1	50		50		100*	200
Regular 2		250	50			300
Overhead 2		50		150		200
Regular 3			100			100
Overhead 3				50		50
Demand	200	300	200	200	100	1000

* Only 100 of the 200 available overhead 1 hours are used. Note that the increase of $0.50 in carrying costs and decrease of $0.50 in overtime costs changes the optimal mix from the solution in 14 above.

17. Regular payroll cost = $C_t W_t = 100 W_t$ for $35 \le W_t \le 45$. This is a linear function so tabulate the end and middle points:

W_t	35	40	45
Cost	3500	4000	4500

19. To evaluate the cost function, must establish the proportionality between P_t and W_t, i.e., $P_t = k W_t$. Assume that $k = 10$ here. Expected cost of overtime $= C_3 (P_t - C_4 W_t) + C_5 P_t - C_6 W_t = 20 (P_t - 100 [0.1] P_t)^2 + 20 P_t - 100[0.1] P_t = 20 (-9 P_t)^2 + 20 P_t - 10 P_t = 1620 P_t^2 + 10 P_t$ for $350 \le P_t \le 450$.

P_t	350	375	400	425	450
Cost in Millions	198	227	259	293	328

Note that the quadratic term dominates here.

Chapter 14

1. 8.5 gallons.
3. 9.3 gallons.
5. Order 2,820 jars when inventory level hits 200 jars.
7. The perpetual system is always better in all aspects. The buffer stock size here is much smaller (200 to 400) resulting in storage and opportunity cost savings and the inventory fluctuations are much smaller.
9. 250 units.
11. Buy new equipment since the direct cost savings of $4 per unit saves about 40% in total costs since other costs here are insignificant.

13. Buy in lots of 500 for c = $8.

15. Unit costs dictate the decision; buy in lots of 600 from competitor for c = $7.

17. The implied carrying cost percentage is unrealistically low at 0.7%. Obtain a reasonable estimate of this percentage and recalculate the optimal lot size.

19. They can all be run on the same equipment if the sum of the run times is less than the shortest cycle time. The cycle time + = Q/d and is the last column in the table in the preceeding problem. Here, the sum of the run times = 17.4 + 11.6 + 6.4 + 13.3 = 48.7 > 37.5, the shortest cycle time. Hence, all items cannot run on the same machine.

21. A carrying rate of 10% is imputed, which is reasonable.

23. For Q = 20,000, TVC = $6.60, which is a cost increase of 0.46% for a decrease in Q of 8.7%. The percentage change in the cost is 20 times less than the percentage change in Q. This tells us that we need only be in the vicinity of Q* since the cost sensitivity is low.

25. Take the discount.

Chapter 15

1.

| | | | End Week | | | | | |
	1	2	3	4	5	6	7	8
Gross Req.	400	200	200	300	500	100	600	400
Stock-on-hand	700	300	100					
Net Req.			100	300	500	100	600	400
PO Receipt			100	300	500	100	600	400
PO Release	100	300	500	100	600	400		

3. Using the average weekly demand of 337.5, Q* = 1039. You order this lot size whenever the net requirements are positive. Here, your first positive net requirement is 100 for week 3, your planned order release for week 1 would be for 1039, which would be carried in inventory appropriately.

5.

| | | | | End Week | | | | | | |
	0	1	2	3	4	5	6	7	8	9	10
Gross Req.			3400	4200	5200	6300	7500	3100	4600	5400	6600	7800
Net Req.			3400	4200	5200	6300	7500	3100	4600	5400	6600	7800
PO Receipt			3400	4200	5200	6300	7500	3100	4600	5400	6600	7800
PO Release		3400	4200	5200	6300	7500	3100	4600	5400	6600	7800	

7. Using the average weekly demand of 5410, Q* = 7656. You order this lot size whenever the net requirements are positive. Here, your first positive net requirement is 3400 for week 1, your planned order release for week 1 would be for 7656, which would be carried in inventory appropriately.

9.

| | | | End Week | | | | | | |
	1	2	3	4	5	6	7	8	9
Gross Req.		240				320			160
Net Req.		240				320			160
PO Receipt		240				320			160
PO Release	240				320			160	

11.

| | | | End Week | | | | | | | |
	1	2	3	4	5	6	7	8	9	10
Gross Req.		150				60				180
Net Req.		150				60				180
PO Receipt		150				60				180
PO Release	150			60				180		

13.

| | | | End Week | | | | | | | |
	1	2	3	4	5	6	7	8	9	10
Gross Req.		60				100			40	60
SOH	80	20	20	20	20		10	10		
Net Req.						80			30	
PO Receipt						90			90	
PO Release					90			90		

15. 5.

17.

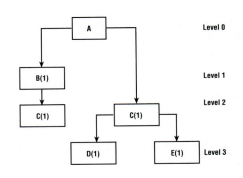

19. 5.

21.

Weeks Ahead	1	2	3	4	5	6	7	8	9
Gross Req.		20				45			60
PO Release	40				90			120	

Chapter 16

1. 1B, 2C, 3A.
3. 7B, 8C, 9A.
5. 1A, 2B, 3C, 4D, 5E.
7.

	A	B	C
M1	$5.40	$2.80	$4.50
M2	$11.40	$6.30	$12.00
M3	$8.16	$3.92	$4.80
MD	$0	$0	$0

9. $1171.48, an improvement of $270.72.
11. 24 hours on M1 for 84 B units.
    12 hours on M1 for 180 C units.
    54 hours on M2 for 1620 C units.
    62.5 hours on M3 for 300 A units.
    17.5 hours on M3 for 98 B units.
13. The best sequence for each department individually is the SPT rule. For A, the sequence would be IV, II, III, I and for B, the sequence would be IV, II, III, I.
15. I, III, II, IV.
17. They are all different.
19. LPT sequence: e, d, f, c, a, b. Total flow time = 181. Mean flow time = 30.17.
21. The SPT sequence would now be b, f, d, a, c, e.
23. Priority-oriented SPT sequence: a, d, c, e, f, b. Total flow time = 155. Mean flow time = 25.83. The flow times here are, of course, longer than the true SPT times.
25. SPT sequence: a, b, e, d, c. Total flow time = 141. Mean flow time = 28.2.
27. None, the bottlenecks are at C, D and assembly.
29. The $50,000 six-unit per hour belt cannot be fully utilized since the output rate is presently only five units per hour with multiple bottlenecks at C, D and assembly. The appropriate time and stock buffers must be set for these at the five-unit rate. Each buffer will have to be individually monitored to keep the system running smoothly since problems can arise only affecting some operations.

Chapter 17

1.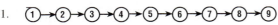

Legend:
1 Remove hub cap (if applicable)
2 With electric lug wrench remove four to six lugs per tire
3 Remove wheel from auto

4 Use tire machine to remove old tire from wheel
5 Take away old tire
6 Use tire machine to mount new replacement tire on wheel
7 Replace wheel on auto
8 With electric lug wrench replace four to six lugs per tire
9 Replace hub cap (if applicable)

3. .8.
5. The number of stations can be reduced to 3.
7.

n	1	2	3	4	5	6
orders/hr.	4.3	8.7	13.0	17.4	21.7	26.1

9. Unfortunately, this gives 5 workstations: I = {1,2} for 2.9, II = {3,4} for 3.3, III = {6,8} for 3.1, IV = {5,7} for 3.1 and V = {9} for 1.4. This gives a cycle time of 3.1 for productivity of 18.2 with balance delay of 16%.
11. I = {1,2,3} for 1.5, II = {4,7,8} for 1.4 and III = {5,6} for 1.5. This gives a cycle time of 1.5 for productivity of 40 with balance delay of 2%.
13. I = {1,2,3,4,8} for 24 and II = {5,6,7,9} for 24.
15. Use two for perfect balance. Four is very inefficient.
17. To achieve a rate of 40, all operations except 8 would have to be speeded up, which would probably be too expensive.
19. a. 6   b. 20%.
21. $L = 4$, $W = 0.5$.
23. Choose WIP storage of 15 cars for a 1% chance of overload. It should be located as close to the painting booth as possible.
25. In 15 minutes, get queue of 150 rolls, which would require a large storage space. Quality would be affected; for example, the rolls would be cold as the queue built up.
27. Adjust down the number in pull-five by word of mouth or a visual signal. Working in parallel would only make sense in a pull-one system. The maximum demand rate would still be limited by the dressing rate of 10/minute.

Chapter 18

1. Note that all the steps are strictly sequential, i.e., the one bar begins where the previous one ends. The chart is obvious.
3. Unless the projects can be partitioned for sequencing, there really is no evident critical-path representation that captures the timing of the Gantt chart. The critical-path model is concerned with absolute times first. There is no direct translation unless the Gantt chart is drawn as a network. It is necessary that some kind of assumptions be made. The diagram can be improved when real information is obtained about milestones.
5. The materials to build A are on hand.

7.

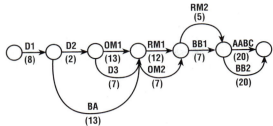

* D = Design, BA = Build A, OM = Order
Materials, RM = Receive Materials, BB = Build B,
AABC = Assemble ABC
* D1 and D2 are the appropriate partition of D, etc.
* (#) = # Time units for activity (approximate times)

9.

Activity	1-2	1-3	2-4	3-4	3-5	4-5	2-7	7-5	5-6	4-7
ES	0	0	30	15	15	33	30	55	63	33
EF	30	15	33	30	35	63	55	59	83	53

83 days to complete project.

11. a.

Activity	1-2	1-3	2-4	3-4	3-5	4-5	2-7	7-5	5-6
Slack	0	3	0	3	28	0	4	4	0

b. Materials for product and package are ordered
while they are still being test marketed. The notion
of separate tests markets for product and package
runs counter to good systems practice. Also, almost
no slack makes the project tough to manage.

13.

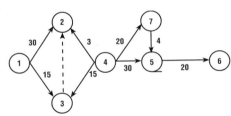

15.

Activity	1-2*	1-3	2-4*	3-4	3-5	4-5*	2-7	7-5	5-6*
LS	0	3	30	18	30	33	39	59	63
LF	30	18	33	33	30	63	59	63	83

Remember that 3-2 is a dummy activity. Critical path
activities are starred (*).

17. A, E, F

19. a.

	A	B	C	D	E	F	G	H	I
Normal Activity Slack	0	3	7	11	0	0	7	3	3
Crash Activity Slack	6	0	5	7	6	6	5	0	0

b. Normal cost is $57, crash cost is $106 (if don't
speed up activities with slacks appropriately, save
$30.8 and cost is $75.2).

21.

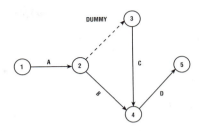

23. The two-person shift decreases the time of A1 from
23 to 21. The critical path which includes A1 is still
the same so the time to completion reduces from 48
to 46 and the time adjustments are proportional.

Chapter 19

1. 63 > 40 so accept change.

3. Benchmarking business school performance should
include measures of the timeliness of courses, quality
of teachers, ratings of companies, positions filled and
student evaluations. Distinctions should be drawn
between large and small schools and the individual
missions of the business schools.

5. Using hypothetical data, where large numbers are
good, the ratings for business schools (I = 1, 2, 3)
are developed.

BS_i Rating = $R_{i,1} \times R_{i,2} \times R_{i,3} \ldots \times R_{i,n}$
BS_1 Rating = $R_{1,1} \times R_{1,2} \times R_{1,3} \ldots \times R_{1,n}$ = $5 \times 4 \times 6 \times 7 = 840$
BS_2 Rating = $R_{2,1} \times R_{2,2} \times R_{2,3} \ldots \times R_{2,n}$ = $3 \times 6 \times 8 \times 2 = 288$
BS_3 Rating = $R_{3,1} \times R_{3,2} \times R_{3,3} \ldots \times R_{3,n}$ = $9 \times 1 \times 5 \times 9 = 405$

The Schools are ranked 1 first, 3 second and 2 third.
More complex scoring systems weight the ratings
before multiplying them.

7. a. Estimated total gross margin = $12,000,000.
b. 15% increase to 45%.
c. 30 % increase to 60%.

9. a. 11.4 days
b. Lattice organization.

11. The chart indicating the appropriate efficiency
frontiers and the points for each project are on the
attached chart. The preference rank order would be:
First place: E with R = 4
Tied for second place: C and D with R = 3
Tied for third place: A and B with R = 2

13.

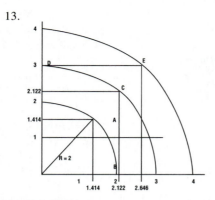

a. 76%.
b. $1,824.
c. $1,824,000.
d. Those requiring a great deal of traveling

GLOSSARY

20:80 rule 20% of the total number of problems and complaints occur 80% of the time.

ABC Classification Concept that some materials are more important than others, and do what is important before doing what is less important.

Acceptance sampling methods Inspection of a production lot sample. Detailing is performed to separate good from bad.

Acceptance sampling Uses statistical sampling theory in conjunction with SQC to check conformance to agreed upon standards; process stability is assumed.

Action learning Trading jobs for a period of time; encourages cooperation and team building.

Activity-based cost (ABC) systems Actual resource expenses are subtracted from revenues to yield the short-term contribution margin (STCM); general expenses are subtracted separately from STCM to yield operating profit.

Actual Capacity Greatest output rate achieved with the existing configuration of resources and the accepted product or service mix plans.

Aggregate planning (AP) Aggregate planning, an internal production management function, matches classes of jobs with the supportive resources needed to produce each of them; requires forecasting.

Analytic process Progressive disassembly; reverse of synthetic process as it breaks up one thing into many things.

Artificial intelligence (AI) Systems which have the capability of sensing and learning; analogous to the brain and its functioning.

Assemble-to-order (ATO) environment End items are built from subassemblies made available to meet demand within promised due dates.

Assignment method Provides production scheduling with a decision model for matching jobs with facilities in an optimal way.

Average performance Employees are grouped above and below an average performance measurement.

Backward pass rule When two or more arrows converge at a single node, choose the smallest value of LS among the converging arrows for the LF value of the preceding arrow.

Balance delay Percent of unproductive time at the stations.

Baldrige Award U.S. award for companies which lead in quality accomplishments.

Bar chart Graphical method for presenting statistics captured by data check sheets; data differentiated by bars.

Benchmarking Procedure of comparing one's own process against the best.

Best location Related to the function of the facility and the characteristics of its products and services.

Best practice Process standard with strong roots in the systems approach; striving to make one's own process the best.

Bidding Buyer requests competing companies to specify how much they will charge for their product; can involve costs, quality, and delivery among others.

Bill of material (BOM) A listing of all subassemblies, intermediates, parts and raw materials that go into a parent assembly showing the quantity of each required to make an assembly; types of BOMs include modular, matrix, cost, and transient or phantom; second major input to the MRP computer system.

Bill of resources (BOR) Listing of the key resources needed to make one unit of the selected items; also called product load profile or bill of capacity.

Bottlenecks (BN) The slowest machine sets the pace of the process flow; when the pace is exceeded by the demand, waiting lines grow.

Breakpoint business process redesign Focuses on creating strategic-level competitive advantages.

Buffer stock Stock carried to prevent outages when demand exceeds expectations; also called reserve and/or safety stock.

c-chart Shows the number of defectives in successive samples.

Capacity Maximum containment and/or maximum sustained output.

Capacity requirements planning (CRP) Function of establishing, measuring, and adjusting limits or levels of capacity to determine how much labor and machine resources are required to accomplish production tasks.

Capital productivity Partial measures of productivity are the number of units of output per dollar of invested capital. Also, a ratio based on dollars of output per dollar of invested capital.

Cause and effect charts Organize and depict the results of analyses concerning the determination of the causes of quality problems; include fishbone (or Ishikawa) charts and scatter diagrams.

Certification of suppliers Process for grading suppliers to ensure that the supplier organizations conform to standards which are essential for meeting the buyer's needs; rated on quality, price, and delivery.

Change management Managers reach strategic decisions which help organizations adapt to significant changes in the competitive realm.

Change management projects— called a Require planning, designing, redesigning, and implementing the system' domain of project managers.

Changeovers Needed steps to prepare equipment and people to do a new job; special form of operations.

Closed-loop MRP System built around MRP which includes the additional planning, manufacturing control and execution functions as well as feedback thereby ensuring planning validity at all times.

Computer-aided design (CAD) Design software which communicates specification to CAM software which translates, instructs, and controls production machinery.

Computer-aided manufacturing (CAM) Software used together with CAD software to determine the feasibility of manufacturing a new design or suggest alternatives.

Concurrent engineering (CE) Multifunctional parallel-path project management.

Continuity Complete one project and then move to another; use of global project teams enabling work to progress following the sun.

Continuous improvement (CI) Constant effort to find better ways.

Continuous processes Work configuration where the product flows through pipes continuously as it is being treated.

Control chart Graphic means of plotting points which should fall within specific upper- and lower-bound limits if the process is stable.

Control limits   Thresholds designed statistically to signal a process is not stable.

Control management processes— called *b*   Entail running the system and shielding it from external disturbances; domain of process managers.

Cost trade-off analysis   Comparisons using only the variable portions of total costs; variable costs are those costs which differ between alternative choices.

Cost-benefit analysis   Model analogous to a ledger with dollar costs and dollar equivalents for the benefits listed which are summed and net benefit is the difference.

Costs of quality   Three basic quality costs which include prevention, appraisal, and failure.

Crash   Each activity will be done in the least possible or minimum time; entails additions to the project budget.

CPM   Critical-Path Method.

Critical path   The unique path that accounts for the longest time sequence; at least one path.

Criticality   Crucial to performance or dangerous to use; reflects costs of failures.

Cross-docking   Transfer of goods from incoming trucks at the receiving dock to outgoing trucks at the shipping docks; goods never enter the warehouse; saves time and money.

Custom shop   Work configuration in which products or services are "made-to-order."

Cybernetics   Study of control systems; employs similarities between human brains and electronic systems.

Cyborg   Technological pairing of people and machines to achieve results neither could otherwise obtain.

Cycle time   Interval that elapses between successive units of finished goods coming off the production line; also length of time a unit spends at each workstation.

Cycles   Composed of four discrete stages: introduction, growth, maturation, and decline.

Days of inventory (DOI)   Measure equivalent to dividing total stock on-hand by the average demand per day; expected length of time before running out of stock

if no further stock is added; inventory backlog to be worked off.

Deming Prize   JUSE worldwide award for companies who have achieved outstanding quality performance.

Design for assembly (DFA)   Starts with the goal of feasible assembly; compliments DFM; objective of minimizing the assembled system's variability and ease of assembly.

Design for manufacturing (DFM) Starts with the goal of feasible fabrication; compliments DFA; objective of minimum variability.

Deterministic line balancing   Each total process takes an exact amount of time which is known and not variable.

Dimensions of quality   Descriptors examined to determine a product's quality.

Distribution requirements planning A time-phased order point approach for determining the needs to replenish inventory at branch warehouses.

Drift-decay quality phenomenon A product's performance can gradually diminish as a result of age and wear.

Due date   Number of days remaining before promised delivery.

Dynamic inventory models   Within the order repetition class of inventory situations; these models place orders repetitively over long periods of time.

Economic lot size (ELS)   The fixed amount to produce; a type of OPP

Economic order quantity (EOQ)   The fixed amount to order; a type of OPP; also called square root models.

Economies of scale   Reductions in variable costs directly related to increasing volumes of production output.

Economies of scope   Set-up cost reductions allow more frequent set ups and thereby, smaller lot sizes can be run.

Efficiency frontiers   Separate the regions by their combined power to deliver improved capabilities.

Elasticity   Rate-of-change measure expressing degree to which demand grows or shrinks in response to a price change.

Empirical testing   Experimental testing which can simulate worst cases of the kind of handling the package design can be expected to undergo; examines alternatives.

Ends   Objectives; part of a means-end model which connects strategies with tactics.

End products   Also called parent products; considered to have independent demand systems where demand is driven by the marketplace.

Ergonomics   Systematic study of how people physically interact with the working environment, as well as their equipment, facilities and products; also called human factors, human engineering, or biomechanics.

Expediting   Following up on orders and doing whatever can be done to move materials faster.

Experience curve   Average unit costs decrease with increasing experience and therefore volume; known as economies of scale.

Expert systems   Computerized sequences for smart procedures developed by copying the way experts solve the problems.

Extrapolation   Process of moving from observed data (past and present) to unknown values of future points; main function of forecasting.

Facilities planning   Composed of the location, structure and specific site selection, equipment choice, and layout of the plant and office within which OM does its work.

FIFO   "First-in, first-out" sequencing rule; the first jobs into the shop get worked on first; sometimes referred to as "first-come, first-serve" (FCFS).

Finite scheduling   Production is scheduled in batches of finite size for a variety of jobs in the shop; feed the bottleneck and protect it with days of buffer stock.

Fixed Costs = FC   Overhead costs paid regardless of production quantity; nonattributable to specific products or services.

Flexible manufacturing systems (FMS)   Individual machines and/or groups of machines (called cells) programmed to produce a menu of different products.

Flexible office systems (FOS) Outgrowth of the new technological capabilities applied to office systems.

Flexible process system (FPS)   Used like flow shops but produce variety due to computer programming and computer-driven equipment operating together to change set up in instants; include flexible manufacturing system (FMS) and flexible office system (FOS); and associated with economics of scope.

Flexible service systems (FSS) Outgrowth of the new technological capabilities applied to service systems.

Floor plan model Graphic two- or three-dimensional method of trial and error used to evaluate the seven layout criteria of capacity, balance, investment and operating costs, flexibility, work in process (WIP), distance parts move, and storage for WIP.

Flow shops (FS) Serialized work configurations where one work unit is at various stages of completion at sequential work stations located along the production line.

Forecasting To calculate or predict events. Types include: base series modification, moving averages, weighted moving averages (WMA), regression analysis, correlation, multiple, Delphi method, exponential smoothing, and the use of forecasting errors.

Forward pass rule When two or more arrows converge at a single node, choose the largest value of EF among the converging arrows for the ES value of the next arrow.

Functional field approach A tactical approach concentrating on specifics to achieve an end goal.

Fundamental theorem of linear programming There cannot be more active variables than real constraints.

Gantt charts Provide a graphic system that is easy to follow by visualizing the progress of jobs and the load on departments; method for trying to optimize assignments.

Group technology cellular manufacturing systems Application of FMS; employs stations connected and coordinated to work together to efficiently produce families of parts.

Hazard Analysis Critical Control Emphasis documentation and updates changes in conditions.

Heuristic line-balancing methods Employ "rules of thumb," common sense, logical thinking, and experience to find near-optimal solutions for large problems.

Heuristics (rules of thumb) Two apply to plant or office layout improvements: assign large work flow rate centers to close locations, and assign small work flow rate centers to distant locations.

Histograms Bar charts measure frequencies with which each of all possible outcomes occurs.

HMMS Model which addresses four issues:
• What are nonlinear costs?
• How complex is HMMS?
• How much of a difference will it make?
• How much is it used?

House of Quality (HOQ) A mapping procedure which analyzes design and process interactions and can identify WHATs, HOWs, WHYs, and HOW MUCHes.

I/O profit model Derived from the costs and revenues of the traditional model based on a specific time period.

Information systems Provide customer data for operations management to supply required services.

Intelligent information system Provides unsolicited data that is timely for decision making; prompts decisions, detects errors, provides a rapid-access, multifunctional systems database.

Interchangeable parts Concept that allows batches of parts to be made, any one of which will fit into the assembled product.

Internal MM system Requires control of production materials which are process flows.

Inventory record files Third major input to the MRP computer system; includes up-to-date stock on-hand levels and ordering information including suppliers' lead times.

Inventory Those stocks or items used to support production, supporting activities, and customer service; storage of all materials used or made by anyone in the organization for the direct or indirect purposes of offering finished products or end services to customers.

ISO 9000 series Widely-known quality standards established by ISO—the international organization for quality standardization.

ISO 14001 Standards for conformity assessment.

Japanese Industrial Standard Z8101 A quality standard system utilized by the Japanese resembling those quality specifications developed in the U.S.

Job shop (JS) Work configuration for processing work in batches or lots.

Just-in-case Components (parts or information) are on hand and brought to where they are needed before they are needed to avoid delivery slip-ups.

Just-in-time Components (parts or information) are on hand and brought to where they are needed as they are needed.

Kanban Any method that calls attention to the fact that the station wishes to receive a JIT delivery.

Labor productivity Partial measure of productivity; ratio of sales dollars to labor cost dollars.

Lateral transferability Flexibility to move operations sideways and into different stations (with respect to precedence diagrams).

Layout A three-dimensional situation which lends itself to creative solutions for alleviating problems; a good layout interacts with a good job design; leads to productivity increases and quality improvements.

Layout or sequencing chart Reserves specific times on the various facilities for the actual assigned jobs; Gantt-type chart.

Lead time (LT) Interval between order placement and receipt.

Lean production systems Combine quality with speed and waste avoidance.

Learning curve Repeated practice for groups or individuals decreases the time required to do the job.

Leveling factors Used to convert average time to adjusted or normal time.

LIFO "Last-in, first-out" rule. Usually true on an elevator.

Line balancing (LB) Effort to nearly equalize the output rate capabilities of successive workstations on the production line; also effort to reduce and match excess idle time at the workstations; three LB classes include: deterministic, heuristic, and stochastic.

Linear programming (LP) Modeling technique used by OM to achieve optimal use of resources.

Loading Takes actual orders and assigns them to designated facilities; decides which department is going to do what work.

Logistics Strategies and tactics of materials and product distribution.

Lot-for-lot ordering Planned-order receipt is equal to the net requirements.

Low-level coding Organizing rule for

computing material requirements; lowest part position determines where the part is located for computer scanning.

Machine-Operator Chart (MOC) Provides a means for visualizing the way in which workers can tend machines over time.

Make or buy decisions One of the seven classes of inventory situations; the decision of whether inventory should be purchased externally or made internally; based on cost and quality considerations.

Make-to-order (MTO) inventory environment No finished goods inventory to call upon; end items are produced to customers' orders.

Make-to-stock (MTS) environment Emphasis is on having sufficient stock on-hand to provide excellent service with minimum delivery times.

Management of technology (MOT) Component of the input-output systems transformation functioning at the societal, company, and international levels.

Manufacturing Fabrication and assembly of goods.

Master production schedule (MPS) One of three inputs to the MRP information system—other two are bill of material (BOM) and inventory stock-level reports; time-phased plan which indicates product specifications and production quantity.

Materials management (MM) Organizing and coordinating all management functions responsible for every aspect of materials movements and transformations.

Materials management information system (MMIS) Provides online information about stock levels for all materials and parts; other data as well.

Materials management system (MMS) System used in materials management which is triggered by demands.

Mean flow time Average amount of time required to complete each job in the group; average wait-to-start and processing times for every job in the group.

Mean time between failures (MTBF) A measurement of the product's expected lifetime.

Means Process components designed to achieve the ends (or objectives); part of a means-end model.

Methods analysis Systematic examination of all operations in any process in search of a better way of doing things;

used to work smarter not harder.

Minimum cost assignment schedule The equivalent to identifying the production policy for the minimum cost point.

Models Developed by operations researchers, management scientists, and location analysts for P/OM's use in facilities planning; types include: location decision, transportation, scoring, center of gravity, plant layout, among others.

MRP II Entails total manufacturing resource planning as part of the strategic business plan.

Multifactor productivity Total productivity; difference between change in output and change in labor and capital inputs engaged in the production of the output.

Multiple-channel stations Total systems utilization must be less than one; the single-line forms in front of M stations waiting for a free station to provide service; more complex than a single-channel system.

Net change MRP MRP is continually retained in the computer; whenever a change is needed, a partial explosion and netting is made only for affected parts.

Net Present Value (NPV) Financial management tool which discounts future earnings by prevailing expected interest rates.

Neural networks Mechanisms that are analogous to human nervous systems designed to enable computers to emulate functions associated with human intelligence.

Normal time Adjusted average time.

Northwest Corner (NWC) method A supply and demand matrix used to determine transportation costs which dominate the plant location decision; used for an initial feasible solution.

Northwest Corner (NWC) method A supply and demand matrix used to determine transportation costs which dominate the plant location decision; used for an initial feasible solution.

Objectives of operations Part of a means-end model which connects strategies with tactics.

Olympics Represent a systems symbol for "going global."

OM system Incorporates all factors with an effect on the purposes and goals of the system.

Operating characteristic curves (O.C. curves) A unique sampling plan which shows what happens to the probability of accepting a lot as the actual percent defective in the lot goes from 0 to 1.

Operations management Systematic planning and control of operations; a methodical, purposeful process.

Operations Purposeful actions methodically conducted as a work plan to achieve practical ends.

Opportunity costs Costs of doing less than best; cost to be paid is the difference in net benefits.

OPT "Optimized production technology"; a production process scheduling system highly sensitive to bottlenecks.

Order point policies (OPP) Define the stock level at which an order will be placed and the amount to be ordered.

Order promising The process of making a delivery commitment; involves a check of uncommitted material and availability of capacity.

p-chart Shows percent defective in successive samples.

Pareto analysis Goal is to determine which problems occur most frequently and arranges them in rank order of relative frequency of occurrence.

Payback period Estimated number of years required to produce enough profits or savings to offset expenditures completely. Also called breakeven time and cost recovery period.

Periodic inventory system One type of OPP where on consecutive specific dates separated by the same amount of time, the item record is called up and an order is placed for a variable number of units.

Permutability of columns Flexibility with respect to the order with which operations are completed at a station (with respect to precedence diagrams).

Perpetual inventory system One type of OPP which continuously records inventory receipts and withdrawals.

PERT Program Evaluation and Review Technique.

Pilot studies Scaled-down models of new methods run in parallel to existing processes.

Planning horizon Used in the first level of production planning (which is aggregate planning); the forecast $\geq$ the longest lead time.

Precedence diagrams Graphic displays which show operations that must be done before others.

Principle of modularity Design, develop, and produce the minimum number of parts (or operations) that can be combined in the maximum number of ways to offer the greatest number of products or services.

Process A series of goal-oriented activities or steps; a detailed description of the input-output transformation sequence.

Process analysis Examines the purposes of the process and attempts to find the best possible way for goal achievement.

Process charts Maps of specific operations, transports, quality checks, storages and delays used by present, or proposed, systems.

Process quality control methods Seven methods used to analyze and improve process quality include: data check sheets, bar charts, histograms, Pareto analysis, cause and effect charts, statistical quality control charts, and run charts.

Process steps Operations or stages within the manufacturing cycle required to transform components into intermediates or finished goods.

Product callbacks Rework involving labor and material costs as well as legal claims and customer dissatisfaction.

Production Physical work which produces a material product.

Production management Planning and decision making for the manufacture of goods.

Production scheduling Assigns actual jobs to designated facilities with unambiguous stipulations for completion at specific times.

Production standards Criteria specifying the amount of work to be accomplished in a given period of time; used to compute real costs of production.

Productivity Overall measure of the ability to produce a good or a service; OM views productivity as the ratio of output over input.

Project management A P/OM responsibility which provides a significant competitive advantage for P/OM.

Projects Special work configurations designed to accomplish singular or nearly singular goals

Pull-production systems Production systems that signal for delivery from upstream stations (including outside suppliers); materials are pulled as needed.

Purchasing agents (PAs) Bring into the organization the needed supplies.

Pure number A number that has no dimensions.

Push-production systems Earlier era production systems which push materials on the downstream stations; when not carefully controlled, they can impede productivity.

Put-away In process analysis, the clean-up activity follows set up and production.

Quality Circles (QCs) Groups of workers, organized around products, who focus on quality enhancements.

Quality Function Deployment (QFD) Comprehensive program for quality extension company-wide.

Quality integrity An ethical standard to deliver only safe and healthy products or services.

R-chart Type of statistical quality control (SQC) chart; monitors the stability of the range (variability).

Redesign Changing the process methods/systems being used; needed when cost of failure is high, the probability of breakdown is great, and it cannot be lowered by preventive maintenance or better practice.

Reengineering (REE) Also called business process redesign; seeks radical improvements over a short period of time; opposite of continuous improvement.

Regeneration MRP MPS is totally reexploded down through all bills of material, to maintain valid priorities; new requirements and planned orders are then recalculated.

Relative productivity Compares the performance of competitive processes for which OM is accountable.

Resource leveling Project design concept which balances resource commitment among activities over time.

Rough-cut capacity planning (RCCP) Uses approximation to determine whether there is sufficient capacity to accomplish objectives; precursor to CRP.

Run Count on a control chart, successive values that fall either above or below the mean line.

Scatter diagrams Help determine causality; type of cause and effect chart composed of data points associated with each other (i.e., height and weight).

Scientific management Numerical measurement and analysis of the way work should be done.

Scoring model Permits simultaneous evaluation of tangible and intangible costs; means for organizing and combining estimates and hard numbers.

Sensitivity analysis Procedures using high and low estimates to test the equation's sensitivity to the range of estimation.

Sequence Used in the third level of production scheduling; establishes the order in which work will be done by departments and people working on jobs.

Sequenced assembly Allows assembly to be a continuous flow shop process; timing must be perfect.

Sequencing Decides the order in which the work will be done; specifies precise timing of assignments thereby allowing customers' notification of delivery dates.

Service Generates revenues either independently of goods, or to help the user of those goods.

Set up Applied to the start-up of a new job, which means the cleanup from the old one.

Shop floor control Part of sequencing; communicates the status of orders and the productivity of workstations; assigns priorities that determine order with which work will be done.

Simulations "Pretend" runs of what might occur using models of processes.

Single-channel system Rho is the average percent of time that the station is providing service.

Slack Unused departmental capacity.

Slope Reflects the rate at which profit increases as additional units of output (percent of capacity) are sold.

Smart benchmarks Capture significant competitive factors.

SMG Wheel World class performance must be supported by a totally interconnected information system.

SPT rule Sequencing jobs according to their shortest processing times (SPT); helps to minimize average delivery lateness and maximize fulfillment of delivery promises.

Standard units The common denominator for aggregated units in the production scheduling problem.

Static inventory models   Within the order repetition class of inventory situations; these models have no repetition.

Static inventory problem   Deciding how much to buy when there is only a one time purchase.

Statistical quality control (SQC)   Producer's ability to control the process's variability of part qualities within specified tolerance limits.

Stochastic line balancing   Exists when the assumption of fixed and deterministic times for operations and total process time is not correct.

Strategy   Comprehensive and overall planning for the organization's future; a continuous, ongoing process formulated by top management and appropriate specialists.

Synchronization   Requires control of the timing of flows; includes how much is made at one time and transferred to the next station.

Synergy   Action of separate agents that produces a greater effect than the sum of their individual actions.

Synthetic or predetermined time standards   Created from a subset of elements common to all jobs.

Synthetic process   Process combining a variety of components to form a single product; employs synthesis to achieve a smooth process flow.

Systems approach   Integrates OM decisions with business functions as a team approach to problem resolution.

Systems utilization factor, ρ (rho)   Average number of orders being serviced over time for single-channel systems.

Syzygy   Teamwork is pulling together with the analogy of maximum tidal pull exerted because the planets and the moon have lined up to support each other's gravitational forces.

Tactics   Procedures used by the organization's line managers to carry out strategic plans; include production scheduling, materials planning, and quality control.

Take-back laws   Laws which compel manufacturers to take back used product; results in products designed for quick disassembly and reuse.

Team building   Training groups of people to have a common focus; replaces internal competition with cooperation.

Team learning   Emphasizes interdependencies that exist between people working toward a common goal; also called cooperative learning; requires application of the systems approach.

Technology   Consists of equipment employed in a manner to bring about specific transformations.

Time and motion study   Study of the motions of which the job is comprised; based on sample observations.

Time series   Stream of data that represents past measurements.

Time series analysis   Uses statistical methods including trend and cycle analysis to predict future values based on past history.

Time study   Study of how long it takes to do specific operations.

Time-based management (TBM)   Cycle-time management and production scheduling management are facets of TBM; units are produced in an improved way by a faster process.

Tolerance limits   Define the range of acceptable product.

Tolerance matching   Measure of how well two parts fit together.

Total Costs = TC = FC + TVC   Sum of the Fixed and Total Variable Costs.

Total flow time   Cumulative time required to complete a group of jobs; composed of the sum of the complete-to-ship times for each job in the group.

Total or cumulative lead time (LT)   Time between recognition of the need for an order and the receipt of goods. Total time to do what needs to be done to get the order delivered.

Total productive maintenance (TPM)   Systematic program to prevent process breakdowns, failures and stoppages; strategies include preventive, remedial, and partial preventive maintenance (PPM).

Total Quality Management (TQM)   Called "The Prime Directive." TQM must be a company-wide systems approach to quality.

Total Revenue = $TR = (p)V$ Volume (V) multiplied by price per unit p.

Total Variable Costs = TVC = $(vc)V$ Result of multiplying the variable costs per unit (vc) by the number of units or Volume (V); also direct costs.

Trade-off models   Needed to evaluate and balance the disadvantages against the benefits.

Transformation   Added-value alteration of materials and components into desired goods.

Turnover   Measured as the number of inventory turns (number of times warehouse is emptied and refilled per time period); good turnover brings large rewards and a reputation for purchasing and sales success.

Value analysis (VA)   Analysis of the value of alternative materials; includes consideration of improvements, opportunities and materials innovations.

Variable costs per unit = vc   Inputs that tend to be fully chargeable and directly attributable to the product; also called direct costs.

Variety   The number of product alternatives a producer offers customers.

Vendor releasing   A large amount of material is purchased to be delivered in small quantities—JIT, or nearly so.

Work centers   Locating specific locations on the plant floor to do specific kinds of jobs.

Work configuration   Physical set up (activity flow) used to make the product; six types: custom shops, job shops, flow shops, flexible manufacturing systems, continuous flow processing, and projects.

Work sampling   Also called operations sampling; method to determine with reasonable accuracy, the percentage of time workers engage in different tasks before and after job design.

x-chart   Type of statistical quality control (SQC) chart; uses classification by variables to detect and discern causes of process quality problems.

Zero breakdowns   Specifies everything possible will be done to eliminate malfunctions and failures.

BUSINESS AND ORGANIZATION INDEX

A

A.G. Simpson Co., 595
Acorn Structures Inc., 429
Agile Manufacturing Enterprises, 66
Air Products and Chemical, Inc., 744
Airbus Industries, 22
AlliedSignal, 516
Alpine Technologies, 873
American Airlines, 22, 33
American Express, 9, 143
American National Standards Institute (ANSI), 125-126, 886
American Production and Inventory Control Society (APICS), 42, 590, 623, 653-654
American Society for Quality Control (ASQC), 125, 143, 886
American Society of Mechanical Engineers (ASME), 215, 220
Ames Rubber Corp., 144
Andersen Windows, 326
Anheuser-Busch, 855, 856
Arthur Andersen & Co., 884
Ashland, Inc., 139
Association of American Railroads (AAR), 322
Association for Robotics in Hazardous Environemnts (RHE), 883
AT&T, 24, 46, 840, 883
AT&T Consumer Communications Services, 142, 144
AT&T Global Business Communications Systems (GBCS), 782
AT&T Transmission Systems, 144
AT&T Universal Card Services, 143, 144
AT&T's Little Rock Repair and Distribution Center, 782-791
AT&T's Multimedia Group, 782
Atlanta Committee for the Olympic Games (ACOG), 573
AutoAlliance International, Inc., 174
Avery Dennison Corporation, 597

B

Baird Information Systems, 605
Bandai Co., 89
Bayer Corporation, 391
Bayway Refining Company, 551
BE&K, 801
Bechtel, 801
Bell Labs, 784
Bellcore TEC, 417
Beta Control, 873
Better Beer Company (simulation), 409
Blackfeet Indian Writing Company, 553
BMW, 871
Boeing Co., 22, 564, 754, 840

Booz Allen & Hamilton, Inc., 806, 860, 861, 885, 894-895
Boston Consulting Group (BCG), 57, 431, 445
Boutwell Owens and Co., Inc., 172
boyd & fraser publishing company, 553, 824
Breck's, 260
Bristol Meyers, 33
Burger King, 350, 8

C

Cadillac Motor Car Co., 143-144
Caldors, 9
Campbell's Soup Company, 621-622
Caterpillar, Inc., 481, 690, 872
Centers for Disease Control and Prevention (CDCP), 875
Central Intelligence Agency (CIA) 887
Century Theaters, 31
Charles Schwab & Co., 861
Chembank, 82, 859
Chiat/Day Inc., 884
Chicago Board of Trade, 180, 213
Chocolate Manufacturer's Association, 177
Chrysler Corporation, 28, 123, 209, 319, 320, 524, 543, 840, 852, 871
Cinergy Corp., 430
Cisco Systems, 708
Citibank, 208
Citicorp, 28, 352, 477
Club Med Sales, Inc., 30, 453
Coca-Cola USA, 55
Coleco Industries Inc., 89
COMB, 9
Commerce Department, 887
Computer and Automated Systems Association (CASA), 888, 895
Computer Science Corporation, 801
Computervision Corp., 85
ConAgra, Inc., 353
Cooper Industries, Inc., 313, 402, 600
Copley Pharmaceutical Inc., 266
Corning Incorporated, 315
Cummins Engine, 481

D

Dana Corporation, 381
Data General, 801
Dean Foods Company, 557
Dean Witter Reynolds, 477
Del Laboratories, 160
Delta Airlines, 22
Department of Health and Human Services, 894
DHL, 9
Digital Equipment Corporation, 50, 370, 514, 519, 531, 695, 729, 840

Disney. See Walt Disney Co.
Disney Development Co., 465
Dougherty, K. S., 893
Dow Chemical Company, 22
Duke Construction
Dun & Bradstreet Corp., 370, 884
DuPont, 323, 497, 531, 622, 806, 867
DynCorp, 280

E

E.D.S., 801
Eastman Chemical Company, 144, 530
Eastman Kodak, 840, 872
Eddie Bauer, 9
EG&G, 801
Elizabeth Arden, 737
Emerson Electric Co., 602
Ernst & Young, 884
European Network of Quality System Assessment and Certification, 127
Export-Import Bank of the United States, 887
Exxon Corp., 28

F

FACS (Financial and Credit Services), 450
FastMAN Software Systems Inc., 651, 669
Federal Express Corp., 9, 144, 211, 323, 622,
Federated Department Stores, 450
Federated Systems Group, 450
Fidelity Investments, 861
Florida Power and Light (FPL), 142, 146
FMC, 517
Food and Drug Administration (FDA), 875, 879
Ford Motor Co,
Ford Motor Company, 28, 50, 123, 139, 174, 226, 254, 333, 356, 514, 543, 840, 852, 855, 856, 871

G

GAF, 55
Gap, The, 31, 9
GE Aircraft Engines, 620
General Electric Co., 28, 33, 738, 840
General Electric-Fanuc, 839
General Motors Corporation, 50, 254, 348, 352, 353, 383, 481, 543, 840, 852, 856, 871, 872, 900
Gensler & Associates, 704
Gensym Corporation, 878
Gillette, 55
Glasstech, Inc., 320, 738
Globe Metallurgical, Inc., 144
GM Hughes Corporation, 83

Goodyear Tire and Rubber Company, 116, 47
Gore, W. L. & Associates, 864, 867, 870
Grand Island Contract Carriers, 409
Granite Rock Co., 144
GTE, 172
GTE Directories Corp., 144
Guardian Industries, 317

H
Harley-Davidson, inc., 191
Help/Systems, Inc., 127
Henry Dreyfuss Associates, 390
Hershey Foods Corporation, 717
Hewlett-Packard Corporation, 26, 352, 353, 524, 543, 651, 769, 840, 856, 858, 872
HNTB Corporation, 475, 799
Honda of America Manufacturing, Inc., 18, 95, 353, 840
Honda Motor Company, 17, 352, 543
Hughes Aircraft Company, 237

I
Iacocca Institute, 50
IBM. See International Business Machines Corporation
IBM Rochester, 144
Inco Ltd., 538
Ingersoll-Rand Company, 187, 840
Intel Corporation, 188, 338, 370
International Business Machines Corporation (IBM), 14, 17, 50, 143, 233, 352, 370, 477, 543, 840, 855, 872, 884
International Olympic Committee (IOC), 573
International Standards Organization (ISO), 125-126, 886
International TechneGroup Incorporated, 137

J
Japanese Union of Scientists and Engineers (JUSE), 141
John Deere, 481
Johnson Controls, 595
Johnson and Johnson, 160

K
Kmart, 9, 172
Kodak, 872
Komatsu, 840
Kraft General Foods, 55

L
L.L. Bean, 9
Lands' End, Inc., 9, 105, 209, 855
Last Best Place, The, 742
LBS Capital Management, Inc., 314
Legal Sea Foods, 118
Levi Strauss, 65
Lexmark, 651

Libbey-Owens-Ford (LOF) Co., 317
Limited, The, 9, 525, 527-528
Litton Industries, Inc., 65
LTV Steel, 405

M
McDonald's, 8-9, 31, 33
McDonnell Douglas Corporation, 354, 466
McKesson Corporation, 46
Macola Software, 652, 666
Marcam Corp., 55
Marlow Industries, Inc., 144
Marriott, 497
Martin Marietta Corporation, 283, 379
Maytag Corporation, 449
Mazda, 855, 856
Mazda North America (MANA), 174
Measurex, 651
Megatest Corporation, 322
Mellon Bank Corp., 314
Merck & Co., Inc., 697
Merrill Lynch, Pierce, Fenner & Smith, Inc., 895
Methanex Corporation, 129
Microsoft Corp., 651, 824
Milliken and Co., 144
Milton Bradley, 172
Mitel Corporation, 669
Mitron Corp., 695, 729
Mitsubishi, 139
Motorola, Inc., 144, 213, 338, 352, 370, 401-402, 543, 56, 64, 651
Mrs Baird's Bakeries, Inc., 605
MTM Association for Standards and Research, 376-378

N
NASA, 685, 708, 754, 799, 806, 876
National Institute of Standards and Technology (NIST), 115, 143
National Semiconductor, 183
National Tooling and Machining Association (NTMA), 812
NCR, 24
New United Manufacturing Motors Inc. (NUMMI), 348-351
New York Mercantile Exchange, 180
Nikon, 124
Nissan, 595, 840
Norm Thompson, 9
Northrop, 840
NUMMI (New United Manufacturing Motors Inc.), 348-351

O
Occupational Safety and Health Administraiton, 389
Olympics, 886
Oticon Holding S/A, 864, 870
Oversears Private Investment Corporation (OPIC), 887
Oxko Corp., 331

P
Panasonic, 651
Parma Community General Hospital, 570
Paul Revere Insurance Company, 208, 209
Pennzoil Products Company, 231
Pepsi-Cola, N.A., 598
Pilkinton, 316
Pillsbury Company, 875
Polaroid, 172
PPG Industries, 317
Price Waterhouse, 832, 884

Q
Qualiticare, Inc., 521
Quality Control Color, 766
Quasar, 401

R
Radio Shack, 172
Red Cross, The, 21
Red Pepper Software, 708, 729
Remington Rand, 806
Ritz-Carlton Hotel Company, 30, 144, 434
Robotic Industries Association (RIA), 883
Rockwell Internaitonal Corporation, 651, 840
Rolls Royce, 735
Rolls-Royce Aerospace Group, 85
Rosenbluth International, 495-506
Royal Bank of Canada, 859
RPS, 9

S
Samsung Group, 260
Saturn Corporation, 351, 353, 872, 898-907
Satyam Software, 370
Sears, 9, 50
Sharper Image, The, 9
Sherwin-Williams Company, 552
Siemens AG Company, 126, 840
Siemens Medical Systems, Inc., 185
Skoda, 595
Society of Manufacturing Engineers (SME), 888, 895
Solectron Corp., 144
Sony Electronics, 338, 837
South Coast Air Quality Management District (SCAQMD), 878
Southwest Airlines, 23
Spiegel, 9
Sporty's, 9
Sprague Electric Company, 445
Steuben, 316
Sun Microsystems Inc., 708
Supercuts, 704
System Software Associates, Inc. (SSA), 646, 670

T
Takenaka Komuten, 142
Tandem, 524
Target Stores, 89
Texas Instruments, 82, 142, 543,

Texas Instruments Defense Systems &
        Electronics Group, 144
Tom's of Maine, 152-161
Toshiba, 543
Toyota, 139, 348, 352, 595, 840
Toyota Motor Manufacturing U.S.A. Inc.,
        215, 222, 231, 258
Toys "R" Us, 89
Transitions Research Corporation, 881-882
Travelers Insurance Company, 884
TRW, Inc., 753
Tyco International Ltd., 559

U
U.S. Air, 735
U.S. Air Force, 840
U.S. Department of Commerce, 143, 887
U.S. Labor Department, 44
U.S. Navy, 226, 806, 825
U.S. Postal Service, 9, 280
U.S. Public Health Service (PHS), 875
U.S. War Department, 254
Union Bank of Switzerland, 885
Union Camp Corporation, 737
Union Electric Company, 169
United Auto Workers (UAW), 383, 481
United Parcel Service, 9, 323

V
Valendrawers, 659
Vehicle Recycling Development Center, 871
Victoria's Secret, 9
Vitreous Environmental Group Inc., 873
Volvo, 202, 383, 481

W
W. L. Gore & Associates, 864, 867, 870
W. R. Grace & Co., 536
Wainwright Industries, Inc., 144
Wallace Co., Inc., 144
Wal-Mart, 172, 46, 527, 531, 89, 9
Walt Disney Co., 21, 172, 209, 465
Waterjet Systems, Inc., 871
Wendy's, 21, 31
Western Electric Company, 284, 383
Westin Hotels and Resorts, 360
Westinghouse, 55
Westinghouse Commercial Nuclear Fuel
        Division, 94, 144
Westvaco Corporation, 60
Weyerhauser Company, 592
Wilson Sporting Goods, 49
Woodstock '94, 828
Worlds of Wonder Inc., 89

X
Xerox Corp., 7, 15, 17, 143, 144, 209,
        855, 872, 873

Z
Zytec Corp., 144

NAME INDEX

A

Abernathy, William J., 36, 203
Adams, M., 149, 150, 300
Ahmadi, Reza H., 775
Altman, Y., 494
Anderson, Jon, 894
Arcus, A.L., 780
Arnold, Horace Lucien, 203
Ashby, W. Ross, 341
Attaran, M., 585

B

Baird, Jerry, 605
Baker, K.R., 680
Baldrige, Malcolm, 142-146, 150
Baloff, Nicholas, 440, 445
Barnett, Carole K., 893
Barringer, Felicity, 507
Baumol, William J., 69, 73
Beer, Stafford, 341, 343
Bennet, James, 392
Bennett, Michael, 907
Berry, William L., 203, 585, 638, 680, 683, 776
Bhame, C.D., 109
Birnie, D.P., 757, 780
Birou, Laura M., 547, 637
Blackman, Sue Anne, 69, 73
Blackstone, John H., 73, 109, 440, 638, 680, 683
Bleustein, Jeffrey, 191
Bogard, Ken, 558
Boisseau, Charles, 507
Borel, Emile, 389, 392
Boudrie, Jack, 812
Bounds, G., 149, 150, 300
Bowman, E.H., 585-586
Box, G.E.P., 109
Bramhall, D.F., 494
Bridges, William, 894
Bridgman, Paul W., 494
Briscoe, Dennis R., 388
Brisson, Mary, 506
Britton, Andrew, 69
Burke, R., 846
Burnham, John M., 494
Burt, D. N., 547
Buzacott, J.A., 680
Bylinsky, Gene, 894
Byrne, John A., 254

C

Cahill, Gerry, 440
Cameron, Brandt, 895
Campbell, Laurel, 49
Capek, Karel, 895
Carr, D.K., 892, 893
Cashion, Patrick, 873
Cellini, Benvenuto, 59

Certo, Sam, 388
Champy, James, 244, 246, 859, 892-893
Chappell, Kate, 152
Chappell, Matthew, 159
Chappell, Tom, 152, 160, 161
Chase, Richard B., 26, 36, 39
Cho, Fujio, 215
Chow, W.M., 775
Christnsen, Fiona, 538
Christopher, Martin, 440
Cirillo, Mary, 477
Clark, K.B., 203, 36
Clausing, Don, 136, 138, 149
Conners, Ron, 417
Conway, R.W., 729
Cooper, R.B., 389, 392, 775
Cordray, Richard, 409
Cornell, Ed, 213
Cousteau, Jacques, 336
Cox, James F., 680, 683
Cox, Jeff, 729
Cox, John F., 73, 109, 203, 205
Crosby, P.E., 300
Crowden, D.J., 300
Cypress, Harold, 494

D

Dannenbring, David G., 780
Davies, John, 754
Davies, R.L., 494
Deaton, Larry, 14
DeLurgio, S.A., 109
DeMeyer, Arnoud, 893
Deming, W. Edwards, 63, 69, 74, 140-142, 149-150, 252, 254, 300, 304
Dickson, W.J., 389, 392
Diesenhouse, Susan, 895
DiMasi, J.A., 894
Dinslage, Pat, 409
Dobler, D.W., 547
Dodge, Harold F., 149, 300, 304
Dougherty, K.S., 893
Drucker, Peter, 109, 887, 895
Dunlop, J.B., 222

E

Eckert Vickie, 322
Eisner, Michael, 465
Eli Whitney, 59
Ellram, Lisa M., 547, 637
Engelberger, Joe, 882
Epstein, L. Ivan, 494
Evans, Philip, 26, 39, 860, 893, 894

F

Fabricant, Soloman, 69, 73
Farley, Maggie, 370
Farrell, P.V., 547
Fauber, John, 191

Faurote, Fay Leone, 203
Fayol, Henri, 62
Fearon, H., 547
Feder, Barnaby J., 46
Feigenbaum, A.V., 300
Flynn, Anna E., 547, 638
Fogarty, D.W., 638
Ford, Henry, 59, 80, 190, 232,
Forrester, Jay W., 69, 73, 389, 392, 411, 440
Forster, N., 494
Fox, R.E., 203, 205, 730
Fraser, Allan, 695

G

Gagnon, R., 775
Galvin, Robert, 401
Gantt, Henry L., 62, 688, 802
Garrett, Dennis J., 521
Ghosh, S., 775
Gilbert, J., 547
Gilbreth, Frank B., 62, 375, 389, 392
Gilbreth, Lillian, 62
Giunipero, L., 547
Gleicher, David B., 854
Godfrey, 300
Goldfield, Robert, 873
Goldratt, Eliyahu M., 184, 203, 205, 716, 729-730
Goldstein, Michael, 89
Gopal, Christopher, 440, 494
Gore, W.L., 892
Grant, Eugene L., 300
Graves, S.C., 775
Greenbury, 494
Greene, Alice H., 878
Greene, Richard Tabor, 300, 893
Gross, D., 775, 780
Gryna, F.M. Jr., 69

H

Haag, Scott, 430
Hall, Edward T., 547
Hammer, Michael, 244, 246, 859, 893
Hammesfahr, R.D.J., 776
Harris, C.H., 775, 780
Hartley, J.R., 846, 847
Hauser, John R., 136, 138, 149, 494
Hax, Arnoldo C., 907
Hayes, Robert H., 26, 36-37, 39, 203, 205
Hazlitt, William, 401
Heady, R.B., 680, 683
Heath, David, 595
Heekin, Lucinda, 742
Heiner Muller-Merbach, 11
Heinritz, S.F., 547
Helgeson, W.B., 757, 780
Henderson, Rex, 314
Hessel, M.P., 341, 345

Hickman, Thomas K., 547
Hickman, William M., 638
Hildrebrand, Carol, 14
Hill, Terry, 203
Hirano, Hiroyuki, 775
Hodgetts, Richard M., 143
Hoffherr, G., 149
Hoffmann, T.R., 638
Hofstadter, Douglas, 341, 896
Holt, Charles C., 577, 585-586
Horne, Martin, 651
Horovitz, Jacques, 36, 39
Howard, Kevin, 494, 858

I
Ignall, Edward J., 759, 775, 780
Illingworth, Montieth M., 897
Imai, Masaaki, 244, 893
Ishikawa, K., 300, 304
Ishiwata, Junichi, 244
Israel, Donna J., 506

J
Jacobson, Richard, 127
Jansson, Brian, 172
Jenkins, G.M., 109
Johansson, Henry J., 494, 892-893
Johnson, Michael, 465
Johnson, S.M., 730
Jones, Billy Mac, 341
Juran, J.M., 63, 69, 74, 300

K
Kamauff, John W., 547, 638
Kantrow, A.M., 36, 203
Kaplan, Robert S., 389, 392
Karaska, G.J., 494
Karolefski, John, 46
Katzenbach, Jon R., 389, 392
Kelly, Wendell M., 521
Kelton, W.D., 776
Kerzner, H., 846
Kidder, Tracy, 801, 847
Kilbridge, M.D., 754, 757, 780
Killen, Kenneth H., 547, 638
King, Bob, 149
King, R.A., 892, 893
Kinsman, Susan E., 417
Klarich, Mary Jane, Jr., 570
Klein, Don, 812
Kleinberg, Eliot, 417
Koelsch, James R., 494
Kolchin, M., 547
Kolind, Lars, 866
Kunkel, Karl, 659

L
LaBarre, Polly, 894
Lane, Polly, 754
Lasdon, L., 585
Law, A.M., 776

Leavenworth, Richard S., 300
LeBlanc, Nicolas, 61
Lee, Consella A., 331
Lee, L., Jr., 547
Leenders, Michiel R., 547, 638
LeFauve, Richard G., 907
Leibfried, Kathleen H.J., 209, 244,
    246, 895
Leonard, George, 440, 445
Letterman, David, 692
Lettieri, Richard, 832
Levering, R., 895
Levitt, Theodore, 37, 39
Lewandowski, Jim, 907
Lilien, Gary L., 494
Lipin, Steven, 28
Littauer, S.B., 300
Lu, D.J., 776

M
MacArthur, Douglas, 140
McAteer, Brett, 651, 669
McCright, John S., 85
McDougall, Doug, 695
McGee, V.E., 109
McGrath, Michael E., 341, 893
McHugh, Patrick, 494, 893, 893
Mackey, Paul, 477
McLaughlin, Ward, 172
McNair, C.J., 209, 244, 246, 895
McNamara, Jennifer, 506
Madeley, John, 897
Magaziner, I., 109
Mahadera, Kumon, 370
Main, Jeremy, 300
Makridakis, S., 109
Maldutis, Julius, 895
Mantel, S.J., 846
Margolis, Paul, 55
Marsh, S., 149
Martin, Andre, 680
Maskell, Brian H., 389
May, Lyn, 573
Meisol, Patricia, 521
Meridith, J.R., 846
Miles, L.D., 547
Miller, Bill, 903, 906
Miller, Jeffrey G., 893
Modigliani, Franco, 577, 585-586
Mooney, M., 341, 345
Moore, James M., 494
Moran, D.E., 892, 895
Moran, J.W., 149
Morris, C.R., 847
Moskowitz, M., 895
Muller-Merbach, Heiner, 37
Munton, A.G., 494
Murnane, William, 172
Murray, Charles J., 907
Muth, J.F., 577, 585-586

N
Nakajima, Seiichi, 245
Nakane, Jinichiro, 893
Nakui, S., 149
Nemhauser, G.L., 585
Newell, A., 753, 780
Normann, R., 203
Nulty, Peter, 254

O
Obershaw, David, 708
Ohno, Taiichi, 69, 74, 231, 245
Olson, Jane, 450
Olson, Thomas, 832
O'Neill, Nancie, 360
Orlicky, Joseph, 643, 680, 683
O'Toole, Jack, 907
Owen, Jean V., 494

P
Pacelle, Mitchell, 896
Paquin, Glenda, 669
Parkinson, C. Northcote, 37, 440, 445
Paton, Scott Madison, 115
Pendlebury, A. John, 494, 896
Pereira, Joseph, 89
Perlez, Jane, 595
Peters, Tom, 37
Plane, Donald r., 585, 638, 776
Pollack, Michael G., 341, 345
Powers, William F., 465
Prather, Catherine Porter, 215
Pritsker, A.A.B., 776
Pterson, R., 638
Puchek, Vladimir, 893

R
Raedels, Alan R., 547, 638
Ranii, David, 55
Ranney, G., 149, 150, 300
Rao, C. Radhakrishna, 494
Rautenstrauch, Walter, 419, 440, 445
Reed, John, 28
Reed, Ruddell, Jr., 494
Reich, R., 109
Reimann, Curt W., 115
Ritterhaus, Gary, 158-159
Roberts, Edward B., 341
Robinson, Alan, 245
Roethlisberger, F.J., 389, 392
Rogers, D., 494
Roger, Jerry, 538
Romig, Harry G., 149, 300, 304
Rosenbluth, Hal, 495-505, 506

S
Sanger, David E., 897
Sarafin, Raymond, 907
Schmit, Julie, 507
Scholtes, P.R., 300, 389, 846
Schonberger, R., 547

Schrage, L., 585
Scism, Jack, 766
Scott, Al, 49
Senge, Peter M., 440, 445
Servan-Schreiber, Jean-Jacques, 69, 74
Shanthikumar, J.G., 680
Shaughnessy, H., 846
Shenkar, Oded, 389
Shewhart, Walter A., 59, 63, 69, 74,
       253-254, 300, 304, 371
Shingo, Shigeo, 215, 231-232, 245-246,
       252, 300, 304, 341-345
Shulman, Lawrence E., 26, 39, 860, 893, 895
Sigal, C.E., 776
Silver, E.A., 638
Simon, Herbert A., 577, 585-586, 753, 780
Simone, Thomas, 46
Smith, Adam, 58, 59, 61, 69, 74
Smith, Doug, 85
Smith, Douglas K., 389, 392
Smith, Jack, 856
Smith, Wayne, 538
Spencer, Michael S., 73, 109, 680, 683
Sprinkles, Jim, 766
Srikarth, M.L., 729
Stahl, Walter R., 494
Stalder, James, 832
Stalk, George, 26, 39, 860, 893
Starr, Martin K., 780
Stein, Charles, 570
Stephens, K., 300
Summers, David A., 893
Surro, Jennifer R., 233

T
Tabit, George E., 331
Tascarella, Patty, 314
Taylor, Frederick W., 59, 61-62
Tersine, R.J., 547, 638
Thomson, R.W., 222
Tichy, Noel, 893
Tilley, A.R., 389, 392
Tippett, L.H.C., 373, 389, 392
Tonge, Fred M., 780
Tooley, F.V., 341
Turner, Richard, 465

U
Umble, M. Michael, 729

V
VanderVeen, Don, 812
Vasilash, Gary S., 906-907
Villers, R., 440, 445
Vincke, P., 494
Vollman, Thomas E., 203, 585, 638, 680,
       683, 776

W
Wadsworth, H., 300
Walker, Tom, 824

Waltman, Les, 620
Waren, A., 585
Warren, Alex M., Jr., 215
Waterman, R.H., Jr., 37
Watts, Thomas G., 506
Weiman, Carl F.R., 896
Wenbin, Yan, 897
West, Tom, 801
Wester, L., 754, 757, 780
Wetzel, Jay, 898
Wheeler, William A. III, 494, 896
Wheelwright, Steven C., 26, 37, 39, 109,
       203, 205
Whitney, Eli, 61
Whybark, D. Clay, 203, 585, 638, 680,
       683, 776
Wight, Oliver W., 680, 683
Wilbur, Donn K., 659
Wingard, Donald P., 73
Wolff, Edward N., 69, 73
Wolsey, L.A., 585
Womack, James P., 907
Wurgaft, Hernan, 775

Y
Yorks, L., 149, 300

Z
Zeleny, M., 341, 345
Zhu, Z., 680, 683
Zuboff, Shoshana, 341, 345
Zuckerman, Laurence, 897
Zweben, Monte, 708

SUBJECT INDEX

A

ABC inventory classification, 539
Acceptance sampling, 283-291
    defined, 132, 283
Action learning, 140, 352
Activities-on-arrows network, 808
Activity cycles, in PERT networks, 810
Activity symbols, process charts with
    sequenced, 216-222
Activity-based costing (ABC) accounting
    systems, 19, 414-415
    job design and, 358-359
Activity-on-nodes network, 808
Actual capacity, 396
Adaptive systems, 332
Aesthetics, as quality dimension,
    119-120
Agents, purchasing, 525
Aggregate demand, 199, 560
Aggregate planning, 548-586
    basic model for, 559
    converting schedule to master
        production schedule, 647
    defined, 550
    linear programming for, 569-574
    nonlinear cost model for, 577-580
    policies for, 561-569
    transportation model for, 574-577
Agricultural business, global, 885
Air traffic control, expert systems for, 335
Alternative materials analysis. *See* Value
    analysis
Allocation rule, for transportation
    costs, 461
Analytic processes, 176-178
    chocolate products, 177
    defined, 176
    marketing products of, 181
    mixed synthetic and, 182-184
    sourcing for, 179-180
AOQL. *See* Average outgoing
    quality limits
Appraisal, cost of, 132
Apprentices, artisans and trainees, 60-61
Artificial intelligence, 332, 336
Assemble-to-order (ATO) inventory
    environment, 652
Assembly
    design for, 853
    line balancing, 736
    sequenced, 59, 62-63
    technology of design for, 328
Assembly chart, many parts to one, 173
Assignment method
    defined, 691
    for loading, 691-696
Assignment schedule, minimum cost, 569
Attributes, control charts for, 275

B

Average
    moving, for short-term trends, 96
    weighted moving, 97
Average outgoing quality limits (AOQL),
    289-290
Average performance, 356

Backorders, 574-575
Backward pass rule, 816, 818
Balance
    line. *See* Line balancing
    perfect, 751
Balance delay
    defined, 751
    perfect balance with zero, 751
Baldrige National Quality Award,
    142-146, 322, 530
    1995 award examination criteria
        categories, 145
    award-winning companies, 144
    criteria framework, 145
    defined, 142
Bar charts, 253, 256
Bar code technology, 787-790
Batch delivery, EOQ models for, 605-612
Batch processing, career success in, 30
Batch size, in compensating for
    defectives, 251
Benchmarks
    HMMS model as, 579
    smart, 856
Benchmarking, 209
    in change management, 855-857
    defined, 182
Best location, 452
Best practice
    adopting, 84-85
    defined, 182
Bidding, 532
Bill of material (BOM), 653-655
    defined, 653
    for job shop, 186
Bill of resources (BOR), 668
Binomial distribution, 291-293
Biomechanics, 389
Bottlenecks
    capacity and, 407
    finite scheduling of, 715-721
    permanent, 408
Breakdowns, zero, 240
Breakeven
    costs, 416
    decision model, 441-445
    profit and loss, 420
Breakeven capacity
    related to work configurations, 424
    interfunctional planning for, 418-419

Breakeven chart
    construction of, 419-421
    linear, 418-419
    nonlinear, 421
    profit and loss, 422
Breakeven equations, linear, 424
Breakeven model
    analysis of linear, 421
    use of in equipment selection, 474
Buffer stock
    defined, 593
    for perpetual inventory system, 624-625
Building, cost determinants of, 467
Bureaucracy, as inhibitor of flexibility and
    productivity, 49
Business unit, strategic, 80
Buyer
    negotiation between supplier and, 287
    risks of, 525-526
Buying, cost determinants in facility
    planning, 467

C

c-chart, 278
CAD/CAM technology, 65, 314, 337
Callbacks, product, 133
Canadian quality standard (CSAZ229.1), 128
Capacity
    actual, 396
    bottlenecks and, 407
    breakeven costs, revenues and, 416
    definitions of, 396-400
    design for, 427
    dynamic adjustment of maximum, 404
    growth adjustments for maximum, 406
    as layout criteria, 482
    learning increases, 431
    linear programming and, 425
    margin contribution and, 425
    maximum rated, 398-402
    qualitative aspects of, 401-402
    service, 400
    supply and demand relationships with,
        399
    systems approach and, 435
Capacity crises, contingency planning for,
    410-411
Capacity management, 394-440
Capacity planning
    cost drivers for, 414-415
    forecasting and, 415
    interfunctional breakeven, 418-419
    for public utilities, 430
    rough-cut, 668
    upscaling and downsizing in, 405-406
Capacity requirements planning, (CRP),
    404, 666-668
    defined, 667

Capital productivity, 45
Capricious demand, 404
Career opportunities in OM, 29-34
Carrying cost, inventory, 597-599
Cause and effect charts, 253, 261-265
    defined, 261
Causes
    assignable, 266
    chance, 266
Cellular layout, 480
Cellular manufacturing systems, 193-195
Centralized global information system, 523
Certification of suppliers, defined, 543
Chance causes, 266
Change, factors that allow organizational,
        854-855
Change management, 848-907
    continuous improvement for, 857-859
    defined, 851
    OM's part in, 853-854
    reengineering for, 859-870
Change management projects, defined, 801
Changeovers
    costs of, 596
    defined, 230
    process, 230-233
    technologies for, 331
Changes, adapting to external, 851-852
Chart
    bar, 256
    breakeven, 418-419
    c-, 278
    cause and effect, 261-265
    control, 267
    layout or sequencing, 703
    machine-operator, 234
    p-, 275
    process. See Process charts
    R-, 273
    x̄-, 267
Check sheets, data, 255-256
Chocolate products, analytic process for, 177
Chunnel, 799
Closed-loop MRP, 671-672
    defined, 671
Coding, for MRP, 656-658
Coefficient
    correlation, 100
    of determination, 101
Columns, permutability of, 750
Commercial diplomacy, 887
Commodities, used in analytic processes,
        179-180
Common good capitalism, 153
Communication flows, systems, 524-525
Community attitude, as factor in location
        planning, 469
Company, stages, 27
Competition, global, 59, 66
Completion time, distribution of project,
        827-829

Complex machine
    carrying parts for, 534
    failure cost matrix for, 534
Components, purchased materials as, 516
Computer-aided design (CAD), 64. See
        also CAD/CAM
Computer-aided manufacturing (CAM), 65
        See also CAD/CAM
Computer-aided process planning (CAPP),
        220
Computer-based training, 354
Computers, as substitute for stopwatches
        in job observation, 362-366
Computer-user interfaces, technology, 337
Concurrent engineering (CE), 838-840
Conformance, as quality dimension, 119
Construction, 12
Consulting businesses, global, 885
Contingency planning, for capacity crises,
        410-411
Continuity, 839
Continuous flow processing, 169
Continuous improvement (CI), 26, 34, 85,
        208-209
    for change management, 857-859
    job design programs for, 381
Continuous job design improvement
        programs, 381
Continuous processes
    defined, 188
    no bottlenecks in, 715
Continuous production, 613
    from EOQ through ELS to, 618-619
Continuous project-development model,
        867-870
Control, statistical quality, 59, 63
Control chart
    for attributes, 275-278
    defined, 267
    for ranges, 274
    for statistical process control, 265-267
    for variables, 274
Control limits
    defined, 269
    interpreting, 274
    upper and lower, 269-274
Control management processes, 801
Controlled storage, 227
    process chart symbol for, 217
Cooperative storage, of inventory, 598
Correction factor, rest and delay, 369-371
Correlation
    in forecasting life cycle stages, 93
    scatter diagram implying strong, 264
Correlation coefficient, as forecasting
        method, 100
Cost
    of appraisal, 132
    estimates of for project management, 823
    expected, of well-designed job, 371
    of failure, 133

    fixed, 416
    flow, 487-488
    with input-output transformation
        model, 23
    inspection, 288
    opportunity, 692
    per unit, 190
    prevention, 130-132, 239
    of quality, 130-132
    standards, 357
    total, 417
    total variable, 417, 57
    trade-offs between time and,
        832-837
    variable, per unit, 417
Cost curve
    discontinuity of, 632
    total, 133
Cost-benefit analysis, 467
CPM, 806
Critical path, 806
Critical path computation, for PERT
        network, 811
Criticality, 540
Critical-path method (CPM), 805-811
Cross-docking, 531
Cumulative (or total) lead time, 647
Curve
    experience, 432
    operating characteristic, 285-287
Custom shop, 169
    adding the, 198
    defined, 184-185
Customer demand, forecasted, 551
Cybernetics, 332
Cyborgs, 336
Cycle time
    analysis of, 362
    constraints on, 746-749
    defined, 734
    management, 732-780
    number of stations and, 745-746
    reduction of, 361-362
    shortest common, 366
Cycles
    defined, 86
    extrapolation of trend lines, step
        functions and, 91-92
    historical forecasting with seasonal, 93
    job, 361
    as time series, 92

D
Data check sheets, 253, 255-256
Days of inventory
    defined, 526
    as performance measurement, 526
Decentralized global information
        system, 523
Decision matrix, 37-39
    equivalent, 72

for two-stage problem, 72
Decision models, 37-39
    breakeven, 441-445
Decision trees, 70-73
    expert system for marine underwriter, 334
Defectives
    compensating for, 251
    percent, versus prevention costs, 130
Delay, rest and, 753
Delivery
    contrasting batch and continuous, 613
    just-in-time, 453, 605, 765-768
    technology of packaging and, 324-325
Delphi method of forecasting, 102
Demand
    aggregating, 199
    capricious, 404
    continuity, 594
    distribution, 593-594
    independent or dependent, 594
    peak and off-peak, 400-405
    when greater than supply, 212
Deming Prize, 141
Deming's 14 points for quality
        achievement, 141
Department supervisor, 32
Dependent demand systems, as
        characteristic of MRP, 643-646
Depreciation, 420
Design
    computer-aided, 64
    use of simulation for modification, 329
Design for assembly (DFA), 853
    defined, 328
    technology of, 328
Design for capacity (DFC), 427
Design for disassembly (DFD), 871-873
Design for manufacturing (DFM), 853
    defined, 328
    technology of, 328
    variance and volume in, 475
Detailing, 132
Determination, coefficient of, 101
Deterministic estimates, for constructing
        PERT networks, 806-807
Deterministic line balancing, 738
Diagnostic production studies, 361
Diagram, scatter, 263
Dies, single-minute exchange of, 231-233
Direct costs, with input-output
        transformation model, 23
Disassembly, design for, 871-873
Discounts, costs of, 599
Distance learning, 354-355
Distribution
    binomial, 291-293
    failure, 403
    normal, [Appendix]
    Poisson, 294, [Appendix]
    Weibull, 403
Distribution problem, 37-39

Distribution resource planning (DRP),
        668-670
Division of labor, interchangeable parts
        and, 59
Downsizing, upscaling and, 405-406
Drift-decay phenomenon, 122
Drum-buffer-rope model, OPT
        system's, 719
Due date, 712
Durability, as quality dimension, 119-120
Dynamic inventory models, 592

E
Economic lot size (ELS)
    defined, 604
    models for, 612-623
Economic order quantity (EOQ)
    defined, 604
    models for, 605-612
Economies
    of scale, 57, 190-192, 431
    of scope, 192, 233
Efficiency frontiers, 862
Elasticity
    defined, 51
    price-demand, productivity and, 51-57
    quality, 55-57
Electronic data interchange (EDI), 315,
        521, 524, 595
Electronic transfer of funds (ETF), 20
Empirical testing, 326
End products, 643
Engineering, concurrent, 838-840
Environment
    greening the, 870
    industry perspective on, 900-907
Environmental impacts, 852
Equipment selection, 474-476
    for process technologies, 449
Ergonomics, 389
Estimates, deterministic, 806-807
Ethics, 874-880
    in change management, 853
    of purchasing, 529
Expected values (EV), 38
Expediting
    to control lead time, 621-622
    defined, 524
Experience curve, defined, 432
Expert system
    defined, 332
    examples, 878
    for marine underwriter, 334
Exponential smoothing methodology,
        109-111
Extraspection, 11
External changes, adapting to, 851-852
Extrapolation, 91

F
Facilities planning, 446-495

decisions, models for, 451
    layout of, 450
    location of, 448
    using scoring models, 468-474
    who does, 451
    for work configuration requirements, 467
Fail-safe systems, 389
Failure
    costs of, 133
    definition critical to quality
        evaluation, 123
    mean time between, 123
    probability of, as function of cost of
        prevention, 239
Failure model, for determining spare parts
        inventory, 533
Features, as quality dimension, 119
Financial services, global, 885
Finished goods, 517
Finite scheduling, 408
    defined, 718
    of bottlenecks, 715-721
First-in, first-out (FIFO), 524, 702
Fishbone chart, 261-263. See also Cause
        and effect charts
Fixed costs
    defined, 416
    with input-output transformation
        model, 23
Flexible manufacturing systems (FMS),
        169, 193-195, 332, 425
    changeovers in, 596
    defined, 332
    inventory costs in, 603
    role of, 198
Flexible office systems, (FOS), 425
    defined, 332
Flexible process systems (FPS), 332, 425
    career success in, 30
    defined, 193-195
    inventory costs in, 603
Flexible processes, 332
Flexible production systems, (FPS) 59
    defined, 64
Flexible service systems (FSS), 332
Flexible-purpose equipment, 425
Float process, 316
Floor plan models, defined, 483
Flowcharts, process, 209-210
Flow costs, 487-488
Flow shop (FS), 169
    benefits and constraints of, 190
    career success in, 30
    comparing job shops and, 189
    defined, 187-190
    intermittent, 192-193
    inventory costs in, 601-602
    job enrichment in, 382-383
    no bottlenecks in, 715
    supply consumption in, 521
    total productive maintenance for, 240

Flow time
  mean, 705
  total, 705
Food and Drug Administration (FDA),
      879-880
Food and drug testing, 879-880
Forecasting
  based on customer demand, 551
  capacity planning and, 415
  considerations for MRP, 645
  defined, 90
  errors, 103-104
  historical, with seasonal cycle, 93, 95
  importance of, 556
  life cycle stages, 78-111
  multiple, 102
  perspectives on, 90
  planning horizon for, 556
Forecasting methods, 95-104
  base series modification of
      historical, 95
  coefficient of determination, 101
  comparative errors for different, 102-
      103
  correlation coefficient, 100
  Delphi method, 102
  exponential smoothing methodology,
      109-111
  moving average for short-term
      trends, 96
  regression analysis, 98
  weighted moving averages, 97
Forward pass rule, defined, 816
Functional field approach, to OM, 10
Fundamental theorem of linear
      programming, 572

G
Game, decision trees for, 71
Gantt charts, 688
  for project-planning, 802-805
  layout, 703
  load, 688
General purpose equipment (GPE), 185
Glass processing technologies, 315-324
Global activity, changing, 852
Global aspects of career paths, 31
Global competition, 59, 66
Global consulting businesses, 885
Global information system, 523-524
Global operations, POM's responsibility
      for successful, 885-888
Global workforce, 370
Goods, differentiating between services
      and, 15-17
Group technology
  cellular manufacturing systems,
      193-195
  families, 431
  layout, 480
  modular production and, 427

H
HACCP system, 875-878
Hawthorne study, 383
Hazard control, preventive system for.
      See HACCP
Health care industry, planning services
      in, 570
Hedging, how purchasing agents use, 204
Heuristic
  defined, 488
    Helgeson and Birnie's, 757-759
    Kilbridge and Wester's, 754-757
    line balancing, 738, 753
    methods, 689
Histograms, 253, 256-258
Historical forecasting
  base series modification of, 95
  with seasonal cycle, 93
History of P/OM, 60-66
HMMS model, 577
Home shopping, online processing
      for, 195
Hotel management, human resources
      function in, 360
House of quality (HOQ), 136-140
  defined, 137
  scoring model for, 149-151
Hub and spoke diagram, 81
Human factors, 389
Human resources management (HRM),
      348-349
Hypergeometric O.C. curves, 301-304

I
Identifiable systems causes, 267
Improvements, 227
Incremental improvement planning, 221
Indirect costs, with input-output
      transformation model, 23
Industrial Revolution, 60-61
Industry, recognition for, 140-146
Inelasticity, perfect, 52
Information, pooling, 102
Information processing, process flow
      layout for, 227-230
Information system
  defined, 18-19
  global, 523-524

  intelligent, 523
  for materials management, 523-525
Inherent systems causes, 266
Innovations, technological, 317
Input-output model
  illustration, 20
  quality and, 133
  profit, 24-25
  with associated costs and revenues,
      21-23
  See also transformation model
Inspection, 227, 282-283

100% versus sampling, 282
  cost of, 132, 288
  process chart symbol for, 217
  receiving, storage and, 531
  See also appraisal
Intelligent information system,
      defined, 523
Interchangeable parts (IP)
  defined, 61
  division of labor and, 59
Intermittent flow shop (IFS), 192-193
  line-balancing in, 736-737
  model, economic lot size for, 614-617
Internal materials management
      system, 515
International producer standards. See
      ISO 9000 and ISO 14000
Internet, 315
Introspection, 11
Inventory
  costs of, 598-603
  days of, 526
  defined, 590
  management, 588-638
  static, 535
Inventory models, static versus dynamic,
      591-592
Inventory record files, 655
Inventory system
  ABC classification, 539
  costs of running, 600-601
  periodic, 603-604, 628-630
  perpetual, 603, 623-628
Ishikawa chart, 261-263. See also Cause
      and effect charts
ISO 9000
  defined, 126
  quality assurance certificate, 126
  standards, 126-129
ISO 9001-9004, 128-129
ISO 14000/14001, 886

J
Japanese Industrial Standard Z8101, 129
Japanese industry, after MacArthur
      initiative, 425
Japanese quality standards, 129
JIT. See Just-in-time delivery
Job cycle, 361
Job design
  activity-based cost systems and,
      358-359
  continuous improvement programs
      for, 381
  enrichment and, 382-383
  evaluation, improvement and, 380
  expected productivity and, 371
  layout of workplace and, 477-480
  planning teamwork and, 346-393
  work sampling and, 373-375
  work simplification and, 381

Job improvement, evaluation, design and, 380
Job observation
  corrections for, 369-371
  sample size for, 371-373
  in setting time standards, 361
  study sheet, 362-363
  training for, 368
Job shop (JS), 169
  bill of materials for, 186
  bottlenecks in, 715
  comparing flow shops and, 189
  defined, 185
  inventory costs in, 602
  materials requirements in, 521
  process layouts, 480
Just-in-case delivery, 175
Just-in-time (JIT) delivery, 453, 605, 765-768
  in auto industry, 595
  defined, 175

K
Kaizen, 215, 221
Kanban
  defined, 765
  dual-bin, 789

L
Labor, division of, 57-58, 59
Labor productivity, defined, 44
Last-in, first-out (LIFO), 524
  defined, 702
Lateral transferability, defined, 750
Layout
  assignments, 485
  cellular, 480
  criteria for, 482
  defined, 478
  developing station, 749
  of facility, 450
  group technology, 480
  heuristics to improve, 488
  interactions within, 476
  job shop process, 480
  models, 482
  opportunity costs of improvement of, 478
  product-oriented, 480
  types of, 480-482
  workplace, job design and, 477-480
Layout chart, 703
  process, 223-225
Lead time
  defined, 591
  determination of, 619-621
  distributions for, 594
  expediting to control, 621-622
  variability of, 622-623
Leaders, project managers as, 800
Lean production systems (LPS), 59
  defined, 63

Learning
  action, 352
  distance, 354-355
  increases capacity, 431
  service, 434
  team, 352
Learning curve
  average time, 432
  cumulative time, 432
  defined, 432
  effects of, 369
Learning organization, 21st century, 525
Least-squares method, in regression analysis, 99
Legal factors, affecting location decision, 453-454
Leveling
  factors, 368
  resource, 830-831
Life cycle stages
  as considerations in equipment selection, 476
  evolution of, 87
  forecasting, 78-111
  product versus process, 196
  understanding for OM action, 86-88
Limits
  control, 269-274
  range of quality tolerance, 131
  tolerance, 131
Line balancing (LB)
  basic concepts for, 739-743
  defined, 736
  deterministic, 738
  heuristic, 738, 753
  methods of, 736-753
  queuing aspects of, 760
  stochastic, 738-739, 759-765
Line function, process management is, 211
Linear breakeven equations, 424
Linear breakeven model, analysis of 421
Linear programming (LP)
  for aggregate planning, 569-574
  capacity and, 425
  defined, 425
  fundamental theorem of, 572
Load chart, Gantt, 688
Loading, 687-688
  defined, 556
  See also Shop loading
Location
  to enhance service contact, 452
  factors affecting decisions for, 453-454
  factors for comparing alternative, 471
Location decisions
  public relations considerations in, 465
  qualitative factors in, 452
  using transportation model, 454-465
Logistics
  defined, 189, 515
  packaging, 858

shop floor, 789
Loss, profit, breakeven and, 420
Lot sizing
  for MRP, 664-665
  part-period policy, 680-683
Lot-for-lot ordering, defined, 660
Low-level coding, defined, 657
Loyalty to the firm, 350

M
Machine-operator chart, 234
Machinery, global, 885
Mail-order business, 742
Maintenance, total productive, 236
Maintenance strategy
  preventive, 239
  remedial, 239
Make or buy decision, 592
Make or buy model, 894
Make-to-order (MTO) inventory environment, 652
Make-to-stock (MTS) inventory environment, 652
Malcolm Baldrige Award. See Baldrige National Quality Award
Management
  capacity, 394-440
  human resources, 348-349
  materials, 512-547
  operations. See Operations management
  production, 17
  project, 27
  scientific, 59, 61-62
  time-based, 211
Management of technology (MOT), 308
Manager
  OM, 32
  project, 800
Manufacturability, design for, 853
Manufacturing
  applications, of OM, 8
  computer-aided, 65
  defined, 15
  enterprise wheel, SME, 888
  information systems in, 18-19
  synchronized, 718-720
  technology of design for, 328
Manufacturing resource planning. See MRP II
Manufacturing systems
  flexible, 332
  group technology cellular, 193-195
Mapping, for quality systems, 135
Marketing models, for predicting sales, 90
Master production schedule (MPS), 646
  inputs and outputs for, 650
Material, bill of, 653-655
Material criticality, 539
Material requirements planning (MRP), 640-683
  background for, 643

basic calculations and concepts for, 658
inputs and outputs of system, 656
operation of, 655-658
See also MRP
Materials
cost leverage of savings in, 518-519
criticality of, 539
dollar value of, 540
penalties of errors in ordering, 519-520
taxonomy of, 515-518
Materials management, 512-547
of critical parts, 533
defined, 514
differences by work configurations, 520-523
information system in, 523-525
systems perspective for, 514-515
Materials shortages, 537
Mathematical degeneracy, 461
Mature product, protection of, 88
Mean, standard error of the, 272
Mean flow time, defined, 705
Mean time between failures (MTBF), 123
Methods analysis, 535
Miniaturization, 338
Minimum cost assignment schedule, 569
Mixed analytic and synthetic processes, 182-184
Mobile robots, 880
Model
aggregate planning, 559
continuous project-development, 867-870
decision, 37-39
failure, 533
facilities planning, 451
floor plan, 483
input-output, quality and, 133
intermittent flow shop, 614-617
layout load, 483
make or buy, 896
marketing, 90
net present value, 896
OPT system's drum-buffer-rope, 719
quality, 119-121
quality control feedback, 134
quantity discount, 630
scoring, 470-472
simulation of queuing, 776
trade-off, 520
trade-off with workforce adjustment, 569
transformation, 19-23
use of by OM, 6-7
Modular production, group technology and, 427
Modularity, principle of, 428
Monte Carlo number assignments, 777
Motivation, employee, 383
Moving averages
for short-term trends, 96
weighted, 97

MRP
closed-loop, 671-672
net change, 665-666
regeneration, 665-666
weaknesses of, 670-671
MRP II, 673
MRP. *See* Material requirements planning
Multifactor productivity
average annual percent changes in, 44-45
defined, 44
Multiple-channel stations, defined, 764
Multiple forecasts, pooling information and, 102
Multiple-sampling plans, 290

N
Negotiation between supplier and buyer, 287
Net change MRP, 665-666
Net present value (NPV)
defined, 310
model, 893
Neural networks, 332
defined, 336
programming and, 314
New orders, job shop process flowcharts for, 216
New products, 86-88
Nonfunctional quality dimension, 124-125
Nonlinear cost model, for aggregate planning, 577-580
Normal time, defined, 368
Northwest Corner method (NWC)
defined, 460
for determining transportation costs, 460

O
O.C. curve
binomial, 292-293
defined, 285
hypergeometric, 301-304
Poisson, 294
Object-oriented software, 477, 708
Observation, job. *See* Job observation
Obsolescence, of inventory, 598
Occupational Safety and Health Act (OSHA), 877
Office, virtual, 883
Office systems, flexible, 332
Olympic games, 1996 Centennial, 573
Olympic perspective, for quality achievement, 251
Olympics, as systems symbol, 886
OM
developments, timeline of, 59
system, 12
transformations, 58-66
See also Operations management
Operating characteristic curves, 285-287.
See also O.C. curves

Operation, process chart symbol for, 217
Operations
analysis of, 225-227
defined, 7
working definition of, 7-9
Operations management
career paths in, 31
as contrasted with production management, 17-19
defined, 7, 17
explaining, 6-9
functional field approach, 10
manufacturing applications, 8
overview and introduction, 3-75
positions and career opportunities in, 29-34
role in developing strategies, 81-88
role in productivity attainment, 40-74
as scarce resource, 85
service applications, 8-9
strategies for, 80
systems approach and, 4-39, 10-13
use of models by, 6-7
Operations managers, workers and, 350-351
Operations sheet, bill of materials and, for job shop, 186
Opportunity costs
defined, 478, 692
for layout improvement, 478
OPT, 717
bottlenecks with, 408
Optical scanners, 327
Optimized production technology (OPT), 717
defined, 184
Optimum production technique, 717.
See also OPT
Order errors, 519-520
Order point policies (OPP), 603
Order promising
changes in, 651-652
defined, 646
Ordering, costs of inventory, 596
Organization chart
for quality deployment, 139
system mapping across traditional, 13
traditional, 10
Organizational change, factors that allow, 854-855
Organizational productivity, operational measures of, 46-47
Outgoing quality limits, average, 289-290
Out-of-stock costs, 600
Output
converting to dollars, 44
quality, 134-135
Outsourcing, in auto industry, 595
Overhead, 414

P
p-chart, defined, 275

P/OM
  development, stages of, 25-28
  history of, 60-66
Packaging
  logistics, 858
  technology of delivery and, 324-325
Pareto analysis, 253, 258-261
  defined, 258
  procedures, 259-261
Part, c-charts for number of defects per, 278
Part-period lot-sizing policy, 680-683
Parts
  interchangeable, 59, 61
  materials management of critical, 533
  value analysis for purchased, 226
Payback period, 310
Payoffs, calculating for distribution
    problem, 37-39
Peak demand, off-peak and, 400-405
Perceived quality, as quality dimension,
    119-120
Perfect inelasticity, 52
Performance
  evaluation, 355-357
  improvement manager, 34
  as quality dimension, 119
Periodic inventory system, 603, 628-630
Permutability of columns, defined, 750
Perpetual inventory system, 623-628
  defined, 603
  two-bin, 627-628
PERT
  constructing networks, 806
  defined, 806
Pilferage, of inventory, 598
Pilot studies
  defined, 865
  for testing designs, 329
Planning
  capacity. See Capacity planning
  capacity requirements, 404, 666-668
  distribution requirements, 668
  facilities, 446-495
  incremental improvement, 221
  material requirements, 640-683
  rough-cut capacity, 667
  teamwork and job design, 346-393
  of transitions, 859
  zero-based, 221
Planning and control system, total
    manufacturing, 648-650
Planning horizon, 556
Plant
  availability, as factor affecting location
      decision, 453-454
  cost differentials, 462
  location, decisions for, 455-456
Poisson distribution, 294, Appendix
Poka-yoke system, 342
Positions, organizational, in OM, 29-34
Precedence diagram, defined, 743

Predetermined time standards, 375
Present value, net, 310
Prevention, costs of, 130-132
Preventive maintenance strategy, 239
Price-demand elasticity, productivity and,
    51-57
Principle of modularity, defined, 428
Probabilistic estimates, for constructing
    PERT networks, 807
Process, 168-171
  configuration strategies, 166-203
  defined, 168
  improvement and adaptation of, 208-209
  simulations, 170
  steps, 168
  synchronization, 234-236
Process analysis
  background for using, 211-216
  defined, 211
  process redesign and, 206-247
Process chart symbols, 217
Process charts
  creating, 220
  defined, 209
  for electrical panel box, 210
  flowcharts, 209-210
  illustrated, 218-219
  purpose of, 213
  with sequenced activity symbols, 216-222
  use of, 214-216
Process flow, 163-508
  layout diagrams, 225
  seven points for consideration in design
      of, 212
  types of, 173-176
Process inputs, as factor affecting location
    decision, 453-454
Process layout charts, 223-225
Process life cycle stages, 196
Process management, as line function, 211

Process quality control methods
  defined, 253
  seven, 253-283
Process redesign, process analysis and,
    206-247
Process systems, flexible, 193-195, 332
Process types, expanded matrix of,
    196-198
Processes
  analytic, 176-178
  career success and types of, 29-31
  classifying, 170
  as contrasted with projects, 801
  continuous, 188
  control management, 801
  strategic aspects of, 172-176
  synthetic, 173-176
  types of, career success in, 29-31
Processing technologies, glass, 315-324

Producer standards, international, 125-130
Product
  new, 86-88
  protection of established, 88
Product callbacks, defined, 133
Product life cycle stages, process life
    cycle stages versus, 196
Production
  continuous, 613
  defined, 17
  modular, 427
  working definition of, 7-9
Production management
  as contrasted with operations
      management, 17-19
  defined, 17
Production schedule, 684-730
  aggregate planning for, 550
  defined, 686
  master, 646
Production standards, 357
Production systems
  flexible, 59, 64
  lean, 59, 63
  push and pull, 766-770
Production theory, seven steps of, 60-66
Productive capacity, 432
Productivity
  attainment, 40-74
  capital, 45
  defined, 42
  expected, of well-designed job, 371
  as global systems measure, 48
  labor, 44
  measures of, 42-51
  multifactor, 44
  operational measures of, 46-47
  price-demand elasticity and, 51-57
  rates, in aggregate planning, 555
  relative, 45
  standards, 367-368

  as systems measure, 47
  OM's key role in, 40-74
Product-oriented layout, 480
Products
  end, 643
  specification of parts used in, 428
Profit, breakeven, loss and, 420
Profit and loss breakeven chart, 422
Profit model, input-output, 24-25
Program evaluation and review technique.
    See PERT
Programming
  computer, 337
  industry in India, 370
Progress reports, project, 829
Project, continuous model for
    development of, 867-870
Project completion times, distribution of,
    827-829

Project management
   basic rules for, 801-802
   defined, 27
   rapid response, 796-847
Project manager, 33
Project managers, 800
Projects, 169
   career success in, 29
   change management, 801
   as contrasted with processes, 801
   defined, 798
   designing, 829-831
   inventory costs in, 603
   managing, 800
Proportional sampling, O.C. curves and, 285-287
Pull-production systems, 766
Purchasing
   ethics of, 529
   function, 525
   requiring bids before, 532-533
Purchasing agents, 527-529
   defined, 525
   hedging used by, 204
Pure number, 473
Purpose utilities, as quality dimension, 121
Push-production systems, 766
Put-away, 230

Q
Quality
   cost of, 117
   costs of, 130
   defined, 116
   defined and measured, 122
   definition of failure critical to evaluation of, 123
   dimensions of, 118-125
   director of, 33
   drift-decay phenomenon, 122
   elasticity of, 55-57
   house of, 136-140
   industry perspective on, 152-161
   input-output model and, 133
   integrity of, 875
   models of, 119-121
   output, 134-135
   teamwork for, 140
   three costs of, 132
   TQM is systems management of, 135
   why is better quality important? 114
Quality achievement
   attitude and mind-set for, 250-252
   Deming's 14 points for, 141
Quality award, Baldrige. See Baldrige National Quality Award
Quality circles (QC), 140
Quality control
   feedback model, 134
   methodology for, 251-252
   methods for, 248-304

statistical, 59, 63
   technology, 327-328
Quality correction, 134
Quality cycle, Deming's, 252
Quality dimension
   extended taxonomy of, 120-121
   Garvin's, 119-120
   nonfunctional, 124-125
   other "ilities", 121
   purpose utilities, 121
Quality evaluation, definition of failure critical to, 123
Quality function deployment (QFD), 138-140
   defined, 138
Quality integrity, defined, 875
Quality limits, average outgoing, 289-290
Quality management. See total quality management (TQM)
Quality standards
   Canadian, 128
   database of, 130
   Japanese, 129
   U.S., 125
Quality systems, mapping, 135
Quality tolerance limits, range of, 131
Quantity discount model, 630
Queue control, 716
Queuing models, simulation of, 776

R
Random numbers, abbreviated table of, 777-778
Random variation, 266
Raw materials, 516
R-charts, 273
   defined, 274
Receiving, inspection, storage and, 531
Recycling technology, 873
Redesign, 237
Reengineering (REE)
   for change management, 859-870
   defined, 28, 221, 859
   example, 783
   redesigning work processes with, 221
Regeneration MRP, 665-666
Regression analysis
   as forecasting method, 98
   scatter diagram for, 101
Relative productivity, defined, 45
Reliability, as quality dimension, 119
Remedial maintenance strategy, 239
Renting, cost determinants of in facility planning, 467
Reorder point, in perpetual inventory system, 624-625
Repairability, as service function of quality, 124
Resource leveling, 830-831
Research and development (R&D), 50, 122, 879
Rest and delay correction factor, 369-371

Revenue, total, 417
Robots, 336
   mobile, 880
   service, 882
Rough-cut capacity planning (RCCP), 667
Rule, 20:80, 259
Rules of thumb, 488. See also heuristics
Run charts, 253
Runs
   analysis of statistical, 280-281
   defined, 280

S
Sales patterns, 90
Sampling
   acceptance, 283-291
   versus 100% inspection, 282
Sampling methods, acceptance, 132
Satellite tracking system, 409
Scale, economies of, 57-58, 190-192
Scatter diagram
   defined, 263
   for regression analysis, 101
Scheduling
   control of, 829
   finite, 718
   production, 684-731
Scientific management (SM), 59
   defined, 61-62
Scientific method, cycle of, 252
Scoring model
   defined, 469
   developing the, 470-472
   facility selection using, 468-474
Seasonal cycle, historical forecasting with, 93
Sensitivity analysis, defined, 479, 577
Sequence, defined, 556
Sequence rule
   first-in, first-out (FIFO), 702
   last-in, first-out (LIFO), 702
Sequenced assembly, (SA), 59, 62-63
Sequencing, 687
   chart, 703
   operations, 702
   with more than one facility, 713-715
Sequential processes, 173
Serialized process, precedence diagram of, 743
Service, industrial perspective on, 496-507
Service applications, of OM, 8
Service capacity, 400
Service function of quality, 124
Service learning, 434
Service robots, 882
Service systems, flexible, 332
Service workforce, expanding, 351
Serviceability, as quality dimension, 119-120
Services
   defined, 15

differentiating between goods and, 15-17
information systems in, 18-19
location to enhance, 452
matrix of process types applied to,
198-199
value analysis for purchased, 226
Set up, 230-233
costs of, 596
defined, 230
Shipments, trial and error pattern of, 457
Shipping costs
factory to customers, 455-458
suppliers to factory, 454-455
Shipping decision, decision tree for, 70
Shop
comparing job and flow, 189
custom, 184-185, 169, 198
flow, 169
intermittent flow, 192-193
job, 169
Shop floor
control, 687
logistics, example, 789
workers, 351
Shop loading
transportation method for, 696-702
See also Loading
Shortages, materials, 537
Shortest processing time (SPT), 707
Short-term trends, moving average for, 96
Simulation
computer, 170
computer-based training, 354-355
defined, 170
process, 170
testing by, 326
use of for design modification, 329
Single-channel system, defined, 761
Single-minute exchange of dies (SMED),
231-233
Site, selection of structure and, 466-468
Site availability, as factor affecting
location decision, 453-454
Site selection, 449
Slack, in PERT chart, 820
Slope, 422
Smart benchmarks, 856
Smart technology, industry perspective on,
782-791
Special purpose equipment (SPE), 186
Spoilage, of inventory, 598
SPT rule
defined, 707
priority-modified, 712-713
Stage I-IV company, 27
Stages
forecasting life cycle, 78-111
of P/OM development, 25-28
Standard error of the mean, 272
Standard hours, 397
Standard units, 551

Standards
database of quality, 130
international producer, 125-130
Japanese quality, 129
productivity, 367
quality, 125
using production and cost, 357
Static inventory models, 591
Static inventory problem, 535
Stations
cycle time and number of, 745-746
developing layouts for, 749
multiple-channel, 764
Statistical process control (SPC), 265-267
Statistical quality control (SQC), 59, 265
defined, 63
Statistical quality control charts, 253
Statistical runs, analysis of, 280-281
Step functions, extrapolation of cycles,
trend lines and, 91-92
Stochastic line balancing, 738-739, 759-765
Stock, buffer, 593
Stock on-hand, patterns of, 645
Stock-outage costs, in perpetual inventory
system, 625-626
Stopwatches, computers as substitute for
in job observation, 362-366
Storage
controlled, process chart symbol for, 217
receiving, inspection and, 531
uncontrolled, process chart symbol for, 217
Strategic business units, 80
Strategic planning, hub and spoke diagram
for, 81
Strategy
defined, 80
OM's role in developing, 81-88
for operations management, 80
participants in planning, 81-88
process configuration, 166-203
tactics vs, 80
who carries out? 82-84
Structure selection
affect of work configurations on, 466
site and, 466-468
Subassemblies, purchased materials as, 516
Suggestion systems, 381
Supplier, 514
certification of, 543
chosen based on location, 517
negotiation between buyer and, 287
outside or self, 592
Supply chain, 189
delay deteriorates performance in, 409-416
game, 410-414
technology, 320-323
Symbols, process charts with sequenced
activity, 216-222
Synchronization, 735
defined, 718
process, 234-236

Synchronized manufacturing, 408, 718-720
Synergy, defined, 84
Synthetic processes, 173-176
automobile assembly, 174
defined, 173
marketing differences between analytic
and, 181
mixed analytic and, 182-184
sourcing for, 178-179
Synthetic time standards
defined, 376
evaluation of applicability of, 376
System
defining the, 12-13
materials management, 514
OM, 12
representation of, 12
total manufacturing planning and
control, 648-650
Systemic approach, 11-12
System mapping, 13
Systematic-constructive systems approach,
11-12
Systems
flexible manufacturing, 332
flexible office, 332
flexible process, 332
flexible production, 59, 64
flexible service, 332
information, 18-19
lean production, 59, 63
Systems approach
capacity and, 435
defined, 11
examples of 14-15
operations management and, 4-39
structure of, 13-14
systematic-constructive, 11-12
taxonomy of, 11
why required? 12
Systems causes
identifiable, 267
inherent, 266
Systems communication flows, 524-525
Systems management, TQM and, 135
Systems utilization factor, 761
Syzygy, 84

T
Tactics
defined, 83
process configurations link strategy
with, 172
strategies vs, 80
Take-back laws, 871
Taxes, as factor affecting location
decision, 454
Team building, 353
Team learning, 352
Teamwork
planning job design and, 346-393

for quality, 140
training and, 348-349
Technology
art and science of, 311-315
of computer-user interfaces, 337
defined, 308
group, 427
management of OM, 306-345
new, 852
of packaging and delivery, 324-325
quality control, 327-328
readying the organization for future,
880-884
recycling, 873
supply chain, 320-323
of testing, 325-327
timing of, 310
training and, 349-350
Technology trap, 309-319
Teleconferencing, distance learning
through, 354-355
Telephony, new era of, 417
Testing
empirical, 326
by simulation, 326
technology of, 325-327
Therbligs, 375
Time
estimates for project management, 823
trade-offs between costs and, 832-837
Time and motion study
defined, 361
synthetic time standards, 376-379
Time series
analysis, 92
defined, 91
extrapolation of, 92
Time standards, 361
synthetic or predetermined, 375
Time study, 361
Time values, for classifications of
motions, 377
Time-based management (TBM)
defined, 211, 735
for rapid response projects, 837-840
Timeline, of OM developments, 59
Tolerance limits, 131
Tolerance matching, 329
Toothpaste, production process for, 152-161
Total (or cumulative) lead time, 647
Total cost
curve, 132
defined, 417
Total flow time, 705
Total manufacturing planning and control
system, 648-650
Total productive maintenance (TPM)
defined, 236
model for determining, 245
Total quality management (TQM), 34, 85,
114-116, 136
design for manufacturing for, 475

inspecting and testing technology of,
327-328
Total revenue, 417
Total transport, minimizing, 214-215
Total variable costs, 57
defined, 416
in EOQ model, 608
TQM, 135. See also Total quality management
Trade, as factor affecting location
decision, 453-454
Trade-off
cost and time, 832-837
models, 520
Trainees, apprentices, artisans and, 60-61
Training
computer-based, 354
effectiveness of program, 351-352
methods, 351-352
teamwork and, 348-349
technology and, 349-350
Transformation model
basic OM, 21-23. See also input-output
model
illustration, 20
technology component of, 308
Transition management, 793-907. See also
Change management
Transitions, planning, 859
Transportation
minimizing, 214-215
process chart symbol for, 217
Transportation analysis, data for, 457
Transportation costs
obtaining, 458-459
total per unit, 458
Transportation method, for shop loading,
696-702
Transportation model
for aggregate planning, 574-577
flexibility of, 464
limitations of, 464
location decisions using, 454-465
Trees, decision, 70-73
Trend lines, step functions and, 91-92
Turnover
examples of, 527
as performance measurement, 526
Twenty-eighty (20:80) rule, 259
Two-bin perpetual inventory control
system, 627-628
Type I and II errors, in decision theory, 288

U
U.S. steel industry, 351
Uncontrolled storage, process chart
symbol for, 217
Units of work, standard, 553
Universal product codes, 327
Updating, for MRP, 665
Upscaling, in capacity planning, 406-406

V
Value analysis (VA), 226, 535

Variable costs, 417
with input-output transformation
model, 22
per unit, 416
total, 57
Variance, volume and, in design for
manufacturing, 475
Variation, random, 266
Variety
defined, 125
of parts and products, 428
VAT, 184
Vendor rationalization process, 538
Vendor releasing, of inventory, 596
Vendors. See Suppliers
Videoconferencing, distance-learning
through, 354-355
Virtual office, 883
Virtual warehouse, 791

W
Wages
determination of, 380
setting based on normal time, 368
Warranty policies, 123-124
Waterjetting, 871
Weibull distribution, 403
Weighted moving averages, 97
WIP. See Work in process
Work, standard units of, 553
Work centers, 484
Work configuration
affect on structure selection, 466
breakeven capacity related to, 424
defined, 168-169
differences in materials management
by, 520-523
moving from one stage to another, 195-198
Work flows, total daily, 486
Work in process (WIP), 517
storage for, 482
Work process configurations, 184
Work sampling
defined, 373-375
job design and, 373-375
size of sample for, 374
Work simplification, job design and, 381
Work system, managing, 379-383
Worker performance, evaluation of, 356
Workers
operations managers and, 350-351
shop floor, 351
Workplace, layout for job design, 477-480

X
$\bar{x}$-chart, 267

Z
Zero breakdowns, 240
Zero-based planning, 221
Zero-error mindset, for quality
achievement, 250-252